ECOIMMUNOLOGY

ECOIMMUNOLOGY

Edited by Gregory E. Demas

AND

Randy J. Nelson

OXFORD
UNIVERSITY PRESS

OXFORD
UNIVERSITY PRESS

Oxford University Press, Inc., publishes works that further
Oxford University's objective of excellence
in research, scholarship, and education.

Oxford New York
Auckland Cape Town Dar es Salaam Hong Kong Karachi
Kuala Lumpur Madrid Melbourne Mexico City Nairobi
New Delhi Shanghai Taipei Toronto

With offices in
Argentina Austria Brazil Chile Czech Republic France Greece
Guatemala Hungary Italy Japan Poland Portugal Singapore
South Korea Switzerland Thailand Turkey Ukraine Vietnam

Published by Oxford University Press, Inc.
198 Madison Avenue, New York, New York 10016
www.oup.com

Library of Congress Cataloging-in-Publication Data
Ecoimmunology / edited by Gregory E. Demas and Randy J. Nelson.
p. ; cm.
Includes bibliographical references and index.
ISBN 978-0-19-973734-5
1. Immune system. 2. Disease susceptibility.
3. Psychoneuroimmunology. I. Demas, Gregory E. II. Nelson, Randy Joe.
[DNLM: 1. Immune System—immunology. 2. Disease Susceptibility.
3. Ecosystem. 4. Immunity, Innate—immunology.
5. Psychoneuroimmunology—methods. QW 504]
QR181.E26 2011
616.07'9—dc23 2011015733

CONTENTS

CONTRIBUTORS

Shelley A. Adamo
Department of Psychology and Neuroscience
Dalhousie University
Halifax, Nova Scotia, Canada

Raise Ahmad
Department of Zoology
Banaras Hindu University
Varanasi, India

Noah T. Ashley
Department of Biological Sciences
University of Alaska, Anchorage
Anchorage, Alaska

Arnaud Aubert
Department of Neurosciences
Department of Psychology
François Rabelais University
Tours, France

Michael T. Bailey
Division of Oral Biology
College of Dentistry
The Ohio State University
Columbus, Ohio

Emily Chester
Department of Biology
Indiana University
Bloomington, Indiana

Christopher L. Coe
Harlow Center for Biological Psychology
University of Wisconsin
Madison, Wisconsin

Robert Dantzer
Department of Animal Sciences, College of ACES
Department of Pathology, College of Medicine
University of Illinois at Urbana-Champaign
Urbana, Illinois

Gregory E. Demas
Department of Biology
Indiana University
Bloomington, Indiana

Bonnie Fairbanks
Department of Biological Sciences
Virginia Tech
Blacksburg, Virginia

Susannah French
Department of Biology
Utah State University
Logan, Utah

Timothy Greives
Department of Migration and Immuno-Ecology
Max Planck Institute for Ornithology
Radolfzell, Germany

Sameer Gupta
Department of Zoology
Banaras Hindu University
Varanasi, India

Chandana Haldar
Department of Zoology
Banaras Hindu University
Varanasi, India

Dennis Hasselquist
Department of Biology
Lund University
Lund, Sweden

Dana M. Hawley
Department of Biological Sciences
Virginia Tech
Blacksburg, Virginia

Anne C. Jacobs
Department of Biology
University of California, Riverside
Riverside, California

Keith W. Kelley
Department of Animal Sciences, College of ACES
Department of Pathology, College of Medicine
University of Illinois at Urbana-Champaign
Urbana, Illinois

Ellen D. Ketterson
Department of Biology
Indiana University
Bloomington, Indiana

Sabra L. Klein
The W. Harry Feinstone Department of Molecular Microbiology and Immunology
Department of Biochemistry and Molecular Biology
Johns Hopkins Bloomberg School of Public Health
Baltimore, Maryland

Jason L. Kubinak
Department of Biology
University of Utah
Salt Lake City, Utah

Jacqueline W. Mays
Laboratory of Viral Diseases, CBS/VIS
NIAID, NIH
Bethesda, Maryland

Adam C. Nelson
Department of Biology
University of Utah
Salt Lake City, Utah

Randy J. Nelson
Department of Neuroscience
Ohio State University
Columbus, Ohio

Jan-Åke Nilsson
Department of Biology
Lund University
Lund, Sweden

Dawn M. O'Neal
Department of Biology
Indiana University
Bloomington, Indiana

Wayne K. Potts
Department of Biology
University of Utah
Salt Lake City, Utah

Nicole D. Powell
Division of Oral Biology
College of Dentistry
The Ohio State University
Columbus, Ohio

E. Rhiannon Pursall
Department of Plant and Animal Sciences
University of Sheffield
Sheffield, United Kingdom

Lars Råberg
Department of Biology
Lund University
Lund, Sweden

Seema Rai
Department of Zoology
Banaras Hindu University
Varanasi, India

Dionne P. Robinson
The W. Harry Feinstone Department of Molecular Microbiology and Immunology
Johns Hopkins Bloomberg School of Public Health
Baltimore, Maryland

Jens Rolff
Department of Plant and Animal Sciences
University of Sheffield
Sheffield, United Kingdom

Louise A. Rollins-Smith
Departments of Microbiology and Immunology, Pediatrics, and Biological Sciences Vanderbilt University Medical Center
Nashville, Tennessee

James S. Ruff
Department of Biology
University of Utah
Salt Lake City, Utah

Paul Schmid-Hempel
Institute of Integrative Biology (IBZ)
ETH Zurich
Switzerland

John F. Sheridan
Division of Oral Biology
College of Dentistry
The Ohio State University
Columbus, Ohio

Martin Stjernman
Department of Biology
Lund University
Lund, Sweden

Michael Tobler
Department of Biology
Lund University
Lund, Sweden
School of Biological Sciences
University of Sydney
Sydney, Australia

Zachary M. Weil
Departments of Neuroscience and Psychology
The Ohio State University
Columbus, Ohio

John C. Wingfield
Department of Neurobiology, Physiology, and Behavior
University of California, Davis
Davis, California

Douglas C. Woodhams
Institute of Evolutionary Biology and Environmental Studies
University of Zurich
Zurich, Switzerland

Rajesh Yadav
Department of Zoology
Banaras Hindu University
Varanasi, India

Marlene Zuk
Department of Biology
University of California, Riverside
Riverside, California

ECOIMMUNOLOGY

CHAPTER 1

INTRODUCTION TO ECOIMMUNOLOGY

Gregory E. Demas and Randy J. Nelson

Susceptibility to a wide variety of infectious diseases, from old "classics" such as influenza and malaria to newly emerging diseases including West Nile virus and severe acute respiratory syndrome (SARS), displays pronounced temporal and geographical variation, driven in part by differences in environmental conditions. Although some of this variation is undoubtedly due to fluctuations in disease prevalence *per se*, changes within the host's own immune system likely contribute to changes in disease susceptibility as well. Within the past decade alone, the role of parasites and pathogens in the evolution of life-history traits has become of increasing interest to both ecologists and evolutionary biologists, giving rise to the field of ecological immunology (ecoimmunology). At the same time, the immune system has become an important area of proximate investigation among animal physiologists and physiological ecologists as a means for understanding changes in disease susceptibility and the neural and neuroendocrine mechanisms that mediate these changes. Indeed, seminal studies with respect to mechanistic causes of changes in immunity and thus disease susceptibility gave rise to the related field of psychoneuroimmunology.

Despite these two important areas of investigation, the majority of research has traditionally focused on only one level of analysis, either ultimate (evolution and ecology of parasites and their effects on life histories) or proximate (neural and endocrine factors that interact with immunity and disease). The primary goals of ecoimmunology are to understand the extrinsic and intrinsic factors leading to changes in immune system function and how these changes contribute to disease susceptibility in both field and laboratory settings and across a wide range of animal species, including humans (Fig. 1.1). To accomplish this, ecoimmunologists themselves must also become broadly integrative, cross-disciplinary scientists willing to pursue diverse conceptual and technical approaches ranging from evolution, ecology, and life-history theory to endocrinology, neuroscience, molecular biology, and behavior. Although this field of study is still in its infancy, the amount of research conducted in this area has virtually exploded over the past 10 years. Surprisingly, however, there has not been a single edited volume on this topic published to date, despite this burgeoning research field.

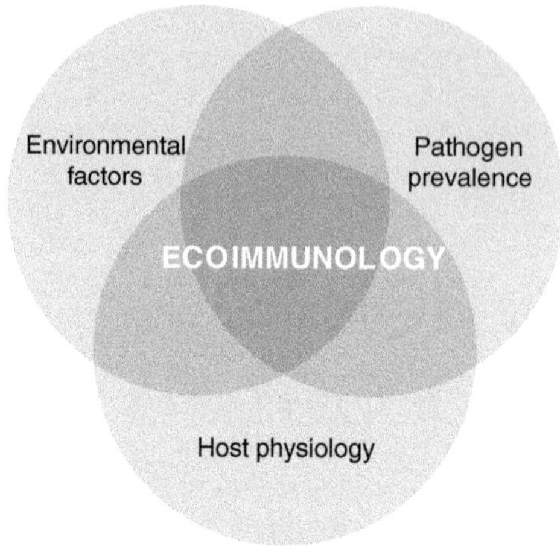

FIGURE 1.1. Ecoimmunology is a newly emergent, interdisciplinary research field that examines interactions among host physiology (e.g., immune function) and disease ecology (e.g., pathogen prevalence) in a wide range of environmentally relevant contexts.

The overarching goal of this book is to critically review recent advances in the discipline of ecoimmunology, and also to attempt to integrate both the proximate and ultimate perspectives into a common theoretical framework. It is our belief that such a framework will help clear up some of the apparent discrepancies within the field while also guiding future research endeavors in this area of investigation. To accomplish this goal, we have brought together many of leading experts within ecoimmunology and provided them with a forum to review the latest groundbreaking research in their respective subfields, as well as point to future directions in this field. The chapters that make up this volume cover a diverse mix of research in ecoimmunology; some chapters focus on the range of environmental factors that can affect host immune function, whereas other chapters emphasize the complex dynamics among host immunity, pathogen prevalence, and disease susceptibility. Still others examine the physiological (genetic, neural, and endocrine) mechanisms that lead to coordinated and possibly adaptive changes in immune responses. The chapters were also chosen to reflect the taxonomic breadth that encompasses the field of ecoimmunology; topics cover research from both field and laboratory studies and across taxonomic study systems, including birds, mammals, amphibians, reptiles, and insects. This book was written for both the novice ecoimmunologist who is interested in ecoimmunology but is not well versed in the literature, as well as the more experienced investigator already engaged in eco-immunological research. In both cases, we hope the reader gains a more integrative, comprehensive understanding of environment–immune system interactions. For readers less familiar with the immune system and its many intricacies, we have provided a brief overview of the vertebrate immune system (below) that we hope will serve as a primer to enhance the comprehension of the findings reviewed in the subsequent chapters. While this review focuses on vertebrate immunity, excellent overviews of invertebrate immunity are presented in Chapters 13 and 17 of this volume. It is truly an exciting time to be working within the field of ecoimmunology; we hope readers of this volume share our excitement for this field.

VERTEBRATE IMMUNE SYSTEM

Virtually all organisms are exposed to a wide range of potential pathogens within their environments and thus have evolved complex physiological responses, including innate and acquired (adaptive) immunity, to resist or eliminate pathogens. Broadly speaking, innate responses are relatively rapid, nonspecific, germline-encoded responses that serve as the initial line of defense against invading pathogens. Such defenses include anatomical barriers (mucous membranes, skin), resident flora (nonpathogenic bacteria), humoral factors (lysozyme, complement, and other acute phase proteins), and cellular responses (phagocytic cells like neutrophils, monocytes, and macrophages; inflammatory mediators produced by basophils, mast cells, and eosinophils; and natural killer cells). The complement system includes enzymes that function to eliminate microorganisms by promoting inflammatory responses, such as changes in local vascular permeability and entry of immune cells into infection sites. Complement also functions in the lysis of foreign cells through the formation of membrane attack complexes, and mediation of phagocytosis through the coating (opsonization) of pathogens and infected cells. Complement is also important for stimulating acquired (adaptive) immune responses. Acquired immune responses are typically slower, pathogen-specific responses that require activation and serve as an additional line of defense. In this case, effector mechanisms allow fast, secondary responses during subsequent exposures. Collectively, the immune system comprises a collection of specialized cells whose integrated function is the differentiation of "self" from "non-self" and the elimination of foreign "non-self" substances, specifically those capable of damaging the host.

The tissues and structures of the immune system, referred to as lymphoid organs, are connected to one another via a type of circulatory system, the lymphatic system, which carries lymphocyte-containing lymph (fluid) throughout the body where these cells encounter a pathogen. Among vertebrates, there are two main types of lymphoid organs, primary and secondary. Primary lymphoid organs include the bursa of Fabricius in birds and bone marrow in mammals and other vertebrates. These tissues are the source of stem cells, progenitor cells for all lymphoid tissues, and are where a specific subtype of lymphocytes, B cells, mature. The second primary lymphoid tissue is the thymus. It is within the thymus that lymphoid cells that have migrated from the bone marrow develop into a second type of lymphocyte, T cells. In addition to these primary lymphoid tissues, most vertebrates possess a host of secondary lymphoid tissues, including lymph nodes, spleen, appendix, tonsils, and Peyer's patches lining parts of the intestinal tract, which further serve to identify, sequester, and eliminate pathogens via a host of immune responses.

Cells of the immune system, called white blood cells (WBCs), all derive from common embryonic hematopoietic stem cells responsible for the production of most blood cell types. Among these pluripotent stem cells, some cells will undergo further development into lymphocytes, whereas others will differentiate into polymorphonuclear leukocytes (PMNs) or monocytes. Lymphocyte progenitor cells subsequently develop into B cells, T cells, or natural killer (NK) cells. B cells produce antibodies that neutralize pathogens and their products, block binding of parasites to host cells, induce complement activation, promote cellular migration to sites of infection, and enhance phagocytosis, among other actions. T cells do not produce antibodies; rather, they require direct contact with an infected cell to eliminate it. T cells can be further divided into different subtypes, including cytotoxic T cells (killer T cells; Tc), T helper cells (Th), and suppressor or regulatory T cells (Treg). Cytotoxic T cells destroy infected host cells by physical and chemical lysis, making T cells effective at destroying intracellular pathogens (e.g., viruses). In contrast, helper T cells serve a complementary function by secreting cytokines, as well as interacting with B cells to coordinate the production of antibodies. Th cells are further differentiated into several subtypes, most notably Th1 and Th2, depending on the specific patterns of cytokine production and their associated functions Cytokines are glycoproteins that perform a variety of functions, such as regulation of cell growth and development Th-1 cytokines include, among others, interferon-gamma (IFNγ), tumor necrosis factor alpha and beta (TNFα, β), and various interleukins (e.g., IL-1β, IL-2, IL-3, and IL-12). These cytokines activate macrophages, neutrophils, and NK cells, mediate inflammatory responses and cellular immunity (T cells), promote cytotoxicity toward tumor cells, and enhance chemotaxis of leukocytes. The Th-2, anti-inflammatory cytokines include many interleukins (e.g., IL-4, IL-5, and IL-10) that induce humoral immunity and antibody production (B cells), eosinophil activation, mast cell degranulation, goblet cell hyperplasia, mucin production, and intestinal mastocytosis (resulting in histamine release). This cytokine phenotype is important for protection against intestinal infections as well as the facilitation of allergic reactions. Although Th-1 and Th-2 cytokines act antagonistically to one another, both are usually present within the host at any given time, although during infection one phenotype usually predominates.

Collectively, B and T cells coordinate an organism's acquired (adaptive) immune responses. NK cells, in contrast, are part of an innate immune response, and nonspecifically attack and lyse infected cells. Along with NK cells, PMNs serve as part of the innate immune response. PMNs consist of three basic cell types: basophils, eosinophils, and heterophils (called neutrophils in humans). Heterophils serve as phagocytes, targeting and killing cells that have been tagged with antibodies and complement proteins. Basophils and eosinophils also participate in innate immune responses, primarily by secreting proteins such as toxins, prostaglandins, and histamine, as well as serving as antigen-presenting cells. Lastly, monocytes develop into macrophages, which are large leukocytes of the innate immune system that provide nonspecific defense against pathogens by phagocytosing pathogen-infected cells and activating B and T lymphocytes. Macrophages, along with lymphocytes, are also responsible for the production of cytokines (e.g., interleukins, tumor necrosis factors, and interferons), soluble signaling molecules that coordinate a wide range of immunological and inflammatory responses to infection. Lastly, the body fluids of most vertebrates possess a wide range of soluble factors consisting of antimicrobial peptides, lysozyme, complement proteins, transferrin, and lectins that serve as part of the innate humoral immune response. As this section suggests, the immune system is a complex, multifaceted system that works in a coordinated fashion to destroy pathogens. Thus, it must be emphasized that single measures of immunity are rarely sufficient to capture the complexity of an immune response. As will be obvious in the following chapters, the field of ecoimmunology has matured from the approach of examining a single immunological endpoint.

CHAPTER 2

LIFE-HISTORY EVOLUTION, HORMONES, AND AVIAN IMMUNE FUNCTION

Dawn M. O'Neal and Ellen D. Ketterson

INTRODUCTION

LIFE-HISTORY THEORY

Life-history theory seeks to explain associations among demographic traits and variation in life-history stages, including development, reproduction, and senescence, by investigating how these associations and trade-offs among them are related to fitness (Fig. 2.1). Life-history theory can also be applied to stages of the annual cycle in intermittent breeders, including molt, migration, and periods of non-breeding. The sequence and duration of life-history stages are affected by the environment to which organisms are adapted, as reflected in phenotypic attributes of morphology, behavior, and physiology (Ricklefs and Wikelski, 2002). The morphological, physiological, and behavioral phenotype that emerges, and that reflects the organism's fit to its environment, is limited by available resources and the relative costs and benefits of one or another phenotypic traits (Stearns, 1989; Zera and Harshman, 2001).

In birds, an excellent example of a trade-off between competing stages of the annual cycle is the trade-off between the timing of molt and reproduction. Molt, the replacement of flight and body feathers in preparation for migration or to refresh secondary sex characteristics, is nutritionally costly and can impair flight, thus impeding foraging and evading predators. Reproduction is also nutritionally as well as energetically costly, requiring the distribution of resources towards the production and care of offspring and away from self-maintenance. Thus, the costs associated with molt and reproduction can be in direct conflict with each other. Parents may be limited in their ability to direct energy towards new feather growth (self-maintenance) and away from offspring

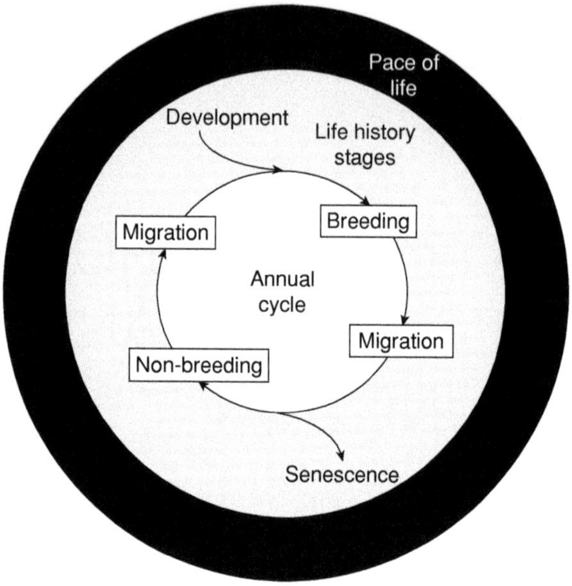

FIGURE 2.1: Schematic of how life-history stages (development, breeding and non-breeding seasons, molt/migration, and senescence) fit within the context of the annual cycle and within the context of a species pace of life. Pace of life is shown as encompassing both the stages of life history and the natural occurrence of these stages through the annual cycle.

care (parental effort) and still maintain maximum fitness benefits via survival and reproduction. As a result, molt and reproduction typically do not overlap in time, and if they do, each may proceed in the face of diminished allocation of resources to the other.

Similarly, selection can shape allocation to survival and reproduction. Most notably expressed in terms of an organism's pace of life, trade-offs can occur between survival and fecundity according to the speed at which an organism lives, fast-paced (short life span) versus slow-paced (long life span) (Ricklefs and Wikelski, 2002). Because fast-paced, short-lived species are limited in the number of breeding episodes, they tend to allocate resources to current reproduction over adult survival or future reproductive opportunities. Conversely, slow-paced species favor future reproduction (adult survival) at the expense of current reproductive effort. As a result of these trade-offs, there are marked differences in physiology, behavior, and morphology between fast- and slow-paced species. Slow-paced species, for example, not only have smaller clutch sizes (Skutch, 1949), longer development time, and longer duration of parental care (Ricklefs and Wikelski, 2002; Stutchbury and Morton, 2008) compared to fast-paced species, but they also often have lower basal metabolic rates (Wikelski et al., 2003a; Wiersma et al., 2007), lower testosterone levels during breeding (Wingfield et al., 2001; Hau et al., 2008), increased corticosterone levels in response to a stressor (Bókony et al., 2009; Malisch and Breuner, 2010), and more robust immune responses (Martin et al., 2004; Tieleman et al., 2005; Martin et al., 2006; Addison et al., 2009), which collectively reflect greater investment in self-maintenance in contrast to reproduction.

HORMONES AND IMMUNE FUNCTION IN THE CONTEXT OF LIFE-HISTORY THEORY

Endocrine and immune responses can have direct and indirect effects on various life-history traits, as well as reproductive success and survival and vice versa (e.g., offspring quality and survival [Grindstaff et al., 2003; Rubolini et al., 2005a; O'Neal et al., 2008], reproductive success [Bonneaud et al., 2003; Bókony et al., 2009], and adult survival [Ricklefs and Wikelski, 2002; Wingfield and Kitaysky, 2002; Lee, 2006; Reed et al., 2006]). Hormones can be an important mediator of life-history trade-offs because they respond to environmental signals such as climate (Wingfield and Kitaysky, 2002; Wingfield, 2003), social interactions (Wingfield et al., 2001; Veiga et al., 2004; Goymann et al., 2007), and habitat quality (Marra and Holberton, 1998; Wingfield, 2003; Martin and Rubenstein, 2008), and thus act to coordinate numerous aspects of the phenotype, including behavior (Soma et al., 2000; Hau et al., 2004; Wingfield, 2005; Zysling et al., 2006), secondary sexual characteristics (Evans et al., 2000; Rubolini et al., 2005b), other hormones (Ketterson et al., 1991; Owen-Ashley et al., 2004), and immune function (Duffy et al., 2000; Casto et al., 2001; Owen-Ashley et al., 2004; Ashley et al., 2009; Roberts et al., 2009), via altered gene expression in both brain and periphery (Ketterson et al., 1992; Finch and Rose, 1995). For example, American redstarts (*Setophaga ruticilla*) wintering on poor winter habitat have greater baseline and stress-induced corticosterone levels than redstarts wintering in richer habitat (Marra and Holberton, 1998; Studds and Marra, 2005), and the difference is reflected in arrival time on the breeding grounds (and thus the transition from non-breeding to breeding) (Marra et al., 1998) and in subsequent reproductive success (a proxy for fitness) (Smith and Moore, 2005). Interestingly, hormonal variation related to trade-offs is also found within fast- and slow-paced species depending on their age. The common tern (*Sterna hirundo*), for example, is a long-lived species, but younger adults are more responsive to stressors in terms of elevation of corticosterone than are older adults, a difference that has been attributed to the greater number of reproductive attempts likely remaining to a younger adult; that is, the "pace of life" of young adults is slower than that of elderly adults of the same species (Heidinger et al., 2006). Immune mechanisms have similar attributes to hormonal mechanisms in that they are also sensitive to environmental variation in climate, habitat quality, and social interactions (Ruiz et al., 2002; Lindström et al., 2005a; Altizer et al., 2006; Lee, 2006; Hawley et al., 2007; Lee et al., 2008; Martin et al., 2008). Immune mechanisms can also interact with hormonal mechanisms to create a complex unit on which selection can act (Zuk and Stoehr, 2002; Hasselquist, 2007), and this interaction is discussed in the following sections.

ECOIMMUNOLOGY IN AVES

Immune responses like hormones are sensitive to environmental variation and can be an important indicator of physical condition in free-living individuals (Siegel, 1980; Nelson and Demas, 1996; Lahti, 1998; Marra and Holberton, 1998; Shephard and Shek, 1998; Ruiz et al., 2002; Wingfield and Hunt, 2002; Wingfield and Kitaysky, 2002; Comendant et al., 2003; Boonstra, 2004; Lindström et al., 2005a; Matson et al., 2005; Tieleman et al., 2005; Hawley, 2006; Martin et al., 2006a; Wikelski and Cooke, 2006). Many techniques have been used to measure immune responses in birds. Table 2.1 summarizes the most common techniques, their function, and what immune component they assess.

Of the main components of the immune system, adaptive (or induced) immunity is the more difficult to measure in the field, as it typically requires challenges with foreign antigens and repeated capture to measure the development of pathogen-specific antibodies. Thus, field studies often measure constitutive (or innate) immune defenses. Field studies also generally use a single technique for measuring immune function, although criticism of this approach has been building (Norris and Evans, 2000; Martin et al., 2006b; Martin et al., 2008). Recent studies have indicated that the costs involved in activating different immune components are variable; for example, antibody production is thought to require more energy redistribution than the development of innate components, and there are high costs associated with the systemic inflammatory responses induced by cell-mediated and innate responses (reviewed in Lee, 2006). Together these observations indicate possible trade-offs between immune components and the importance of taking multiple measures of the immune system in order to accurately evaluate an organism's ability to mount an immune response—that is, its immunocompetence (Norris and Evans, 2000; Martin et al., 2006b; Matson et al., 2006a).

In the sections to follow, we consider immune function across multiple life-history stages and in conjunction with other environmental, behavioral, and physiological mechanisms that play significant roles in mediating immune function. In particular, we focus on the most notable hormonal mechanisms involved in the regulation of avian immunity, including melatonin, prolactin, growth hormone, and several neuroendocrine hormones. We will provide examples of how maternally derived hormones during development can act as immunosuppressants, how breeding hormones can act in concert with immune function to signal mate quality, and how some non-breeding hormones can actually enhance immune function. We also provide examples of the cost and benefits associated with hormone–immune interactions and the resulting trade-offs that can arise between life-history traits and stages.

DEVELOPMENT/GROWTH

IMMUNE FUNCTION DURING DEVELOPMENT AND MATERNAL ANTIBODIES

Environmental variation plays a critical role in organismal growth and development and can influence both immune responses and hormonal signals. Mothers are a significant source of environmental variability, and female birds are known to influence embryonic and fetal development via hormones and/or antibodies they deposit in the egg (Boulinier and Staszewski, 2008; Gil, 2008; Navara and Mendonca, 2008; Hasselquist and Nilsson, 2009) (see Chapter 6). Unlike mammalian systems, in which the maternal endocrine and immune environment can continue to influence development during the embryonic and fetal stages, in avian systems the maternal influence on the composition of the embryo's endocrine and immune environments has been determined

TABLE 2.1: Common techniques used to measure immune function in birds

Immune defense measured	Assay	Function	Citation
Constitutive innate			
	Bacteria killing	Assess complement, natural antibodies, acute phase proteins and cellular components; indicative of an individual's susceptibility to bacterial infection	(Tieleman et al., 2005; Matson et al., 2006b; Millet et al., 2007)
	Hemolytic complement assay	Assesses ability of complement to lyse foreign blood cells	(Matson et al., 2005; Greives et al., 2006)
	Differential cell counts	Assesses concentrations of phagocytes and lymphocytes; indicative of circulating cellular immunity	(Norris and Evans, 2000; Campbell and Ellis, 2007)
Induced innate			
	Lipopolysaccharide (LPS)	Measurement of acute phase response or sickness behavior mediated by pro-inflammatory cytokines	(Owen-Ashley et al., 2006; Owen-Ashley and Wingfield, 2006)
Constitutive adaptive			
	Total immunoglobulin G (total IgG; nonspecific)	Nonspecific antibody response; indicative of specific antibody response	(Bonneaud et al., 2003; Greives et al., 2006)
	Hemagglutination assay	Assessment of natural antibody-mediated complement activity; approximates specific antibody response	(Matson et al., 2005)
Induced cell-mediated			
	Phytohemagglutinin (PHA)	Cytokine production/T-cell activity; indicative of resistance to intracellular infection	(Casto et al., 2001; Martin et al., 2006a)
	Concanavalin A		(Fitzgerald et al., 1992)
Induced humoral			
	Key-hole limpet hemocyanin (KLH)	Antibody response; indicative of resistance to extracellular infection	(Hasselquist et al., 1999)
	Sheep red blood cells (SRBC)/ Chukar (*Alectoris chukar*) red blood cells (CRBC)		(Deerenberg et al., 1997; Fair and Ricklefs, 2002; Smits and Baos, 2005)
	Newcastle disease virus (NDV)		(Fair and Ricklefs, 2002)
	Diphtheria–tetanus vaccine		(Svensson et al., 1998; Ilmonen et al., 2000)

by the time the egg is laid. Interestingly, females may prepare their offspring for current climate, social, disease, or habitat environments by adjusting levels of hormone and immune (antibody) deposition in egg yolk. For example, female great tits (*Parus major*) exposed to ectoparasites prior to laying transfer more maternal antibodies to eggs than non-exposed females, thereby providing chicks with needed defenses upon hatching (Buechler et al., 2002). Severe environmental perturbations (e.g., drought, high predation, or severe weather events), however, can reduce the ability of females to deposit antibodies in eggs, presumably due to trade-offs between self-maintenance and reproductive effort (Groothuis et al., 2005b; Rubolini et al., 2005a; Moreno et al., 2008). Poor nutritive environments, for example, can reduce condition in female pied flycatchers (*Ficedula hypoleuca*) and also reduce maternal antibody deposition, thus reducing nestling growth (Moreno et al., 2008). This maternal preparation of offspring for environmental conditions they may experience has even been related to pace of life (Addison et al., 2009). In species with longer development times and longer lives (i.e., a slower pace of life), individuals may experience repeated infections and encounter a greater variety of parasites. Females of such species are known to invest more in the maternal transfer of antibodies compared to species with shorter development times and a faster pace of life (reviewed in Tella et al., 2002; Lee et al., 2006; Lee et al., 2008).

Early in life, developing birds have limited ability to synthesize antibodies endogenously, so the immune defenses of chicks primarily consist of the innate immune response and any maternally derived antibodies (Lawrence et al., 1981; Grindstaff et al., 2003). These maternally transferred antibodies can reflect the lifetime exposure of the mother to pathogens, including those pathogens that are represented in the local disease environment at the time of reproduction (Gasparini et al., 2001; Lemke et al., 2003). Thus, offspring susceptibility to disease can be reduced by prior maternal antibody responses that are transmitted in the egg (Grindstaff et al., 2003). Immunoglobulin Y (IgY), the equivalent of IgG found in mammals, is produced maternally in response to immune challenge, and it is the primary antibody found in avian eggs (Addison et al., 2009). This transfer of antibodies is thought to reduce the need of offspring to generate an endogenous humoral response that might require the redistribution of energy away from growth and general development in order to support immune responses such as sickness behavior (anorexia, fever, nutrient redistribution) and potentially costly nonspecific inflammatory immune responses (Pihlaja et al., 2006; Grindstaff, 2008; Hasselquist and Nilsson, 2009). For example, in Japanese quail (*Coturnix japonica*), immunization with lipopolysaccharide ([LPS)]an antigen derived from *Salmonella typhimurium* that does not proliferate but induces sickness behavior, suppresses growth in developing chicks (Grindstaff, 2008). However, immunizing the mother with LPS prior to egg laying induces the synthesis of antibodies, which when transferred to offspring dampen the growth-suppressive effects of chick immunization with LPS (Grindstaff, 2008). Maternal antibodies can be particularly effective at suppressing endogenous humoral responses in offspring during early development, when they would presumably be too costly for offspring to generate directly. Later, when the cost is more manageable, offspring immune function is believed to be activated in response to the specific pathogenic environment of the offspring's generation (Gasparini et al., 2009). The age at which nestlings begin to produce their own antibodies differs among bird species, ranging from 10 to 14 days post-hatching (Grindstaff et al., 2006; Pihlaja et al., 2006; reviewed by Hasselquist and Nilsson, 2009). However, despite the brief exposure offspring have to maternal antibodies, these antibodies have long-term protective effects, influencing the strength of offspring response to challenges years after maternal exposure (Grindstaff et al., 2003; Grindstaff et al., 2006; Reid et al., 2006; Boulinier and Staszewski, 2008).

MATERNAL HORMONES: CORTICOSTERONE

Traditionally, corticosterone has been at the forefront of hormones associated with the mediation of immune function. In adult birds, this glucocorticoid is generally associated with the liberation of glucose for energy and can have enhancing or suppressive effects on immune responses depending on the length of corticosterone exposure and the type of immune parameter measured (reviewed by Martin, 2009). Due to its sensitivity to environmental perturbations, corticosterone is often used to evaluate how stressful the environment is on physiology and behavior (Wingfield, 2003). For example, an increase in the amount of corticosterone deposited in eggs is often seen as a negative effect of a stressful environment on females and their offspring. Maternally elevated corticosterone levels in European starling (*Sturnus vulgaris*) embryos generally delay hatching, and post-hatching, this hormone often suppresses development and growth (Love et al., 2005). Exogenous corticosterone has similar effects on hatchling house sparrows (*Passer domesticus*), depressing immune function (response to phytohemagglutinin [PHA]) and growth, in addition to mediating parental effort (Loiseau et al., 2008). Experimental manipulations of corticosterone in developing eggs have also implicated corticosterone in reducing the cell-mediated immune response (Love et al., 2005; Rubolini et al., 2005a), but at least in yellow-legged gulls (*Larus michahellis*), exogenous corticosterone has no effect on humoral immunity (Rubolini et al., 2005a). It should be noted, however, that the effects of corticosterone are not always negative. It has been associated with increases in begging displays with a positive effect on growth in developing chicks, with normal nest departure, and with post-hatching dispersal (Heath, 1997; Kitaysky et al., 2003). Thus, whether a response to corticosterone should be viewed as positive or negative may depend on the environment or an individual's condition.

MATERNAL HORMONES: TESTOSTERONE

Like corticosterone, maternally deposited testosterone can also have significant effects on chick growth, development, and immune function (Groothuis et al., 2005a; Gil, 2008). Maternally derived testosterone has been associated with enhanced growth in developing chicks via elevated begging behavior and activity levels, which may effectively enhance the ability of chicks to stimulate parental feeding and their ability to compete with siblings for resources (reviewed by Groothuis et al., 2005a). Exposure to maternal testosterone has also been associated with behavioral measures of aggression and boldness, and it has even been linked to social status (reviewed in Gil, 2008). Furthermore, studies to date suggest that maternally deposited testosterone may play a role in mediating immune function during early development. Generally, maternally deposited and experimentally manipulated yolk testosterone have significant immunosuppressive effects (Gil, 2008), with higher concentrations suppressing both cell-mediated response (PHA) (Groothuis et al., 2005a; Navara et al., 2005; but see Tschirren et al., 2005; Rubolini et al., 2006; and Pitala et al., 2009) and humoral immunity (Muller et al., 2005; Sandell et al., 2009) in developing chicks of many species. Nestling jackdaws (*Corvus monedula*), for example, hatching from eggs treated with testosterone and androstenedione (a precursor to testosterone) display lower antibody titers against challenge with LPS at 6 days of age, and lower antibody responses to diphtheria–tetanus antigens as well as cell-mediated response to PHA at 8 days of age, as compared to control chicks (Sandell et al., 2009). The immunosuppressive effects of testosterone in chicks have largely been linked to trade-offs between growth and immune system development and the costs associated with maintaining and mounting an energetically costly immune defense (Groothuis et al., 2005a). This immunosuppressive nature of maternally deposited testosterone has also been shown to have long-term effects. Zebra finch (*Taeniopygia guttata*) fledglings from eggs treated with testosterone did not initially display differences in humoral (diphtheria–tetanus challenge) or cell-mediated (PHA challenge) responses compared to control birds, but did show enhanced humoral responses at 5 and 7 months (Tobler et al., 2009). Interestingly, females may adjust the concentration of deposited testosterone depending on the local parasite environment, which can have significant effects on dispersal behavior. Great tit (*Parus major*) females exposed to hen fleas, for example, reduce the amount of testosterone and androstenedione deposited into eggs during laying (Tschirren et al., 2004). Furthermore, nestlings from infested nests express shorter dispersal distances compared to those from un-infested nests, presumably as an adaptive host response to the local parasite community, or conversely because of a lowered ability of parasite-infested birds to deal with new environments (Tschirren et al., 2007).

DEVELOPMENTAL HORMONES: NON-MATERNALLY TRANSFERRED HORMONES AND THE IMMUNE SYSTEM

Several hormones generally associated with immune development are not known to be maternally transferred in birds (Fig. 2.2). Growth hormone, for example, is associated with overall nestling growth and also stimulates the growth of lymphoid cells and tissues and receptors found in thymus, bursa, and spleen of newly hatched chicken chicks (Hull et al., 1996; Luna et al., 2005). The avian thymus is involved in the development, proliferation, and differentiation of T lymphocytes, whereas the bursa (bursa of Fabricius) is responsible for the maturation of B cells involved in humoral responses (Nieuwenhuis, 1996). In addition, the avian spleen is responsible for systemic immunity in the form of lymphocyte recirculation and differentiation, antibody production, phagocytosis, and the development of macrophages and parts of the complement system (John, 1994). The observation that these immune organs are stimulated by growth hormone during development suggests a role for this hormone in macrophage profile (Dicarlo et al., 1996) activity and at least indirectly in lymphocyte maturation, two of the foremost cell types in initiating immune responses (Hull et al., 1996; Luna et al., 2005).

Prolactin also plays a role in thymic and lymphatic development in developing chicks, as does thyroid hormone with its enhancing effects on lymphocyte proliferation and humoral immunity (Skwarlo-Sonta, 1992; Moreno et al., 1994; Kelley et al., 2007). Finally, melatonin, the main neuro-hormone synthesized and released by the pineal gland, is also known to play an important role in the development of the immune system in birds, accelerating the development of both cellular

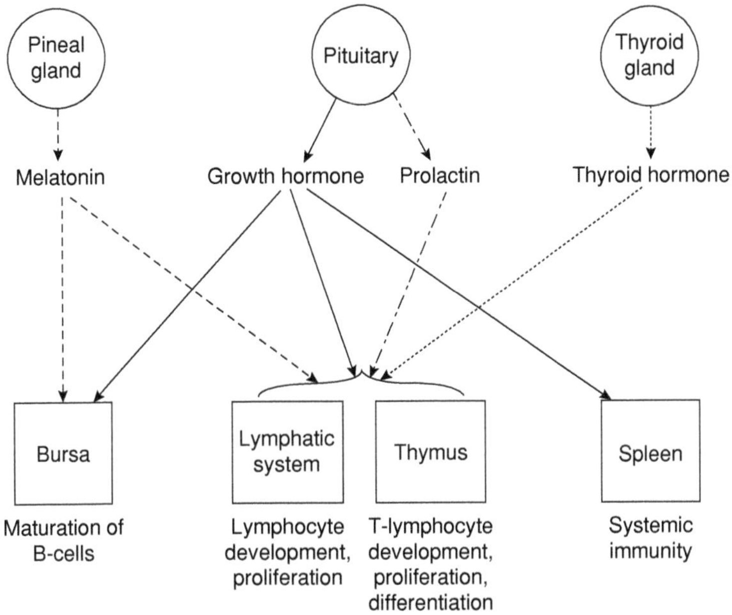

FIGURE 2.2: Non-maternally transferred hormones, the tissues they originate from, and the immune tissues and cells they affect. In general, these hormones are involved in the enhancement, development, proliferation, and differentiation of these cells and tissues.

and humoral immune responses (Moore and Siopes, 2005). Chicken embryos (*Gallus gallus domesticus*) whose pineal gland was removed exhibit retarded development of the lymphoid glands, the thymus, and bursa of Fabricius, leading to reductions in immune responses (Skwarlo-Sonta, 2002). Similarly, pinealectomized neonatal chickens exhibit diminished immune responses, which were rescued with exogenous melatonin treatment (Skwarlo-Sonta, 1996).

BREEDING/REPRODUCTION

BREEDING IMMUNE FUNCTION

The benefits associated with increased investment in current reproduction can be offset by reductions in survival or future fecundity, and these costs have been well documented in numerous avian species (Bell, 1980; Clutton-Brock and Godfray, 1991). Although the mechanisms underlying these trade-offs are not well understood, immune function and hormones have received considerable attention as possible mediators (reviewed in Harshman and Zera, 2007). Both steroid hormones and to some extent immune function fluctuate throughout the breeding season but are generally depressed during offspring care (Bonneaud et al., 2003; Ardia, 2005a; Greenman et al., 2005). Hormones, because of their pleiotropic effects, which include the redistribution of energy and interference with parental care behaviors (Zera and Harshman, 2001), and immune function, because of its potential to redistribute energy away from provisioning efforts, are likely mediators of tradeoffs between reproduction, survival, and future fecundity.

The mediation of individual immune function during breeding is thought to be rooted in trade-offs between immune function and reproductive effort, and evidence of these trade-offs has been recently documented in birds (Martin et al., 2006; Hasselquist, 2007; Lee et al., 2008; Stutchbury and Morton, 2008). A comparative study of 70 neotropical birds demonstrated positive correlations between natural antibody titers and incubation time, suggesting that the strength of the adaptive immune system may covary with development times (Lee et al., 2008). However, species with large clutches (generally fast-paced species) displayed stronger complement activity, suggesting that species undergoing short development times may depend more on innate defenses, which are thought to be less costly to develop than adaptive immunity (Lee et al., 2008). Furthermore, studies manipulating parental effort through the experimental increase of clutch size followed by challenges with pathogens have shown that increased clutch size can reduce parental immune responses (reviewed in Knowles et al., 2009). For example, female common eiders (*Somateria mollissima*) incubating experimentally increased clutches showed lower lymphocyte levels and suppressed humoral immune responsiveness against diphtheria–tetanus challenge (Hanssen et al., 2005). Immune and parasite challenges of parents during parental care can also have adverse effects, reducing provisioning and brooding rates (Bonneaud et al., 2003; Tomás et al., 2007; Gallizzi et al., 2008). Female blue tits (*Cyanistes caeruleus*), for example, treated for blood parasites with primaquine, an antimalarial drug, increased their provisioning rates (Tomás et al., 2007), indicating that infection may have been suppressing parental behavior prior to treatment. Similar effects were reported in house sparrows challenged with LPS (Bonneaud et al., 2003).

GEOGRAPHIC AND SEX VARIATION IN BREEDING IMMUNE FUNCTION

Sex and geography are also known to influence the immune responses of breeding birds, although their effects are often dependent on which measure of immune function is being considered (Ardia, 2005b; Martin et al., 2006b; Owen-Ashley et al., 2008). Geographic variation in immune function is particularly evident when considering a species' pace of life, in part because pace of life generally varies on a latitudinal scale, with slower-living species found most often at low latitudes (Wikelski et al., 2003a). Furthermore, various aspects of disease also display latitudinal variation. In general, disease prevalence increases with decreasing latitude as climate becomes milder and biodiversity increases, fostering pathogen abundance and transmission (Stromberg, 1997; Harvell et al., 2002; Lafferty, 2009). As such, immune function varies latitudinally, with more robust individual immune responses at low latitudes compared to responses at high latitudes (Martin et al., 2004; Ardia, 2007; Hasselquist, 2007). Thus, breeding immune responses are often much lower in fast-paced, short-lived species compared to slow-paced, long-lived species (Martin et al., 2004; Ardia, 2005b; Owen-Ashley et al., 2008). Again, however, this variation is not absolute and depends significantly on what component of the immune system is being measured (Martin et al., 2006). Slow-living, tropical house sparrows, for example, display higher secondary humoral responses compared to their fast-living, temperate counterparts, but temperate-zone house sparrows have stronger cell-mediated responses by comparison (Martin et al., 2006). Similarly, humoral responses to diphtheria–tetanus vaccination are significantly higher in high-latitude breeding white-crowned sparrows (*Zonotrichia l. gambelii*) compared to lower-latitude breeding sparrows (*Zonotrichia l. pugetensis*) (Owen-Ashley et al., 2008). Subspecies did not differ in cell-mediated responses to PHA.

With respect to sex, females generally display more robust immune responses than males owing to physiological and ecological/social mechanisms that vary with the life history of each sex (Zuk and McKean, 1996; Hasselquist, 2007; Nunn et al., 2009). Sex differences in immune function have been attributed to differences in mating systems, with the idea that in species in which variance in reproductive success differs by sex (generally polygynous species), the sex with the higher variance in success should invest relatively more in mate attraction and less in immune function (Klein, 2000; Rolff, 2002; Klein, 2004). Conversely, in monogamous species the sexes should have similar levels of immune function. Indeed, studies suggest that females of socially polygynous species invest more in immunity than males (Zuk and McKean, 1996; Hasselquist et al., 1999). In addition to mating systems, male-typical behaviors of aggression, dispersal, and flocking have also been suggested as mechanisms that lead to increased parasitism and reduced immune function in males (Zuk and McKean, 1996). The hormonal mechanisms mediating sex differences in immune function have most often been related to the immunosuppressive effects of testosterone, discussed in more detail below, with the result that males, the sex that generally displays higher levels of testosterone, show reductions in immune response compared to females (Folstad and Karter, 1992; Klein, 2004). Recently, however, the role of testosterone in the mediation of sex differences in immune function has been seen as indirect, with testosterone influencing immune function more by contributing to the elevation of circulating corticosterone (Owen-Ashley et al., 2004; Hau, 2007).

BREEDING HORMONES: TESTOSTERONE AND THE IMMUNOCOMPETENCE HANDICAP HYPOTHESIS

Research regarding immune function and mate choice has most notably been linked with sex hormones, which are involved in many behavioral and morphological aspects of reproduction. Also, many sex hormones are elevated during the breeding season when immune function is normally suppressed, suggesting a role for these hormones as immunosuppressive agents. Of the sex hormones, testosterone has received much consideration as a possible mediator of immune function during breeding. Testosterone has been extensively studied in male birds and more recently in females (Ball and Balthazart, 2004; Clotfelter et al., 2004; Garamszegi et al., 2005; Ketterson et al., 2005; Hirschenhauser and Oliveira, 2006; Zysling et al., 2006; Sandell, 2007; O'Neal et al., 2008; Ketterson et al., 2009). Testosterone mediates aggressive interactions, as well as reproductive and parental care behaviors (Ketterson and Nolan, 1994; Wingfield, 2005; Hirschenhauser and Oliveira, 2006). Testosterone is most notably studied within the context of immune function under the immunocompetence handicap hypothesis, which proposes that testosterone mediates trade-offs between sexual signaling (i.e., secondary sex characteristics such as plumage), social status, and immunocompetence (Folstad and Karter, 1992) (see Chapter 15). The hypothesis states that testosterone enhances the expression of some sexual signals, but also decreases survival due to its immunosuppressive nature. Thus, only high-quality males, those that can withstand pathogen or parasite infection presumably due to their superior genetic makeup or their condition, can fully express testosterone-mediated secondary sex traits despite the immunological cost (Folstad and Karter, 1992; reviewed in Roberts et al., 2004). Evidence for this hypothesis has been inconclusive. Whereas several studies have indicated immunosuppressive effects of testosterone (Duffy et al., 2000; Casto et al., 2001; Greives et al., 2006; Zysling et al., 2006; Ashley et al., 2009), others have found immunoenhancing effects, no effect, or differential effects depending on the immune component measured (Hasselquist et al., 1999; Lindström et al., 2001; Roberts and Peters, 2009). Furthermore, several studies have suggested that testosterone suppresses immune function in a context-dependent manner such that only individuals in poorer condition experience immunosuppression owing to elevated testosterone, while those in good condition may not (Roberts and Peters, 2009).

BREEDING HORMONES: TESTOSTERONE AND OTHER IMMUNOMODULATORS

Carotenoids, which are associated with red and yellow pigments in secondary sex characteristics, have also been suggested as possible buffers for the immunosuppressive effects of testosterone (McGraw and Ardia, 2007). Many studies have provided evidence supporting carotenoid-based plumage coloration as an indicator of parasite resistance (measured as clearance of infection or parasite count) (Piersma et al., 2001; Hill and Farmer, 2005; Dawson and Bortolotti, 2006; Baeta et al., 2008) and immunocompetence (Lindström and Lundstrom, 2000; McGraw and Ardia, 2003; Saks et al., 2003; Horak et al., 2006; Maney et al., 2008). Research has also revealed reductions in plasma carotenoid levels with immune challenge, suggesting that the immunomodulatory effect of carotenoids may come at a cost to the expression of colored sexual signals (but see Hill, 1999; McGraw and Ardia, 2003; Biard et al., 2009). In other words, use of carotenoids to foster expression of sexual signals may divert their use for immune maintenance, and, as a result, sexually selected traits may act as an indicator to potential mates of parasite resistance and thus physiological/genetic/physical quality (Hamilton and Zuk, 1982; Folstad and Karter, 1992).

In addition to carotenoids, recent work has also suggested that the immunosuppressive nature of testosterone may be mediated by access to resources via dominance interactions (Evans et al., 2000; Poiani et al., 2000), leptin (Alonso-Alvarez et al., 2007), and even corticosterone (Evans et al., 2000; Casto et al., 2001; Ashley et al., 2009; but see Roberts et al., 2009). For example, dominant birds generally have high levels of testosterone, often reflected in sexually selected characteristics, and as a result tend to have greater access to food resources, leading to better body condition and more effective immune responses (Poiani et al., 2000). Similarly, individuals able to maintain high leptin levels, and thus large lipid reserves in conjunction with high testosterone levels, might be able to develop secondary sexual characteristics without incurring the immunosuppressive costs of testosterone (Alonso-Alvarez et al., 2007). Indeed, exogenous testosterone coupled with treatment with leptin rescues cell-mediated immune response to PHA in zebra finches (*Taeniopygia guttata*) (Alonso-Alvarez et al., 2007).

BREEDING HORMONES: CORTICOSTERONE

In common with testosterone, corticosterone responses can mediate phenotypic traits that have direct and indirect effects on various fitness components (e.g., reproductive success, offspring quality and survival, and adult survival) (Martin and Rubenstein, 2008; Bonier et al., 2009), and it may be the most important hormone in the mediation of environmental effects on physiology and behavior (Wingfield and Kitaysky, 2002; Wingfield, 2003). Most often corticosterone is studied in the context of the stress response—that is, increases in plasma corticosterone concentrations in response to demanding activities such as breeding or migration, or environmental stressors such as unpredictable changes in climate, aggressive encounters, or exposure to predators (Sapolsky et al., 2000). During breeding, the stress response is often dampened or suppressed, presumably because of the negative effects corticosterone can have on parental care behaviors, reducing offspring care or inducing nest abandonment (Wingfield and Kitaysky, 2002; Bókony et al., 2009). With respect to immune function, the effects of stressors on immunity are highly dependent on the duration of the stress response (i.e., corticosterone secretion) as well as the type of immune response measured (reviewed in Martin, 2009). Acute corticosterone secretion (magnitude of minutes to days) can result in immunoenhancement, with individuals displaying more robust immune responses (Buehler et al., 2008), or no changes in immune response (Martin et al., 2005; Bourgeon et al., 2006). In red knots (*Calidris canutus*), for example, bacteria killing (*Staphylococcus aureus* and *Candida albicans*) increases after 60 or 90 min of handling stress (Buehler et al., 2008). Enhancement of the immune system in response to acute stress is thought to be a preparation for wound healing or to ward off potential infection as a result of stressors such as failed predation events or aggressive interactions (Martin, 2009). It should be noted, however, that in this same species, leukocyte and lymphocyte concentrations decrease with handling stress, suggesting that acute corticosterone can also stimulate the redistribution of immune cells and resources rather than just suppressing or enhancing immune function (Dhabhar et al., 1995; Braude et al., 1999; Dhabhar, 2002). Thus, cells involved in the stress response such as lymphocytes may be redirected from the blood to the lymph system, while other immune cells such as heterophils, phagocytes that play a role in bacterial killing, may be increased (Dhabhar et al., 1995). Indeed, several studies have measured the redistribution of immune cells, increases in heterophils and decreases in lymphocytes, to assess the stress response (reviewed in Davis et al., 2008). Chronically elevated corticosterone, on the other hand, can act as a mechanism by which immune function is suppressed, increasing disease susceptibility and possibly reducing survival (Svensson et al., 1998; Bourgeon and Raclot, 2006; Butler et al., 2009). However, these immunosuppressive effects of chronic corticosterone secretion can also be beneficial by preventing tissue damage associated with inflammatory immune responses and the impairment of other metabolic functions; it can also increase the ability of individuals to react to unpredictable environmental perturbations (Wingfield et al., 1998; Sapolsky et al., 2000; Martin, 2009).

BREEDING HORMONES: MELATONIN AND PROLACTIN

Recently, melatonin has also been reported to play a fundamental role in immunomodulation, by "rescuing" immune function from the immunosuppressive effects of reproductive activities (Nelson et al., 1995; Bentley, 2001; Moore and Siopes, 2003; Majewski et al., 2005). In birds, melatonin is released by the pineal gland and plays a critical role in the modulation of circadian activity, facilitating hatching, foraging, and migratory activities (reviewed in Bentley, 2001). With respect to the immune system in particular, studies have indicated that when individuals are photostimulated and the reproductive system has recrudesced, treatment with melatonin can enhance both cell-mediated and humoral immune function (Skwarlo-Sonta, 1996; Bentley, 2001; Moore and Siopes, 2003). Melatonin has also been implicated in the circadian regulation of immune function (Nelson and Demas, 2004; Singh and Haldar, 2007a; Siopes and Underwood, 2008) (see Chapters 9 and 10). In bush quail (*Perdicula asiatica*), for example, immune activity (total leukocyte and lymphocyte count) peaks in parallel with nighttime increases in melatonin and decreases in testosterone, presumably in anticipation of pathogenic encounters during the day (Singh and Haldar, 2007a). Similarly, cell-mediated immunity and humoral immunity were found to vary diurnally in Japanese quail (*Coturnix japonica*) but in contrasting ways. Cell-mediated responses (PHA) were higher during the day when melatonin was low, and humoral responses (to chukar red blood cells) were elevated at night when melatonin secretion was high (Siopes and Underwood, 2008).

Prolactin, which is associated with egg-laying, incubation, and parental care behaviors during breeding, also has immunomodulating properties (Skwarlo-Sonta, 1992). Many immune cells express prolactin receptors and some types of lymphocytes produce and secrete prolactin, suggesting that prolactin may act as an autocrine or paracrine modulator of immune activity (Matera, 1996). In superb starlings (*Lamprotornis superbus*), for example, prolactin has been shown to be positively correlated with bacteria killing ability (Rubenstein et al., 2008). Similarly, in ring doves (*Streptopelia risoria*) high concentrations of prolactin (exogenous or naturally high levels associated with incubation) increase phagocytic function (Rodriguez et al., 1999).

NON-BREEDING/WINTER

NON-BREEDING IMMUNE FUNCTION

In general, non-breeding/winter immune function in birds is vastly understudied, particularly with respect to free-living individuals. Of the few studies that have been conducted, some have suggested that immune function is generally more robust during non-breeding compared to breeding (Nelson and Demas, 1996; Bentley et al., 1998; Martin et al., 2008). Hypotheses to explain higher immunocompetence during the non-breeding season postulate that immune function is elevated because (1) this period is less demanding than the breeding season and individuals are freed from reproductive activities that can be both energetically and physiologically immunosuppressive (Lee, 2006; Martin et al., 2008), or (2) individuals need to prepare for harsh, and potentially novel, winter climates by upregulating the immune system to buffer against the immunosuppressive stressors of reduced food availability and unpredictable climate (Nelson and Demas, 1996). Furthermore, this enhancement or higher immunocompetence should be especially notable in long-lived species that invest more in survival (Lee, 2006). The few studies that have investigated non-breeding immune function in birds, whether in captive or free-living individuals, have indicated surprising and somewhat contradictory results. Temperate-zone house sparrows (*Passer domesticus*), for example, display low immune function (response to challenge with PHA) during early breeding and high immune function during late breeding in comparison to intermediate responses during non-breeding periods (Martin et al., 2004; Greenman et al., 2005). Similarly, PHA responses in ruffs (*Philomachus pugnax*) are higher during non-breeding compared to breeding, though there are no sex differences in responses during non-breeding (Lozano and Lank, 2003). Leukocyte counts are also higher in non-breeding than breeding red jungle fowl (*Gallus gallus*), but response to PHA challenge does not differ between red jungle fowl in the non-breeding and breeding condition (Zuk and Johnsen, 1998). Cell-mediated immune function (response to PHA) is also relatively stable between breeding and non-breeding tropical house sparrows (Martin et al., 2004) and Japanese quail (*Coturnix japonica*) (Moore and Siopes, 2000). Furthermore, in Japanese quail, humoral responses to chukar red blood cells also do not vary between seasons. On the other hand, non-breeding compared to breeding song sparrows (*Melospiza melodia*) actually display a reduction in sickness behavior in response to challenge with LPS (Owen-Ashley and Wingfield, 2006), and European starlings (*Sturnus vulgaris*) show reduced humoral immunity in response to the T-cell mitogen concanavalin A (Bentley et al., 1998). These results caution against making broad generalizations about the mechanisms mediating non-breeding immune function and reveal the need for more studies investigating seasonal differences in immune function, particularly in an environmental context and with respect to free-living individuals.

Recent studies in wintering birds have shown immune function to be sensitive to variation in temperature, as well as habitat and social interactions (Nelson and Demas, 1996; Shephard and Shek, 1998; Ruiz et al., 2002; Lindström et al., 2005a; Hawley, 2006). Increases in social competition in house finches (*Carpodacus mexicanus*), for example, decrease antibody response (Hawley et al., 2006). Dominant individuals, however, display weaker humoral responses to challenge with sheep red blood cells (SRBCs) but stronger cell-mediated responses to challenge with PHA and reduced symptom severity to infection with *Mycoplasma gallisepticum* (Hawley et al., 2007). Wintering dark-eyed juncos (*Junco h. hyemalis*) exhibit latitudinal variation in immune function (complement activity and levels of total IgG), with more robust responses in birds wintering in the milder climate of the southern as opposed to northern United States (O'Neal et al., unpublished data). This sensitivity of immune function to environmental factors is thought to be a result of trade-offs between thermoregulation and foraging activity/access to food and the associated costs of mounting an immune response, which include reductions in food intake, higher metabolic rate, and weight loss (Shephard and Shek, 1998; Svensson et al., 1998; Ots et al., 2001; Bonneaud et al., 2003; Martin et al., 2003; Sandland and Minchella, 2003; Segerstrom, 2007). For example, great tits (*Parus major*) challenged with SRBCs have higher metabolic rates and lose 3% of their body mass subsequent to immune challenge (Ots et al., 2001). Although much work has been conducted on environmental factors influencing non-breeding immune function, there is a limited amount of research investigating hormonal mechanisms outside of melatonin and corticosterone.

NON-BREEDING: MELATONIN AND THE IMMUNOENHANCEMENT HYPOTHESIS

Many of the studies investigating hormonal mediation of non-breeding immune function have focused on photoperiodic mediation of immune function by comparing the impact of day length on immune development and responses (Bentley et al., 1998; Moore and Siopes, 2000; Owen-Ashley et al., 2006). Photoperiodic effects on immune function are thought to be mediated via melatonin secretion. The increase in melatonin secretion associated with decreasing day length in autumn is predicted to enhance immune function in preparation for harsh wintering conditions (Nelson and Demas, 1996). This prediction has been best studied in mammals and is known as the immunoenhancement hypothesis, which states that day length-induced variation in melatonin prepares temperate-zone animals for immunologically challenging conditions by increasing lymphatic tissue development and thus immune function directly (Nelson and Demas, 1996; Sinclair and Lochmiller, 2000). In birds, however, the link between melatonin and annual variation in day length is much weaker, and the role of melatonin in mediating many seasonal processes, particularly outside of the laboratory, is still unclear (Bentley, 2001; Buehler et al., 2009). Nonetheless, several studies have demonstrated that melatonin can enhance avian immune function during winter. Wintering bush quail (*Perdicula asiatica*), for example, display increased levels of circulating melatonin compared to breeding individuals, and this increase in melatonin is associated with enhanced immune function (lymphoid organ weight, total leukocyte count, lymphocyte count and percentage stimulation ratio) (Singh and Haldar, 2007b). Meanwhile, exogenous melatonin in Japanese quail held under short day lengths enhances both cellular (response to PHA) and humoral (response to chukar red blood cells) immune function (Moore and Siopes, 2002). Similarly, photorefractory European starlings implanted with melatonin display greater splenocyte proliferation in response to concanavalin A than controls (Bentley et al., 1998). Conversely, a recent study in red knots (*Calidris canutus*), for which winter is not the most demanding time of year, demonstrated that non-breeding individuals do not display correlated immunoenhancement despite increases in melatonin during winter/non-breeding (Buehler et al., 2009). The authors posit that these results suggest that stressors such as climate and food availability may counteract immunoenhancement in natural conditions (Buehler et al., 2009).

NON-BREEDING: CORTICOSTERONE

Corticosterone, usually measured in the context of the stress response, is often considered as a mediator of environmental stressors because of its sensitivity to variations in food and refuge availability, social interactions, and climate (Siegel, 1980; Marra and Holberton, 1998; Wingfield and Hunt, 2002; Wingfield and Kitaysky, 2002; Boonstra, 2004; Lindström et al., 2005a). Generally, the sensitivity of corticosterone to these potential stressors is thought to allow for rapid physiological adjustment (i.e., immune function) in response to unpredictable food availability and more extreme climatic and environmental conditions (Rogers et al., 1993; Holberton and Able, 2000). Several studies have indicated declines in non-breeding immune function in conjunction with increases in corticosterone secretion. In free-living house finches, for example, corticosterone increases over winter, peaking during the harshest months along with peak prevalence in *Mycoplasma gallisepticum* (Lindström et al., 2005b). Similarly, free-living wintering song sparrows (*Melospiza melodia*) display increased sickness behavior in response to LPS compared to breeding individuals, a difference that appears to be mediated by lowered baseline corticosterone levels in non-breeding birds (Owen-Ashley and Wingfield, 2006). Temperate-zone house sparrows implanted with corticosterone also display suppressed cell-mediated immunity (PHA), but, interestingly, tropical house sparrows do not, suggesting that tropical birds may be insensitive to corticosterone (Martin et al., 2005). In a different approach to the role of corticosterone in non-breeding immune function, experimental studies manipulating testosterone in non-breeding European starlings demonstrated declines in cell-mediated (response to PHA) and humoral (response to key-hole limpet hemocyanin [KLH]) immune function with concurrent increases in plasma corticosterone levels, suggesting that corticosterone may be mediating the immunosuppressive effects of testosterone in non-breeding individuals (Duffy et al., 2000). It should be noted, however, that despite reduced baseline corticosterone in free-living dark-eyed juncos wintering in suburban compared to rural habitats, no differences in immune response (complement activity) were observed between habitats (O'Neal and French, unpublished data).

MIGRATION

IMMUNE FUNCTION DURING MIGRATION

Migration is the most energetically demanding activity in a bird's life history (Wikelski et al., 2003b), with the potential for dramatic increases in basal metabolic rate (Mckechnie, 2008), declines in body mass and breast muscle (Lindström et al., 2000), and reduction in the mass of several organs (Battley et al., 2000). Consequently, we would expect trade-offs between migration and a multitude of other life-history stages, particularly reproduction because it is bookended by migratory activity and is also energetically costly. Indeed, the relationship between migration and immune function is quite complex. Migratory activity can be immunosuppressive; for example, migrating thrushes at their stopover sites display depressed immune function on arrival (Owen and Moore, 2008), and Bewick's swans (*Cygnus bewickii*) infected with avian influenza A are poor migrators, delaying departure from wintering grounds, traveling shorter distances, and foraging at reduced rates (Van Gils et al., 2007).

These data raise interesting questions about winter immunoenhancement and how it pertains to migratory species. In migrating birds, winter immunoenhancement theoretically would coincide with preparation for autumn migration, which includes hyperactivity, molt, an increase in fat stores, and spleen development (Fange and Silverin, 1985; John, 1994; Silverin et al., 1999), which are all energetically costly in their own right. Thus, one would expect tradeoffs to occur between enhancement of the immune system and migratory preparation, perhaps in favor of migratory preparation, because these two events are competing for the same resources and individuals would be unable to migrate or would be delayed in migration without this physiological preparation. Enhancement of the immune system, however, would prepare migratory species for the energetically costly migration journey during which upregulating the immune system is unlikely, as well as introduction to vectors and diseases at stopover sites and end destinations. Also, postponing immunoenhancement until arrival on winter grounds, during the winter months, can be costly if food is not readily available for lymphatic and fat store development (Lochmiller and Deerenberg, 2000; Nelson and Demas, 2004). However, as several studies have suggested, birds in poor physiological condition, which may be immunosuppressed, may not take on the stress of migratory activity (Hasselquist et al., 2007; Weber and Stilianakis, 2007). Red knots (*Calidris canutus*), for example, exercised in a wind tunnel for the equivalent of a natural migratory journey, display no differences in cell-mediated or humoral immune function compared to non-migrating controls (Hasselquist et al., 2007). Furthermore, birds that have weaker responses to challenge with diphtheria–tetanus refuse to migrate at all, suggesting perhaps that only birds in superior condition, those that have cleared infections prior to migration and perhaps by extension prepared their immune systems for migration, actually migrate.

MIGRATORY HORMONES

A variety of hormones have been proposed to be involved in the regulation of migratory behavior (*Zugunruhe*) and physiology (hyperphagia and fattening), though this is an unresolved area of understanding (reviewed in Wingfield et al., 1990). Interestingly, the endocrine mechanisms underlying spring migration are thought to be different from those underlying autumn migration, such that spring migration is influenced by gonadal steroids whereas autumn migration is not (Wingfield et al., 1990). Though the exact hormonal mechanisms may differ between these migratory periods, in general, increases in hormones such as corticosterone, testosterone, prolactin, growth hormone, and thyroid hormones stimulate hyperphagia, fattening, and *Zugunruhe* and may even play a role in migratory direction (Wingfield et al., 1990; Hintz, 2000; Piersma et al., 2000; Landys-Ciannelli et al., 2002). The relationship among these hormones and immune function, however, is not known at this time. Considering that many of these hormones suppress immune function during other life-history stages and that migration itself is energetically and nutritionally costly, it seems likely that increases in these hormones may play a role in mediating trade-offs between migration and immune function at this time.

SENESCENCE

IMMUNOSENESCENCE

According to the terminal investment hypothesis, as an individual senesces it should redirect energy from self-maintenance to current reproductive effort/success (Trivers, 1972). Indeed, experimental manipulations of short-term physiological costs via immune challenge increase individuals' perceived health risk and increase reproductive effort, particularly in aged individuals (Bonneaud et al., 2004; Hanssen, 2006; Velando et al., 2006; Marzal et al., 2008). For example, treatment with LPS before laying in male blue-footed boobies (*Sula nebouxii*) results in older males more than doubling their reproductive effort compared to younger males (Velando et al., 2006).

Similarly, life-history theory predicts that as an individual ages, it should experience a decline in immune and hormone responses that may interfere with reproduction as the chances for future reproductive bouts become more uncertain (Lavoie, 2005; Ottinger and Lavoie, 2007; Vleck et al., 2007). Immunosenescence has been demonstrated in a multitude of species, including several birds (Cichon et al., 2003; Lozano and Lank, 2003; Haussmann et al., 2005; Lavoie, 2005;

Martin et al., 2006b; Lavoie et al., 2007). Five- to six-year-old female collared flycatchers (*Ficedula albicollis*), for example, display lower antibody titers in response to *in vitro* challenge with sheep red blood cells (hemagglutination assay) compared to young (1-year-old) and middle-aged (3-year-old) females (Cichon et al., 2003). Similar age-related declines in humoral immune response were found with respect to *in vivo* Newcastle disease virus (NDV) challenge in barn swallows (*Hirundo rustica*) (Saino et al., 2003). Likewise, studies taking a comparative approach to cell-mediated responses to challenge with PHA demonstrated declines in immune response with age in several avian species (Tella et al., 2002; Haussmann et al., 2005).

Interestingly, studies involving more than one immune measure have demonstrated trade-offs between immune components as individuals age, such that older individuals have stronger humoral immune responses and weaker cell-mediated response compared to younger individuals (Lavoie et al., 2007; Palacios et al., 2007). In Japanese quail (*Coturnix japonica*), for instance, aged birds (28 months) display reduced cell-mediated response to PHA compared to juvenile (6 weeks) and adult (10 months) birds, but are not significantly different from younger adults with respect to humoral responses to chukar red blood cells or avian (H9N2) influenza virus (Lavoie et al., 2007). Similarly, cell-mediated response to *in vitro* stimulation with PHA and ConA (concanavalin A, a T-cell mitogen) declined with age in nesting tree swallows (*Tachycineta bicolor*), but there were no effects of age on humoral responses to *in vitro* challenge with LPS or on measures of natural antibodies (Palacios et al., 2007). Differences in age-related depression of immune components could be a result of the cost associated with using these components. Cell-mediated response can have high energetic and nutritive costs due to the secretion of pro-inflammatory cytokines (reviewed in Lee, 2006); as a result, individuals may be favored if they implement the less costly humoral response when faced with infection. To date, however, hypotheses put forth to explain differences between immune components in the rate of senescence have centered on species differences in pace of life.

Within pace-of-life theory, immunosenescence has received consideration under the hypothesis that species with longer and slower life spans, as well as a reduced extrinsic mortality, should invest more in immune function over a longer period of their life span, and thus rates of immunosenescence should be slower in these species (Tella et al., 2002; Haussmann et al., 2005; Lee, 2006; Palacios et al., 2007). Certainly, there is some support for this hypothesis, with rates of cell-mediated immunosenescence shown to be higher in "fast-paced" zebra finches (*Taeniopygia guttata*) compared to tree swallows (*Tachycineta bicolor*; "medium-paced") and Leach's storm petrels (*Oceanodroma leucorhoa*; "slow-paced") (Tella et al., 2002). Furthermore, this hypothesis may explain results with respect to humoral measures and age-related declines in antibody titers in short-lived barn swallows (*Hirundo rustica*) (Saino et al., 2003) and collared flycatchers (*Ficedula albicollis*) (Cichon et al., 2003), but not in long-lived tree swallows (Palacios et al., 2007). Much work, however, is needed to find further support for hypotheses surrounding immunosenescence and theories on pace of life (reviewed in Lee, 2006).

IMMUNOSENESCENCE AND HORMONES

Despite several studies suggesting that hormonal signals, like immune function, may experience degrees of senescence owing to declines in signal strength or number of signal receptors/cells

(Nisbet et al., 1999; Tarlow et al., 2003; Angelier et al., 2007; Ottinger, 2007; Ottinger and Lavoie, 2007), few studies have investigated the relationships among immune function, hormone signals, and senescence in birds. Those that have investigated these relationships have discovered links between the immunosuppressive effects of testosterone and age in male red-legged partridges (*Alectoris rufa*) indicating that the immunosuppressive effects of testosterone on cell-mediated responses to PHA may be a factor in aged males only where the combined cost of immunosuppression by high testosterone levels and overall immunosenescence is elevated (Alonso-Alvarez et al., 2009). Corticosterone, however, seems to have no relationship with cell-mediated immunosenescence in tree swallows (Palacios et al., 2007), whereas treatment with melatonin and its amino acid precursor tryptophan can rescue age-related immunodepression as measured by immunoglobulin titers, bactericidal activity, and hemolytic activity (Terron et al., 2009). For the most part, though, senescence of various physiological functions and signals has been studied in isolation, leaving huge potential for research regarding aging in birds.

CONCLUSIONS

ECOIMMUNOLOGY AND LIFE-HISTORY THEORY

In the context of life-history evolution, ecoimmunology attempts to explain individual and species variation in immune function as it relates to associations among demographic traits and variation in life-history stages in an ecological context (Martin et al., 2006b; French et al., 2009). In other words, considering the costs of immune responses, how and when does immune function trade off with other physiological, morphological, and behavioral traits to maximize reproductive success and survival? In this review, we considered these trade-offs with respect to endocrine mechanisms across life-history stages. Understanding the relationship between hormones and immune function will obviously require more research, but we can learn much from current data about the evolution of endocrine and immune systems and how they are able to orchestrate suites of traits. The collected research demonstrates that both endocrine and immune responses may vary in their strength and timing of implementation. Furthermore, the sensitivity of immune tissues and cells to hormonal signals can also vary, becoming more or less sensitive over life-history stages and the annual cycle. Thus, immune function can directly affect self-maintenance and reproductive success or act indirectly via hormonal mediation. There is clearly a need for greater understanding of how the immune and endocrine systems are integrated. Phenotypic integration refers to the relationships among behavioral, physiological, and morphological traits that are tightly connected, which can constrain or enable rapid evolution in response to environmental change. Phenotypic independence refers to connections that are readily uncoupled during evolution (Pigliucci, 2003; Ketterson et al., 2009). In the context of the above research, hormones tend to play an important role in coordinating immune responses, suggesting a high level of integration (Table 2.2). It should be noted, however, that these physiological correlations are highly dependent on environmental influences. Increases in access to resources, for example, have been shown to rescue hormone-mediated immune suppression (Poiani et al., 2000; Alonso-Alvarez et al., 2007; Hawley et al., 2007; Baeta et al., 2008).

TABLE 2.2: Well studied hormone–immune interactions across life-history stages

Corticosterone			
Life-history stage	**Species**	**Effect on immune system**	**Reference**
Development			
	Yellow-legged gulls (*Larus michahellis*); European starlings (*Sturnus vulgaris*)	EEC suppressed cell-mediated immune function	(Love et al., 2005; Rubolini et al., 2005a)
	American kestrel	EEC elevated cell-mediated immune function	(Butler et al., 2009)
Breeding			
	House sparrows (*Passer domesticus*)	EEC suppressed cell-mediated immune response in temperate sparrows	(Martin et al., 2005)
	Multiple species	Handling stress suppressed bacteria killing ability	(Matson et al., 2006b)
	Red knots (*Calidris canutus*)	Acute corticosterone response to handling stress enhanced bacteria killing	(Buehler et al., 2008)
	Common eiders (*Somateria mollissima*)	EEC suppresses total immunoglobulin levels	(Bourgeon and Raclot, 2006)
Non-breeding/ winter			
	House finch (*Carpodacus mexicanus*)	Stress-induced corticosterone correlated with *Mycoplasma gallisepticum* symptoms	(Lindström et al., 2005b)
	House sparrows (*Passer domesticus*)	EEC suppressed cell-mediated response in temperate birds but no effect on tropical	(Martin et al., 2005)
	Dark-eyed junco (*Junco hyemalis hyemalis*)	No relationship between baseline or stress-induced corticosterone and complement activity	(O'Neal and French, unpublished data)
Senescence			
	Tree swallow (*Tachycineta bicolor*)	No relationship between corticosterone and cell-mediated immunosenescence	(Palacios et al., 2007)

TABLE 2.2: (Continued)

Melatonin			
Life-history stage	Species	Effect on immune system	Reference
Development			
	Chickens (*Gallus gallus domesticus*)	Exogenous melatonin rescues immune function in pinealectomized individuals	(Skwarlo-Sonta, 1996)
	Chickens (*Gallus gallus domesticus*)	Exogenous melatonin inhibits lymphocyte proliferation in response to PHA	(Markowska et al., 2001)
Breeding			
	Japanese quail (*Coturnix japonica*), European starlings (*Sturnus vulgaris*)	Exogenous melatonin enhances cell-mediated and humoral immune function	(Bentley, 2001; Moore and Siopes, 2003)
	Japanese quail (*Coturnix japonica*), Bush quail (*Perdicula asiatica*)	Diurnal elevations in nighttime melatonin associated with enhanced leukocyte/lymphocyte counts and cell-mediated immune function	(Singh and Haldar, 2007a; Siopes and Underwood, 2008)
Non-breeding/ winter			
	Bush quail (*Perdicula asiatica*)	Elevated circulating levels of melatonin associated with enhanced immune function	(Singh and Haldar, 2007b)
	European starlings (*Sturnus vulgaris*), Japanese quail (*Coturnix japonica*)	Exogenous melatonin enhances cell-mediated and humoral immune function	(Bentley et al., 1998; Moore and Siopes, 2002)
	Red knots (*Calidris canutus*)	No associations with increased melatonin and immunoenhancement	(Buehler et al., 2009)
Senescence			
	Ring Doves (*Streptopelia risoria*)	EEM rescues age-related immunodepression	(Terron et al., 2009)

Testosterone			
Life-history stage	Species	Effect on immune system	Reference

TABLE 2.2: (Continued)

Development			
	Great tit (*Parus major*)	No effect of EET on ectoparasite infection or cell-mediated immune response	(Tschirren et al., 2005)
			(Tobler 2009)
	Jackdaw (*Corvus monedula*); black-headed gull (*Chroicocephalus ridibundus*)	EE-androgens (T or androstenedione) suppressed humoral response (LPS & diphtheria–tetanus) and cell-mediated response	(Muller et al., 2005; Sandell et al., 2009)
	Collared flycatcher (*Ficedula albicollis*)	EET and androstenedione had no effect on cell-mediated immune response	(Pitala et al., 2009)
	Black-headed gull (*Chroicocephalus ridibundus*)	Decreases in immunoglobulin deposition but increases in testosterone deposition with laying order	(Groothuis et al., 2006)
	Eastern bluebird (*Sialia sialis*)	EET suppressed cell-mediated (PHA) immune response	(Navara et al., 2005)
Breeding			
	Dark-eyed junco (*Junco hyemalis*); European starling (*Sturnus vulgaris*)	EET suppressed humoral (KLH), cell-mediated (PHA), and innate (total IgG and complement) immune function	(Duffy et al., 2000; Casto et al., 2001; Greives et al., 2006; Zysling et al., 2006)
	White-crowned sparrows (*Zonotrichia leucophrys gambelii*)	EET suppressed sickness behavior (response to LPS)	(Ashley et al., 2009)
	Japanese quail (*Coturnix coturnix*)	High-T males (intact) had lower cell-mediated response compared to low-T (castrated) males	(Boughton et al., 2007)
	Zebra finch (*Taeniopygia guttata*)	EET in individuals selected for high stress response displayed enhanced secondary antibody response to diphtheria–tetanus vaccine	(Roberts et al., 2007)
	Zebra finch (*Taeniopygia guttata*)	Carotenoids buffer the immunosuppressive effects (cell-mediated) of EET	(Mcgraw and Ardia, 2007)
	Zebra finches (*Taeniopygia guttata*)	Leptin buffers the immunosuppressive effects (cell-mediated, PHA) of EET	(Alonso-Alvarez et al., 2007)
	Red-winged blackbird (*Agelaius phoeniceus*)	EET had no effect on humoral immune response (KLH)	(Hasselquist et al., 1999)

TABLE 2.2: (Continued)

Testosterone			
Life-history stage	Species	Effect on immune system	Reference
Molt/migration			
	Blue tits (*Cyanistes caeruleus*)	EET enhanced humoral and cell-mediated (PHA) immune function (in males in good condition)	(Roberts and Peters, 2009)
Non-breeding/winter			
	European starlings (*Sturnus vulgaris*)	EET via corticosterone suppresses cell-mediated and humoral immune function	(Duffy et al., 2000)
Senescence			
	Red-legged partridge (*Alectoris rufa*)	EET suppressed cell-mediated immune function in old males but not middle-aged males	(Alonso-Alvarez et al., 2009)

EEC = experimentally elevated corticosterone; EET = experimentally elevated testosterone

Given the considerable plasticity in both endocrine and immune biology, we might anticipate rapid responses to changes in selection. On the other hand, given the powerful predictive nature of pace of life, we may find that endocrine–immune relationships have similar deep connections that are resistant to uncoupling.

CORTICOSTERONE: THE IMMUNOMODULATORY HORMONE?

Equally important is the consideration that corticosterone may be a universal hormone in the mediation of immune responses across life-history stages. Research has indicated that immunosuppression associated with increases in testosterone and decreases in melatonin, the two most studied hormones outside of corticosterone itself, are associated with increases in corticosterone and potential immunosuppression (Duffy et al., 2000; Casto et al., 2001; Rodríguez et al., 2001; Guerrero and Reiter, 2002; Zysling et al., 2006; Ashley et al., 2009). In other wordss, in many cases in which testosterone is elevated or when melatonin is in decline, such as during breeding, corticosterone is elevated and immune function tends to be depressed (but see Tschirren et al., 2005; Roberts et al., 2009, for notable exceptions). Because corticosterone is sensitive to many more types of environmental variation (i.e., climate, habitat, resources) in addition to those that tend to trigger variation in melatonin (i.e., photoperiod) and testosterone (i.e., social interactions), this hormone could indeed be *the* immunomodulatory hormone.

FUTURE RESEARCH

The study of avian immunology in an ecological and evolutionary context is a relatively new field, and much research still needs to be accomplished. In particular, studies examining developing avian immune function should investigate how non-maternally transferred hormones such as melatonin, prolactin, thyroid, and growth hormone work in concert with maternally derived hormones, which often reflect current environmental conditions, to influence the development of the immune system. Additionally, current research with respect to hormones and immune development has focused mainly on the relationship between testosterone and corticosterone, ignoring the potential effects of androstenedione, estradiol, and progesterone as well as other maternally derived factors such as diet-derived antioxidants, which can counteract the immunosuppressive effects of hormones (Royle et al., 2001). Progesterone, estradiol, and to some extent prolactin (but see Skwarlo-Sonta, 1992; Rodriguez et al., 1999; Rubenstein et al., 2008) have also been overlooked with respect to their potential immunomodulatory effects in breeding individuals. Research on testosterone during breeding, however, is far from complete, as studies testing the immunocompetence handicap hypothesis have produced mixed results. More needs to be known about the mechanisms by which testosterone might interact with immune function, such as the role immune cell androgen receptors may play in mediating cell expression and proliferation and much remains to be learned about the role of carotenoids, food resources, and other factors that may modulate hormonal effects on breeding immune function (Evans et al., 2000; Poiani et al., 2000; Alonso-Alvarez et al., 2007). Of the completed research on non-breeding and senescence, questions remain regarding melatonin-mediated immunoenhancement during non-breeding (see Buehler et al., 2009) and senescence (Terron et al., 2009), and the role corticosterone (Palacios et al., 2007) and testosterone (Alonso-Alvarez et al., 2009) may play in immunosuppression. For the most part, however, whether due to the downregulation of hormones during non-breeding and migration or the complex nature of measuring age effects on physiological measures heavily influenced by the environment, hormone–immune interactions during non-breeding, migration, and senescence remain perhaps the largest area requiring future research.

There are also several areas of more general research that remain to be explored. Across life-history stages and in the context of geographic variation, pace of life, and sex, there has been little research looking at the interactions between hormones and immune function, despite evidence that both hormones and immune function vary geographically (Martin et al., 2004; Ardia, 2007; Hasselquist, 2007; Owen-Ashely et al., 2008), with life span (Martin et al., 2006; Lee et al., 2008; Knowles et al., 2009), and with sex (Zuk and McKean, 1996; Hasselquist, 2007; Hau, 2007; Nunn et al., 2009). Moreover, much of current research has focused on individuals held in captivity and has been beneficial to understanding how hormones and immune function interact at the basic level, but captivity poses its own constraints with respect to stress and unnatural resource conditions, which can have their own effects on endocrine and immune responses. Perhaps more important than ecological context, however, is the need for more research using comprehensive measures of both endocrine and immune systems, as well as consideration of seasonality and how it links physiology across the annual cycle, particularly when considering hormone–immune interactions across life-history stages.

REFERENCES

Addison, B, Klasing, KC, Robinson, WD, Austin, SH and Ricklefs, RE (2009). Ecological and life-history factors influencing the evolution of maternal antibody allocation: A phylogenetic comparison. *Proceedings of the Royal Society B-Biological Sciences* 276, 3979–3987.

Alonso-Alvarez, C, Bertrand, S and Sorci, G (2007). Energetic reserves, leptin and testosterone: A refinement of the immunocompetence handicap hypothesis. *Biology Letters* 3, 271–274.

Alonso-Alvarez, C, Perez-Rodriguez, L, Garcia, JT and Vinuela, J (2009). Testosterone-mediated trade-offs in the old age: A new approach to the immunocompetence handicap and carotenoid-based sexual signalling. *Proceedings of the Royal Society B-Biological Sciences* 276, 2093–2101.

Altizer, S, Dobson, A, Hosseini, P, Hudson, P, Pascual, M and Rohani, P (2006). Seasonality and the dynamics of infectious diseases. *Ecology Letters* 9, 467–484.

Angelier, F, ric, Weimerskirch, H, Dano, S and Chastel, O (2007). Age, experience and reproductive performance in a long-lived bird: A hormonal perspective. *Behavioral Ecology and Sociobiology* 61, 611–621.

Ardia, DR (2005a). Individual quality mediates trade-offs between reproductive effort and immune function in tree swallows. *Journal of Animal Ecology* 74, 517–524.

Ardia, DR (2005b). Tree swallows trade off immune function and reproductive effort differently across their range. *Ecology* 86, 2040–2046.

Ardia, DR (2007). The ability to mount multiple immune responses simultaneously varies across the range of the tree swallow. *Ecography* 30, 23–30.

Ashley, NT, Hays, QR, Bentley, GE and Wingfield, JC (2009). Testosterone treatment diminishes sickness behavior in male songbirds. *Hormones and Behavior* 56, 169–176.

Baeta, R, Faivre, B, Motreuil, S, Gaillard, M and Moreau, J (2008). Carotenoid trade-off between parasitic resistance and sexual display: An experimental study in the blackbird (*Turdus merula*). *Proceedings of the Royal Society B-Biological Sciences* 275, 427–434.

Ball, GF and Balthazart, J (2004). Hormonal regulation of brain circuits mediating male sexual behavior in birds. *Physiology & Behavior* 83, 329–346.

Battley, PF, Piersma, T, Dietz, MW, Tang, SX, Dekinga, A and Hulsman, K (2000). Empirical evidence for differential organ reductions during trans-oceanic bird flight. *Proceedings of the Royal Society of London Series B-Biological Sciences* 267, 191–195.

Bell, G (1980). The costs of reproduction and their consequences. *The American Naturalist* 116, 45–76.

Bentley, GE (2001). Unraveling the enigma: The role of melatonin in seasonal processes in birds. *Microscopy Research and Technique* 53, 63–71.

Bentley, GE, Demas, GE and Nelson, RJ (1998). Melatonin, immunity and cost of reproductive state in male European starlings. *Proceedings of the Royal Society of London. Series B: Biological Sciences* 265, 1191–1195.

Biard, C, Hardy, C, Motreuil, S and Moreau, J (2009). Dynamics of PHA-induced immune response and plasma carotenoids in birds: Should we have a closer look? *Journal of Experimental Biology* 212, 1336–1343.

Bókony, V, Lendvai, ÁZ, Liker, A, Angelier, F, Wingfield, JC and Olivier, C (2009). Stress response and the value of reproduction: Are birds prudent parents? *The American Naturalist* 173, 589–598.

Bonier, F, Martin, PR, Moore, IT and Wingfield, JC (2009). Do baseline glucocorticoids predict fitness? *Trends in Ecology & Evolution* 24, 634–642.

Bonneaud, C, Mazuc, J, Chastel, O, Westerdahl, H and Sorci, G (2004). Terminal investment induced by immune challenge and fitness traits associated with major histocompatibility complex in the house sparrow. *Evolution* 58, 2823–2830.

Bonneaud, C, Mazuc, J, Gonzalez, G, Haussy, C, Chastel, O, Faivre, B and Sorci, G (2003a). Assessing the cost of mounting an immune response. *American Naturalist* 161, 367.

Boonstra, R (2004). Coping with changing northern environments: The role of the stress axis in birds and mammals. *Integrative and Comparative Biology* 44, 95.

Boughton, RK, Bridge, ES and Schoech, SJ (2007). Energetic trade-offs between immunity and reproduction in male Japanese quail (*Coturnix coturnix*). *Journal of Experimental Zoology Part A-Ecological Genetics and Physiology* 307A, 479–487.

Boulinier, T and Staszewski, V (2008). Maternal transfer of antibodies: Raising immuno-ecology issues. *Trends in Ecology & Evolution* 23, 282–288.

Bourgeon, S, Criscuolo, Fo, Le Maho, Y and Raclot, T (2006). Phytohemagglutinin response and immunoglobulin index decrease during incubation fasting in female common eiders. *Physiological and Biochemical Zoology* 79, 793–800.

Bourgeon, S and Raclot, T (2006). Corticosterone selectively decreases humoral immunity in female eiders during incubation. *Journal of Experimental Biology* 209, 4957–4965.

Braude, S, Tang-Martinez, Z and Taylor, GT (1999). Stress, testosterone, and the immunoredistribution hypothesis. *Behavioral Ecology* 10, 345–350.

Buechler, K, Fitze, PS, Gottstein, B, Jacot, A and Richner, H (2002). Parasite-induced maternal response in a natural bird population. *Journal of Animal Ecology* 71, 247–252.

Buehler, Deborah M, Bhola, N, Barjaktarov, D, Goymann, W, Schwabl, I, Tieleman, BI and Piersma, T (2008). Constitutive immune function responds more slowly to handling stress than corticosterone in a shorebird. *Physiological and Biochemical Zoology* 81, 673–681.

Buehler, DM, Koolhaas, A, Van't Hof, TJ, Schwabl, I, Dekinga, A, Piersma, T and Tieleman, BI (2009). No evidence for melatonin-linked immunoenhancement over the annual cycle of an avian species. *Journal of Comparative Physiology A-Neuroethology Sensory Neural and Behavioral Physiology* 195, 445–451.

Butler, MW, Leppert, LL and Dufty, AM (2009). Effects of small increases in corticosterone levels on morphology, immune function, and feather development. *Physiological and Biochemical Zoology* 83, 78–86.

Campbell, T and Ellis, C (2007). *Avian and Exotic Animal Hematology and Cytology*. Wiley, Ames.

Casto, JM, Nolan, V and Ketterson, ED (2001). Steroid hormones and immune function: Experimental studies in wild and captive dark-eyed juncos (*Junco hyemalis*). *American Naturalist* 157, 408–420.

Cichon, M, Sendecka, J and Gustafsson, L (2003). Age-related decline in humoral immune function in collared flycatchers. *Journal of Evolutionary Biology* 16, 1205–1210.

Clotfelter, ED, O'Neal, DM, Gaudioso, JM, Casto, JM, Parker-Renga, IM, Snajdr, EA, Duffy, DL, Nolan, V and Ketterson, ED (2004). Consequences of elevating plasma testosterone in females of a socially monogamous songbird: Evidence of constraints on male evolution? *Hormones and Behavior* 46, 171–178.

Clutton-Brock, TH and Godfray, C (1991). Parental investment. In Krebs, JR, and Davies, NB, eds. *Behavioral Ecology: An Evolutionary Approach*, pp. 234–262. Blackwell Publishing, Boston.

Comendant, T, Sinervo, B, Svensson, EI and Wingfield, J (2003). Social competition, corticosterone and survival in female lizard morphs. *Journal of Evolutionary Biology* 16, 948–955.

Davis, AK, Maney, DL and Maerz, JC (2008). The use of leukocyte profiles to measure stress in vertebrates: A review for ecologists. *Functional Ecology* 22, 760–772.

Dawson, RD and Bortolotti, GR (2006). Carotenoid-dependent coloration of male American kestrels predicts ability to reduce parasitic infections. *Naturwissenschaften* 93, 597–602.

Deerenberg, C, Arpanius, V, Daan, S and Bos, N (1997). Reproductive effort decreases antibody responsiveness. *Proceedings of the Royal Society of London. Series B: Biological Sciences* 264, 1021–1029.

Dhabhar, F, Miller, A, McEwen, B and Spencer, R (1995). Effects of stress on immune cell distribution. Dynamics and hormonal mechanisms. *Journal of Immunology* 154, 5511–5527.

Dhabhar, FS (2002). Stress-induced augmentation of immune function—the role of stress hormones, leukocyte trafficking, and cytokines. *Brain, Behavior, and Immunity* 16, 785–798.

DiCarlo, R, Meli, R, Florio, S, Raso, GM, Gualillo, O and Pagnini, G (1996). The effect of age and sex on the expression of prolactin binding activity in the chicken bursa of Fabricius. *Life Sciences* 59, 1803–1808.

Duffy, DL, Bentley, GE, Drazen, DL and Ball, GF (2000). Effects of testosterone on cell-mediated and humoral immunity in non-breeding adult European starlings. *Behavioral Ecology* 11, 654–662.

Evans, MR, Goldsmith, AR and Norris, SRA (2000). The effects of testosterone on antibody production and plumage coloration in male house sparrows (*Passer domesticus*). *Behavioral Ecology and Sociobiology* 47, 156–163.

Fair, JM and Ricklefs, RE (2002). Physiological, growth, and immune responses of Japanese quail chicks to the multiple stressors of immunological challenge and lead shot. *Archives of Environmental Contamination and Toxicology* 42, 77–87.

Fange, R and Silverin, B (1985). Variation of lymphoid activity in the spleen of a migratory bird, the pied flycatcher (*Ficedula-hypoleuca* – aves, passeriformes). *Journal of Morphology* 184, 33.

Finch, CE and Rose, MR (1995). Hormones and the physiological architecture of life-history evolution. *Quarterly Review of Biology* 70, 1–52.

Fitzgerald, SD, Fitzgerald, AL, Reed, WM and Burnstein, T (1992). Immune function in pheasants experimentally infected with marble spleen disease virus. *Avian Diseases* 36, 410–414.

Folstad, I and Karter, AJ (1992). Parasites, bright males, and the immunocompetence handicap. *American Naturalist* 139, 603–622.

French, SS, Moore, MC and Demas, GE (2009). Ecological immunology: The organism in context. *Integrative and Comparative Biology* 49, 246–253.

Gallizzi, K, Alloitteau, O, Harrang, E and Richner, H (2008). Fleas, parental care, and transgenerational effects on tick load in the great tit. *Behavioral Ecology* 19, 1225–1234.

Garamszegi, LZ, Eens, M, Hurtrez-Bousses, S and Moller, AP (2005). Testosterone, testes size, and mating success in birds: A comparative study. *Hormones and Behavior* 47, 389–409.

Gasparini, J, McCoy, KD, Haussy, C, Tveraa, T and Boulinier, T (2001). Induced maternal response to the Lyme disease *spirochaete Borrelia burgdorferi sensu lato* in a colonial seabird, the *kittiwake rissa tridactyla*. *Proceedings of the Royal Society of London Series B-Biological Sciences* 268, 647–650.

Gasparini, J, Piault, R, Bize, P and Roulin, A (2009). Pre-hatching maternal effects inhibit nestling humoral immune response in the tawny owl *Strix aluco*. *Journal of Avian Biology* 40, 271–278.

Gil, D (2008). Hormones in avian eggs: Physiology, ecology and behavior. *Advances in the Study of Behavior* 38, 337–398.

Goymann, W, Landys, MM and Wingfield, JC (2007). Distinguishing seasonal androgen responses from male-male androgen responsiveness: revisiting the challenge hypothesis. *Hormones and Behavior* 51, 463–476.

Greenman, CG, Martin, LB and Hau, M (2005). Reproductive state, but not testosterone, reduces immune function in male house sparrows (*Passer domesticus*). *Physiological and Biochemical Zoology* 78, 60–68.

Greives, TJ, McGlothlin, JW, Jawor, JM, Demas, GE and Ketterson, ED (2006). Testosterone and innate immune function inversely covary in a wild population of breeding dark-eyed juncos (*Junco hyemalis*). *Functional Ecology* 20, 812–818.

Grindstaff, JL (2008). Maternal antibodies reduce costs of an immune response during development. *Journal of Experimental Biology* 211, 654–660.

Grindstaff, JL, Brodie, ED and Ketterson, ED (2003). Immune function across generations: Integrating mechanism and evolutionary process in maternal antibody transmission. *Proceedings of the Royal Society of London Series B-Biological Sciences* 270, 2309–2319.

Grindstaff, JL, Hasselquist, D, Nilsson, JA, Sandell, M, Smith, HG and Stjernman, M (2006). Transgenerational priming of immunity: Maternal exposure to a bacterial antigen enhances offspring humoral immunity. *Proceedings of the Royal Society B-Biological Sciences* 273, 2551–2557.

Groothuis, TGG, Eising, CM, Blount, JD, Surai, P, Apanius, V, Dijkstra, C and Muller, W (2006). Multiple pathways of maternal effects in black-headed gull eggs: Constraint and adaptive compensatory adjustment. *Journal of Evolutionary Biology* 19, 1304–1313.

Groothuis, TGG, Eising, CM, Dijkstra, C and Muller, W (2005a). Balancing between costs and benefits of maternal hormone deposition in avian eggs. *Biology Letters* 1, 78.

Groothuis, TGG, Müller, W, von Engelhardt, N, Carere, C and Eising, C (2005b). Maternal hormones as a tool to adjust offspring phenotype in avian species. *Neuroscience & Biobehavioral Reviews* 29, 329–352.

Guerrero, JM and Reiter, RJ (2002). Melatonin-immune system relationships. *Current Topics in Medicinal Chemistry* 2, 167.

Hamilton, WD and Zuk, M (1982). Heritable true fitness and bright birds: a role for parasites. *Science* 218, 384–387.

Hanssen, SA (2006). Costs of an immune challenge and terminal investment in a long-lived bird. *Ecology* 87, 2440–2446.

Hanssen, SA, Hasselquist, D, Folstad, I and Erikstad, KE (2005). Cost of reproduction in a long-lived bird: Incubation effort reduces immune function and future reproduction. *Proceedings of the Royal Society B: Biological Sciences* 272, 1039–1046.

Harshman, LG and Zera, AJ (2007). The cost of reproduction: The devil in the details. *Trends in Ecology & Evolution* 22, 80–86.

Harvell, CD, Mitchell, CE, Ward, JR, Altizer, S, Dobson, AP, Ostfeld, RS and Samuel, MD (2002). Ecology: climate warming and disease risks for terrestrial and marine biota. *Science* 296, 2158.

Hasselquist, D (2007). Comparative immunoecology in birds: Hypotheses and tests. *Journal of Ornithology* 148, S571–S582.

Hasselquist, D, Lindström, A, Jenni-Eiermann, S, Koolhaas, A and Piersma, T (2007). Long flights do not influence immune responses of a long-distance migrant bird: A wind-tunnel experiment. *Journal of Experimental Biology* 210, 1123–1131.

Hasselquist, D, Marsh, JA, Sherman, PW and Wingfield, JC (1999). Is avian humoral immunocompetence suppressed by testosterone? *Behavioral Ecology and Sociobiology* 45, 167–175.

Hasselquist, D and Nilsson, JA (2009). Maternal transfer of antibodies in vertebrates: Trans-generational effects on offspring immunity. *Philosophical Transactions of the Royal Society B-Biological Sciences* 364, 51–60.

Hau, M (2007). Regulation of male traits by testosterone: Implications for the evolution of vertebrate life histories. *Bioessays* 29, 133.

Hau, M, Gill, SA and Goymann, W (2008). Tropical field endocrinology: Ecology and evolution of testosterone concentrations in male birds. *General and Comparative Endocrinology* 157, 241–248.

Hau, M, Stoddard, ST and Soma, KK (2004). Territorial aggression and hormones during the non-breeding season in a tropical bird. *Hormones and Behavior* 45, 40–49.

Haussmann, MF, Winkler, DW, Huntington, CE, Vleck, D, Sanneman, CE, Hanley, D and Vleck, CM (2005). Cell-mediated immunosenescence in birds. *Oecologia* 145, 270–275.

Hawley, DM (2006). Asymmetric effects of experimental manipulations of social status on individual immune response. *Animal Behaviour* 71, 1431–1438.

Hawley, DM, Jennelle, CS, Sydenstricker, KV and Dhondt, AA (2007). Pathogen resistance and immunocompetence covary with social status in house finches (*Carpodacus mexicanus*). *Functional Ecology* 21, 520–527.

Hawley, DM, Lindström, K and Wikelski, M (2006). Experimentally increased social competition compromises humoral immune responses in house finches. *Hormones and Behavior* 49, 417–424.

Heath, J (1997). Corticosterone levels during nest departure of juvenile American kestrels. *Condor* 99, 806–811.

Heidinger, BJ, Nisbet, ICT and Ketterson, ED (2006). Older parents are less responsive to a stressor in a long-lived seabird: A mechanism for increased reproductive performance with age? *Proceedings of the Royal Society B: Biological Sciences* 273, 2227–2231.

Hill, GE (1999). Is there an immunological cost to carotenoids based ornamental coloration? *The American Naturalist* 154, 589–595.

Hill, GE and Farmer, KL (2005). Carotenoid-based plumage coloration predicts resistance to a novel parasite in the house finch. *Naturwissenschaften* 92, 30–34.

Hintz, JV (2000). The hormonal regulation of premigratory fat deposition and winter fattening in red-winged blackbirds. *Comparative Biochemistry and Physiology, Part A: Molecular & Integrative Physiology* 125, 239–249.

Hirschenhauser, K and Oliveira, RF (2006). Social modulation of androgens in male vertebrates: Meta-analyses of the challenge hypothesis. *Animal Behaviour* 71, 265.

Holberton, RL and Able, KP (2000). Differential migration and an endocrine response to stress in wintering dark-eyed juncos (*Junco hyemalis*). *Proceedings of the Royal Society of London Series B-Biological Sciences* 267, 1889–1896.

Horak, P, Zilmer, M, Saks, L, Ots, I, Karu, U and Zilmer, K (2006). Antioxidant protection, carotenoids and the costs of immune challenge in greenfinches. *Journal of Experimental Biology* 209, 4329–4338.

Hull, KL, Thiagarajah, A and Harvey, S (1996). Cellular localization of growth hormone receptors binding proteins in immune tissues. *Cell and Tissue Research* 286, 69–80.

Ilmonen, P, Taarna, T and Hasselquist, D (2000). Experimentally activated immune defence in female pied flycatchers results in reduced breeding success. *Proceedings: Biological Sciences* 267, 665–670.

John, JL (1994). The avian spleen: a neglected organ. *Quarterly Review of Biology* 69, 327.

Kelley, KW, Weigent, DA and Kooijman, R (2007). Protein hormones and immunity. *Brain, Behavior, and Immunity* 21, 384–392.

Ketterson, ED, Atwell, JW and McGlothlin, JW (2009). Phenotypic integration and independence: Hormones, performance, and response to environmental change. *Integrative and Comparative Biology* 49, 365–379.

Ketterson, ED and Nolan, V (1994). Male parental behavior in birds. *Annual Review of Ecology and Systematics* 25, 601–628.

Ketterson, ED, Nolan, V and Sandell, M (2005). Testosterone in females: Mediator of adaptive traits, constraint on sexual dimorphism, or both? *American Naturalist* 166, S85–S98.

Ketterson, ED, Nolan, V, Wolf, L and Ziegenfus, C (1992). Testosterone and avian life histories: Effects of experimentally elevated testosterone on behavior and correlates of fitness in the dark-eyed junco (*Junco hyemalis*). *American Naturalist* 140, 980–999.

Ketterson, ED, Nolan, V, Wolf, L, Ziegenfus, C, Dufty, AM, Ball, GF and Johnsen, TS (1991). Testosterone and avian life histories: the effect of experimentally elevated testosterone on corticosterone and body-mass in dark-eyed juncos. *Hormones and Behavior* 25, 489–503.

Kitaysky, AS, Kitaiskaia, E, Piatt, J and Wingfield, JC (2003). Benefits and costs of increased levels of corticosterone in seabird chicks. *Hormones and Behavior* 43, 140–149.

Klein, SL (2000). Hormones and mating system affect sex and species differences in immune function among vertebrates. *Behavioural Processes* 51, 149–166.

Klein, SL (2004). Hormonal and immunological mechanisms mediating sex differences in parasite infection. *Parasite Immunology* 26, 247–264.

Knowles, SCL, Nakagawa, S and Sheldon, BC (2009). Elevated reproductive effort increases blood parasitaemia and decreases immune function in birds: A meta-regression approach. *Functional Ecology* 23, 405–415.

Lafferty, KD (2009). The ecology of climate change and infectious diseases. *Ecology* 90, 888–900.

Lahti, K (1998). Social dominance and survival in flocking passerine birds: A review with an emphasis on the willow tit *Parus montanus*. *Ornis Fennica* 75, 1.

Landys-Ciannelli, MM, Ramenofsky, M, Piersma, T, Jukema, J, Group, CR and Wingfield, JC (2002). Baseline and stress-induced plasma corticosterone during long-distance migration in the bar-tailed godwit, *Limosa lapponica*. *Physiological and Biochemical Zoology* 75, 101–110.

Lavoie, ET (2005). Avian immunosenescence. *Age* 27, 281–285.

Lavoie, ET, Sorrell, EM, Perez, DR and Ottinger, MA (2007). Immunosenescence and age-related susceptibility to influenza virus in Japanese quail. *Developmental & Comparative Immunology* 31, 407–414.

Lawrence, EC, Arnaudbattandier, F, Grayson, J, Koski, IR, Dooley, NJ, Muchmore, AV and Blaese, RM (1981). Ontogeny of humoral immune function in normal chickens: a comparison of immunoglobulin-secreting cells in bone-marrow, spleen, lungs and intestine. *Clinical and Experimental Immunology* 43, 450–457.

Lee, KA (2006). Linking immune defenses and life history at the levels of the individual and the species. *Integrative and Comparative Biology* 46, 1000–1015.

Lee, KA, Martin, LB, Hasselquist, D, Ricklefs, RE and Wikelski, M (2006). Contrasting adaptive immune defenses and blood parasite prevalence in closely related passer sparrows. *Oecologia* 150, 383–392.

Lee, KA, Wikelski, M, Robinson, WD, Robinson, TR and Klasing, KC (2008). Constitutive immune defences correlate with life-history variables in tropical birds. *Journal of Animal Ecology* 77, 356–363.

Lemke, H, Hansen, H and Lange, H (2003). Non-genetic inheritable potential of maternal antibodies. *Vaccine* 21, 3428–3431.

Lindström, A, Kvist, A, Piersma, T, Dekinga, A and Dietz, MW (2000). Avian pectoral muscle size rapidly tracks body mass changes during flight, fasting and fuelling. *Journal of Experimental Biology* 203, 913–919.

Lindström, K and Lundstrom, J (2000). Male greenfinches (*Carduelis chloris*) with brighter ornaments have higher virus infection clearance rate. *Behavioral Ecology and Sociobiology* 48, 44–51.

Lindström, KM, Hasselquist, D and Wikelski, M (2005a). House sparrows (*Passer domesticus*) adjust their social status position to their physiological costs. *Hormones and Behavior* 48, 311–320.

Lindström, KM, Hawley, DM, Davis, AK and Wikelski, M (2005b). Stress responses and disease in three wintering house finch (*Carpodacus mexicanus*) populations along a latitudinal gradient. *General and Comparative Enocrinology* 143, 231–239.

Lindström, KM, Krakower, D, Lundstrom, JO and Silverin, B (2001). The effects of testosterone on a viral infection in greenfinches (*Carduelis chloris*): An experimental test of the immunocompetence-handicap hypothesis. *Proceedings of the Royal Society of London Series B-Biological Sciences* 268, 207–211.

Lochmiller, RL and Deerenberg, C (2000). Trade-offs in evolutionary immunology: Just what is the cost of immunity? *Oikos* 88, 87–98.

Loiseau, C, Sorci, G, Dano, S and Chastel, O (2008). Effects of experimental increase of corticosterone levels on begging behavior, immunity and parental provisioning rate in house sparrows. *General and Comparative Endocrinology* 155, 101–108.

Love, OP, Chin, EH, Wynne-Edwards, KE and Williams, TD (2005). Stress hormones: A link between maternal condition and sex biases reproductive investment. *The American Naturalist* 166, 751–766.

Lozano, GA and Lank, DB (2003). Seasonal trade-offs in cell-mediated immunosenescence in ruffs (*Philomachus pugnax*). *Proceedings of the Royal Society of London Series B-Biological Sciences* 270, 1203–1208.

Luna, M, Barraza, N, Berumen, L, Carranza, M, Pedernera, E, Harvey, S and Aramburo, C (2005). Heterogeneity of growth hormone immunoreactivity in lymphoid tissues and changes during ontogeny in domestic fowl. *General and Comparative Endocrinology* 144, 28–37.

Majewski, P, Adamska, I, Pawlak, J, Baranska, A and Skwarlo-Sonta, K (2005). Seasonality of pineal gland activity and immune functions in chickens. *Journal of Pineal Research* 39, 66–72.

Malisch, JL and Breuner, CW (2010). Steroid-binding proteins and free steroids in birds. *Molecular and Cellular Endocrinology* 316, 42–52.

Maney, DL, Davis, AK, Goode, CT, Reid, A and Showalter, C (2008). Carotenoid-based plumage coloration predicts leukocyte parameters during the breeding season in Northern cardinals (*Cardinalis cardinalis*). *Ethology* 114, 369–380.

Markowska, M, Waloch, M and Skwarlo-Sonta, K (2001). Melatonin inhibits PHA-stimulated chicken lymphocyte proliferation in vitro. *Journal of Pineal Research* 30, 220–226.

Marra, PP, Hobson, KA and Holmes, RT (1998). Linking winter and summer events in a migratory bird by using stable-carbon isotopes. *Science* 282, 1884–1886.

Marra, PP and Holberton, RL (1998). Corticosterone levels as indicators of habitat quality: Effects of habitat segregation in a migratory bird during the non-breeding season. *Oecologia* 116, 284–292.

Martin, LB, Hasselquist, D and Wikelski, M (2006). Investment in immune defense is linked to pace of life in house sparrows. *Oecologia* 147, 565.

Martin, LB (2009). Stress and immunity in wild vertebrates: Timing is everything. *General and Comparative Endocrinology* 163, 70–76.

Martin, LB, Gilliam, J, Han, P, Lee, K and Wikelski, M (2005). Corticosterone suppresses cutaneous immune function in temperate but not tropical house sparrows, *Passer domesticus*. *General and Comparative Endocrinology* 140, 126.

Martin, LB, Han, P, Lewittes, J, Kuhlman, JR, Klasing, KC and Wikelski, M (2006a). Phytohemagglutinin-induced skin swelling in birds: Histological support for a classic immunoecological technique. *Functional Ecology* 20, 290–299.

Martin, LB, Weil, ZM and Nelson, RJ (2006b). Refining approaches and diversifying directions in ecoimmunology. *Integrative and Comparative Biology* 46, 1030–1039.

Martin, LB, Pless, M, Svoboda, J and Wikelski, M (2004). Immune activity in temperate and tropical house sparrows: A common-garden experiment. *Ecology* 85, 2323–2331.

Martin, LB and Rubenstein, DR (2008). Stress hormones in tropical birds: Patterns and future directions. *Ornitologia Neotropical* 19, 207–218.

Martin, LB, Scheuerlein, A and Wikelski, M (2003). Immune activity elevates energy expenditure of house sparrows: A link between direct and indirect costs? *Proceedings of the Royal Society of London Series B-Biological Sciences* 270, 153.

Martin, LB, Weil, ZM and Nelson, RJ (2008). Seasonal changes in vertebrate immune activity: Mediation by physiological trade-offs. *Philosophical Transactions of the Royal Society B-Biological Sciences* 363, 321–339.

Marzal, A, Bencsch, S, Revieriego, M, Balbontin, J and de Lope, F (2008). Effects of malaria double infection in birds: One plus one is not two. *Journal of Evolutionary Biology* 21, 979–987.

Matera, L (1996). Endocrine, paracrine and autocrine actions of prolactin on immune cells. *Life Sciences* 59, 599–614.

Matson, KD, Cohen, AA, Klasing, KC, Ricklefs, RE and Scheuerlein, A (2006a). No simple answers for ecological immunology: Relationships among immune indices at the individual level break down at the species level in waterfowl. *Proceedings of the Royal Society B-Biological Sciences* 273, 815–822.

Matson, KD, Ricklefs, RE and Klasing, KC (2005). A hemolysis-hemagglutination assay for characterizing constitutive innate humoral immunity in wild and domestic birds. *Developmental and Comparative Immunology* 29, 275–286.

Matson, KD, Tieleman, BI and Klasing, KC (2006b). Capture stress and the bactericidal competence of blood and plasma in five species of tropical birds. *Physiological and Biochemical Zoology* 79, 556.

McGraw, KJ and Ardia, DR (2003). Carotenoids, immunocompetence, and the information content of sexual colors: An experimental test. *The American Naturalist* 162, 704–712.

McGraw, KJ and Ardia, DR (2007). Do carotenoids buffer testosterone-induced immunosuppression? An experimental test in a colourful songbird. *Biology Letters* 3, 375–378.

McKechnie, AE (2008). Phenotypic flexibility in basal metabolic rate and the changing view of avian physiological diversity: A review. *Journal of Comparative Physiology B-Biochemical Systemic and Environmental Physiology* 178, 235–247.

Millet, S, Bennett, J, Lee, KA, Hau, M and Klasing, KC (2007). Quantifying and comparing constitutive immunity across avian species. *Developmental and Comparative Immunology* 31, 188–201.

Moore, CB and Siopes, TD (2000). Effects of lighting conditions and melatonin supplementation on the cellular and humoral immune responses in Japanese quail *Coturnix coturnix japonica*. *General and Comparative Endocrinology* 119, 95–104.

Moore, CB and Siopes, TD (2002). Melatonin can produce immunoenhancement in Japanese quail (*Coturnix coturnix japonica*) without prior immunosuppression. *General and Comparative Endocrinology* 129, 122–126.

Moore, CB and Siopes, TD (2003). Melatonin enhances cellular and humoral immune responses in the Japanese quail (*Coturnix coturnix japonica*) via an opiatergic mechanism. *General and Comparative Endocrinology* 131, 258–263.

Moore, CB and Siopes, TD (2005). Enhancement of cellular and humoral immunity following embryonic exposure to melatonin in turkeys (*Meleagris gallopavo*). *General and Comparative Endocrinology* 143, 178–183.

Moreno, J, Lobato, E, Morales, J, Merino, S, Martinez-De La Puente, J and Tomàs, G (2008). Pre-laying nutrition mediates maternal effects on offspring immune capacity and growth in the pied flycatcher. *Oecologia* 156, 727–735.

Moreno, J, Vicente, A, Heijnen, I and Zapata, AG (1994). Prolactin and early T-cell development in embryonic chicken. *Immunology Today* 15, 524–526.

Muller, W, Groothuis, TGG, Kasprzik, A, Dijkstra, C, Alatalo, RV and Siitari, H (2005). Prenatal androgen exposure modulates cellular and humoral immune function of black-headed gull chicks. *Proceedings of the Royal Society B-Biological Sciences* 272, 1971.

Navara, KJ, Hill, GE and Mendonca, MT (2005). Variable effects of yolk androgens on growth, survival, and immunity in Eastern bluebird nestlings. *Physiological and Biochemical Zoology* 78, 570–578.

Navara, KJ and Mendonca, MT (2008). Yolk androgens as pleiotropic mediators of physiological processes: A mechanistic review. *Comparative Biochemistry and Physiology A-Molecular & Integrative Physiology* 150, 378–386.

Nelson, RJ and Demas, GE (1996). Seasonal changes in immune function. *Quarterly Review of Biology* 71, 511–548.

Nelson, RJ and Demas, GE (2004). Seasonal patterns of stress, disease, and sickness responses. *Current Directions in Psychological Science* 13, 198–201.

Nelson, RJ, Demas, GE, Klein, SL and Kriegsfeld, LJ (1995). The influence of season, photoperiod, and pineal melatonin on immune function. *Journal of Pineal Research* 19, 149–165.

Nieuwenhuis, P (1996). Histophysiology of the lymphoid system: The thymus and T cells. In Marsh, JA and Kendall MD, eds. *The Physiology of Immunity*, pp. 3–32. CRC Press, New York.

Nisbet, ICT, Finch, CE, Thompson, N, Russek-Cohen, E, Proudman, JA and Ottinger, MA (1999). Endocrine patterns during aging in the common tern (*Sterna hirundo*). *General and Comparative Endocrinology* 114, 279–286.

Norris, K and Evans, MR (2000). Ecological immunology: Life history trade-offs and immune defense in birds. *Behavioral Ecology* 11, 19–26.

Nunn, CL, Lindenfors, P, Pursall, ER and Rolff, J (2009). On sexual dimorphism in immune function. *Philosophical Transactions of the Royal Society B-Biological Sciences* 364, 61–69.

O'Neal, DM, Reichard, DG, Pavilis, K and Ketterson, ED (2008). Experimentally-elevated testosterone, female parental care, and reproductive success in a songbird, the dark-eyed junco (*Junco hyemalis*). *Hormones and Behavior* 54, 571–578.

Ots, I, Kerimov, AB, Ivankina, EV, Ilyina, TA and Horak, P (2001). Immune challenge affects basal metabolic activity in wintering great tits. *Proceedings of the Royal Society of London Series B- Biological Sciences* 268, 1175–1181.

Ottinger, MA (2007). Neuroendocrine aging in birds: Comparing lifespan differences and conserved mechanisms. *Ageing Research Reviews* 6, 46–53.

Ottinger, MA and Lavoie, E (2007). Neuroendocrine and immune characteristics of aging in avian species. *Cytogenetic and Genome Research* 117, 352–357.

Owen-Ashley, NT, Hasselquist, D, Raberg, L and Wingfield, JC (2008). Latitudinal variation of immune defense and sickness behavior in the white-crowned sparrow (*Zonotrichia leucophrys*). *Brain Behavior and Immunity* 22, 614–625.

Owen-Ashley, NT, Hasselquist, D and Wingfield, JC (2004). Androgens and the immunocompetence handicap hypothesis: Unraveling direct and indirect pathways of immunosuppression in song sparrows. *American Naturalist* 164, 490–505.

Owen-Ashley, NT, Turner, M, Hahn, TP and Wingfield, JC (2006). Hormonal, behavioral, and thermoregulatory responses to bacterial lipopolysaccharide in captive and free-living white-crowned sparrows (*Zonotrichia leucophrys gambelii*). *Hormones and Behavior* 49, 15–29.

Owen-Ashley, NT and Wingfield, JC (2006). Seasonal modulation of sickness behavior in free-living northwestern song sparrows (*Melospiza melodia morphna*). *Journal of Experimental Biology* 209, 3062–3070.

Owen, JC and Moore, FR (2008). Relationship between energetic condition and indicators of immune function in thrushes during spring migration. *Canadian Journal of Zoology-Revue Canadienne De Zoologie* 86, 638–647.

Palacios, MG, Cunnick, JE, Winkler, DW and Vleck, CM (2007). Immunosenescence in some but not all immune components in a free-living vertebrate, the tree swallow. *Proceedings of the Royal Society B-Biological Sciences* 274, 951–957.

Piersma, T, Mendes, L, Hennekens, J, Ratiarison, S, Groenewold, S and Jukema, J (2001). Breeding plumage honestly signals likelihood of tapeworm infestation in females of a long-distance migrating shorebird, the bar-tailed godwit. *Zoology-Analysis of Complex Systems* 104, 41–48.

Piersma, T, Reneerkens, J and Ramenofsky, M (2000). Baseline corticosterone peaks in shorebirds with maximal energy stores for migration: A general preparatory mechanism for rapid behavioral and metabolic transitions? *General and Comparative Endocrinology* 120, 118–126.

Pigliucci, M (2003). Phenotypic integration: Studying the ecology and evolution of complex phenotypes. *Ecology Letters* 6, 265–272.

Pihlaja, M, Siitari, H and Alatalo, RV (2006). Maternal antibodies in a wild altricial bird: Effects on offspring immunity, growth and survival. *Journal of Animal Ecology* 75, 1154–1164.

Pitala, N, Ruuskanen, S, Laaksonen, T, Doligez, B, Tschirren, B and Gustafsson, L (2009). The effects of experimentally manipulated yolk androgens on growth and immune function of male and female nestling collared flycatchers *Ficedula albicollis*. *Journal of Avian Biology* 40, 225–230.

Poiani, A, Goldsmith, AR and Evans, MR (2000). Ectoparasites of house sparrows (*Passer domesticus*): An experimental test of the immunocompetence handicap hypothesis and a new model. *Behavioral Ecology and Sociobiology* 47, 230–242.

Reed, WL, Clark, ME, Parker, PG, Raouf, SA, Arguedas, N, Monk, DS, Snajdr, E, Nolan Jr., V and Ketterson, ED (2006). Physiological effects on demography: A long-term experimental study of testosterone's effects on fitness. *American Naturalist* 167, 667–683.

Reid, JM, Arcese, P, Keller, LF and Hasselquist, D (2006). Long-term maternal effect on offspring immune response in song sparrows *Melospiza melodia*. *Biology Letters* 2, 573–576.

Ricklefs, RE and Wikelski, M (2002). The physiology/life-history nexus. *Trends in Ecology & Evolution* 17, 462–468.

Roberts, M and Peters, A (2009). Is testosterone immunosuppressive in a condition-dependent manner? An experimental test in blue tits. *Journal of Experimental Biology* 212, 1811–1818.

Roberts, ML, Buchanan, KL and Evans, MR (2004). Testing the immunocompetence handicap hypothesis: A review of the evidence. *Animal Behaviour* 68, 227–239.

Roberts, ML, Buchanan, KL, Evans, MR, Marin, RH and Satterlee, DG (2009). The effects of testosterone on immune function in quail selected for divergent plasma corticosterone response. *Journal of Experimental Biology* 212, 3125–3131.

Roberts, ML, Buchanan, KL, Hasselquist, D and Evans, MR (2007). Effects of testosterone and corticosterone on immunocompetence in the zebra finch. *Hormones and Behavior* 51, 126–134.

Rodriguez, AB, Barriga, C and Lea, RW (1999). Effect of prolactin, in vivo and in vitro, upon heterophil phagocytic function in the ring dove (*Streptopelia risoria*). *Developmental & Comparative Immunology* 20, 451–457.

Rodríguez, AB, Terrón, MP, Durán, J, Ortega, E and Barriga, C (2001). Physiological concentrations of melatonin and corticosterone affect phagocytosis and oxidative metabolism of ring dove heterophils. *Journal of Pineal Research* 31, 31–38.

Rogers, CM, Ramenofsky, M, Ketterson, ED, Nolan, V and Wingfield, JC (1993). Plasma-corticosterone, adrenal mass, winter weather, and season in nonbreeding populations of dark-eyed juncos (*Junco hyemalis hyemalis*). *Auk* 110, 279–285.

Rolff, J (2002). Bateman's principle and immunity. *Proceedings of the Royal Society of London. Series B: Biological Sciences* 269, 867–872.

Royle, NJ, Surai, PF, and Hartley, IR (2001). Maternally derived androgens and antioxidants in bird eggs: complementary but opposing effects? *Behavioral Ecology* 12, 381–385.

Rubenstein, DR, Parlow, AF, Hutch, CR and Martin, LB (2008). Environmental and hormonal correlates of immune activity in a cooperatively breeding tropical bird. *General and Comparative Endocrinology* 159, 10–15.

Rubolini, D, Romano, M, Boncoraglio, G, Ferrari, RP, Martinelli, R, Galeotti, P, Fasola, M and Saino, N (2005a). Effects of elevated egg corticosterone levels on behavior, growth, and immunity of yellow-legged gull (*Larus michahellis*) chicks. *Hormones and Behavior* 47, 592–605.

Rubolini, D, Romano, M, Martinelli, R, Leoni, B and Saino, N (2005b). Effects of prenatal yolk androgens on armaments and ornaments of the ring-necked pheasant. *Behavioral Ecology and Sociobiology* 59, 549–560.

Rubolini, D, Romano, M, Martinelli, R and Saino, N (2006). Effects of elevated yolk testosterone levels on survival, growth and immunity of male and female yellow-legged gull chicks. *Behavioral Ecology and Sociobiology* 59, 344–352.

Ruiz, G, Rosenmann, M, Novoa, FF and Sabat, P (2002). Hematological parameters and stress index in rufous-collared sparrows dwelling in urban environments. *Condor* 104, 162–166.

Saino, N, Ferrari, RP, Romano, M, Rubolini, D and Møller, AP (2003). Humoral immune response in relation to senescence, sex and sexual ornamentation in the barn swallow (*Hirundo rustica*). *Journal of Evolutionary Biology* 16, 1127–1134.

Saks, L, Ots, I and Horak, P (2003). Carotenoid-based plumage coloration of male greenfinches reflects health and immunocompetence. *Oecologia* 134, 301–307.

Sandell, MI (2007). Exogenous testosterone increases female aggression in the European starling (*Sturnus vulgaris*). *Behavioral Ecology and Sociobiology* 62, 255–262.

Sandell, MI, Tobler, M and Hasselquist, D (2009). Yolk androgens and the development of avian immunity: An experiment in jackdaws (*Corvus monedula*). *Journal of Experimental Biology* 212, 815–822.

Sandland, GJ and Minchella, DJ (2003). Costs of immune defense: An enigma wrapped in an environmental cloak? *Trends in Parasitology* 19, 571–574.

Sapolsky, RM, Romero, LM and Munck, AU (2000). How do glucocorticoids influence stress responses? Integrating permissive, suppressive, stimulatory, and preparative actions. *Endocrine Reviews* 21, 55–89.

Segerstrom, SC (2007). Stress, energy, and immunity: An ecological view. *Current Directions in Psychological Science* 16, 326–330.

Shephard, RJ and Shek, PN (1998). Cold exposure and immune function. *Canadian Journal of Physiology and Pharmacology* 76, 828–836.

Siegel, HS (1980). Physiological stress in birds. *Bioscience* 30, 529.

Silverin, B, Fange, R, Viebke, PA and Westin, J (1999). Seasonal changes in mass and histology of the spleen in willow tits *Parus montanus*. *Journal of Avian Biology* 30, 255.

Sinclair, JA and Lochmiller, RL (2000). The winter immunoenhancement hypothesis: Associations among immunity, density, and survival in prairie vole (*Microtus ochrogaster*) populations. *Canadian Journal of Zoology* 78, 254–264.

Singh, SS and Haldar, C (2007a). Biological significance of daily variation in immunity of perdicula asiatica: Role of melatonin and testosterone. *Biological Rhythm Research* 38, 95–106.

Singh, SS and Haldar, C (2007b). Peripheral melatonin modulates seasonal immunity and reproduction of Indian tropical male bird *Perdicula asiatica*. *Comparative Biochemistry and Physiology A-Molecular & Integrative Physiology* 146, 446–450.

Siopes, TD and Underwood, HA (2008). Diurnal variation in the cellular and humoral immune responses of Japanese quail: Role of melatonin. *General and Comparative Endocrinology* 158, 245–249.

Skutch., AF (1949). Do tropical birds rear as many young as they can nourish? *Ibis* 91, 430–455.

Skwarlo-Sonta, K (1992). Prolactin as an immunoregulatory hormone in mammals and birds. *Immunology Letters* 33, 105–121.

Skwarlo-Sonta, K (1996). Functional connections between the pineal gland and immune system. *Acta Neurobiologiae Experimentalis* 56, 341–357.

Skwarlo-Sonta, K (2002). Melatonin in immunity: Comparative aspects. *Neuroendocrinology Letters* 23, 61–66.

Skwarlo-Sonta, K (1996). Functional connections between the pineal gland and immune system. *Acta Neurobiologiae Experimentalis* 56, 341–357.

Smith, RJ and Moore, FR (2005). Arrival timing and seasonal reproductive performance in a long-distance migratory landbird. *Behavioral Ecology and Sociobiology* 57, 231–239.

Smits, JEG and Baos, R (2005). Evaluation of the antibody mediated immune response in nestling american kestrels (*Falco sparverius*). *Developmental and Comparative Immunology* 29, 161–170.

Soma, KK, Tramontin, AD and Wingfield, JC (2000). Oestrogen regulates male aggression in the non-breeding season. *Proceedings of the Royal Society of London Series B-Biological Sciences* 267, 1089–1096.

Stearns, SC (1989). Trade-offs in life-history evolution. *Functional Ecology* 3, 259–268.

Stromberg, BE (1997). Environmental factors influencing transmission. *Veterinary Parasitology* 72, 247–256.

Studds, CE and Marra, PP (2005). Nonbreeding habitat occupancy and population processes: An upgrade experiment with a migratory bird. *Ecology* 86, 2380–2385.

Stutchbury, BJM and Morton, ES (2008). Recent advances in the behavioral ecology of tropical birds *Wilson Journal of Ornithology* 120, 26–37.

Svensson, E, Raberg, L, Koch, C and Hasselquist, D (1998). Energetic stress, immunosuppression and the costs of an antibody response. *Functional Ecology* 12, 912–919.

Tarlow, EM, Hau, M, Anderson, DJ and Wikelski, M (2003). Diel changes in plasma melatonin and corticosterone concentrations in tropical Nazca boobies (*Sula granti*) in relation to moon phase and age. *General and Comparative Endocrinology* 133, 297–304.

Tella, JL, Scheuerlein, A and Ricklefs, RE (2002). Is cell-mediated immunity related to the evolution of life-history strategies in birds? *Proceedings of the Royal Society of London Series B-Biological Sciences* 269, 1059–1066.

Terron, MP, Delgado, J, Paredes, SD, Barriga, C, Reiter, RJ and Rodriguez, AB (2009). Effect of melatonin and tryptophan on humoral immunity in young and old ringdoves (*Streptopelia risoria*). *Experimental Gerontology* 44, 653–658.

Tieleman, BI, Williams, JB, Ricklefs, RE and Klasing, KC (2005). Constitutive innate immunity is a component of the pace-of-life syndrome in tropical birds. *Proceedings of the Royal Society B-Biological Sciences* 272, 1715.

Tobler, M, Hasselquist, D, Smith, HG and Sandell, MI (2009). Short- and long-term consequences of prenatal testosterone for immune function: An experimental study in the zebra finch. *Behavioral Ecology and Sociobiology*, online.

Tomàs, G, Merino, S, Moreno, J, Morales, J and Puente, JM-DL (2007). Impact of blood parasites on immunoglobulin level and parental effort: A medication field experiment on a wild passerine. *Functional Ecology* 21, 125–133.

Trivers, RL (1972). Parental investment and sexual selection. In: Campbell, B, ed. *Sexual Selection and the Descent of Man*, pp. 136–179. Aldine, Chicago.

Tschirren, B, Fitze, Patrick, S and Richner, H (2007). Maternal modulation of natal dispersal in a passerine bird: An adaptive strategy to cope with parasitism? *The American Naturalist* 169, 87–93.

Tschirren, B, Richner, H and Schwabl, H (2004). Ectoparasite-modulated deposition of maternal androgens in great tit eggs. *Proceedings of the Royal Society of London Series B-Biological Sciences* 271, 1371–1375.

Tschirren, B, Saladin, V, Fitze, PS, Schwabl, H and Richner, H (2005). Maternal yolk testosterone does not modulate parasite susceptibility or immune function in great tit nestlings. *Journal of Animal Ecology* 74, 675.

van Gils, JA, Munster, VJ, Radersma, R, Liefhebber, D, Fouchier, RAM and Klaassen, M (2007). Hampered foraging and migratory performance in swans infected with low-pathogenic avian influenza a virus. *PLos One* 2, 6.

Veiga, JP, Vinuela, J, Cordero, PJ, Aparicio, JM and Polo, V (2004). Experimentally increased testosterone affects social rank and primary sex ratio in the spotless starling. *Hormones and Behavior* 46, 47–53.

Velando, A, Drummond, H and Torres, R (2006). Senescent birds redouble reproductive effort when ill: Confirmation of the terminal investment hypothesis. *Proceedings of the Royal Society B: Biological Sciences* 273, 1443–1448.

Vleck, CM, Haussmann, MF and Vleck, D (2007). Avian senescence: Underlying mechanisms. *Journal of Ornithology* 148, S611–S624.

Weber, TP and Stilianakis, NI (2007). Ecologic immunology of avian influenza (H5N1) in migratory birds. *Emerging Infectious Diseases* 13, 1139–1143.

Wiersma, P, Munoz-Garcia, A, Walker, A and Williams, JB (2007). Tropical birds have a slow pace of life. *Proceedings of the National Academy of Sciences of the United States of America* 104, 9340–9345.

Wikelski, M and Cooke, SJ (2006). Conservation physiology. *Trends in Ecology & Evolution* 21, 38–46.

Wikelski, M, Spinney, L, Schelsky, W, Scheuerlein, A and Gwinner, E (2003a). Slow pace of life in tropical sedentary birds: A common-garden experiment on four stonechat populations from different latitudes. *Proceedings of the Royal Society of London. Series B: Biological Sciences* 270, 2383–2388.

Wikelski, M, Tarlow, EM, Raim, A, Diehl, RH, Larkin, RP and Visser, GH (2003b). Costs of migration in free-flying songbirds. *Nature* 423, 704–704.

Wingfield, JC (2003). Control of behavioural strategies for capricious environments. *Animal Behaviour* 66, 807.

Wingfield, JC (2005). A continuing saga: The role of testosterone in aggression. *Hormones and Behavior* 48, 253–255.

Wingfield, JC and Hunt, KE (2002). Arctic spring: Hormone-behavior interactions in a severe environment. *Comparative Biochemistry And Physiology B-Biochemistry & Molecular Biology* 132, 275.

Wingfield, JC and Kitaysky, AS (2002). Endocrine responses to unpredictable environmental events: Stress or anti-stress hormones? *Integrative and Comparative Biology* 42, 600.

Wingfield, JC, Lynn, SE and Soma, KK (2001). Avoiding the "costs" of testosterone: Ecological bases of hormone-behavior interactions. *Brain Behavior and Evolution* 57, 239–251.

Wingfield, JC, Maney, DL, Breuner, CW, Jacobs, JD, Lynn, S, Ramenofsky, M and Richardson, RD (1998). Ecological bases of hormone-behavior interactions: The "emergency life history stage." *American Zoologist* 38, 191–206.

Wingfield, JC, Schwabl, H and Mattocks Jr., PW (1990). Endocrine mechanisms of migration. In: Gwinner, E, ed. *Bird Migration: Physiology and Ecophysiology*, pp. 232–256. Springer-Verlag, Berlin.

Zera, AJ and Harshman, LG (2001). The physiology of life history trade-offs in animals. *Annual Review of Ecology and Systematics* 32, 95–126.

Zuk, M and Johnsen, TS (1998). Seasonal changes in the relationship between ornamentation and immune response in red jungle fowl. *Proceedings of the Royal Society of London. Series B: Biological Sciences* 265, 1631–1635.

Zuk, M and McKean, KA (1996). Sex differences in parasite infections: Patterns and processes. *International Journal for Parasitology* 26, 1009–1023.

Zuk, M and Stoehr, AM (2002). Immune defense and host life history. *American Naturalist* 160, S9–S22.

Zysling, DA, Greives, TJ, Breuner, CW, Casto, JM, Dernas, GE and Ketterson, ED (2006). Behavioral and physiological responses to experimentally elevated testosterone in female dark-eyed juncos (*Junco hyemalis carolinensis*). *Hormones and Behavior* 50, 200–207.

CHAPTER 3

SICKNESS BEHAVIOR IN VERTEBRATES

Allostasis, Life-History Modulation, and Hormonal Regulation

Noah T. Ashley and John C. Wingfield

INTRODUCTION

Over the past several decades, the rapidly growing field of ecological immunology has yielded important insights for understanding the impact that parasites and disease have upon host populations, life-histories, and sexual selection. For example, there is accumulating evidence that anti-parasite defenses of hosts are energetically costly to produce and may therefore compete for resources with other life-history functions, such as reproduction, growth, and development of particular sexual ornaments (Sheldon and Verhulst, 1996; Lochmiller and Deerenberg, 2000; Norris and Evans, 2000; Zuk and Stoehr, 2002; Schmid-Hempel and Ebert, 2003). There are also survival costs associated with immune system activation (Moret and Schmid-Hempel, 2000; Hanssen et al., 2004; Eraud et al., 2009). Furthermore, on a mechanistic level, much progress has been made to develop techniques and assays that quantify immunological performance, or "immunocompetence," in both vertebrate and invertebrate taxa. However, a continual challenge is linking these various measures with patterns of host susceptibility to specific pathogens, and ultimately host fitness (Adamo, 2004; Viney et al., 2005).

There is now an increased precedent for ecologists and evolutionary biologists to explore host–parasite interactions from psychoneuroimmunological (Ader and Cohen, 1995) and neuroimmunoendocrinological (Blalock, 1992) perspectives (Demas, 2004; Ottaviani et al., 2008). A large body of evidence has documented multidirectional communication among the immune, nervous, and endocrine systems, which regulate host defenses while maintaining essential homeostatic mechanisms. Although the field of psychoneuroimmunology has explored some of these interrelationships largely from a clinical standpoint (Raison et al., 2002), researchers are now starting to examine how immune-endocrine-behavior circuits are useful in predicting a host's susceptibility to parasitism within life-history and evolutionary contexts (Zuk, 1996; Ricklefs and Wikelski, 2002; Demas, 2004; Owen-Ashley and Wingfield, 2007; Adelman and Martin, 2009). A key finding is that some of these circuits exhibit plasticity according to energy demands and life-history stage of the host.

One such immune-neuroendocrine circuit that has exhibited modulation involves the activation of the acute phase response (APR) and accompanying nonspecific behavioral symptoms that occur during the onset of infection. Collectively known as "sickness behavior," these behavioral alterations include anorexia and adipsia (reduced appetite and thirst, respectively), reduced activity, soporific behavior, increased slow-wave sleep, anhedonia (inability to experience pleasure), hyperalgesia (decreased threshold to perception of pain), a general withdrawal from social activities (e.g., territorial aggression, parental care, sexual behavior), reduced exploratory behavior, decreased libido, malaise, and a reduction in grooming behavior (Hart, 1988; Kent et al., 1992; Dantzer et al., 1993; Exton, 1997; Dantzer et al., 1998; Dunn and Swiergiel, 1998; Maier and Watkins, 1998; Kavaliers et al., 2000; Langhans, 2000; Dantzer, 2001; Konsman et al., 2002; Tizard, 2008). Impaired learning and memory can also occur. Sick animals will often shiver to increase heat production and seek warmer environments and adopt certain postures (e.g., curled position, puffed out) to promote heat conservation (Hart, 1988), which reduces the energetic costs associated with maintaining a fever (a rise in the thermoregulatory set-point). Taken together, these behavioral modifications, along with fever, inflammation, hormonal changes, reduced intestinal motility, and alteration of acute phase proteins (APPs) released from the liver (Andus et al., 1991; Baumann and Gauldie, 1994), constitute the generalized APR in vertebrates. Note that many of the basic components of APRs are homologous in invertebrates (e.g., Toll-like receptors on immune cells that recognize immunogenic moieties on pathogens), suggesting that the response is evolutionarily conserved.

Sickness behavior and fever were traditionally viewed as a weakened physiological state of the host that was induced by pathogenesis. When the field of Darwinian or evolutionary medicine emerged in the 1980s, one of its earliest tenets was that symptoms of disease can be functionally categorized into three groups: manifestations of infectious agents, byproducts of infection, or actual host defenses shaped by natural selection (Ewald, 1980; Hart, 1988; Williams and Neese, 1991). The familiar adage "treating the symptoms doesn't cure the disease" comes readily to mind as a lesson learned by medical students when treating patients. That some symptoms confer a net benefit to the host was first documented experimentally by pioneering research of Matt Kluger and colleagues in the 1970s. Kluger and others demonstrated that fever in an ectothermic animal, desert iguanas (*Dipsosaurus dorsalis*), has adaptive value. Ectotherms can raise their body temperature only through behavioral means. Iguanas experimentally infected with bacteria tended to seek warmer environments through behavioral thermoregulation (Vaughn et al., 1974). Furthermore, survival was enhanced in infected iguanas kept in these warm environments compared to those kept in cool conditions (Kluger et al., 1975).

A parallel finding was observed in endotherms. Rabbits (*Oryctolagus cuniculus*) infected with live bacteria were more likely to die when treated with antipyretics than controls that were allowed to develop fever (Kluger and Vaughn, 1978). Fever is now thought to have two major benefits for facilitating a host's ability to combat infection: (1) enhancement of innate and specific immunological defenses through temperature-dependent and independent mechanisms (Kluger, 1991), and (2) direct inhibition of some pathogenic bacteria and viruses that tend to exhibit optimal growth at temperatures at or below the normal body temperature of hosts (Kluger, 1979). However, fever also has some well-known negative effects upon host fitness, such as increased energy requirements and potential for self-damage, which can have lethal effects. As long as the cost–benefit ratio is skewed towards the benefits, fever should be selected for, even if it leads to morbidity and mortality in rare circumstances (Kluger et al., 1998).

Benjamin Hart extended this line of adaptationist reasoning to the study of sickness behavior in his seminal work (Hart, 1988). He argued that sickness behavior is not a maladaptive response due to pathogen-induced debilitation *per se*, but rather a highly organized and adaptive host behavioral strategy that redirects energy to fever and immune defenses at the expense of other energy-consuming activities, such as foraging, reproduction, growth, and social interactions. More specifically, Hart postulated that anorexia and adipsia are behavioral responses designed to minimize energy spent searching for food and water while reducing intake of micronutrients, such as iron, zinc, and copper, which are rate-limiting to some microorganisms for replication and growth. Reduced activity, by definition, would curb energy consumption while preventing convective heat loss associated with movement, as well as decrease the risk of predation. Hart also viewed reduced grooming behavior in sick animals as another means to save energy, and to minimize evaporative water loss for hosts that employ licking behavior as an anti-parasite defense, such as in mammals. It should be noted that several aspects of Hart's theory have found support from empirical studies, while others have been met with some criticism (e.g., benefits of reduced energy intake). These issues are discussed later in this chapter.

Although there is a wealth of information regarding the general immune-neuroendocrine mechanisms that underlie sickness behavior, particularly in mammals, and some excellent reviews on the subject (Kent et al., 1992; Exton, 1997; Dantzer et al., 1998; Dunn and Swiergiel, 1998; Aubert, 1999; Maier and Watkins, 1999; Dantzer, 2001; Johnson, 2002; Konsman et al., 2002; Dantzer and Kelley, 2007; Tizard, 2008), much less is known about how animals in the wild cope behaviorally with an infection (Hasselquist, 2007; Owen-Ashley and Wingfield, 2007; Adelman and Martin, 2009; French et al., 2009) . The major goals of this chapter are to provide a synoptic overview of sickness behavior using the framework of life-history theory and allostasis, to review the costs and benefits of this response, and to summarize the body of work that has demonstrated modulation of the sickness response in a number of vertebrate taxa. Lastly, the proximate factors underlying the modulation of sickness behavior are discussed.

ACTIVATION OF THE ACUTE PHASE RESPONSE (APR) AND SICKNESS BEHAVIOR

Before consideration of sickness behavior from an ecological perspective, it is necessary to provide a brief mechanistic overview of how the sickness response and the underlying APR are activated. This immune-neuroendocrine cascade begins when circulating immune cells, such as macrophages and dendritic cells, detect invading microbes and/or tissue damage in the periphery (Fig. 3.1). Recognition occurs when specific Toll-like receptors (TLR) on phagocytic cells bind to

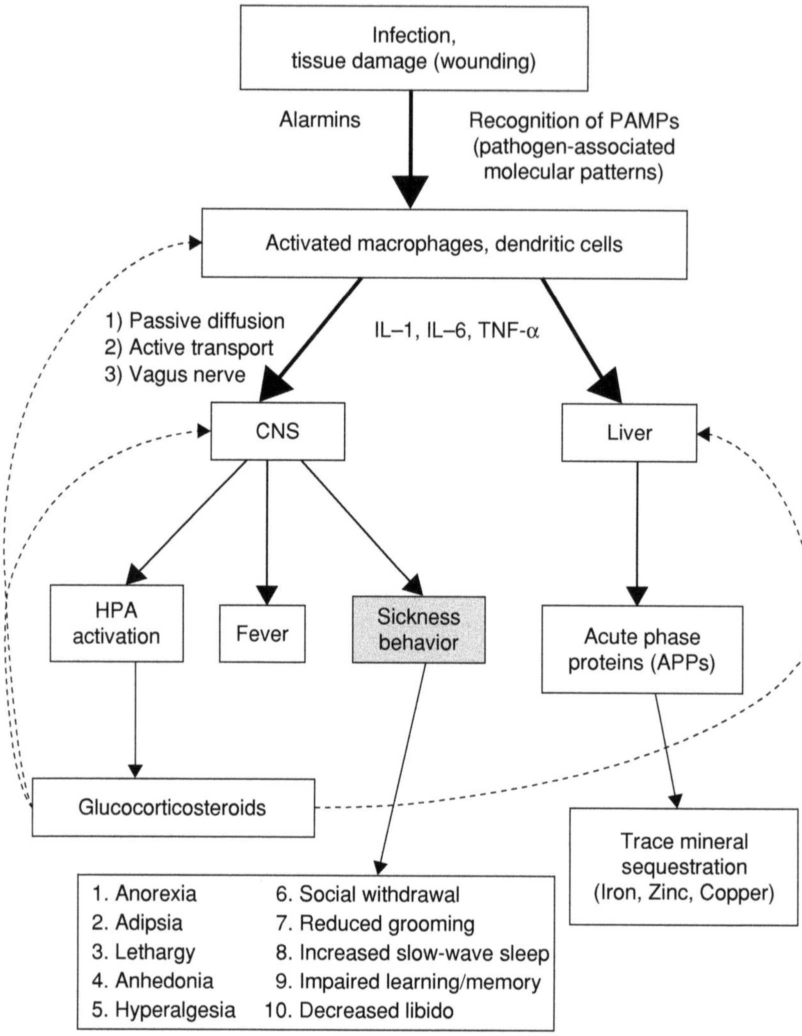

FIGURE 3.1: Schematic of the neuroendocrine-immune circuit for activating sickness behavior. Recognition of infection or injury triggers the release of cytokines, which activates a suite of physiological and behavioral responses that include sickness behavior, fever, acute phase protein (APP) production from the liver, and activation of the hypothalamo-pituitary-adrenal (HPA) axis and subsequent release of glucocorticosteroids. These hormones will initially stimulate immune function but later provide negative feedback at multiple levels to the overall circuit. Dotted and continuous lines represent inhibitory and stimulatory actions, respectively.

pathogen-associated molecular patterns (PAMPs) that are conserved on the surface of various microbes. Alarmins, which are released from destroyed cells as a result of tissue damage, are also recognized by these sentinel immune cells (Tizard, 2008).

Once immune cells are activated, macrophages then engulf and destroy foreign invaders, but also synthesize and release pro-inflammatory cytokines, which mainly involve interleukin 1-beta (IL-1β), IL-6, and tumor necrosis factor-alpha (TNF-α). Cytokines are well known to have pleiotropic functions: they act in an autocrine and/or paracrine fashion to stimulate a secondary wave of cytokine release and to recruit additional immune cells to the site of infection or injury, while signaling the brain and liver through endocrine and neural pathways that infection or injury is occurring (Fig. 3.1).

During an APR, cytokines (primarily IL-6) alter liver function with hepatocytes increasing production of positive APPs such as C-reactive protein and haptoglobin, while reducing negative APPs, such as albumin and carrier proteins (Baumann and Gauldie, 1994). Some of these proteins have the ability to opsonize bacteria, activate complement, enhance phagocytosis, and sequester minerals from the bloodstream, such as iron, copper, zinc, that are limiting for bacterial growth (Weinberg, 1974; Kluger and Rothenberg, 1979; Klasing, 1984; Tufft et al., 1988; Kontoghiorghes and Weinberg, 1995).

Upon entering the general circulation, cytokines reach the brain using several pathways: (1) passive diffusion through semi-permeable regions in the blood–brain barrier (BBB) such as the circumventricular organs (Konsman et al., 2002), (2) active transport through the BBB (Banks, 2001), or (3) stimulation of afferent neural inputs, such as the vagus nerve (Laye et al., 1995; Bluthe et al., 1996). These cytokines then bind to specific receptors in the brain to trigger metabolic, behavioral, and hormonal changes, such as fever, sickness behavior, activation of the hypothalamic-pituitary-adrenal (HPA) axis, and inhibition of the hypothalamic-pituitary-gonadal (HPG), somatotropic, and thyrotropic axes (Besedovsky et al., 1986; Rivier, 1990; Rivier and Vale, 1990; Turnbull and Rivier, 1995; Kondo et al., 1997; Soto et al., 1998). Some physiological effects of cytokines noted above (e.g., fever, anorexia) are also regulated by prostaglandins, such as PGE2. Glucocorticosteroids, the end products of HPA axis activation, initially stimulate immunological function in the short term (Dhabhar, 2002) but eventually provide negative feedback and effectively close the immune-neuroendocrine loop at multiple levels upstream to prevent the APR from causing potential damage to the body from an over-reactive immune system (Munck et al., 1984; Besedovsky et al., 1986; Besedovsky and del Ray, 1996) (Fig. 3.1).

In experimental studies, pro-inflammatory cytokines (e.g., IL-1) or lipopolysaccharide (LPS), an immunogenic component of the outer membrane of gram-negative bacteria, are commonly employed to activate sickness behavior. Peptidoglycan and lipoteichoic acid from gram-positive bacteria are other potent endotoxins that can activate the APR. The TLRs that recognize these endotoxins are evolutionarily conserved throughout the animal kingdom. Even sponges (phylum Porifera), the phylogenetically oldest metazoa, have TLRs on cell surfaces that recognize LPS (Wiens et al., 2005; Wiens et al., 2007).

From a methodological standpoint, comparative studies have relied upon LPS for stimulating sickness behavior. Most of the cytokines that are commercially available are produced for use in humans and mice, and the structural homology of cytokines, in general, is variable among taxa. Thus, human and murine cytokines may be less effective to activate sickness behaviors in other species. Because of the evolutionarily conserved TLRs as discussed above, most animals will readily respond to LPS. Furthermore, unlike most cytokines, LPS is stable at different temperatures and dissolves readily in solution, making it ideal for field studies. Effects of LPS are also dose-dependent, although the dose–response curve and maximum effective dose are typically species-specific. Ideally, a dose–response curve should be constructed for each species examined. In addition, it should be noted that the physiological and behavioral responses to LPS vary depending upon the route of injection (peripheral vs. central). Typically, LPS and cytokines are more potent when injected centrally than via peripheral routes (Kent et al., 1992). Caution is therefore warranted in comparing those studies that use different injection types and doses. Lastly, response to LPS is relevant to understanding host–parasite interactions, given that LPS is a component of an infectious agent and hosts respond, at least initially, to bacterial endotoxin as if they were infected with live, replicating organisms.

THE EMERGENCY LIFE-HISTORY STAGE AND SICKNESS BEHAVIOR

Most animals at some point in their life cycle are confronted with unanticipated perturbations in their environment. Such perturbations might include an attack from a predator, which would require a rapid flight-or-flight response, or a more gradual threat, such as inclement weather, pollution/endocrine disruption, or a change in social status. These unpredictable events, termed labile perturbation factors (LPFs), are typically transitory, although human disturbance and pollution can often result in more permanent effects. LPFs can further be divided into two categories (Wingfield et al., 1998). Indirect LPFs involve a loss of offspring or deterioration of habitat, whereas direct LPFs affect the individual directly by reducing availability of food resources, by restricting access to resources through disturbance of optimal habitat, and/or by increasing energy demands (e.g., escape from a predator, adverse weather conditions, or disease).

Although animals use predictable events in the environment, such as changing day length, temperature, and rainfall, to anticipate future events and to adjust life-history stages accordingly, responses to LPFs generally require a rapid response, which has been collectively termed the emergency life-history stage (ELHS) (Wingfield et al., 1998). This stage involves redirecting the animal away from normal life-history stages into a physiological and behavioral state that increases the likelihood of survival. A key feature of this response in vertebrates is the activation of the HPA axis, which triggers the release of glucocorticosteroids from the adrenals. These steroid hormones have a number of effects that promote fitness in the short term, including increased gluconeogenesis, suppression of reproductive behavior without inhibiting the reproductive system, immune system regulation, increased foraging behavior, and promotion of escape behavior during the day and restfulness during the night (Wingfield and Ramenofsky, 1999; Sapolsky et al., 2000). Once the LPF passes, normal life-history functions can resume. If an individual is unable to acclimate or move away from the LPF, then chronic symptoms of stress may result from prolonged elevation of high levels of glucocorticosteroids, and eventually death will occur. The chronic effects of glucocorticosteroids are also well described and can include complete inhibition of reproductive function, increased susceptibility to disease from suppression of the immune system, neuronal cell damage (death), severe protein loss, disruption of the arachidonic acid cascade, and inhibition of growth and metamorphosis (Sapolsky et al., 2000). It is suggested that the short-term effects of glucocorticosteroids and activation of the ELHS are adaptive mechanisms to avoid the severe, and in some cases deadly, effects of chronically elevated glucocorticosteroid levels.

There is now compelling evidence that host response to infection can disrupt various life-history stages and functions in both laboratory and field settings. The inhibitory effects of APR activation on the HPG axis are well documented, which can lead to a temporary deactivation of reproductive capacity and effort (Besedovsky and del Ray, 1996, for review). Sexual activity is also suppressed in female laboratory rats administered either LPS or IL-1, but not in males (Yirmiya et al., 1995; Avistur and Yirmiya, 1999). In free-living birds, LPS decreases male territorial aggression and song in the white-crowned sparrow (*Zonotrichia leucophrys gambelii*) during the breeding season (Owen-Ashley et al., 2006) and in song sparrows (*Melospiza melodia morphna*) during the non-breeding, winter season (Owen-Ashley and Wingfield, 2006). Parental care is also reduced in female house sparrows (*Passer domesticus*) after LPS treatment, with some females even opting to abandon their entire clutch (Bonneaud et al., 2003).

The concept of sickness behavior and fever as unique components of the generalized ELHS (Fig. 3.2) was suggested by Wingfield (2003). Expression of sickness behavior and APRs can be triggered either directly by invading microorganisms or in combination with other substages where wounding/infection might occur (Fig. 3.2). Briefly, these additional ELHS strategies include (1) proactive/reactive coping styles (Koolhaas et al., 1999) for responding to psychosocial stressors, predation, etc., which are regulated neurally and by both HPA and HPG axes, (2) the fight-or-flight response such as anti-predator aggression, which is mediated by the sympathetic arm of the autonomic nervous system, and (3) facultative behavioral and physiological responses, which are mediated by the HPA axis and associated neuropeptides.

The APR and sickness behavior are considered a fourth coping mechanism that is mediated by cytokines, prostaglandins, and HPA activation. Although the interactions between these substages

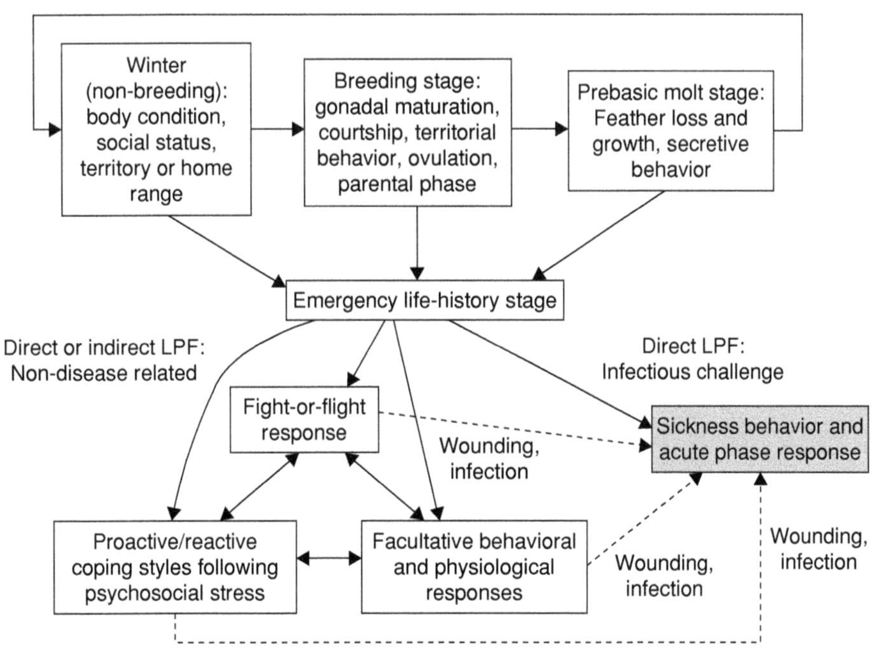

FIGURE 3.2: The emergency life-history stage (ELHS) and sickness behavior (modified from Wingfield, 2003). There are four major substages that may be expressed in varying combinations according to context and circumstances. LPFs are labile perturbation factors in the environment that disrupt the normal life cycle.

are not well understood, it logically follows that an APR and sickness behavior would be activated, for example, when an individual suffers injury from escaping a predator during a flight-or-flight response. If the animal successfully escapes and does not experience other LPFs, then APRs and sickness behavior can be triggered if the wounds become infected. Another possibility is that these different ELHS responses would be expressed sequentially, rather than simultaneously. Short-term increases in glucocorticosteroids would also enhance immunological function and regulate the APR (Dhabhar and McEwen, 1999). Return to a normal life-history stage would occur only once the individual has fully recovered from infection and other LPFs, or at least recovered to an extent that would permit other life-history stages to be expressed.

Presenting sickness behavior in the framework of the ELHS allows us to make some predictions about how an animal might respond to multiple perturbations in the environment. In addition, the idea that sickness behavior has a temporal or "opportunity" cost associated with it becomes apparent because normal life-history stages are usually deactivated in favor of expressing sickness behavior and the APR, although there is some modulation depending upon life-history stage and level of reproductive investment (see below). This viewpoint is a departure from past ecoimmunological studies that have typically focused upon interpreting the regulation of the immune system as a function of energetic or immunopathological costs. The inclusion of temporal costs incurred during a sickness response provides another dimension for assessing trade-offs between host defenses and other fitness components.

ALLOSTASIS AND SICKNESS BEHAVIOR

The process by which organisms actively modify their behavior and physiology to accommodate both predictable and unpredictable changes in the environment is known as *allostasis*, or stability through change (McEwen, 1998; McEwen and Wingfield, 2003). The primary mediators of allostasis are catecholamines, cytokines, and glucocorticosteroids. They orchestrate behavioral and physiological alterations in response to changing environments and LPFs. Various patterns and combinations of these primary mediators are known as an allostatic state or reactive scope and vary according to the type of perturbation and potential for stress (McEwen, 1998; Romero et al., 2009). These states can be sustained for periods of time if food and/or energy reserves (e.g., fat stores) are accessible to fuel homeostatic processes. However, when energy demand outweighs supply, then glucocorticoid levels increase further, activating an ELHS, and symptoms of allostatic overload can appear (McEwen and Wingfield, 2003; Wingfield, 2005; Fig. 3. 3).

The basic existence of an organism requires a minimum energy threshold (Ee) that maintains essential homeostatic functions necessary for survival. The energy required to adjust and respond to predictable changes in the environment and to forage, obtain, and process food (Ei) is regulated allostasis (Fig. 3.3). During reproduction, energy demands increase above Ei but alternatively may decline during hibernation or torpor, for example. In all cases, the sum of energy output should be normally below the energy available that can be gained in the environment (Eg) and that the individual can consume. If an LPF in the environment occurs, such as a storm or changes in habitat, then the energy required to forage and obtain food, Eo, increases and the HPA axis is triggered. If the summation of Ee + Ei + Eo is less than Eg, then allostatic load increases, but the individual still

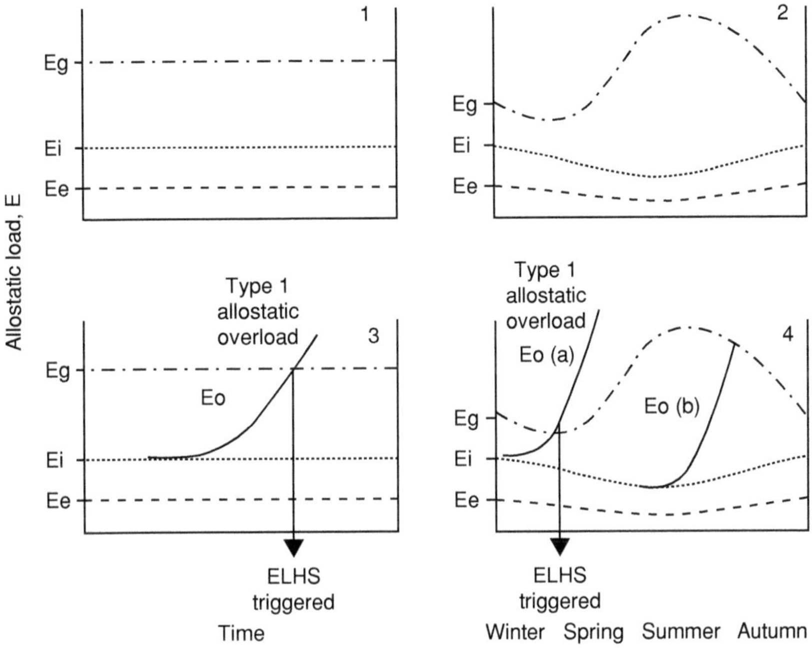

FIGURE 3.3: The effects of labile perturbation factors (LPFs; the frequency, intensity, and duration of storms, predators, human disturbance, etc.) on allostatic load (E; energy levels) in free-living organisms. Ee is the existence energy simply to sustain life and Ei is the energy required to go about daily routines such as foraging, maintaining a territory, etc. In panel 1, Eg is held constant, but as environmental conditions deteriorate the energy required to go out, obtain food, and process and assimilate it increases and a new line, Eo, can be inserted (panel 3). As the LPF persists (panel 3), then Eo increases until Ee + Ei + Eo > Eg. At this point the individual goes into negative energy balance (allostatic overload type 1) and an emergency life-history stage (ELHS) is triggered. Secretion of glucocorticosteroids increases dramatically, thus driving the transition to an ELHS (see Wingfield et al., 1998). In panel 2, a more realistic scenario based on seasons is presented in which Ee and Ei decrease in summer because of higher ambient temperatures, etc., and increase in winter. Eg tends to increase in spring due to primary productivity in the environment and then decreases through autumn and winter. If we then insert a LPF, Eo (a), in winter allostatic overload is achieved very quickly and an ELHS is triggered. However, if the same LPF occurs in summer, Eo (b), then allostatic overload is not reached and the individual can maintain its normal life-history stage throughout this perturbation (panel 4). In this case, the same LPF would result in the same rise in Eo, but an ELHS would be triggered in one case but not the other. In this way, the energy model can explain individual differences in coping with the environment but keeping the potential mechanisms on a common framework. Note also that when in negative energy balance, the triggering of an ELHS is designed to reduce allostatic load (Ee + Ei + Eo) so that the individual can regain positive energy balance and survive the LPF in the best condition possible. If allostatic load is so high that even an ELHS cannot reduce it sufficiently to gain positive energy balance, then chronic stress will ensue. From McEwen and Wingfield (2003).

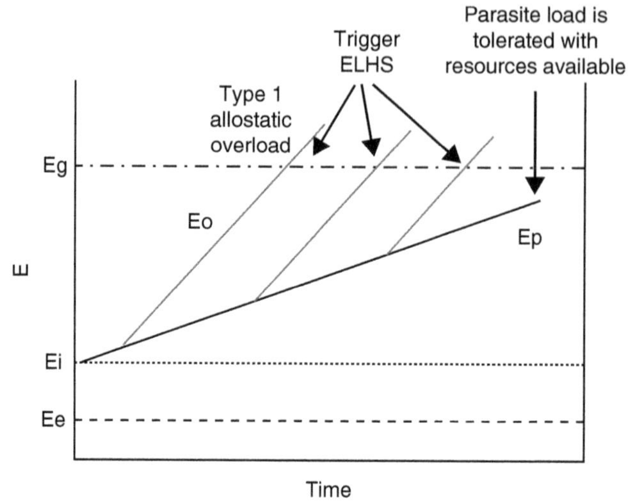

FIGURE 3.4: Using the allostasis model we can insert a line, Ep, which represents the additional energy required to fuel increasing parasite infections. As long as Ep remains below Eg, then the parasite load can be tolerated, assuming that no other pathological effects occur. However, such an individual is susceptible to further perturbations of the environment (Eo, in light gray). Note that as the parasite load (Ep) increases, then the same perturbation (Eo) will exceed Eg and trigger an emergency life-history stage (ELHS) more quickly. Note that increasing infection will also limit how much energy can be allocated to other life-history stages such as breeding.

remains in its normal life-history cycle. However, if Ee + Ei + Eo surpasses Eg, then type I allostatic overload occurs, which activates the ELHS following further release of glucocorticoids (Fig. 3.3).

Using the allostasis model we can insert a line, Ep (Fig. 3.4), which represents the additional energy required to fuel increasing ecto- and/or endo-parasite infections. As long as Ep remains below Eg, then the parasite load can be tolerated as long as no other pathological effects occur. However, such an individual is susceptible to further perturbations of the environment (Eo, in light gray). Note that as the parasite load (Ep) increases, then the same perturbation (Eo) will exceed Eg and trigger an ELHS more quickly (Fig. 3.4). Also note that increasing infection will limit how much energy can be allocated to other life-history stages such as breeding.

Allostatic load refers to the cumulative cost to the body from homeostatic processes involved in daily and seasonal routines as well as additional costs of the predictable life cycle (e.g., breeding, migration) and any unpredictable perturbations (McEwen, 1998; McEwen and Wingfield, 2003). *Allostatic overload* represents a state in which the sum of the energetic costs of daily routines, life-history stage, and any perturbations exceeds the energy available to fuel these costs (see Fig. 3.3). If allostatic overload persists, then serious pathophysiology can occur. There are two types of allostatic overload.

Type 1 occurs when energy demand exceeds supply, resulting in activation of an ELHS. The primary function of the ELHS is to decrease allostatic load so that the individual resumes positive energy balance (see Fig. 3.3). Type 2 allostatic overload is initiated when there is sufficient or even excess energy available, but the effects of social dysfunction, presence of predators, infection and/or injury, and other long-term changes in the environment increase Ee and Ei permanently but not

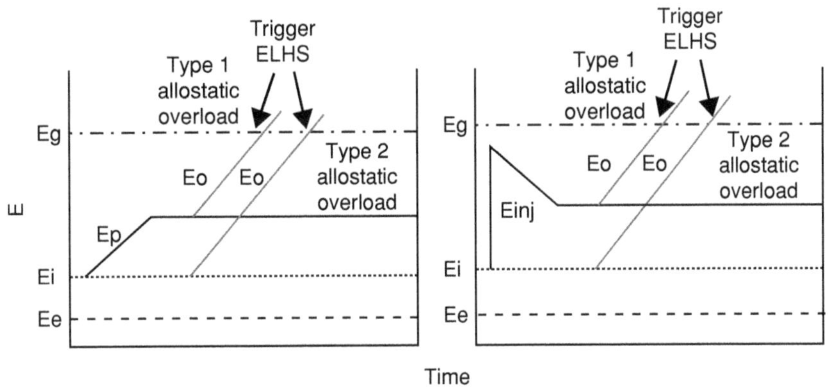

FIGURE 3.5: Examples of allostatic overload type 2. An individual infected with parasites will incur an increased allostatic load (Ep in left-hand panel). If it is assumed that this infection reaches a stable point, then the increased allostatic load becomes long term, consistent with allostatic overload type 2. Note that Eg has not been exceeded, but mediators of allostasis, such as glucocorticoids and cytokines, may be high. In another example (right-hand panel), an injury may suddenly increase allostatic load (Einj), which then declines as the injury heals. If the injury does not completely heal or impairs daily routines, then the allostatic load may become permanent, consistent with allostatic overload type 2. Note that in both panels a further perturbation (Eo in light gray) will exceed Eg and trigger an emergency life-history stage (ELHS) more quickly in the individual with allostatic overload type 2. Note also that in both panels, allostatic overload type 2 will also limit how much energy can be allocated to other life-history stages such as breeding.

sufficiently to exceed Eg (Fig. 3.5). There are many examples of this pathophysiological response in human societies and animals in captivity, and the long-term consequences (e.g., atherosclerosis, inflammatory diseases) can be debilitating even though food supply is still sufficient to fuel such overload. Type 2 allostatic overload does not trigger a fight-or-flight response but does make an individual more susceptible to perturbations of the environment and thus allostatic overload type 1 (Fig. 3.5). Allostatic overload type 2 may be long term (such as in cases of infection or injury; Fig. 3.5) and can be mitigated only through learning and changes in social behavior and structure (McEwen and Wingfield, 2003).

Again, energy requirements (E) are broadly defined as energy needs, but also include limiting factors such as minerals, vitamins, essential amino acids, and so forth. Allostatic overload type 2 may be a frequent consequence of infection and injury, as illustrated in Figure 3.5. An individual infected with ecto- and/or endo-parasites will incur an increased allostatic load (Ep in left-hand panel, Fig. 3.5). If it is assumed that this infection reaches a stable point, then the increased allostatic load becomes long term, consistent with allostatic overload type 2. Note that Eg has not been exceeded, but mediators of allostasis such as glucocorticoids and cytokines may be high. In another example (right-hand panel, Fig. 3.5), an injury may suddenly increase allostatic load (Einj) that then declines as the injury heals. If this injury does not completely heal or impairs normal daily routines, then the allostatic load may become permanent, consistent with allostatic overload type 2. Note that in both panels a further perturbation (Eo in light gray) will exceed Eg and trigger an ELHS more quickly in the individual with allostatic overload type 2. Note also that in both panels, allostatic overload type 2 will also limit how much energy can be allocated to other life-history stages such as breeding. This type of response will not be discussed further.

Activation of sickness behavior and the APR is actually a special case of allostatic overload type 1. The interesting aspect of the sickness response is that food intake is drastically reduced by anorexia at a time when the energetic demands of activating fever and other components of the APR, such as protein accretion from the liver, are quite high. This energetic bottleneck is offset by decreased activity but still leads to a negative energy balance, an increase in glucocorticosteroid levels, and potentially rapid entry into an ELHS. To compare sickness responses with classic allostatic overload type 1, it is useful to modify the simple energy models that have been used previously to understand both the ELHS and allostasis (see Fig. 3.3).

The energetic dynamics of the sickness response, including the mediators of allostasis such as glucocorticoids and cytokines of the immune system, are different compared to classic examples of allostatic overload types 1 and 2 outlined above. First, despite available energy in the environment (Eg), food intake essentially drops to zero (Fig. 3.6B). Thus, we must insert a new line, Ef (energy from food intake) that is independent of Eg, that is not changing and is removed from the graph (Fig. 3.6B). We will now assume that Eg is always sufficient and will exceed any increase in Ef. Because Ef is less than Ee (Fig. 3.6B) in an anorexic state, endogenous energy reserves (Er) must be drawn upon (Fig. 3.6C and D). Next, we need to derive an additional term. Eo, the additional energy needed for finding food during an LPF, is replaced with the additional energy needed to produce an APR (Eapr) to fight infection.

Energy reserves in the form of protein, carbohydrates, and some lipids (Er) are necessary to fuel the APR and other immunological defenses and will decline over time, culminating in body mass loss. Note that as Eapr increases, Er decreases, and the point at which Er intersects Ee + Ei represents the depletion of energy available to activate and maintain the APR. Refeeding should occur at this time (increased Ef) or the individual will likely die. A further inspection of Figure 3.6 reveals that an individual with high energy stores, depicted as Figure 3.6C, will tend to produce a stronger APR of longer duration than an animal with low energy reserves (Fig. 3.6D). This observation lays the foundation for explaining how sickness behavior is primarily modulated by energy limitation later on in this chapter. In addition, both high Er and low Er will decline until reaching Ee + Ei, after which a minimum body mass threshold is strictly defended. This threshold has been demonstrated in a number of experimental studies (Bilbo et al., 2002; Owen-Ashley et al., 2006; Owen-Ashley and Wingfield, 2006; Owen-Ashley et al., 2008; Buehler et al., 2009). Note that a further reduction in Er typically does not occur during an acute infection, as individuals will need energy to forage and search for food once they have recovered (remember Eg has remained sufficiently high throughout) and need to regain lost body mass (Fig. 3.6A). The exception to this rule is when a chronic, debilitating disease induces cachexia and muscle wastage. In these cases, body mass cannot be reversed through nutritional means, and the probability of death from the underlying condition increases substantially. In humans, cachexia is seen in patients with cancer, AIDS, and congestive heart failure, and can be viewed as a chronic sequela of allostatic overload type 1.

Energy requirements for most animals typically vary on a seasonal basis as food availability (Eg) changes over the annual cycle (see Fig. 3.3). For instance, a defining characteristic of high-latitude and temperate environments is that Eg increases in the spring and summer and declines in autumn and winter (see Fig. 3.3). Existence energy (Ee) and the energy needed to go out, find,

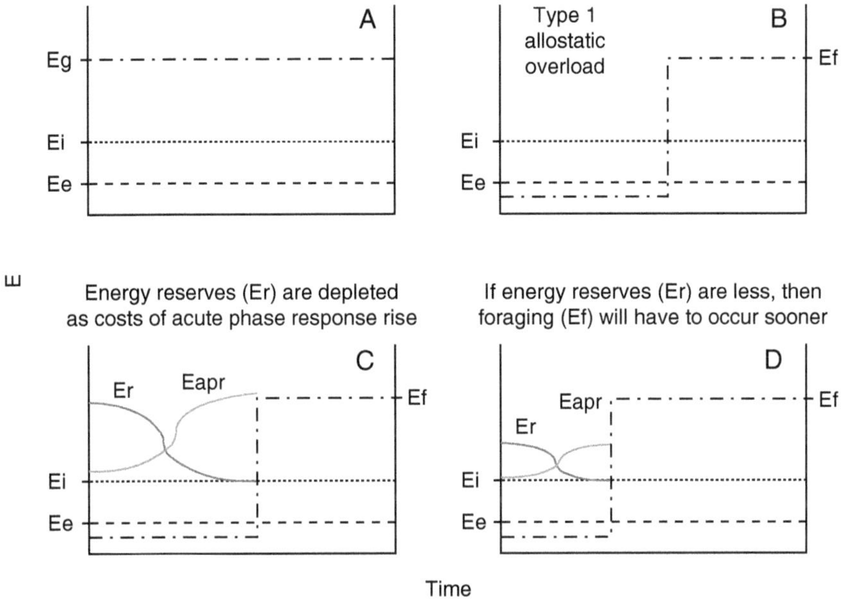

FIGURE 3.6: The concept of allostasis allows the modeling of sickness behavior using allostatic load as a metric of energetic costs (panel A) to mount an immune response in relation to availability (Eg, the energy to be gained from the environment) and the energy required to go about daily and seasonal routines (Ee, existence energy, and Ei, energy required to go out, forage, maintain a territory, etc.). When the sum of Ee, Ei, and Eo exceeds Eg, then the individual is in negative energy balance (called allostatic overload type 1, see Fig. 3.3). During activation of the acute phase response following infection and/or injury, anorexia is frequently a result. Thus, we remove Eg from panels B, C, and D, although it should be borne in mind that food (Eg) is probably always available. In panel B, we introduce a new term Ef (energy gained from food intake). Thus anorexia (very low Ef) results in immediate allostatic overload type 1 (Panel B). At this point energy reserves (Er) must be tapped (dark gray line in panel C) to fuel the rising energetic costs of mounting an APR (light gray line in panel C). Once Er is depleted (intersection with Ee + Ei), then Ef must increase (panel C). If an individual has less Er, then the ability to mount an APR will be lessened and of shorter duration (panel D). Thus Ef will have to increase earlier. This model provides a basic framework to make predictions about how individuals should respond to infection and wounding depending upon circumstances and energy reserves.

and process food (Ei) will correspondingly decline in the spring and summer and increase again in the autumn and winter (see Fig. 3.3). Because of these differences in seasonal energy requirements, allostatic overload type 1 will occur rapidly during the winter (see Fig. 3.3) compared to the summer. In contrast, activation of the APR and the duration of allostatic overload type 1 is not directly dependent upon seasonal shifts in energy availability in the environment (Fig. 3.6). Instead, differences in energy reserves (Er) are still the primary factor regulating allostasis in this case. However, it is likely that fat stores will vary over the year and coincide with seasonal fluctuations in food availability (Figs. 3.3 and 3.6).

These theoretical energy models provide an important tool for making predictions about how environmental conditions, social status, and disease interact to regulate allostatic load and how and when individuals might become vulnerable to allostatic overload type 1. These models also depict some of the costs (both in terms of time and energy) that are associated with expressing sickness behavior.

COSTS OF SICKNESS BEHAVIOR

A recurrent principle of ecological immunology is that immunological defenses are optimized and balanced against their costs to activate and maintain at the expense of competing physiological processes and their relative performance in mitigating and neutralizing continual challenges from a variety of pathogens and parasites. It is clear that pathogenesis can cause extensive damage to the host, including lethal effects, in immunocompromised individuals. Alternatively, there are also costs to excessive immunological defenses that lead to inflammatory diseases and autoimmunological disorders (Råberg et al., 1998; Graham et al., 2005) and in some cases septic shock and death. These opposing selective forces presumably optimize the strength, duration, efficacy, and rate of particular host defenses.

ENERGY AND NUTRITION

The costs of exhibiting sickness behavior and accompanying APRs can be substantial. Much of the energetic costs associated with this nonspecific defense are attributed to fever and alterations in nutrient use to different organs, especially the liver (Klasing and Austic, 1994a, b; Lochmiller and Deerenberg, 2000; Klasing, 2004). Although broadly effective at controlling infection, the APR is thought to offer less protection to the host and exacts a higher cost-to-benefit ratio than specific lymphocyte-mediated defenses (T-cell activation, antibody production) (Klasing, 2004). Nevertheless, the APR may be favored in cases where an immediate and rapid response is required or when rapid development precludes investment in adaptive immunity (Klasing, 2004). There are also nutritional costs associated with APR activation, especially in terms of amino acid use and turnover. For example, accretion of APPs from the liver accounts for 4.6% of the lysine intake in a growing chick (Klasing, 2004). Glucose utilization can also increase up to 68% during an APR (Klasing, 1998). Thus, some nutrients that would otherwise be utilized for reproduction or growth are diverted to the APR, which creates allocation constraints for resources. Reduced locomotion as a result of sickness behavior should ameliorate some of these costs (Hart, 1988), but there is still an energy deficit (Fig. 3.6).

HOST DAMAGE

Sickness behavior inflicts damage to the host if it leads to cachexia, which is a pathophysiological response to chronic disease (e.g., cancer, AIDS) and is caused by metabolic abnormalities from tumor products and cytokines. Cachexia involves severe weight loss, lipolysis, wasting of muscle and visceral protein, chronic nausea, and weakness. Fever is commonly viewed as inflicting collateral damage to the host during an infection. A high fever (>41°C in humans), although rare, is clearly detrimental to fitness and triggers seizures, cerebral ischemia, impairments in brain metabolism and secondary messenger systems, cytoskeletal degradation, as well as eventual total organ failure. For lower fevers, depending upon duration, increased metabolic demands may exceed the metabolic or cardiopulmonary capacities of the host, which can cause ischemic injury to certain tissues. In addition, increased production of reactive nitrogen and oxygen species from activated phagocytes during inflammation and fever may facilitate the neutralization of invading pathogens but may also cause lipid peroxidation and DNA damage to non-infected host cells (Bertrand et al., 2006; Sorci and Faivre, 2009). Some of this damage is prevented by antioxidant protection (Finkel and Holbrook, 2000). Kluger (1998) speculated that fever typically becomes maladaptive only when cytokines and other inflammatory mediators are overproduced. This leads to exaggerated responses, which include high fevers and potentially harmful effects from vascular leakage (Baluna and Vitetta, 1997), potentially leading to stroke.

TEMPORAL COSTS

The activation of the APR and sickness behavior in vertebrate hosts is a component of the ELHS as discussed above, while temporarily suspending other life-history stages and behaviors, such as breeding, sexual behavior and displays, parental care, and territorial aggression. These "opportunity" costs involve disruption of normal activities not only *during* an APR, but also when an individual is recovering from an infection and attempting to recoup some of the energy reserves that were lost from illness-induced anorexia (Owen-Ashley and Wingfield, 2007). Importantly, the time devoted to combating and recovering from an infection can impart serious fitness consequences to hosts if such opportunity costs are high (e.g., foregone opportunities to breed and care for young). Thus, it is predicted that hosts will modulate the expression of APRs and sickness behavior depending upon the intrinsic value of the activities to be sacrificed, which vary in response to ecological conditions (Owen-Ashley and Wingfield, 2007; Adelman and Martin, 2009).

Evidence for such modulation exists in both field and laboratory studies. For example, lactating mice injected with LPS decreased nest-building behavior and delayed retrieval of pups at an ambient temperature of 22°C (Aubert et al., 1997). However, LPS-injected females exhibited nest-building behavior and pup-retrieval rates that were comparable to saline-injected females when the ambient temperature was decreased to 6°C (Aubert et al., 1997). At these lower temperatures, pup survival is considerably reduced without attendant care from the mother. In another study, IL-1 administration reduced sexual behavior in female rats but failed to disrupt mating behavior in males (Yirmiya et al., 1995). Importantly, the suppressive effects of IL-1 on locomotion were comparable between males and females (Yirmiya et al., 1995), suggesting that these sexual differences observed are localized to mating. Given the divergent reproductive strategies between the sexes, male rats are likely mating while sick to increase fitness returns, whereas females are opting to delay conception when infected, thereby decreasing the chance of spontaneous abortion or abnormal development of offspring (Avistur and Yirmiya, 1999).

In free-living house sparrows, breeding females treated with LPS exhibited a smaller decline in maternal feeding rates when clutch size was experimentally increased (Bonneaud et al., 2003), implying modulation of the sickness response when fitness returns are increased.

Similarly, in another field study, male song sparrows (*Melospiza melodia morphna*) injected with LPS during the winter (non-breeding season) reduced territorial aggression and became anorexic, while breeding males in the spring showed little or no response to an identical dose of endotoxin (Owen-Ashley and Wingfield, 2006). Since territorial establishment early in the breeding season is critical for maximizing reproductive success in this species (Arcese, 1989; Hiebert et al., 1989), males responding strongly to LPS would likely suffer net losses in fitness because of either territory loss or cuckolding by intruding males.

These studies demonstrate that the expression of sickness behavior and APRs is dependent upon trade-offs with other competing functions (reproduction, locomotion, growth, etc.) that involve energy, nutrition, self-damage, and time as the limiting factors (Owen-Ashley and Wingfield, 2007). Presented another way, expression of various components of sickness behavior may depend upon whether a host has a low or high residual reproductive value (RVV; Adelman and Martin 2009). Hosts with high RRV increase fitness by delaying reproduction, while hosts with low RRV decrease fitness when reproduction is delayed (Clutton-Brock, 1984). The prediction is that hosts with low RRV would invest heavily in reproduction over expression of sickness behaviors (the terminal investment hypothesis), while the opposite relationship would be true for hosts with high RRV (Adelman and Martin, 2009). A similar relationship has been delineated between reproductive value and the vertebrate stress response: when RRV is low, a reduced stress response is expected to enable successful breeding and to maximize fitness (Bókony et al., 2009).

BENEFITS OF SICKNESS BEHAVIOR

From Hart's influential paper (1988) discussed above, sickness behavior is proposed to constitute an adaptive strategy that involves energy-saving measures such as decreased locomotion, increased slow-wave sleep, postural changes, and general disinterest in pleasurable activities (anhedonia), social interactions, and sexual behavior. In addition, activating the ELHS precludes the individual from engaging in energetically demanding functions that could compete with immune function, while compensating for the high costs of fever through conservation of heat and allowing repair of tissues if injured (Hart, 1988; Hart, 1990). These behaviors might also decrease the risk of predation, although vigilance is likely diminished (Hart, 1988).

THE PARADOX OF ANOREXIA

The loss of appetite, or anorexia, is often associated with many bacterial, viral, and parasitic infectious diseases. It is somewhat paradoxical, then, that anorexia is beneficial to the host given that energy is consumed by (1) the high costs of an APR, and (2) parasites/pathogens imposing both nutritional and metabolic demands. Taking into account these costs and the central axiom surrounding ecological immunology that allocation constraints produce trade-offs between immune functioning and other competing functions when resources are limited (Sheldon and Verhulst, 1996; Norris and Evans, 2000), it is puzzling that food intake would be suppressed during this time. Although often described as a side effect of disease (Scrimshaw, 1977; Theologides, 1977), anorexia is a robust phenomenon that is observed in both invertebrates (Dunn et al., 1994; Adamo, 2006; Adamo et al., 2007) and vertebrates (Kent et al., 1992; Exton, 1997), and many of the same cytokines that initiate fever in vertebrates also activate anorexia, suggesting that there is an adaptive component to this response (Exton, 1997). Several compelling experimental studies seem to support this notion. Infected mice that are force-fed to levels seen in healthy individuals were more likely to suffer increased mortality than mice that were not force-fed (Murray and Murray, 1979). No signs of mortality were observed in mice that were force-fed but not infected (Murray and Murray, 1979). Conversely, mice that were acutely starved had higher survival rates after a bacterial infection than mice provided with food *ad libitum* (Wing and Young, 1980). In caterpillars (*Manduca sexta* larvae), increased mortality occurred only in infected caterpillars that were force-fed a high-lipid diet (Adamo et al., 2007). Diets consisting of water or sucrose were ineffective at reducing survival (Adamo et al., 2007). Further investigations have identified an interesting trade-off between lipid transport and immune function in crickets (*Gryllus texensis*), with subjects eating less food and preferring foods that contained less fat after an immune challenge (Adamo et al., 2010). These results suggest that anorexia may provide a means to reduce lipid transport to maximize immune responses. It is unknown whether such a trade-off exists in vertebrates, and this represents an exciting area of future study.

Kyriazakis and others (1998) methodically reviewed five hypotheses that could provide a functional basis for anorexia: (1) induction of the response for the parasite's benefit, (2) reduced food intake starves parasites, (3) reduction in energetic efficiency leads to anorexia, (4) anorexia enhances effective immune responses, and (5) anorexia leads to increased diet selectivity during infection. The first three hypotheses failed to demonstrate consistency with the physiological and behavioral alterations that occur during an infection and the generality of the response being exhibited in both protozoan and helminth infections (Kyriazakis et al., 1998). To support the idea that anorexia enhances immune responses (hypothesis #4), several studies have demonstrated that food restriction enhances several components of immune function (Nakamura et al., 1990; Zysling et al., 2009). However, the research must be viewed with caution for several reasons. First, food restriction does not necessarily mimic an anorexic response from infection, and there are different mechanisms controlling the two responses (Konsman and Dantzer, 2001). Second, it is generally recognized that food restriction suppresses immune function, although these differences are likely influenced by the duration and degree of restriction and whether glucocorticoids are elevated, which are potent immunomodulators. The last hypothesis, that anorexia increases diet selectivity, assumes that a reduction in food intake improves an animal's ability to select foods that either reduce the chance of infection or possess certain anti-infectious properties. There is some support that animals alter their preferences for certain foods when infected, such as the case of conditioned taste aversion where animals exposed to a novel taste before an illness subsequently avoid this food item during infection (Dantzer and Kelley, 2007). Protein intake is also voluntarily reduced following treatment with LPS or IL-1 compared to carbohydrate- and fat-rich foods, presumably because protein is more likely to contain iron (Aubert et al., 1995). Nevertheless, it is still unclear why increased selectivity of certain food types would necessitate a reduction in food intake.

Anorexic responses to pathogens are thought to facilitate the reduction of plasma levels of trace metals, such as iron, zinc, and copper, which are essential elements required for many bacteria and other pathogens to grow and replicate (Weinberg, 1974; Kontoghiorghes and Weinberg, 1995). For example, low levels of these elements do not by themselves impair bacterial growth, but when reduced levels of iron, for example, are coupled with fever, the proliferation of bacteria is inhibited (Kluger and Rothenberg, 1979). However, Kyriazakis et al. (1998) argued that anorexia itself is independent of host nutrition and is an ineffective defense against parasites since anorexia typically involves a gradual rather than an abrupt loss of nutrients from the host. Furthermore, it is unclear how the pathogen could suffer more harm than the host during this response.

A final hypothesis for why anorexia might benefit host survival involves apoptosis, or cell suicide. It is hypothesized that anorexia could lead to premature death of infected cells through enhanced apoptosis, thereby preventing infection from other non-infected cells (if they survive; LeGrand, 2000). Food restriction induces apoptosis in the liver and constitutes a defense mechanism for combating cancerous cells and some pathogens (Grasi-Kraupp et al., 1994). Nonetheless, it remains to be seen whether apoptosis is increased with infection-induced anorexia (LeGrand, 2000).

Given these various explanations for understanding the functional significance of illness-induced anorexia, there is no clear resolution to the debate and other studies are needed to pinpoint what the actual benefits of the response are and how they might outweigh the costs.

MODULATION OF SICKNESS BEHAVIOR

If the behavioral changes that accompany infection are part of a coordinated strategy to increase the likelihood of survival during an infectious challenge, then it seems reasonable that modulation of sickness behavior may occur with respect to underlying fitness consequences, and the internal and external constraints that are experienced throughout the life cycle of the host (Aubert, 1999). This behavioral flexibility is often termed *motivation*, which is defined as a central organized state that coordinates both perception and action (Bolles, 1967). A key feature of a motivational state is that by its very nature of being expressed, it competes with other types of motivational states for behavioral output (Aubert, 1999), thereby exacting an opportunity cost as discussed above. For example, a male vertebrate foraging for food will find it challenging and likely ineffective to defend a breeding territory at the same time. Furthermore, many competing behaviors are simply incompatible, such as the simultaneous expression of parental care and territorial aggression (Wingfield et al., 1990), or migratory behavior and nesting. Environmental and social stimuli are thought to adjust the expression of these competing behaviors into a discrete life-history stage or substage, where motivation for one set of behavioral output prevails (Aubert, 1999; Wingfield, 2003). This type of behavioral hierarchy seems to apply to the expression of sickness behaviors as well. Even though responding to an infectious insult is critical to an animal's survival, the time window to overcome an infection is typically measured in days to weeks, which, in theory, leaves opportunity for switching behavioral phenotypes to accommodate more urgent needs (e.g., predator aversion, sexual behavior, care of offspring) that might optimize fitness. We next review studies that have demonstrated the plasticity of sickness behavior in a number of vertebrate taxa.

EARLY MOTIVATIONAL STUDIES

Neal Miller and colleagues in the 1960s conducted the initial studies to demonstrate that sickness behavior is, in actuality, a motivational state (Holmes and Miller, 1963; Miller, 1964). Using operant conditioning, they reported that rats treated with endotoxin would stop pressing a bar to receive a water reward, but would drink freely from a water cup when placed in their cage (although to a lesser extent than normal). Furthermore, when placed on a running wheel, rats injected with endotoxin would press a bar at a higher rate compared to controls to temporarily stop the wheel, thereby obtaining more time to rest (Miller, 1964). These studies indicate that rats were motivated to display sickness behavior by either increasing or decreasing their operant response. Unfortunately, these important experiments were not fully appreciated until the field of psychoneuroimmunology prompted a resurgence in the study of sickness behavior several decades later (Aubert, 1999; Dantzer and Kelley, 2007).

SEASONAL MODULATION

There is now abundant evidence that immune function varies on a seasonal basis in several vertebrate species (for review see Nelson and Demas 1996; Martin et al., 2008; and Chapter 7 in this volume). Many animals seasonally reproduce in the spring and summer to take advantage of predictable seasonal increases in food supply and warmer temperatures at mid- and high latitudes. In contrast, winter is often considered an energetically demanding time for animals that are sedentary (e.g., rodents). These animals must adjust their physiology and behavior to cope with the lower ambient temperatures, reduced food supply, and increased thermogenic requirements that are encountered during the winter months. Migratory species, on the other hand, are able to lessen the impact of seasonal energy deficits by wintering in areas that have warmer temperatures and a relatively abundant supply of resources (e.g., tropics), albeit parasite pressure is typically greater than higher latitudes (Piersma, 1997).

Given seasonal changes in energy availability, it is predicted that the partitioning of resources allocated to immunological defenses and other competing functions also shifts accordingly over the annual cycle. Empirical studies suggest that immunocompetence varies on a seasonal basis, but in no clear-cut direction. For example, in rodents, immune function is typically diminished in winter when food is limited and thermogenesis is increased to adjust to colder temperatures (Nelson and Demas, 1996; Nelson et al., 2002; Nelson, 2004). In captivity, the effect is reversed and short day lengths enhance components of immune function to supposedly anticipate these seasonal energy deficits incurred during the winter, often termed the "winter immunoenhancement hypothesis" (Nelson and Demas, 1996). However, this is not a universal phenomenon, as some captive rodent species exhibit short-day–induced immunosuppression (Nelson and Demas, 1996) or even fail to display any photoperiodic changes in immune function (Demas and Nelson, 2003). Furthermore, similar experiments in birds have also yielded mixed results (Hasselquist, 2007). Some studies demonstrate heightened adaptive immune function in short-day (non-breeding) conditions (Bentley et al., 1998; Zuk and Johnsen, 1998; Lozano and Lank, 2003; Martin et al., 2004; Greenman et al., 2005) or after melatonin treatment (Bentley et al., 1998; Moore and Siopes, 2000), whereas others have identified enhancement during the breeding season

(Hasselquist et al., 1999; Møller et al., 2003) or simply a lack of seasonal modulation (Gonzalez et al., 1999; Moore and Siopes, 2000; Owen-Ashley et al., 2008).

Many of these discrepancies, at least in rodents, have been rectified by establishing the connection between immune function and seasonal fluctuations in energy stores (Demas et al., 2003; Demas, 2004). Species that increase body fat during the winter tend to exhibit immunoenhancement, whereas animals that lose fat on short days show reduced immune function (Demas, 2004). Some of these effects are also independent of immunomodulatory properties of gonadal steroids and melatonin (Bentley et al., 1998; Wen et al., 2007; Prendergast et al., 2008), which also vary on a seasonal basis. Furthermore, experimental reductions in total body fat of several rodent species housed on long days suppressed humoral immune function (Demas et al., 2003), suggesting that immunological responses are coupled with seasonal changes in energy reserves. The apparent link between energy availability and immune function has important implications for examining seasonal changes in the expression of sickness behavior and is described for rodents in Chapter 9.

In captive birds, similar seasonal variation in sickness behavior has been identified. Captive male Gambel's white-crowned sparrows (*Zonotrichia leucophrys gambelii*) held on short days (nonbreeding conditions) exhibited reduced sickness responses after LPS challenge compared to males exposed to long days (breeding). This variation corresponded with long-day males being heavier and having larger fat depots than short-day males (Owen-Ashley et al., 2006). Interestingly, body mass and fat stores in female sparrows did not increase upon exposure to long days, which is consistent with females failing to exhibit seasonal changes in LPS-induced sickness responses in this particular experiment (Owen-Ashley et al., 2006). In captivity, females tend to be fatter all the time, whereas in the field they do fatten in spring during migration and arrival and prior to incubating.

In another study, seasonal variation in sickness behavior was compared between two subspecies of white-crowned sparrow that seasonally breed at different latitudes and have disparate migratory strategies (Owen-Ashley et al., 2008). Gambel's white-crowned sparrow (*Z. l. gambelii*) migrates long distances from wintering grounds in the southern United States and Mexico to reach high-latitude breeding areas in Alaska and British Columbia (47–68°N), while the Puget Sound white-crowned sparrow (*Z. l. pugetensis*) is a short-distance migrant that breeds in temperate regions from southern Oregon to southern British Columbia (40–49°N) and winters in central California. When exposed to long photoperiod treatments (either 16L:8D or 20L:4D), male Gambel's white-crowned sparrow gained body mass and fat stores and exhibited stronger sickness responses after LPS challenge than males on short days (8L:16D), as previously described (Owen-Ashley et al., 2006). These fat reserves are critical for fueling the vernal migratory bouts to high-latitude breeding grounds, and birds rapidly gain body mass by exhibiting premigratory hyperphagia upon long-day stimulation (King, 1961). This seasonal variation in sickness behavior and energy reserves was in stark contrast to the response of Puget Sound white-crowned sparrows. Body mass or fat stores did not increase in males after long-day photostimulation, and LPS-induced sickness responses failed to vary seasonally (Owen-Ashley et al., 2008). Given that Puget Sound white-crowned sparrows migrate only short distances in the spring to reach breeding areas, there is little need to accumulate large energy reserves for migration. These differences in the response of short- versus long-distance migratory songbirds to LPS challenge exemplify the variable responses that can occur during a single life-history stage and demonstrate the importance of possessing sufficient energy stores during an infectious challenge that individuals *can afford to lose* when they become anorexic during an acute illness.

Such experiments on captive animals have a distinct advantage relative to field studies by permitting the study of behavioral responses in the laboratory that would either be difficult or impossible to measure in free-living animals. However, recent improvements in animal tracking technology have vastly improved over the past decade, making it possible to examine fine-scale alterations in movement as well as various physiological parameters, such as body temperature and heart rate (Wikelski et al., 2007; Hetem et al., 2008; Bisson et al., 2009; Adelman et al., 2010b). Furthermore, a drawback of captive studies is that conditions are not very representative of those experienced in the wild, where food supply is typically limited and ambient temperatures are variable.

Several studies have examined sickness responses in wild animals through experimental (Muchlinski et al., 2000; Bonneaud et al., 2003; Owen-Ashley et al., 2006; Owen-Ashley and Wingfield, 2006; Eraud et al., 2009; Adelman et al., 2010b) or opportunistic (Hetem et al., 2008) methods, and seasonal variation has been reported. Rufous song sparrows (*Melospiza melodia morphna*) of western Washington State are sedentary and males defend breeding territories throughout the year, except when undergoing pre-basic molt from August to September. To examine seasonal patterns in sickness responses, free-living males were injected with LPS or saline (vehicle), banded with a unique color combination, and then immediately released back onto their territories in either January (non-breeding) or April (early breeding) (Owen-Ashley and Wingfield, 2006). Approximately 24 hours later, territorial aggressive behavior of focal males was measured in response to a caged male decoy (simulated territorial intrusion), and a proportion of males was recaptured to evaluate changes in body mass. LPS treatment decreased territorial aggression in males during the winter (non-breeding), but not in the spring (breeding), compared to saline-injected birds (Owen-Ashley and Wingfield, 2006). In addition, LPS males that were recaptured 25 hours later experienced anorexia and body mass loss in the winter, but not in the spring (relative to controls), indicating relative insensitivity to endotoxin in breeding males (Owen-Ashley and Wingfield, 2006). At face value, these results contradict the captive studies that demonstrate attenuation of sickness behavior on short day lengths. However, it should be emphasized that free-living male song sparrows exhibit opposing patterns of seasonal energy accumulation compared to captive white-crowned sparrows and Siberian hamsters. Male song sparrows are actually heavier and have larger fat stores during the winter compared to the spring, when energy reserves are minimal. Winter fattening (King et al., 1961, 1970) is a common strategy employed by many seasonally breeding vertebrates to compensate for longer periods of overnight fasting during the darker months of winter. In contrast, captive Gambel's white-crowned sparrows accumulate substantial energy stores when exposed to long days as a result of premigratory hyperphagia (Wingfield et al., 1996) while Siberian hamsters undergo significant reductions in body mass and white adipose tissue during the winter (Bartness et al., 1989). Thus, this opposing pattern of seasonal energy regulation is likely the driving factor behind why seasonal modulation of sickness behavior is different between captive and free-living studies (Owen-Ashley and Wingfield, 2007).

SEXUAL MODULATION

Sexual differences in immune function are common (Poulin, 1996; Klein, 2000; Nunn et al., 2009; Chapter 16 of this volume), but less attention has focused upon whether the behavioral symptoms of infection vary between the sexes. Epidemiological studies involving humans suggest that males are more susceptible to sepsis than females (Schroder et al., 1998; Offner et al., 1999; Wichmann et al., 2000). Sepsis represents a whole-body inflammatory response to infection, which also involves sickness behavior (Annane et al., 2006). A caveat of these studies is that sepsis typically occurs in elderly populations, which introduces a confounding interaction between gender and age. Female rats also tend to produce lower fevers than male rats (Murakami and Ono, 1987). Sex hormones or sex-related gene polymorphisms are thought to figure prominently into the regulation of these gender-based differences. For example, female mice in proestrus tolerate sepsis better than males (Zellweger et al., 1997), and flutamide (an androgen receptor blocker) improves survival in male mice after hemorrhagic shock and sepsis (Angele et al., 1997). There is also a sexual dimorphism of the HPA axis response to endotoxin, with females producing a heightened response (Frederic et al., 1993; Spinedi et al., 2002). These responses are generally enhanced by castration and attenuated by androgen or estrogen replacement in mammals (Spinedi et al., 1992).

Reproductive strategies between the sexes differ, with the variance in reproductive success being larger in males than in females. Typically, male fitness is limited by the number of mates fertilized, while female fitness is constrained by the number of offspring produced (pregnancy for mammals and the production of eggs in birds and reptiles) (Andersson, 1994). Thus, expression of sickness behavior is likely balanced according to both current reproductive prospects and the expectation of future fitness benefits of each sex. For example, male rats are less sensitive to the effects of IL-1 on mating behavior than female rats (Yirmiya et al., 1995). Males that suppress symptoms of infection when presented with the chance to mate will likely gain an immediate fitness advantage, especially if mating opportunities are limited. Similarly, females that decrease mating behavior when ill are reducing the risk of spontaneous abortion of the fetus during infection (Avistur and Yirmiya, 1999), and thereby minimizing their future fitness losses. Attenuation of sickness behavior in females is more likely to occur when a reduction in parental care from illness has deleterious repercussions upon offspring survival (Aubert et al., 1997; Bonneaud et al., 2003; Weil et al., 2006a).

TRADE-OFFS WITH REPRODUCTION

Life-history trade-offs between immune defense and reproduction are well described, and owing to the costly nature of the APR and sickness behavior, it is expected that this response will covary negatively with reproductive effort of hosts (Avistur and Yirmiya, 1999). Evidence for such trade-offs has been identified in both field and laboratory experiments. Wild female house sparrows when challenged with LPS lost body mass and fed nestlings less than saline-injected controls. However, experimental increases in clutch size (as a proxy for elevated reproductive effort) attenuated the sickness response in females (Bonneaud et al., 2003). Female mice were less sensitive to the suppressive effects of LPS on parental care (Aubert et al., 1997) and maternal aggression (Weil et al., 2006a) when survival of pups was threatened. Febrile responses to endotoxin are suppressed in rats during the later part of pregnancy and immediately after birth compared to virgin and lactating rats (Martin et al., 1995). Besides direct energetic trade-offs between fever and parturition, it is hypothesized that suppression of the APR minimizes the risk of anaphylaxis when maternal and fetal blood mix during delivery, while ensuring that sickness behaviors do not interfere with maternal pup care during the early postpartum period (Martin et al., 1995). Treatment with LPS delayed short-day gonadal regression in male Siberian hamsters (Weil et al., 2006c) and increased egg-laying in female crickets (Adamo, 1999). These studies provide support for the terminal investment hypothesis, which states that organisms should invest in current reproduction if the probability of the next reproductive event is low (Clutton-Brock, 1984). The physiological mechanisms that dictate this life-history strategy during illness are largely unexplored and represent an exciting area of future research.

In male songbirds, LPS treatment reduced territorial aggression and song in non-breeding song sparrows relative to saline-injected controls as previously described, but males that were establishing and defending early breeding territories in the spring were relatively insensitive to the same dose of endotoxin (Owen-Ashley and Wingfield, 2006). In contrast, LPS injection in free-living male Gambel's white-crowned sparrows dramatically decreased both aggression and song during the breeding season compared to saline-injected controls (Owen-Ashley et al., 2006). However, males were captured during the parental phase when subjects were feeding nestlings. In birds that breed at high latitudes, such as Gambel's white-crowned sparrow, the transition between the territorial and parental phase is abrupt, and aggressive behavior can decline rapidly (Meddle et al., 2002; Wingfield and Hunt, 2002). Expression of sickness behavior might be feasible for males during the parental period if breeding females compensate for decreased paternal care, although this hypothesis needs to be explicitly tested.

SOCIAL EFFECTS

Increasing evidence suggests that sickness behaviors can be modulated according to different social stimuli (Avitsur et al., 1998; Avistur and Yirmiya, 1999; Konsman et al., 2002). As previously described, IL-1 administration decreases mating behavior in female but not male rats, even though both sexes experience a general decline in activity levels (Yirmiya et al., 1995). Presence of a sexually receptive female has been implicated as a motivational factor for males to suppress sickness behavior (Aubert, 1999; Avistur and Yirmiya, 1999), although this effect was not apparent in mice (Weil et al., 2006b) or white-crowned sparrows (Owen-Ashley, 2004). Interestingly, in male mice, exposure to a receptive female induced an unexpected increase in hypothalamic IL-1 and TNF-α gene expression as well as a heightened anhedonic response compared to isolated males, suggesting that social interactions actually exacerbate sickness responses in this case (Weil et al., 2006b).

Social dominance hierarchies can also influence the expression of sickness behavior in male mice. Dominant mice showed expected reductions in activity and aggressive behavior after LPS treatment, but submissive males exhibited defensive behaviors, as well as social exploratory behavior (Cohn and de Sá-Rocha, 2006). These results imply that dominant males can afford to prioritize recuperative behavior due to their high social ranking, while submissive males are forced to display defensive behaviors when housed with dominant males. It is proposed that high glucocorticoid levels in submissive mice are responsible for suppressed responses to LPS (although these levels were not directly measured). This laboratory study sets the stage for future comparative studies to examine if modulation of sickness behaviors occurs in social hierarchal systems of wild vertebrates.

The absence of social contact also has a profound impact upon activation of the APR and sickness behavior, especially in young animals that have been deprived of parental care. For example, early social isolation increased LPS-induced sickness responses while impairing suckling behavior in domestic piglets (Tuchscherer et al., 2006). Mouse pups that have been separated from their dams exhibited a similar effect (Avitsur and Sheridan, 2009), suggesting that the early postnatal environment can influence how individuals cope with disease later on in adulthood. Studies showing modulation of the sickness responses in adults caring for young (Aubert et al., 1997; Bonneaud et al., 2003) imply that the presence or absence of offspring could influence the expression of sickness behavior, although this has not been directly tested.

LATITUDINAL/PACE-OF-LIFE EFFECTS

Species with long life spans tend to produce small numbers of offspring that develop slowly and mature relatively late in life. These organisms are often referred to as "slow-living" or K-selected species. At the opposite end of this spectrum are organisms that have short life spans and generate large numbers of offspring that develop rapidly and reach sexual maturity at an early age, generally termed "fast-living" or r-selected species (Promislow and Harvey, 1990). It is hypothesized that higher-quality immunological defenses will be favored in species that have increased investment in self-maintenance over reproduction (Lochmiller and Deerenberg, 2000; Ricklefs and Wikelski, 2002; Klasing, 2004; Tieleman et al., 2005).

As tropical species tend to exhibit a slower pace of life than congeners inhabiting higher latitudes (Wiersma et al., 2007), pace-of-life syndromes are also biased by geography. This association is likely influenced by corresponding latitudinal gradients in parasite abundance and diversity, which is greater at lower latitudes (Piersma, 1997; Harvell et al., 2002). Thus, for comparative studies, it is difficult to specifically attribute intraspecific and interspecific variation of immunological defenses to pace-of-life strategy, parasite-mediated selection, or both without conducting additional studies (e.g., selective breeding experiments).

For costly APRs and sickness behavior, it is predicted that fast-paced species will attenuate responses if they interfere with rapid growth and/or high reproductive potential, although APRs could be favored in cases where rapid response to infection is required or when investments in adaptive immunity are precluded (Klasing, 2004; Lee, 2006). For example, young chickens that have been selected for rapid growth exhibited diminished febrile responses to LPS compared to chicks selected for high egg production in adulthood (Leshchinsky and Klasing, 2001). In contrast, house sparrows lay smaller clutches than closely related European tree sparrows (*Passer montanus*) but are relatively insensitive to a dose of LPS that significantly reduces activity in the latter, faster-paced species during the non-breeding season (Lee et al., 2005). In congeneric rodents (*Peromyscus* spp.), pace of life did not influence LPS-induced sickness behavior, but species that mounted high fevers tended to display minimal sickness behavior (Martin et al., 2008), implying a physiological trade-off between different components of the APR among congeneric rodents. This interesting effect requires further study in other taxa.

Variation in sickness behavior is also expected to occur in populations inhabiting different latitudes. For example, in high-latitude breeding birds, reproductive prospects are limited to one clutch because of the short polar summers, and any disruption in nesting can lead to decreased reproductive success (Hahn et al., 1995; Wingfield and Hunt, 2002). Allocation of resources towards reproduction should be favored in lieu of immune defense, especially if parasites are presumably depauperate in high-latitude habitats (Piersma, 1997). Sickness behavior was diminished in free-living song sparrows breeding in Washington State compared to birds breeding in California (Adelman et al., 2010b), and this difference persisted in controlled captive conditions (Adelman et al., 2010a). Thus, latitudinal differences in sickness behavior are likely not influenced by environmental effects, but rather genetic, maternal, or early-life factors. Contrary to this hypothesis, captive Gambel's white-crowned sparrow, a high-latitude breeder, exhibited stronger sickness responses to LPS than conspecific Puget Sound white-crowned sparrows that breed in temperate regions when exposed to long days (Owen-Ashley et al., 2008). However, these differential responses were highly correlated with body condition, which is a result of subspecies differences in the expression of premigratory hyperphagia after exposure to long days. Future studies that examine latitudinal variation in sickness responses should take into account body condition and other photoperiodic-dependent effects.

AGE EFFECTS AND IMMUNOSENESCENCE

It is well established that immunosenescence, or age-related deterioration of the immune system, occurs in humans and laboratory animals (Miller, 1996). There is now accumulating evidence that free-living populations also show age-dependent declines in immune function (Adamo et al., 2001; Lozano and Lank, 2003; Saino et al., 2003; Haussman et al., 2005; Palacios et al., 2007). Contrary to this relationship, sickness behavior seems to be exacerbated at older ages. For example, aged laboratory mice produced prolonged anorectic responses, leading to increased weight loss, following LPS administration compared to younger cohorts (Godbout et al., 2005). LPS-induced decreases in activity levels and social behavior persisted later in aged mice than adults. Furthermore, aged mice had a different pattern of gene expression in the brain after LPS injection and exaggerated neuroinflammatory and oxidative stress levels in the brain (Godbout et al., 2005). This study suggests that older animals that have experienced senescence of the adaptive immune system switch to more costly and damaging APRs, which can eventually lead to sepsis and death if the infection does not resolve. Alternatively, in rats, older subjects exhibited diminished febrile responses compared to young rats when exposed to lower ambient temperatures (Buchanan et al., 2003; Peloso et al., 2003). However, old rats developed febrile responses that were equivalent to young rats when the ambient temperature was raised. These data suggest that there are temperature-dependent effects upon the management of fever in older individuals, which could be related to lower metabolic rates in aged subjects (Buchanan et al., 2003; Peloso et al., 2003). Senescence of the APR in free-living animals has yet to be examined.

PROXIMATE MECHANISMS REGULATING MODULATION OF SICKNESS BEHAVIOR

We next examine proximate mechanisms that could be responsible for mediating variation in sickness behavior according to different life-history stages. Because the APR is intimately connected with different arms of the vertebrate immune system (innate and adaptive immunity), the same mechanisms that proximately regulate these components will likely extend to the APR and sickness behavior. The interplay between the neuroendocrine and immune systems is well established. Increasing evidence indicates that many hormones are potent immunoregulatory molecules that participate in various aspects of immune function.

ENERGY RESERVES

A key factor that regulates the extent and duration of sickness behavior is energy reserves. Figure 3.6 depicts this relationship on a theoretical level using the principles of allostasis. Because anorexia is a central component of sickness behavior and the energetic costs of mounting an APR are high, an energetic bottleneck is created. Without energy from food intake, animals are forced to rely upon energy reserves (e.g., fat, protein, and carbohydrate stores) to fuel the high costs of the APR. These reserves will continue to dwindle until a minimum body mass threshold is reached (Fig. 3.6). In healthy individuals, this threshold is typically defended, and animals will immediately resume feeding. However, if body mass falls below this minimum set-point, then a cachexic state ensues, which typically results in morbidity and mortality (Plata-Salamán, 1996).

This phenomenon has been termed the *energy limitation hypothesis* (Owen-Ashley and Wingfield, 2007) and has received support from both field and laboratory studies. Figure 3.7 presents a selection of studies from the primary literature that has documented variation of sickness behavior in different life-history contexts. The first study (Fig. 3.7A) demonstrates that Siberian hamsters attenuate LPS-induced sickness behaviors on short days (Bilbo et al., 2002), whereas the second study (Fig. 3.7B) shows an opposite seasonal effect where male song sparrows diminish sickness responses to LPS during the early breeding season (long days) compared to winter (short days), as measured by a percentage reduction in body mass compared to saline (SAL)-injected controls (Owen-Ashley and Wingfield, 2006). However, these contrasting seasonal relationships coincide with seasonal changes in body mass, with lower body mass being associated with diminished sickness responses (Fig. 3.7).

Migration and the preparatory life-history activities that precede it (e.g., hyperphagia) can also have an impact upon the expression of sickness behavior (Fig. 3.7C). As discussed above, when exposed to long vernal day lengths, Gambel's white-crowned sparrows, a long-distance migrant that breeds in the Arctic, will gain body mass and accumulate fat stores (not shown in figure) to a greater extent than conspecific Puget Sound white-crowned sparrows, which migrate shorter distances (Wingfield et al., 1996; Wingfield et al., 1997). This difference likely explains why the Puget Sound white-crowned sparrow mounts a greater sickness response compared to Gambel's white-crowned sparrow (Owen-Ashley et al., 2008) on long day lengths. Note that anorexic responses to LPS, as measured by a percentage decrease in body mass, did not significantly differ between subspecies on short days, which is consistent with both subspecies having similar body masses during this time (Fig. 3.7C).

A final demonstration of the energy limitation hypothesis involves the effect of food access and restriction upon sickness responses (Fig. 3.7D). Red knots (*Calidris canutus*) are molluscivorous shorebirds that primarily feed in the intertidal zone and can experience food shortages on a daily basis (van Gils et al., 2006). Tidal surges make food unavailable twice a day, but these periods of food deprivation are predictable. When coupled with unpredictable adverse weather, food availability may drop precipitously and birds may endure bouts of fasting for several days (Zwarts et al., 1996; Vézina et al., 2009), and body mass can fluctuate substantially. Captive red knots that were experimentally deprived of food for 18 hours per day failed to respond to LPS treatment compared to unmanipulated controls on the same feeding schedule (Buehler et al., 2009). In contrast, birds receiving 22 hours of food access a day and injected with LPS exhibited anorexia relative to controls, as evidenced by a significant interaction between injection treatment and feeding regime (Buehler et al., 2009). Red knots that experienced food shortage had significantly lower body mass than birds given nearly unlimited access to food (Fig. 3.7D), implying an inability to activate an APR and sickness behavior simply because of the high costs of this response.

Other studies using food restriction or genetic mutation to alter body mass have reported similar findings. For example, mice restricted to 88% and 94% of their normal body weight exhibited reduced sickness responses to LPS compared to rats given food *ad libitum* (Lennie, 1998). Zucker rats genetically predisposed towards obesity (fa/fa) exhibit greater cytokine-induced anorexic responses than lean (+/+) wild-type rats (Plata-Salamán et al., 1997). Fever amplitude and duration after LPS treatment were reduced in Siberian hamsters when deprived of food for 24 hours compared to controls fed *ad libitum* (Bilbo and Nelson, 2002). However, a third group of hamsters was subjected to a restricted diet that involved a 26% reduction in body mass, but did not exhibit reduced fever compared to controls even though body mass changes were similar to food-deprived hamsters. These differences in sickness response between food-restricted versus food-deprived individuals were likely driven by the higher cortisol levels reported in food-restricted hamsters compared to hamsters on restricted or *ad libitum* diets (Bilbo and Nelson, 2002). Higher levels of cortisol likely suppress the fever response. However, in female golden hamsters, food restriction alone or in combination with exercise suppressed certain components of the APR (Conn et al., 1995). Thus, these interrelationships between body condition and expression of the APR are complex and presumably involve intermediary changes in hormones, such as glucocorticoids, sex steroids, and leptin, that have immunoregulatory functions (see below).

It should be noted that the dependence of the APR on energy reserves is not limited to endotherms. For example, juvenile green iguanas (*Iguana iguana*) in prime condition when challenged with LPS tend to select warmer temperatures to behaviorally regulate body temperature. Otherwise, individuals will actually resort to behavioral hypothermia (selection of colder temperatures) if energy reserves are not sufficient to sustain metabolism associated with higher temperatures and behavioral fever (Deen and Hutchinson, 2001).

LEPTIN AND GHRELIN

Leptin is a cytokine-like hormone that is produced in adipocytes and plays an important role in regulating energy balance by having an anorexigenic effect upon appetite. This hormone is actively transported across the BBB, where it inhibits the stimulatory effects of neuropeptide Y on feeding behavior in the hypothalamus (Myers et al., 2008). Because leptin levels are positively correlated with energy stores (Campfield et al., 1995; Pellymounter et al., 1995), this hormone provides a physiological signal of energy availability that allows for appropriate regulation of food intake, metabolism, and other physiological systems, such as immune function. For example, as described

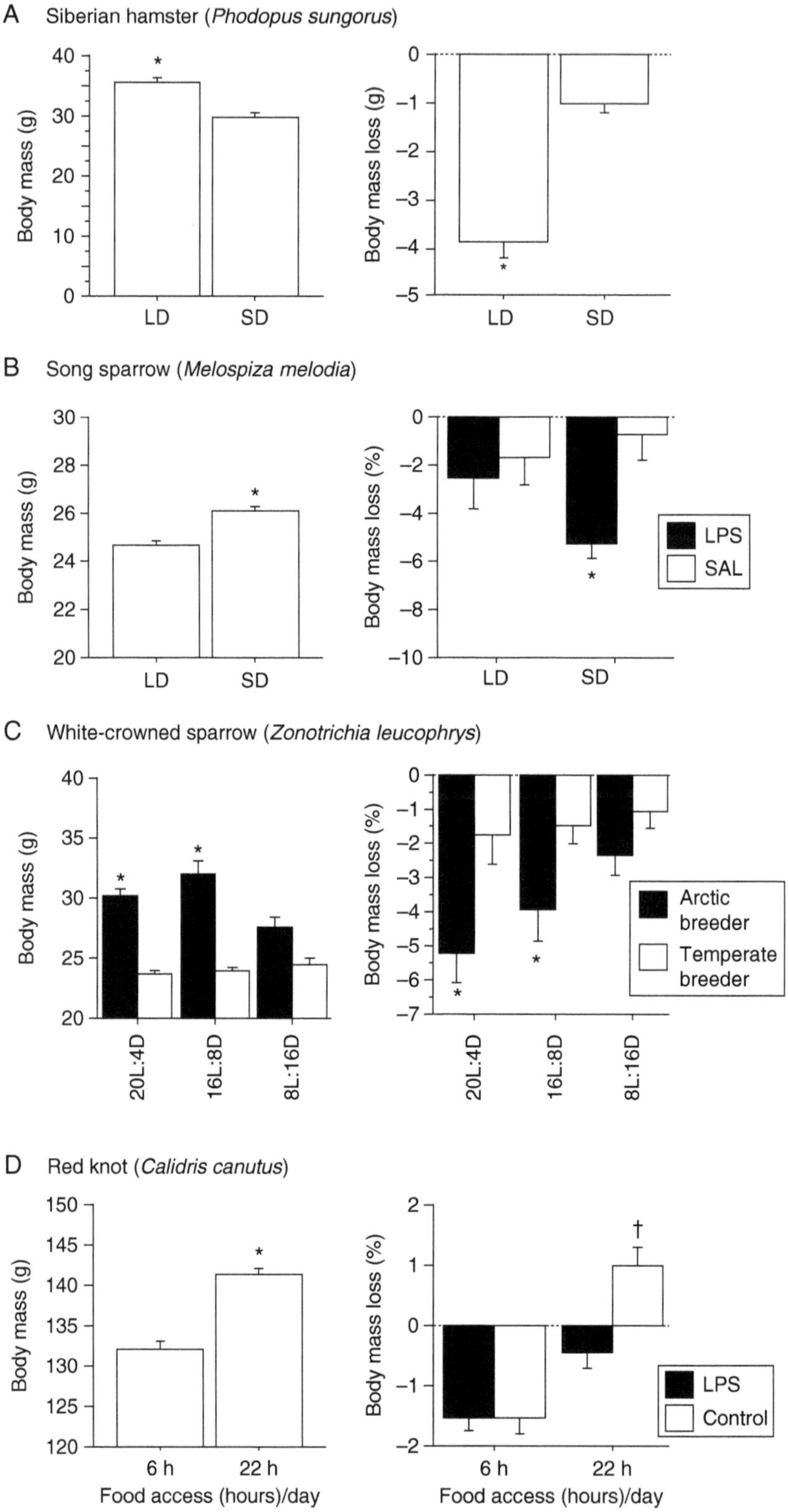

FIGURE 3.7: A selection of studies that demonstrates the energy limitation hypothesis in regulating variation of sickness behavior in different life-history contexts. (A) Captive male Siberian hamsters

in Chapter 8, fat reduction in Siberian hamsters decreased humoral immune responses compared to shams, and exogenous leptin treatment attenuated this immunosuppressive effect (Demas and Sakaria, 2005), implying a neuroendocrine link between energy availability and immune function.

Since the magnitude of expression of LPS-induced anorexic behavior is positively associated with energy stores (see above), leptin is an obvious candidate for mediating this response. As evidence, *in vitro* studies show that leptin stimulates the release of pro-inflammatory cells from immune cells (Dixit et al., 2004). LPS and cytokines acutely increased leptin gene expression in adipose tissue and circulating levels of leptin (Grunfeld et al., 1996; Sarraf et al., 1996). At first glance, these results suggest that leptin mediates the anorexic effect of sickness behavior. However, additional research is less straightforward. For example, obese leptin-deficient (ob/ob) mice exhibited more pronounced LPS-induced anorexia than lean littermates, whereas leptin receptor-deficient (db/db) mice developed less severe anorexia compared to lean controls (Faggioni et al., 1997). Interestingly, other measures of sickness behavior, such as reduced social exploration, were exacerbated in db/db mice (O'Connor et al., 2005). However, in mice and rats, neutralization of the bioactivity of leptin reversed LPS-induced anorexia, suggesting at least an indirect role in affecting sickness behavior. Thus high levels of leptin (and thus high energy reserves) are expected to promote increased anorexia during sickness.

Ghrelin, the counterpart to leptin, is a polypeptide that is produced in the stomach and intestine and has orexigenic effects upon metabolism and appetite. In addition, ghrelin has been shown to inhibit inflammatory cytokine production and promote thymic function (Taub, 2007). Ghrelin

FIGURE 3.7 (Continued)

(*Phodopus sungorus*) lose body mass on short days and attenuate LPS-induced sickness responses as measured by a change in body mass 48 hours post-injection compared to long days. Note that pre-injection body mass is lower in short-day males compared to long-day subjects. Modified from Bilbo et al. (2002). (B) Free-living male song sparrows (*Melospiza morphna*) exhibit an opposing seasonal pattern that involves suppression of sickness behavior during the breeding season (long days). Males injected with LPS (black bars) lost body mass 25 hours later during the winter (short days) compared to saline (SAL)-injected controls (white bars) in the same season. Males in the spring when defending breeding territories did not lose body mass compared to controls, suggesting insensitivity to LPS or an inability to mount a sickness response. These results are consistent with wintering sparrows being significantly heavier than breeding birds. Modified from Owen-Ashley and Wingfield (2006).
(C) Differences in sickness behaviors responses (as measured by a 24-h percentage change in body mass) in two subspecies of white-crowned sparrows that have different migratory strategies. The Arctic-breeding subspecies (black bars) is a long-distance migrant and accumulates more body mass (g) and fat stores (not shown) upon vernal exposure to long days (16L:8D or 20L:4D) than the temperate-breeding subspecies (white bars). These changes in body mass parallel the magnitude of anorexic responses to LPS. Saline-injected controls are not shown in this figure. Modified from Owen-Ashley et al. (2008). (D) Red knots (*Calidris canutus*) given restricted access to food (6 h of food access) fail to respond to LPS challenge compared to birds given 22 h of food access as measured by a 24-h percentage change in body mass. Restricted access to food decreased body mass (g), which is consistent with birds failing to mount a response. Adapted from Buehler et al. (2009) and Vézina et al. (2009). For all graphs, bars show means ± 1 SEM. Asterisk denotes a significant ($p < 0.05$) pair-wise difference. Dagger denotes a significant ($p < 0.05$) interaction in the overall ANOVA model, but no significant pair-wise difference.

administration exerts protective effects on the survival of rats during endotoxemia and attenuates responses to sepsis (Chang et al., 2003) and cachexia (Hataya et al., 2003). High levels of ghrelin correspond with low body condition, providing a mechanism for diminished or absent sickness responses in animals in poor condition. Thus, the energy limitation hypothesis is likely regulated by the antagonistic interplay of ghrelin and leptin, which signals energy availability. However, more work in comparative species is needed to establish how levels are regulated in the wild and how they might affect behavioral responses to infection.

GLUCOCORTICOSTEROIDS

There is overwhelming evidence that glucocorticosteroids, steroid hormones released from the adrenal cortex of vertebrates, regulate nearly every aspect of immunological function (Munck et al., 1984; Besedovsky and del Ray, 1996; McEwen et al., 1997; Dhabhar, 2002). The aim of this section is not to revisit these studies, but instead to focus upon how these hormones might influence the expression of APRs and sickness behavior in relation to both acute and chronic stress. Referring back to Figure 3.1, glucocorticosteroids have an essential function of providing negative feedback to the APR such that an overactive immune response is prevented (Besedovsky et al., 1986; Bateman et al., 1989; Baumann and Gauldie, 1994). However, at the onset of infection or when exposed to acute stress, IL-1 and other cytokines can stimulate the HPA axis to release glucocorticosteroids, which can have an enhancing effect upon immunological function through early mobilization of immune cells and their redistribution (Dhabhar and McEwen, 1999; Dhabhar, 2002). It is hypothesized that this upregulation is adaptive, allowing animals to prepare for potential challenges to the immune system, such as wounding or infection that may occur during a response to an LPF (see Fig. 3.2).

After sustained elevation, glucocorticoids tend to suppress immune function. The arachidonic acid cascade that mediates inflammation in suppressed (Floman and Zor, 1976; Lewis et al., 1986), and cytokine production and function are inhibited through interaction of glucocorticoid receptors with transcription factors, which in turn regulate cytokine gene expression (Lee et al., 1988; Auphan et al., 1995). Removal of the adrenal gland sensitizes animals to endotoxic shock (Bertini et al., 1988; Ruzek et al., 1999), increases the febrile response (Coehlo et al., 1992; Morrow et al., 1993), and further exacerbates sickness behavior (Goujon et al., 1995a, b; Johnson et al., 1996; Pezeshki et al., 1996) compared to sham-adrenalectomized subjects when treated with LPS. Excessive and unregulated production of TNF-α (cachectin) is the main mediator of these abnormal responses. Replacement with glucocorticosteroids alleviates many of these effects (Goujon et al., 1995a, b; Johnson et al., 1996; Pezeshki et al., 1996). Thus, activation of the HPA axis and release of glucocorticosteroids plays an important role in regulating the APR.

MELATONIN

Melatonin, an indoleamine, is derived from the amino acid tryptophan and is synthesized primarily by the pineal gland. Biosynthesis and secretion of pineal melatonin are stimulated by darkness and inhibited by light. As a result, the duration of nightly melatonin secretion accurately encodes photoperiodic information, allowing seasonally breeding mammals to time reproduction and other life-history activities over the annual cycle. In birds, the association between melatonin and photoperiodic time measurement is less apparent. Instead, it is thought that photoperiodic responses involve direct interaction between deep encephalic photoreceptors and GnRH neurons (Dawson et al., 2001).

Like reproduction, immune function also varies on a seasonal basis, and melatonin has been shown to mediate many of these effects (Maestroni and Conti, 1993; Nelson and Demas, 1996; Nakamura et al., 1998; Nelson et al., 2002; Nelson, 2004). Melatonin receptors have been isolated on lymphocytes (Nelson and Demas, 1996), which suggests that melatonin exerts some effects directly through a receptor-mediated pathway. Generally speaking, melatonin is immunoenhancing, although there are cases of this hormone suppressing the immune system as well.

One such example of immunosuppression by melatonin involves APR-related physiological and behavioral responses. Melatonin treatment counteracts LPS-induced adipsia, hyperalgesia, and fever in laboratory rodents (Nava et al., 1997; Raghavendra et al., 2000; Bruno et al., 2005). Chronic daily injections of melatonin for 6 weeks in Siberian hamsters housed on long days attenuated febrile responses to LPS, but cortisol levels were also elevated from melatonin treatment, making it difficult to attribute this suppressive effect to one or both hormones (Bilbo and Nelson, 2002). Pinealectomy in the same species abolished the attenuation of sickness behavior by LPS, but not fever, on short days compared to long days, which suggests that the pineal gland is important for regulating at least sickness behavior (Wen et al., 2007). However, this counteracting effect could also be explained by the energy limitation hypothesis because pinealectomy altered body mass regulation of hamsters in a photoperiod-dependent manner such that there were no seasonal differences in body mass between pinealectomized subjects compared to intact hamsters. Furthermore, if melatonin does indeed act upon the APR and sickness behavior in a suppressive fashion, then it would explain the seasonal difference found in rodents (Bilbo et al., 2002), but not in birds, where long-day suppression has been reported (Owen-Ashley and Wingfield, 2006). Clearly other proximate factors might be involved, but studies are needed to clarify the role of melatonin in directly modulating sickness behavior by controlling the input of other possible mediators, such as energy stores and glucocorticoids.

ANDROGENS

Androgens, including testosterone, regulate male reproductive behavior while promoting spermatogenesis and development of some morphological traits important for female choice. Produced in the male testis, the steroid hormone testosterone also influences aggressive interactions in a variety of vertebrate taxa (Wingfield et al., 1994). However, there is evidence that elevated levels of testosterone also exact costs, such as suppression of the immune system (Grossman, 1984; Alexander and Stimson, 1988; Zuk, 1990; Hillgarth and Wingfield, 1997). These bidirectional effects of testosterone upon male fitness led to the development of the *immunocompetence handicap hypothesis,* which proposes that honesty of androgen-dependent traits is enforced by the obligatory suppressive effects of testosterone upon immune function (Folstad and Karter, 1992). Thus, only high-quality males with genetic resistance or tolerance to parasites can withstand the effects of immunosuppression and develop and maintain androgen-dependent ornaments.

However, empirical support for this hypothesis is lacking because testosterone does not always promote immunosuppression (Hasselquist et al., 1999; Roberts et al., 2004) and other hormones, such as glucocorticosteroids, may be indirectly involved in mediating these testosterone-induced effects upon immune function (Owen-Ashley et al., 2004).

Immunosuppression by testosterone has traditionally been regarded as an evolutionary constraint, or handicap, on male fitness rather than an adaptation. However, it has been hypothesized that suppression of sickness behavior by testosterone could be adaptive if sickness behavior interferes with certain testosterone-mediated behaviors that are important for reproductive success (Ashley et al., 2009). For example, in seasonally breeding songbirds, high levels of testosterone during the breeding season are associated with male–male territorial aggression and mate-guarding (Wingfield, 1985; Wingfield et al., 1990). Exhibiting sickness behavior from an injury and/or infection during male contests exacts an opportunity cost and could detrimentally affect mating chances in the future. The same line of reasoning could be applied to sexual behavior in males (Avistur and Yirmiya, 1999).

Experimental studies have generally supported the hypothesis that testosterone suppresses sickness behavior and accompanying APRs. In rats, castrated males were more sensitive to the depressive effect of IL-1β on social exploration than intact males (Dantzer et al., 1991). This effect is mediated by androgen-dependent vasopressinergic neurons in the brain, which seemed to antagonize the neural effects of IL-1β (Dantzer et al., 1991). Similarly, in mice, castration stimulated adrenal and immunological responses to LPS compared to shams, and testosterone replacement counteracted this effect (Spinedi et al., 1992; Seale et al., 2004). In contrast, castrated Siberian hamsters attenuated sickness behaviors to LPS compared to gonad-intact males (Prendergast et al., 2008). In birds, male white-crowned sparrows that were castrated and implanted with testosterone exhibited reduced sickness behavior to LPS compared to castrated and intact controls with empty implants, which did not differ in sickness responses (Ashley et al., 2009). This suggests that only high levels of testosterone typically associated with social instability have the capacity to suppress sickness behavior in this species. Alternatively, because testosterone implants elevated baseline plasma levels of corticosterone in this species compared to other groups, this suppressive effect could be mediated by a stress-induced mechanism. Nonetheless, this modulation is consistent with the seasonal pattern of behavioral symptoms of infection documented in free-living songbirds, where males suppress sickness responses when testosterone levels are elevated from defending breeding territories compared to wintering males (Owen-Ashley and Wingfield, 2006). Thus, the costs of suppressing sickness behavior are likely transient, but they probably contribute to the general trend that males, on average, exhibit increased susceptibility to infection compared to females.

Besides testosterone, other androgen metabolites have been shown to mediate levels of pro-inflammatory cytokines and inflammatory responses. Dihydrotestosterone, an androgen metabolite important for development of masculine characteristics, decreased the release of pro-inflammatory cytokines, IL-1β and IL-6, in splenic macrophages, while increasing levels of IL-10, an anti-inflammatory cytokine (Angele et al., 1999; Angele et al., 2001). Dehydroepiandrosterone (DHEA) is a steroid precursor that is secreted predominantly from the adrenal cortex. DHEA and its direct metabolite, androstenediol (AED), have stimulatory effects upon immune function in mammals (Loria et al., 1996; Svec and Porter, 1998; Hazeldine et al., 2010). Both DHEA and AED treatment protect mice against lethal bacterial infections and septic shock from high doses of

LPS (Ben-Nathan et al., 1999). The sulfated form of DHEA does not have the same effect (Chen and Johnson, 2004). Because DHEA secretion decreases with age, a phenomenon known as "adrenopause," it is speculated that this decline could contribute to patterns of immunosenescence in some mammal populations, namely humans (Hazeldine et al., 2010). In song sparrows, DHEA is the only detectable androgen in the plasma during the non-breeding season (Soma and Wingfield, 2001) and has been shown to regulate territorial behavior (e.g., song) during the winter (Soma et al., 2002). Because DHEA is not immunosuppressive compared to testosterone (Owen-Ashley et al., 2004) and winter is an energetically demanding time, it is hypothesized that song sparrows escape the high costs of testosterone that are typically unavoidable during the breeding season (Wingfield et al., 2001).

ESTROGENS AND PROGESTERONE

Estrogens and progesterone, the primary steroids that regulate many aspects of female reproduction, also have immunomodulatory properties. Estrogens are predominantly synthesized in the follicles of the ovary, the corpus luteum, and placenta, and are generally immunoenhancing, whereas progesterone is produced in the corpus luteum, placenta, and adrenals and is immunosuppressive. Because progesterone plays an integral role in supporting pregnancy in placental mammals, it is hypothesized that its suppressive actions upon maternal immune response during conception and gestation increase the probability that pregnancy will be accepted and maintained. Likewise, the immunoenhancing qualities of estrogens are thought to decrease susceptibility to disease in females, while increasing the likelihood of acquiring autoimmunological disorders (e.g., rheumatoid arthritis), which are more prevalent in females.

Sickness behavior seems to be influenced by both of these hormones. Female rats in estrus exhibited more severe anorexic responses to LPS than females during diestrus or males (Geary et al., 2004). Furthermore, in ovariectomized rats, estradiol treatment enhanced the anorectic effect of IL-1β (Geary et al., 2004), while progesterone treatment diminished sickness behavior to an inflammatory challenge (Lennie, 2004). These studies provide a proximate mechanism for females to modulate the expression of sickness behavior according to stage of reproduction. During estrus, a strong response to a pathogenic challenge is favored because illness could lead to spontaneous abortion and wasted reproductive effort (Avistur and Yirmiya, 1999). However, during pregnancy, a severe response to infection could trigger excessive production of cytokines, which could lead to rejection of the fetus through autoimmune processes. Thus, suppression of sickness behavior is favored in these circumstances. Taken together, the antagonistic effects of these hormones are likely tied to a female's current and future reproductive prospects.

PROLACTIN

Prolactin is a polypeptide hormone that is synthesized and secreted from the anterior pituitary and has over 300 known biological functions (Freeman et al., 2000). It is widely recognized for stimulating milk production in mammals and crop milk in some birds (e.g., doves), promoting brood patch formation in birds (in conjunction with estrogen), and influencing parental behavior

(Höhn and Cheng, 1965; Freeman et al., 2000). Prolactin is also known to have potent stimulatory effects upon humoral and cell-mediated components of immune system (Spangelo et al., 1987; Matera, 1996; Weigent, 1996). It serves as a growth factor for lymphocytes and increases T-cell–mediated responses.

There is also a role for prolactin in buffering stress-induced immunosuppression (Dorshkind and Horseman, 2001). Similar to estrogen, prolactin has protective effects against inflammation derived from severe trauma. Prolactin treatment reduces plasma levels of corticosterone and IL-6, while enhancing other components of immunological function during trauma and improving survival of animals during septic shock (Zellweger et al., 1996).

Prolactin inhibits IL-6–induced inflammation in female reproductive tissues (Deb et al., 1999) but promotes inflammation in the prostrate (Tangbanluekal and Robinette, 1993). These studies suggest that prolactin can have either enhancing or suppressive effects upon inflammation, depending upon tissue type and/or gender. If prolactin is important in mediating parental behavior, then a suppressive effect upon pro-inflammatory cytokines, and therefore sickness behavior, should be favored, especially if expression of sickness behavior would disrupt parental behavior such that survival of offspring is threatened. However, this is merely conjecture, as studies directly examining prolactin and sickness behavior have not been conducted.

CONCLUSIONS

Modulation of sickness behavior is a prime example of a life-history trade-off between host defenses and other competing physiological processes, and this phenotypic variation seems to follow the same rule of most life-history traits, which is to optimize fitness. Given the evolutionarily conserved nature of the APR in animals, one might argue that there is little selection pressure for these responses to vary in natural contexts, especially if they are nonspecific towards pathogens. However, natural selection should set a limit upon the optimal levels of resources that can be dedicated to particular immune defenses. Using the concept of life-history theory and allostatic load, we demonstrated substantial energetic and temporal costs of activating sickness behavior and APRs, suggesting that this response would be modulated if the costs outweigh the benefits of responding to infection. Evidence to date suggests that season, sex, age, reproductive state, social environment, and pace of life contribute to modulation of the APR and sickness behavior. Energy reserves and hormones, such as glucocorticosteroids, sex steroids, leptin, ghrelin, and prolactin, play a role in regulating this response, but more experimental studies that manipulate energy levels, nutrition, and hormone levels are certainly needed.

FUTURE DIRECTIONS

Much remains to be learned about the costs and benefits of the APR and sickness behavior. Chief among these is identifying the fitness consequences associated with APR modulation. For example, the long-term costs of suppressing sickness behavior are unexplored and could involve decreases in survivorship or fecundity through increased susceptibility to disease or costs incurred at other times of the year (e.g., molt, migration). Only one study to date has identified survival costs of exhibiting sickness behavior in free-living animals (Eraud et al., 2009), and more are needed. Transgenerational and perinatal effects of LPS challenge in wild vertebrates would also be interesting to pursue. More generally, most of the studies described in this chapter occurred in a laboratory setting where temperatures and food supply are held constant. More comparative work is needed in wild animals where environmental conditions and food resources are less predictable to understand the ecological factors that regulate this response.

Experimental research to validate the energy limitation hypothesis would also be helpful. One method is to experimentally remove fat stores (Demas et al., 2003) to establish the role of energy reserves in regulating the sickness response. Evaluating the role of hormones that regulate food intake, such as leptin and grhelin, would also prove useful. The adaptive significance of illness-induced anorexia is still an enigma. Although progress has been made to understand the function of anorexia in insects (Adamo et al., 2010), a satisfactory explanation is lacking in vertebrates. If anorexia reduces the energetic trade-off between lipid transport and immune function, as seen in insects, studies that measure energy expenditure in conjunction with fatty acid metabolism after LPS challenge would be particularly valuable.

More research is needed to understand why APRs are highly conserved and to what extent these defenses regulate host resistance to particular pathogens. Virtually nothing is known about how parasites evolve in response to behavioral changes of the host, except in special cases where particular pathogens manipulate host behavior (e.g., rabies) (Klein, 2003). Conversely, the relative effectiveness of APRs in controlling different types of infectious diseases remains elusive. Elucidating how previous exposure to parasites and the interactions between the genetic makeup of individuals and the natural environment regulate expression of sickness behavior is also an emerging challenge. Lastly, a better understanding of cytokine-related genes and how they vary within and between populations could improve our understanding of the co-evolutionary forces that shape host defenses in the wild.

ACKNOWLEDGMENTS

N. T. A. was supported by a National Science Foundation Polar Postdoctoral Fellowship. J.C.W. is grateful for support from the National Science Foundation (grant number IOS-0750540) and for the Endowed Chair in Physiology from the University of California, Davis.

REFERENCES

Adamo, S. A. (1999). Evidence for adaptive changes in egg laying in crickets exposed to bacteria and parasites. Anim. Behav. 57, 117–124.

Adamo, S. A. (2004). How should behavioral ecologists interpret measures of immunity. Anim. Behav. 68, 1443–1449.

Adamo, S. A. (2006). Comparative psychoneuroimmunology: evidence from the insects. Behav. Cogn. Neurosci. Rev. 5, 128–140.

Adamo, S. A., Bartlett, A., Le, J., Spencer, N., Sullivan, K. (2010). Illness-induced anorexia may reduce trade-offs between digestion and immune function. Anim. Behav. 79, 3–10.

Adamo, S. A., Fidler, T. L., Forestell, C. A. (2007). Illness-induced anorexia and its possible function in the caterpillar, *Manduca sexta*. Brain Behav. Immun. 21, 293–300.

Adamo, S. A., Jensen, M., Younger, M. (2001). Changes in lifetime immunocompetence in male and female *Gryllus texensis* (formerly *G-integer*): trade-offs between immunity and reproduction. Anim. Behav. 62, 417–425.

Adelman, J. S., Bentley, G. E., Wingfield, J. C., Martin, L. B., Hau, M. (2010a). Population differences in fever and sickness behaviors in a wild passerine: a role for cytokines. J. Exp. Biol. 213, 4099–4109.

Adelman, J. S., Córdoba- Córdoba, K., Spoelstra, K., Wikelski, M., Hau, M. (2010b). Radiotelemetry reveals variation in fever and sickness behaviours with latitude in a free-living passerine. Funct. Ecol. 24, 813–823.

Adelman, J. S., Martin, L. B. (2009). Vertebrate sickness behaviors: adaptive and integrated neuroendocrine immune responses. Integ. Comp. Biol. 49, 202–214.

Ader, R., Cohen, N. (1995). Psychoneuroimmunology: interactions between the nervous system and the immune system. Lancet. 349, 99–103.

Alexander, J., Stimson, W. H. (1988). Sex hormones and the course of parasitic infection. Parasitol. Today. 4, 1891–1893.

Andersson, M. (1994). Sexual Selection. Princeton University Press, Princeton, New Jersey.

Andus, T., Bauer, J., Gerok, W. (1991). Effect of cytokines on the liver. Hepatology. 13, 364–375.

Angele, M. K., Knöferi, M. W., Ayala, A., Bland, K. I., Chaudry, I. H. (2001). Testosterone and estrogen differently effect Th1 and Th2 cytokine release folowing trauma-haemorrhage. Cytokine. 16, 22–30.

Angele, M. K., Knöferi, M. W., Schwacha, M. G., Ayala, A., Cioffi, W. G., Bland, K. I., Chaudry, I. H. (1999). Sex steroids regulate pro- and ant-inflammatory cytokine release by macrophages after trauma-hemorrhage. Am. J. Physiol. 277, C35–C42.

Angele, M. K., Wichmann, M. W., Ayala, A., Cioffi, W. G., Chaudry, I. H. (1997). Testosterone receptor blockade after hemorrhage in males. Arch. Surg. 132, 1207–1214.

Annane, D., Bellissant, E., Cavaillon, J. (2006). Septic shock. Lancet. 365, 63–78.

Arcese, P. (1989). Territory acquisition and loss in male song sparrows. Anim. Behav. 37, 45–55.

Ashley, N. T., Hays, Q. R., Bentley, G. E., Wingfield, J. C. (2009). Testosterone treatment diminishes sickness behavior in male songbirds. Horm. Behav. 56, 169–176.

Aubert, A. (1999). Sickness and behaviour in animals: a motivational perspective. Neurosci. Biobehav. Rev. 23, 1029–1036.

Aubert, A., Goodall, G., Dantzer, R. (1995). Compared effects of cold ambient temperature and cytokines on macronutrient intake in rats. Physiol. Behav. 57, 869–873.

Aubert, A., Goodall, G., Dantzer, R., Gheusi, G. (1997). Differential effects of lipopolysaccharide on pup retrieving and nest building in lactating mice. Brain. Behav. Immun. 11, 107–118.

Auphan, N., DiDonato, J., Rosettte, C., Helmburg, I., Karin, M. (1995). Immunosuppression by glucocorticoids: inhibition of NK-kappa B activity through induction of I kappa B synthesis. Science. 270, 286–290.

Avistur, R., Yirmiya, R. (1999). The immunobiology of sexual behavior: gender differences in the suppression of sexual activity during illness. Pharm. Biochem. Behav. 64, 787–796.

Avitsur, R., Cohen, E., Yirmiya, R. (1998). Effects of interleukin-1 on sexual attractivity in a model of sickness behavior. Physiol. Behav. 63, 25–30.

Avitsur, R., Sheridan, J. F. (2009). Neonatal stress modulates sickness behavior. Brain Behav. Immun. 2009, 977–985.

Baluna, R., Vitetta, E. S. (1997). Vascular leak syndrome: a side effect of immunotherapy. Immunopharmacology. 37, 117–132.

Banks, W. A. (2001). Anorectic effects of circulating cytokines: role of the vascular blood-brain barrier. Nutrition. 17, 434–437.

Bartness, T. J., Elliott, J. A., Goldman, B. D. (1989). Control of torpor and body weight patterns by a seasonal timer in Siberian hamsters. Am. J. Physiol. 257, R142–R149.

Bateman, A., Singh, A., Kral, T., Solomon, A. (1989). The immune-hypothalamic-pituitary-adrenal axis. Endocr. Rev. 10, 92–112.

Baumann, H., Gauldie, J. (1994). The acute phase response. Immunol. Today. 15, 74–81.

Ben-Nathan, D., Padgett, D. A., Loria, R. M. (1999). Androstenediol and dehydroepiandrosterone protect mice against lethal bacterial infections and lipopolysaccharide toxicity. J. Med. Microbiol. 48, 425–431.

Bentley, G. E., Demas, G. E., Nelson, R. J., Ball, G. F. (1998). Melatonin, immunity and cost of reproductive state in male European starlings. Proc. R. Soc. Lond. B 265, 1191–1195.

Bertini, R., Bianchi, M., Ghezzi, P. (1988). Adrenalectomy sensitizes mice to the lethal effects of interleukin 1 and tumor necrosis factor. J. Exp. Med. 167, 1708–1712.

Bertrand, S., Criscuolo, F., Faivre, B., Sorci, G. (2006). Immune activation increases susceptibility to oxidative tissue damage in Zebra Finches. Funct. Ecol. 20, 1022–1027.

Besedovsky, H. O., del Ray, A. (1996). Immune-neuro-endocrine interactions: facts and hypothesis. Endocr. Rev. 17, 64–102.

Besedovsky, H. O., Del Ray, A., Sorkin, E., Dinarello, C. A. (1986). Immunoregulatory feedback between interleukin-1 and glucocorticoid hormones. Science. 243, 652–654.

Bilbo, S. D., Drazen, D. L., Quan, N., He, L., Nelson, R. J. (2002). Short day lengths attenuate the symptoms of infection in Siberian hamsters. Proc. R. Soc. Lond. B. 269, 447–454.

Bilbo, S. D., Nelson, R. J. (2002). Melatonin regulates energy balance and attenuates fever in Siberian hamsters. Endocrinology. 143, 2527–2533.

Bisson, I., Butler, L. K., Hayden, T. J., Romero, L. M., Wikelski, M. C. (2009). No energetic cost of anthropogenic disturbance in a songbird. Proc. R. Soc. Lond. B. 276, 961–969.

Blalock, J. E. E. (1992). Neuroimmunoendocrinology. Charger, New York.

Bluthe, R. M., Michaud, B., Kelley, K. W., Dantzer, R. (1996). Vagotomy attenuates behavioural effects of interleukin-1 injected peripherally but not centrally. Neuroreport. 8, 1485–1488.

Bókony, V., Lendvai, A., Liker, A., Angelier, F., Wingfield, J. C., Chastel, O. (2009). Stress response and the value of reproduction: are birds prudent parents? Am. Nat. 173, 589–598.

Bolles, R. C. (1967). Theory of Motivation. Harper, New York.

Bonneaud, C., Mazuc, J., Gonzalez, G., Haussy, C., Chastel, O., Faivre, B., Sorci, G. (2003). Assessing the cost of mounting an immune response. Am. Nat. 161, 367–379.

Bruno, V. A., Scacchi, P. A., Perez-Lloret, S., Esquifino, A. I., Cardinali, D.P., Cutrera, R. A. (2005). Melatonin treatment counteracts the hyperthermic effect of lipopolysaccharide injection in the Syrian hamster. Neurosci. Lett. 389, 169–172.

Buchanan, J. B., Peloso, E., Satinoff, E. (2003). Thermoregulatory and metabolic changes during fever in young and old rats. Am. J. Physiol. Regul. Integr. Comp. Physiol. 285, R1165–R1169.

Buehler, D. M., Encinas-Viso, F., Petit, M., Vèzina, F., Tieleman, B. I., Piersma, T. (2009). Limited access to food and physiological trade-offs in a long distance migrant shorebird. II. Constitutive immune function and the acute-phase response. Physiol. Biochem. Zool. 82, 561–571.

Campfield, L. A., Smith, F. J., Guisez, Y., Devos, R., Burn, P. (1995). Recombinant mouse OB protein: evidence for a peripheral signal linking adiposity and central neural networks. Science. 269, 546–549.

Chang, L., Zhao, J., Yang, J., Zhang, Z., Du, J., Tang, C. (2003). Therapeutic effects of ghrelin on endotoxic shock in rats. Eur. J. Pharmacol. 473, 171–176.

Chen, J., Johnson, R. W. (2004). Dehydroepiandrosterone-sulfate did not mitigate sickness behavior in mice. Physiol. Behav. 82, 713–719.

Clutton-Brock, T. H. (1984). Reproductive effort and terminal investment in iteroparous animals. Am. Nat. 123, 212–229.

Coehlo, M. M., Souza, G. E. P., Pela, I. R. (1992). Endotoxin-induced fever is modulated by endogenous glucocorticoids in rats. Am. J. Physiol. Regul. Integr. Comp. Physiol. 32, R243–R427.

Cohn, D. W. H., de Sá-Rocha, L. C. (2006). Differential effects of lipopolysaccharide in the social behavior of dominant and submissive mice. Physiol. Behav. 87, 932–937.

Conn, C. A., Kozak, W. E., Tooten, P. C., Gruys, E., Borer, K. T., Kluger, M. J. (1995). Effect of voluntary exercise and food restriction in response to lipopolysaccharide in hamsters. J. Appl. Physiol. 78, 466–477.

Dantzer, R. (2001). Cytokine-induced sickness behavior: where do we stand? Brain. Behav. Immun. 15, 7–24.

Dantzer, R., Bluthé, R., Kent, S., Goodall, G. (1993). Behavioral effects of cytokines: an insight into mechanisms of sickness behavior. In: E. G. DeSouza (Ed.), Neurobiology of Cytokines. Academic Press, San Diego.

Dantzer, R., Bluthé, R., Laye, S., Bret-Dibat, J., Parnet, P., Kelley, K. (1998). Cytokines and sickness behavior. Ann. N.Y. Acad. Sci. 840, 586–590.

Dantzer, R., Bluthe, R. M., Kelley, K. W. (1991). Androgen-dependent vasopressinergic neurotransmission attenuates interleukin-1 induced sickness behavior. Brain Res. 557, 115–120.

Dantzer, R., Kelley, K. W. (2007). Twenty years of research on cytokine-induced sickness behavior. Brain Behav. Immunol. 21, 153–160.

Dawson, A., King, V. M., Bentley, G. E., Ball, G. F. (2001). Photoperiodic control of seasonality in birds. J . Biol. Rhythms. 16, 365–380.

Deb, S., Tessier, C., Prigent-Tessier, A., Barkai, U., Ferguson-Gottschall, S., Srivastava, R. K., Faliszek, J., Gilbori, G. (1999). The expression of interleukin-6 (IL-6), IL-6 receptor, and gp130 kDa glycoprotein in the rat decidua and a decidual cell line: regulation by 17ß-estradiol and prolactin. Endocrinology. 140, 4442–4450.

Deen, C. M., Hutchinson, V. H. (2001). Effects of lipopolysaccharide and acclimation temperature on induced behavioral fever in juvenile *Iguana iguana*. J. Therm. Biol. 26, 55–63.

Demas, G. E. (2004). The energetics of immunity: a neuroendocrine link between energy balance and immune function. Horm. Behav. 45, 173–180.

Demas, G. E., Drazen, D. L., Nelson, R. J. (2003). Reductions in total body fat decrease humoral immunity. Proc. R. Soc. Lond. B. 270, 905–911.

Demas, G. E., Nelson, R. J. (2003). Lack of photoperiodic changes in humoral or cell-mediated immunity in a desert-dwelling rodent, *Peromyscus aztecus*. J. Comp. Physiol. B. 173, 171–176.

Demas, G. E., Sakaria, S. (2005). Leptin regulates energetic tradeoffs between body fat and humoral immunity. Proc. R. Soc. Lond. B. 272, 1845–1850.

Dhabhar, F. S. (2002). A hassle a day may keep the doctor away: stress and the augmentation of immune function. Integ. Comp. Biol. 42, 556–564.

Dhabhar, F. S., McEwen, B. S. (1999). Enhancing versus suppressive effects of stress hormones on skin immune function. Proc. Natl. Acad. Sci. USA. 96, 1059–1064.

Dixit, V. D., Schaffer, E. M., Pyle, R. S., Collins, G. D., Sakthivel, S. K., Palaniappan, R., Lillard Jr., J. W., Taub, D. D. (2004). Ghrelin inhibits leptin- and activation-induced proinflammatory cytokine expression by human monocytes and T cells. J. Clin. Invest. 114, 57–66.

Dorshkind, K., Horseman, N. D. (2001). Anterior pituitary hormones, stress, and immune system homeostasis. Bioessays. 23, 288–294.

Dunn, A. J., Swiergiel, A. H. (1998). The role of cytokines in infection-related behavior. Ann. N.Y. Acad. Sci. 840, 577–585.

Dunn, P. E., Bohnert, T. J., Russel, V. (1994). Regulation of antibacterial protein synthesis following infection and during metamorphosis of *Manduca sexta*. Ann. N.Y. Acad. Sci. 712, 117–130.

Eraud, C., Jacquet, A., Fairvre, B. (2009). Survival cost of an early immune soliciting in nature. Evolution. 63, 1036–1043.

Ewald, P. W. (1980). Evolutionary biology and the treatment of signs and symptoms of infectious disease. J. Theor. Biol. 86, 169–176.

Exton, M. S. (1997). Infection-induced anorexia: active host defence strategy. Appetite. 29, 369–383.

Faggioni, R., Fuller, J., Moser, A., Feingold, K. R., Grunfeld, C. (1997). LPS-induced anorexia in leptin-deficient (ob/ob) and leptin receptor-deficient (db/db) mice. Am. J. Physiol. 273, R181–R186.

Finkel, T., Holbrook, N. J. (2000). Oxidants, oxidative stress and biology of ageing. Nature. 408, 239–247.

Floman, Y., Zor, U. (1976). Mechanism of steroid action in inflammation: inhibition of prostaglandin synthesis and release. Prostaglandins. 12, 402–413.

Folstad, I., Karter, A. J. (1992). Parasites, bright males, and the immunocompetence handicap. Am. Nat. 139, 603–622.

Frederic, F., Oliver, C., Wollman, E., Delhaye-Bouchard, N., Mariani, J. (1993). IL-1 and LPS induce a sexually dimorphic response of the hypothalamo-pituitary-adrenal axis in several mouse strains. Eur. Cytokine Netw. 4, 321–329.

Freeman, D. A., Kanyicka, B., Lerant, A., Nagy, G. (2000). Prolactin: structure, function, and regulation of secretion. Physiol. Rev. 80, 1523–1631.

French, S. S., Moore, M. C., Demas, G. E. (2009). Ecological immunology: the organism in context. Integ. Comp. Biol. 49, 246–253.

Geary, N., Asarian, L., Sheahan, J., Langhans, W. (2004). Estradiol-mediated increases in the anorexia induced by intraperitoneal injection of bacterial lipopolysaccharide in female rats. Physiol. Behav. 82, 251–261.

Godbout, J. P., Chen, J., Abraham, J., Richwine, A. F., Berg, B. M., Kelley, K. W., Johnson, R. W. (2005). Exaggerated neuroinflammation and sickness behavior in aged mice after activation of the peripheral innate immune system. FASEB J. 19, 1329–1331.

Gonzalez, G., Sorci, G., de Lope, F. (1999). Seasonal variation in the relationship between cellular immune response and badge size in male house sparrows (*Passer domesticus*). Behav. Ecol. Sociobiol. 46, 117–122.

Goujon, E., Parnet, P., Aubert, A., Goodall, G., Dantzer, R. (1995a). Corticosterone regulates behavioral effects of lipopolysaccharide and interluekin-1β in mice. Am. J. Physiol. 269, R154–R159.

Goujon, E., Parnet, P., Cremona, S., Dantzer, R. (1995b). Endogenous glucocorticoids down regulate central effects of interleukin-1β on body temperature and behaviour in mice. Brain Res. 702, 173–180.

Graham, A.L., Allen, J. E., Read, A. F. (2005). Evolutionary causes and consequences of immunopathology. Annu. Rev. Ecol. Syst. 36, 373–393.

Grasi-Kraupp, B., Bursch, W., Ruttkay-Nedecky, B., Wagner, A., Lauer, B., Shulte-Hermann, R. (1994). Food restriction eliminates preneoplastic cells through apoptosis and antagonizes carcinogenesis in rat liver. Proc. Natl. Acad. Sci. USA. 91, 9995–9999.

Greenman, C. G., Martin, L. B., Hau, M. (2005). Reproductive state, but not testosterone, reduces immune function in male house sparrows (*Passer domesticus*). Physiol. Biochem. Zool. 78, 60–68.

Grossman, C. J. (1984). Regulation of the immune system by sex steroids. Endocrin. Rev. 5, 435–455.

Grunfeld, C., Zhao, C., Fuller, J., Pollock, A., Moser, A., Friedman, J., Feingold, K. (1996). Endotoxin and cytokines induce expression of leptin, the ob gene product in hamsters: a role for leptin in the anorexia of infection. J. Clin. Invest. 97, 2152–2157.

Hahn, T. P., Wingfield, J. C., Mullen, R., Deviche, P. J. (1995). Endocrine bases of spatial and temporal opportunism in arctic-breeding birds. Am. Zool. 35, 259–273.

Hanssen, S. A., Hasselquist, D., Folstad, I., Erikstad, K. E. (2004). Costs of immunity: immune responsiveness reduces survival in a vertebrate. Proc. R. Soc. Lond. B. 271, 925–930.

Hart, B. L. (1988). Biological basis of the behavior of sick animals. Neurosci. Biobehav. Rev. 12, 123–137.

Hart, B. L. (1990). Behavioral adaptations to pathogens and parasites: five strategies. Neurosci. Biobehav. Rev. 14, 273–294.

Harvell, C. D., Mitchell, C. E., Ward, J. R., Altizer, S., Dobson, A. P., Ostfeld, R. S., Samuel, M. D. (2002). Climate warming and disease risk for terrestrial and marine biota. Science. 296, 2158–2162.

Hasselquist, D. (2007). Comparative immunoecology in birds: hypotheses and tests. J. Ornithol. 148 (Suppl 2), 571–582.

Hasselquist, D., Marsh, J. A., Sherman, P. W., Wingfield, J. C. (1999). Is avian humoral immunocompetence suppressed by testosterone? Behav. Ecol. Sociobiol. 45, 167–175.

Hataya, Y., Akamizu, T., Hosada, H., Kanamoto, N., Moriyama, K., Kangawa, K., Takaya, K., Nakao, K. (2003). Alterations of plasma ghrelin levels in rats with lipopolysaccharide-induced wasting syndrome and effects of ghrelin treatment on the syndrome. Endocrinology. 144, 5365–5371.

Haussman, M. F., Winkler, D. W., Huntington, C. E., Vleck, D., Sanneman, C. E., Hanley, D., Vleck, C. M. (2005). Cell-mediated immunosenescence in birds. Oecologia. 145, 270–275.

Hazeldine, J., Arlt, W., Lord, J. M. (2010). Dehydroepiandrosterone as a regulator of immune cell function. J. Steroid. Biochem. Mol. Biol. 120, 127–136.

Hetem, R. S., Mitchel, D., Maloney, S. K., Meyer, L. C. R., Fick, L. G., Kerley, G. I. H., Fuller, A. (2008). Fever and sickness behavior during an opportunistic infection in a free-living antelope, the greater kudu (*Tragelaphus strepsiceros*). Am. J. Physiol. Regul. Integr. Comp. Physiol. 294, R246–R254.

Hiebert, S. M., Stoddard, P. K., Arcese, P. (1989). Repertoire size, territory acquisition and reproductive success in the song sparrow. Anim. Behav. 37, 266–273.

Hillgarth, N., Wingfield, J. C. (1997). Testosterone and immunosuppression in vertebrates: implications for parasite-mediated sexual selection. In: D. H. Clayton, J. Moore, Eds., Host–Parasite Evolution: General Principles and Avian Vodels. Oxford University Press, Oxford, pp. 78–104.

Höhn, E. O., Cheng, S. C. (1965). Prolactin and the incidence of brood patch formation and incubation behaviour of the two sexes in certain birds with special reference to the phalaropes. Nature. 208, 197–198.

Holmes, J. E., Miller, N. E. (1963). Effects of bacterial endotoxin on water intake, food intake, and body temperature in the albino rat. J. Exp. Med. 118, 649–658.

Johnson, R. W. (2002). The concept of sickness behavior: a brief chronological account of four key discoveries. Vet. Imm. Immunopath. 87, 443–450.

Johnson, R. W., Propes, M. J., Shavit, Y. (1996). Corticosterone modulates behavioral and metabolic effects of lipopolysaccharide. Am. J. Physiol. 270, R192–198.

Kavaliers, M., Colwell, D. D., Choleris, E. (2000). Parasites and behaviour: an ethopharmacological perspective. Parasitol. Today. 16, 464–468.

Kent, S., Bluthé, R., Kelley, K. W., Dantzer, R. (1992). Sickness behavior as a new target for drug development. Trends Pharmacol. Sci. 13, 24–28.

King, J. R. (1961). On the regulation of vernal premigratory fattening in the white-crowned sparrow. Physiol. Zool. 34, 145–157.

King, J. R. (1970). Photoregulation of food intake and fat metabolism in relation to avian sexual cycles. In: In J. Benoit and I. Assenmacher Eds., La Photorégulation de la Reproduction chez les Oiseaux et les Mammiferes, pp. 365–385. Coll. Int. C.N.R.S. No. 172. Paris.

Klasing, K. C. (1984). Effect of inflammatory agents and interleukin 1 on iron and zinc metabolism. Am. J. Physiol. 247, R901–R904.

Klasing, K. C. (1998). Nutritional modulation of resistance to infectious diseases. Poult. Sci. 77, 1119–1125.

Klasing, K. C. (2004). The costs of immunity. Acta Zool. Sinica 50, 961–969.

Klasing, K. C., Austic, R. E. (1994a). Changes in protein degradation in chickens due to an inflammatory challenge. Proc. Soc. Exp. Biol. Med. 176, 292–296.

Klasing, K. C., Austic, R. E. (1994b). Changes in protein synthesis due to an inflammatory challenge. Proc. Soc. Exp. Biol. Med. 176, 285–291.

Klein, S. L. (2000). The effects of hormones on sex differences in infection: from genes to behavior. Neurosci. Biobehav. Rev. 24, 627–638.

Klein, S. L. (2003). Parasite manipulation of the proximate mechanisms that mediate social behavior in vertebrates. Physiol. Behav. 79, 441–449.

Kluger, M. J. (1979). Fever. Its biology, Evolution and Function. Princeton University Press, Princeton.

Kluger, M. J. (1991). Fever: role of pyrogens and cryogens. Physiol. Rev. 71, 93–127.

Kluger, M. J., Kozak, W., Conn, C. A., Leon, L. R., Soszynski, D. (1998). Role of fever in disease. Ann. N.Y. Acad. Sci. 856, 224–233.

Kluger, M. J., Ringler, D. H., Anver, M. R. (1975). Fever and survival. Science. 188, 166–168.

Kluger, M. J., Rothenberg, B. A. (1979). Fever and reduced iron: their interaction as a host defense response to bacterial infection. Science. 203, 374–376.

Kluger, M. J., Vaughn, L. K. (1978). Fever and survival in rabbits infected with *Pasteurella multocida*. J. Physiol. 282, 243–251.

Kondo, K., Harbuz, M. S., Levy, A., Lightman, S. L. (1997). Inhibition of the hypothalamic-pituitary-thyroid axis in response to lipopolysaccharide is independent of changes in circulating corticosteroids. Neuroimmunomodulation. 4, 188–194.

Konsman, J. P., Dantzer, R. (2001). How the immune and nervous systems interact during disease-associated anorexia. Nutrition. 17, 664–668.

Konsman, J. P., Parnet, P., Dantzer, R. (2002). Cytokine-induced sickness behavior: mechanisms and implications. Trends Neurosci. 25, 154–159.

Kontoghiorghes, G. J., Weinberg, E. D. (1995). Iron: mammalian defense systems, mechanisms of disease, and chelation therapy approaches. Blood Rev. 9, 33–45.

Koolhaas, J. M., Korte, S. M., Boer, S. F., van der Vegt, B. J., van Renen, C. G., Hopster, H., de Jong, I. C., Ruis, M. A. W., Blokhuis, H. J. (1999). Coping styles in animals: current status in behavior and stress-physiology. Neurosci. Biobehav. Rev. 23, 925–935.

Kyriazakis, I., Tolkamp, B. J., Hutchings, M. R. (1998). Towards a functional explanation for the occurrence of anorexia during parasitic infections. Anim. Behav. 56, 265–274.

Langhans, W. (2000). Anorexia of infection: current prospects. Nutrition. 16, 996–1005.

Laye, S., Bluthe, R. M., Kent, S., Combe, C., Medina, C., Parnet, P., Kelley, K. W., Dantzer, R. (1995). Subdiaphragmatic vagotomy blocks induction of IL-1 beta mRNA in mice brain in response to peripheral LPS. Am. J. Physiol. Regul. Integr. Comp. Physiol. 268, R1327–R1331.

Lee, K. A. (2006). Linking immune defenses and life history at the levels of the individual and the species. Integ. Comp. Biol. 46, 1000–1015.

Lee, K. A., Martin, L. B., Wikelski, M. C. (2005). Responding to inflammatory challenges is less costly for a successful avian invader, the house sparrow (*Passer domesticus*), than its less-invasive congener. Oecologia. 145, 244–251.

Lee, S. W., Tsou, A. P., Chan, H., Thomas, J., Petrie, K., Eugui, E. M., Allison, C. A. (1988). Glucocorticoids selectively inhibit the transcription of the interleukin 1beta gene and decrease the stabiliy of interleukin 1B mRNA. Proc. Natl. Acad. Sci. USA. 85, 1204–1208.

LeGrand, E. K. (2000). Why infection-induced anorexia? The case for enhanced apoptosis of infected cells. Med. Hypotheses. 54, 579–602.

Lennie, T. A. (1998). Relationship of body energy status to inflammation-induced anorexia and weight loss. Physiol. Behav. 64, 475–481.

Lennie, T. A. (2004). Sex differences in severity of inflammation-induced anorexia and weight loss. Biol. Res. Nurs. 5, 255–264.

Leshchinsky, T. V., Klasing, K. C. (2001). Divergence of the inflammatory response in two types of chickens. Dev. Comp. Immunol. 25, 629–638.

Lewis, G. D., Campbell, W. B., Johnson, A. R. (1986). Inhibition of prostaglandin synthesis by glucocorticoids in human endothelial cells. Endocrinology. 119, 62–69.

Lochmiller, R. L., Deerenberg, C. (2000). Trade-offs in evolutionary immunology. Just what is the cost of immunity? Oikos. 88, 87–98.

Loria, R. M., Padgett, D. A., Huynh, P. N. (1996). Regulation of the immune response by dehdyroepiandrosterone and its metabolites. J. Endocrinol. 150, S209–S220.

Lozano, G. A., Lank, D. B. (2003). Seasonal trade-offs in cell-mediated immunosenscence in ruffs (*Philomachus pugnax*). Proc. R. Soc. Lond. B. 270, 1203–1208.

Maestroni, G. J. M., Conti, A. (1993). Melatonin in relation with the immune system. In: Y. Hing-Su, R. J. Reiter, Eds., Melatonin, Biosynthesis, Physiological Effects and Clinical Applications. CRC Press, Boca Raton, pp. 289–311.

Maier, S. F., Watkins, L. R. (1998). Cytokines for psychologists: implications of bi-directional immune-to-brain communication for understanding behavior, mood, and cognition. Psychol. Rev. 105, 83–107.

Maier, S. F., Watkins, L. R. (1999). Bidirectional communication between the brain and the immune system: implications for behavior. Anim. Behav. 57, 741–751.

Martin, L. B., Pless, M., Svoboda, J., Wikelski, M. (2004). Immune activity in temperate and tropical house sparrows: a common-garden experiment. Ecology. 85, 2323–2331.

Martin, L. B., Weil, Z. M., Nelson, R. J. (2008). Fever and sickness behaviour vary among congeneric rodents. Funct. Ecol. 22, 68–77.

Martin, L. B., Weil, Z. M., Nelson, R. J. (2008). Seasonal changes in vertebrate immune activity: mediation by physiological trade-offs. Phil. Trans. R. Soc. Lond. B 363, 321–339.

Martin, S. M., Malkinson, T. J., Veale, W. L., Pittman, Q. J. (1995). Fever in pregnant, parturient, and lactating rats. Am. J. Physiol. Regul. Integr. Comp. Physiol. 268, R919–R293.

Matera, L. (1996). Endocrine, paracrine and autocrine actions of prolactin on immune cells. Life Sci. 59, 599–614.

McEwen, B. S. (1998). Stress, adaptation, and disease. Allostasis and allostatic load. Ann. N.Y. Acad. Sci. 840, 33–44.

McEwen, B. S., Baron, C. A., Brunson, K. W., Bulloch, K., Chambers, W. H., Dhabhar, F. S., Goldfarb, R. H., Kitson, R. P., Miller, A. H., Spencer, R. L., Weiss, J. M. (1997). The role of adrenocorticoids as modulators of immune function in health and disease: neural, endocrine and immune interactions. Brain Res. Rev. 1997, 79–133.

McEwen, B. S., Wingfield, J. C. (2003). The concept of allostasis in biology and biomedicine. Horm. Behav. 43, 2–15.

Meddle, S. L., Romero, L. M., Astheimer, L. B., Buttemer, W. A., Moore, I. T., Wingfield, J. C. (2002). Steroid hormone interrelationships with territorial aggression in an arctic-breeding songbird, Gambel's white-crowned sparrow, *Zonotrichia leucophrys gambelii*. Horm. Behav. 42, 212–221.

Miller, N. E. (1964). Some psychophysiological studies of the motivation and the behavioral effects of illness. Bull. Br. Psychol. Soc. 17, 1–20.

Miller, R. A. (1996). The aging immune system: primer and prospectus. Science. 273, 70–74.

Møller, A. P., Erritzøe, J., Saino, N. (2003). Seasonal changes in immune response and parasite impact on hosts. Am. Nat. 161, 657–671.

Moore, C. B., Siopes, T. D. (2000). Effects of lighting conditions and melatonin supplementation on the cellular and humoral immune responses in Japanese quail *Coturnix coturnix japonica*. Gen. Comp. Endocrinol. 119, 95–104.

Moret, Y., Schmid-Hempel, P. (2000). Survival for immunity: the price of immune system activation for bumblebee workers. Science. 290, 1166–1170.

Morrow, L. E., McClellan, J. L., Conn, C. A., Kluger, M. J. (1993). Glucocorticoids alter fever and IL-6 responses to psychological stress and to lipopolysaccharide. Am. J. Physiol. Regul. Integr. Comp. Physiol. 33, R1010–R1016.

Muchlinski, A. E., Baldwin, B. C., Gramajo, R. (2000). Endotixin elicits a febrile response in laboratory-maintained but not free-living California ground squirrels. J. Mammalogy. 81, 701–708.

Munck, A., Guyre, P. M., Holbrook, N. J. (1984). Physiological functions of glucocorticoids in stress and their relation to pharmacological actions. Endocr. Rev. 5, 25–44.

Murakami, N., Ono, T. (1987). Sex-related differences in fever development of rats. Am. J. Physiol. 252, R284–289.

Murray, M. J., Murray, A. B. (1979). Anorexia of infection as a mechanism of host defense. Am. J. Clin. Nutr. 32, 593–596.

Myers, M. G., Cowley, M. A., Münzberg, H. (2008). Mechanisms of leptin action and leptin resistance. Annu. Rev. Physiol. 70, 537–556.

Nakamura, I., Mitarai, Y., Yoshioka, M., Koizumi, N., Shibahara, T., Nakajima, Y. (1998). Serum levels of interlekin-6, alpha-acid glycoprotein, and corticosterone in two-week-old chickens inoculated with *Escherichia coli* lipopolysaccharide. Poul. Sci. 77, 908–911.

Nakamura, K., Aioke, A., Hosokawa, T., Rokutan, K., Koyama, K., Nishi, Y., Yoshida, A., Kawai, K. (1990). Effect of food restriction on immune response in mice. J. Neuroimmunol. 30, 23–29.

Nava, F., Calapai, G., Facciolà, G., Cuzzocrea, S., Giuliani, G., De Sarro, A., Caputi, A. P. (1997). Melatonin effects on inhibition of thirst and fever induced by lipopolysaccharide in rat. Eur. J. Biochem. 331, 267–274.

Nelson, R. J. (2004). Seasonal immune function and sickness responses. Trends Immunol. 25, 187–192.

Nelson, R. J., Demas, G. E. (1996). Seasonal changes in immune function. Q. Rev. Biol. 71, 511–548.

Nelson, R. J., Demas, G. E., Klein, S. L., Kriegsfeld, L. J. (2002). Seasonal Patterns of Stress, Immune Function, & Disease. Cambridge University Press, Cambridge.

Norris, K., Evans, M. R. (2000). Ecological immunology: Life history trade-offs and immune defense in birds. Behav. Ecol. 11, 19–26.

Nunn, C. L., Lindenfors, P., Pursall, R., Rolff, J. (2009). On sexual dimorphism in immune function. Phil. Trans. R. Soc. B. 364, 61–69.

O'Connor, J. C., Satpahty, A., Hartman, M. E., Horvath, E. M., Kelley, K. W., Dantzer, R., Johnson, R. W., Freund, G. G. (2005). IL-1beta-mediated innate immunity is amplified in the db/db mouse model of type 2 diabetes. J. Immunol. 174, 4991–4997.

Offner, P. J., Moore, E. E., Biffl, W. L. (1999). Male gender is a risk factor for major infections after surgery. Arch. Surg. 1999, 935–938.

Ottaviani, E., Malagoli, D., Capri, M., Franceschi, C. (2008). Ecoimmunology: is there any room for the neuroendocrine system? Bioessays. 30, 868–874.

Owen-Ashley, N. T. (2004). Environmental Regulation of Immune-Endocrine Phenomena in Songbirds. Ph.D. Dissertation. Dept. of Zoology, Univ. of Washington, Seattle, pp. 232.

Owen-Ashley, N. T., Hasselquist, D., Råberg, L., Wingfield, J. C. (2008). Latitudinal variation of immune defense and sickness behavior in the white-crowned sparrow (*Zonotrichia leucophrys*). Brain Behav. Immun. 22, 614–625.

Owen-Ashley, N. T., Hasselquist, D., Wingfield, J. C. (2004). Androgens and the immunocompetence handicap hypothesis: unraveling direct and indirect pathways of immunosuppression in song sparrows. Am. Nat. 164, 490–505.

Owen-Ashley, N. T., Turner, M., Hahn, T. P., Wingfield, J. C. (2006). Hormonal, behavioral, and thermoregulatory responses to bacterial lipopolysaccharide in captive and free-living White-crowned Sparrows (*Zonotrichia leucophrys gambelii*). Horm. Behav. 49, 15–29.

Owen-Ashley, N. T., Wingfield, J. C. (2006). Seasonal modulation of sickness behavior in free-living northwestern song sparrows (*Melospiza melodia morphna*). J. Exp. Biol. 209, 3062–3070.

Owen-Ashley, N. T., Wingfield, J. C. (2007). Acute phase responses of passerines: characterization and seasonal variation. J. Ornithol. 148 (Suppl 2), S583–S591.

Palacios, M. G., Cunnick, J. E., Winker, D. W., Vleck, C. M. (2007). Immunosenescence in some but not all immune components in a free-living vertebrate, the tree swallow. Proc. R. Soc. Lond. B. 274, 951–957.

Pellymounter, M. A., Cullen, M. J., Baker, M. B., Hecht, R., Winters, D., Boone, T., Collins, F. (1995). Effects of the obese gene product on body weight regulation in ob/ob mice. Science. 269, 540–543.

Peloso, E. D., Florez-Duquet, M., Buchanan, J. B., Satinoff, E. (2003). LPS fever in old rats depend on the ambient temperature. Physiol. Behav. 78, 651–654.

Pezeshki, G., Pohl, T., Schöbitz, B. (1996). Corticosterone controls interleukin-1 beta expression and sickness behavior in the rat. J. Neuroendocrinol. 8, 129–135.

Piersma, T. (1997). Do global patterns of habitat use and migration strategies co-evolve with relative investments in immunocompetence due to spatial variation in parasite pressure? Oikos. 80, 623–631.

Plata-Salamán, C. R. (1996). Anorexia during acute and chronic disease. Nutrition. 12, 69–78.

Plata-Salamán, C. R., Vasselli, J. R., Sonti, G. (1997). Differential responsiveness of obese (fa/fa) and lean (Fa/Fa) Zucker rats to cytokine-induced anorexia. Obes. Res. 5, 36–42.

Poulin, R. (1996). Sexual inequalities in helminth infections: a cost of being a male? Am. Nat. 147, 287–295.

Prendergast, B. J., Ballie, S. R., Dhabhar, F. S. (2008). Gonadal hormone-dependent and -independent regulation of immune function by photoperiod in Siberian hamsters. Am. J. Physiol. Regul. Integr. Comp. Physiol. 294, R384–R392.

Promislow, D. E. L., Harvey, P. H. (1990). Living fast and dying young: a comparative analysis of life-history variance among mammals. J. Zool. 220, 417–437.

Råberg, L., Grahn, M., Hasselquist, D., Svensson, E. (1998). On the adaptive significance of stress-induced immunosuppression. Proc. R Soc. Lond. B 265, 1637–1641.

Raghavendra, V., Agrewala, J. N., Kulkarni, S. K. (2000). Melatonin reversal of lipopolysaccharide-induced thermal and behavioral hyperalgesia in mice. Eur. J. Pharmacol. 395, 15–21.

Raison, C. L., Gumnick, J. F., Miller, A. H. (2002). Neuroendocrine-immune interactions: implications for health and behavior. In: D. Pfaff, A. Etgen, S. Fahrbach, R. Rubin, eds, Hormones, Brain and Behavior. Elsevier Science, pp. 209–261.

Ricklefs, R. E., Wikelski, M. (2002). The physiology/life-history nexus. Trends Ecol. Evol. 17, 462–468.

Rivier, C. (1990). Role of endotoxin and interleukin-1 in modulating ACTH, LH and sex steroid secretion. Adv. Exp. Med. Biol. 275, 295–301.

Rivier, C., Vale, W. (1990). Cytokines act within the brain to inhibit luteinizing hormone secretion and ovulation in the rat. Endocrinology. 127, 849–856.

Roberts, M. L., Buchanan, K. L., Evans, M. R. (2004). Testing the immunocompetence handicap hypothesis: a review of the evidence. Anim. Behav. 68, 227–239.

Romero, L. M., Dickens, M. J., Cyr, N. E. (2009). The reactive scope model: a new model integrating homeostasis, allostasis, and stress. Horm. Behav. 55, 375–389.

Ruzek, M. C., Pearce, B. D., Miller, A. H., Biron, C. A. (1999). Endogenous glucocorticoids protect against cytokine-mediated lethargy during viral infection. J. Immunol. 162, 3527–3533.

Saino, N., Ferrari, R. P., Romano, M., Rubolini, D., Møller, A. P. (2003). Humoral immune response in relation to senescence. J. Evol. Biol. 16, 1127–1134.

Sapolsky, R. M., Romero, L. M., Munck, A. U. (2000). How do glucocorticosteroids influence stress responses? Integrating permissive, suppressive, stimulatory and preparative actions. Endocr. Rev. 21, 55–89.

Sarraf, P., Frederich, R. A., Turner, E. M., Ma, G., Jaskowiak, N. T., Rivet III, D. J., Flier, J. S., Lowell, B. B., Fraker, D. L., Alexander, H. R. (1996). Multiple cytokines and acute inflammation raise mouse leptin levels: potential role in inflammatory anorexia. J. Exp. Med. 185, 171–175.

Schmid-Hempel, P., Ebert, D. (2003). On the evolutionary ecology of specific immune defence. Trends Ecol. Evol. 18, 27–32.

Schroder, J., Kahlke, V., Staubach, K. H., Zabel, P., Stuber, F. (1998). Gender differences in human sepsis. Arch. Surg. 133, 1200–1205.

Scrimshaw, N. S. (1977). Effect of infection on nutrient requirements. Am. J. Clin. Nutr. 30, 1536–1544.

Seale, J. V., Wood, S. A., Atkinson, H. C., Harbuz, M. S., Lightman, S. L. (2004). Gonadal steroid replacement reverses gonadectomy-induced changes in the corticosterone pulse profile and stress-induced hypothalamic-pituitary-adrenal axis activity of male and female rats. J. Neuroendocrinol. 16, 989–998.

Sheldon, B. C., Verhulst, S. (1996). Ecological immunology: Costly parasite defences and trade-offs in evolutionary ecology. Trends Ecol. Evol. 11, 317–321.

Soma, K. K., Wingfield, J. C. (2001). Dehydroepiandrosterone in songbird plasma: seasonal regulation and relationship to territorial aggression. Gen. Comp. Endocrinol. 123, 144–155.

Soma, K. K., Wissman, A. M., Brenowitz, E. A., Wingfield, J. C. (2002). Dehydroepiandrosterone (DHEA) increases territorial song and the size of an associated brain region in a male songbird. Horm. Behav. 41, 203–212.

Sorci, G., Faivre, B. (2009). Inflammation and oxidative stress in vertebrate host-parasite systems. Phil. Trans. R. Soc. B. 364, 71–83.

Soto, L., Martín, A. I., Millám, J. R., Vara, E., López-Calderón (1998). Effects of endotoxin lipopolysaccharide administration on the somatotropic axis. J. Endocrinol. 159, 239–246.

Spangelo, B. L., Hall, N. R., Ross, P. C., Goldstein, A. L. (1987). Stimulation of in vivo antibody production and concanavalin-A-induced mouse spleen cell mitogenesis by prolactin. Immunopharmacology. 14, 11–20.

Spinedi, E., Gaillard, R. C., Chisari, A. (2002). Sexual dimorphism of neuroendocrine-immune interactions. In: R. C. Gaillard, Ed., Neuroendocrine-Immune Interactions. Karger, Basel, Switzerland, pp. 91–107.

Spinedi, E., Suescun, M. O., Hadid, R., Daneva, T., Gaillard, R. C. (1992). Effects of gonadectomy and sex hormone therapy on the endotoxin-stimulated hypothalamo-pituitary-adrenal axis: evidence for a neuroendocrine-immunological sexual dimorphism. Endocrinology. 131, 2430–2436.

Svec, F., Porter, J. (1998). The actions of exogenous dehydroepiandrosterone in experimental animals and humans. Proc Soc Exp Biol Med. 218, 174–191.

Tangbanluekal, L., Robinette, C. L. (1993). Prolactin mediates estradiol-induced inflammation in the lateral prostrate of Wistar rats. Endocrinology. 132, 2407–2416.

Taub, D. D. (2007). Novel connections between the neuroendrocrine and immune systems: the ghrelin immunoregulatory network. Vitam. Horm. 77, 325–346.

Theologides, A. (1977). Anorexia-producing intermediary metabolites. Am. J. Clin. Nutr. 29, 552–560.

Tieleman, B. I., Williams, J. B., Ricklefs, R. E., Klasing, K. C. (2005). Constitutive innate immunity is a component of the pace-of-life syndrome in tropical birds. Proc. R. Soc. Lond. B. 272, 1715–1720.

Tizard, I. (2008). Sickness behavior, its mechanisms and significance. Anim. Health Res. Rev. 9, 87–99.

Tuchscherer, M., Kanitz, E., Puppe, B., Tuchscherer, A. (2006). Early social isolation alters behavioral and physiological responses to an endotoxin challenge in piglets. Horm. Behav. 50, 753–761.

Tufft, L. S., Nockels, C. F., Fettman, M. J. (1988). Effects of *Escherichia coli* on iron, copper, and zinc metabolism in chicks. Avian Dis. 32, 779–786.

Turnbull, A. V., Rivier, C. L. (1995). Regulation of the HPA axis by cytokines. Brain Behav. Immun. 9, 253–275.

van Gils, J. A., Spaans, B., Dekinga, A., Piersma, T. (2006). Foraging in a tidally structured environment by red knots (*Calidris canutus*): ideal, but not free. Ecology. 87, 1189–1202.

Vaughn, L. K., Bernheim, H. A., Kluger, M. J. (1974). Fever in the lizard *Dipsosaurus dorsalis*. Nature. 252, 473–474.

Vézina, F., Petit, M., Buehler, D. M., Dekinga, A., Piersma, T. (2009). Limited access to food and physiological trade-offs in a long-distance migrant shorebirds. I. Energy metabolism, behavior, and body-mass regulation. Physiol. Biochem. Zool. 82, 549–560.

Viney, M. E., Riley, E. M., Buchanan, K. L. (2005). Optimal immune responses: immuncompetence revisited. Trends Ecol. Evol. 20, 665–669.

Weigent, D. A. (1996). Immunoregulatory properties of growth hormone and prolactin. Pharmacol. Ther. 69, 237–257.

Weil, Z. M., Bowers, S. L., Dow, E. R., Nelson, R. J. (2006a). Maternal aggression persists following lipopolysaccharide-induced activation of the immune system. Physiol. Behav. 87, 694–699.

Weil, Z. M., Bowers, S. L., Pyter, L. M., Nelson, R. J. (2006b). Social interactions alter pro-inflammatory cytokine gene expression and behavior following endotoxin administration. Brain Behav. Immun. 20, 72–79.

Weil, Z. M., Martin, L. B., Workman, J. L., Nelson, R. J. (2006c). Immune challenge retards seasonal reproductive regression in rodents: evidence for terminal investment. Biol. Lett. 2, 393–396.

Weinberg, E. D. (1974). Iron and susceptibility to infectious disease. Science. 184, 952–955.

Wen, J. C., Dhabhar, F. S., Prendergast, B. J. (2007). Pineal-dependent and -independent effects of photoperiod on immune function in Siberian hamsters (*Phodopus sungorus*). Horm. Behav. 51, 31–39.

Wichmann, M. W., Inthorn, D., Andress, H. J. (2000). Incidence and mortality of severe sepsis in surgical intensive care patients: the influence of patient gender on disease process and outcome. Intensive Care Med. 2000, 167–172.

Wiens, M., Korzhev, M., Krasko, A., Thakur, N. L., Perovic-Ottstadt, S., Breter, H. J., Ushijima, H., Diehl-Seifert, B., Müller, I. M., Müller, W. E. G. (2005). Innate immune defense of the sponge *Suberites domuncula* against bacteria involves a MyD88-dependent signaling pathway. J. Biol. Chem. 280, 27949–27959.

Wiens, M., Korzhev, M., Perovic-Ottstadt, S., Luthringer, B., Brandt, D., Klein, S., Müller, W. E. G. (2007). Toll-like receptors are part of the innate immune defense system of sponges (Demospongiae: Porifera). Mol. Biol. Evol. 24, 792–804.

Wiersma, P., Muñoz-Garcia, A., Walker, A., Williams, J. B. (2007). Tropical birds have a slow pace of life. Proc. Natl. Acad. Sci. USA. 104, 9340–9345.

Wikelski, M., Kays, R. W., Kasdin, N. J., Thorup, K., Smith, J. A., Swenson, G. W. (2007). Going wild: what a global small-animal tracking system could do for experimental biologists. J. Exp. Biol. 210, 181–186.

Williams, G. C., Neese, R. M. (1991). The dawn of Darwinian medicine. Q. Rev. Biol. 66, 1–22.

Wing, E. J., Young, J. B. (1980). Acute starvation protects mice against *Listeria monocytogenes*. Infect. Immun. 28, 771–776.

Wingfield, J. C. (1985). Short-term changes in plasma levels of hormones during establishment and defense of a breeding territory in male song sparrows, *Melospiza melodia*. Horm. Behav. 19, 174–187.

Wingfield, J. C. (2003). Control of behavioural strategies for capricious environments. Anim. Behav. 66, 807–816.

Wingfield, J. C. (2005). The concept of allostasis: coping with a capricious environment. J. Mammalogy. 86, 248–254.

Wingfield, J. C., Hahn, T. P., Wada, M., Astheimer, L. B., Schoech, S. (1996). Interrelationship of day length and temperature on the control of gonadal development, body mass, and fat score in White-crowned Sparrows, *Zonotrichia leucophrrys gambelii*. Gen. Comp. Endocrinol. 101, 242–255.

Wingfield, J. C., Hahn, T. P., Wada, M., Schoech, S. J. (1997). Effects of day length and temperature on gonadal development, body mass, and fat depots in White-crowned Sparrows, *Zonotrichia leucophrys pugetensis*. Gen. Comp. Endocrinol. 107, 44–62.

Wingfield, J. C., Hegner, R. E., Dufty, A. M., Jr, Ball, G. F. (1990). The "challenge hypothesis": theoretical implications for patterns of testosterone secretion, mating systems, and breeding strategies. Am. Nat. 136, 829–846.

Wingfield, J. C., Hunt, K. E. (2002). Arctic spring: hormone behavior interactions in a severe environment. Comp. Biochem. Physiol. B. 132, 275–286.

Wingfield, J. C., Lynn, S. E., Soma, K. K. (2001). Avoiding the "costs" of testosterone: ecological bases of hormone-behavior interactions. Brain Behav Evol. 57, 239–251.

Wingfield, J. C., Maney, D. L., Breuner, C. W., Jacobs, J. D., Lynn, S., Ramenofsky, M., Richardson, R. D. (1998). Ecological bases of hormone-behavior interactions: the "emergency life history stage." Am. Zool. 38, 191–206.

Wingfield, J. C., Ramenofsky, M. (1999). Hormones and the behavioral ecology of stress. In: P. H. M. Balm, Ed., Stress Physiology in Animals. Academic Press, Sheffield.

Wingfield, J. C., Whaling, C. S., Marler, P. R. (1994). Communication in vertebrate aggression and reproduction: the role of hormones. In: E. Knobil, J. D. Neill, Eds., Physiology of Reproduction. Raven Press, New York, pp. 303–342.

Yirmiya, R., Avitsur, R., Donchin, O., Cohen, E. (1995). Interleukin-1 inhibits sexual behavior in female but not in male rats. Brain Behav. Immun. 9, 220–233.

Zellweger, R., Wichmann, M. W., Ayala, A., Stein, S., Demaso, C. M., Chaudry, I. H. (1997). Females in proestrus state maintain splenic immune functions and tolerate sepsis better than males. Crit. Care Med. 25, 106–110.

Zellweger, R., Zhu, X. H., Wichmann, M. W., Ayala, A., Demaso, C. M., Chaudry, I. H. (1996). Prolactin administration following hemorrhagic shock improves macrophage cytokine release capacity and decreases mortality from subsequent sepsis. J. Immunol. 157, 5748–5754.

Zuk, M. (1990). Reproductive strategies and disease susceptibility: an evolutionary viewpoint. Parasitol. Today. 6, 231–233.

Zuk, M. (1996). Disease, endocrine-immune interactions, and sexual selection. Ecology. 77, 1037–1042.

Zuk, M., Johnsen, T. S. (1998). Seasonal changes in the relationship between ornamentation and immune response in red jungle fowl. Proc. R. Soc. Lond. B. 265, 1631–1635.

Zuk, M., Stoehr, A. M. (2002). Immune defense and host life history. Am. Nat. 160, S9–S22.

Zwarts, L., Hulscher, J. B., Koopman, K., Zegers, P. M. (1996). Short-term variation in the body weight of oystercatchers *Haematopus ostralegus*: effect of available feeding time by day and night, temperature and wind force. Ardea. 84A, 357–372.

Zysling, D. A., Garst, A. D., Demas, G. E. (2009). Photoperiod and food restriction differentially affect reproductive and immune responses in Siberian hamsters *Phodopus sungorus*. Funct. Ecol. 23, 979–988.

CHAPTER 4

AMPHIBIAN IMMUNITY

Staying in Tune with the Environment

Louise A. Rollins-Smith and Douglas C. Woodhams

INTRODUCTION

Ecoimmunology is the field of study that attempts to understand the functions of the immune system in the context of the environment of the host. For amphibians, this is a very broad subject that encompasses the effects of temperature, chemicals in the environment, other animals in the community, stress resulting from competition and predation, and the challenges of having distinct embryonic, larval, and adult life stages. All of these variables affect disease susceptibility. Because ecoimmunology is a new field of study, there have been no previous attempts to pull together the threads of information about the interactions of the immune system with the environment in amphibians. Yet, there is a critical need for this kind of comprehensive review because amphibians are facing enormous challenges due to a number of environmental factors interacting with disease. Although there have been no detailed reviews of environmental interactions with immune function in amphibians, amphibian immunity, amphibian disease ecology, immune changes at metamorphosis, and other aspects of this subject have been previously reviewed. Whenever possible, we will cite these reviews and distinguish the primary literature from the reviews.

There are several central facts that define amphibian interactions with the environment. Amphibians are ectotherms, unable to regulate their temperature except by moving to a warmer or colder environment. Thus, temperature affects all physiological processes, including the functions of the immune system. Amphibians are critically dependent on water. Many species must come to water bodies to breed. Tadpoles must metamorphose before the pond dries out. Food is there, but so are competitors and disease organisms. All of these factors have an impact on behavior, growth, and immune defenses. During larval life the immune system is immature, and it is dramatically reorganized at metamorphosis. Thus, during larval life and metamorphosis, amphibians are more vulnerable to some pathogens until the adult-type immune system emerges.

Immunity can be defined as the collective mechanisms used by the host to resist infection but tolerate self. It differs among hosts and changes with environmental conditions and upon challenge by pathogens or parasites. There are costs and trade-offs necessary to maintain an effective immune defense. These may include reduced growth, longevity, or fecundity. Trading increased body growth for more limited development of the lymphocyte repertoire, for example, may leave the host vulnerable to infection and resulting pathology. Alternatively, an overactive immune defense may lead to immunopathology or sap the energy of the host, reducing its ability to seek more favorable conditions and defend against predators and competitors, and may affect reproduction and neurological functions such as learning and memory. There may be optimal immune defenses in a given environment, against a given parasite or pathogen. However, these optima are always shifting in response to natural selection.

Throughout the text, we have indicated amphibian species by their common names and their binomial (genus and species) names. To connect the most current literature with past literature we have retained the use of well-established species names rather than adopt the names recommended by Darrel Frost and others (Frost et al., 2006). The common names, established names, and newly recommended names are shown in Table 4.1.

THE PROBLEM OF GLOBAL AMPHIBIAN DECLINES

Worldwide, approximately 6,700 species of amphibians have been described, and efforts to define the diversity of biological species continue to uncover even more species (Vieties et al., 2009; www.amphibiaweb.org). Separation of cryptic species by DNA barcoding reveals additional new species (Weight et al., 2005; Wang et al., 2008; Crawford et al., 2010). While new species predominantly originate from South America, India, New Guinea, and Madagascar; several recent discoveries have been made in Europe (Temple and Cox, 2009). Sadly, many species are suffering historically unprecedented population declines, even as they are being discovered (Crawford et al., 2010). These losses have led to more than 150 apparent species extinctions (Stuart et al., 2004; reviewed in Skerratt et al., 2007; Wake and Vredenburg, 2008). A number of factors contribute to population declines (reviewed in Collins and Storfer, 2003; Collins and Crump, 2009). They include habitat destruction (Ash and Bruce, 1994; Delis et al., 1996; Hecnar and M'Closkey, 1996, 1998; Marsh and Beckman, 2004), climate change (Pounds and Crump, 1994; Donnelly and Crump, 1998; Pounds et al., 1999; Carey and Alexander, 2003; Daszak et al., 2005; Pounds et al., 2006; Wake, 2007; Whitfield et al., 2007; Lips et al., 2008; Rohr et al., 2008; Rohr and Raffel, 2010), over-harvesting from wild populations (Gorzula, 1996; Schlaepfer et al., 2005; Schloegel et al., 2009), introduction of non-native species (Bradford, 1989; Bradford et al., 1993; Drost and Fellers, 1996; Lawler et al., 1999; Mazzoni et al., 2003; Vredenburg, 2004; Garner et al., 2006; Knapp et al., 2007; Fisher and Garner, 2007), environmental chemicals (Berrill et al., 1997; Bridges and Semlitsch, 2001; Sparling et al., 2001; Relyea and Mills, 2001; Davidson et al., 2002; Blaustein et al., 2003; Relyea, 2004, 2005; Relyea et al., 2005; Davidson et al., 2007; Hayes et al., 2010), and interactions of factors (Alford and Richards, 1999; Stuart et al., 2004; Hayes et al., 2010). Growing evidence links many declines to the emerging infectious diseases caused by ranaviruses and chytridiomycosis caused by the chytrid fungus, *Batrachochytrium dendrobatidis* (Berger et al., 1998; Longcore et al., 1999; Pessier et al., 1999; Lips et al., 2006; reviewed in Carey et al., 1999; Daszak et al., 1999; Chinchar, 2002; Skerratt et al., 2007; Gray et al., 2009). *B. dendrobatidis* occurs on every continent where amphibians exist, and understanding this pathogen is at the top

TABLE 4.1: Amphibian species mentioned in this chapter

Common name	Scientific name	Suggested modified name
African clawed frog	*Xenopus laevis*	
American toad	*Bufo americanus*	*Anaxyrus americanus*
boreal toad	*Bufo boreas*	*Anaxyrus boreas boreas*
Bronze frog	*Rana clamitans*	*Lithobates clamitans*
Bullfrog	*Rana catesbeiana*	*Lithobates catesbeianus*
California newt	*Taricha torosa*	
Cascades frog	*Rana cascadae*	
common toad	*Bufo bufo*	
Cope's gray tree frog	*Hyla chrysoscelis*	
Eastern gray tree frog	*Hyla versicolor*	
Eastern newt	*Notophthalmus viridescens*	
edible frog	*Rana esculenta*	*Pelophylax esculentus*
European common frog	*Rana temporaria*	
foothill yellow-legged frog	*Rana boylii*	
Fowler's toad	*Bufo fowleri*	*Anaxyrus fowleri*
fire-bellied toad	*Bombina bombina*	
green treefrog	*Hyla cinerea*	
green-eyed treefrog	*Litoria genimaculata*	
magnificent treefrog	*Litoria splendida*	
marine toad	*Bufo marinus*	*Rhinella marina*
Mexican axolotl	*Ambystoma mexicanum*	
mountain yellow-legged frog	*Rana muscosa*	
Northern leopard frog	*Rana pipiens*	*Lithobates pipiens*
Northern cricket frog	*Acris crepitans*	
Northern red-legged frog	*Rana aurora*	
orange-eyed treefrog	*Litoria chloris*	
Pacific chorus frog	*Hyla regilla*	*Pseudacris regilla*
Panamanian golden frog	*Atelopus zeteki*	
Peron's tree frog	*Litoria peronii*	
pig frog	*Rana grylio*	
pool frog	*Rana lessonae*	*Pelophylax lessonae*
redback salamander	*Plethodon cinereus*	
Roth's tree frog	*Litoria rothii*	
Sierra Nevada yellow-legged frog	*Rana sierrae*	
Sonoran tiger salamander	*Ambystoma mavortium stebbinsi*	
Southern bell frog	*Litoria raniformis*	
Southern leopard frog	*Rana utricularia*	*Lithobates sphenocephalus utricularius*
Southern toad	*Bufo terrestris*	*Anaxyrus terrestris*
spring peeper	*Pseudacris crucifer*	
tiger frog	*Rana tigrina rugulosa*	
tiger salamander	*Ambystoma tigrinum*	
waterfall frog	*Litoria nannotis*	
wood frog	*Rana sylvatica*	*Lithobates sylvaticus*
Woodhouse's toad	*Bufo woodhousii*	*Anaxyrus woodhousii*

Standard common names, established scientific names, and names modified for the most current scientific relationships are shown as suggested by the Center for North American Herpetology (http://www.cnah.org/, Oct. 27, 2010) and IUCN Red List ofThreatened Species, or AmphibiaWeb (www.amphibiaweb.org, Oct. 27, 2010).

of actions to address the causes of declines and extinctions in the most recent Amphibian Conservation Action Plan (online supplement to Mendelson et al., 2006).

EMERGING INFECTIOUS DISEASES IN AMPHIBIANS

Because of the global association with amphibian declines, both ranaviral disease and chytridiomycosis were listed as internationally notifiable diseases by the World Organization for Animal Health (OIE) in 2008 (Schloegel et al., 2010). Because both diseases are of paramount concern to amphibian biologists, we will briefly summarize our current understanding of each pathogen and the diseases that they cause.

B. dendrobatidis is a unique fungal pathogen. It is primitive in comparison with other fungal pathogens of vertebrates because it retains a swimming zoospore as the infectious stage (Longcore et al., 1999) and is most closely related to other fungi that hold a basal position in fungal evolution before advanced fungi diverged (James et al., 2000; James et al., 2007; Stajich et al., 2009). *B. dendrobatidis* colonizes skin cells of adult amphibians and the mouthparts of tadpoles (Berger et al., 1998; Longcore et al., 1999; Pessier et al., 1999; Berger et al., 2005). Infection occurs when zoospores land on the skin or the mouthparts and encyst (stop movement and settle, resorb the flagellum, and form a cell wall) (Berger et al., 1998; Longcore et al., 1999; Pessier et al., 1999; Berger et al., 2005). Through mechanisms that are not understood, the pathogen moves from the surface to the stratum granulosum of the skin. There it enters healthy cells, and the organism, now called a thallus (or germling), grows in size and develops into a single urn-shaped zoosporangium in which the zoospores develop. This pattern of chytrid development within cells is termed "exogenous development" (Barr, 1980). As infected skin cells move toward the surface, the zoosporangium matures, the discharge papilla opens, and mature zoospores swim out (Berger et al., 1998; Longcore et al., 1999; Pessier et al., 1999; Berger et al., 2005). Zoospores are motile for about 24 hours at 23°C before they encyst or die, but they survive more than 48 hours at 4° to 14°C (Piotrowski et al., 2004; Woodhams et al., 2008). Movement is directed toward nutritional cues (Moss et al., 2008). On the skin, thalli are found in clusters, suggesting zoospore transmission over a short distance. In calm water, zoospores move less than 2 cm (Piotrowski et al., 2004). They may be carried by currents to infect a new host, and many highly susceptible species are stream-dwelling species (Laurance et al., 1996; Lips, 1998, 1999). The mechanism by which this pathogen causes chytridiomycosis and death appears to be by interfering with the ion-transport functions of the skin, leading ultimately to cardiac arrest (Voyles et al., 2007, 2009). Unlike other fungal pathogens that move from the site of infection to other organs, *B. dendrobatidis* remains confined to the skin or tadpole mouth parts.

Viruses in the genus *Ranavirus* are among members of the iridovirus family of DNA viruses that cause pathology in fish, amphibians, and reptiles (reviewed in Chinchar, 2002; Williams et al., 2005; Gray et al., 2009). They have been linked to amphibian mortality events in the United States, Canada, the United Kingdom, Venezuela, Brazil, and Australia (Cunningham et al., 1996; Jancovich et al., 1997; Zupanovic et al., 1998; Bollinger et al., 1999; Cullen and Owens, 2002; Green et al., 2002; Docherty et al., 2003; Greer et al., 2005; Mazzoni et al., 2009). Pathological findings include hemorrhagic and necrotic lesions within internal organs (liver, spleen, kidney), gastrointestinal ulceration, skin ulceration and erosion, and necrosis within hematopoietic cells (Bollinger et al., 1999;

Greer et al., 2005; reviewed in Gray et al., 2009). Transmission occurs by exposure to contaminated water or soil, contact with infected individuals, or ingestion of infected tissue through predation (Jancovich et al., 2001; reviewed in Gray et al., 2009). Viruses infect tadpoles and adults of many species, although adults are more resistant (Robert et al., 2005).

OTHER AMPHIBIAN DISEASES

The causal web of chytridiomycosis (Fig. 4.1) and other amphibian diseases can be quite complex. Infection and disease development may depend on the dose and timing of exposure, the host species, pathogen or parasite, environmental conditions, and host body condition. Therefore, a potentially infectious microorganism is not always parasitic, and a "pathogen" is not always a pathogen.

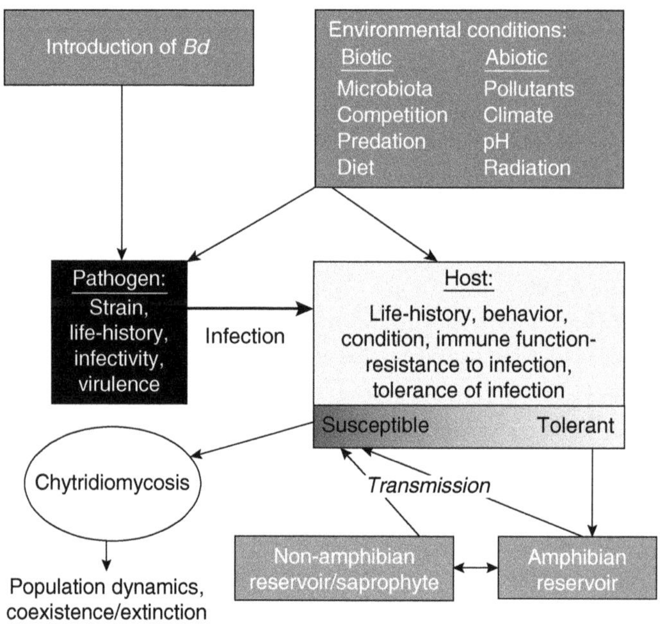

FIGURE 4.1: Causal web of the amphibian disease chytridiomycosis. Upon introduction of *Batrachochytrium dendrobatidis* (*Bd*), infection and disease development in an amphibian host depends on the infectivity, virulence, and life-history of the pathogen strain. Host susceptibility varies among and within species and may depend on the life-history stage of the amphibian (tadpole, new metamorph, adult), its behavioral ecology (microhabitat preferences, behavioral fever, contact rate), its body condition, and immune function. Transmission to new hosts is dependent on the presence of infected amphibians and possibly nonamphibian reservoirs in the environment. Environmental conditions affect pathogen infectivity, growth, and transmission, as well as host ecology and susceptibility. Host–pathogen interactions may change with biotic conditions (other skin microbes, competition, predation, and nutrition) and abiotic conditions, including climate, radiation, pH, and chemical pollutants. Under some conditions, hosts may tolerate infections and coexist with the pathogen. Alternatively, infection may lead to development of the disease chytridiomycosis, population declines, extinctions, and cascading ecosystem alterations.

Under some conditions, a "pathogen" may be a beneficial mutualist. However, the term "pathogen" is regularly used to indicate an organism that can potentially be infectious and cause disease. Koch's postulates for the causation of chytridiomycosis were fulfilled by Nichols et al. (2001). In addition to *B. dendrobatidis,* several other pathogens are associated with disease outbreaks or fulfill Koch's postulates for disease in amphibians (reviewed in Carey et al., 1999; Daszak et al., 1999, 2003; Densmore and Green, 2007). They include gram-negative bacteria such as *Aeromonas hydrophila, Pseudomonas aeruginosa* (reviewed in Taylor et al., 2001), *Klebsiella pneumoniae, Proteus mirabilis* (Schadich and Cole, 2010), other fungi (reviewed in Taylor, 2001; Densmore and Green, 2007), the water mold *Saprolegnia* (Kiesecker et al., 2001), and protozoan parasites (Poyton and Whitaker, 1994; Pascolini et al., 2003; reviewed in Densmore and Green, 2007). Most of these pathogens can influence population dynamics. Even symbiotic microbiota can be facultative or opportunistic pathogens that emerge under stressful host conditions. Thus, given the urgent conservation concerns, disease ecology of amphibians is an important area of study.

OVERVIEW OF AMPHIBIAN IMMUNITY

The immune system of a limited number of amphibian species has been studied in detail. Most of what is known derives from the studies of model species (the South African clawed frog, *Xenopus laevis;* the Northern leopard frog, *Rana pipiens,* and the Mexican axolotl, *Ambystoma mexicanum*). From these studies, we have learned that the immune system of amphibians is very similar to that of all other vertebrate groups (reviewed in Du Pasquier et al., 1989; Carey et al., 1999; Robert and Ohta, 2009). The hematopoietic cells, the organs, and the critical effector molecules (antibodies, complement, cytokines, and chemokines) are very similar to those found in fish, reptiles, birds, and mammals (reviewed in Robert and Ohta, 2009). Amphibians lack true lymph nodes, but in other respects the immune system of amphibians is very much like that of mammals. A very detailed description of the organs of the immune system of amphibians can be found in Manning and Horton (1982). The organs containing lymphoid tissue in each amphibian order and in tadpoles or adults of anuran species are summarized in Table 4.2 (adapted from Manning and Horton, 1982).

Briefly, the major organs of the amphibian immune system are the thymus, spleen, liver, and kidney. There are also accumulations of lymphoid tissue in the gut, mesentery, and gills (Table 4.2). Bone marrow is present in some adult anuran species, where it is involved in lymphopoiesis and granulopoiesis. In the tadpole and adult of all species, the thymus is the source of T lymphocytes. In the tadpole of *R. pipiens,* B lymphocytes develop in the pronephros or mesonephros (Zettergren, 1982). B cells develop in the spleen and liver of larval *X. laevis* (Hadji-Azimi et al., 1982) and in the spleen of larval *A. mexicanum* (Fellah et al., 1989). In adult *X. laevis,* B-cell development continues in the spleen (Hadji-Azimi et al., 1990), whereas it occurs predominately in the bone marrow of *R. pipiens* adults (Eipert et al., 1979; reviewed in Hansen and Zapata, 1998) (Fig. 4.2).

It is somewhat useful to describe the immune system as functionally divided into the adaptive system and the innate system. In fact, the functions overlap. As a guide to ecologists who would like to use assays of immune function in their studies, we have listed a number of commonly used immunological assays with reference to key publications in Table 4.3.

TABLE 4.2: Organs containing lymphoid tissue in amphibians

	Gymnophiona	Caudata	Anura: Primitive anurans (e.g. *Xenopus*) Larva	Anura: Primitive anurans (e.g. *Xenopus*) Adult	Anura: Higher anurans (Ranidae and Bufonidae) Larva	Anura: Higher anurans (Ranidae and Bufonidae) Adult
Thymus	+	+	+	+	+	+
Spleen	+	+	+	+	+	+
Kidney		±	+	+	+	+
Lymphomyeloid nodes (e.g., jugular body, lymph gland	–	–	–	–	–	+
Gut-associated lymphoid tissue		+	+	+	+	+
Bone marrow	–	+ (Plethodontidae)	–	+	–	+
Liver	+	+	+	+	+	±

A plus symbol (+) indicates present. Negative symbol (–) indicates absent, and a plus/minus symbol (±) indicates present but not a major lymphoid organ.
Adapted from Manning and Horton (1982).

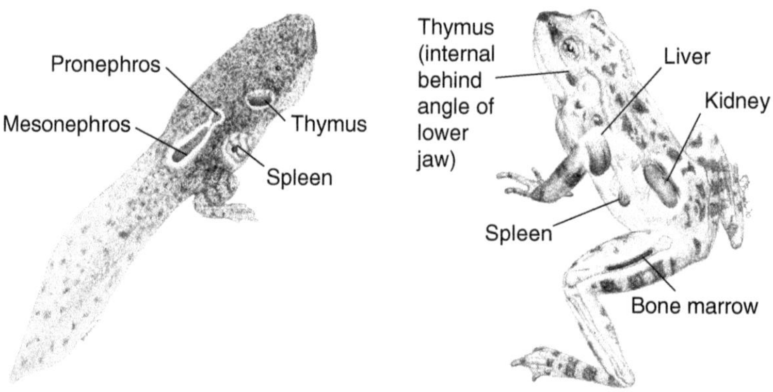

FIGURE 4.2: The major organs of the immune system in a tadpole (left) and an adult anuran amphibian (original artwork by Robert M. Brucker). In the tadpole and adult, the thymus is the source of T lymphocytes. In the tadpole of *Rana pipiens*, B lymphocytes develop in the pronephros or mesonephros (Zettergren, 1982). However, they develop in the spleen and liver of *Xenopus laevis* (Hadji-Azimi et al., 1982) tadpoles, and in the spleen of larval *Ambystoma mexicanum* (Fellah et al., 1989). In adult *X. laevis*, B-cell development continues in the spleen (Hadji-Azimi et al., 1990), whereas it occurs predominately in the bone marrow of *R. pipiens* adults (Eipert et al., 1979).

TABLE 4.3: Common immune assays for amphibians

Immune assay	Purpose	Some Key References
Antigen-specific antibody response	Measure of B-lymphocyte activation and adaptive humoral immunity	Cone and Marchalonis, 1972; Turner and Manning, 1974; Du Pasquier et al., 1985; Charlemagne, 1987; Rollins-Smith et al., 1990; Gilbertson et al., 2003; Maniero et al., 2006 ; Ramsey et al., 2010
In vitro lymphocyte proliferation	Lymphocyte proliferation in response to allogeneic cells, specific antigens, mitogens, or bacterial antigens	Du Pasquier and Weiss, 1973; Green (Donelly) and Cohen, 1979; Bleicher et al., 1983; Rollins-Smith et al., 1984; Rollins-Smith et al., 1990; Maniero and Carey, 1997; Morales et al., 2003
T-cell activity in the skin	T-lymphocyte proliferation in response to a mitogen, phytohemagglutinin (PHA), measured as swelling at the site of injection	Gilbertson et al., 2003; Gervasi and Foufopolous, 2008
Serum complement activity	Measure of innate defense in blood serum or plasma	Green and Cohen, 1977; Jensen and Koch, 1991; Maniero and Carey, 1997
Phagocytosis activity and oxidative burst examined by chemiluminescence (CL)	Measure of innate defense in blood, including production of oxygen radicals by circulating neutrophils	Christin et al., 2003, 2004; Gilbertson et al., 2003; Brodkin et al., 2007; Zhang et al., 2008
Microbial growth inhibition caused by antimicrobial skin peptides	Measure of innate skin defense	Rollins-Smith et al., 2002a, 2002b; Woodhams et al., 2006a; Sheafor et al., 2008
Relative intensities of skin defense peptides	Measure of innate skin defense	Woodhams et al., 2006a, 2006b, 2007b, 2010
Lysozyme activity	Measure of innate defense in gut or skin mucosa or in blood serum against gram-positive bacteria	Raffel et al., 2006; Zhao et al., 2006
Differential leukocyte counts	Measure the absolute numbers or relative proportions of circulating blood cells important for innate and adaptive immune defense	Hadji-Azimi et al., 1987; Maniero and Carey, 1997; Belden and Kiesecker, 2005; Raffel et al., 2006; Woodhams et al., 2007b; Gervasi and Foufopolous, 2008; Zhang et al., 2008

THE INNATE IMMUNE SYSTEM

The innate immune system (reviewed in Carey et al., 1999; Rollins-Smith, 2001; Robert and Ohta, 2009) is the rapidly responding but less specific and regulated arm of the immune system. It is considered to be the first line of defense against pathogens and may directly remove them or slow down disease development until the adaptive immune system can become activated. Many of the same innate defenses that have been described in other vertebrates are present in amphibians. Phagocytic cells capable of directly killing pathogens include macrophages, neutrophils, basophils, and eosinophils (Hadji-Azimi et al., 1987). Cells with dendritic morphology that patrol the skin and tissues and present antigens to lymphocytes have been described (Carrillo et al., 1990; Du Pasquier and Flajnik, 1990; Castell-Rodriguez et al., 1999; Mescher et al., 2007). Natural killer cells capable of directly killing virus-infected cells have also been described (Horton et al., 2000). Other essential components of innate immunity are complement proteins (Green and Cohen, 1977; Jensen and Koch, 1991; Kato et al., 1994), lysozyme (Zhao et al., 2006), and antimicrobial peptides (reviewed in Rollins-Smith and Conlon, 2005; Rollins-Smith, 2009). Genes for Toll-like receptors (TLRs) involved in innate recognition of pathogen-associated molecular patterns (PAMPs) in mammals have been identified in *Xenopus* (Roach et al., 2005; Ishii et al., 2007). Thus, many of the central components of innate immunity are present in amphibians.

THE ADAPTIVE IMMUNE SYSTEM

The adaptive immune system in amphibians is generally a slowly responding, but highly specific and regulated arm of the immune defense system. It is characterized by T lymphocytes (developing in the thymus) and B lymphocytes (developing in the liver and spleen or bone marrow) expressing rearranging immunoglobulin (Schwager et al., 1988, 1991a, 1991b) and T-cell (Chrétien et al., 1997; Haire et al., 2002) receptors that enable the lymphocytes to bind to a great variety of molecules. Central to the cooperative function of the adaptive immune system is the major histocompatibility complex (MHC) (reviewed in Flajnik and Du Pasquier, 1990). B cells can recognize their antigens directly, but T cells recognize only fragments of antigens (peptides) displayed by antigen-presenting cells (macrophages, dendritic cells, and B lymphocytes) in association with MHC molecules. B cells need to interact with T cells for development of effective antibody responses (Turner and Manning, 1974; Blomberg et al., 1980). The MHC of all vertebrates has three regions. Class I genes encode a set of molecules that form heterodimers with beta-2 microglobulin and display antigenic peptides on the surface of all nucleated cells. These molecules are critical for self-recognition and presentation of viral antigens and other intracellular antigens. Class II genes encode molecules that associate as heterodimers and display antigens collected from outside of the cell by professional antigen-presenting cells. The MHC also encodes molecules involved in processing of protein antigens in the proteosome (low-molecular-weight proteins [LMP] 2 and 7), endoplasmic reticulum transporters (transporters associated with antigen processing [TAP] 1 and 2), and the so-called class III molecules (complement proteins and some cytokines). The *Xenopus* MHC encodes the same components as all other jawed vertebrates except that the arrangement of the genes is probably in a primordial configuration, more like that of fish and birds (Nonaka et al., 1997; reviewed in Flajnik and Kasahara, 2001).

Many amphibian pathogens attack the skin. The skin of vertebrates is protected by innate defenses and the adaptive immune system. Antigens from the skin can be transported to the spleen, where an immune response may develop involving both T cells and B cells. The immune response is characterized by classical first- and second-set kinetics and is dependent on the incubation temperature of the hosts (reviewed in Cohen et al., 1985; Robert and Ohta, 2009).

WHICH COMPONENTS OF THE IMMUNE SYSTEM ARE IMPORTANT FOR PROTECTION FROM B. DENDROBATIDIS AND OTHER SKIN PATHOGENS?

The mucous layer covering amphibian skin is an ideal niche for many opportunistic pathogens because it is moist and provides a rich nutrient source. Therefore, defense of the skin is critical for protection from many amphibian pathogens, and involves innate immune defenses such as epithelial barriers, phagocytic cells, lysozyme, and antimicrobial peptides (reviewed in Rollins-Smith and Conlon, 2005; Rollins-Smith, 2009). Because *B. dendrobatidis* is confined to the outer keratinized epithelial layers and can quickly lead to host mortality, it has generally been thought that the adaptive immune system responds too slowly to adequately protect the host (Berger et al., 1998, Pessier et al., 1999). Therefore, innate defenses, such as antimicrobial peptides, would be critical for survival. As described below, skin microbes may also contribute to protection of the skin. In addition to innate defenses, the adaptive immune system is important for defense of the skin. In *X. laevis,* a host that is relatively resistant to *B. dendrobatidis,* both innate immune mechanisms (antimicrobial peptides in the mucus) and the adaptive immune system (antibodies in the mucus) are involved in protection (Ramsey et al., 2010). There is a paradox here: If amphibians have robust immune defenses, why are so many species susceptible to death from chytridiomycosis? The answer may be that the pathogen itself can inhibit development of cell-mediated immune responses in the skin. Current studies show that factors secreted by *B. dendrobatidis* can inhibit lymphocyte proliferation by induction of apoptosis (J. Ramsey et al., unpublished). This may be a classic example of pathogen–host co-evolution. The pathogen needs to avoid immune destruction and has evolved a mechanism to inhibit immune defenses. Our current view of the immune defenses of the skin is shown in Figure 4.3.

AMPHIBIAN ANTIMICROBIAL PEPTIDES

Mucous glands and granular glands (also called poison or serous glands) are prominent features in the dermal layer of amphibian skin (Noble and Noble, 1944; Bovbjerg, 1963; Sjoberg and Flock, 1976; Mills and Prum, 1984). The mucous glands produce a material rich in mucopolysaccharides that keeps the skin moist (Duellman and Treub, 1986; Schumacher et al., 1994; Goniakowska-Witalinska and Kubiczek, 1998). Granular glands produce bioactive peptides including neuropeptides and antimicrobial peptides that are thought to play a role in defense against vertebrate predators as well as microbes (reviewed in Erspamer, 1994; Daly, 1995; Nicolas and Mor, 1995; Simmaco et al., 1998; Zasloff, 2002; Rinaldi, 2002; Conlon et al., 2004; Apponyi et al., 2004; see also the special issue on amphibian antimicrobial peptides of *Biochimica et Biophysica Acta,* 2009,

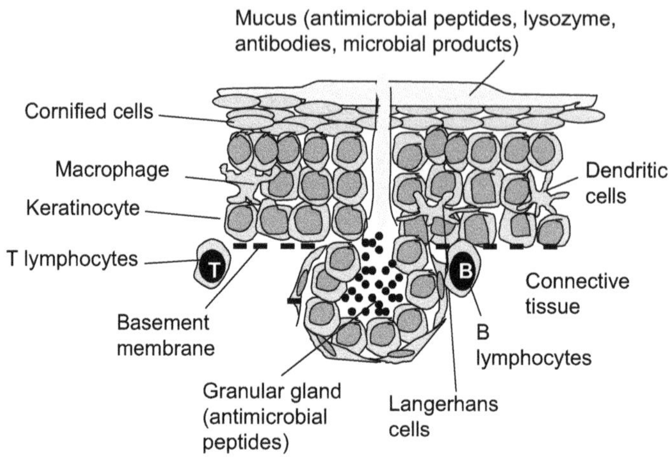

FIGURE 4.3: Diagram of amphibian skin and immune cells. Mucus containing antimicrobial peptides, lysozyme, antibodies, and microbial metabolites covers the epidermis. These factors in the mucus may inhibit *B. dendrobatidis* zoospore survival and colonization. Granular glands in the dermis synthesize and release active antimicrobial peptides. Keratinocytes in the epidermis are susceptible to infection by *B. dendrobatidis*, and infected cells may be recognized by macrophages, dendritic cells, or Langerhans cells to activate adaptive immunity. These antigen-presenting cells should recruit lymphocytes to limit the infection. However, products produced by fungal cells may inhibit activation of the antigen-presenting cells or lymphocytes.

Vol. 1788: 1535–1694). Both mucous and granular glands are composed of a syncytium of epithelial cells surrounding a secretory compartment (Dockray and Hopkins, 1975; Mills and Prum, 1984). In granular glands, the center of the gland is filled with granules packed with active peptides (Gibson et al., 1986; Giovannini et al., 1987). Granular glands are surrounded by a layer of myoepithelial cells with sympathetic axons terminating between the contractile elements (Sjoberg and Flock, 1976). The myoepithelial cells possess α-adrenoreceptors, and epinephrine or norepinephrine induces contraction and release of granular contents by a holocrine mechanism (Benson and Hadley, 1969; Dockray and Hopkins, 1975; Holmes and Balls, 1978). A schematic diagram of a granular gland is shown in Figure 4.4. Holocrine secretion involves loss of most of the contents of the gland; however, the multiple nuclei of the syncytial gland remain, and a new gland regenerates from remaining epithelial cells (Neuwirth et al., 1979; Toledo and Jared, 1995). Granular glands can be found all over the body, with the largest ones in the dorsolateral skin (dermal plicae) and behind the eyes (paratoid glands) (Noble and Noble, 1944; Bovbjerg, 1963; Sjoberg and Flock, 1976; Neuwirth, et al., 1979; Toledo and Jared, 1995).

An extensive literature characterizes the amino acid sequences and activity of a large number of amphibian antimicrobial peptides ranging in size from 10 to 50 amino acids. They are active against gram-positive and gram-negative bacteria, fungi, protozoa, and viruses (reviewed in Nicolas and Mor, 1995; Simmaco et al., 1998; Zasloff, 2002; Rinaldi, 2002; Conlon et al., 2004; Apponyi et al., 2004). (See also the special issue on amphibian antimicrobial peptides of *Biochimica et Biophysica Acta*, 2009, Vol. 1788: 1535–1694.) Although families of peptides are shared by related species, there is little or no overlap in the peptides from one species to another (reviewed in Conlon et al., 2004; Nicolas and El Amri, 2009). There is no consensus amino-acid sequence

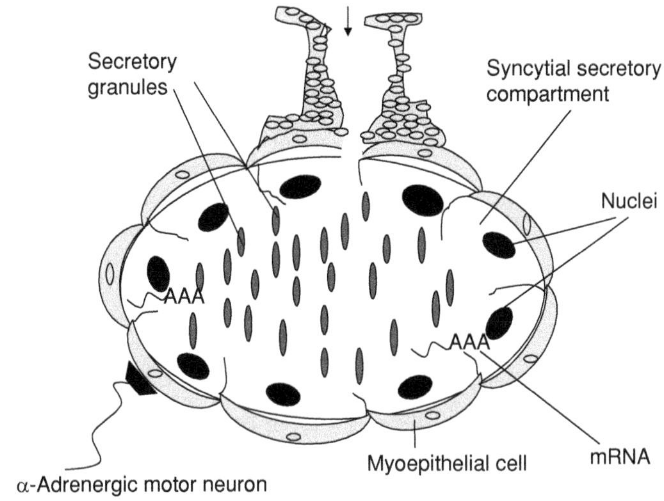

FIGURE 4.4: Schematic diagram of a granular gland after Dockray and Hopkins (1975). Amphibian granular glands are composed of glandular epithelial cells that fuse into a syncytium with multiple nuclei visible. The gland is surrounded by myoepithelial cells innervated by α-adrenergic motor neurons. Defensive peptides are stored in membrane-limited granules within the lumen of the gland. When the neurotransmitters epinephrine and norepinephrine are released, the myoepithelial cells contract, forcing the contents of the gland onto the surface of the skin. The lumen of the gland also contains messenger RNA.

associated with biological activity, but the peptides are usually cationic and relatively hydrophobic and have the ability to form an amphipathic α-helix in a membrane-mimetic environment (reviewed in Yeaman and Yount, 2003). This structure provides them with an ability to disturb biological membranes, and this seems to be the main mechanism of killing of their targets (reviewed in Nicolas and Mor, 1995; Simmaco et al., 1998; Rinaldi, 2002; Zasloff, 2002; Apponyi et al., 2004; Conlon et al., 2004).

Accumulated evidence shows that antimicrobial peptides secreted into the skin mucus of amphibians play an important role in protection from infection by skin pathogens such as *B. dendrobatidis* or spread of the pathogen to other areas of the skin (reviewed in Rollins-Smith and Conlon, 2005; Rollins-Smith, 2009). Both *X. laevis* and *R. pipiens* produce and secrete small amounts of antimicrobial peptides that are available on the skin of a resting frog to deter infection. This amount is dramatically increased when frogs are chased, simulating a predator attack (Ramsey et al., 2010; J. Pask, unpublished data), and the concentrations are sufficient to inhibit chytrid growth. Thus, the concentrations present on the skin of normally active amphibians appear to be sufficient to inhibit colonization by an infectious organism such as *B. dendrobatidis*. This evidence supports the general hypothesis that antimicrobial peptides are an important component of the innate defense against skin pathogens. In addition to these *in vitro* effects on *B. dendrobatidis,* there is increasing evidence that the suite of antimicrobial peptides expressed by each species may predict whether they survive or decline when confronted with natural *B. dendrobatidis* infections in the wild. Species expressing a skin peptide mixture that potently inhibits *B. dendrobatidis* growth *in vitro* are among the more common, apparently resistant, species. In contrast, declining or endangered species expressed peptides with poor *in vitro* activity against *B. dendrobatidis* (Woodhams et al., 2006a). Among species in Panama that have experienced serious declines, very few species expressed skin peptides with potent anti-*Bd* activity. These few species appear to persist in geographic areas that have experienced declines (Woodhams et al., 2006b). Among four species of Australian frogs that were experimentally infected with *B. dendrobatidis,* those with more potent antimicrobial peptides were more resistant to experimental infection (Woodhams et al., 2007a).

DEFENSES DUE TO SYMBIOTIC SKIN MICROBES

Normal microbial communities associated with mucosal surfaces, including the human digestive and reproductive systems, are essential to health and disease resistance (reviewed in Wilson, 2005). Because many pathogens must first encounter mucosal surfaces before invasion, innate defenses found there are critical. Symbiotic microbiota naturally integrated in association with the epithelia may function to limit disease risk through production of antimicrobial metabolites, competition and exclusion of exogenous pathogens, or enhancement of host immunity. Thus, amphibian skin is a biomedical model of host–pathogen interactions similar to those that occur in humans and other vertebrates (reviewed in Barra and Simmaco, 1995; Zasloff, 2002; Belden and Harris, 2007). Understanding the interactions of host defenses with symbiotic microorganisms is critical for elucidating mechanisms of infection and the ecoimmunology of amphibian disease.

Community ecology theory predicts that ecosystem integrity, including stability, productivity, and disease resistance, may be a function of species diversity and interactions (Robinson and Valentine, 1979; Case, 1990; Tilman et al., 1997, 2006; Ostfeld and Keesing, 2000; Hector and Bagchi, 2007; Johnson and Thieltges, 2010). Certain species or communities of natural microbes may help maintain health by preventing infection by pathogenic organisms or preventing other members of the microbial community from proliferating and causing disease (reviewed in Bäkhed et al., 2005; Belden and Harris, 2007; Zilber-Rosenberg and Rosenberg, 2008). Because polyparisitism is common in nature (Pederson and Fenton, 2007; Bordes and Morand, 2009), eco-immunologists should note that co-infections may alter disease risk by producing synergistic or additive effects, by mediating virulence, or by reducing the impact of otherwise harmful pathogens (Massey et al., 2004; Harrison et al., 2006; Cattadori et al., 2008). Examples of beneficial bacterial symbionts in various taxa are accumulating. They occur in the cnidarian *Hydra,* fish skin, locusts, honey bees, human intestines, wheat rhizosphere, fungus-growing ants, and others (Spanggard et al., 2001; Dillon et al., 2005; Matos et al., 2005; Evans and Armstrong, 2006; Currie et al., 2006; Lacava et al., 2007; reviewed in Dethlefsen et al., 2006; Vine et al., 2006; Fraune et al., 2009). Recent evidence suggests that natural microbiota may be important in preventing disease in amphibians as well (Austin, 2000; Harris et al., 2006, 2009a, 2009b; Lauer et al., 2007, 2008; Woodhams et al., 2007b; Becker et al., 2009; Becker and Harris, 2010). If mutualistic microbes are an extension of the innate immune system of amphibian hosts, rather than representing chance colonization from the host's environment, the following conditions should apply:

1. Resident microbes on amphibian skin should inhibit potential pathogens *in vitro*. This has been demonstrated for bacteria isolated from salamander or frog skin. They inhibit

B. dendrobatidis and the egg-killing fungi *Mariannaea elegans* and *Rhizomucor variabilis* (Harris et al., 2006; Lauer et al., 2007, 2008; Banning et al., 2008). Many skin-associated bacteria of mountain yellow-legged frogs (*Rana muscosa* and *Rana sierrae*) inhibited *B. dendrobatidis* in co-culture challenge assays (Woodhams et al., 2007b; Lam et al., 2010).

2. Bio-augmentation, or probiotic addition of mutualistic microbes to amphibians or their habitat, should reduce susceptibility to pathogen colonization. In mountain yellow-legged frogs (*R. muscosa*) inoculated with the bacterium *Janthinobacterium lividum*, higher concentrations of antimicrobial metabolites were present on the skin, thus reducing infection risk (Becker et al., 2009; Harris et al., 2009a). Symbiotic microbial communities may also function to exclude exogenous pathogens through competition for space or resources, or indirectly through stimulating host immune defenses.
3. Bio-augmentation should decrease morbidity and increase survival of pathogen-infected hosts by inhibiting pathogens *in vivo*. Treatment of endangered mountain yellow-legged frogs (*R. muscosa*) with the bacterium *J. lividum*, which naturally occurs on some host frogs, decreased morbidity associated with chytridiomycosis and increased survival (Harris et al., 2009a). Non-lethal effects of *B. dendrobatidis* infection on red-backed salamanders (*Plethodon cinereus*) were ameliorated by addition of the mutualistic bacterium *Pseudomonas fluorescens* (Harris et al., 2009b).
4. Disease-resistant host populations should have a higher prevalence of pathogen-inhibiting microbiota than susceptible populations. Field surveys of a few populations of *R. muscosa* in the Sierra Nevada Mountains of California suggest this trend against emerging chytridiomycosis (Woodhams et al., 2007b; Lam et al., 2010).
5. Host amphibians may have adaptations to select for and enhance mutualistic microbiota, such as specialized epidermal structures, mucosal microenvironment, or antimicrobial skin peptides that selectively reduce pathogens or stimulate mutualistic microbiota. Cutaneous bacteria that inhibit *B. dendrobatidis* have been found that are resistant to high concentrations of host antimicrobial skin peptides (Woodhams et al., 2007c). Behavioral adaptations to transmit beneficial microbiota should also be suspected, although there is currently no information about these potential adaptations.
6. After removal of symbiotic microbiota, hosts should be more susceptible to infection and disease. Although this prediction needs to be more thoroughly investigated, there is some evidence that removal of host skin microbes allows for a more adverse disease outcome in *P. cinereus*. When the community of skin microbes was reduced prior to exposure to *B. dendrobatidis*, the animals exhibited several signs of morbidity (weight loss and modified behavior) that suggested they were more severely affected by the infection than controls with skin microbes intact (Becker and Harris, 2010).

MECHANISMS OF DISEASE PROTECTION CONFERRED BY SYMBIOTIC MICROBIOTA

There are a variety of ways by which symbiotic microbes could provide protection from disease. They may prevent infection through competitive interactions such as interfering with pathogen adherence or fitness (discussed in Wilson, 2005). Biofilms may be associated with the epidermal mucus layer and form a barrier to pathogens (Reid et al., 2001). Essential nutrients may not be easily available to arriving or opportunistic pathogens. Alternatively, natural antibiotics such as violacein or 2,4-diacetylphloroglucinol from symbiotic bacteria may actively suppress disease organisms (Bettin and Greven, 1986; Brucker et al., 2008a, 2008b; Becker et al., 2009). If biotherapy, or probiotic treatment of amphibians with beneficial microbes does not improve resistance to infection, the stimulation of host immune defenses may improve host tolerance of infection and prevent the development of disease.

The bioaugmentation approach is new for ecologists, but it is not new in agricultural systems (Table 4.4). However, few studies have examined the ecological impacts of augmenting the bacteria in soil (Yang et al., 2001; Dechesne et al., 2005; de Castro et al., 2010), and the concept is unknown in wildlife disease management.

For understanding disease dynamics, experimental bioaugmentation studies may provide a wealth of information. However, the application of this method for disease control or treatment of natural systems raises several concerns. For example, some common microbial symbionts such as *Aeromonas hydrophila* can be transmitted through the aquatic environment. Vertical transmission of protective microbiota may occur in amphibians that exhibit parental care (Banning et al., 2008). The modes of transmission and persistence for most other amphibian skin microbes are unknown. To prevent opportunistic microbes from becoming pathogenic and to prevent unwanted ecosystem cascades, the impacts of augmenting microbes needs further study. However, the possibility of conferring disease resistance by way of microbial symbionts that persist on the host and offer long-term and even heritable protection deserves careful consideration. Natural selection acting on the host–microbe symbiosis in coordination with changing environments may allow a greater adaptive potential than for host immunity alone (Zilber-Rosenberg and Rosenberg, 2008).

TABLE 4.4: Examples of the 92 specific microbials recorded by the California Department of Pesticide Regulation and total pounds of active ingredient used in California in 2005

Use	Name	Chemical code	Total lbs
Fungicide	*Agrobacterium radiobacter*	1984	0.29
Fungicide	*Bacillus subtilis* GBO3	3945	14.79
Insecticide	*Bacillus thuringiensis*, subsp. Kurstaki, strain HD-1	4023	3120.48
Insecticide	*Beauveria bassiana* strain GHA	3993	491.10
Fungicide	*Gliocladium virens* GL-21 (spores)	3854	15.96
Insecticide, Nematicide	*Myrothecium verrucaria*, dried fermentation solids & solubles	3966	17527.96
Fungicide	*Pseudomonas fluorescens* A506	2842	887.04
Fungicide	QST 713 strain of dried *Bacillus subtilis*	5447	2348.95
Insecticide	Spinosad	3983	73062.22
Fungicide	*Streptomyces griseoviridis* strain K61	3937	0.23
Fungicide	*Trichoderma harzianum rifai* strain KRL-AG2	3977	11.75

Data provided by Dr. Carlos Davidson, San Francisco State University. The Pesticide Action Network's pesticide database provides use, toxicity, and regulatory information for 228 microbial pesticides (http://www.pesticideinfo.org/, March 22, 2010).
Over-the-counter garden products containing live microbes are not tracked in the Pesticide Use Report.

ONTOGENY OF IMMUNITY

Understanding interactions of the immune system with the environment requires some knowledge about the effectiveness of the immune system at each stage of amphibian life. Thus, we will briefly review what is known about ontogeny of immunity in amphibians, the changes that occur at metamorphosis, and differences in disease susceptibility at each stage. The organs that are involved in immunity are listed in Table 4.2. This information will help to explain life-history trade-offs with immunity.

DEVELOPMENT OF IMMUNE SYSTEM IN ANURAN AND URODELAN AMPHIBIANS

The ontogeny of the amphibian immune system has been described in a series of excellent reviews (Du Pasquier, 1973; Cohen et al., 1985; Du Pasquier et al., 1986; Rollins-Smith, 1998; Robert and Ohta, 2009). Several important aspects of immune system development should be reiterated here. In anuran amphibians, the immune system develops in two phases. In phase one, the organs of the immune system (thymus, spleen, kidney, liver, and gut-associated lymphoid tissue) develop in the tadpole. Although both B-cell–mediated antibody responses and T-cell–mediated rejection responses emerge, the antibody and T-cell repertoire is very limited. Tadpoles reject skin grafts (Hildemann and Haas, 1959; Bovbjerg, 1966; Horton, 1969; Baculi and Cooper, 1970; Chardonnens and Du Pasquier, 1973; Du Pasquier and Chardonnens, 1975; Roux and Volpe, 1975; Plytycz, 1981; DiMarzo and Cohen, 1982a, 1982b; Obara et al., 1983; Rollins-Smith et al., 1988) or tumor cells (Robert et al., 1994, 1995) from donors known or assumed to express different MHC antigens, but they fail to reject grafts differing by minor histocompatibility antigens (Chardonnens and Du Pasquier, 1973; Du Pasquier and Chardonnens, 1975; DiMarzo and Cohen, 1982b; Obara et al., 1983; Rollins-Smith et al., 1988; Robert et al., 1994, 1995). Tadpoles develop specific IgM and IgY antibody responses (Moticka et al., 1973; Du Pasquier and Haimovich, 1974, 1976; Pross and Rowlands, 1976; Du Pasquier et al., 1979; Hsu and Du Pasquier, 1984a, 1984b), but the antibodies to a specific small antigen (dinitrophenol-conjugated to a larger protein) are less diverse and of lower affinity (Du Pasquier and Haimovich, 1976; Du Pasquier et al., 1979; Hsu and Du Pasquier, 1984a, 1984b, 1992) than those of genetically identical adults. In the tadpole, conventional MHC class Ia molecules are expressed relatively late and at lower density on a per-cell basis than on adult cells (Flajnik et al., 1986; Flajnik and Du Pasquier, 1988; Rollins-Smith et al., 1997a). Tadpoles express MHC class II antigens on B cells and antigen-presenting cells (Du Pasquier and Flajnik, 1990; Rollins-Smith and Blair, 1990; Gravenor et al., 1995). In phase two of immune system development, the adult-type pattern of immune responses emerges. The adult pattern includes acute (rapid) rejection of allografts differing by minor histocompatibility antigens or MHC (Simnett, 1964; Horton, 1969; Horton and Manning, 1972; Chardonnens and Du Pasquier, 1973; Tochinai and Katagiri, 1975; DiMarzo and Cohen, 1982a, 1982b; Obara et al., 1983; Rollins-Smith et al., 1988), the expression of diverse higher-affinity antibodies (Hsu and Du Pasquier, 1984b; Du Pasquier et al., 1979), the robust expression of MHC class I on all cells (Flajnik et al., 1986; Flajnik and Du Pasquier, 1988; Rollins-Smith et al., 1997a;

Salter-Cid et al., 1998), and expression of class II antigens on antigen-presenting cells and B and T cells (Du Pasquier and Flajnik, 1990; Rollins-Smith and Blair, 1990; Gravenor et al., 1995; Rollins-Smith et al., 1996). Metamorphosis is characterized by a striking loss of lymphocytes from liver, thymus, and spleen (Du Pasquier and Weiss, 1973; Rollins-Smith et al., 1984; Rollins-Smith et al., 1988). If a triploid thymus is implanted into a diploid host before metamorphosis, it is colonized by host cells that completely replace the tadpole cells around the time of metamorphosis (Turpen and Smith, 1989; Rollins-Smith et al., 1992). B cells also increase and decrease in two waves (Hadji-Azimi et al., 1982; Hadji-Azimi et al., 1990). These observations and the observation that the B- and T-cell–recognition repertoires change after metamorphosis suggest that one set of lymphocytes is lost at metamorphosis and replaced by a second set in the adult period (reviewed in Rollins-Smith, 1998).

Much less is known about the development of the immune system in urodele amphibians. In general, the emergence of immune system functions (skin graft rejection, antibody production) is a gradual age-dependent process (reviewed in Rollins-Smith, 1998).

THE ROLE OF CORTICOSTEROID HORMONES IN IMMUNE SYSTEM REORGANIZATION AT METAMORPHOSIS

Both thyroid hormones and corticosteroid hormones increase dramatically during climax of metamorphosis (Leloup and Buscaglia, 1977; Miyauchi et al., 1977; Regard et al., 1978; Mondou and Kaltenback, 1979; Jaffe, 1981; Suzuki and Suzuki, 1981; Krug et al., 1983; Jaudet and Hatey, 1984; Kikuyama et al., 1986). Short-term culture of amphibian lymphocytes in physiologically relevant concentrations of corticosterone or aldosterone inhibited lymphocyte proliferation and induced cell death due to apoptosis (Rollins-Smith and Blair, 1993; Rollins-Smith et al., 1997b; Barker et al., 1997). These effects on apoptosis and proliferation were reversed at metamorphosis in the presence of the corticosteroid hormone receptor antagonist RU486 (Rollins-Smith et al., 1997b; Barker et al., 1997). In contrast, thyroid hormones did not induce consistent inhibition of lymphocyte proliferation or loss of lymphocyte viability (Rollins-Smith and Blair, 1993). These observations were interpreted as evidence that the natural elevation of corticosteroid hormones at metamorphosis leads to the loss of a population of larval lymphocytes that is replaced in the post-metamorphic period (reviewed in Rollins-Smith, 1998).

DEVELOPMENTAL STAGES AND DISEASE SUSCEPTIBILITY

There are clear differences in disease susceptibility at different life stages. Ranaviruses were associated with mass mortality episodes affecting pig frog (*Rana grylio*) tadpoles (Zhang et al., 2001), tiger frog (*Rana tigrina rugulosa*) tadpoles (Weng et al., 2002), and bullfrog (*Rana catesbeiana*) tadpoles (Majji et al., 2006). The high susceptibility of larval stages to ranaviruses has also been established experimentally in *Xenopus*. Adults can be infected, especially if they are immunosuppressed, but tadpoles are much more sensitive to the lethal effects of ranavirus infections (Gantress et al., 2003; Robert et al., 2005). Others have reported that late-stage tadpoles and new metamorphs of the wood frog (*Rana sylvatica*) and *R. pipiens* are more susceptible to ranavirus infections than younger tadpoles (Green et al., 2002; Greer et al., 2005). The greater susceptibility of tadpoles to virus infections is likely due to the immaturity of the immune defenses.

The reorganization of the amphibian immune system at metamorphosis may also cause frogs to become more susceptible to some diseases during this period of development (reviewed in Carey et al., 1999; Rollins-Smith, 1998). Tiger salamanders (*Ambystoma tigrinum*) in Colorado died at metamorphosis due to an unknown disease (Carey and Bryant, 1995). Newly metamorphosed frogs become susceptible and die of chytridiomycosis, while adults and tadpoles are less susceptible (Berger et al., 1998; Bosch et al., 2001; Green et al., 2002; Rachowicz and Vredenburg, 2004; Rachowicz et al., 2006; Andre et al., 2008; Walker et al., 2010; Tobler and Schmidt, 2010). A survey of 64 morbidity and mortality events in the United States revealed that ranaviruses caused disease in late-stage larvae and recent metamorphs (Green et al., 2002). Juvenile frogs appear to be more susceptible to disease caused by the Bohle iridovirus than either tadpole or adult stages (Cullen et al., 1995; Cullen and Owens, 2002; Mazzoni et al., 2003). In developing Southern toads (*Bufo terrestris*) and spring peepers (*Pseudacris crucifer*), enteric bacterial loads were higher during metamorphosis (Fedewa, 2006).

LIFE-HISTORY TRADE-OFFS WITH IMMUNITY

It is generally thought that immunity comes at a cost to the host (reviewed in Sheldon and Verhulst, 1996; Lochmiller and Deerenberg, 2000). Increasingly, amphibian immune defenses are considered among traits that are traded off for advantages in other life-history traits such as reproduction or growth. Trade-offs may be viewed as negative functional interactions between traits and can be studied to understand the variation and evolution of life-history traits (reviewed in Zera and Harshman, 2001). Immune defenses, therefore, can be studied with the understanding that limited host resources must be allocated to various physiological processes, and the optimal allocation of resources depends on ecological context (reviewed in Sadd and Schmid-Hempel, 2009).

In amphibians, physiological trade-offs have been described between immune responses and reproductive effort or development (Rollins-Smith et al., 1988; McCallum and Trauth, 2007; Gervasi and Foufopoulos, 2008; Garner et al., 2009). *R. sylvatica* develop in temporary ponds at variable rates. In quickly drying ponds, tadpoles are able to metamorphose earlier, at a smaller size, and at the expense of decreased total leukocyte counts and a cellular immune response (skin swelling in response to injection of a standard T-cell mitogen, phytohemagglutinin) (Gervasi and Foufopoulos, 2008). Precocious metamorphosis limited lymphocyte development and impaired allograft-rejection responses in *X. laevis* (Rollins-Smith et al., 1988). Immunologically challenged male Northern cricket frogs (*Acris crepitans*) grew at the same rate as non-challenged frogs but appeared to divert investment from reproduction. That is, spermatic cyst diameter, germinal epithelium depth, and gonadosomatic index were significantly reduced (McCallum and Trauth, 2007). It should be noted that these frogs were immunized weekly for about 6 weeks with foreign red blood cells, and thus the immune system was probably stressed by this heavy antigen burden. As described above, the immune system of anurans involutes at metamorphosis. This may be due to the heavy demands for energy needed for remodeling of all organ systems, or it may be necessary to avoid immune recognition and destruction of self due to the emergence of new organ-specific antigens in the developing adult organs (discussed in Rollins-Smith, 1998).

Although immune investment was not measured directly, Garner et al. (2009) showed that mortality of common toads (*Bufo bufo*) due to exposure to the amphibian chytrid fungus *B. dendrobatidis* depended on life-history stage, host condition (body size), and pathogen dose. Exposure to *B. dendrobatidis* as tadpoles significantly reduced mass at metamorphosis. Toads exposed to *B. dendrobatidis* that survived through metamorphosis had significantly reduced chances of survival after metamorphosis in comparison with unexposed controls (Garner et al., 2009). Infection with *B. dendrobatidis* reduced developmental rates and body size of some amphibian larvae of Cope's gray tree frog (*Hyla chrysoscelis*) and Fowler's toads (*Bufo fowleri*) by reducing foraging activity and efficiency (Venesky et al., 2009, 2010). Although exposure to *B. dendrobatidis* shortly after metamorphosis did not cause death, pathogen-exposed foothill yellow-legged frogs (*Rana boylii*) had significantly reduced growth (Davidson et al., 2007) in comparison with controls. Thus, diverting resources to immunological defense against even low doses of *B. dendrobatidis* zoospores and resolving infections appears to come at the cost of reduced growth and/or reduced survival.

ENVIRONMENTAL CONDITIONS AFFECTING IMMUNE DEFENSES AND DISEASE

There are many environmental factors that affect the amphibian immune system and thus the ability of amphibians to resist pathogens. Because it is impossible to fully cover all aspects of this subject, we have chosen to limit our discussion to the effects of temperature on immunity, the effects of other animals in the community, some discussion of stress resulting from competition and predation, and some discussion of effects of chemicals in the environment. The subject of amphibian immunotoxicology was briefly reviewed in Rollins-Smith and Smits (2005) and Rollins-Smith et al. (2007).

TEMPERATURE EFFECTS ON IMMUNE DEFENSES

There is a fairly extensive literature describing the effects of cold on the immune system or immune responses in amphibians dating from the 1970s. It is probably useful to distinguish the effects of natural changes that occur with hibernation in cold-adapted species such as *R. pipiens* or the European common frog (*Rana temporaria*) from those due to temporary temperature fluctuations under natural conditions or in the laboratory. *R. pipiens* have been shown to lose hematopoietic cells from the blood, spleen, thymus, jugular bodies, and bone marrow during hibernation (Cooper et al., 1992). All hematopoietic cells in the blood and organs of the immune system were restored within 30 days of the end of hibernation (Cooper et al., 1992). This was accompanied by a spike in leukocyte traffic in the blood during spring in natural conditions (Harris, 1972). Post-hibernation antibody responses were reduced in magnitude, and secondary responses were slightly delayed in *R. pipiens* (Cooper et al., 1992). Likewise, the thymus has been shown to shrink in size in the winter and recover in the summer in *R. temporaria* (Plytycz and Bigaj, 1983; Plytycz et al., 1991; Miodoński et al., 1996). This natural involution is thought to be due to a decrease in lymphopoiesis in the major hematopoietic organs rather than lymphocyte migration (Cooper et al., 1992) or active apoptosis of lymphocytes due to corticosteroid hormones. Interrenal activity and corticosteroid hormone production are minimal during winter (Leboulenger et al., 1979). The Eastern newt (*Notophthalmus viridescens*) also appears to lose lymphocytes and eosinophils during winter. Counts of these two cell types were negatively correlated with temperature (Raffel et al., 2006). In this study, neutrophil numbers did not change with different temperature regimes (Raffel et al., 2006). The authors attempted to separate seasonal effects from temperature effects, and their analysis does suggest some seasonal effects independent of temperature (Raffel et al., 2006). However, the central data for leukocyte cell counts are based on the number of each leukocyte type per 5,000 erythrocytes in a blood smear collected in the field. While this is a convenient measure of blood cells for a field study, the interpretation of the results could be distorted if erythrocyte numbers rise or fall in a pattern that differs from leukocyte numbers, as was observed in *R. pipiens* by Harris (1972). Whenever possible, it is best to obtain absolute leukocyte counts independent of erythrocyte counts.

Other measures of immune system function were also reduced by cold temperature. The activity of complement factors in the serum of *R. pipiens* was greatly reduced when frogs were exposed to hibernation temperatures (Green and Cohen, 1977; Maniero and Carey, 1997). Complement activity recovered quickly when animals were returned to warmer temperatures (22°C). *In vitro* lymphocyte proliferation in response to the T-cell mitogen, PHA, was also impaired if the lymphocytes came from *R. pipiens* held in hibernation conditions for 2 to 5 months (Maniero and Carey, 1997). Microbial population reductions and altered community composition of gut microbiota in response to hibernation may also be linked to immune function in *R. pipiens* (Gossling et al., 1982a, 1982b; Banas et al., 1988). Forbes et al. (2004) suggest that immunosuppression upon emergence from hibernation could explain the initially high and subsequently declining prevalence of *Aeromonas hydrophila* infections in breeding *R. catesbeiana, R. clamitans,* and *R. pipiens*. The latter species had the highest infection prevalence, perhaps due to an earlier and contracted breeding season.

These changes in the immune system organs and functions due to the reduced temperatures of hibernation conditions might be considered to be an example of an immune system accommodation or trade-off. It is energetically expensive to continue to produce lymphocytes in winter when metabolism is reduced to a minimum. Pathogens would also be slowed by the cold and would be less likely to kill their hosts during hibernation. The downside of reducing resources devoted to the immune system during hibernation is that there is a lag of weeks before the immune system is restored in the spring. Whether this temporary disturbance results in greater susceptibility to disease in post-hibernation amphibians is unknown. It is unclear whether the immune systems of species that do not experience true hibernation also have annual cycles with reduced immune system activity during winter. Future studies should examine this question in additional species.

Changes in immune system function due to hibernation should be distinguished from those that are strictly due to temporary fluctuations in temperature. The antibody response of the marine toad (*Bufo marinus*) was delayed and antibody titers were reduced when the toads were kept at 15°C rather than the optimal 25°C. Toads immunized at 25°C and transferred to 15°C had a delayed switch to low-molecular-weight antibodies (Lin and Rowlands, 1973). In a separate study of antibody responses in marine toads, toads kept at 20°C had significantly reduced antibody titers against horse red blood cells in comparison with toads at 37°C (Cone and Marchalonis, 1972). Although most antibody responses are T-cell dependent, another classic measure of T-cell function in amphibians is skin allograft rejection. Skin graft rejection is markedly delayed by colder temperatures in a number of amphibian species. Cohen (1966) examined skin allograft rejection in the Eastern newt, *N. viridescens*, and observed no differences in the kinetics of rejection at 23°C or 30°C. However, at 20°C rejection was delayed, and at 10° to 15°C, rejection was halted. When cold newts were transferred to warmer temperatures, rejection occurred within 36 days (Cohen, 1966). In adults of several species (the edible frog, *Rana esculenta*; the fire-bellied toad, *Bombina bombina*; and the common toad, *B. bufo*), there was a marked delay in the timing of skin allograft rejection at 10°C in comparison with 22°C (Jozkowicz and Plytycz, 1998). In *X. laevis*, even a small (3°C) difference in temperature significantly delayed rejection by young post-metamorphic frogs. At 24°C, rejection occurred within 15 days, but it was delayed to about 18 to 19 days at 21°C (DiMarzo, 1980; reviewed in Cohen et al., 1985). Thus, classic B-cell and T-cell responses are impaired when temperatures are reduced from the optimal temperature for the host.

CHYTRIDIOMYCOSIS AND TEMPERATURE

Because amphibian immune defenses are impaired by cold temperatures, it might be expected that some disease outbreaks would be associated with colder temperatures. Chytridiomycosis has been associated with high elevation and cooler temperatures in several studies (Bradley et al., 2002; Berger et al., 2004; Drew et al., 2006; McDonald et al., 2005; Kriger et al., 2007; reviewed in Fisher et al., 2009). Furthermore, *B. dendrobatidis* thrives at cool temperatures (more zoospores shed, longer period of activity) (Woodhams et al., 2008). Thus, prevalence of infection can increase in cooler seasons (McDonald et al., 2005; Woodhams and Alford, 2005). Whether or not chytridiomycosis (the disease) develops probably depends on multiple factors, including the ability to produce defensive peptides in the skin. A recent analysis of the pattern of declines of members of the genus *Atelopus* in Central America (generally thought to be due to chytridiomycosis) attributes the pattern to extreme temperature variations that may have impaired immunity (Rohr and Raffel, 2010). Because *B. dendrobatidis* thrives at cool temperatures, the balance between immunity and disease may shift when temperatures fall.

ENVIRONMENTAL MOISTURE AND CHYTRIDIOMYCOSIS

The infectious dose of a pathogen or parasite required for successful infection varies widely among different pathogens. Different strategies and mechanisms of infection may be important considerations in disease ecology studies. To infect a host cell, pathogens with direct-attachment mechanisms of infection may require fewer infectious particles than pathogens that must modify the host cell environment (reviewed in Schmidt-Hempel, 2008). Molecular studies on the pathogenicity of the emerging amphibian chytrid fungus are beginning to unravel the genetic and virulence factors involved with disease (Rosenblum et al., 2008). In experimental studies with boreal toads (*Bufo boreas*) the infectious dose of *B. dendrobatidis* zoospores and the duration of exposure has a large impact on subsequent patterns of infection and disease (Carey et al., 2006; Murphy et al., 2009). This dose-effect may help to explain the observation that when *B. dendrobatidis*-exposed toads are given access to dry refuges, infection intensity is reduced (C. Carey and L. Livo, unpublished). Similarly, juvenile orange-eyed tree frogs (*Litoria chloris*) infected with *B. dendrobatidis* survived longer under experimental conditions of dry air or rain than under conditions of constantly high humidity (Woodhams, 2003; and R.A. Alford, unpublished). Disease or infection may be prevented when the number of infecting cells is not adequate to initiate infection or overcome host immunity.

ANTIMICROBIAL PEPTIDES AND ENVIRONMENTAL TEMPERATURE, SEASON, AND POPULATION

Antimicrobial peptides are thought to be secreted in response to alarm or injury. It is not clear whether exposure to either pathogens or symbiotic microbes induces this response. However, edible frogs (*Rana esculenta*) treated with antibiotics (tetracycline and erythromycin) did not regenerate granular glands depleted by electrostimulation and produce antimicrobial peptides, whereas control frogs exposed to natural flora recovered normal gland morphology and antimicrobial activity (Mangoni et al., 2001). The role of environmental temperature on amphibian skin peptide production and expression is currently being studied. Peptide secretions were sampled from larval green-eyed tree frogs (*Litoria genimaculata*) from streams of various temperatures (Woodhams, 2003). Skin peptide recovery was positively correlated with stream water temperature (Fig. 4.5).

In concert with temperature, seasonal effects on skin peptide composition have been observed in Australian frogs. There is increased production of the pheromone splendipherin in male magnificent tree frogs (*Litoria splendida*) during the breeding season (Wabnitz et al., 2000). In Peron's tree frog (*Litoria peronii*) from southeast Australia, there are distinct differences in the pattern of winter and summer skin peptides, with production of two antimicrobial peptides reduced in the winter (Bilusich et al., 2009). Similarly, Roth's tree frog (*Litoria rothii*) from tropical Australia appears to produce antimicrobial peptides only in the summer (Sherman et al., 2009). These studies would suggest a winter depression in the production of antimicrobial peptides. It is tempting to speculate that skin pathogen challenges would generally be reduced in winter and the "cost" of maintaining production of antimicrobial peptides would consume unnecessary energy. However, since *B. dendrobatidis* thrives in the cold (Woodhams et al., 2008), this pathogen would have a greater survival advantage in winter. Additional studies with Northern leopard frogs (*R. pipiens*) (L.A. Rollins-Smith, unpublished) and pool frogs (*Rana lessonae*) (J. Hall and D.C. Woodhams, unpublished) suggest that production of skin peptides is inhibited during hibernation. In hibernating *R. sylvatica*, antimicrobial peptides were undetectable, but a single brevinin-1 family peptide was detected after housing the frogs at 30°C for 3 weeks (Mattute et al., 2000).

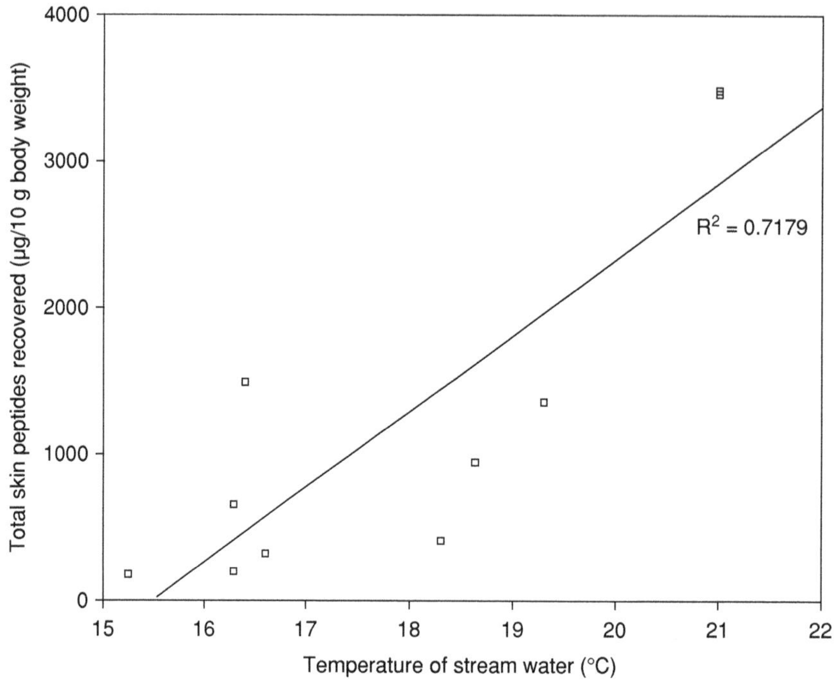

FIGURE 4.5: Temperature effects on larval green-eyed treefrog (*Litoria genimaculata*) peptide production. Skin peptide recovery was positively correlated with stream water temperature (linear regression, F1,8 = 20.3582, P = 0.0020).

Genetic differentiation also influences the skin peptide repertoire. In Northern leopard frogs (*R. pipiens*) from three populations that were acclimated to laboratory conditions for several weeks before sampling, the profile of all peptides and known antimicrobial peptides varied among populations (Fig. 4.6; Tennessen et al., 2009). In less geographically disparate populations of *L. genimaculata* from the wet tropics region of Australia, high-elevation populations had skin peptides with less effectiveness against the amphibian chytrid fungus compared to low-elevation populations, and this corresponded to higher infection intensities in upland frogs (Woodhams et al., 2010).

ENVIRONMENTAL STRESSORS AND IMMUNE DEFENSE

Nutrition may be an important variable in immune system development and resistance to disease. Tadpoles provided with less-than-optimal nutrition metamorphosed later and at a smaller size (Skelly and Werner, 1990; Newman, 1994; Kupferberg et al., 1994; Martinez et al., 1996). Amphibians metamorphosing at smaller-than-optimal body sizes have reduced numbers of total lymphocytes and thus may be more susceptible to disease (Rollins-Smith et al., 1988; Gervasi and Foufopoulos, 2008). This view appears to be supported by two studies of toad species infected with *B. dendrobatidis* soon after metamorphosis. *B. boreas* exposed as young of the year to a

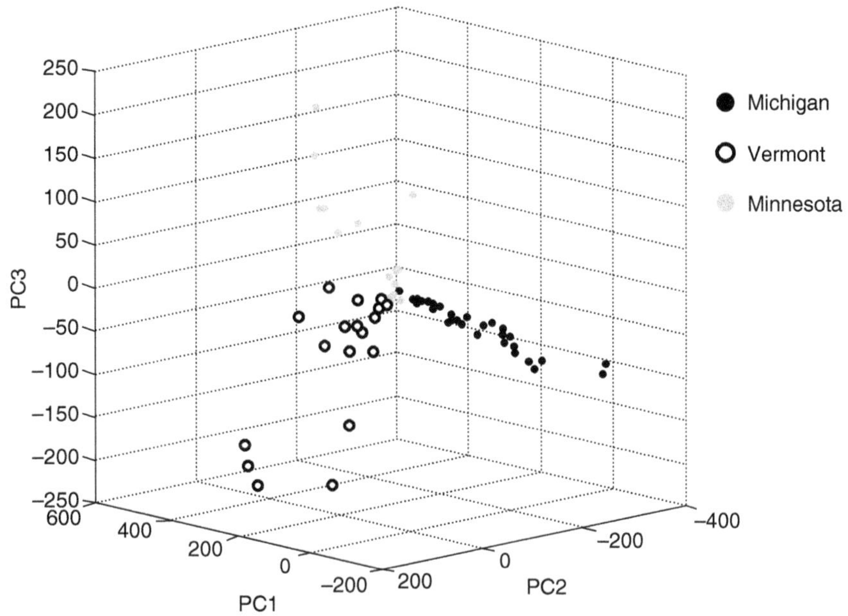

FIGURE 4.6: Principal components analysis of antimicrobial skin peptides from three populations of Northern leopard frogs based on intensities of known antimicrobial peptides. Populations are readily distinguishable by skin peptide profiles analyzed by MALDI-TOF mass spectrometry. Figure produced by Pierre Chaurand from data presented in Tennessen et al. (2009).

relatively high dose of *B. dendrobatidis* survived longer than small toads (Carey et al., 2006). Likewise, larger-than-average toads of *B. bufo* survived low-dose infections that universally killed toads of small body size (Garner et al., 2009). These studies suggest that natural stressors such as competition for food resources can inhibit development of the immune system or development of an immune response to a specific pathogen.

From ecotoxicology studies another type of stressful interaction has been demonstrated. Although the impact on the immune system was not examined in these studies, predatory stress (exposure to predator cues) on tadpoles caused an increase in mortality from exposure to several common pesticides (carbaryl, malathion, and Roundup) in some amphibian species (Relyea and Mills, 2001; Relyea, 2004, 2005). Agrochemical pollutants may be additive stressors, potentially leading to death. Increased stress hormones during amphibian development may reduce pesticide tolerance or cause immunosuppression. Skin peptide defenses were significantly reduced after carbaryl exposure in *R. boylii* or Southern bell frogs (*Litoria raniformis*) (Davidson et al., 2007; Schadich, 2008). These cryptic sublethal effects of pesticides and other lag effects occur in addition to the immediate trade-off between survival under poor conditions and growth or size at metamorphosis (reviewed in Rollins-Smith 1998, 2001; Denver, 2009).

Lyte and Ernst (1992) showed that catecholamines such as norepinephrine, epinephrine, dopamine, and L-dopa induce the growth of gram-negative bacteria in culture, and thus these stress hormones may potentiate the development of infections, or perhaps boost symbiotic bacteria. Stress hormones may also trigger innate immune defenses such as premature release of antimicrobial peptides from the skin of amphibians. Some agrochemicals such as the widely used cholinesterase-inhibiting pesticide carbaryl may mimic the molecular structure of catecholamines, and lead to a reduction in antimicrobial peptide defenses and increase the disease susceptibility of exposed amphibians (Davidson et al., 2007; Schadich, 2008). Likewise, atrazine exposure at a concentration considered to be environmentally relevant (400 ppb) significantly downregulated expression of several antimicrobial peptide genes expressed in the skin (magainin II, levitide A, preprocaerulein, and skin granule protein) of metamorphosing *X. laevis* females (Langerveld et al., 2009). If skin peptide defenses are impaired, skin pathogens such as *B. dendrobatidis* can more easily gain a foothold, as shown when skin peptides were depleted by norepinephrine treatment (Ramsey et al., 2010).

Pesticide exposure suppressed circulating eosinophils and led to increased trematode infections and increased limb malformations in *R. sylvatica* (Kiesecker, 2002). A mixture of pesticides representative of wetland environments in southwest Quebec suppressed some functions of the immune system of African clawed frogs, *X. laevis*, and Northern leopard frogs, *R. pipiens* (Christin et al., 2003, 2004). There are many similar reports of long-term immunosuppression in *R. pipiens* exposed to environmental chemicals (Gendron et al., 2003; Gilbertson et al., 2003; Albert et al., 2007; Rohr et al., 2008; reviewed in Rohr and McCoy, 2010). Few studies have directly tested the effects of a pesticide on disease susceptibility in amphibians. Taylor et al. (1999) showed that exposure to malathion increased the susceptibility of adult Woodhouse's toads (*Bufo woodhousii*) to the lethal effects of the bacterium *Aeromonas hydrophila*. Thus, pesticides can increase disease susceptibility in amphibians. Carey et al. (1999), Rollins-Smith (2001), and Belden and Harris (2007) hypothesized that environmental stress on amphibians may suppress immune defenses such as antimicrobial skin peptide production or maintenance of symbiotic microbiota. Widespread exposure to such stressors, in this view, will hypothetically lead to immunosuppression and subsequent disease-related amphibian population declines.

AMPHIBIAN BEHAVIOR AND IMMUNE DEFENSES

Amphibian behavior can dramatically affect the function of the immune system. Although amphibians cannot alter their body temperature metabolically as mammals do, they can seek warmer or cooler environments that may favor or inhibit pathogen growth. Here we discuss amphibian thermal preferences, behavioral fever and other sickness behaviors, and community interactions that affect immune defenses.

THERMAL PREFERENCES AND BEHAVIORAL FEVER

Many studies have examined the preferred temperatures of adult and larval amphibians using thermal gradients. They have found variation in preferred temperatures with daily cycles and seasons (Hutchison and Spriestersbach, 1986), with acclimation temperatures (Lucas and Reynolds,

1967; Lillywhite, 1971; Hutchison and Hill, 1976), between species from different habitats (Claussen, 1973), and between individuals and social groups (Brattstrom, 1963; Gatten and Hill, 1984). Variation in preferred temperatures might correlate with nutritional status, age, photoperiod, biotic interactions, and other factors (reviewed in Reynolds and Casterlin, 1979; Crawshaw, 1980; Hutchison and Dupré, 1992). Thermal preferences, therefore, appear to reflect complex compromises between physiological processes that may include metabolism, hydroregulation, and other pressures (Licht and Brown, 1967; Bundy and Tracy, 1977; reviewed in Brattstrom, 1979; Crawshaw, 1980). Disease is one such pressure that can increase the preferred temperature, leading to behavioral fever.

Fever can be described as a core body temperature elevated above normal caused by an increase in the thermoregulatory set-point. In mammals, this set-point is changed when exogenous pyrogens such as microbial cells or products trigger cytokine receptors or Toll-like receptors (reviewed in Kluger, 1991, 1992; Dinarello, 2004). Fever is considered to be a phylogenetically conserved adaptive host response (reviewed in Kluger, 1990, 1991). Febrile temperatures are thought to enhance immunological responses such as migration and cytotoxic activity of neutrophils and macrophages, heat shock transcription factor activation (Singh and Aballay, 2006), optimization of cytokine expression, and T-cell proliferation, as well as to decrease growth of microorganisms in an environment of reduced plasma iron concentration (reviewed in Kluger, 1991; Hasday et al., 2000). In ectotherms, behavioral fever is a response that involves a higher preferred and selected environmental temperature, resulting in an increased body temperature. Many ectotherms exhibit behavioral fever in response to infection, leading to an enhanced host immune response and perhaps reduced pathogenicity (Casterlin and Reynolds, 1977; Kluger, 1977; Myhre et al., 1977; Lefcort and Eiger, 1993; Lefcort and Blaustein, 1995; reviewed in Kluger, 1990; Kluger et al., 1996). Adult amphibians and tadpoles move to warmer temperatures when injected with killed bacteria or when naturally infected. Adult green tree frogs (*Hyla cinerea*) injected with killed bacteria (*A. hydrophila*) moved to a warm area of the aquarium and elevated their body temperature by greater than 2°C (Kluger, 1977), while control frogs did not change temperature. Similarly, adult *R. esculenta* injected with killed pathogenic bacteria (*Mycobacterium xenopi* or *Mycobacterium ranae*) preferred warmer temperatures, whereas injections with a nonpathogenic bacterial species (*Mycobacterium aquae*) or saline caused no change (Myhre et al., 1977). *R. catesbeiana* and *R. pipiens* tadpoles injected with bacteria or naturally infected with a fungus (*Candida humicola*) sought warmer water (Casterlin and Reynolds, 1977; Lefcort and Eiger, 1993; Lefcort and Blaustein, 1995). For some diseases, heat (or elevated body temperature) can be applied as a simple treatment or prophylaxis; this was described for juvenile *L. chloris* with chytridiomycosis (Woodhams et al., 2003).

In addition to behavioral fever, a generalized set of behavioral changes may act to promote recovery from infection. In mammals, these sickness behaviors are induced by pro-inflammatory cytokines acting on the motivational centers of the brain to increase the priority of recuperative behaviors such as sleep, immobility, inappetance, and asociality (reviewed in Aubert, 1999; Kelley et al., 2003). As part of an acute phase response or general stress, fever is often associated with altered metabolism, endocrine activity, and immune responses, including upregulation of heat shock proteins (Jiang et al., 2010; reviewed in Kushner, 1982; Heinrich et al., 1990; Robert, 2003; Heikkila, 2010). Almost nothing is known about a possible acute phase response in amphibians. A C-reactive protein has been isolated from *X. laevis*, but it does not appear to have the same function as that of mammalian species (Liu and Liu, 1993). Complement proteins with the ability to directly kill bacteria have been described (Jensen and Koch, 1991; Kato et al., 1994; Green and Cohen, 1977). Key mediators of an acute phase response in mammals and fish, interleukin 1-β (IL-1β) and a tumor necrosis factor-α (TNF-α) homolog, have been described (Zou et al., 2000; Mawaribuchi et al., 2008). Finally, peritoneal leukocytes from *X. laevis* rapidly upregulated expression of TNF-α and IL-1β following infection of the animals with frog virus 3 (Morales et al., 2010). These observations suggest that amphibians can respond systemically to infection with some version of an acute phase response.

In amphibians, injection of pyrogens (fever-inducing substances) usually produces behavioral fever after a short latency period and for a short duration (from minutes to hours), and the temperature increase ranges from 0.5° to 6.5°C (Kluger, 1977; Myhre et al., 1977; Lefcort and Eiger, 1993; Lefcort and Blaustein, 1995). Natural infections may cause prolonged fever with the potential for physiological costs such as risk of tissue damage, dehydration, increased oxygen consumption and metabolic rate, and ecological consequences. However, a few studies have shown that infected ectotherms can maintain prolonged behavioral fever that is beneficial. Two studies of reptiles showed that elevated temperature over seven days was protective, and prevention of the fever response with an antipyretic drug could be lethal (Kluger et al., 1975; Bernheim and Kluger 1976). Another study of goldfish (*Carassius auratus*) showed that elevated temperature over three days was protective for the fish that were able to seek the higher temperature, and most fish forced to remain at a lower temperature died (Covert and Reynolds, 1977). Prolonged behavioral fever may result from chronic infections or pathogen exposure. Sherman (2008) described a behavioral fever of greater than 11°C in *N. viridescens* infected with an *Icthyophonus*-like fungus. Infected newts did not have an increased thermal tolerance, and it is unclear whether the prolonged behavioral fever response benefited the host or rather pathogen transmission. In contrast, Panamanian golden frogs (*Atelopus zeteki*) elevated their body temperature above the normal set-point, and the odds of infection with the amphibian chytrid fungus were reduced for these warmer frogs during an epizootic event. This suggests they were able to develop a prolonged behavioral fever that was somewhat protective (Richards-Zawacki, 2010).

CAN FEVER BE MANIPULATED BY PATHOGENS?

There are many examples of parasites influencing behavior (reviewed in Moore, 2002; Mouritsen and Poulin, 2002). Behavioral fever of amphibians in response to infection may normally help alleviate infection. However, this response may also be manipulated by pathogens to their advantage. Changes in host behavior can alter disease development, or lead to changes in the probability of disease transmission, predation risk, or survival (discussed in Lefévre and Thomas, 2008). Since fever in amphibians is associated with changes in habitat preference, social interactions, activity, and feeding, host preference for warmer microhabitats, for example, may allow increased pathogen growth or transmission. Thus, ecological trade-offs are produced between the potentially positive immunological effect of high temperature and the risks associated with altered behavior of seeking a warmer environment. For example, Lefcort and Blaustein (1995) found that larval red-legged frogs (*Rana aurora*) infected with the yeast *Candida humicola* tended to move toward warmer water and were also less responsive to chemical cues from a predator. Infected tadpoles suffered greater predation than uninfected tadpoles (Lefcort and Blaustein, 1995). Similarly, the behavior of larval *R. catesbeiana* was altered by injection of alcohol-killed bacteria. Tadpoles injected with bacteria moved to warmer water, tended to be more lethargic, and were more vulnerable to predation by California newts (*Taricha torosa*) (Lefcort and Eiger, 1993). This observation indicates that physiological changes due to bacterial or fungal infections may induce a temporary behavioral fever, reduce host fitness, and perhaps increase the risk of transmission for some pathogens with multiple host life cycles that are transmitted through predation of intermediate hosts (Lefcort and Eiger, 1993). In a one-host system, *Ambystoma tigrinum* virus infections led to a behavioral fever, but behavioral responses did not lead to greater predation of Sonoran tiger salamander (*Ambystoma mavortium stebbinsi*) larvae by larval dragonflies, *Anax junius* (Parris et al., 2004). The capacity of infected salamanders to behaviorally reduce predation risk and increase activity in the absence of predators may facilitate viral transmission (Parris et al., 2004).

Rather than a generalized response to infection, some evidence indicates that behavioral fever may be triggered by injection of a pathogenic bacterium but not by a nonpathogenic bacterium (Myhre et al., 1977). In this study, both the pathogenic and the nonpathogenic bacteria were in the genus *Mycobacteria*. Thus, it is not clear how the bacterial species differed. Clearly, there was a difference in the innate immune system recognition of each bacterial type that resulted in behavioral fever in response to one but not the other.

COMMUNITY INTERACTIONS AND BEHAVIORAL IMMUNE DEFENSES

Associated with thermoregulation and habitat preference is aggregation behavior in some amphibians (Wilson-Rich et al., 2009). Amphibian larvae often form aggregations and can increase local temperature while basking in groups. Dense social aggregations do not necessarily increase disease risk (discussed in Elliot and Hart, 2010); however, the direct mode of transmission for many amphibian pathogens suggests that exposure, and thus infection rate and disease incidence, increases with density (discussed in McCallum et al., 2001; Rachowicz and Briggs, 2007). Han et al. (2008) found that unexposed larval cascades frogs (*Rana cascadae*) tended to aggregate with tadpoles that were infected with *B. dendrobatidis*, thereby increasing the risk of infection. Larval *B. boreas* did not show this tendency, and the larvae of these two species, as well as larvae of the Pacific chorus frog (*Hyla regilla*) and the red-legged frog (*R. aurora*), did not demonstrate behavioral fever upon exposure to *B. dendrobatidis*. It seems likely that behavioral fever was not induced in tadpoles exposed to *B. dendrobatidis* because the infection remains confined to the mouthparts, and the immune response is immature at this stage of life. In a study of adult frog behavior, Rowley and Alford (2007) found that species-specific behavioral variation in the probability of *B. dendrobatidis* transmission matched susceptibility to chytridiomycosis-related population declines. The most susceptible species in their study, the waterfall frog (*Litoria nannotis*) was found most frequently in contact with other frogs and with water. In addition to the behavior of larvae and adults, some amphibians vary placement of egg masses depending on temperature and aggregate them in the cold. This may affect pathogen transmission under some circumstances (Ruthig, 2008).

In amphibian husbandry, population density is a critical factor affecting growth, development, and survival of amphibians. Crowding is often thought to stress animals by increasing competition for resources, aggressive encounters, and the microbial burden associated with soiled living conditions. Richards (1958) suggested that the effects of high density on amphibian larvae can be mediated by pathogens (interference competition) in addition to direct competition for resources (exploitative competition). Three pathogens have been implicated as mediators of competition either among tadpoles of different anuran species, between tadpoles and mosquito larvae, or between tadpoles of different size classes. These include two fungal species, *Candida humicola* (Steinwascher, 1979) and *Rhodotorula glutinis* (Mokany and Shine, 2003), and the protozoan *Anurofeca* (or *Prototheca*) *richardsi* (Beebee, 1991; Petranka, 1995; Bardsley and Beebee, 1998, 2001; Baker et al., 1999). For example, in *B. bufo*, the as-yet-unculturable protozoan *A. richardsi* can be found in the intestine and feces of tadpoles. Small tadpoles may be diverted to a low-quality diet and coprophagy (consumption of feces), leading to starvation (Beebee, 1991). The effects of this pathogen interference on tadpole growth are regulated by environmental conditions. At low temperatures similar to field breeding conditions, *Anurofeca*-induced growth inhibition was not observed in larval *R. sylvatica* (Biesterfeldt et al., 1993).

Between and within some amphibian species, competition for recruitment, growth, development, and survival may be mediated by pathogens. Larval *H. chrysoscelis* developed more slowly upon exposure to amphibian chytrid fungus, *B. dendrobatidis*, but only when exposed to predatory newts (*N. viridescens*) (Parris and Beaudoin, 2004). *B. dendrobatidis* also caused developmental stress and affected competitive interactions between larval *H. chrysoscelis* and *B. fowleri* (Parris and Cornelius, 2004). In the presence of water the mold *Saprolegnia ferax*, the usual outcome of competition between *R. cascadae* and *H. regilla* was reversed—that is, without the pathogen, the presence of *R. cascadae* tadpoles inhibited growth development and survival of *H. regilla* tadpoles. When *Saprolegnia* was present, survival of *R. cascadae* tadpoles was reduced more than that of the H. regilla tadpoles. However, *R. cascadae* survivors grew faster and developed to be larger than frogs that were not exposed. Thus, in the presence of a pathogen, competitve interactions may be drastically altered (Kiesecker and Blaustein, 1999). In system with *H. cinerea* and *R. utricularia*, chemical interference and aggressive behaviors from *R. utricularia* tadpoles against the *H. cinerea* tadpoles inhibited growth and increased mortality of *H. cinerea*. This observation was interpreted to indicate that avoiding interactions may improve survival and metamorphosis of *Hyla* (Faragher and Jaeger, 1998). Social aggregation or avoidance behaviors may benefit one species over another if chemical or pathogen interference is involved in the competitive interaction. Although the immunological implications of these interactions were not studied, it is likely that aggressive competition that disadvantages development and growth of one species would likely alter development of the immune system when resources are limited.

Reducing host density below a threshold may alleviate disease pressure by reducing pathogen exposure and density-dependent transmission (Rachowicz and Briggs, 2007; Briggs et al., 2010). Behavioral reductions in contact rate or pathogen exposure will alleviate disease pressure no matter what the transmission function (i.e., frequency- or density-dependent transmission). Thus, the timing and spatial distribution of egg deposition and the degree of social aggregation in larval and adult amphibians may be influenced by disease pressures (Faragher and Jaeger, 1998; Ruthig, 2008) and should be considered potential behavioral immune defenses.

COMMUNITY COMPOSITION AND DEFENSE AGAINST PATHOGENS OR PARASITES

Community composition and structure may also contribute to the defense of amphibians against disease. If several host amphibian species do not directly compete for resources because of niche partitioning or other reasons, shared pathogens may still lead to apparent competition (reviewed in Lefévre et al., 2009). Increased prevalence of a pathogen in a disease-resistant carrier species could negatively affect a disease-susceptible species by amplifying disease risk or even drive it to extinction. Pathogens that can be transmitted to a series of hosts are of great concern for small endangered populations of mammals (Cleaveland et al., 2001). Alternatively, an infection-incompetent species may dilute or reduce the chances for infection in a disease-susceptible species (reviewed in Keesing et al., 2006) For example, when American toads (*Bufo americanus*) were experimentally exposed to cercariae of *Ribeiroia ondatrae*, mortality and malformation were high. However, when Eastern gray tree frogs (*Hyla versicolor*) were also present as less susceptible decoy hosts, infection of toads was reduced (Johnson et al., 2008). For multiple pathogen systems, co-infection may alter disease risk by producing synergistic or additive effects. Virulence may be reduced through processes that enhance immunity such as cross-reactivity to related parasites. Alternatively, susceptibility to one pathogen may be increased if the first infection is immunosuppressive (Cattadori et al., 2008). Virulence may also be reduced through microbial competitive interactions (Massey et al., 2004; Harrison et al., 2006). In nature, polyparasitism is common and interactive effects are important for understanding parasite pressure, particularly in changing environments (Pederson and Fenton, 2007). Mammalian species that were exposed to a greater number of helminthic parasites had higher numbers of circulating lymphocytes. This was interpreted as evidence that they invested more in immune defenses than species encountering fewer parasites (Bordes and Morand, 2009). Thus the biotic community is an important component of disease risk. While choosing from a range of biotic communities to which to belong may be outside the behavioral scope for many amphibians, such responses to disease pressure certainly occur at an evolutionary time scale and contribute to patterns of community diversity, distributions of species, and the stability and resilience of food webs (Johnson and Hartson, 2009; Bordes and Morand, 2009; reviewed in Keesing et al., 2006; Zilber-Rosenberg and Rosenberg, 2008; Lafferty et al., 2008). This stretches the definition of immune defense from the individual to the ecological level.

Indeed, social groups, such as social insects, may be advantaged by the capacity for behavioral immunity, given its relatively lower cost than physiological immune defenses (Eliot and Hart, 2010). In social insects, behavioral immunity includes border, soma, and germline defenses. It may be prophylactic and inducible, and shows memory (Walker and Hughes, 2009; reviewed in Cremer and Sixt, 2009). Behavioral defenses in amphibians may not be as elaborate as in social insects; however, community interactions have an important role in immunity.

CONCLUDING REMARKS

In this review chapter, we have tried to draw together the findings of a number of amphibian ecologists and amphibian immunologists. All would likely agree that the environment can dramatically affect the immune system in amphibians and modify disease outcomes. Our best view of the possible costs of amphibian survival responses are summarized in Figure 4.7. Various environmental stressors may induce host stress responses through the hypothalamic–pituitary-adrenal (HPA) axis, resulting in elevation of corticosteroid hormones that suppress immunity. Concurrently, pathogen recognition can induce life-history trade-offs to enhance immunity and subsequent survival from pathogen challenge.

Although disease is one potential outcome of infection, other outcomes, including immune defense, can also have fitness costs. The increased survival benefits of immune responses come

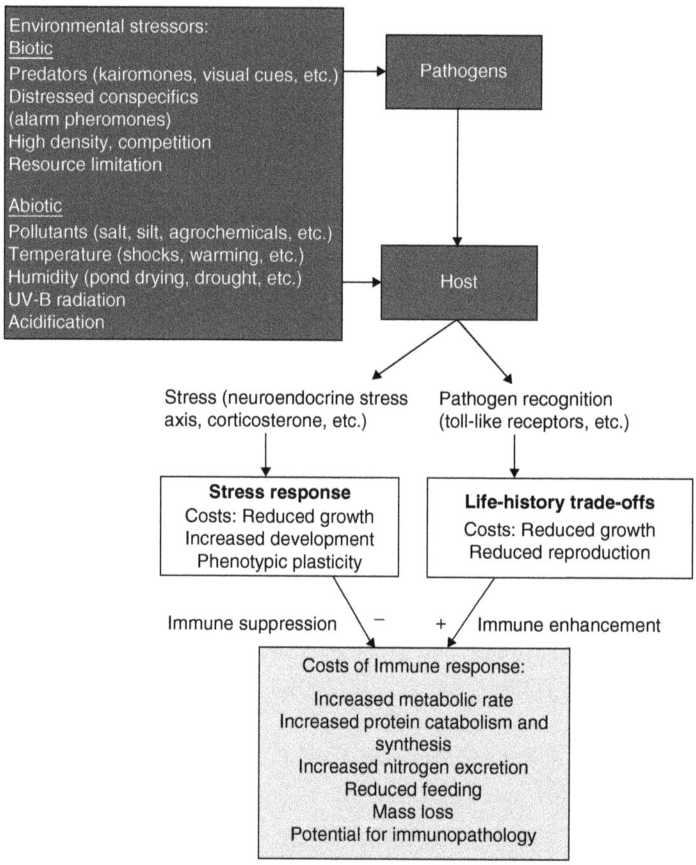

FIGURE 4.7: Possible costs of amphibian survival responses. Various environmental stressors may induce host stress responses that suppress immunity. Concurrently, pathogen recognition can induce life-history trade-offs to enhance immunity and subsequent survival from pathogen challenge. Although disease is one potential outcome of infection, other outcomes, including immune defense, can also have substantial fitness costs. The increased survival benefits of these responses come with associated costs, some of which are indicated. An acute survival response of amphibians may be reduced activity, including feeding, reproduction, and dispersal activities, sometimes leading to altered hydration, thermoregulation, and predation risk. Chronic responses to stress in amphibians may come with the cost of reduced growth. In amphibian larvae, chronic stress may result in precocious metamorphosis, with possible carryover or lag effects in later developmental stages.

with associated costs such as increased metabolic rate and catabolism. An acute survival response of amphibians may be reduced activity, including feeding, reproduction, and dispersal activities, sometimes leading to altered hydration, thermoregulation, and predation risk. Chronic responses to stress in amphibians may come with the cost of reduced growth. In amphibian larvae, chronic stress may result in precocious metamorphosis with possible carryover or lag effects in later developmental stages. Future research that combines the efforts of immunologists and ecologists may provide insights that will suggest better management strategies for threatened or endangered species.

ACKNOWLEDGMENTS

Recent work from the Rollins-Smith laboratory was supported by NSF grants IOS-0520847, IOS-0619536, and IOS-0843207. D.C.W. was supported by the Claraz Foundation, and the Swiss National Science Foundation (31–125099). Earlier studies from the Rollins-Smith laboratory developed as collaborations within NSF Integrated Research Challenges in Environmental Biology (IRCEB) grants IBN-9977063 and DEB-0213851 (James P. Collins, P.I.).

REFERENCES

Albert, A, Drouillard, K, Haffner, GD, and Dixon, B (2007). Dietary exposure to low pesticide doses causes long-term immunosuppression in the leopard frog (*Rana pipiens*). Environ. Toxicol. Chem. 26:1179–1185.

Alford, RA and Richards, SJ (1999). Global amphibian declines: A problem in applied ecology. Ann. Rev. Ecol. Sys. 30:133–165.

AmphibiaWeb 2000–2010. Information on amphibian biology and conservation. Accessed October 27, 2010. http://amphibiaweb.org.

Andre, SE, Parker, J, and Briggs, CJ (2008). Effect of temperature on host response to *Batrachochytrium dendrobatidis* infection in the mountain yellow-legged frog (*Rana muscosa*). J. Wildl. Dis. 44:716–720.

Apponyi, MA, Pukala, TL, Brinkworth CS, Vaselli, VM, Bowie, JH, Tyler MJ, Booker GW, Wallace, JC, Carver, JA, Separovic, F, Doyle, J, and Llewellyn, L (2004). Host-defence peptides of Australian anurans: structure, mechanisms of action and evolutionary significance. Peptides 25:1035–1054.

Ash, AN and Bruce RC (1994). Impacts of timber harvesting on salamanders. Conserv. Biol. 8:300–301.

Aubert, A (1999). Sickness and behavior in animals: a motivational perspective. Neurosci. Biobehav. Rev. 23:1029–1036.

Austin Jr, RM (2000). Cutaneous microbial flora and antibiosis in *Plethodon ventralis*. In: Bruce RC, Jaegar RG, and Houck LD (Eds). The Biology of Plethodontid Salamanders. Kluwer Academic/Plenum Publishers, New York, pp. 451–463.

Bäckhed, F, Ley RE, Sonnenburg, JL, Peterson, DA, and Gordon, JI (2005). Host–bacterial mutualism in the human intestine. Science 307:1915–1920.

Baculi, BS and Cooper, EL (1970). Histopathology of skin allograft rejection in larval *Rana catesbeiana*. J. Exp. Zool. 173:329–340.

Baker, GC, Beebee, TJC, and Ragan, MA (1999). *Prototheca reichardsi*, a pathogen of anuran larvae, is related to a clade of protistan parasites near the animal–fungal divergence. Microbiology 145:1777–1784.

Banning, JL, Weddle, AL, Wahl III, GW, Simon, MA, Lauer, A, Walters, RL, Harris, R (2008). Antifungal skin bacteria, embryonic survival, and communal nesting in four-toed salamanders, *Hemidactylium scutatum*. Oecologia. 156:423–429.

Banas, JA, Loesche, WJ, and Nace, GW (1988). Classification and distribution of large intestinal bacterial in nonhibernating and hibernating leopard frogs (*Rana pipiens*). App. Environ. Microbiol. 54:2305–2310.

Bardsley, L and Beebee, TJC (1998). Interspecific competition between *Bufo* larvae under conditions of community transition. Ecology 79:1751–1760.

Bardsley, L and Beebee, TJC (2001). Non-behavioural interference competition between anuran larvae under semi natural conditions. Oecologia 128:360–367.

Barker, KS, Davis, AT, Li, B, and Rollins-Smith, LA (1997). In vitro studies of spontaneous and corticosteroid-induced apoptosis of lymphocytes populations from metamorphosing frogs/RU486 inhibition. Brain Behav. Immun. 11:119–131.

Barr, DJS (1980). An outline for the reclassification of the *Chytridiales*, and for a new order, the *Spizellomycetales*. Can. J. Bot. 58:2380–2394.

Barra, D, and Simmaco, M (1995). Amphibian skin: a promising resource for antimicrobial peptides. Trends Biotechnol. 13:205–209.

Becker, MH, and Harris, RN (2010). Cutaneous bacteria of the redback salamander prevent morbidity associated with a lethal disease. PLoS One 5:e10957.

Becker, MH, Brucker, RM, Schwantes, CR, Harris, RN, and Minbiole, KPC (2009). The bacterially produced metabolite violacein is associated with survival of amphibians infected with a lethal fungus. Appl. Environ. Microbiol. 75:6635–6638.

Beebee, TJC (1991). Purification of an agent causing growth inhibition in anuran larvae and its identification as a unicellular alga. Can. J. Zool. 69:2146–2153.

Belden, LK, and Harris, RN (2007). Infectious diseases in wildlife: the community ecology context. Front. Ecol. Environ. 5:533–539.

Belden, LK and Kiesecker, JM (2005). Glucocorticoid hormone treatment of larval treefrogs increases infection by *Alaria* sp. trematode cercariae. J. Parasitol. 91:686–688.

Benson, BJ and Hadley, ME (1969). In vitro characterization of adrenergic receptors controlling skin gland secretion in two anurans *Rana pipiens* and *Xenopus laevis*. Comp. Biochem. Physiol. 30:857–864.

Berger, L, Speare, R, Daszak, P, Green, DE, Cunningham, AA, Goggin, CL, Slocombe, R, Ragan MA, Hyatt, AD, McDonald, KR, Hines, HB, Lips, DR, Marantelli, G, and Parkes H (1998). Chytridiomycosis causes amphibian mortality associated with population declines in the rain forests of Australia and Central America. Proc. Natl. Acad. Sci. USA 95:9031–9036.

Berger, L, Speare, R, Hines, HT, Marantelli, G, Hyatt, AD, McDonald, KR, Skerratt, LF, Olsen, V, Clarke, JM, Gillespie, G, Mahony, M, Sheppard, N, Williams, C and, Tyler, M (2004). Effect of season and temperature on mortality in amphibians due to chytridiomycosis. Aust. Vet. J. 7:434–439.

Berger, L., Hyatt, AD, Speare, R, and Longcore, JE (2005). Life cycle stages of the amphibian chytrid *Batrachochytrium dendrobatidis*. Dis. Aquat. Org. 68:51–63.

Bernheim, HA and Kluger, MJ (1976). Fever: effect of drug-induced antipyresis on survival. Science 193: 237–239.

Berrill, M, Bertran, S, and Pauli, B (1997). Effects of pesticides on amphibian embryos and larvae. Herpetolog. Conserv. 1:233–245.

Bettin, C and Greven, H (1986). Bacteria on the skin of *Salamandra salamandra* (L.) (Amphibia, Urodela) with notes on their possible significance. Zool. Anz. 216:267–270.

Biesterfeldt, JM, Petranka, JW, and Sherbondy, S (1993). Prevalence of chemical interference competition in natural populations of wood frogs, *Rana sylvatica*. Copeia 1993: 688–695.

Bilusich, D, Jackway, RJ, Musgrave, IF, Tyler, MJ, and Bowie, JH (2009). The host-defence skin peptide profiles of Peron's tree frog *Litoria peroni* in winter and summer. Sequence determination by electrospray mass spectrometry and activities of the peptides. Rapid Commun. Mass Spectrom. 23:2628–2636.

Blaustein, AR, Romansic, JM, Kiesecker, JM, and Hatch, AC (2003). Ultraviolet radiation, toxic chemicals and amphibian population declines. Divers. Distrib. 9:123–140.

Bleicher, PA, Rollins-Smith, LA, Jacobs, DM, and Cohen, N (1983). Mitogenic responses of frog lymphocytes to crude and purified preparations of bacterial lipopolysaccharide (LPS). Dev. Comp. Immunol. 7: 483–496.

Blomberg, B, Bernard, CCA, and Du Pasquier, L (1980). In vitro evidence for T-B lymphocyte collaboration in the clawed toad, *Xenopus*. Eur. J. Immunol. 10:869–876.

Bollinger, TK, Mao, J, Schock, D, Brigham, RM, and Chinchar, VG (1999). Pathology, isolation, and preliminary molecular characterization of a novel iridovirus from tiger salamanders in Saskatchewan. J. Wildl. Dis. 35:413–429.

Bordes, F and Morand, S (2009). Coevolution between multiple helminth infestations and basal immune investment in mammals: cumulative effects of polyparasitism. Parisitol. Res. 106:33–37.

Bosch, J, Martínez-Solano, I, and García-París, M (2001). Evidence of a chytrid fungus I nfection involved in the decline of the common midwife toad (*Alytes obstetricans*) in protected areas of central Spain. Biol. Conserv. 97:331–337.

Bovbjerg, AM (1963). Development of the glands of the dermal plicae in *Rana pipiens*. J. Morph. 113: 321–243.

Bovbjerg, AM (1966). Rejections of skin homografts in larvae of *Rana pipiens*. J. Exp. Zool. 162:69–80.

Bradford, DF (1989). Allotopic distribution of native frogs and introduced fishes in high Sierra Nevada lakes of California: implication of the negative effect of fish introductions. Copeia 1989: 775–778.

Bradford, DF, Tabatabai, F, and Graber, DM (1993). Isolation of remaining populations of the native frog, *Rana muscosa*, by introduced fishes in Sequoia and Kings Canyon National Parks, California. Cons. Biol. 7:882–888.

Bradley, GA, Rosen, PC, Sredl, MJ, Jones, TR, and Longcore, JE (2002). Chytridiomycosis in native Arizona frogs. J. Wildl. Dis. 38:206–212.

Brattstrom, BH (1963). A preliminary review of the thermal requirements of amphibians. Ecology 44: 238–255.

Brattstrom, BH (1979). Amphibian temperature regulation studies in the field and laboratory. Integr. Comp. Biol. 19:345–356.

Bridges, CM and Semlitsch, RD (2001). Genetic variation in insecticide tolerance in a population of southern leopard frogs (*Rana sphenocephala*): Implications for amphibian conservation. Copeia 2001:7–13.

Briggs, CJ, Knapp, RA, and Vredenburg, VT (2010). Enzootic and epizootic dynamics of the chytrid fungal pathogens of amphibians. Proc. Natl. Acad. Sci. USA 107:9695–9700.

Brodkin, MA, Madhoun, H, Rameswaran, M, and Vatnick I (2007). Atrazine is an immune disruptor in adult northern leopard frogs (*Rana pipiens*). Environ. Toxicol. Chem. 26:80–84.

Brucker, RM, Baylor, CM, Walters, RL, Lauer, A, Harris, RN, and Minbiole, KPC (2008a). The identification of 2,4-diacetylphloroglucinol as an antifungal metabolite produced by cutaneous bacteria of the salamander *Plethodon cinereus*. J. Chem. Ecol. 34:39–43.

Brucker, RM, Harris, RN, Schwantes, CR, Gallaher, TN, Flaherty, DC, Lam, BA, and Minbiole, KPC (2008b). Amphibian chemical defense: antifungal metabolites of the microsymbiont *Janthinobacterium lividum* on the salamander *Plethodon cinereus*. J. Chem. Ecol. 34:1422–1429.

Bundy, C and Tracy, CR (1977). Behavioral responses of American toads (*Bufo americanus*) to stressful thermal and hydric environments. Herpetologica 33:455–458.

Carey, C and Alexander, MA (2003). Climate change and amphibian declines: is there a link? Divers. Distrib. 9:111–121.

Carey, C and Bryant CJ (1995). Possible interrelations among environmental toxicants, amphibian development, and decline of amphibian populations. Environ. Health Perspect. 103 (Suppl 4):13–17.

Carey, C, Bruzgul, JE, Livo, LJ, Walling, ML, Kuehl, KA, Dixon, BF, Pessier, AP, Alford, RA, and Rogers, KB (2006). Experimental exposures of boreal toads (*Bufo boreas*) to a pathogenic chytrid fungus (*Batrachochytrium dendrobatidis*). EcoHealth 3:5–21.

Carey, C, Cohen, N, and Rollins-Smith, L (1999). Amphibian declines: an immunological perspective. Dev. Comp. Immunol. 23:459–472.

Carrillo, J, Castel, A, Perez, A, and Rondan, A (1990). Langerhans-like cells in amphibian epidermis. J. Anat. 172:39–46.

Case, TJ (1990). Invasion resistance arises in strongly interacting species-rich model competition communities. Proc. Natl. Acad. Sci. USA. 87:9610–9614.

Castell-Rodriguez, AE, Hernandez-Penaloza, A, Sampedro-Carrillo, WA, Herrera-Enriquez, MA, Alvarez-Perez, SJU, and Rondan-Zarate, A (1999). ATPase and MHC class II molecules co-expression in *Rana pipiens* dendritic cells. Dev. Comp. Immunol. 23:473–485.

Casterlin, ME and Reynolds, WW (1977). Behavioral fever in anuran amphibian larvae. Life Sci. 20: 593–596.

Cattadori, IM, Boag, B, and Hudson, PJ (2008). Parasite co-infection and interaction as drivers of host heterogeneity. Int. J. Parasitol. 38:371–380.

Chardonnens, X, and Du Pasquier, L (1973). Induction of skin allograft tolerance during metamorphosis of the toad *Xenopus laevis*: a possible model for studying generation of self tolerance to histocompatibility antigens. Eur. J. Immunol 3:569–573.

Charlemagne, J (1987). Antibody diversity in amphibians. Non-inbred axolotls used the same unique heavy chain and a limited number of light chains for their anti-DNP antibody response. Eur. J. Immunol 17:421–424.

Chinchar, VG (2002). Ranaviruses (family *Iridoviridae*): emerging cold-blooded killers. Arch. Virol. 147: 447–470.

Chrétien, I, Marcuz, A, Fellah, J, Charlemagne, J, and Du Pasquier, L (1997). The T cell receptor beta genes of *Xenopus*. Eur. J. Immunol. 27:763–771.

Christin, M-S, Gendron, AD, Brousseau, P, Ménard, L, Marcogliese, DJ, Cyr, D, Ruby, S, and Fournier, M (2003). Effects if agricultural pesticides on the immune system of *Rana pipiens* and on resistance to parasitic infection. Environ. Toxicol. Chem. 22:1127–1133.

Christin, MS, Ménard, L, Gendron, AD, Ruby, S, Cyr, D, Marcogliese, DJ, Rollins-Smith, L, and Fournier, M (2004). Effects of agricultural pesticides on the immune system of *Xenopus laevis* and *Rana pipiens*. Aquatic Toxicol. 30: 33–43.

Claussen, DL (1973). The thermal relations of the tailed frog, *Ascaphus truei*, and the Pacific treefrog, *Hyla regilla*. Comp. Biochem. Physiol. A 44:137–153.

Cleaveland, S, Laurenson, MK, and Taylor, LH (2001). Diseases of humans and their domestic mammals: pathogen characteristics, host range and risk of emergence. Philos. Trans. R. Soc. Lond. B Biol. Sci. 356:991–999.

Cohen, N (1966). Tissue transplantation immunity in the adult newt, *Diemictylus viridescens*. III. The effects of X-irradiation and temperature on the allograft reaction. J. Exp. Zool. 163: 231–240.

Cohen, N, DiMarzo, S, Rollins-Smith, L, Barlow, E, and Vanderschmidt-Parsons, S (1985). The ontogeny of allo-tolerance and self-tolerance in larval *Xenopus laevis*. In: Balls M, Bownes M, eds. Metamorphosis. Oxford: Oxford University Press. pp. 388–419.

Collins, JP and Crump ML (2009). Extinction in Our Times: Global Amphibian Decline. Oxford University Press, New York, pp. 55–58.

Collins, JP and Storfer, A (2003). Global amphibian declines: Sorting the hypotheses. Divers. Distrib. 9: 89–98.

Cone, RE and Marchalonis, JJ (1972). Cellular and humoral aspects of the influence of environmental temperature on the immune response of poikilothermic vertebrates. J. Immunol. 108:952–957.

Conlon, JM, Kolodziejek, J, and Nowotny, N (2004). Antimicrobial peptides from ranid frogs: taxonomic and phylogenetic markers and a potential source of new therapeutic agents. Biochim. Biophys. Acta. 1696: 1–14.

Cooper, EL, Wright, RK, Klempau, AE, and Smith, CT (1992). Hibernation alters the frog's immune system. Cryobiology 29:616–631.

Covert, JB and Reynolds, WW (1977). Survival value of fever in fish. Nature 267:43–45.

Crawford, AJ, Lips, KR, and Bermingham, E (2010). Epidemic disease decimates amphibian abundance, species diversity, and evolutionary history in the highlands of central Panama. Proc. Natl. Acad. Sci. USA 107:13777–13782.

Crawshaw, LI (1980). Temperature regulation in vertebrates. Annu. Rev. Physiol. 42:473–491.

Cremer, S and Sixt, M (2009). Analogies in the evolution of individual and social immunity. Phil. Trans. R Soc. Lond. B Biol. Sci. 364:129–142.

Cullen, BR, and Owens, L (2002). Experimental challenge and clinical cases of Bohle iridovirus (BIV) in native Australian anurans. Dis. Aquat. Org. 49:83–92.

Cullen, BR, Owens, L, and Whittington, RJ (1995). Experimental infection of Australian anurans (*Limnodynastes terraereginae* and *Litoria latopalmata*) with Bohle iridovirus. Dis. Aquat. Org. 23:83–92.

Cunningham, AA, Langton, TES, Bennett, PM, Lewin, JF, Drury, SEN, Gough, RE, and MacGregor, SK (1996). Pathological and microbiological findings from incidents of unusual mortality of the common frog (*Rana temporaria*). Phil. Trans. R. Soc. London B. 351:1539–1557.

Currie, CR, Poulsen, M, Mendenhall, J, Boomsma, JJ, and Billen, J (2006). Coevolved crypts and exocrine glands support mutualistic bacteria in fungus-growing ants. Science 311:81–83.

Daly, JW (1995). The chemistry of poisons in amphibian skin. Proc. Natl. Acad. Sci. USA 92:9–13.

Davidson, C, Benard, MF, Shaffer, HB, Parker, J, O'Leary, C, Conlon, JM, and Rollins-Smith, LA (2007). Effects of chytrid and carbaryl exposure on survival, growth, and skin peptide defenses in foothill yellow-legged frogs. Environ. Sci. Technol. 41:1771–1776.

Davidson, C, Shaffer, HB, and Jennings, MR (2002). Spatial tests of the pesticide drift, habitat destruction, UV-B and climate change hypotheses for California amphibian declines. Conserv. Biol. 16:1588–1601.

Daszak, P, Berger, L, Cunningham, AA, Hyatt, AD, Green, DE, Speare, R (1999). Emerging infectious diseases and amphibian population declines. Emerg. Infect. Dis. 5:735–748.

Daszak, P, Cunningham, AA, and Hyatt, AD (2003). Infectious disease and amphibian population declines. Divers. Distrib. 9:141–150.

Daszak, P, Scott, DE, Kilpatrick, AM, Faggioni C, Gibbons, JW, and Porter, D (2005). Amphibian population declines at Savannah River site are linked to climate, not chytridiomycosis. Ecology 86:3232–3237.

de Castro, VL, Johsson, CM, Silva, CM, and de Holanda Nunes Maia, A (2010) Assessing the safety of *Pseudomonas putida* introduction in the environment: An overview of ecotoxicological tests. Regul. Toxicol. Pharmacol. 56:300–305.

Dechesne, A, Pallud C, Bertolla, F, and Grundmann, GL (2005). Impact of the microscale distribution of a *Pseudomonas* strain introduced into soil on potential contacts with indigenous bacteria. Appl. Environ. Microbiol. 71:8123–8131.

Delis, PR, Mushinsky, HR, and McCoy, ED (1996). Decline of some west-central Florida anuran populations in response to habitat degradation. Biodiversity Conserv. 5:1579–1595.

Densmore, CL and Green, DE (2007). Diseases of amphibians. ILAR J. 48:235–254.

Denver, RJ (2009). Stress hormones mediate environment-genotype interactions during amphibian development. Gen. Comp. Endocrinol. 164:20–31.

Dethlefsen, L, Eckburg, PB, Bik, EM, and Relman, DA (2006). Assembly of the human intestinal microbiota. Trends Ecol. Evol. 21:517–523.

Dillon, RJ, Vennard, CT, Buckling, A, and Charnley, AK (2005). Diversity of locust gut bacteria protects against pathogen invasion. Ecol. Lett. 8:1291–1298.

DiMarzo, SJ (1980). Ontogeny of alloreactivity in the toad, *Xenopus laevis*. Ph.D. Thesis. University of Rochester, Rochester, New York.

DiMarzo, SJ and Cohen, N (1982a). An in vivo study of the ontogeny of alloreactivity in the frog, *Xenopus laevis*. Immunology 45:39–48.

DiMarzo, SJ and Cohen, N (1982b). Immunogenetic aspects of in vivo allotolerance induction during ontogeny of *Xenopus laevis*. Immunogenetics 16:103–116.

Dinarello, CA (2004). Review: infection, fever, and exogenous and endogenous pyrogens: some concepts have changed. Innate Immun. 10:201–222.

Docherty, DE, Meteyer, CU, Wang, J, Mao, J, Case, S, and Chinchar, GD (2003). Diagnostic and molecular evaluation of three iridovirus-associated salamander mortality events. J. Wildl. Dis. 39:556–566.

Dockray, GJ and Hopkins, CR (1975). Caerulein secretion by dermal glands in *Xenopus laevis*. J. Cell Biol. 64:724–733.

Donnelly, MA, and Crump, ML (1998). Potential effects of climate change on two neotropical amphibian assemblages. Climate Change 39:541–561.

Drew, A, Allen, EJ, and Allen, LJS (2006). Analysis of climatic and geographic factors affecting the presence of chytridiomycosis in Australia. Dis. Aquat. Organ. 68:245–250.

Drost, CA, and Fellers, GM (1996). Collapse of regional frog fauna in the Yosemite area of the California Sierra Nevada. Cons. Biol. 10:414–425.

Du Pasquier, L (1973). Ontogeny of the immune response in cold-blooded vertebrates. Curr. Top. Microimmunol. 61:37–88.

Du Pasquier, L and Chardonnens, X (1975). Genetic aspects of the tolerance to allografts induced at metamorphosis in the toad *Xenopus laevis*. Immunogenetics 2:431–440.

Du Pasquier, L and Flajnik, MF (1990). Expression of MHC class II antigens during *Xenopus* development. Dev. Immunol. 1:85–95.

Du Pasquier, L and Haimovich J (1974). Changes in affinity of IgM antibodies in amphibian larvae. Eur. J. Immunol. 4:580–583.

Du Pasquier, L and Haimovich J (1976). The antibody response during amphibian ontogeny. Immunogenetics 3:381–391.

Du Pasquier, L and Weiss, N (1973). They thymus during the ontogeny of the toad *Xenopus laevis*: growth membrane-bound immunoglobulins and mixed lymphocyte reaction. Eur. J. Immunol. 3:773–777.

Du Pasquier, L, Blomberg, B, and Bernard, CCA (1979). Ontogeny of immunity in amphibians: changes in antibody repertoires and appearance of adult major histocompatibility antigens in *Xenopus*. Eur. J. Immunol. 9:900–906.

Du Pasquier, L, Flajnik, MF, Guiet, C, and Hsu, E (1985). Methods used to study the immune system of *Xenopus* (Amphibia, Anura). In: Lefkovits I, Pernis V, eds. Immunological Methods, Vol. III. Elsevier Science and Technology Books, Orlando, FL, pp. 425–465.

Du Pasquier, L, Flajnik, MF, Hsu, E, and Kaufman, JF (1986). Ontogeny of the immune system in anuran amphibians. Prog. Immunol. 6:1079–1088.

Du Pasquier, L, Schwager, J, and Flajnik MF (1989). The immune system of *Xenopus*. Ann. Rev. Immunol. 7:251–275.

Duellman, WE and Treub, L (1986). Biology of Amphibians. Johns Hopkins University Press, pp. 369–370.

Eipert, E, Klempau, AE, Lallone, RA, and Cooper, EL (1979). Bone marrow as a major lymphoid organ in *Rana*. Cell Immunol. 46:275–280.

Elliot, SL and Hart, AG (2010). Density-dependent prophylactic immunity reconsidered in the light of host group living and social behavior. Ecology 91:65–72.

Erspamer, V (1994). Bioactive secretions of the amphibian integument. In: Heatwole, H, Barthalmus, GT, and Heatwole, AY (eds). Amphibian Biology. Volume 1: The Integument. Surrey Beatty and Sons, Chipping Norton, pp. 178–350.

Evans, JD and Armstrong, TN (2006). Antagonistic interactions between honey bee bacterial symbionts and implications for disease. BMC Ecol. 6:4.

Faragher, SG and Jaeger, RG (1998). Tadpole bullies: examining mechanisms of competition in a community of larval anurans. Can. J. Zool. 76:144–153.

Fedewa, LA (2006). Fluctuating gram-negative microflora in developing anurans. J. Herp. 40:131–135.

Fellah, JS, Vaulot, D, Tournefier, A, and Charlemange, J (1989). Ontogeny of immunoglobulin expression in the Mexican axolotl. Development 107:253–263.

Fisher, MC and Garner, TWJ (2007). The relationship between the emergence of *Batrachochytrium dendrobatidis*, the international trade in amphibians and introduced amphibian species. Fung. Biol. Rev. 21:2–9.

Fisher, MC, Garner, TWJ, and Walker, SF (2009). Global emergence of *Batrachochytrium dendrobatidis* and amphibian chytridiomycosis in space, time, and host. Annu. Rev. Microbiol. 63:291–310.

Flajnik, MF and Du Pasquier, L (1988). MHC class I antigens as surface markers of adult erythrocytes during the metamorphosis of *Xenopus*. Dev. Biol. 128:198–206.

Flajnik, MF and Du Pasquier, L (1990). The major histocompatibility complex of frogs. Immunol. Rev. 113:47–63.

Flajnik, MF and Kasahara, M (2001). Comparative genomics of the MHC: Glimpses into the evolution of the adaptive immune system. Immunity 15:351–362.

Flajnik, MF, Kaufman, JF, Hsu, E, Manes, M, Parisot, R, and Du Pasquier, L (1986). Major histocompatibility complex-encoded class I molecules are absent in immunologically competent *Xenopus* before metamorphosis. J. Immunol. 137:3891–3899.

Forbes, MR, McRuer, DL, and Rutherford, PL (2004). Prevalence of *Aeromonas hydrophila* in relation to timing and duration of breeding in three species of ranid frogs. EcoScience. 11:282–285.

Fraune, S, Augustin, R, and Bosch, TCG (2009). Exploring host–microbe interactions in hydra. Microbe 4:457–462.

Frost, DR, Grant T, Faivovich, J, Bain, RH, Haas, A, Haddad, CFB, de Sá, RO, Channing A, Wilkinson, M, Donnellan, SC, Raxworthy, CJ, Campbell, JA, Blotto, BL, Moler, P, Drewes, RC, Nussbaum, RA, Lynch, JD, Green, DM, and Wheeler, WC (2006). The amphibian tree of life. Bull. Am. Mus. Nat. Hist. 297:1–371.

Gantress, J., Maniero, GD, Cohen, N, and Robert, J (2003). Development and characterization of a model system to study amphibian immune responses to iridoviruses. Virology 311:254–262.

Garner, TWJ, Perkins, MS, Govindarajulu, P, Seglie, D, Walker, S, Cunningham, AA, and Fisher, MC. (2006). The emerging amphibian pathogen *Batrachochytrium dendrobatidis* globally infects introduced populations of the North American bullfrog, *Rana catesbeiana*. Biol. Lett. 2:455–459.

Garner, TWJ, Walker, S, Bosch, J, Leech, S, Rowcliffe, LJ, Cunningham, AA, and Fisher, MC. (2009). Life history tradeoffs influence mortality associated with the amphibian pathogen *Batrachochytrium dendrobatidis*. Oikos 118:783–791.

Gatten, RE Jr and Hill, CJ (1984). Social influence on thermal selection by *Hyla crucifer*. J. Herpetol. 18: 87–88.

Gendron, AD, Marcogliese, DJ, Barbeau, S, Christin, M-S, Brousseau, P, Ruby, S, Cyr, D, and Fournier, M (2003). Exposure of leopard frogs to a pesticide mixture affects life history characteristics of the lungworm *Rhabdias ranae*. Oecologia 135:469–476.

Gervasi, SS and Foufopoulos, J (2008). Costs of plasticity: responses to desiccation decrease post-metamorphic immune function in a pond-breeding amphibian. Funct. Ecol. 22:100–108.

Gibson, BW, Poulter, L, Williams, DH, and Maggio, JE (1986). Novel peptide fragments originating from PGLa and the caerulein and xenopsin precursors from *Xenopus laevis* J. Biol. Chem. 261: 5341–5349.

Gilbertson, M-K, Haffner, GD, Drouillard, KG, Albert, A, and Dixon, B (2003). Immunosuppression in the northern leopard frog (*Rana pipiens*) induced by pesticide exposure. Environ. Toxicol. Chem. 22:101–110.

Giovannini, MG, Poulter, L, Gibson, BW, and Williams, DH (1987). Biosynthesis and degradation of peptides derived from *Xenopus laevis* prohormones. Biochem. J. 243:113–120.

Goniakowska-Witalinska, L and Kubiczek, U (1998). The structure of the skin of the tree frog (*Hyla arborea arborea* L.). Ann. Anat. 180:237–246.

Gorzula, S (1996). The trade in dendrobatid frogs from 1987 to 1993. Herp. Rev. 27:116–123.

Gossling, J, Loesche, WJ, and Nace, GW (1982a). Response of intestinal flora of laboratory-reared leopard frogs (*Rana pipiens*) to cold and fasting. App. Environ. Microbiol. 44:67–71.

Gossling, J, Loesche, WJ, and Nace, GW (1982b). Large intestine bacterial flora of nonhibernating and hibernating leopard frogs (*Rana pipiens*). App. Environ. Microbiol. 44:59–66.

Gravenor, I, Horton, TL, Ritchie, P, Flint, E, and Horton, JD (1995). Ontogeny and thymus-dependence of T cell surface antigens in *Xenopus*: flow cytometric studies on monoclonal antibody-stained thymus and spleen. Dev. Comp. Immunol. 19:507–523.

Gray, MJ, Miller, DL, and Hoverman, JT (2009). Ecology and pathology of amphibian Ranaviruses. Dis. Aquat. Organ. 87:243–266.

Green, DE, Converse, KA, and Schrader, AK (2002). Epizootiology of sixty-four amphibian morbidity and mortality events in the USA, 1996–2001. Ann. NY. Acad. Sci. 969:323–339.

Green, N and Cohen, N (1977). Effect of temperature on serum complement levels in the leopard frog, *Rana pipiens*. Dev. Comp. Immunol 1:59–64.

Green (Donelly), N and Cohen, N (1979). Phylogeny of immunocompetent cells: III. Mitogen response characteristics of lymphocyte subpopulations from normal and thymectomized frogs (*Xenopus laevis*). Cell Immunol 48:59–70.

Greer, AL, Berrill, M, and Wilson, PJ (2005). Five amphibian mortality events associated with ranavirus infection in south central Ontario, Canada. Dis. Aquat. Organ. 67:9–14.

Hadji-Azimi, I, Coosemans, V, and Canicatti, C (1987). Atlas of adult *Xenopus laevis laevis* hematology. Dev. Comp. Immunol. 11:807–874.

Hadji-Azimi, I, Coosemans, V, and Canicatti, C (1990). B-lymphocyte populations in *Xenopus laevis*. Dev. Comp. Immunol. 14:69–84.

Hadji-Azimi, I, Schwager, J, and Thiebaud, C (1982). B-lymphocyte differentiation in *Xenopus laevis* larvae. Dev. Biol. 90:253–258.

Haire, RN, Kitzen Haindfeld, MK, Turpen, JB, and Litman GW (2002). Structure and diversity of T-lymphocyte antigen receptors alpha and gamma in *Xenopus*. Immunogenetics 54:431–438.

Han, BA, Bradley, PW, and Blaustein, AR (2008). Ancient behaviors of larval amphibians in response to an emerging fungal pathogen, *Batrachochytrium dendrobatidis*. Behav. Ecol. Sociobiol. 63:241–250.

Hansen, JD and Zapata AG (1998). Lymphocyte development in fish and amphibians. Immunol. Rev. 166:199–220.

Harris, JA (1972). Seasonal variation in some hematological characteristics of *Rana pipiens*. Comp. Biochem. Physiol. A Comp. Physiol. 43:975–989.

Harris, RN, Brucker, RM, Walke, JB, Becker, MH, Schwantes, CR, Flaherty, DC, Lam, BA, Woodhams, DC, Briggs, CJ, Vredenburg, VT, and Minbiole, KPC (2009a). Skin microbes on frogs prevent morbidity and mortality caused by a lethal skin fungus. ISME J. 3:818–824.

Harris, RN, James, TY, Lauer, A, Simon, MA, and Patel, A (2006). Amphibian pathogen *Batrachochytrium dendrobatidis* is inhibited by the cutaneous bacteria of amphibian species. EcoHealth. 3:53–56.

Harris, RN, Lauer, A, Simon, MA, Banning, JL, and Alford, RA (2009b). Addition of antifungal skin bacteria to salamanders ameliorates the effects of chytridiomycosis. Dis. Aquat. Org. 83:11–16.

Harrison, F, Browning, LE, Vos, M, and Buckling, A (2006). Cooperation and virulence in acute *Pseudomonas aeruginosa infections.* BMC Biol. 4:21.

Hasday, JD, Fairchild, KD, and Shanholtz, C (2000). The role of fever in the infected host. Microbes Infect. 15:1891–1904.

Hayes, TB, Falso, P, Gallipeau, S, and Stice, M (2010). The cause of global amphibian declines: a developmental endocrinologist's perspective. J. Exp. Biol. 213:921–923.

Hecnar, SJ and M'Closkey, RT (1996). Regional dynamics and the status of amphibians. Ecology 77: 2091–2097.

Hecnar, SJ and M'Closkey, RT (1998). Species richness pattern of amphibians in southwestern Ontario ponds. J. Biogeography 25:763–772.

Hector, A and Bagchi, R (2007). Biodiversity and ecosystem multifunctionality. Nature. 448: 188–190.

Heikkila, JJ (2010). Heat shock protein gene expression and function in amphibian model systems. Comp. Biochem. Physiol. A. 156:19–33.

Heinrich, PC, Castell, JV, and Andus, T (1990). Interleukin-6 and the acute phase response. Biochem. J. 265:621–636.

Hildemann, WH and Haas, R (1959) Homotransplantation immunity and tolerance in the bullfrog. J. Immunol. 83:478–485.

Holmes, C and Balls, M (1978). In vitro studies on the control of myoepithelial cell contractions in the granular glands of *Xenopus laevis* skin. Gen. Comp. Endocrinol. 36:255–263.

Horton, JD (1969). Ontogeny of the immune response to skin allografts in relation to lymphoid organ development in the amphibian *Xenopus laevis* Daudin. J. Exp. Zool. 170:449–466.

Horton, JD and Manning, MJ (1972). Response to skin allografts in *Xenopus laevis* following thymectomy at early stages of lymphoid organ maturation. Transplantation 14:141–154.

Horton, TL, Minter, R, Stewart, R, Ritchie, P, Watson, MD, and Horton, JD (2000). *Xenopus* NK cells identified by novel monoclonal antibodies. Eur. J. Immunol. 30:604–613.

Hsu, E and Du Pasquier, L (1984a). Ontogeny of the immune system in *Xenopus*. I. Larval immune response. Differentiation 28:109–115.

Hsu, E and Du Pasquier, L (1984b). Ontogeny of the immune system in Xenopus. II. Antibody repertoire differences between larvae and adults. Differentiation 28:116–122.

Hsu, E and Du Pasquier, L (1992). Changes in the amphibian antibody repertoire are correlated with metamorphosis and not with age or size. Dev. Immunol. 2:1–6.

Hutchison, VH and Dupré, RK (1992). Thermoregulation. In: Feder, ME and Burggren, WW (eds). Environmental Physiology of the Amphibians. University of Chicago Press, Chicago, pp. 206–249.

Hutchison, VH and Hill, LG (1976). Thermal selection in the hellbender, *Cryptobranchus alleganiensis*, and the mudpuppy *Necturus maculosus*. Herpetologica 32:327–331.

Hutchison, VH and Spriestersbach, KK (1986). Diel and seasonal cycles of activity and behavioral thermoregulation in the salamander *Necturus maculosus*. Copeia 1986:612–618.

Ishii, A, Kawasaki, M, Matsumoto, M, Tochinai, S, and Seya, T (2007). Phylogenetic and expression analysis of amphibian *Xenopus* toll-like receptors. Immunogenetics 59:281–293.

Jaffe, RC (1981). Plasma concentrations of corticosterone during *Rana catesbeiana* tadpole metamorphosis. Gen. Comp. Endocrinol. 44:314–318.

James, TY, Letcher, PM, Longcore JE, Mozley-Standridge, SE, Porter, D, Powell, JJ, Griffith, GW, and Vilgalys, R. (2007). A molecular phylogeny of the flagellated fungi (Chytridiomycota) and description of anew phylum (Blastocladiomycota). Mycologia 98:860–871.

James, TY, Porter, D, Leander, CA, Vilaglys, R, and Longcore, JE (2000). Molecular phylogenetics of the chytridiomycota supports the utility of ultrastructural data in chytrid systematics. Can. J. Bot. 78: 336–350.

Jancovich, JK, Davidson, EW, Morado, JF, Jacobs, BL, and Collins, JP (1997). Isolation of a lethal virus from the endangered tiger salamander *Ambystoma tigrinum stebbinsi.* Dis. Aquat. Organ. 31:161–167.

Jancovich, JK, Davidson, EW, Seiler, A, Jacobs, BL, and Collins, JP (2001). Transmission of the *Ambystoma tigrinum* virus to alternative hosts. Dis. Aquat. Organ. 46: 159–163.

Jaudet, GJ and Hatey, JH (1984). Variations in aldosterone and corticosterone plasma levels during metamorphosis in *Xenopus laevis* tadpoles. Gen. Comp. Endocrinol. 56:59–65.

Jensen, LB and Koch, C (1991). An assay for complement factor B in species at different levels of evolution. Dev. Comp. Immunol. 15:173–179.

Jiang, Z, Schatzmayr, G, Mohnl, M, and Applegate, TJ (2010). Net effect of an acute phase response: partial alleviation with probiotic supplementation. Poult. Sci. 89:28–33.

Johnson, PT and Hartson, RB (2009). All hosts are not equal: explaining differential patterns of malformations in an amphibian community. J. Anim. Ecol. 78:191–201.

Johnson, PTJ and Thieltges, DW (2010). Diversity, decoys and the dilution effect: how ecological communities affect disease risk. J. Exp. Biol. 213:961–970.

Johnson, PTJ, Hartson, RB, Larson, DJ, and Sutherland, DR (2008). Diversity and disease: community structure drives parasite transmission and host fitness. Ecol. Lett. 10:1017–1026.

Jozkowicz, A and Plytycz, B (1998). Temperature but not season affects the transplantation immunity of anuran amphibians. J. Exp. Zool. 281:58–64.

Kato, Y, Salter-Cid, L, Flajnik, MF, Kasahara, M, Namikawa, C, Sasaki, M, and Nonaka, M. (1994). Isolation of the *Xenopus* complement factor B complementary DNA and linkage of the gene to the frog MHC. J. Immunol. 143:4536–4554.

Keesing, F, Holt, RD, and Ostfeld, RS (2006). Effects of species diversity on disease risk. Ecol. Lett. 9: 485–498.

Kelley, KW, Bluthé, R-M, Dantzer, R, Zhou, JH, Johnson, RW, Shen, WH, and Broussard, SR (2003). Cytokine-induced sickness behavior. Brain Behav. Immun. 17, Suppl 1:112–118.

Kiesecker, JM (2002). Synergism between trematode infection and pesticide exposure: A link to amphibian limb deformities in nature? Proc. Natl. Acad. Sci. USA 99:9900–9904.

Kiesecker, JM and Blaustein, AR (1999). Pathogen reverses competition between larval amphibians. Ecology 80:2442–2448.

Kiesecker, JM, Blaustein, AR, and Belden, LK (2001). Complex causes of amphibian population declines. Nature 410:681–684.

Kikuyama, S, Suzuki, MR, and Iwamuro, S (1986). Elevation of plasma aldosterone levels in tadpoles at metamorphic climax. Gen. Comp. Endocrinol. 63:186–190.

Kluger, MJ (1977). Fever in the frog *Hyla cinerea.* J. Therm. Biol. 2:79–81.

Kluger, MJ (1990). The febrile response. In: Morimoto, RI, Tissières, A, and Georgopoulos C (eds). Stress Proteins in Biology and Medicine. Cold Spring Harbor Laboratory Press, Cold Spring Harbor, NY, pp. 61–78.

Kluger, MJ (1991). Fever: role of pyrogens and cryogens. Physiol. Rev. 71:93–127.

Kluger, MJ (1992). Fever revisited. Pediatrics 90:846–850.

Kluger, MJ, Kozak, W, Corm, CA, Leon, LR, and Soszynski, D (1996). The adaptive value of fever. Infect. Dis. Clin. North Am. 10:1–20.

Kluger, MJ, Ringler, DH, and Anver, MR. (1975). Fever and survival. Science 188:166–168.

Knapp, RA, Boiano, DM, Vredenburg, VT (2007). Removal of nonnative fish results in population expansion of a declining amphibian (mountain yellow-legged frog, *Rana muscosa*). Biol. Conserv. 135:11–20.

Kriger, KM, Pereoglou, F, and Hero, J-M (2007). Latitudinal variation in the prevalence and intensity of chytrid (*Batrachochytrium dendrobatidis*) infection in eastern Australia. Conserv. Biol. 21:1280–1290.

Krug, EC, Horn, KV, Battista, J, and Nicoll, CS (1983). Corticosteroids in serum of *Rana catesbeiana* during development and metamorphosis. Gen. Comp. Endocrinol. 52:232–241.

Kupferberg, SJ, Marks, JC, and Power, ME (1994). Effects of variation in natural algal and detrital diets on larval anuran (*Hyla regilla*) life-history traits. Copeia 1994:446–457.

Kushner, I (1982). The phenomenon of the acute phase response. Ann. NY. Acad. Sci. 389:39–48.

Lacava, PT, Parker, J, Andreote, FD, Dini-Andreote, F, Ramirez, JL, and Miller, TA (2007). Analysis of the bacterial community in glassy-winged sharpshooter heads. Entomological Res. 37:261–266.

Lafferty, KD, Allesina, S, Arim, M, Briggs, CJ, De Leo, G, Dobson, AP, Dunne, JA, Johnson, PTJ, Kuris, AM, Marcogliese, DJ, Martinez, ND, Memmott, J, Marquet, PA, McLaughlin, JP, Mordecai, EA, Pascual, M, Poulin, R, and Thieltges, DW (2008). Parasites in food webs: the ultimate missing links. Ecol. Lett. 11: 533–546.

Lam, BA, Walke, JB, Vredenburg, VT, and Harris, RN (2010). Proportion of individuals with anti-*Batrachochytrium dendrobatidis* skin bacteria is associated with population persistence in the frog *Rana muscosa.* Biol. Cons. 143:529–531.

Langerveld, AJ, Celestine, R, Zaya, R, Mihalko, D, and Ide, CF (2009). Chronic exposure to high levels of atrazine alters expression of genes that regulate immune and growth-related functions in developing *Xenopus laevis* tadpoles. Environ. Res. 109:379–389.

Lauer, A, Simon, MA, Banning, JL, Andre, E, Duncan, K, and Harris, RN (2007). Common cutaneous bacteria from the eastern red-backed salamander can inhibit pathogenic fungi. Copeia. 2007:630–640.

Lauer, A, Simon, MA, Banning JL, Lam, BA, and Harris, RN (2008). Diversity of cutaneous bacteria with antifungal activity isolated from female four-toed salamanders. ISME J. 2:145–157.

Laurance, WF, McDonald, KR, and Speare, R (1996). Epidemic disease and catastrophic decline of Australian rain forest frogs. Cons. Biol. 10: 406–413.

Lawler, SP, Dritz, D, Strange, T, and Holyoak, M (1999). Effects of introduced mosquitofish and bullfrogs on the threatened California red-legged frog. Cons. Biol. 13:613–622.

Leboulenger, F, Delarue, D, Tono, MC, Jegou, S, Leroux, P, and Vaudry, H (1979). Seasonal study of the interrenal function of the European green frog, in vivo and in vitro. Gen. Comp. Endocrinol. 39: 388–396.

Lefcort, H and Eiger, SM (1993). Antipredatory behaviour of feverish tadpoles: implications for pathogen transmission. Behaviour 126:13–27.

Lefcort, H and Blaustein, AR (1995). Disease, predator avoidance, and vulnerability to predation in tadpoles. Oikos 74:469–474.

Lefévre, T and Thomas, F (2008). Behind the scene, something else is pulling the strings: Emphasizing parasitic manipulation in vector-borne diseases. Infect. Genet. Evol. 8:504–519.

Lefévre, T, Lebarbenchon, C, Gauthier-Clerc, M, Missé, D, Poulin, R, and Thomas, F (2009). The ecological significance of manipulative parasites. Trends Ecol. Evol. 24:41–48.

Leloup, J and Buscaglia, M (1977). La triiodothyronine, hormone de la métamorphose des Amphibians. CR Acad Sci. Paris 284D:2261–2263.

Licht, P and Brown, AG (1967). Behavioral thermoregulation and its role in the ecology of the red-bellied newt, *Taricha rivularis.* Ecology 48:598–611.

Lillywhite, HB (1971). Temperature selection by the bullfrog, *Rana catesbeiana.* Comp. Biochem. Physiol. A 40:213–227.

Lin, HH and Rowlands, DT Jr (1973). Thermal regulation of the immune response in South American toads (*Bufo marinus*). Immunology 24:129–133.

Lips, KR (1998). Decline of tropical montane amphibian fauna. Cons. Biol. 12:106–117.

Lips, KR (1999). Mass mortality and population declines of anurans at an upland site in western Panama. Cons. Biol. 13:117–125.

Lips, KR, Brem, F, Brenes, R, Reeve, JD, Alford, RA, Voyles, J, Carey, C, Livo, L, Pessier, AP, and Collins, J. (2006). Emerging infectious disease and the loss of biodiversity in a Neotropical amphibian community. Proc. Natl. Acad Sci. USA 103:3165–3170.

Lips, KR, Diffendorfer, J, Mendelson, JR III, and Sears, MW (2008). Riding the wave: Reconciling the roles of disease and climate change in amphibian declines. PLOS Biol. 6:441–454.

Liu, L and Liu, TY (1993). Isolation and characterization of C-reactive protein (CRP) cDNA and genomic DNA from *Xenopus laevis*. A species representing an intermediate stage in CRP evolution. J. Biol. Chem. 268:6809–6815.

Lochmiller, RL and Deerenberg, C (2000). Trade-offs in evolutionary immunology: just what is the cost of immunity? Oikos 88:87–98.

Longcore, JE, Pessier, AP, and Nichols, DK (1999). *Batrachochytrium dendrobatidis* gen. et sp. nov., a chytrid pathogenic to amphibians. Mycologia 91:219–227.

Lucas, EA and Reynolds, WA (1967). Temperature selection by amphibian larvae. Physiol. Zool. 40: 159–171.

Lyte, M and Ernst, S (1992). Catecholamine induced growth of gram-negative bacteria. Life Sci. 3:203–212.

Majji, S, LaPatra, S, Long, SM, Sample, R, Bryan, L, Sinning, A, and Chinchar, VG. (2006). *Rana catesbeiana* virus Z (RCV-Z): a novel pathogenic ranavirus. Dis. Aquat. Organ. 73:1–11.

Mangoni, ML, Miele, R, Renda, TG, Barra D, and Simmaco, M (2001). The synthesis of antimicrobial peptides in the skin of *Rana esculenta* is stimulated by microorganisms. FASEB J. 15:1431–1432.

Maniero, GD and Carey C (1997). Changes in selected aspects of immune function in the leopard frog, *Rana pipiens*, associated with exposure to cold. J. Comp. Physiol. B. 167:256–263.

Maniero, GD, Morales, H, Gantress, J, and Robert, J (2006). Generation of a long-lasting, protective, and neutralizing antibody response to the ranavirus FV3 by the frog *Xenopus*. Dev. Comp. Immunol. 30:649–657.

Manning, MJ and Horton, JD (1982). RES structure and function of the amphibian. In: Cohen, N and Sigel, MM (eds). The Reticuloendothelial System, Volume 3, Phylogeny and Ontogeny. Plenum Press, New York, pp. 423–459.

Marsh, DM and Beckman NG (2004). Effects of forest roads on the abundance and activity of terrestrial salamanders. Ecol. Appl. 14:1882–1891.

Martinez, IP, Alvarez, R, and Paz-Herraez, M (1996). Growth and metamorphosis of *Rana perezi* larvae in culture: effects of larval density. Aquaculture 142:163–170.

Massey, RC, Buckling, A, and ffrench-Constrant, R (2004). Interference competition and parasite virulence. Proc. R. Soc. Lond. B 271:785–788.

Matos, A, Kerkhof, L, and Garland, JL (2005). Effects of microbial community diversity on the survival of *Pseudomonas aeruginosa* in the wheat rhizosphere. Microb. Ecol. 49:257–264.

Matutte, B, Storey, KB, Knoop, FC, and Conlon, JM (2000). Induction of synthesis of an antimicrobial peptide in the skin of the freeze-tolerant frog, *Rana sylvatica,* in response to environmental stimuli. FEBS Lett. 483:135–138.

Mawaribuchi, S, Tamura, K, Okano, S, Takayama, S, Yaoita, Y, Shiba, T, Takamatsu, N, and Ito, M (2008). Tumor necrosis factor-alpha attenuates thyroid home-induced apoptosis in vascular endothelial cell line XLgoo established from *Xenopus* tadpole tails. Endocrinology 149:3379–3389.

Mazzoni, R, Cunningham, AA, Daszak, P, Apolo, A, Perdomo, E, and Speranza G (2003). Emerging pathogen of wild amphibians in frogs (*Rana catesbeiana*) farmed for international trade. Emerg. Infect. Dis. 9: 995–998.

Mazzoni, R, de Mesquita, AJ, Fleury, LFF, et al (2009). Mass mortality associated with a frog virus 3-like Ranavirus infection in farmed tadpoles *Rana catesbeiana* from Brazil. Dis. Aquat. Org. 86:181–191.

McCallum, H, Barlow, N, and Hone, J (2001). How should pathogen transmission be modeled? Trends Ecol. Evol. 16:295–300.

McCallum, ML and Trauth, SE (2007). Physiological tradeoffs between immunity and reproduction in the Northern cricket frog. Herpetologica 63:269–274.

McDonald, KR, Mendez, D, Muller, R, Freeman, AB, and Speare, R (2005). Decline in the prevalence of chytridiomycosis in upland frog populations in North Queensland, Australia. Pacific Conserv. Biol. 11:114–120.

Mendelson, JR III, Lips, KR, Gagliardo, RW, Rabb, GB, Collins, JP, Diffendorfer, JE, Daszak, P, Ibanez, R, Zippel, KC, Lawson, DP, Wright, KM, Stuart, SN, Gascon C, da Silva, HR, Burrowes, PA, Joglar, RL, La Marca, E, Lotters, S, du Preez, LH, Weldon, C, Hyatt, A, Rodriguez-Mahecha, JV, Hunt, S, Robertson, H, Lock, B, Raxworthy, CJ, Frost, DR, Lacy, RC, Alford, RA, Campbell, JA, Parra-Olea, G, Bolanos, F, Domingo, JJC, Halliday, T, Murphy, JB, Wake, MH, Coloma, LA, Kuzmin, SL, Price, MS, Howell, KM, Lau, M, Pethiyagoda, R, Boone, M, Lannoo, MJ, Blaustein, AR, Dobson, A, Griffiths, RA, Crump, ML, Wake, DB, and Brodie, ED Jr. (2006). Confronting amphibian declines and extinctions. Science 313:48.

Mescher, AL, Wolf, WL, Moseman, EA, Hartman, B, Harrison, C, Nguyen, E, and Neff, A (2007). Cells of cutaneous immunity in *Xenopus*: studies during larval development and limb regeneration. Dev. Comp. Immunol. 31:383–393.

Mills, JW and Prum, BE (1984). Morphology of the exocrine glands of the frog skin. Am. J. Anat. 171:91–106.

Miodonski, AJ, Bagij, J, Mika, J, and Plytycz, B (1996). Season-specific Thymic architecture in the frog. *Rana temporaria*: SEM studies. Dev. Comp. Immunol. 20:129–137.

Miyauchi, H, LaRochelle, FT Jr, Suzuki, M, Freeman, M, and Frieden, E (1977). Studies on thyroid hormones and their binding in bullfrog tadpole plasma during metamorphosis. Gen. Comp. Endocrinol. 33: 254–266.

Mokany, A and Shine, R (2003). Biological warfare in the garden pond: tadpoles suppress the growth of mosquito larvae. Ecol. Entomol. 28:102–108.

Mondou, PM and Kaltenbach, JC (1979). Thyroxine concentration in blood serum and peritoneal fluid of metamorphosing tadpoles and of adult frogs. Gen. Comp. Endocrinol. 39:343–349.

Moore, J (2002). Parasites and the Behaviour of Animals. Oxford: Oxford University Press.

Morales, HD, Abramowitz, L, Gertz, J, Sowa, J, Vogel, A, and Robert, J (2010). Innate immune responses and permissiveness to ranavirus infection of peritoneal leukocytes in the frog *Xenopus laevis*. J. Virol. 84:4912–4922.

Morales, H, Muharemagic, A, Gantress, J, Cohen, N, and Robert, J (2003). Bacterial stimulation upregulates the surface expression of the stress protein gp96 on B cells in the frog *Xenopus*. Cell Stress Chaparones 8:265–271.

Moss, AS, Reddy, NS, Dortag, IM, and San Franciso, MJ (2008). Chemotaxis of the amphibian pathogen *Batrachochytrium dendrobatidis* and its response to a variety of attractants. Mycologia 100:1–5.

Moticka, EJ, Brown, BA, and Cooper, EL (1973). Immunoglobulin synthesis in bullfrog larvae. J. Immunol. 110:855–861.

Mouritsen, KN and Poulin, R (2002). Parasitism, community structure and biodiversity in intertidal ecosystem. Parasitology 124 suppl:S101–117.

Murphy, PJ, St-Hilaire, S, Bruer, S, Corn, PS, and Peterson, CR (2009). Distribution and Pathogenicity of *Batrachochytrium dendrobatidis* in boreal toads from the Grand Teton area of western Wyoming. EcoHealth 6:109–120.

Myhre, K, Cabanac, M, and Myhre, G (1977). Fever and behavioural temperature regulation in the frog *Rana esculenta*. Acta Physiol. Scand. 101:219–229.

Neuwirth, M, Daly, JW, Myers, CW, and Tice LW (1979). Morphology of the granular secretory glands in skin of poison-dart frogs (*Dendrobatidae*). Tissue Cell 11:755–771.

Newman, RA (1994). Effects of changing density and food level on metamorphosis of a desert amphibian, *Scaphiopus couchii*. Ecology 75:1085–1096.

Nicolas, P and El Amri, C (2009). The dermaseptin superfamily: A gene-based combinatorial library of antimicrobial peptides. Biochim. Biophys. Acta 1788:1537–1550.

Nicolas, P and Mor, A (1995). Peptides as weapons against microorganisms in the chemical defense system of vertebrates. Annu. Rev. Microbiol. 49: 277–304.

Nichols, DK, Lamirande, EW, Pessier, AP, and Longcore, JE (2001). Experimental transmission of cutaneous chytridiomycosis in dendrobatid frogs. J. Wildl. Dis. 37:1–11.

Noble, GA and Noble, ER (1944). On the histology of frog skin glands. Trans. Am. Microsco. Soc. 63: 254–263.

Nonaka, M, Namikawa, C, Kat, Y, Sasaki, M, Salter-Cid, L, and Flajnik, MF (1997). Major histocompatibility complex gene mapping in the amphibian *Xenopus* implies a primordial organization. Proc. Natl. Acad. Sci. USA 94:5789–5791.

Obara, N, Kawahara, H, and Katagiri, C (1983). Response to skin grafts exchanged among siblings of larval and adult gynogenetic diploids in *Xenopus laevis*. Transplantation 36:91–95.

Ostfeld, RS and Keesing, F (2000). Biodiversity and disease risk: the case of Lyme disease. Cons. Biol. 14: 722–728.

Parris, MJ and Beaudoin, JG (2004). Chytridiomycosis impacts predator-prey interactions in larval amphibian communities. Oecologia 4:626–632.

Parris, MJ and Cornelius, TO (2004). Fungal pathogen causes competitive and developmental stress in larval amphibian communities. Ecology 85:3385–3395.

Parris, MJ, Davis, A, and Collins, JP (2004). Single-host pathogen effects on mortality and behavioral responses to predators in salamanders (Urodela: *Ambystomatidae*). Can. J. Zool. 82:1477–1483.

Pascolini, R, Daszak, P, Cunningham, AA, Tei, S, Vagnitti, D, Bucci, S, Fagotti, A, and Di Rosa, I (2003). Parasitism by *Dermocystidium ranae* in a population of *Rana esculenta* complex in central Italy and description of *Amphibiocystidium* n.g. Dis. Aquat. Organ. 56:65–74.

Pedersen, AB and Fenton, A (2007). Emphasizing the ecology in parasite community ecology. Trends Ecol. Evol. 22:133–139.

Pessier, AP, Nichols, DK, Longcore, JE, and Fuller, MS (1999). Cutaneous chytridiomycosis in poison dart frogs (*Dendrobates* spp.) and White's tree frogs (*Litoria caerulea*). J. Vet. Diagn. Invest. 11:194–199.

Petranka, JW (1995). Interference competition in tadpoles: Are multiple agents involved? Herp. J. 5:206–207.

Piotrowski, JS, Annis, LS, and Longcore, JE (2004). Physiology of *Batrachochytrium dendrobatidis,* a chytrid pathogen of amphibians. Mycologia 96:9–15.

Plytycz, B (1981). Ontogeny of transplantation immunity in the common frog, *Rana temporaria* L. Differentiation 20: 71–76.

Plytycz, B and Bigaj, J (1983). Seasonal cyclic changes in the thymus gland of the adult frog, *Rana temporaria*. Thymus 5:327–344.

Plytycz, B, Dulak, J, Bigaj, J, and Janeczko, K (1991). Annual cycle of mitotic activity in the thymus of the frog, *Rana temporaria*. Herpetopathologia 2:23–29.

Pounds, JA and Crump, ML (1994). Amphibian declines and climate disturbance: The case of the golden toad and the harlequin frog. Cons. Biol. 8:72–85.

Pounds, JA, Bustamante, MR, Coloma, LA, Consuegra, JA, Fogden, MPL, Foster, PN, La Marca, E, Masters, KL., Merino-Viteri, A, Puschendorf, R, Santiago, RR, Sanchez-Azofeifa, GA, still, CJ, and Young, BE (2006). Widespread amphibian extinctions from epidemic disease driven by global warming. Nature. 439:161–167.

Pounds, JA, Fogden, MPL, and Campbell, JH (1999). Biological response to climate change on a tropical mountain. Nature 398:611–615.

Poynton, SL and Whitaker, BR (1994). Protozoa in poison dart frogs (*Dendrobatidae*): Clinical assessment and identification. J. Zoo. Wildl. Med. 25:29–39.

Pross, SH and Rowlands DT Jr (1976). Humoral immune responses in *Rana catesbiana* frogs and tadpoles. J. Exp. Zool. 197:97–104.

Rachowicz, LJ and Briggs, CJ (2007). Quantifying the disease transmission function: effects of density on *Batrachochytrium dendrobatidis* transmission in the mountain yellow-legged frog *Rana muscosa*. J. Anim. Ecol. 76:711–721.

Rachowicz, LJ and Vredenburg, VT (2004). Transmission of *Batrachochytrium dendrobatidis* within and between amphibian life stages. Dis. Aquat. Organ. 61:75–83.

Rachowicz LJ, Knapp, RA, Morgan, JA, Stice, MJ, Vredenburg, VT, Parker, JM and Briggs, CJ (2006). Emerging infectious disease as a proximate cause of amphibian mass mortality. Ecology 87:1671–1683.

Raffel, TR, Rohr, JR, Kiesecher, JM, and Hudson, PJ (2006). Negative effects of changing temperature on amphibian immunity under field conditions. Funct. Ecol. 20:819–828.

Ramsey, JP, Reinert, LK, Harper, LK, Woodhams, DC, and Rollins-Smith, L (2010). Innate and adaptive immune defenses against a fungus linked to global amphibian declines in the South African clawed frog, *Xenopus laevis*. Infect. Immun. 78:3981–3992.

Regard, E, Taurog, A, and Nakashima, T (1978). Plasma thyroxine and thriiodothyronine levels in spontaneously metamorphosing *Rana catesbeiana* tadpoles and in adult anuran amphibia. Endocrinology 102:674–684.

Reid, G, Howard, J, and Gan, BS (2001). Can bacterial interference prevent infection? Trends Microbiol. 9:424–428.

Relyea, RA (2004). Synergistic impacts of malathion and predatory stress on six species of North American tadpoles. Environ. Toxicol. Chem. 23:1080–1084.

Relyea, RA (2005). The lethal impacts of Roundup and predatory stress on six species of North American tadpoles. Arch. Environ. Contam. Toxicol. 48:351–357.

Relyea, RA and Mills, N (2001). Predator-induced stress makes the pesticide carbaryl more deadly to gray treefrog tadpoles (*Hyla versicolor*). Proc. Natl. Acad. Sci. USA 98:2491–2496.

Relyea, R, Schoeppner, NM, and Hoverman, JT (2005). Pesticides and amphibians: The importance of community context. Ecol. Appl. 15:1125–1134.

Reynolds, WW and Casterlin, ME (1979). Behavioral thermoregulation and the "final preferendum" paradigm. Integr. Comp. Biol. 19:211–224.

Richards, CM (1958). The inhibition of growth in crowded *Rana pipiens* tadpoles. Physiol. Zool. 31:138–151.

Richards-Zawacki, CL (2010). Thermoregulatory behavior affects prevalence of chytrid fungal infection in a wild population of Panamanian golden frogs. Proc. R. Soc. B 277:519–528.

Rinaldi, AC (2002). Antimicrobial peptides from amphibian skin: an expanding scenario. Curr. Opin. Chem. Biol. 6:799–804.

Roach, JC, Glusman, G, Rowen, L, Kaur, A, Purcell, MK, Smith, KD, Hood, LE, and Aderem, A (2005). The evolution of vertebrate toll-like receptors. Proc. Natl. Acad. Sci. USA 102:9577–9582.

Robert, J (2003). Evolution of heat shock protein and immunity. Dev. Comp. Immunol. 27:449–464.

Robert, J and Ohta, Y (2009). Comparative and developmental study of the immune system in *Xenopus*. Dev. Dyn. 238:1249–1270.

Robert, J, Guiet, C, and Du Pasquier, L (1994). Lymphoid tumors of *Xenopus laevis* with different capacities for growth in larvae and adults. Dev. Immunol 3:297–307.

Robert, J, Guiet, C, and Du Pasquier, L (1995). Ontogeny of the alloimmune response against a transplanted tumor in *Xenopus laevis*. Differentiation 59:135–144.

Robert, J., Morales, H, Buck, W, Cohen, N, Marr, S, and Gantress, J (2005). Adaptive immunity and histopathology in frog virus 3-infected *Xenopus*. Virology 332:667–675.

Robinson, JV and Valentine, WD (1979). The concepts of elasticity, invulnerability and invadability. J. Theor. Biol. 81:91–104.

Rohr, JR and McCoy, KA (2010). A qualitative meta-analysis reveals consistent effects of atrazine on freshwater fish and amphibians. Environ. Health Perspect. 118:20–32.

Rohr, JR and Raffel, TR (2010). Linking global climate and temperature variability to widespread amphibian declines putatively caused by disease. Proc. Natl. Acad. Sci. USA 107:8269–8274.

Rohr, JR, Raffel, TR, Romansic, JM, McCallum, H, and Hudson, PJ (2008). Evaluating the links between climate, disease spread, and amphibian declines. Proc. Natl. Acad. Sci. USA 105:17436–17441.

Rohr, JR, Schotthoefer, AM, Raffel, TR, Carrick, HJ, Halstead, N, Hoverman, JT, Johnson, CM, Johnson, LB, Lieske, C, Piwoni, MD, Schoff, PK, and Beasley, V. (2008). Agrochemicals increase trematode infections in a declining amphibian species. Nature. 455:a1235–1239.

Rollins-Smith, LA (1998). Metamorphosis and the amphibian immune system. Immunol. Rev. 166: 221–230.

Rollins-Smith, LA (2001). Neuroendocrine-immune system interactions in amphibians: Implications for understanding global amphibian declines. Immunologic Res. 23(2&3):273–280.

Rollins-Smith, LA (2009). The role of amphibian antimicrobial peptides in protection of amphibians from pathogens linked to global amphibian declines. Biochim. Biophys. Acta. 1788:1593–1599.

Rollins-Smith, LA and Blair, P (1990). Expression of class II major histocompatibility complex antigens on adult T cells in *Xenopus* is metamorphosis-dependent. Dev. Immunol 1:97–104.

Rollins-Smith, LA and Blair, P (1993). The effects of corticosteroid hormones and thyroid hormones on lymphocyte viability and proliferation during development and metamorphosis of *Xenopus laevis*. Differentiation 54:155–160.

Rollins-Smith, LA and Conlon, JM (2005). Antimicrobial peptide defenses against chytridiomycosis, an emerging infectious disease of amphibian populations. Dev. Comp. Immunol. 29: 589–598.

Rollins-Smith, LA and Smits, JEG (2005). Amphibian models and approaches to immunotoxicology. In: Tryphonas, H, Fournier, M, Blakley, B, Brousseau, P, and Smits, JEG (eds.). Investigative Immunotoxicology. CRC Press, Boca Raton, FL, pp. 77–90.

Rollins-Smith, LA, Barker, KS, and Davis, AT (1997b). Involvement of glucocorticoids in the reorganization of the amphibian immune system at metamorphosis. Dev. Immunol. 5:145–152.

Rollins-Smith, LA, Blair, PJ, and Davis, AT (1992). Thymus ontogeny in frogs: T cell renewal at metamorphosis. Dev. Immunol. 2:207–213.

Rollins-Smith, LA, Carey, C, Longcore, J, Doersam, JK, Boutte, A, Bruzgal, JE, and Conlon, JM (2002b). Activity of antimicrobial skin peptides from ranid frogs against *Batrachochytrium dendrobatidis*, the chytrid fungus associated with global amphibian declines. Dev. Comp. Immunol. 26:471–479.

Rollins-Smith, LA, Doersam, JK, Longcore, JE, Taylor SK, Shamblin, JC, Carey, C, and Zasloff, MA (2002a). Antimicrobial peptide defenses against pathogens associated with global amphibian declines. Dev. Comp. Immunol. 26:63–72.

Rollins-Smith, LA, Flajnik, MF, Blair, PJ, Davis, AT, and Green, WF (1997a). Involvement of thyroid hormones in the expression of MHC class I antigens during ontogeny in *Xenopus*. Dev. Immunol. 5: 133–144.

Rollins-Smith, LA, Needham DAP, Davis, AT, and Blair, PJ (1996). Late thymectomy in *Xenopus* tadpoles reveals a population of T cells that persists through metamorphosis. Dev. Comp. Immunol 20:165–174.

Rollins-Smith, LA, Parsons, SCV, and Cohen, N (1984). During frog ontogeny, PHA and con A responsiveness of splenocytes precedes that of thymocytes. Immunology 52:491–500.

Rollins-Smith, LA, Parsons, SCV, and Cohen, N (1988). Effects of thyroxine-driven precocious metamorphosis on maturation of adult-type allograft rejection responses in early thyroidectomized frogs. Differentiation 37:180–185.

Rollins-Smith, LA, Parsons, SCV, and Cohen, N (1990). Immune responses of intact and embryonically enucleated frogs to self-lens antigens. J. Immunol. 145:3262–3267.

Rollins-Smith, LA, Rice, CD, and Grasman, KA (2007). Immunotoxicology in amphibians, fish, and birds. In: Luebke, RW, House, RV, and Kimber, I (eds.). Immunotoxicology and Immunopharmacology, 3rd ed. CRC Press Taylor and Francis, Boca Raton, FL, pp. 385–402.

Rosenblum EB, Stajich JE, Maddox N, and Eisen MB (2008). Global expression profiles for life stages of the deadly amphibian pathogen *Batrachochytrium dendrobatidis*. Proc. Natl. Acad. Sci. USA 105: 107034–107039.

Roux, KH and Volpe, EP. (1975). Evidence for a major histocompatibility complex in the leopard frog. Immunogenetics 2:577–589.

Rowley, JJL and Alford, RA (2007). Behavior of Australian rainforest stream frogs may affect transmission of chytridiomycosis. Dis. Aquat. Organ. 77:1–9.

Ruthig, G (2008). The influence of temperature and spatial distribution on the susceptibility of southern leopard frog eggs to disease. Oecologia 156:895–903.

Sadd, BM and Schmid-Hempel, P (2009). Principles of ecological immunology. Evol. Appl. 2:113–121.

Salter-Cid, L, Nonaka, M, and Flajnik, MF (1998). Expression of MHC class Ia and class Ib during ontogeny: high expression in epithelia and coregulation of class Ia and *lmp7* genes. J. Immunol. 160: 2853–2861.

Schadich, E (2008). Skin peptide defences of African clawed frogs (*Xenopus laevis*) and New Zealand Litoria frogs against bacterial dermatosepticemia. Ph.D. Thesis. University of Canterbury, Christchurch, New Zealand.

Schadich, E and Cole, AL (2010). Pathogenicity of *Aeromonas hydrophila, Klebsiella pneumoniae,* and *Proteus mirabilis* to brown tree frogs (*Litoria ewingii*). Comp. Med. 60:114–117.

Schlaepfer, MA, Hoover, C, and Dodd, CK Jr (2005). Challenges in evaluating the impact of the trade in amphibians and reptiles on wild populations. BioScience 55: 256–264.

Schloegel, LM, Daszak, R, Cunningham, AA, Speare, R, and Hill, B (2010). Two amphibian diseases, chytridiomycosis and ranaviral disease, are now globally notifiable to the World Organization for Animal Health (OIE): an assessment. Dis. Aquat. Organ. Jan. 13, 2010 [Epub ahead of print].

Schloegel, LM, Picco, AM, Kilpatrick, AM, Davies, AJ, Hyatt, AD, and Daszak, P (2009). Magnitude of US trade in amphibians and presence of *Batrachochytrium dendrobatidis* and ranavirus infestation in imported North American bullfrogs (*Rana catesbeiana*). Biol. Conserv. 142:1420–1426.

Schmidt-Hempel, P (2008). Parasite immune evasion: a momentous molecular war. Trends Ecol. Evol. 23: 318–326.

Schumacher, U, Adam, W, Hauser, F, Probst, JC, and Hoffmann, W (1994). Molecular anatomy of a skin gland: Histochemical and biochemical investigations on the mucous glands of *Xenopus laevis*. J. Histochem. Cytochem. 42:57–65.

Schwager, J, Bürckert, N, Courtet, M, and Du Pasquier, L (1991b). The ontogeny of diversification of the immunoglobulin heavy chain locus in *Xenopus*. EMBO J. 10: 2461–2470.

Schwager, J, Bürckert, N, Schwager, M, and Wilson, M (1991a). Evolution of immunoglobulin light chain gene: analysis of *Xenopus* IgL isotypes and their contribution to antibody diversity. EMBO J. 10: 505–511.

Schwager, J, Mikoryak, CA, and Steiner, LA (1988). Amino acid sequence of heavy change from *Xenopus laevis* IgM deduced from cDNA sequence: implications for evolution of immunoglobulin domains. Proc. Natl. Acad. Sci. USA 85:2245–2249.

Sheafor, BA, Davidson, EW, Parr, L, and Rollins-Smith, L (2008). Antimicrobial peptide defenses in the salamander, *Ambystoma tigrinum*, against emerging amphibian pathogens. J. Wildl. Dis. 44:226–236.

Sheldon, BC and Verhulst, S (1996). Ecological immunology: costly parasite defences and trade-offs in evolutionary ecology. Trends Ecol. Evol. 11: 317–321.

Sherman, E (2008). Thermal biology of newts (*Notophthalmus viridescens*) chronically infected with a naturally occurring pathogen. J. Therm. Biol. 33:27–31.

Sherman, PJ, Jackway, RJ, Nicholson, E, Musgrave, KF, Boonethung, P, and Bowie, JH (2009). Activities of seasonably variable caerulein and rothein skin peptides form the tree frogs *Litoria splendida* and *Litoria rothii*. Toxicon 54:828–835.

Simmaco, M, Mignogna, G, and Barra, D (1998). Antimicrobial peptides from amphibian skin: What do they tell us? Biopolymers (Peptide Science) 47:435–450.

Simnett, JF (1964). Histocompatibility in the platanna, *Xenopus laevis* (Daudin) following nuclear transplantation. Exp. Cell Res. 33:232–239.

Singh, V and Aballay, A (2006). Heat-shock transcription factor (HSF)-1 pathway required for *Caenorhabditis elegans* immunity. Proc. Natl. Acad. Sci. USA 103:13092–13097.

Sjoberg, E and Flock, A (1976). Innervation of skin glands in the frog. Cell Tiss. Res. 172:81–91.

Skelly, DK and Werner, EE (1990). Behavioral and life historical responses of larval American toads to an odonate predator. Ecology 71:2313–2322.

Skerratt, LF, Berger, L, Speare, R, Cashins, S, McDonald, KR, Phillott, AD, Hines, HB, and Kenyon, N (2007). Spread of chytridiomycosis has caused the rapid global decline and extinction of frogs. EcoHealth 4: 125–134.

Spanggaard, B, Huber, I, Nielsen, J, Sick, EB, Pipper, CB, Martinussen, T, Slierendrecht, WJ, and Gram, L (2001). The probiotic potential against vibriosis of the indigenous microflora of rainbow trout. Environ. Microbiol. 3:755–765.

Sparling, DW, Fellers, GM, and McConnell, LL (2001). Pesticides and amphibian population declines in California, USA. Environ. Toxicol. Chem. 20:1591–1595.

Stajich, JE, Berbee, ML, Blackwell, M, Hibbett, DS, James, TY, Spatafora, JW, and Taylor, JW. (2009). The fungi. Curr. Biol. 19:R840–R845.

Steinwascher, K (1979). Competitive interactions among tadpoles: responses to resource level. Ecology 60:1172–1183.

Stuart, SN, Chanson, JS, Cox, NA, Young, BE, Rodrigues, AS, Fischman, DL, and Waller, RW (2004). Status and trends of amphibian declines and extinctions worldwide. Science 306:1783–1786.

Suzuki, S and Suzuki, M (1981). Changes in thyroidal and plasma iodine compounds during and after metamorphosis of the bullfrog, *Rana catesbeiana*. Gen. Comp. Endocrinol. 45:74–81.

Taylor, SK (2001). Mycotic diseases. In: Wright, K and Whitaker, B (eds.). Amphibian Medicine and Captive Husbandry. Krieger Publications, Melbourne, FL, pp. 181–191.

Taylor, SK, Williams, ES, and Mills, KW (1999). Effects of malathion on disease susceptibility in Woodhouse's toads. J. Wildl. Dis. 35:536–541.

Taylor, SK, Green, DE, Wright, KM, and Whitaker, BR (2001). Bacterial diseases. In: Wright, K and Whitaker, B (eds.). Amphibian Medicine and Captive Husbandry. Krieger Publications, Melbourne, FL, pp. 159–179.

Temple, HJ and Cox, N (2009). European red list of amphibians. Office for Official Publications of the European Communities, Luxembourg, pp. 1–32.

Tennessen, JA, Woodhams, DC, Chaurand, P, Reinert, LK, Billheimer, D, Shyr, Y, Caprioli, RM, Blouin, MS, and Rollins-Smith, LA (2009). Variations in the expressed antimicrobial peptide repertoire of northern leopard frog (*Rana pipiens*) populations suggest intraspecies differences in resistance to pathogens. Dev. Comp. Immunol. 33:1247–1257.

Tilman, D, Lehman, CL, and Thomson, KT (1997). Plant diversity and ecosystem productivity: Theoretical considerations. Proc. Natl. Acad. Sci. USA. 94:1857–1861.

Tilman, D, Reich, PB, and Knops, JMH (2006). Biodiversity and ecosystem stability in a decade-long grassland experiment. Nature 441:629–632.

Tobler, U and Schmidt, BR (2010). Within-and among-population variation in chytridiomycosis-induced mortality in the toad *Alytes obstetricans*. PLoS One 5(6):e10927.

Tochinai, S and Katagiri, C (1975). Complete abrogation of immune response to skin allografts and rabbit erythrocytes in the early thymectomized *Xenopus*. Dev. Growth Diff. 17:383–394.

Toledo, RC and Jared C (1995). Cutaneous granular glands and amphibian venoms. Comp. Biochem. Physiol. 111A:1–29.

Turner, RJ and Manning, MJ (1974). Thymic dependence of amphibian antibody responses. Eur. J. Immunol. 4:343–346.

Turpen, JB and Smith, PB (1989). Precursor immigration and thymocyte succession during larval development and metamorphosis in *Xenopus*. J. Immunol. 142:41–47.

Venesky, MD, Parris, MJ, and Storfer, A (2009). Impacts of *Batrachochytrium dendrobatidis* infection on tadpole foraging performance. EcoHealth 6:565–575.

Venesky, MD, Wassersug, RJ, and Parris, MJ (2010). Fungal pathogen changes feeding kinematics of larval anurans. J. Parasitol. 96:522–557.

Vieites, DR, Wollenberg, KC, Andreone, F, Kohler, J, Glaw, F, and Vences, M (2009). Vast underestimation of Madagascar's biodiversity evidence by an integrative amphibian inventory. Proc. Natl. Acad. Sci. USA 106:8267–8272.

Vine, NG, Leukes, WD, and Kaiser, H (2006). Probiotics in marine larviculture. FEMS Microbiol. Rev. 30:404–427.

Voyles, J, Berger, L, Young, S, Speare, R, Webb, R, Warner, J, Rudd, D, Campbell, R, and Skerratt, LF (2007). Electrolyte depletion and osmotic imbalance in amphibians with chytridiomycosis. Dis. Aquat. Org. 77:113–118.

Voyles, J, Young, S, Berger, L, Campbell, C, Voyles, WF, Dinudom, A, Cook, D, Webb, R, Alford, RA, Skerratt, LF, and Speare, R (2009). Pathogenesis of chytridiomycosis, a cause of catastrophic amphibian declines. Science 326:582–585.

Vredenburg, VT (2004). Reversing introduced species effects: Experimental removal of introduced fish leads to rapid recovery of a declining frog. Proc. Natl. Acad. Sci. USA 101:7646–7650.

Wabnitz, PA, Bowie, JH, Tyler, MH, Wallace, JC, and Smith, BP (2000). Differences in the skin peptides of the male and female Australian tree frog *Litoria splendida*—the discovery of the aquatic male sex pheromone splendipherin together with Phe8 caerulein and a new antibiotic peptide caerin 1.10. Eur. J. Biochem. 267: 269–275.

Wake, DB (2007). Climate change implicated in amphibian and lizard declines. Proc. Natl. Acad. Sci. USA 104:8201–8202.

Wake, DB and Vredenburg, VT (2008). Are we in the midst of the sixth mass extinction? A view of the world of amphibians. Proc. Natl. Acad. Sci. USA 105:11466–11473.

Walker, SF, Bosch, J, Gomez, V, Garner, TWJ, Cunningham, AA, Schmeller, DS, Ninyerola, M, Henk, DA, Ginestet, C, Authur, C-P, and Fisher, MC (2010). Factors driving pathogenicity vs. prevalence of amphibian panzootic chytridiomycosis in Iberia. Ecol. Lett. 13:372–382.

Walker, TN and Hughes, WHO (2009). Adaptive social immunity in leaf-cutting ants. Biol. Lett. 5:446–448.

Wang, IJ, Crawford, AJ, and Bermingham, E (2008). Phylogeography of the pygmy rain frog (*Pristimantis ridens*) across the lowland wet forests of isthmian Central America. Mol. Phylogenet. Evol. 47:992–1004.

Weight, LA, Crawford, AJ, Rand, AS, and Ryan, MJ (2005). Biogeography of the tungara frog, *Phsalaemus pustulosus*: a molecular perspective. Mol. Ecol. 14:3858–3876.

Weng, SP, He, JG, Wang, XH, Lu, L, Deng, M, and Chan, S-M (2002). Outbreaks of an iridovirus disease in cultured tiger frog, *Rana tigrina rugulosa*, in southern China. J. Fish Dis. 25:423–427.

Whitfield, SM, Bell, KE, Philippi, T, Sasa, M, Bolanos, F, Chaves, G, Savage, J, and Donnelly, MA (2007). Amphibian and reptile declines over 35 years at La Selva, Costa Rica. Proc. Natl. Acad. Sci. USA 104:8352–8356.

Williams, T, Barbosa-Solomieu, V and Chinchar, VG (2005). A decade of advances in iridovirus research. Adv. Virus Res. 65:173–248.

Wilson, M (2005) Microbial Inhabitants of Humans: Their Ecology and Role in Health and Disease. Cambridge University Press, Cambridge, UK.

Wilson-Rich, N, Spivak, M, Fefferman, NH, and Starks, PT (2009). Genetic, individual, and group facilitation of disease resistance in insect societies. Annu. Rev. Entomol. 54:405–423.

Woodhams, DC (2003). The ecology of chytridiomycosis, an emerging infectious disease of Australian rainforest frogs. Ph.D. thesis. James Cook University, Townsville, Australia.

Woodhams, DC and Alford, RA (2005). Ecology of chytridiomycosis in rainforest stream frog assemblages of tropical Queensland. Conserv. Biol. 19:1449–1459.

Woodhams, DC, Alford, RA, and Marantelli, G (2003). Emerging disease cured by elevated body temperature. Dis. Aquat. Organ. 55:65–67.

Woodhams, DC, Alford, RA, Briggs, CJ, Johnson, M, and Rollins-Smith, LA (2008). Life-history trade-offs influence disease in changing climates: strategies of an amphibian pathogen. *Ecology.* 89:1627–1639.

Woodhams, DC, Ardipradja, K, Alford, RA, Marantelli, G, Reinert, LK, and Rollins-Smith, LA (2007a). Resistance to chytridiomycosis varies by amphibian species and is correlated with skin peptide defenses. Anim. Conserv. 10:409–417.

Woodhams, DC, Kenyon, N, Bell, S, Alford, RA, Chen, S, Billheimer, D, Shyr, Y, and Rollins-Smith, LA (2010). Adaptations of skin peptide defences and possible response to the amphibian chytrid fungus in populations of Australian green-eyed treefrogs, *Litoria genimaculata.* Diversity Distrib. 16:703–712.

Woodhams, DC, Rollins-Smith, LA, Alford, RA, Simon, MA, and Harris, RN (2007c). Innate immune defenses of amphibian skin: antimicrobial peptides and more. Anim. Conserv. 10:425–428.

Woodhams, DC, Rollins-Smith, LA, Briggs, CJ, Vredenburg, VT, Simon, MA, Billheimer, D, Shakhtour, B, Shyr, Y, and Harris, RN (2007b). Symbiotic bacteria contribute to innate immune defenses of the threatened mountain yellow-legged frog, *Rana muscosa.* Biol. Cons. 138:390–398.

Woodhams, DC, Rollins-Smith, LA, Carey, C, Reinert, L, Tyler, MJ, and Alford, R (2006a). Population trends associated with antimicrobial peptide defenses against chytridiomycosis in Australian frogs. Oecologia 46:531–540.

Woodhams, DC, Voyles, J, Lips, KR, Carey, C, and Rollins-Smith, LA (2006b). Predicted disease susceptibility in a Panamanian amphibian assemblage based on skin peptide defenses. J. Wildl. Dis. 42:207–218.

Yang, C, Crowley, DE, and Menge, JA (2001). 16S rDNA fingerprinting of rhizosphere bacterial communities associated with healthy and Phytophthora-infected avocado roots. FEMS Microbiol Ecol. 35:129–136.

Yeaman, MR and Yount, NY (2003). Mechanisms of antimicrobial peptide action and resistance. Pharmacol. Rev. 55:27–55.

Zera, AJ and Harshman, LG (2001). The physiology of life history trade-offs in animals. Annu. Rev. Ecol. Syst. 32:95–126.

Zasloff, M (2002). Antimicrobial peptides of multicellular organisms. Nature 415:389–395.

Zettergren, LD (1982). Ontogeny and distribution of cells in B lineage in the American leopard frog, *Rana pipiens.* Dev. Comp. Immunol. 6:311–320.

Zhang, J, Wehzheng, Z, and Yan, Q (2008). Non-specific immune response of bullfrog *Rana catesbeiana* to intraperitoneal injection of bacterium *Aeromonas hydrophila.* Chin. J. Oceanol. Limnol. 26:248–255.

Zhang, Q-Y, Xiao, F, Li, Z-Q, Gui, J-F, Mai, J, and Chinchar, VG (2001). Characterization of an iridovirus from the cultured pig frog *Rana grylio* with lethal syndrome. Dis. Aquat. Organ 48:27–36.

Zhao, Y, Jin, Y, Lee, WH, and Zhang, Y (2006). Purification of lysozyme from skin secretions of *Bufo andrewsi.* Comp. Biochem Physiol C Toxicol Pharmacol. 142:46–52.

Zilber-Rosenberg, I and Rosenberg, E (2008). Role of microorganisms in the evolution of animals and plants: the hologenome theory of evolution. FEMS Microbiol. Rev. 32:723–735.

Zou, J, Bird, S, Minter, R, Horton, J, Cunningham, C, and Secombes, CJ (2000). Molecular cloning of the gene for interleukin-1β from *Xenopus laevis* and analysis of expression *in vivo* and *in vitro*. Immunogenetics 51:332–338.

Zupanovic, Z, Musso, C, Lopez, C, Louriero, CL, Hyatt, AD, Hengstberger, S, and Robinson, AJ (1998). Isolation and characterization of iridoviruses from the giant toad *Bufo marinus* in Venezuela. Dis Aquat. Organ. 33:1–9.

CHAPTER 5

IMMUNITY IN PRIMATES WITHIN A PSYCHOBIOLOGICAL PERSPECTIVE

Christopher L. Coe

Primates, like other animals, are critically dependent upon their immune systems for an effective defense against pathogens and the maintenance of health. This statement is certainly true in natural settings where most primate species have to coexist with enteric pathogens and ectopic parasites, control bacterial infections of wounds, and occasionally contend with a traumatic, life-threatening injury. Immune competence continues to be of importance in the more hygienic conditions of the laboratory or zoo, where there may still be some contagion of infectious illness or a periodic bite wound from an aggressive encounter with another animal. As we will discuss in this chapter, the immune system also carries out a number of other functions essential to successful reproduction and infant development, which makes this topic germane to almost everyone, regardless of whether interested in pathology or the normal physiology of the healthy individual.

The immune system typically does such an effective job that it is largely unnoticed by the casual observer. Adult animals in their prime just seem to be healthy and rarely show overt signs of infectious disease. The myriad cells and functions that account for their robust appearance are not readily evident until one begins to examine the underlying cellular processes that ensure this immune competence and vigor. The absence of pathology is determined not only by the host's ability to prevent and control infection; it also requires that the immune system be continuously vigilant in its surveillance of the internal milieu for abnormal cells, including for errant lymphocytes that might react incorrectly against healthy tissue to initiate autoimmune reactions. In older animals, specific immune cells, such as the cytolytic T lymphocytes and natural killer cells also monitor for neoplastic malignancies, which must be eliminated when found. The surveillance and clearance functions of the immune system are often overlooked, but the day-to-day importance of these activities becomes very evident if an individual is immune-suppressed for a sustained period.

Studies on primates have contributed to our current awareness that immune responses can be significantly affected by many extrinsic and intrinsic factors that are not specifically a part of the immune system. Immunity does not just function in an autonomous manner, targeting pathogens only in a reactive way. The cellular reactions required for host defense can be performed over and over again, they are malleable and dynamically responsive to many features of the environment. For example, immune responses are influenced by numerous noninfectious factors, including diet, day/night biorhythms, climatic changes across seasons, and sleep duration. In addition, changes in hormone levels can markedly affect the number and types of cells in circulation, and their ability to proliferate or synthesize antibody. The fact that an animal's or a person's ability to mount an effective immune response can shift across different conditions makes the topic of immune modulation of special relevance to the new field of ecoimmunology. In addition, this chapter reviews some of the prenatal and postnatal factors that can shape the development of immune responses in infant primates (Coe and Lubach, 2008). The maturation of immunity should be viewed as a "learning system" responsive to many aspects of the early rearing environment, including antigen exposure and maternal factors. Stressful events can further influence immune processes at any point in the life span, but are of particular significance in the young and old primate (Coe and Laudenslager, 2007).

An important corollary of taking this type of developmental perspective is that one quickly realizes that the beginning and end of the life span are periods of heightened vulnerability to disease. Immune responses are not yet fully mature and competent in the young infant. Moreover, the immature animal frequently encounters a pathogen for the first time without the benefits of immunological memory and the more robust secondary type of antibody response. Ultimately, old animals and elderly people eventually undergo a progressive deterioration in a number of immune processes, which makes them more frail and vulnerable to infection. Beyond this weakening of immune control over infection, which in natural settings is often the cause of ultimate mortality, the process of immune senescence contributes to an overall physiological dysregulation in the aged individual. In humans, it is manifest clinically by an increased prevalence of inflammatory disorders, autoimmune conditions, and the initiation of cancerous growths (Bruunsgaard et al., 2001). In older monkeys, immune competence starts to decline when they reach 20 years of age, and they are likely to die within the next few years. Their longevity can be extended for several more years in modern laboratory facilities and zoos by veterinary care and the prevention of infectious disease behind protective barriers. But then one sees an increased incidence of cancer and other age-related ailments, similar to the ones found in elderly humans when our life span is extended (Uno, 1997).

A FOCUS ON PRIMATES ACROSS THE LIFE SPAN

Our review starts at the very beginning. First, we provide a brief description of some immune processes involved in implantation, reproductive success, and the maintenance of pregnancy. As an exemplary process of importance to primates, one immune-related aspect of the communication between the mother and fetal primate is discussed: the placental transfer of maternal antibody. Because of the significance for the care of primates in captivity, a series of studies on how early rearing and housing conditions influence immune responses in young monkeys are then reviewed. Those findings are also germane to juvenile monkeys in the natural setting, because this developmental stage is frequently a time of high mortality. The recently weaned infant monkey starts to move away from the mother and may experience aggressive encounters and bite wounds for the first time. It is further challenged by the need to obtain food independently. Coincidentally, this demanding social transition occurs just as the juveniles are losing some of the immune protection garnered through a process known as *passive immunity*—from prenatally derived antibody provided by the mother. Moreover, they have simultaneously lost one of the immune benefits afforded by nursing: a stimulation of mucosal immune defenses in the gut by breast milk. Within another year or two, depending upon the primate species, either the male or female offspring, or both sexes, will have to emigrate from the natal group upon reaching puberty. They have to relocate to a new troop, which makes their adolescence another time of heightened morbidity and mortality due to poor nutrition, injury, and infection. Finally, the last section of this chapter briefly reviews some of our research on immunological aging in the old monkey. That topic serves as the capstone for our "cradle-to-grave" survey of immunity in the nonhuman primate.

One caveat should be given. Depending upon the taxonomic source you check, there are up to 625 species in the Order Primates. They range from the small, one-pound prosimian species of Madagascar and Africa, which are more similar to the Paleocene ancestors of all primates, to the dozens of different monkeys in the forests of the New and Old World countries (Table 5.1). The Order also includes the three genera of great apes, among which is the largest primate, the 400-pound male gorilla. Thus, one cannot readily generalize findings to all primates based solely on the more commonly studied monkeys in the laboratory.

Much of our knowledge has been based on one species, the rhesus monkey (*Macaca mulatta*), which has been the primate used most frequently used in biomedical research. Experiments with rhesus monkeys enabled the development of the polio vaccine, and their Rh initials still label one of the important protein factors found on human red blood cells. A lot of what we have learned from rhesus monkeys will carry over to other Old World monkeys, including the snow monkeys of

TABLE 5.1: Representative species of the four major primate taxonomic groups, including some descriptive characteristics about their reproduction, development, and life span

	Prosimian	New World monkey	Old World monkey	Ape
Representative species	galago, lemur	squirrel monkey, capuchin	rhesus monkey, baboon, colobus	chimpanzee, gorilla
Pregnancy	2–4 mo	5–7 mo	6 mo	8 mo
Infancy	4–6 mo	6 mo	6–12 mo	3–4 yr
Puberty	1–2 yr	3–4 yr	3–4 yr	9–12 yr
Life span	6–15 yr	15–30 yr	20–30 yr	40–50 yr

Most immune studies have used Old World monkeys, such as the rhesus macaque, although considerable viral and cancer research has also been conducted on several New World monkey species.

Japan and the baboons of Africa. But even within this closely related taxonomic family (Cercopithecoidea), there can be dramatic differences. For example, rhesus monkeys are omnivorous, whereas the diet of quite a few African and Asian monkeys is more folivorous (i.e., the Colobinae). Given this dietary variation and the toxins present in many plants and leaves, the gut microbiota, mucosal immune responses, and gut parasites in the vegetarian colobus monkeys are likely to be very different than in the rhesus monkey (Gillespie et al., 2005).

Another example of the marked species variation can be found in the many different types of retroviruses that infect primates. The first retroviral infections in primates go back millions of years to the earliest prosimians (Gifford et al., 2008). But the originating source of the human immunodeficiency virus (HIV) that causes AIDS is found naturally only in African monkeys, not in Asian primates like the rhesus monkey (Apetrei et al., 2004). That is why the rhesus monkey can be used as the experimental model for simian acquired immunodeficiency syndrome (SAIDS). It has not evolved the natural protective immunity against the strains of simian immunodeficiency virus (SIV) that occur naturally in the mangabey and vervet monkeys of Africa. Thus, when infected with SIV, the rhesus monkey succumbs with an immunosuppressive disease as it starts to lose more and more virally infected T lymphocytes (Schmitz et al., 2010).

This example of specific host–pathogen relationships is just one of many that could be cited to convey the important species differences across such a diverse and very successful Order of animals. Monkeys from South America are equally distinct. A number of infectious viruses in Old World monkeys are not contagious to South American monkeys (Family Ceboidea), including the infamous herpes B virus of the rhesus monkey (renowned because of its potential lethality in humans) (Huff and Barry, 2003). Instead the Ceboid monkeys have their own species-specific herpes viruses, such as the herpes saimiri of the squirrel monkey (*Saimiri sciureus*), which can infect only some other South American monkeys, but not any Old World monkeys (Fickenscher and Fleckenstein, 2001). At the same time, there are a number of some zoonotic pathogens that do readily cross the species barriers among primates. Of clinical significance for us, it is known that primate populations are a natural reservoir for several diseases that can be transmitted to humans, including yellow fever and the Ebola and Marburg virus-induced hemorrhagic diseases (Schou and Hansen, 2000).

A comprehensive review of the numerous viral and bacterial pathogens and parasites that can infect primates is beyond the scope of this chapter. Field researchers have generated an increasingly impressive and encyclopedic array of information about the parasites and infectious diseases with which primates must contend (Muehlenbein, 2006). The National Institutes of Health financially supports some of this research because of a growing awareness that new diseases emerge as humans and animals come into contact due to shrinking tropical rainforests and the deeper encroachment of farms and ranches into previously pristine habitats (Greger, 2007; Keesing et al., 2010). Over time, even the *E. coli* that is a component of the gut microbiome in primates begins to be shared with farm animals and also with the farmers who enter the disturbed and fragmented forest areas (Goldberg et al., 2008; Rwego et al., 2009). Our review takes a more narrow focus. It is designed to provide the interested reader with a primer on some important issues and principles germane to the study of immunity in primates. A secondary take-home message is that social factors, housing, and rearing conditions can significantly affect immune responses in primates, which is really a conclusion that is equally relevant to most other types of animals.

IMMUNITY AND PREGNANCY

The importance of immunity goes back before the fetus even has an immune system, literally to the initiation of pregnancy. While reproduction and pregnancy are usually discussed in endocrine terms, immune-related cells in the reproductive tract actively participate in the establishment and maintenance of pregnancy. One of the first life-sustaining tasks for the embryo implanting into the uterine endometrium is to inhibit some of the mother's white blood cells from rejecting it, while at the same time stimulating others to help engulf it (Fig. 5.1). Sophisticated histological and molecular biology studies have demonstrated that the leading edge of the trophoblast cells, which will soon become the placenta, induces these changes in the leukocytes present in the uterus. Normally these T-cell–like and natural killer cells in the reproductive tract protect against bacterial and fungal infections, which are a potential concern on any exposed mucosal surface. Now they have to be manipulated to permit a successful implantation. Given the complexity of what must occur, it makes sense that many gynecological conditions, such as bacterial vaginosis and endometriosis, can interfere with fertility and the success of implantation.

Specifically, in the rhesus monkey, it has been shown that antigenic signals from the fetus related to its major histocompatibility complex evoke the required inhibition and facilitation of different maternal leukocytes in the uterine endometrium (Golos et al., 2006). In addition, the same type of trophoblast signaling will stimulate monocytic cells present in the uterine tissue to initiate angiogenesis, leading to the formation and sculpting of the mother's life-sustaining blood supply for the placenta. Much is known about these cellular and molecular processes in the field of reproductive immunology (Laird et al., 2003). But it is surprising that virtually no research has

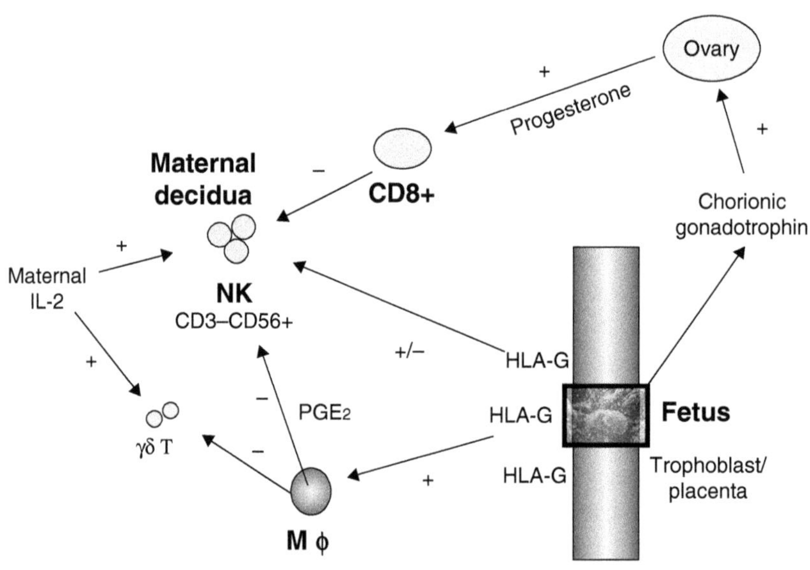

FIGURE 5.1: Fetal and placental modulation of leukocytes in the female uterus is required for successful implantation in primates. The trophoblasts, which create the placenta, downregulate maternal lymphocytes to prevent rejection. Simultaneously, they stimulate macrophages to promote angiogenesis in order to create the maternal blood supply for the placenta (modified from Coe and Lubach, 2007).

been conducted in primates to determine whether environmental factors or psychosocial processes influence these first immune-related steps required for the initiation of life.

Immune processes continue to be important throughout pregnancy. The growing fetus and placenta must strike a delicate balance in higher primates, because the maternal immune system is increasingly exposed to fetally derived proteins and even to some fetal cells that transit across the placenta. As primates evolved, the epithetheliochorial placenta of the prosimian species, which is more like that found in farm animals, switched over to the more penetrating type of hemochorial placenta seen in the monkeys and apes (Wildman et al., 2006). The benefits of having a placenta that is more proximal to the maternal blood supply created a new challenge for the mother's immune system. Because fetal cells will now present paternal antigens to the mother, it can potentially result in immune rejection, comparable to what is seen with a tissue graft or organ transplant. The evolutionary solution for higher primates was to selectively inhibit certain aspects of the primate mother's immune armamentarium while maintaining and even enhancing other responses to protect her during pregnancy. In gravid female monkeys and pregnant women, one can readily measure the progressive inhibition of certain cellular immune responses across pregnancy, while others related to humoral immunity and the cellular generation of antibody may actually be enhanced (Fig. 5.2).

A cascade of hormones and immune proteins are involved in mediating these immune changes in the mother. Rising levels of progesterone and cortisol in the mother's blood drive some of these effects on the mother's immune system, as do some fetal-derived proteins (e.g., alpha fetoprotein). Evidence of these immune-related changes can be monitored by assessing shifts in circulating levels of cytokines, including several of the interleukins. For example, in women, tumor necrosis factor (TNF-alpha) levels in the bloodstream should decline in the third trimester (Rigo et al., 2004). If they do not and TNF-alpha levels actually rise, then it may result in a premature delivery. In fact, a large surge of pro-inflammatory cytokines due to a bacterial infection during pregnancy is known to be associated with miscarriage early in pregnancy and, later on, a foreshortened

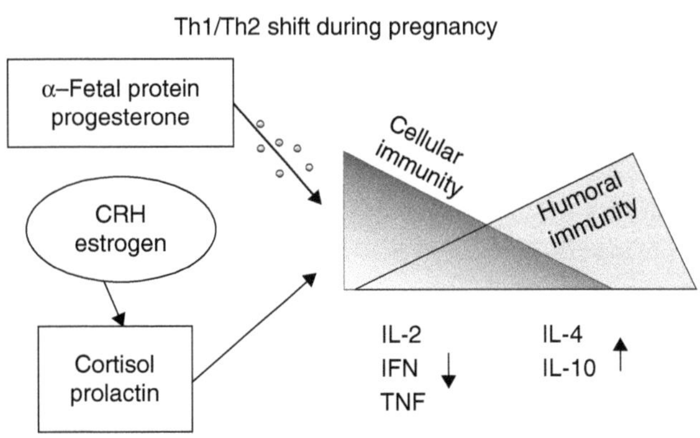

FIGURE 5.2: Many immune changes are required to maintain pregnancy and to prevent rejection of the developing baby later in gestation (modified from Coe and Lubach, 2007). Hormone alterations shift the immune balance from cellular toward humoral responses in the gravid female primate, which is reflected by the cytokine levels in the systemic circulation and the release of cytokines in assays when activating cells with *in vitro* stimulants.

gestation length. Here too it is somewhat surprising that there has not been more research in primates to determine whether psychosocial factors or ecological pressures alter this aspect of pregnancy physiology. In contrast, it is a commonly studied topic in the human obstetrical literature. Moreover, we already know from research on women that psychological stress during pregnancy can induce a more pro-inflammatory state, which in turn contributes to a greater risk for premature delivery (Culhane et al., 2001).

If not already evident, then it should be reiterated that the fetus is an active participant in this physiological dialogue. Certainly, the placenta, which is derived from fetal tissue, is an active contributor to the alterations taking place in the maternal compartment. The placenta initiates many of the hormone changes and alterations in blood volume that occur in a gravid female. It is also the conduit of proteins and other soluble substances between the maternal and fetal compartments. Some cytokines, such as interleukin-6, can cross the placenta and, when elevated in the pregnant female, will reach the amniotic fluid and even make it all the way into the fetal bloodstream (Urakubo et al., 2001). In our research on primates, we became fascinated by one unique aspect of this duet as pregnancy progressed: the transfer of beneficial maternal antibodies to the fetus via the placenta.

PLACENTAL TRANSFER OF MATERNAL ANTIBODY AND PASSIVE IMMUNITY

In higher primate species, the maternal transition toward a more humoral immune bias as pregnancy progresses will serve another function. As one looks across the Order Primates from the prosimians to the monkeys and apes, we see an evolutionary trend toward an increased reliance on a prenatal rather than postnatal transfer of maternal antibody to the infant. We investigated this process by collecting blood samples from new mothers and neonates in five different primate species, and assessing antibody of the G class (IgG) (Coe et al., 1994). As can be seen in Figure 5.3, the amount of IgG was relatively low in newborn galago, which is typical for an animal that will provide maternal antibody later in breast milk. In the squirrel monkey, representing South American monkeys, the neonatal IgG levels were approximately 40% of maternal levels. In contrast, newborn IgG levels in the rhesus monkey and chimpanzee were similar to adult values, indicating that they reached an equilibrium with the mother's antibody titers. But these values were still far below the nearly 150% to 200% of maternal IgG levels that are commonly seen in human newborns.

This multi-species comparison demonstrated that across the Order Primates, there had been a progressive shift toward the more active prenatal transport of antibody seen in humans. It requires the presence of a receptor for IgG on the placenta, which can bind and transfer the maternal antibody. We also know from other experiments in the rhesus monkey that much of this maternal antibody is transferred in the last month of pregnancy (Coe et al., 1993). In that way, a large bolus of antibody will be present in the newborn infant, which ensures a longer period of "passive immunity" against viruses and bacteria to which the mother had been exposed at prior points in her life. Given the half-life of antibody, which is usually about 3 weeks, the maternally derived antibody will typically last for 3 to 6 months after birth. Then, the infant must start to generate its own

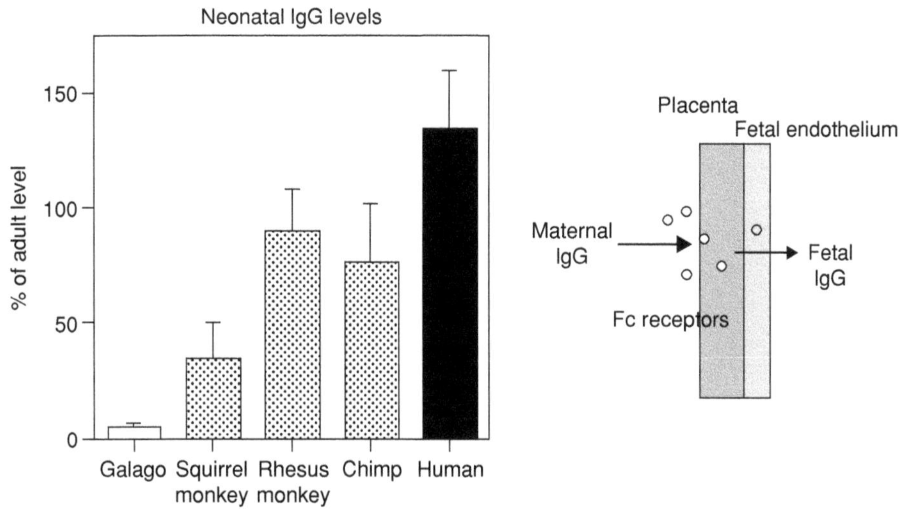

FIGURE 5.3: Levels of maternal antibody in the newborn infants of five different primate species, including humans, presented as a percent of adult levels. Across the Primate Order there has been a progressive evolution of an active transport process, which requires a receptor for the Fc region of the IgG on the placenta. The galago, a prosimian species, appears to rely primarily on the postnatal transfer of IgG in breast milk, which is similar to many other mammals with altricial young.

antibody, which coincides with the maturation of more efficient antigen-presenting cells, and the more mature capability of its own B lymphocytes to synthesize antibody.

A complementary aspect of this species variation in antibody transmission is that the maturity of the young infant's gut at term must match whether the maternal antibody is transferred before or after birth. Mammals that provide more maternal IgG primarily in breast milk typically have altricial offspring with permeable intestines that will permit the translocation of the consumed maternal antibody to reach systemic circulation. As described above, however, in the higher primate species, including the Old World monkeys and great apes, the transfer of IgG is largely a prenatal event. In these species, the antibody present in the mother's milk after birth is predominantly of the IgA class. It is designed instead to coat the exposed surface of the oral cavity and gut lining with neutralizing antibody against bacterial pathogens.

There are a number of important health-related ramifications of the passive immunity seen in infant monkeys, apes, and humans. Typically, there is a respite from infections for several months after birth, due to this beneficial protection conferred by the maternal antibody. In human infants, the first upper respiratory or ear infections usually do not occur until 3 months postpartum. It is also the reason why in clinical practice, one normally delays most pediatric immunizations until after the 3 month age point. Otherwise, the presence of maternal antibody may inhibit the infant from being able to generate its own antibody to the vaccines. In other experiments with monkeys we showed that the amount of maternal IgG reaching the baby can be affected by a number of factors. The most disruptive event is a premature birth, because it preempts the large transfer of antibody during the final weeks of pregnancy. In research on squirrel monkeys, we demonstrated further that the social stress of re-housing females several times during pregnancy would affect the amount of antibody reaching the developing fetus (Coe and Crispen, 2000). It was of particular interest that the antibody transfer was differentially affected by whether the fetus was a male or female. Male infants born after a socially stressed pregnancy received less antibody from the mother than normal. In contrast, female infants from similarly stressed pregnancies actually had higher levels of maternal antibody at birth than did the comparison group of undisturbed infants. The maternal IgG present at birth will then influence the duration and amount of the prentally acquired antibody that is sustained over time (see also Chapter 6).

MICROBIOME OF THE INFANT GASTROINTESTINAL TRACT

We also investigated another novel means that allows the mother to confer some additional protection for her young infant. The gut of the newborn infant primate is sterile at birth. Yet, within a few days after parturition, literally millions of bacteria will become established within the gastrointestinal tract, especially in the large intestine and colon. Over time these bacteria become the resident microbiota. The first of these commensal bacterial species are acquired from the mother, literally by the infant ingesting them during the birth process, and from contact with her skin and through suckling. We tracked the numbers and progression of two beneficial species, Lactobacilli and Bifidobacteria, across the first 6 months of life in rhesus monkeys (Bailey et al., 2004). Both types of bacteria are promoted by consumption of breast milk, and they serve as a protective layer within the gut, a lawn-like barrier that interferes with the attachment and growth of enteric bacterial pathogens that cause diarrheic illnesses. We found that if the concentrations of Lactobacilli and Bifidobacteria were low, then the infant monkey was more prone to an opportunistic infection with Shigella and Campylobacter. We were also interested in whether stressful pregnancy conditions might affect the microbiota profile that will emerge in each infant after birth. If the mother had experienced a 6-week period of daily acute stress early in her 24-week gestation, then the infant had lower levels of Lactobacilli. When the 6-week period of maternal stress period occurred late in gestation, it affected both the Lactobacilli and Bifidobacteria concentrations across the whole nursing stage in infants across until 6 months of age (Fig. 5.4).

The capacity of prenatal conditions to affect the gut microbiota and likelihood of enteric infections is of special significance to human infants in the non-industrialized world. Poor families in these regions are often unable to obtain clean water, and the likelihood of exposure to disease-causing pathogens is high. In these countries, diarrhea remains a leading cause of morbidity and mortality in early infancy. Our findings on the increased risk for gastro-intestinal disease in the infant monkeys from stressed pregnancies are also relevant to veterinary medicine. Farm animals, and many other infant mammals, are very vulnerable to pathogens that enter by the mouth in the course of eating and oral exploration of the world. In subsequent experiments, we showed further that the gut microbiota continue to be sensitive to psychological disturbance even in older infants and juveniles. For husbandry reasons in our monkey facility, we typically wean the 6- to 8-month-old monkey infants into small peer groups. Their bacterial profiles were tracked over the first week with fecal swabs, and revealed that the weaning events caused a transient decrease in the levels of Lactobacilli (Bailey and Coe, 1999). Typically, this decline was maximal on the third day after the

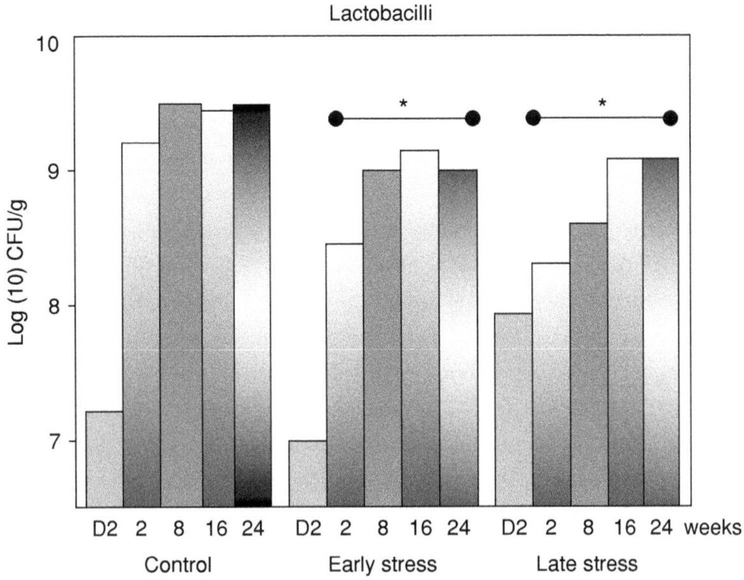

FIGURE 5.4: Fecal concentrations of Lactobacilli in infant rhesus monkeys on Day 2 after birth and across the first 24 weeks of life. Compared to infants from undisturbed pregnancies, maternal stress during the prenatal period significantly reduced the numbers of this beneficial bacteria in the gut microbiota. Lower levels of Lactobacilli and Bifidobacteria were associated with increased likelihood of infection with enteric pathogens, such as Shigella and Campylobacter.

re-housing. In some animals, it coincided with the emergence of diarrheic symptoms due to a flare-up of Campylobacter. While its a type of infection that can be readily treated with antibiotics, we had discerned one mediating factor for the diarrhea that emerges in some stressed weanlings. This clinical observation is probably not too surprising to veterinarians and colony managers. One can surmise that similar events must occur in nature. In the seasonally breeding monkeys that mate annually, the mothers will actively wean their infants at about 6 months of age, so they can reproduce again. The abruptness of the transition can engender a period of heightened mortality among infants in some species. Because assessments of the gut microbiota and parasites can be readily performed in a noninvasive manner from fecal specimens, it would certainly be a topic amenable to investigation in a field setting (Gillespie, 2006).

MATERNAL INFLUENCES ON THE MATURATION OF INFANT IMMUNITY

Given the two preceding sections on the transfer of maternal antibody and maternal priming of the infant's gut bacteria, it is probably not too surprising that mothers continue to influence the development of their offspring's immune responses in other ways. For over 50 years, we have known that nurturant maternal care in mammals, especially in primates, is essential for the promotion of normal infant behavior and emotional development. Today, it is equally common to emphasize how important maternal care is for stimulating the maturation of the brain, promoting neurogenesis and facilitating the establishment of the normal regulatory set-points for the monoamine neurotransmitters. Yet, when we first began to investigate this type of maternal and rearing influence on immune maturation in primates back in the 1980s, relatively little was known. For example, there had not been any systematic research on how the human rearing of infant monkeys in a nursery might affect their immune development as compared to ones reared naturally by the biological mothers. The issue was of greater husbandry significance at that time because many zoos and research facilities were still raising primates in this way. Even today, for humane reasons, it is sometimes necessary to hand-raise an infant monkey if an inexperienced primiparous female rejects it or the parturient female has had delivery complications at birth.

Initially, we assessed how the lymphocytes of 1- to 2-year-old rhesus monkeys would respond in stimulated cultures (Coe et al., 1989). Small blood samples were collected from animals that had been reared either by the natural mother or by humans. All were socially living with other animals at the time of assessment. When the mononuclear cells (MNCs) were stimulated to proliferate by the addition of one of three plant proteins as a mitogenic stimulus (PHA, Con A, or PWM), the MNCs of the human-reared animals were found to be more reactive. The finding was replicated with a different set of monkeys, and we became used to seeing what appeared to be larger proliferative responses (Coe et al., 1992). Ultimately, in a prospective study where we followed young monkeys across the first 2 years of life, it was determined that the differences should be viewed as a type of dysregulation, due to an abnormal ratio of T lymphocytes (CD4+/CD8+ cells) in the human-reared infant monkey (Lubach et al., 1995). Because of the sustained differences in lymphocyte subsets in circulation, the ratios of T cells aliquoted in each well of the culture plate were not the same. Another striking aspect of these immune alterations in nursery-reared monkeys was that they proved to be resistant to recovery (Fig. 5.5). In one experiment all of the

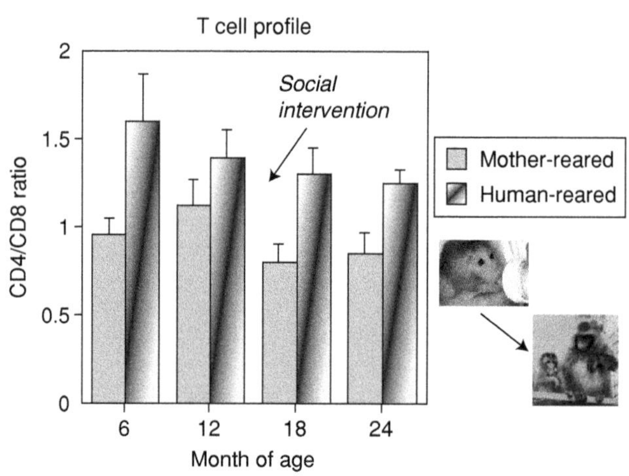

FIGURE 5.5: T-cell profiles of mother-reared and human-reared rhesus monkeys at 6-month intervals across the first 2 years of life. Infants reared in a nursery had a skewed CD4/CD8 ratio, due primarily to elevated numbers of CD8+ cells, which was associated with an exaggerated proliferative response to mitogen stimulation. This immune dysregulation did not return to normal even after the monkeys were housed socially with an adult female at 1 year of age.

young monkeys in one experiment were housed socially in peer groups after being differentially reared during the first year of life. Each large pen contained an adult female monkey to serve as a role model and support provider. The behavior of the human-reared monkeys improved markedly over the next year in this social group setting. Stereotypic behaviors and anxious emotional reactivity, which are common when primates are reared this way, were now less evident. Nevertheless, the differences in the lymphocyte proliferative responses persisted. We concluded that the absence of normal maternal influences across the formative first year of life had exerted an indelible effect on the immune system that could not readily be restored.

IMMUNE EFFECTS OF MATERNAL SEPARATION AND OTHER DISTURBANCES OF SOCIAL RELATIONSHIPS

This conclusion that a maternal influence on immune responses continues to linger long after the nursing phase concurs with a large body of prior research on farm animals and laboratory rodents (Solomon et al., 1968; Ader, 1983). Animals raised in the germ-free environment of a gnotobiotic laboratory are very different than ones raised in natural conditions where they are exposed to pathogens. In addition, numerous papers had also reported that even acute disruptions of early rearing, including brief separations form the mother, can affect many aspects of immunity (Ackerman et al., 1988). As the field of psychoneuroimmunology (PNI) coalesced into a discipline during the 1970s and early 1980s, these developmental models provided compelling evidence for the potency of psychological influences on immune responses. Studies in several different species of monkeys demonstrated that when young infants were separated from the mother or removed from their familiar social groups, pervasive changes in their immune responses could be easily detected (Laudenslager et al., 1982; Coe, 1993; Coe and Hall, 1996). In keeping with what was known about immune alterations in stressful situations, there was a decline in lymphocyte numbers in the bloodstream, and a skewing of the ratio of CD4+ to CD8+ lymphocytes (Table 5.2). In addition, for 1 to 2 weeks after the change in housing conditions, the ability of the separated infant monkey to mount an antibody response was reduced (Coe et al., 1987). If the cells of the stressed monkey were stimulated *in vitro*, then they evinced smaller proliferative responses. When the cells were examined in cytolytic assays, there was a reduced capacity to lyse cancer cells as well as to kill virally infected cell targets (Coe and Erickson, 1997).

As other components of immunity were assessed, however, it became equally evident that some responses were actually enhanced (Table 5.2). In particular, immune processes typically categorized under the heading of innate immunity seemed to be stimulated (Coe et al., 1988a). The number of neutrophils in the bloodstream increased in parallel with the decrease in lymphocytes, sometimes to such a degree that the total white cell count was now elevated by the neutrophil surge. Some functional responses of neutrophils, including the production of superoxides, which is an ancient reaction to protect against bacterial infections, were increased in the young

TABLE 5.2: Altered immune responses in infant and juvenile monkeys evoked by acute separation from the mother or after removal from the social group and re-housing into new cages

Decreased
Lymphocyte number, with a lower CD4+/CD8+ T-cell ratio
Lymphocyte proliferation responses (e.g., to mitogens such as PHA, Con A)
Natural killer cell lysis of target cells (e.g., K562, Raji, Daudi cells)
Antibody response to vaccines and antigens (e.g., KLH, bacteriophage)
Concentrations of Lactobacilli in the gut
Increased
Neutrophil number, which may be sufficient to increase total white count
Neutrophil and monocyte superoxide production
Hemolytic complement activity
Delayed hypersensitivity responses (e.g., to recall antigens such as DNCB)

monkey during the social separation period. The release of superoxides by monocyte/macrophages was similarly increased in the stress monkeys for several weeks. Moreover, the complement system was activated (Coe et al., 1988b). *In vitro* assays that used the lysing of sheep red blood cells as the indicator demonstrated that hemolytic complement activity was markedly increased in socially separated infants and juveniles, a change that persisted for more than 2 weeks. These findings in young primates contributed to a growing awareness among researchers studying other animals and humans that the effects of stress on the immune system were not unidimensional. While stressful events inhibit some aspects of cellular immunity, other components of innate immunity are often turned on simultaneously.

There has been a growing awareness that a diverse array of inflammatory processes are often stimulated by psychological disturbance (Danese et al., 2007). But back in the 1980s, it was not immediately clear why a separated or re-housed animal would show an exaggerated skin reaction after a test of cutaneous hypersensitivity. To obtain this measure, animals were first sensitized in an undisturbed state by exposure to an allergenic stimulus or chemical irritant, such as to chemical dinitrochlorobenzene (DNCB). Then, at a later date, either while either undisturbed or following a social separation, the animal was re-exposed to the DNCB to elicit the recall response on the skin. When this type of test was conducted in a psychologically aroused state, a much larger, redder, and more inflamed reaction was elicited over the next 3 to 7 days. From today's perspective, knowledgeable readers will immediately interpret this differential effect to mean that stressed individuals are evincing a "pro-inflammatory bias." However, that interpretation was not readily apparent back when our studies were first conducted in primates. Even now, it is still common to see many introductory textbooks describe the state of stress as being just one that involves a global suppression of the immune system. The author failed to delineate that while many lymphocyte processes are inhibited in the stressed state, other immune responses are simultaneously enhanced. The latter enhancement of inflammatory biology becomes especially important if interested in the long-term consequences of chronically stressful conditions (Kiecolt-Glaser et al., 2003). In fact, when it comes to research on the biology of aging, we often see that it is the negative consequences of pro-inflammatory responses that ultimately result in the most harm, especially with respect to secondary, cumulative effects on other tissues, such as the cardiovascular system.

IMMUNE MODULATION AND SENESCENCE IN THE AGED PRIMATE

If interested in reflecting on the truly long-term effects of environmental and social factors on the immune system, one is led to consider the biology of aging. Social stressors, poor nutrition, and disruptions in biorhythms can affect immune responses at any point in the life span, but there is considerable redundancy and resilience built into the immune system in the middle-aged, healthy adult. It is actually quite rare for stress alone to evoke an immune-related disease in an adult animal or human. In contrast, similar research on older animals and elderly humans has demonstrated that the immune changes tend to be larger and more long-lasting. In older humans, stressful events can significantly slow down the healing of wounds. There may be a full-blown, symptomatic reactivation of a herpes virus, such as in the clinical case of shingles due to the emergence of herpes varicella-zoster from the latent state (Kiecolt-Glaser et al., 1987, 1995).

In a series of studies on old rhesus monkeys, we characterized some other immune changes that can occur during aging (Coe and Ershler, 2001; Coe, 2004). These evaluations documented further that even routine husbandry events could affect their immune responses. Normally, the immune responses of monkeys in the laboratory setting or a zoo appear robust and quite stable through 20 years of age. Thereafter, one begins to see signs of what can be labeled as a progressive "immune senescence," or the "going to sleep" of the immune system (Ershler et al., 1988). When the lymphocytes of an old monkey are stimulated in culture with mitogens, the proliferative responses tend to be smaller than seen in middle-aged animals. If vaccinated against influenza virus or tetanus toxoid, the typically robust antibody responses are often reduced in the older animals. It is probably not coincidental that these immune declines first emerge as monkeys cross a critical age threshold that roughly corresponds to the end of the normal life span in nature. Here it is also noteworthy that to meet the old age criteria set by the National Institutes on Aging, one should not investigate immunological aging in monkeys until they cross this chronological age. At that point, they usually look and act old in many ways. They may show signs of osteoarthritis when moving; some will have cataracts. If one evaluate even older monkeys at 30 years of age, approximately one third of these primate Methuselahs will have a cancerous malignancy, which may ultimately prove to be the cause of their death (Uno, 1997).

We were fortunate to be able to evaluate many elderly rhesus monkeys living in different housing conditions. In one project, we investigated whether we could index their health and predict their life span using immune measures. Assays of natural killer cell activity were conducted on 14 female monkeys as they crossed over the important threshold of 20 years into old age (Coe, 2004). They were not subjected to further experiments and allowed to live undisturbed. If their lymphocytes had showed high levels of cytolytic activity, they were destined to live for many more years, some for another decade to 30 years of age. Conversely, if their *in vitro* killing of cell targets was already at a low ebb at 20 years, then they had a foreshortened life span of just a few years.

The ready inhibition of immune responses in elderly monkeys was also evaluated in response to routine clinical procedures, such as after an abdominal laparoscopic examination to check for endometriosis (Lemieux et al., 1996). Immune responses were evaluated before and after changes in housing conditions, including cage relocation and after introduction to a new social companion. We were especially interested in evaluating the influence of social factors on immunity in order to discern the optimal housing conditions in a geriatric monkey, at least in terms of promoting maximal immune responses. Lymphocyte responses were examined when old animals lived alone, if housed with one juvenile or one other aged adult, and while they lived in larger social groups with many juvenile monkeys (Coe et al., 1992). Perhaps it is not surprising that complex social settings tended to reduce immune responses, but even the addition of just one juvenile monkey into the cage of an older animal also elicited a negative immune change (Fig. 5.6). For up to 1 month, decreases in proliferative responses were evident. We eventually learned that it was possible to ameliorate this negative effect of excessive social stimulation by constructing double-compartment cages with a tunnel between the two sides. This unique cage design allowed the old monkeys to control the juvenile's access, and the negative immune effects were no longer observed.

Given the prevailing pressures today from regulatory agencies and animal welfare groups to encourage the housing of monkeys in more complex social settings, it is also worth mentioning what proved to be the optimal housing condition for an old rhesus monkey. It was not a large social troop, even if they lived in larger pens. The best social housing condition was living quietly with just one other old companion, at least using the criterion of an environmental setting that promoted maximal lymphocyte responses. Obviously, this discussion about housing conditions is of most relevance to laboratory facilities or zoos. In nature, most animals of this age would already have succumbed to predation, undernutrition, infection, or other lethal provocations.

As a final footnote to this section on aging and longevity, it should be noted that the life span of monkeys can be extended even further with certain dietary manipulations. There has not yet been a comprehensive assessment of immunity in these special monkeys, but we have recently

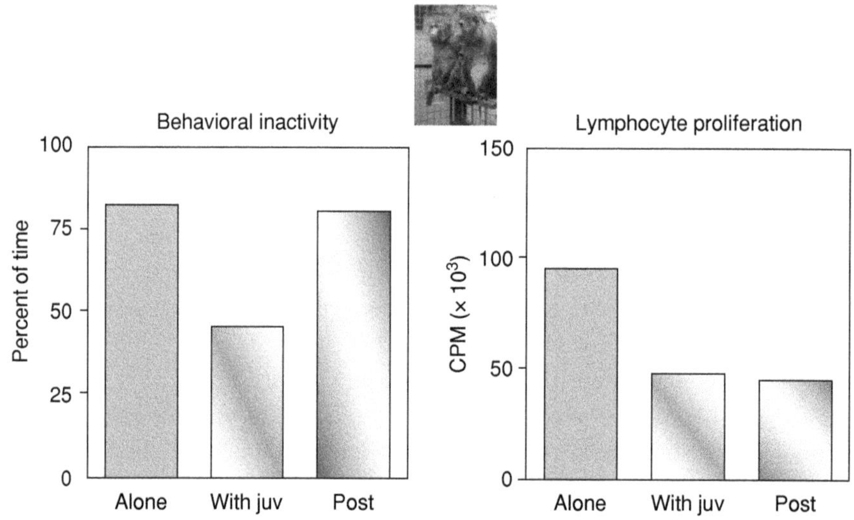

FIGURE 5.6: Immune responses of old rhesus monkeys are very sensitive to environmental change. Even adding one juvenile to the cage of an old female may induce a reduction in lymphocyte proliferative responses for 1 month, and the decrease persisted for another month after removal of the younger monkey. Increased motoric activity in the presence of a companion animal may look like a beneficial behavioral improvement but appears to be taxing for an old monkey. The optimal housing condition proved to be either living with just one other old monkey, or when housed in special double cages with a tunnel to permit solitude as desired.

reported on some of the salubrious benefits for immune/brain relationships that were attained by consuming a low-calorie diet. For example, rhesus monkeys eating a calorie-restricted diet since middle age were not as likely to show the typical age-related association between high IL-6 levels and cortical brain abnormalities in old age (Willette et al., 2010). Monkeys on this unique diet that provides adequate nutrition, but with fewer calories, were also not as likely to evince the age-related deposition of iron in the striatal regions of the brain, which is a marker of neural pathology in humans with dementia and Parkinson's disease (Kastman et al., 2010).

CONCLUDING COMMENTS ON IMMUNODULATION IN PRIMATES

Although the findings discussed in this review may seem most germane to primates in captive settings, many of the general conclusions do apply more broadly to animals living in nature. For example, from an epidemiological perspective, when do we see the most infectious disease and mortality? Infancy and old age are both distinct periods of heightened vulnerability, and the changes in immunity that occur during these two periods contribute to the observed increases in morbidity (Barry et al., 2006). Young animals are not yet fully immune competent, even though some cellular responses in neonatal monkeys, especially innate immunity, are already quite good at birth. In the older animal, there will eventually be a progressive decline in immune vigor. Those decreases in immunity tend to occur at what would normally be the typical end of the life span. It is as if the immune system was designed to function well up to a certain age, and then its natural warranty expires.

The functions of the immune system extend beyond just protection against infection. Leukocytes in the uterine endometrium play a critical role in the initiation of life by helping to engulf the implanting embryo. This aspect of reproductive immunology is of considerable importance if one is interested in the causes of infertility. Similarly, many immune changes must occur in the gravid female in order to maintain and sustain the pregnancy. They are actually induced by the fetus and placenta, partly to prevent rejection, but also to foster some immune-related processes that continue to benefit the infant after birth. In primates, one dramatic example is the evolutionary improvement in the prenatal transfer of maternal IgG across the placenta. As we look from the prosimians to the monkeys and apes, there is a clear transition from the postnatal transfer of maternal antibody in breast milk to becoming largely a prenatal event in the higher primates. Maternal antibody of the G class is transferred via the placenta, whereas the subsequent bolus of beneficial antibody in breast milk is primarily of the secretory IgA class. The latter is really designed to protect the oral cavity and gastrointestinal tract, not to reach the infant's blood.

In fact, both types of antibody are important and contribute to the phenomenon of "passive immunity." Primates are K-selected species, and make significant investments in each infant. Their long gestations, more invasive hemochorial placenta, and extended periods of nursing and maternal care are symbolic of this investment. Therefore, it should not be surprising that maternal behavior and the early rearing environment also have major effects on immune development. Many research papers have clearly demonstrated that maternal infections and stress during pregnancy affect the immune and brain development in infant monkeys, which would make them even more susceptible to the effects of adverse rearing conditions (Barry et al., 2006; Coe et al., 2002; Short et al., 2010). There is a lot we still need to learn about the prenatal programming of the immune system during fetal development.

In this chapter we described the immune effects of one extreme deviation from the normal rearing condition. Lymphocyte responses in rhesus monkeys raised by humans were found to be very different than in the infants reared by the biological mothers. Relatively little is known about the comparable development of cellular immune responses in natural settings. One can assume that intestinal worms and parasite load are a significant burden for a young primate in nature, but we don't have the quantitative evaluations yet to truly gauge the cost. Most of what we know has been acquired in the more hygienic setting of the laboratory, where natural disease is largely prevented through sanitation. If illness does occur, it is usually controlled by quarantine procedures and antibiotic treatments. Here it is important to remember to think of immunity as a "learning system." Immune responses in young animals will be markedly affected by the rearing environment and the relative level of antigen exposure. Thus, they will certainly be different in the laboratory or zoo setting than in the field. There has also been considerable pediatric interest in whether the prevention of infectious disease in early childhood hasn't had the inadvertent consequence of diverting immune development toward the atopic phenotype that underlies allergies and asthma, the so-called "hygiene hypothesis" (Wills-Karp et al., 2001).

In our own research, we have continued to be interested in the gut microbiota, as both as a promoter of health and a potential cause of disease. The gut microbiome is dramatic in many ways. There are over 500 species of commensal bacteria that live in the gut. It is estimated that there are more bacteria that live in us and on us (about 10^{14}) than there are cells in the body (a mere 10^{13}). Yet, we still know relatively little about this important symbiotic relationship in primates, and even less about how it varies across different environmental settings. Fortunately, the techniques of assessment are becoming easier to use and apply. Whereas in the past one had to actually culture and identify all of the different bacteria species by their appearance and characteristics, now it is possible to apply molecular biology approaches to screen for the presence of non-mammilian RNA from the diverse microbiota.

Given the biological closeness of humans and other primates, the topic of zoonosis and disease transmission is of obvious importance to consider. While many pathogens are uniquely attuned to their host, and some viruses cannot readily cross between primate species, others can infect disparate species. In fact, when a pathogenic virus or bacteria is able to make that species jump, it often encounters a host without an immune strategy to respond successfully. The discovery of how retroviruses in monkeys mutated to become contagious among the great apes, and ultimately bridged the species gap and infected humans, is a remarkable one (Worobey et al., 2010). Who could have imagined 20 to 30 years ago how many species of African monkeys are infected with different strains of retroviruses? We have already learned a lot about the unique cellular strategies permitting a host to tolerate these infections from those monkeys that have coexisted with retroviruses for millennia. Other zoonotic pathogens of major concern for both humans and other primates are the viruses that cause hemorrhagic disease. In some cases, humans may actually be the vector for infection back to animals, such as during the recent Ebola epidemic in the lowland gorillas of West Africa (Le Gouar et al., 2009).

Here we see common ground between the interests and expertise of laboratory and field researchers. Knowledge garnered by years of hard work at the bench can improve our understanding of health and disease in natural populations of animals. In a complementary manner, studies of pathogens in nature will continue to inform us about risks for newly emerging diseases of concern for humans. Ideally, new research on the successful immune defenses employed by primates living in the pathogen-rich and parasite-laden tropical regions will also reveal novel approaches to treatment. It is striking how few studies have been undertaken to compare the immunity of any domesticated primate species to their counterparts in nature. That information will be invaluable for promoting the well-being of animals when they are under our care in breeding colonies and in zoos. Similarly, that type of lab/field comparison will provide new insights into how animals maintain their health in the context of so many pathogens and parasites in nature.

FUTURE DIRECTIONS

Listed below are 10 possible topics for future research that are logical sequelae to the information reviewed in this chapter:

1. Discerning the impact of pathogen load and infection on reproductive success in adult females and males
2. Determining the role of maternal health during pregnancy and fetal programming for the development of postnatal immunity
3. Investigating the influence of nursing and breast milk on infant immunity, both on mucosal and systemic immune responses
4. Assessing immune competence during the often treacherous adolescent transition, when maturing primates typically emigrate from their natal group to a new social group
5. Identifying differences in the immune responses and gut microbiota of folivorous and omnivorous primates
6. Quantifying the effects of dietary factors, including plant phytoestrogens and toxins, on the gut microbiota and host immune responses
7. Describing annual variation in immunity in both the seasonally breeding prosimians and monkeys and the less-seasonal, long-lived apes
8. Correlating measures of immune aging in primates with other biomarkers of aging
9. Characterizing the effects of viral load, including the many latent herpes viruses, on the age-related declines in immunity in primates at the end of the life span
10. Refining better indicators of immunity that can be obtained noninvasively from fecal, urinary, or salivary specimens but still reflect systemic immunity and overall immune competence

ACKNOWLEDGMENTS

The research reviewed in this chapter was supported in part by grants from the National Institutes of Health to CLC (AI067518, HD39386, HD057064).

REFERENCES

Ackerman, SH, Keller, SE, Schleifer, SJ, Shindledecker, RD, Camerino, M, Hofer, M, Weiner, H, and Stein, M (1988). Premature maternal separation and lymphocyte function. Brain Behav. Immun. 2(2):161–165.

Ader, R. (1983). Developmental psychoneuroimmunology. Devel. Psychobiol. 16:251–267.

Apetrei, C, Robertson, DL, and Marx, PA (2004). The history of SIVS and AIDS: epidemiology, phylogeny and biology of isolates from naturally SIV infected non-human primates (NHP) in Africa. Front. Biosci. 9:225–254.

Bailey, MT and Coe, CL (1999). Maternal separation disrupts indigenous microflora of infant monkeys. Dev. Psychobiol. 35(2):146–155.

Bailey, MT, Lubach, GR, and Coe, CL (2004). Prenatal conditions alter the bacterial colonization of the gut in the infant monkey. J. Pediatr. Clin. Gastroenterol. 38:414–421.

Bruunsgaard, H, Pedersen, M, and Pedersen, BK (2001). Aging and proinflammatory cytokines. Curr. Opin. Hematol. 8:131–136.

Coe, CL (1993). Psychosocial factors and immunity in nonhuman primates: A review. Psychosomat. Med. 55:298–308.

Coe, CL (2004). Biological and social predictors of immune senescence in the aged primate. Mech. Aging Dev. 125(2):95–98.

Coe, CL and Crispen, H (2000). Social stress in pregnant monkeys differentially affects placental transfer of maternal antibody to male and female infants. Health Psych. 19(6):554–559.

Coe, CL and Erickson, C (1997). Stress decreases natural killer cell activity in the young monkey even after blockade of steroid and opiate hormone receptors. Dev. Psychobiol. 30(1):1–10.

Coe, CL and Ershler, WB (2001). Intrinsic and environmental influences on immune senescence in the aged monkey. Physiol. Behav. 73:379–384.

Coe, CL, Ershler, WB, Champoux, M, and Olson, J (1992). Psychosocial factors and immune senescence in the aged primate. Ann. NY Acad. Sci. 650:276–282.

Coe, CL and Hall, NR (1996). Psychological disturbance alters thymic and adrenal secretion in a parallel but independent manner. Psychoeneuroendocrinology 21:237–247.

Coe, CL, Kemnitz, JW, and Schneider, M. (1993). Vulnerability of placental antibody transfer and fetal complement synthesis to disturbance of the pregnant monkey. J. Med. Primatol. 22(5):294–300.

Coe, CL, Kramer, M, Kirschbaum, C, Netter, P, and Fuchs, E (2002). Prenatal stress diminishes the cytokine response of leukocytes to endotoxin stimulation in juvenile monkeys. J. Clin. Endocrinol. Metabol. 87:675–681.

Coe, CL and Lubach, GR (2007). Mother–infant interactions and the development of immunity from conception through weaning. In: Ader, R and Irwin, M (eds.). Psychoneuroimmunology. Elsevier, San Diego, pp. 415–434.

Coe, CL and Lubach GR (2008). Fetal programming: prenatal origins of health and illness. Curr. Dir. Psych. Sci. 17(1):36–41.

Coe, CL, Lubach, GR, Ershler, WB, and Klopp, RG (1989). Effect of early rearing on lymphocyte proliferation responses in rhesus monkeys. Brain Behav. Immun. 3:47–60.

Coe, CL, Lubach, GR, and Izard, KM (1994). Progressive improvement in the transfer of maternal antibody across the Order Primates. Am. J. Primatol. 32(1):51–55.

Coe, CL, Lubach, GR, Karaszewski, J, and Ershler, WB (1996). Prenatal endocrine activation influences the postnatal development of immunity in the infant monkey. Brain Behav. Immun. 10: 221–234.

Coe, CL, Lubach, GR, Schneider, ML, Dierschke, DJ, and Ershler, WB (1992). Early rearing conditions alter immune responses in the developing infant primate. Pediatrics. 90(3):505–509.

Coe, CL and Laudenslager, M (2007). Psychosocial influences on immunity, including effects on immune maturation and senescence. Brain Behav. Immun. 21:1000–1008.

Coe, CL, Rosenberg, LT, and Levine, S (1987). Psychological factors capable of preventing the inhibition of antibody responses in separated primate infants. Child Devel. 58:1420–1430.

Coe, CL, Rosenberg, LT, and Levine, S (1988a). Prolonged effect of psychological disturbance on macrophage chemiluminescence in the squirrel monkey. Brain Behav. Immun. 2(2):151–160.

Coe, CL, Rosenberg, LT, and Levine, S (1988b). Effect of maternal separation on the complement system and antibody responses in infant primates. Intern. J. Neurosci. 40:289–302.

Culhane, JF, Rauh, V, McCollum, KF, Hogan, VF, Agnew, K, and Wadhwa, PD (2001). Maternal stress is associated with bacterial vaginosis. Maternal Child Health 5(2):127–134.

Danese, A, Pariante, CM, Caspi, A, Taylor, A, and Poulton, R (2007). Childhood maltreatment predicts adult inflammation in a life-course study. PNAS 104:1319–1324.

Ershler, WB, Coe, CL, Gravenstein, S, Klopp, RG, Meyer, M, and Houser, WD (1988). Aging and immunity in nonhuman primates. I. Effects of age and gender on cellular immune function in rhesus monkeys (*Macaca mulatta*). Am. J. Primatol. 15:181–188.

Fickenscher, H and Fleckenstein, B (2001). Herpesvirus saimiri. Philos. Trans. Roy. Soc. Lond. B Biol. Sci. 356(1408):545–567.

Gifford, RJ, Katzourakis, A, Tristem, M, Pybus, OG, Winters, M, and Shafer, RW (2008). A transitional endogenous lentivirus from the genome of a basal primate and implications for lentivirus evolution. PNAS 105(51):20362–20367.

Gillespie, TR, Greiner, EC, and Chapman, CA (2005). Gastrointestinal parasites of the colobus monkeys of Uganda. J. Parasitol. 91(3):569–573.

Gillespie, TR (2006). Noninvasive assessment of gastrointestinal parasites in free-ranging primates. Intern. J. Primatol. 27(4):1129–1143.

Goldberg, T, Gillespie, IB, Rwego, EE, Estoff, and Chapman, CA (2008). Forest fragmentation as cause of bacterial transmission among primates, humans and livestock. Emerging Infectious Dis. 14(9):1375–1382.

Golos, TG, Bondarenko, GI, Breburda, EE, Dambaeva, SV, Durning, M, and Slukvin, I (2006). Immune and trophoblast cells at the rhesus monkey maternal-fetal interface. Methods Molecular Med. 122(3):93–108.

Greger, M (2007) The human/animal interface: Emergence and resurgence of zoonotic infectious diseases. Crit. Rev. Microbiol. 33:243–299.

Huff, J and Barry, P (2003). B-virus (Cercopithecine Herpesvirus 1) infection in humans and macaques: potential for zoonotic disease. Emerging Infectious Dis. 9:246–250.

Kastman, EK, Willette, AA, Coe, CL, Bendlin, B, Kosmatka, K, McLaren, D, Xu, G, Canu, E, Alexander, A, Voytko, ML, Beasley, MT, Colman, R, Weindruch, R, and Sterling, SJ (2010). A calorie restricted diet decreases brain iron accumulation and preserves motor performance in old rhesus monkeys. J. Neurosci. 20:7940–7947.

Keesing, F, Belden LK, Daszak, P, Dobson, A, Harvel, CD, Holt, RD, Hudson, P, Jolles, A, Jones, KE, Mitchell, CE, Myers, SS, Bogich, and Ostfeld, RS. (2010) Impacts of biodiversity on the emergence and transmission of infectious diseases. Nature 468:647–652.

Kiecolt-Glaser, JK, Glaser, R, Dyer, C, Suttleworth, E, Ogrocki, P, and Speicher, CE. (1987). Chronic stress and immunity in family caregivers of Alzheimer's disease victims. Psychosomat. Med. 49:523–535.

Kiecolt-Glaser, JK, Marucha, PT, Malarkey, WB, Mercado, AM, and Glaser, R (1995). Slowing of wound healing by psychological stress. Lancet 346:1194–1196.

Kiecolt-Glaser, JK, Preacher, KJ, MacCallum, RC, Atkinson, C, Malarkey, WB, and Glaser, R (2003). Chronic stress and age-related increases in the proinflammatory cytokine IL-6. PNAS 100:9090–9095.

Laird, SM, Tuckerman, EM, Cork, BA, Linjawi, S, Blaekmore, AIF, and Li, TC (2003). A review of immune cells and molecules in women with recurrent miscarriage. Human Reprod. Update 9(2):163–174.

Laudenslager, ML, Reite, M, and Harbeck, RJ (1982). Suppressed immune response in infant monkeys associated with maternal separation. Behav. Neural Biol. 36:40–48.

Le Gouar, PJ, Vallet, D, David, L, Bermejo, M, Gatt, S, Levrero, F, Petit, EJ, and Menard, N (2009) How Ebola impacts genetics of Western lowland gorilla populations. PLoS ONE 4(12):e8375.

Lemieux, A, Coe, CL, and Ershler, WB (1996). Surgical and psychological stress differentially affect cytolytic responses in the aged female monkey. Brain Behav. Immun. 10:27–43.

Lubach, GR, Coe, CL, and Ershler, WB (1995). Effects of early rearing on immune responses in infant rhesus monkeys. Brain Behav. Immun. 9:31–46.

Muehlenbein, MP (2006). Intestinal parasite infections and fecal steroid levels in wild chimpanzees. Am. J. Phys. Anthro. 130:546–550.

Rigo, JJR, Szelenyl, J, Selmeczy, Z, Papp, Z, and Vizi, ES (2004). Endotoxin-induced TNF production changes inversely to its plasma level during pregnancy. Eur. J. Obstet. Gynecol. 114(2):236–238.

Rwego, IB, Isabire-Basuta, G, Gillespie, TR, and Goldberg, TL (2009). Bacterial exchange between gorillas, humans and livestock in Bwindi. Gorilla J. 38:16–18.

Schmitz, JE, Zahn, RC, Brown, CR, Rett, MD, Li, M, Tang, H, Pryputniewicz, S, Byrum, RA, Kaur, A, Monefiori, DC, Allan, JS, Goldstein, S, and Hirsch, V. (2010). Inhibition of adaptive immune responses to a fatal clinical outcome in SIV-infected pigtail macaques but not vervet African green monkeys. PLOS Pathogens 5(12):e1000691.

Schou, S and Hansen, AK (2000). Marburg and Ebola virus infections in laboratory nonhuman primates: A literature review. Comp. Med. 50(2):108–123.

Short, SJ, Lubach, GR, Karasin, AI, Olsen, CW, Styner, M, Knickmeyer, RC, Gilmore, JH, and Coe, CL (2010). Maternal influenza infection during pregnancy impacts postnatal brain development in the rhesus monkey. Biol. Psychiat. 67:965–973.

Solomon, GF, Levine, S, and Kraft, JK (1968). Early experience and immunity. Nature 220:821–822.

Uno, H (1997). Age-related pathology and biosenescent markers in captive rhesus macaques. Age 20:1–3.

Urakubo, A, Jarskog, LF, Lieberman, JA, and Gilmore, JH (2001). Prenatal exposure to maternal infection alters cytokine expression in the placenta, amniotic fluid, and fetal brain. Schizophrenia Res, 47:27–36.

Wildman, DE, Chen, C, Erez, O, Grossman, LI, Goodman, M, and Romero R (2006). Evolution of the mammalian placenta revealed by phylogenetic analysis. PNAS 103(6):3203–3208.

Willette, AA, Bendlin, BB, McLaren, DG, Canu, E, Kastman, EK, Kosmatka, KJ, Xu, G, Field, AS, Alexander, AL, Colman, RJ, Weindruch, RH, Coe, CL, and Johnson, SC (2010). Age-related changes in neural volume and microstructure associated with interleukin-6 are ameliorated by a calorie-restricted diet in rhesus monkeys. Neuroimage 51:987–984.

Wills-Karp, M, Santeliz, J, and Karp, CL (2001). The germless theory of allergic disease: revisiting the hygiene hypothesis. Nat. Rev. Immunol. 1:69–75.

Worobey, M, Telfer, P, Souquiere, S, Hunter, M, Coleman, CA, Metzger, MJ, Reed, P, Makuwa, M, Hearn, G, Honarvar, S, Roques, P, Apetrei, C, and Marx, P (2010). Island biogeography reveals the deep history of SIV. Science 329(2):1487.

CHAPTER 6

MATERNAL MODULATION OF OFFSPRING IMMUNE FUNCTION IN VERTEBRATES

Dennis Hasselquist, Michael Tobler, and Jan-Åke Nilsson

A maternal effect occurs when a mother's phenotype, including her environmental experiences, has a direct modulating effect on the phenotype of her offspring (Mousseau and Fox, 1998). Maternal effects are a type of epigenetic effect—that is, an effect where the appearance and inheritance of phenotypic traits are altered without changes in the underlying DNA sequence (Jablonka and Lamb, 1999; Poulin and Thomas, 2008). Such effects can be adaptive if they have positive consequences for maternal and offspring fitness. Previous studies have identified a number of factors that can be agents for maternal effects, ranging from social status and maternal provisioning to physiological transfer of nutrients, immunoglobulins, antioxidants, and hormones. Recent research has shown that epigenetic influences in the form of maternal effects can have a profound impact on offspring physiology and fitness (Mousseau and Fox, 1998; Surai, 2002; Groothuis et al., 2005b; Boulinier and Staszewski, 2008; Gil, 2008; Hasselquist and Nilsson, 2009).

In the present review, we will focus on three important components of maternal effects that can affect offspring immune function: maternal transfer of (1) antibodies, (2) hormones (sex steroids and glucocorticoids), and (3) antioxidants (in particular carotenoids). We will first briefly outline the importance of the different maternal factors, as well as their timing and route of transmission in different types of vertebrates. For each of these maternal effects, we will then outline hypotheses and review available data to evaluate them. We further analyze short-term (neonatal stage) and long-term (matured, adult stage) effects on offspring. We will also evaluate costs and benefits and discuss effects of potential trade-offs between the different maternal effects.

MATERNAL TRANSFER OF ANTIBODIES AND OFFSPRING IMMUNE FUNCTION

Transgenerational modulation of parasite defense has been observed in plants as well as in invertebrate and vertebrate animals (Agrawal et al., 1999). In vertebrates, transfer of maternal antibodies (matAb) is a well-known process in model organisms. For example, in domestic chickens it has been used to boost broiler production, and it is also commercially used for the production of specific antibodies in human medicine research and vaccine production (Davison et al., 2008; Schijns et al., 2008). Also in humans, matAb transfer has been studied for decades partly because they may interfere with vaccination of human infants (e.g., Nicoara et al., 1999; Siegrist, 2003). Hence, there is ample information about matAb transfer in humans and other vertebrate immunology model organisms, including mice and rats (e.g., Anderson, 1995; Lundin et al., 1999; Siegrist, 2003; Lemke et al., 2004, 2009). However, studies of matAb transfer in non-model vertebrates are still scarce, although interest has increased considerably recently. This is not surprising, given the potential impact of matAb transfer on offspring fitness (Grindstaff et al., 2003; Boulinier and Stazewski, 2008; Hasselquist and Nilsson, 2009). In this chapter, we will first present information about routes and timing of matAb transmission in vertebrates, and this part will necessarily include extensive information from humans and model organisms. We will then evaluate strategies and factors that can affect matAb transfer, with an emphasis on non-model organism studies. Next, we will present information on for how long matAb is present in the neonate offspring, as well as short- and long-term effects of matAb transfer on offspring physiology and fitness. Finally, we outline a number of hypotheses that have been proposed to explain the transfer of matAb to offspring in vertebrates, and evaluate these hypotheses based on the available data.

ROUTE AND TIMING OF matAb TRANSMISSION

There has been a large body of work describing routes for and timing of matAb transmission in different types of vertebrates, in particular in humans and model organisms such as rat, mice, and chicken. Hence, much of the available information on these proximate factors, and thus the basis for many hypotheses regarding matAb transfer, is based on data from a few species living under non-natural circumstances. Still, this knowledge is useful also for wild vertebrates, although one has to keep in mind the limitations and risks for misinterpretations when extrapolating patterns between such species.

MAMMALS

In humans, transfer of matAb over the placenta begins 3 months before birth and increases rapidly. At birth, the IgG level in serum of the fetus can be 1.2 to 1.8 times higher than in the mother (Brugha et al., 1996; Nicoara et al., 1999). Moreover, the neonate is given a final boost of matAb (IgG) through the colostrum that is breastfed by the mother to the infant just after birth. In rodents, matAb (IgG) is transferred to offspring via the placenta before birth as well as in breast milk during lactation (Grindstaff et al., 2003).

BIRDS

In birds, only IgG (or IgY, as this more ancient form of IgG sometimes is called) antibodies are transferred over the egg sac into the egg yolk, whereas IgM cannot pass over the yolk sac (possibly due to its larger size) and therefore stays in the albumen (Hamal et al., 2006; Davison et al., 2008). This has been confirmed also for wild altricial birds, as plasma of newborn magpie (*Pica pica*) chicks (0 to 24 h) contained IgG only, whereas plasma of 10-day-old chicks contained both IgM and IgG, confirming that intrinsic production of immunoglobulins had begun in the nestlings (Pihlaja et al., 2006).

OTHER VERTEBRATES

In fish it is well documented that both innate (e.g., complement factors, lectins) and adaptive (Ig/antibodies) immune components can be transferred from mother via the egg yolk to offspring (Swain and Nayak, 2009). The main type of antibody in fish is IgM, and transfer of matAb seems to involve reduced or breakdown products of this Ig type (as well as chimeras between parts of IgM and other Ig types) (e.g., Bly et al., 1986; Mor and Avtalion, 1990; Swain and Nayak, 2009). The persistence of matAb in offspring circulation is usually short in fish, with low levels already at the completion of the yolk absorption and complete disappearance at the larval stage (Takemura and Takano, 1997). However, given that young fish do not develop a well-functioning acquired immune system until an age of 2 to several weeks, the transfer of matAb is likely to be important for early immune function (Zapata et al., 2006).

In reptiles matAb transfer occurs via the egg yolk to offspring, for example as demonstrated in a study of the desert tortoise (Schumacher et al., 1999). However, virtually nothing is known about how matAbs influence the immune system of offspring in reptilian species.

IS matAb TRANSFER AN ACTIVE OR A PASSIVE PROCESS?

In mammals, there are convincing data for matAb transfer being an active process. A first line of evidence for active maAb transfer is that levels of matAb in the embryo often exceed that of the mother. For example, in humans anti-measles Ab levels are 1.2 to 1.8 times higher in the fetus than in the mother's circulation (Nicoara et al., 1999). Moreover, in mice uptake of IgG from the gut into the bloodstream is an active process that ends after about 2 weeks—that is, 1 week before end of lactation (Anderson, 1995). In birds, the pattern is less clear. For example, the level of IgG in egg yolk relative to that in the mother's plasma showed high variation between different chicken lines (range of egg yolk IgG/mother plasma IgG = 0.67 to 1.34; Abdel-Mouniem and Abel-Gawad, 2006), and the egg yolk to mother plasma IgG ratio was found to be 0.63 in collared flycatchers (Hargitai et al., 2006). A positive correlation between antigen-specific IgG levels in the mother's plasma and the levels in the egg yolk has been reported (Apanius, 1998; Blount et al., 2002b; Gasparini et al., 2002; Saino et al., 2002a; Grindstaff et al., 2003, 2005; Morales et al., 2006). This could reflect passive leakage of the mother's intrinsic level of plasma IgG into the egg yolk, but it could also reflect an active process if mothers of high immune system quality are able to invest more in reproduction and therefore transfer more IgG into their egg yolk. However, the fact that females might transfer different levels of matAb to specific eggs, for example, over the laying sequence or in relation to ecological or social factors (Table 6.1a), also indicates that females have a mechanism for altering the magnitude of matAb deposition. Furthermore, at the receptor level in chickens, matAb transfer over the yolk sac is an active process (Kowalczyk et al., 1985; Davison et al., 2008) involving specific receptors (Fc RY; West et al., 2004). Hence, females may be able to allocate matAb differentially among their offspring in potentially adaptive ways. In the next section we discuss these aspects, identify potential factors that may affect matAb transfer, outline hypotheses proposed to explain how such factors may influence matAb transfer, and evaluate the available data.

TABLE 6.1A: Factors that could be affected by/related to maternal antibody (matAb) transfer

Species	matAb affected by food given to mother[1]	matAb affected by mother condition or age[2]	matAb affected by antigen/ pathogen exposure[3]	matAb related to egg-laying order	matAb affected by male quality	Reference
Mammals						
Mice (*Mus musculus*)			+			Lemke and Lange, 1999; Lundin et al., 1999
Rat (*Rattus norvegicus*)	+ Fm					Chandra, 1975; Michalek et al., 1975
White-tailed deer (*Odocoileus virginianus*)	0 Pm					Sams et al., 1995
Iberian red deer (*Cervuselaphus hispanicus*)	−Ca					Landete-Castillejos et al., 2002
Birds						
Common tern (*Sterna hirundo*)		0 a				Apanius and Nisbet, 2003
Chicken (*Gallus gallus*)		+ a	+ Ie			Smith et al., 1994 Heller et al., 1990
Japanese quail (*Coturnix japonica*)	0 Pa	0 c		0		Grindstaff et al., 2005
Lesser black-backed gull (*Larus fuscus*)	− C			–		Blount et al., 2002b
Kittiwake (*Rissa tridactyla*)	−fF		+ Ib			Gasparini et al., 2001; 2007
Black-headed gull (*Larus ridibundus*)				–		Müller et al., 2004; Groothius et al., 2006
Ural owl (*Strix uralensis*)	+ cF					Karell et al., 2008

TABLE 6.1A: (Continued)

Species	matAb affected by food given to mother[1]	matAb affected by mother condition or age[2]	matAb affected by antigen/ pathogen exposure[3]	matAb related to egg-laying order	matAb affected by male quality	Reference
Magpie (*Pica pica*)	+ eF			+		Pihlaja et al., 2006
Pied flycatcher (*Ficedula hypoleuca*)	(+) mF[4]	+ c − c[5]				Moreno et al, 2008 Morales et al., 2006 Kilpimaa et al., 2006
Collared flycatcher (*Ficedula albicollis*)		+ c	0 S	+	0	Hargitai et al., 2006
Great tit (*Parus major*)			+ If + If	0		Buechler et al., 2002; Gallizzi et al., 2008a,b Tschirren et al., 2009
Barn swallow (*Hirundo rustica*)				0	+	Saino et al., 2002a, 2003a

[1]Food manipulation of mother: Fm = food malnourishment, Ca = calorie restriction 50%, Pa = protein content manipulated in food given *ad libitum*, Pm = protein malnourishment, C = carotenoid supplementation, Food supplementation (eF = raw egg, cF = dead chicken, mF = meal worms, fF = fish [capelin, *Mallotusvillosus*]).
[2] c = female condition, a = female age.
[3]Antigen exposure of mother: S = SRBCs, Pathogen exposure of mother: If = infection caused by fleas; Ib = infection caused by tick-transmitted Borrelia, Ie = infection caused by *Echerichia coli* parasites.
[4]Mothers supplemented with mealworms but matAb not measured—only total IgG in 12-day-old chicks that had been cross-fostered. Hence, effects on offspring may depend on any change in egg composition, not necessarily egg yolk matAb content.
[5]Experimental weight put on females pre-laying—effect on matAb transfer measured in 7-day-old nestlings.

WHICH FACTORS AFFECT FEMALE matAb TRANSFER?

A number of factors have been suggested to have the potential to affect the mother's transmission of matAb to offspring. These are female (1) intrinsic factors, (2) diet and condition, (3) stress levels, (4) infectious experience, (5) strategic decisions, and (6) quality of mate.

FEMALE INTRINSIC FACTORS

Intrinsic factors of the female that could influence matAb transfer can be both genetically and phenotypically determined. Of these factors, the genetic influence on matAb transfer is the hardest to approach—simply because the very nature of matAb transfer is epigenetic, which would confound relationships established with conventional heritability analyses. However, one way to overcome this problem is to investigate selection lines differing in humoral immunocompetence (e.g., Verhulst et al., 1999). In studies of chickens, selection lines have been found to differ, both in levels of Ab in mothers as well as in matAb transfer efficiency and matAb titers in 1-day-old neonates (Abdel-Moniem and Abel-Gawad, 2006). Moreover, in a chicken line selected for high Ab-mediated immunity, transfer of high levels of matAb into egg yolk was found compared with lines selected for low Ab-mediated immunity and control lines (Parmentier et al., 1996). However, in these kind of selection line studies, it is unknown if the relationship found is a response to selection resulting in higher circulating Ab level in the mother followed by a passive transmission of matAb to the egg yolk, or if matAb transfer increased as a response to selection of factors increasing matAb transfer into egg yolk irrespective of the circulating Ab levels in the mother.

Another way of investigating the effects of female intrinsic factors on matAb transfer is to conduct repeatability analysis. High repeatability of matAb transfer occurs in kittiwakes exposed to natural levels of *Borrelia* infections, both within eggs of the same clutch and also between first and replacement clutches during the same season (Gasparini et al., 2001). However, these results also can be either caused by passive transfer of circulating Ab that reflects the previous *Borrelia* exposure, or a result of genetically determined individual levels of matAb transfer. To test the latter possibility we would need to compare data of matAb transfer to eggs/offspring between mothers and their daughters, although such analyses will not be straightforward to interpret because they may be confounded by the potential long-term phenotypic effects of the matAb transfer *per se* (which may even last for two generations; Lundin et al., 1999). Hence, when searching for underlying genetic effects on matAb transfer, heritability analyses have to account for potentially confounding epigenetic (phenotypically determined) effects. One possibility is to compare mothers and daughters both with reference to (1) transfer of total IgG, which may be more genetically determined, and (2) transfer of antigen-specific matAb, which is more epigenetically affected. We know of no study that has compared matAb transfer between mothers and their daughters, at least not in non-domesticated animals.

FEMALE DIET AND CONDITION

Female diet can affect matAb transfer. Female magpies provided with extra food (raw chicken eggs) before egg laying produced offspring with higher IgG plasma levels at hatching (day 0) compared to unsupplemented control females; however, this difference had disappeared at 10 days of age (Pihlaja et al., 2006). Most interestingly, as there were no effects of food supplementation on clutch size, hatching success, or egg weight, this study shows that mothers can invest specifically in the immune-related quality of their eggs and newborn chicks. In contrast, in a study of captive Japanese quails that were fed *ad libitum* diets of high or low protein content, treatment groups did not differ in plasma Ab or egg yolk matAb levels (Grindstaff et al., 2005). In Table 6.1a, we have summarized information on how female diet may influence matAb transfer to eggs and neonates. The pattern is very mixed, with negative, no, or positive relationships with diet improvements. However, different components of the diet have been altered in these studies, and it is therefore difficult to draw any general conclusions besides the fact that diet can affect matAb transfer. More studies are clearly needed in non-model organisms.

Also, female condition may influence matAb transfer (Table 6.1a). For example, pied flycatcher (*Ficedula hypoleuca*) females in better condition produced eggs of higher blue-green color intensity and the egg yolk of these eggs contained a higher concentration of matAb, resulting in a higher fledging success (Morales et al., 2006).

FEMALE STRESS LEVEL

The only study of which we are aware that tested the effect of stress on matAb transfer reported female stress levels to be negatively related to matAb levels in egg yolk (Hargitai et al., 2006). Stress may influence matAb transfer, either directly due to the immunosuppressive effect of prolonged stress and high corticosterone levels (Råberg et al., 1998; Owen-Ashley et al., 2004), or indirectly through its negative effect on overall condition, which in turn could suppress matAb transfer (Hargitai et al., 2006).

FEMALE INFECTIOUS EXPERIENCE

It is intuitively reasonable to predict that female infectious experience could influence matAb transfer. Such relationships have been found in invertebrates (Moret and Schmid-Hempel, 2001; Little et al., 2003). In vertebrates, the effect of previous infectious experience on matAb transfer has been particularly well studied among cavity-nesting birds parasitized by hen fleas (*Ceratophyllus gallinae*). In studies of great tits, experimental infestation of mothers with hen fleas resulted in higher growth rate, survival, protection against hen fleas in nestlings, and higher reproductive success in matured offspring (Heeb et al., 1998). However, some of the results on the importance of matAb mediating these effects has been questioned because the manipulation of hen flea infestation was done so late that matAb from this infection could not have been incorporated into the eggs included in the study (Tschirren et al., 2009). In line with this, no relationship was reported between matAb in one egg of a clutch and the growth rate of the remaining nestlings hatching from the clutch, independent of whether nests were exposed to hen fleas or not (Tschirren et al., 2009). However, when the time between flea infestation and egg production was sufficiently long to allow for flea-specific matAb transfer, flea-infested mothers produced eggs with higher matAb levels and their offspring had higher body mass when 16 days old (Buechler et al., 2002). Moreover, under such conditions, flea-exposed mothers produced offspring of higher body mass on day 14 if living under benign clutch-competition circumstances without ongoing flea infestation (Gallizzi et al., 2008a), and such offspring were also more protected against a tick ectoparasite (Gallizzi et al., 2008b). Hence, the pattern in the hen flea–great tit system is mixed, and at this stage it is hard to say how much of the positive impact of maternal effects on offspring performance and fitness (e.g., Heeb et al., 1998) can be attributed to matAb transfer and how much depends on other maternal effect factors, such as hormones and antioxidants.

Many other studies indicate that female pathogen exposure is important for matAb transfer (Table 6.1a). For example, in a study comparing 13 local populations of kittiwakes breeding in different cliffs, a positive relationship was found between frequency of birds with tick infestation and the percentage of nests with Borrelia antibody-positive eggs (Gasparini et al., 2001). In a study of antibody responses to pox virus (*Poxvirus avium*) infection in Darwin's finches, it was found that females had higher antibody titers than males despite similar exposure to pox virus infections (Huber et al., 2010). Although no matAb transfer was measured in this study, it is tempting to speculate that females may react to pox virus with a more antibody-mediated response, which could benefit the offspring through matAb transfer.

FEMALE STRATEGIC DECISIONS

Parental investment theory predicts that the mother (and father) should invest differentially in their progeny depending on offspring value, sex, and cost of production (Trivers, 1974). Hence, it can be expected that females make strategic decisions over how much matAb to transfer to different offspring.

Order of offspring in the brood

In birds, the order in which eggs are laid can be an important feature related to offspring value, because progeny from different eggs within the laying sequence face drastically different conditions during their neonatal period. For example, maternally transferred androgens into egg yolk (Schwabl, 1993; Schwabl et al., 1997) and offspring sex ratio (Hasselquist and Kempenaers, 2002) vary with laying order in apparently adaptive ways. Also, matAb transfer varies with laying order in some birds (Table 6.1a), and there are two hypotheses proposed to explain the observed patterns.

The first hypothesis predicts that *matAb allocation should increase with egg-laying order* in species with a "brood survival" reproductive strategy; that is, species producing clutches that hatch asynchronously as a response to environmental factors (Hargitai et al., 2006). Results in favor of this hypothesis are based on studies of the collared flycatcher and the magpie (Table 6.1a). Collared flycatcher females transfer more matAb to last-laid eggs, which is potentially adaptive because late-hatched chicks could then do better in competition with their older siblings, increasing their survival prospects (Hargitai et al., 2006). In magpies, matAb in plasma of newly hatched chicks (0 to 24 h old) also increased with laying order (Pihlaja et al., 2006). In this case, last-laid eggs contained higher testosterone levels, which may suppress immunity (Folstad and Karter, 1992), and it was therefore argued that mothers enhanced transfer of matAb into last-laid eggs as an adaptation to alleviate the immunosuppressive effects of testosterone in the chick (Pihlaja et al., 2006).

The second hypothesis predicts that *matAb allocation should decrease with laying order* in species with a "brood reduction" reproductive strategy; that is, species in which chicks of last-laid eggs have low survival prospects and mainly act as insurance for infertility or other losses of early laid eggs or for buffering parental reproductive success in unpredictable environments (Royle et al., 1999; Blount et al., 2002b). This relationship has been found in two species of gulls (Blount et al., 2002b; Müller et al., 2004), concomitant with reduced egg size and carotenoid levels in the laying sequence, indicating that these gulls are "brood reduction" species.

Hence, both of the hypotheses explaining matAb transfer in relation to egg-laying order have gained some support, although causality is hard to infer from these studies. One of the other factors proposed to explain matAb may covary with egg-laying order and may therefore confound the analysis of how reproductive strategy influences the mother's decision over matAb transfer to offspring.

Offspring sex

Few studies have reported variation in matAb transfer in relation to offspring sex, which may reflect that such relationships are uncommon. However, in a study of barn swallows, it was found that mothers transferred more matAb into the egg yolk of eggs containing a daughter than into eggs containing a son (Saino et al., 2003a), and a possible reason for this was that mothers may have favored daughters, which have higher mortality rates than sons.

QUALITY OF MALE PARTNER

Theory predicts that females should invest more in offspring fathered by high-quality and/or attractive males, because such offspring have potentially higher fitness values (Sheldon, 2000)—the so-called "differential investment" hypothesis. Such differential maternal investment in egg hormone and egg size exists in several bird species (Gil et al., 1999; Cunningham and Russell, 2000; Loyau et al., 2007). In terms of matAb transfer, barn swallow mothers mated to males with experimentally elongated tail feathers (an exaggerated secondary sexual trait) transferred more matAb into egg yolk than females whose partners were controls or had shortened tail feathers, particularly in the first-laid egg (Saino et al., 2002a). However, in a study on collared flycatchers, no allocation of matAb transfer in relation to male qualities was reported (Hargitai et al., 2006). Unfortunately, we have not been able to find any other study presenting data on the effect of male quality on female matAb allocation. We must await more studies before we can evaluate the extent to which females can make adaptive decision on matAb allocation in relation to their mate's quality and attractiveness.

WHAT IS TRANSFERRED: LIFETIME Ab REPERTOIRE OR ONLY Ab FROM ONGOING INFECTIONS?

There are essentially two hypotheses that have been proposed to explain what types of matAb are transferred from mother to offspring.

First, infection at the time of egg laying can be beneficial because it results in transfer of pathogen-specific matAb that protects the neonate. According to this theory, it could even be beneficial for mothers to become infected with common pathogens in order to prepare for their offspring's protection at the neonate stage (Richner et al., 1993; Heeb et al., 1998). An obvious problem with this hypothesis is the cost to the female of an ongoing infection, including negative effects on physiology and feeding rate during the nestling period (e.g., Ilmonen et al., 2000; Råberg et al., 2000).

Second, even antibodies against antigens to which the mother has been exposed to some time ago can be transferred to offspring. This hypothesis states that mothers in principle can transfer their complete immunological experience to their offspring during matAb transmission (Lemke et al., 2003). Support for this hypothesis comes from two studies of birds. In a study on song sparrows (*Melospiza melodia*), females were either vaccinated (inducing primary Ab response against tetanus) or not in autumn (Reid et al., 2006). Next spring (7 to 9 months later), these females produced eggs and raised young that differed in their response to tetanus according to whether their mother had been vaccinated or not the previous autumn (Reid et al., 2006). Thus, this study implies that at least in passerine birds, mothers have a "long-term memory" for antibodies that is transferred to their offspring many months after the mother's own (primary) immune response had ceased. Also, a study of kittiwakes supports a maternal "long-term Ab memory," because vaccination of females in the previous breeding season affected matAb transfer to neonates in the next breeding season (Staszewski et al., 2007). More studies are definitely needed on this exciting topic.

TRADE-OFFS FOR THE MOTHER IN CONNECTION WITH matAb TRANSFER

From an evolutionary perspective, mothers may face trade-offs in relation to matAb transfer, because this process is likely to be physiologically demanding. For example, an egg-producing hen transfers 10% to 20% of her steady-state IgG into the developing egg every day, which results in a 35% to 50% increase in the daily IgG turnover (Kowalczyk et al., 1985). The potential costs for the mother can be several and mainly related to resource allocation. First, making matAbs and transferring them over the egg sac into the egg yolk may require non-negligible levels of energy during the period of egg formation, which has been identified as an energetically demanding period to start with (Nilsson and Råberg, 2001; Williams, 2005) and often takes place early in the breeding season, when conditions in temperate regions are characterized by cold weather and relatively low food abundance (Nager et al., 1997; Royle et al., 2003). Second, transfer of matAb involves a need for high production of antibodies and thus an increased demand for essential nutrients during the period of egg production, resulting in a potential trade-off due to nutrient limitations (Klasing, 1998). Third, transfer of large amounts of matAb into the egg yolk may result in compromised humoral immunity in females. Fourth, if only (or mainly) matAbs from ongoing infections are transferred to offspring, then mothers may pay a cost of infection in order to be able to transfer antigen-specific matAb (Gallizzi et al., 2008a).

The potential benefit for the female is that her offspring may gain fitness benefits from the transferred matAb. These benefits for the mother may be incurred in terms of (1) direct protection of the neonate from pathogens during the first period in its life, when its own intrinsic immune system is not yet fully developed (Grindstaff et al., 2003), and (2) indirect, long-lasting effects priming the offspring's own immunity to the pathogen experience of their mothers (Lemke et al., 2003, 2004). However, the extent to which fitness benefits are dependent mainly on the direct or indirect effects of matAb on offspring immunity is yet not well known.

OFFSPRING UPTAKE OF matAb

There is a broad consensus regarding the potential positive impact of matAb to protect the neonate against pathogens during the first days of life, when the neonate is more or less lacking a functioning humoral and cell-mediated immune defense of its own (Lemke et al., 2009). Studies of model organisms and humans generally support this view, although the length of the period when matAbs have a decisive protective role are likely to differ between organisms due to the route and magnitude of matAb transfer, matAb "half-life" in the offspring, and the timing of development of the offspring's own immune system.

In mammals, where matAb transfer occurs before and at birth (via placenta and colostrums [i.e., the first milk that is breastfed to the newborn]) but also after birth via lactation, matAbs continue to protect the offspring until the end of lactation and beyond, depending on the half-life of matAbs. In humans, passively acquired matAbs before birth have been found to have an exponential clearance rate, with an estimated half-life of 35 to 40 days (Sato et al., 1979; Nicoara et al., 1999; Brugha et al., 1996). matAbs in milk obtained through lactation will only protect the gut (IgA). Still, data from vaccination studies show that in humans, matAbs may continue to protect the infant for up to 9 to 12 months (or more) (Albrecht et al., 1977; Nicoara et al., 1999). In mice, the active uptake of IgG into the bloodstream continues until about 2 weeks of age, although lactation ends after 3 weeks (Anderson, 1995). As the half-life of matAbs in neonate mice is estimated to be 1 to 3 weeks (Takemori and Rajewsky, 1984; Okamoto et al., 1989), matAbs may continue to protect the neonate until an age of 4 to 7 weeks. However, in a detailed study of mice, matAb half-life was 4 to 7 days, and the clearance rate of matAb after the end of lactation (21 days of age) was highly dependent on the antigen dose. Infants of low-dose–infected mothers had cleared matAb completely within a week after weaning, whereas infants of high-dose mothers retained some matAb 3 weeks or more after weaning (Okamoto et al., 1989).

In birds, where matAbs are transferred via the egg yolk and therefore taken up in offspring's circulation before and around hatching, the direct protective role of matAb must be restricted to a rather short period early in life. Studies in the precocial chicken suggest that matAb may be predominant in offspring circulation for up to 14 days of age (Apanius, 1998). However, detailed examination of data from chicken indicates that matAb levels peak at hatching and may be present in the neonate for only 7 to 10 days (Hamal et al., 2006; see also Fig. 8.5 in Apanius, 1998). Furthermore, matAb half-life in chickens is estimated to be 3 to 7 days (Patterson et al., 1962; Davison et al., 2008), also supporting the view of a rather fast decline of matAb in neonate chicken. Moreover, neonate production of IgG can begin at 6 days of age because the IgG-secreting cells are then present in the neonate chicken (Lawrence et al., 1981). In Japanese quail (*Coturnix japonica*), matAbs seemed to be predominant in chicks up to 11 days of age, although titers had dropped by 30% to 50% between day 6 and 11 (Grindstaff, 2008). In pekin ducks, IgG (and IgN) levels in neonates showed a peak at about 0 to 4 days of age and then dropped dramatically over the first 10 days of life, staying low for 1 to 2 months before slowly increasing (Ng and Higgins, 1988; see also Fig. 8.8b in Apanius, 1998). This suggests that matAb levels were high in offspring circulation for only a few days after hatch and were lost from offspring circulation within 10 days of hatching.

Also, in large altricial birds such as gulls, data support a rapid decline of matAb in neonates. In kittiwakes, the level of specific matAb was correlated but lower in 5-day-old offspring than in egg yolk, suggesting a rapid decline of matAb in the neonate (Gasparini et al., 2002). Studies of wild passerine birds also support the view that matAbs are lost from offspring circulation soon after hatching. For example, in nestling pied flycatchers, total Ig levels were low at an age of 5 days and thereafter increased steadily up to an age of 14 days, implying that this increase reflected the endogenous production of Ab in the offspring (Grindstaff et al., 2006). In a study on tree swallows (*Tachycineta bicolor*), no antigen-specific antibodies were found in 10-day-old nestlings of antigen-injected females, suggesting that the matAbs were already lost from offspring circulation (Lozano and Ydenberg, 2002). Similar patterns was found in house sparrows exposed to West Nile virus (WNV), where the half-life of matAb in neonates was estimated to be 3 days and anti-WNV matAb was absent from 9-day-old chicks (Nemeth et al., 2008). Even more surprising is the finding that in the jackdaw (*Corvus monedula*), which is a comparably large passerine bird that allows for plasma sampling at 2 days of age, total antibody levels were at a minimum in 2-day-old chicks and then continuously increased until an age of 25 days, suggesting that matAbs were lost from offspring circulation 1 to 2 days after hatching (M. Eliasson, M. Sandell, J-Å. Nilsson and D. Hasselquist, unpublished). In magpies (a species that is closely related and similar in size to the jackdaw), antibody levels were very low at hatching and increased 30 times by the age of 8 to 10 days, therefore also strongly suggesting that matAbs were lost from offspring circulation very early in life (Pihlaja et al., 2006). Results from the magpie study were corroborated by a recent detailed study on house sparrows in which King et al. (2010) found a very rapid loss of matAb from neonate circulation (within 3 days of hatching), a very short half-life of matAb (2.2 days), and an early onset of offspring intrinsic Ab production (at an age of 3 to 6 days). More studies are needed to investigate the concentration of matAb in neonates during the early development period in order to elucidate how fast matAbs are lost from offspring blood and to determine for how long matAbs may have a direct protective effect on the neonates in different types of birds. However, if the results from the above studies indicating an almost complete loss of matAb in the bloodstream of offspring already at an age of 1 to 5 days are general for passerine birds, then these results question the importance of matAb for direct protection of the neonate except for a few days subsequent to hatching (cf. King et al., 2010).

CONSEQUENCES OF matAb TRANSFER FOR THE OFFSPRING

SHORT-TERM EFFECTS OF matAb ON NEONATE GROWTH AND SURVIVAL

A number of studies on rodents have investigated the effect of female infection/vaccination on offspring growth and survival. In a study of house mice (*Mus musculus*), an experimental group of females was infected with a nematode causing intestinal infection. Offspring of infected mothers had higher birth mass, grew faster, and had larger intestines than offspring of control mothers (Kristan, 2002, 2004). In deer mice (*Peromyscus maniculatus*), females infected with a trematode gave birth to offspring that had increased growth rate and higher survival early in life (Schwanz, 2008a, 2008b). However, offspring to bank vole females with or without Puumala hantavirus (PUUV) infection did not differ in growth rate or survival (Kallio et al., 2006). Still, in several of these cases it cannot be excluded that the infection of the mother induced some kind of "terminal investment" effect so that infected mothers provided more care to their offspring, rather than the benefits gained by offspring being a direct effect of matAb.

Short-term effects of matAb in reptiles are also variable. In a study of an agamid lizard, the Mallee dragon (*Ctenophorus fordi*), lipopolysaccharide (LPS) treatment of the mother had no or even negative effects on neonate growth (Uller et al., 2006). In the common lizard (*Lacerta vivipara*), however, offspring of mite-infested mothers had a higher growth rate than offspring of non-infected mothers (Sorci and Clobert, 1995).

In a study on tree swallows, nestlings of mothers that had been injected with sheep red blood cells (SRBC) about 10 days before egg laying had a significantly higher growth rate than nestlings from control females (Lozano and Ydenberg, 2002). Because broods were cross-fostered to other nests after hatching, the results could not be an effect of injected females investing more in nestling feeding. Instead, the interpretation of these results was either that immune system activation of the mother *per se* resulted in a generally enhanced immune protection in the nestlings, or that the immune challenge in the mothers prompted them to transfer more of other maternal effects, such as androgens (Lozano and Ydenberg, 2002). However, several studies have failed to find any effect of matAb transfer on the growth of bird nestlings (e.g., Grindstaff et al., 2006; Staszewski et al., 2007; Table 6.1b). A particularly interesting case is a study of magpies, where there was a positive relationship between nestling mass and intrinsically produced IgG at day 10 but a negative relationship between mass-controlled growth until day 10 and intrinsically produced IgG at day 10 (Pihlaja et al., 2006). This suggests a trade-off between growth and IgG production in neonates; still, individuals with high body mass could produce stronger IgG responses, making interpretations problematic. In conclusion, studies of the effects of matAb on neonate growth rate show mixed results and generalizations are hard to make at this stage.

Nestling survival can also be affected by matAb transfer. Magpie offspring with higher matAb levels at hatching survived better to an age of 20 days than their nest-mates with lower matAb levels (Pihlaja et al., 2006). However, in this study a between-nest analysis showed that broods in which nestlings had on average higher levels of matAb experienced lower survival over the first 3 weeks of life. Also in a study of kittiwakes, mothers were vaccinated with NDV 1 year before breeding, and then their offspring were vaccinated when 1 day old; offspring of mothers not vaccinated with NDV grew better than offspring of NDV-vaccinated mothers in one out of two years (Staszewski et al., (2007). This result could imply a cost of immunity induced by the NDV antigen during the neonatal period. MatAb may have helped to activate the neonates' own innate immunity, causing a trade-off between immune response and growth rate. However, as the results were not consistent between years, the results have to be interpreted with caution.

SHORT-TERM EFFECTS OF matAb ON NEONATE IMMUNITY

To date only a few studies of wild vertebrates have investigated the effect of matAbs on the neonates' own immunity. A reason for this is that it is hard to determine when the matAbs have been cleared from offspring circulation and the intrinsic production of Ab has begun. However, matAb induced a general enhancement of the immune function of nestling pied flycatchers (Grindstaff et al., 2006). The antigen used in this study was LPS, which induces a nonspecific activation of most B-cell clones, and hence production of a wide variety of antibodies. However, we do not know if this effect was transient over the first few weeks of life in these birds, or if the matAb transfer induced a permanent immune stimulation in the offspring. Furthermore, kittiwake offspring that received naturally high amounts of anti-Borrelia matAb produced higher anti-Borrelia Ab titers when 10 and 20 days old (Gasparini et al., 2006). These data were interpreted as suggesting a "maternal educational" effect. However, at 20 days of age, the offspring's own immunity is not fully matured and its physiology and condition is still heavily dependent on maternal investment, an observation that questions this interpretation. Moreover, this is a correlational study, and thus there are alternative interpretations of these results—for example, that mothers with good genetically based Ab resistance against *Borrelia* produced offspring of similar superior Ab resistance quality, or that maternal effects besides matAb affected offspring Ab resistance. In contrast, in another study of kittiwakes, neonates were NDV vaccinated when 1 day old and when 3 weeks old; offspring differed in their intrinsic anti-NDV Ab levels. Offspring of NDV-vaccinated mothers had nondetectable whereas offspring of non-vaccinated mothers had substantial and increasing levels of anti-NDV Ab titers (Staszewski et al., 2007). Similarly, tawny owl (*Strix aluco*) females with high Ab levels against TETRAVAC vaccine (containing four different disease antigens) produced offspring that when challenged with this antigen at 4 days of age had lower anti-TETRAVAC Ab levels than offspring of mothers with low anti-TETRAVAC Ab levels (Gasparini et al., 2009). These two latter studies support the view that matAb transfer can have negative short-term consequences on offspring intrinsic immunity.

In bank voles, mothers with PUUV hantavirus infections gave birth to offspring that had higher levels of PUUV-specific matAb and were better protected against this pathogen until 3 months after birth compared with offspring of mothers free of PUUV infection (Kallio et al., 2006). In a study of mice (*Mus musculus*), only offspring of previously infected mothers could clear infections of a nematode intestinal parasite (Kristan, 2002). However, if the nematode infection in offspring was not cleared, then infection intensity was higher in the offspring of previously infected versus non-infected mothers.

To conclude, these data seem to fit a scenario where matAbs can have positive effects on the offspring's own immunity during development. However, if the neonate is exposed to a specific antigen that matches its matAb, then this seems instead to have (at least) short-term immunosuppressive effects (cf. Siegrist, 2003; Hasselquist and Nilsson, 2009).

LONG-TERM CONSEQUENCES OF matAb ON MATURE OFFSPRING IMMUNITY

matAbs result in suppressed offspring Ab responses at adult age

It has long been known from human vaccination studies that antigen-specific matAbs can inhibit the infant's own specific Ab response against an antigen, resulting in lack of infant protection after vaccination too early in life (e.g., Anderson, 1995; Carlier and Truyens, 1995; Siegrist, 2003). For example, in studies of human measles vaccination, passively acquired matAbs must reach low levels for the vaccination to be efficient, and this can take 6 to 12 months (Albrechts et al., 1977; Brugha et al., 1996; Nicoara et al., 1999).

Based on such vaccination studies in humans and mice, a number of hypotheses have been proposed to explain the negative effect of matAbs on offspring intrinsic (own) immune responses (Siegrist, 2003). The mechanism that seems to fit the data best at present is based on the idea that matAbs mask epitopes of the antigen, resulting in a lack of B-cell antigen recognition, and hence lack of antigen-specific Ab production in the infant (Siegrist, 2003). This mechanism fits well with available data where matAbs above a threshold level are inhibitory, whereas matAb levels below the threshold allow for antigen-specific Ab production in infants (Campagne et al., 1998; Nicoara et al, 1999). Moreover, vaccine with higher doses of antigen can also overcome the matAb inhibition and activate infant B cells, resulting in a robust intrinsic Ab response in the infants (Siegrist, 2003).

Moreover, the inhibitory effect of matAb may be only transient in mammals, because T-cell reactions seem unaffected by matAb, and hence there will be no matAb inhibition of secondary antibody responses (Siegrist, 2003; see Fig. 1 in Hasselquist and Nilsson, 2009). However, a study on rats suggested that matAb transfer may bias the offspring immune system permanently towards higher Th2 and lower Th1 responses, resulting in suppressed T-cell–mediated immune responses (Lundin et al., 1999). As far as we know, the possibility of a permanent suppression of B- or T-cell–mediated immunity has not been studied in detail and not in organisms besides mice and humans; thus, our knowledge about these long-term processes is limited, and they may differ between different vertebrate groups.

matAbs result in enhanced offspring Ab responses at adult age

Based primarily on studies of mice, a long-term beneficial effect of matAb on offspring intrinsic immunity has been proposed (Lemke and Lange, 1999; Lemke et al., 2004; Hasselquist and Nilsson, 2009). Female mice injected with antigens support this effect, because they produced offspring that had higher antigen-specific Ab titers at adult age (Lange et al., 1999), and in rats this could be detected even in the F2 generation (Lundin et al., 1999). The mechanisms proposed to explain these results are based on matAb transfer to the next generation, where matAbs either act as internal images (Anderson, 1995) or induce idiotypic network interactions (Tanasa et al., 2010) in the developing immune system of the neonate, priming it to the antigen repertoire to which the mother has been exposed (Lemke et al., 2004). Independent of the exact mechanism, it will result in an epigenetic maternal priming of offspring intrinsic immune function (Grindstaff et al., 2003; Boulinier and Staszewski, 2008; Hasselquist and Nilsson, 2009).

Besides studies of mice, two studies of birds support the scenario of long-term enhancement of offspring immunity caused by matAb transfer. In song sparrows, offspring to mothers injected with novel antigens 9 months prior to egg laying had higher antigen-specific Ab responses when 3 to 4 months old compared with offspring of non-injected mothers (Reid et al., 2006). Furthermore, when chicken mothers were vaccinated with *Salmonella typhimurium*, their grown-up offspring had better resistance against wild-type Salmonella strains (Hassan and Curtiss, 1996). These data strongly suggest an epigenetic maternal priming of antigen-specific immune responses in the offspring with long-lasting effects on their own intrinsic Ab responses.

LONG-TERM EFFECTS ON OFFSPRING FITNESS

Few studies have evaluated the long-term effects of maternal infection or matAb transfer on offspring quality and reproductive performance. In deer mice, sons of trematode-infected mothers were heavier when they had matured at 3 months of age, and large body size at this stage is a decisive factor for gaining high reproductive success (Schwanz, 2008b). In the common lizard, offspring of mite-infested mothers experienced higher reproductive success compared with offspring to non-infected mothers (Sorci and Clobert, 1995). In neither of these studies were matAbs measured, so these results may reflect other factors such as differential direct investment in offspring by the mother according to a "terminal investment" theory (see "Short-term effects of matAb on neonate growth and survival" above). Thus, the effect of matAb on offspring fitness is definitely an area open for further studies.

HYPOTHESES PROPOSED TO EXPLAIN MATERNAL ANTIBODY TRANSFER

In this section, we will outline four hypotheses proposed to explain the occurrence of matAb transfer in vertebrates: matAbs (1) have no adaptive function, (2) have positive effects on growth and development of the neonate, (3) provide direct immunoprotection of the neonate, and (4) provide long-term benefits in matured offspring.

"NO ADAPTIVE FUNCTION" HYPOTHESIS

According to this hypothesis, matAb transfer is merely a passive process involving small IgG molecules that pass over the placenta or egg sac membranes into the growing embryo or egg yolk, but without any functional effects on offspring quality (e.g., Tschirren et al., 2009). However, there are several findings that are not compatible with this hypothesis. First, in humans, mice, and chicken, mechanisms for active transfer involving specialized receptors have been identified (e.g., Anderson, 1995; West et al., 2004). Indeed, in birds the level of Ab in the egg yolk may be 20% to 80% higher than in the circulation of the mother (Kowalczyk et al., 1985; Abdel-Moniem and Abel-Gawad, 2006), and in human anti-measles Ab levels have been found to be about 50% higher in the fetus than in the circulation of the mother (Nicoara et al., 1999). Another line of evidence for the adaptive role of matAb transfer to offspring comes from studies of fitness effects of matAb transfer in birds and mammals, where a number of studies show positive effects on components of offspring performance, such as growth rate, immunity, and survival (Table 6.1b). All these observations strongly imply that matAb transfer is not a passive non-adaptive process, but rather a demanding physiological adaptation to the benefit of the embryo/neonate.

"GROWTH AND DEVELOPMENT TRADE-OFF" HYPOTHESIS

According to this hypothesis, nutrients are involved in a trade-off between requirements for mounting immune responses and rapid growth and development in the neonate. Transfer of matAbs could alleviate this trade-off by reducing the need for nutrients directed to the immune system, which instead can be used for rapid growth during the earliest post-hatching period (Klasing and Leshchinsky, 1999; Buechler et al., 2002; Pihlaja et al., 2006). matAbs in general may be important to alleviate the costs of making a broad repertoire of Abs that can protect against low-pathogenic (or even normally commensal) bacteria and virus, as well as avian malaria and gut macroparasites. Under this scenario, we would predict that the development of the offspring's own immune function should be negligible until the end of the rapid growth period (at least 2 to 4 weeks in birds and small mammals). Studies of matAb transfer in altricial birds do not support this prediction, because matAbs are quickly lost (within 2 to 5 days) from circulation. In precocial birds such as chicken and quail, matAbs may be retained for up to 10 to 14 days (see "Offspring uptake of matAb"), and matAbs could then help to alleviate infection costs, facilitating growth. Consistent with this, Japanese quail chicks immunized with the same antigen their mother had been exposed to prior to egg laying grew better than offspring immunized with an antigen to which their mother had not been exposed (Grindstaff, 2008). In humans, matAbs remain in offspring circulation for a much longer period (often 9 to 12 months or longer), potentially alleviating nutritional costs for the offspring. However, the matAb levels in offspring circulation may reach rather low levels already at an offspring age of 6 to 9 months (Nicoara et al., 1999), which somewhat restricts the positive effects on the growth/development trade-off in humans. Furthermore, it can be debated the extent to which the production of immune cells is sufficiently nutrient-demanding compared to growth to be of any importance for nutrient allocation (Klasing, 1998, 2004). Still, there could be other costly effects of antigen exposure. For example, antigens may elicit a costly innate response in the neonate (e.g., Råberg et al., 2002) that retards growth, and under such a scenario matAbs could help to alleviate immune costs and facilitate growth. More studies of different systems are needed to elucidate the growth trade-off function of matAbs.

"DIRECT IMMUNOPROTECTION OF NEONATE" HYPOTHESIS

According to this hypothesis, matAbs protect the neonate against moderate or highly pathogenic parasites during the early post-hatching period, until the newborn has developed its own intrinsic immune system to a stage when it can be protective. The logic behind this hypothesis is that the protection of the neonate against specific parasites encountered by the mother is the adaptive driving force for the epigenetic behavior of transgenerational Ab transfer (e.g., Buechler et al., 2002; Grindstaff, 2008). Hence, we would expect neonates to suffer less from infections and parasites frequently encountered by their mother during the period when the offspring's own intrinsic immune system has not yet matured—about 1 to 2 weeks in altricial and 2 to 5 weeks in precocial birds, 4 to 6 weeks in mice, and at least 12 months in humans (Grindstaff et al., 2003; Hasselquist and Nilsson, 2009). There are some studies that support this hypothesis, in particular the studies on great tits and resistance to flea-induced costs on growth (Heeb et al., 1998; Buechler et al., 2002; Table 6.1b). However, in birds matAbs seem to have vanished from the bloodstream of the neonate within a few days after hatching (see "Offspring uptake of matAb"), leaving several weeks open to attack from diseases before the offspring's own immune system has developed. This seems to be a paradox.

"LONG-TERM BENEFIT" HYPOTHESIS

This hypothesis proposes that a major advantage of matAb transfer is the long-term (potentially lifelong) positive effects on the offspring's own immune system. In this scenario, mothers will prime the intrinsic immune system of their offspring to the locally prevailing parasites (Lemke and

TABLE 6.1B: Effects of maternal antibody transfer on offspring performance.

Species	Neonate growth rate	Neonate survival	Neonate immunity or health[1]	Offspring maturation & reprod[2]	Offspring surviving to adult age	Mature offspring immunity	Treatment	References
Mammals								
House mice (*Mus musculus*)	+	+				+ and −	Natural nematode infection	Kristan, 2002, 2004
	+	+				+		Reuman et al., 1983
							Influenza challenge	Gustafsson et al., 1994
							Virus challenge	Okamoto et al., 1989
Rat (*Rattus norvegicus*)						+	Antigen challenge	Lundin et al., 1999
Deer mice (*Peromyscus maniculatus*)	+	+					Natural trematode infection	Schwanz 2008a, 2008b
Bank vole (*Clethrionomys glareolus*)	−	−	+ h	+ m			Natural hantavirus infection	Kallio et al., 2006
Birds								
Chicken (*Gallus gallus*)		+ [5]	+ i					e.g., Smith et al., 1994; Heller et al., 1990
Rock dove (*Columba livia*)		+				+	Virus challenge	Kissling et al., 1954[6]; Reeves et al., 1954[6]
Kittiwake (*Rissa tridactyla*)	0	(−) [4]						Staszewski et al., 2007
Tawny owl (*Strix aluco*)	0							Gasparini et al., 2009
Magpie (*Pica pica*)	0	(+)	(+) i				Raw egg food suppl.	Pihlaja et al., 2006
Pied flycatcher (*Ficedula hypoleuca*)	(+)[3]		(+)[3]		(+)[3]		Mealworm food suppl.[3]	Moreno et al., 2008
Pied flycatcher	0	0	+ i				Exp antigen inject (dip/tet)	Grindstaff et al., 2006

Pied flycatcher		(+)					Observational data	Morales et al., 2006
Great tit (*Parus major*)	(+) 0	(+)	+ h				Observational data; flea + cross-fostering exp	Heeb et al., 1998; Buechler et al., 2002; Tschirren et al., 2009
Song sparrow (*Melospiza melodia*)						+	Exp antigen injection (dip/tet)	Reid et al., 2006
Tree swallows (*Tachycineta bicolor*)	+						Exp antigen injection (SRBC)	Lozano and Ydenberg, 2002
Lizards								
Common lizard (*Lacerta vivipara*)	+			+ r			Natural mite infection	Sorci and Clobert, 1995
Mallee dragon (*Ctenophorus fordi*)	0 (−)						Exp antigen injection	Uller et al., 2006

[1] h = matAb positive effect on offspring health, i = matAb positive effect on offspring immune response

[2] m = maternal infection/matAb positive for maturation of offspring, r = maternal infection/matAb positive for offspring reproduction

[3] Mothers supplemented with mealworms but matAb not measured—only total IgG in 12-day-old chicks that had been cross-fostered. Hence, effects on offspring may depend on any change in egg composition, not just egg yolk matAb content.

[4] Offspring injected with NDV when 1 day old, and anti-NDV measured when they were 15–25 days old.

[5] Neonate survival measured after injection with *E. coli* strain to which their mother previously had been exposed.

[6] Cited in Grindstaff et al., 2003.

Lange, 1999; Grindstaff et al., 2003; Lemke et al., 2004, 2009; Hasselquist and Nilsson, 2009). Thus, we do not necessarily expect direct positive effects on the neonate physiology or performance, but instead we expect higher endogenously derived immune responses in offspring against parasites/antigens to which their mother previously had been exposed. Support for this hypothesis comes from some studies of birds (Grindstaff et al., 2006; Reid et al., 2006) and small mammals (Lange et al., 1999; Lundin et al., 1999). However, vaccination studies in humans do not seem to support this hypothesis (Siegrist, 2003). Moreover, in experiments on barn owls and kittiwakes, an apparently suppressive effect of matAbs on Ab production in 3-week-old offspring were found (Staszewski et al., 2007; Gasparini et al., 2009), although in these studies the immunosuppression occurred early in life and at the first exposure to the antigen.

CONCLUSION

At present, we know surprisingly little about the sources, timing, and strategies of maternal antibody transfer in non-model vertebrates, and this is also true for the potential short- and long-term effects on the offspring. Thus, there is an ample need for both basic, rather mechanistic, studies of how matAb transfer works in non-model organisms, as well as studies of the direct and long-term epigenetic and evolutionary effects on the offspring. In particular, the long-term effects of matAb transfer on immunity and fitness of mature offspring have hardly been studied in wild vertebrates. However, such studies are essential to evaluate the adaptive value of matAb transfer. Moreover, it would also be very interesting to put matAb transfer into the framework of recent immunoecology and life-history theory—for example, conducting comparative, across-species investigations of matAb transfer in relation to fast/slow pace-of-life (Riklefs and Wikelski, 2002; Hasselquist, 2007) as has already been done for some tropical passerine species (Addison et al., 2009). Another interesting route of investigation would be to analyze matAb transfer in species that differ in migration strategy (sedentary, short- or long-distance migrants), hence facing very different risks of contracting pathogens during their first winter (Hasselquist, 2007).

MATERNALLY TRANSFERRED STEROID HORMONES AND OFFSPRING IMMUNE FUNCTION

Mothers not only transfer antibodies, but they also convey a number of other immunomodulatory substances, such as nutrients or hormones, to their offspring. The transfer of maternally derived steroid hormones in particular has received considerable attention within the field of immunoecology. In this section, we first describe some characteristics of steroid hormones and how they may interact with the immune system. We then review the literature dealing with maternal/prenatal steroids and their effects on the development and functioning of the immune system in different vertebrate species. We sum up this section with some cautionary remarks and future directions.

HORMONAL PROGRAMMING

Steroid hormones are chemical regulators that are effective in minute quantities. They can bind to specific receptor molecules, which in turn may bind to DNA and thereby influence the expression of specific genes and the production of proteins (e.g., Becker et al., 2002). As regulators of gene expression, steroid hormones can have a substantial impact on development and the differentiation of the phenotype. One particular feature of steroid hormones is that they can have so-called "organizational effects" on the phenotype. Exposure to steroid hormones during a short critical period early in the development can induce permanent changes in morphology, physiology, and behavior (e.g., Collaer and Hines, 1995; Martin, 2000; Welberg and Seckl, 2001; Rhen and Crews, 2002; Groothuis et al., 2005b). These changes, as we will see in the next sections, also include alterations of the immune system. Organizational effects can be contrasted with "activational effects," which alter the state of the phenotype temporarily and are thus reversible (e.g., Arnold and Breedlove, 1985), and such activational effects also influence the vertebrate immune system (e.g., Folstad and Karter, 1992). Another important characteristic of steroid hormones is that they typically act on various target tissues and therefore can affect a wide array of phenotypic traits. Hence, the effects of single steroid hormones have been compared to pleiotropic effects of single genes influencing multiple phenotypic traits in the organism (e.g., Ketterson and Nolan, 1992; Groothuis and Schwabl, 2008). In evolutionary terms, this means that selection on a trait that is under the influence of steroid hormones may lead to correlated selection on other traits that are under the influence of the same hormones. Depending on the nature of the effect of the steroids on the different traits, selection may be antagonistic (investment in one trait has to be traded off against investment in another trait) or reinforcing (investment in one trait will also benefit evolution of another) (e.g., Ketterson and Nolan, 1992). Through transfer of steroid hormones, mothers may thus influence their offspring's phenotype in various ways.

Organizational effects as a consequence of maternally derived steroid hormones are often referred to as "prenatal programming," "maternal phenotypic engineering," or "fetal programming" (e.g., Gill et al., 1998; Kapoor et al., 2006; Rettenbacher et al., 2009). Vertebrate mothers, including humans, transfer notable amounts of steroid hormones to their offspring (e.g., Groothuis et al., 2005b). It is now evident that maternal hormone transfer is an important maternal effect that contributes to phenotypic variation in the offspring (e.g., Martin, 2000; Welberg and Seckl, 2001; Rhen and Crews, 2002; Seckl, 2004; Groothuis et al., 2005b; Kapoor et al., 2006; Gil, 2008). However, even though this particular area of research has received a surge of interest during recent years, it becomes increasingly clear that the effects of maternal hormones are complex. Environmental factors interact with maternal hormones to modify phenotypic plasticity in the offspring (e.g., Pilz et al., 2004; Groothuis et al., 2005b; Cucco et al., 2008; Müller et al., 2010; Tobler et al., 2010). To discuss these interactions in detail is beyond this overview, but it should be borne in mind that variable effects of prenatal steroids reported for different species partly derive from this complexity.

Several mechanisms have been proposed to explain how steroid hormones can influence the development of the immune system. Martin (2000) discusses immune modulation by sex steroids (androgens and estrogens) as a consequence of perinatal masculinization/defeminization in mammals, where sexes generally differ in immune function and in sex-steroid levels. This has led to a multitude of studies investigating sex-steroid modulation of the immune system (reviewed in

Schuurs and Verheul, 1990; Martin, 2000). One mechanism that has been proposed to affect prenatal programming of immune function is that sex steroids influence the development of the central nervous system, and differences in immune responses later in life are affected by variation in signaling from the brain to immune cells (Martin, 2000). Immune cells express receptors for both hormones and neurotransmitters (Roszman and Brooks, 1997), which means that the endocrine and the nervous system can interact via the immune system ("cross-talk"; Viltart and Vanbesien-Mailliot, 2007).

Another possible mechanism is that early sex-hormone exposure permanently alters the development of immune organs (thymus, spleen, lymph nodes [mammals], bursa of Fabricius [birds]). Studies on poultry show that the development of the bursa of Fabricius is influenced by prenatal egg hormone manipulation, because low doses have no or a stimulatory effect on bursa growth whereas high doses impair growth (e.g., Glick, 1961; Glick and Sadler, 1961; Norton and Wira, 1977; Verheul et al., 1986; Kondo et al., 2004). The immune cells that develop in these organs (e.g., lymphocytes) would then have different functionalities (or activities) depending on earlier organ development (Martin, 2000). Similar examples of effects of prenatal hormones on immune organ development and function can be found in mammals. For example, administration of testosterone in 2-day-old female rats has been shown to alter thymus development (Leposavic et al., 2009). Prenatally administered glucocorticoids reduce thymus weight in mice (Eishi et al., 1983), and similarly prenatal stress during late gestation (mediated by glucocorticoids) reduces thymus weight in neonatal pigs (Tuchscherer et al., 2002).

A further possibility is that sex steroids act on the hypothalamus-pituitary-gonadal (HPG) axis, the hypothalamus-pituitary-adrenal (HPA) axis, or the hypothalamus-pituitary-somatotrophic axis (HPS) to permanently alter the production or action of hormones, which then directly interact with the immune system. For birds, there is as yet little evidence that yolk androgens affect these systems, although some studies are suggestive (reviewed in Groothuis and Schwabl, 2008). For mammals, there is increasing evidence that prenatal steroid exposure (both sex steroids and glucocorticoids) affect not only the production of steroid hormones later in life, but also the expression and functionality of steroid receptors in immune organs and the brain (e.g., Levitt et al., 1996; Welberg et al., 2001; Kanitz et al., 2006; Recabarren et al., 2009; Guzman et al., 2009).

Apart from direct effects, prenatal steroids may also have indirect effects on the vertebrate immune system. Prenatal steroids affect a wide range of physiological and behavioral characteristics (see above), and hormone-mediated changes in these characteristics may lead to alterations in immune function. It is well known that glucocorticoids have programming effects on behavior in mammals (e.g., Seckl, 2004; Viltart and Vanbesien-Mailliot, 2007) and behavioral alterations may make animals more or less disease-prone. Moreover, accelerated growth, mediated by prenatal hormones (e.g., Schwabl, 1996; McCormick, 1999; Navara et al., 2005; Uller et al., 2007) may result in less energy invested in immune function (e.g., Soler et al., 2003; Uller and Olsson, 2003; Andersson et al., 2004) or in an adaptive downregulation of the immune system to avoid oxidative stress (e.g., von Schantz et al., 1999; see also "Female stress level" and "The effect of antioxidants on the immune system—studies of adults").

The following overview on the effects of maternally derived hormones on the immune function of offspring is divided in two parts: (1) "generally" oviparous species (birds, reptiles, amphibians, and fish, although many species of reptiles, amphibians, and fish are also live-bearing) and

(2) viviparous species (humans and other mammals). The reasoning for this separation is that maternal hormone transfer differs markedly between the two groups. In oviparous species, mothers transfer a preset amount into the egg, but once the egg is laid, no further adjustment of the offspring's prenatal hormone milieu is possible. In viviparous species, in contrast, the embryo develops inside the mother and until birth there is a constant exchange of hormones between mother and offspring. Moreover, in mammals mothers transfer steroids to their offspring even after birth through lactation. Thus, unlike in oviparous species, the amount of maternally transferred compounds to offspring can vary throughout prenatal and neonatal development. Hence, the timing of steroid exposure is important as steroids can have variable effects at different stages of the development. Also, in mammals, hormone transfer is not simply unidirectional (from mother to the offspring) because the fetus can also release its own hormones via the placenta into the mother's bloodstream. Furthermore, the types of steroids that have been investigated and/or are relevant in the context of immune system development differ between the two groups (see "Maternal steroid hormone transfer and offspring immunity in egg-laying vertebrates").

MATERNAL STEROID HORMONE TRANSFER AND OFFSPRING IMMUNITY IN EGG-LAYING VERTEBRATES

Because the majority of studies that have investigated prenatal steroid effects on the immune system from an ecological/evolutionary perspective in oviparous species are avian studies, we will discuss them here in more detail. The first studies investigating the effects of steroid hormones on the development of the avian immune system were done on poultry. These early studies showed that when eggs were dipped in hormone solutions, both the growth of the bursa of Fabricius (an immunological organ in birds) and antibody production were impaired in chicken (e.g., Glick, 1961; Glick and Sadler, 1961; Norton and Wira, 1977; Verheul et al., 1986). The problem with these studies is that they used pharmacological doses of hormones, which makes it difficult to assess how maternally derived hormones (i.e., doses to which the embryos are naturally exposed) affect the avian immune system. With the discovery of systematic intra-clutch variation of yolk testosterone in canary eggs (*Serinus canaria*) (Schwabl, 1993) came the idea that maternal hormone transfer is adaptive, preparing the offspring to the prevailing environmental conditions (Schwabl, 1993; Schwabl et al., 1997). Subsequently, systematic intra-clutch variation of egg androgens was found in many other bird species, which resulted in a rapidly increasing number of studies investigating the effects of egg androgens on a wide range of offspring characteristics (reviewed in Groothuis et al., 2005b; Gil, 2008). The studies on egg-laying species that have investigated egg androgen effects on immune function (or characteristics linked to it) are summarized in Table 6.2.

For birds, there is a representative sample of 14 studies on a total of 12 species. All but one of these studies have focused on androgens (testosterone and androstenedione), which is in contrast to studies on mammalian species, which have more strongly focused on glucocortocoids (cortisol, corticosterone; see "Effects of prenatal glucocorticoids on mammalian offspring immune function"). The reason for this is that while the presence of androgens in bird eggs is now well established (e.g., Schwabl, 1993; Groothuis et al., 2005b; Gil, 2008), the presence of

TABLE 6.2: Studies within the field of evolutionary biology/behavioral ecology on the effects of experimentally elevated hormone levels on immune function in birds and reptiles (no comparable studies on amphibian or fish) T=testosterone; A4= androstenedione; CORT= corticosterone

Species Birds	Treatment	Immune estimate	Life stage	Effect	References
Chinese painted quail (*Coturnix chinensis*)	Elevated T[a]	PHA assay	Chick fast growers	Negative in	Andersson et al., 2004
Black-headed gull (*Larus ridibundus*)	Elevated T, A4	PHA assay	chick	Negative	Groothuis et al., 2005a; Müller et al., 2005
	Elevated T, A4	Total ab levels	chick	Negative	Müller et al., 2005
Eastern bluebird (*Sialia sialis*)	Elevated T[b]	PHA assay	nestling	Neg./no effect	Navara et al., 2005
	Elevated T[b]	Leukocyte count	nestling	No effect	Navara et al., 2005
	Elevated T[b]	Heterophil: lymphocyte ratio	Nestling	No effect	Navara et al., 2005
Grey partridge (*Perdix perdix*)	Elevated T[b]	PHA assay	chick	Neg./pos.	Cucco et al., 2008
		PHA assay	adult	No effect	Cucco et al., 2008
Jackdaw (*Corvus monedula*)	Elevated T, A4	PHA assay	nestling	Negative	Sandell et al., 2009
	Elevated T, A4	Specific antibody production (diphtheria–tetanus)	nestling	Negative	Sandell et al., 2009
Great tit (*Parus major*)	Elevated T	PHA assay	nestling	No effect	Tschirren et al., 2005
Pheasant (*Phasanus colchicus*)	Elevated T	PHA assay	chick	No effect	Rubolini et al., 2006a
Yellow-legged gull (*Larus michahellis*)	Elevated T	PHA assay	chick	No effect	Rubolini et al., 2006b
	Elevated CORT	PHA assay	chick	Negative	Rubolini et al., 2005
	Elevated CORT	Specific antibody production	chick	No effect	Rubolini et al., 2005
Spotless starling (*Sturnus unicolor*)	Elevated T, A4	PHA assay	nestling	No effect	Müller et al., 2007
Collared flycatcher (*Ficedula albicollis*)	Elevated T, A4	PHA assay	nestling	No effect	Pitala et al., 2009

TABLE 6.2: (Continued)

Species Birds	Treatment	Immune estimate	Life stage	Effect	References
House finch (*Carpodacus mexicanus*)	Elevated T	PHA assay	nestling	Positive	Navara et al., 2006a
Zebra finch (*Taeniopygia guttata*)	Elevated T	PHA assay	nestling	(Negative)[c]	Tobler et al., 2010
	Elevated T	PHA assay	adult	No effect	Tobler et al., 2010
	Elevated T	Total antibody levels	nestling	No effect	Tobler et al., 2010
	Elevated T	Total antibody levels	adult	No effect	Tobler et al., 2010
	Elevated T	Specific antibody production (diphtheria–tetanus)	adult [c]	Positive	Tobler et al., 2010
Reptiles					
Common lizard (*Lacerta vivipara*)	Elevated T	Ectoparasite exposure	juvenile	No effect[d]	Uller and Olsson, 2003
	Elevated T	Leukocyte count	juvenile	No effect[d]	Uller and Olsson, 2003

[a]: supraphysiological dose
[b]: both low and high dose used in the same study
[c]: only in small broods
[d]: indirect effect on immunity through growth

glucocorticoids is still under debate (Rettenbacher et al., 2009). Thus, although the study on yellow-legged gull (*Larus michahellis*) chicks shows a negative effect of experimentally elevated yolk corticosterone on cell-mediated immunity (Table 6.2; Rubolini et al., 2005), the relevance of such data will have to be established in the future. Until then, however, it seems safe to assume that maternally derived androgens can influence the development of the avian immune system, even though effects may be species-specific (Table 6.2). The common approach to study egg androgen effects on the avian immune system is to experimentally elevate androgen levels in the eggs and then challenge the offspring that hatch from these eggs with an immune stimulant. To simplify the discussion, we have divided maternal androgen effects on the avian immune system in two groups: short- and long-term effects. Effects of prenatal steroid hormones on immune function (both short and long term) in non-avian species are discussed in the last part of this section.

SHORT-TERM EFFECTS

Most of the avian studies have investigated the effects of egg androgens during the development of the immune system—that is, during the nestling or chick stage (Table 6.2).

CELL-MEDIATED IMMUNITY

All of the studies focusing on the early postnatal life stage used a phytohemagglutinin-A (PHA) assay (a frequently used measure of innate and cell-mediated immunity in avian studies; see, for example, Martin et al., 2006). The effects of egg androgen manipulation on cell-mediated immunity of young birds are very variable, with four studies demonstrating negative effects, one study with a partly negative effect, five studies with no effects, one study with a positive effect, and two studies with negative or positive/no effect (depending on the dose used for manipulation, Table 6.2). Such variation is difficult to interpret but not very surprising, as egg androgen effects are highly variable with respect to a whole range of other offspring characteristics besides immune function (e.g., growth, begging behavior; see Groothuis et al., 2005b; Gil, 2008, for comprehensive reviews on egg androgen effects in birds). One plausible explanation is that the context in which egg androgen effects are studied is important. For example, the negative effect of experimental egg testosterone manipulation on cell-mediated immunity in zebra finch nestlings was apparent only in small, but not large, broods (Tobler et al., 2010). The competitive environment in which the nestlings grow up seems to influence the degree to which egg androgens modulate offspring immunity (see also Müller et al., 2010, for a similar example on offspring growth). It is therefore important to take the context into account before drawing general conclusion about the effect of egg androgens on cell-mediated immunity. Apart from context, other factors such as species or dose (see below) used for manipulation are also likely to be important. Despite the variation in egg androgen effects, we can tentatively conclude that egg androgens have the potential to contribute to variation in cell-mediated immunity during early life stages.

HUMORAL IMMUNITY

Three studies also investigated egg androgen effects on the humoral component of the immune system during the nestling/chick stage. In black-headed gulls (*Larus ridibundus*) and in zebra finches (*Taeniopygia guttata*), experimental elevation of egg androgen levels did not influence total antibody levels in 1-week-old gull and 2.5-week-old finch nestlings, respectively (Müller et al., 2005; Tobler et al., 2010). This may suggest that overall health status is not suppressed by egg androgens. However, when the black-headed gull chicks were challenged with LPS 1 week after hatching, their subsequent total antibody levels (2 days after the challenge) increased in control chicks but not in chicks that had hatched from eggs with elevated androgen levels (Müller et al., 2005). In the third study evaluating humoral immunity, about 1-week-old jackdaw chicks exposed to experimentally elevated prenatal androgen levels had lower naturally occurring LPS-specific antibodies compared to control chicks (Sandell et al., 2009). However, when experimentally challenged with LPS, specific antibodies against LPS were similar in androgen-exposed and control chicks 1 week after the challenge. In this respect, the jackdaw study contrasts somewhat with the one on black-headed gulls. Different timing of post-challenge blood sampling and differences in early life stages (precocial vs. altricial) may have resulted in species-specific responses toward LPS or LPS-like antigens. At 4 weeks of age, jackdaw chicks were challenged with diphtheria–tetanus antigens. Chicks exposed to experimentally elevated egg androgen levels showed lower antibody responses toward diphtheria–tetanus compared to control chicks (Sandell et al., 2009), thus corroborating the negative effects of prenatal androgens on the humoral immunity found in the study on black-headed gulls (Müller et al., 2005). Although further studies are required to assess the effects of prenatal androgens on humoral immunity, there is convincing evidence from two cases in which prenatal androgens were immunosuppressive. Maternal steroids should thus be taken into account as modulators of the humoral immune system during early development.

LONG-TERM EFFECTS

Only two studies so far have investigated whether prenatal androgens have long-lasting consequences for immune function in adulthood.

Cell-mediated immunity

Experimentally elevated egg testosterone levels did not influence cell-mediated immunity in adult grey partridges (*Perdix perdix*) or in adult zebra finches (Cucco et al., 2008; Tobler et al., 2010). Thus, there is currently no evidence that cell-mediated immunity in adulthood is modulated by egg androgens.

Humoral immunity

Tobler and colleagues challenged sexually mature zebra finches that originated from either eggs with experimentally elevated testosterone levels or control eggs repeatedly with diphtheria–tetanus antigens (at about 5 and 7 months of age) (Tobler et al., 2010). Zebra finches that had been exposed to elevated egg testosterone showed higher primary and secondary responses towards the two antigens. Furthermore, there appeared to be a trade-off between adult cell-mediated and humoral immunity in control birds but not in testosterone-exposed birds. In addition, primary and secondary humoral immune responses (toward diphtheria and tetanus) appeared to be more tightly linked in control birds compared to testosterone-exposed birds. This is in contrast with the findings of immunosuppressive effects of prenatal androgens during early development (see above). It appears as if adult zebra finches had enhanced immune function due to prenatal exposure to testosterone. Nevertheless, the findings from the zebra finch study are intriguing as they suggest that prenatal androgens may have a permanent (i.e., organizing) effect on the immune phenotype of avian species. However, further studies are needed to validate these results.

EFFECTS OF MATERNAL STEROIDS ON IMMUNE FUNCTION IN NON-AVIAN SPECIES

For other egg-laying vertebrates (reptiles, amphibians, fish), there are as yet few or no data available on whether and how maternal steroid hormones affect offspring immunity. The only study that is comparable to the bird studies is a study on live-bearing common lizards (*Lacerta vivipara*) (Table 6.2) (Uller and Olsson, 2003). In this study, offspring exposed to experimentally elevated testosterone levels during gestation showed impaired growth in response to ectoparasite infestation (ticks, *I. ricinus*). There was, however, no effect of prenatal testosterone exposure on the number of ectoparasites or the number of leukocytes. Hence, the effect of testosterone on offspring immunity was indirect. Investment in growth was traded off against immune defense toward ectoparasites. Although this is the only study on reptiles and there are as yet no studies on amphibians or fish, it seems likely that maternal steroid hormones may have similar immune-modulating effects on offspring immunity as in birds—particularly so, because experimental manipulations have shown that other phenotypic characters such as growth, behavior, and survival are affected by steroid hormones in these taxa (e.g., McCormick, 1999; Uller et al., 2007; Hu et al., 2008; Gagliano and McCormick, 2009; Uller et al., 2009; but see Radder and Shine, 2007). Thus, as in the study on common lizards, changes in growth or behavior mediated by maternally derived steroid hormones may have to be traded off against changes in immune function.

MATERNAL HORMONE TRANSFER AND OFFSPRING IMMUNITY IN HUMANS AND OTHER MAMMALS

As mentioned, maternal hormone transfer in mammals differs in various ways from maternal steroid hormone transfer in egg-laying species. This is primarily because the embryo develops inside the mother. Mammalian mothers are able to convey hormones to their offspring throughout gestation (i.e., at all stages of the prenatal development). Moreover, even after birth they will transfer steroid hormones to their offspring through lactation. The period of maternal steroid hormone exposure is much more extended compared to egg-laying species (the same reasoning goes for the transfer of maternal antibodies and maternal antioxidants). As a consequence of these differences, the timing of maternal steroid exposure becomes an important factor. Moreover, different experimental approaches are needed to investigate the effects of maternally transferred steroids. Common approaches are supplementation of mothers with additional steroids during gestation or neonatal offspring immediately after birth (e.g., Schlabritz-Loutsevitch et al. 2006; Otten et al., 2007; Leposavic et al. 2009). Compared to egg-laying vertebrates, these approaches encompass several drawbacks that should be kept in mind when interpreting data on early hormone exposure in mammals. First, the amount of hormones supplemented to mothers may not directly translate into the amount transferred to the offspring because mothers may be able to regulate hormone transfer to some degree. Second, supplementation of hormones during gestation may alter the physiology and behavior of mothers, which may indirectly affect offspring development. Such indirect effects may be hard to separate from the effects of the hormones *per se*. Third, supplementation of neonatal offspring with steroids will neglect the prenatal period, which is potentially problematic because certain organizational effects of hormones may be strongest during the prenatal period. Irrespective of these drawbacks, there is convincing evidence that maternal hormones play an important role for the development of the mammalian immune system.

EFFECTS OF PRENATAL ANDROGENS AND ESTROGENS ON MAMMALIAN OFFSPRING IMMUNE FUNCTION

Mammalian prenatal sex steroids have a range of effects on the development of immune function (reviewed by Schuurs and Verheul, 1990; Martin, 2000). However, unlike egg-laying vertebrate species there are, to the best of our knowledge, no studies that have investigated the influence of prenatal sex-steroid exposure on immune function on wild mammals. A correlative study on humans is maybe most comparable to a study on wild mammals. In this study, which included a large birth cohort of nearly 15,000 pregnancies, higher maternal testosterone levels during pregnancy were associated with lower immunoglobulin E (an antibody class that plays an important role in allergy) in 5- to 7-year-old boys (Shaheen et al., 2007). No association was reported in girls (see below for sex-specific action of prenatal steroids on immunity).

The remaining studies investigating the effect of sex steroids on offspring immunity were conducted on laboratory animals (mice, rats, primates). We will use a few of these studies to illustrate the range of prenatal sex-steroid effects. Administration of a single dose of 17β-estradiol in 4-day-old mice increases the cellular immune response and results in resistance to helminth infection (Guzman et al., 2009). Adult female rats showed reduced thymus development and male-like ratios of mature thymocyte subsets when they were exposed to a single dose of testosterone when 2 days old (Leposavic et al., 2009). In a study of male marmosets, treatment with a gonadotropin-releasing hormone receptor antagonist (which reduces androgen production) resulted in increased thymus size (Lunn et al., 1997). Although these studies clearly show that prenatal (and potentially maternally transferred) sex steroids can influence offspring immune function, it is necessary to validate their results in wild mammal populations if we want to understand the consequences of maternal hormone transfer within the context of life-history evolution.

EFFECTS OF PRENATAL GLUCOCORTICOIDS ON MAMMALIAN OFFSPRING IMMUNE FUNCTION

Prenatal glucocorticoids have been found to influence immune function in various vertebrate species (reviewed in Merlot et al., 2008). For livestock, for example, there has been a considerable interest in the influence of prenatal stress on offspring vitality and health. Numerous studies have been performed on pigs, in which pregnant sows have either been socially stressed (which is assumed to induce increased production and maternal transfer of glucocorticoids) or repeatedly administered glucocorticoids (e.g., adrenocorticotrophic hormone, cortisol) to investigate subsequent glucocorticoid-mediated changes in the immune function of their offspring (e.g., Tuchscherer et al., 2002; Kanitz et al., 2004; de Groot et al., 2007; Otten et al., 2007). These studies show that prenatal stress or elevation of glucocorticoid levels in mothers reduces immune function in offspring. For example, prenatal stress has an immunosuppressive effect on lymphocyte proliferation in response to the T-cell mitogen concanavalin (cell-mediated immunity) and the B-cell mitogens LPS and pokeweed mitogen (humoral immunity) at the day of birth (Tuchscherer et al., 2002). These types of data are corroborated by studies on other vertebrates: for instance, in rats (Kay et al., 1998; Gotz et al., 2007), mice (Fonseca et al., 2002; Zahwa et al., 2008), rhesus monkeys (Coe and Lubach, 2005, and references therein), and baboons (Schlabritz-Loutsevitch et al., 2006). It is important to note, however, that prenatal glucocorticoids do not always exert immunosuppressive effects on offspring. The exact effect depends on the timing (e.g., early or late gestation) and the dose administered (e.g., Schlabritz-Loutsevitch et al., 2006; Otten et al., 2007, see below). Although there is as yet little direct evidence that maternal glucocorticoids affect immunity of infants, the available data suggest that prenatal stress/maternal glucocorticoids also play a role in the development of the human immune system (Merlot et al., 2008). Again, as with the sex steroids (see "Effects of prenatal androgens and estrogens on mammalian offspring immune function" above), studies in the wild are needed to assess prenatal glucocorticoid effects from an ecological/evolutionary perspective.

SEX-SPECIFIC EFFECTS OF PRENATAL STEROID HORMONES ON IMMUNITY

In mammals, adult females mount generally stronger immune responses than males (Klein, 2000). Prenatal sex steroids may play an important role as mediators of sex differences in mammalian immune function (Martin, 2000; Chapter 9). In one study, 2- to 4-day-old male Norway rats were gonadectomized and females were injected with testosterone to investigate whether sex-steroid hormones organize adult responses to hanta virus infection (Klein et al., 2002). Untreated males and females were used as controls. Neonatally gonadectomized males had antibody responses against infection and patterns of virus shedding that resembled those of control females. Administration of testosterone in neonatal females had, however, no effect on the same immune characteristics, being similar to those of control females and gonadectomized males. Thus, neonatal steroid hormone manipulation altered responses to infection in male but not in female rats. Another example for sex-specific effects of prenatal sex-steroid exposure in mammals (Shaheen et al., 2007) reported a negative association between maternal testosterone and IgE production in boys but not in girls. For birds, there is as yet little evidence that prenatal sex steroids exert sex-specific effects on the immune system. The only study suggesting such an effect is the study on zebra finches (Tobler et al., 2010). In this study, male zebra finch fledglings showed reduced weight gain 24 hours after challenge with the mitogen PHA compared to female fledglings. This pattern was reversed in fledglings exposed to experimentally elevated yolk testosterone levels. Thus, there was a sex-specific effect of elevated yolk testosterone levels on the ability to cope with the PHA immune challenge during early development. The same study also showed that adult male zebra finches had lower total antibody levels than adult females. However, this sex difference in total antibody levels was irrespective of prenatal testosterone exposure and, thus, does not support the existence of a sex-specific organizing effect of yolk testosterone on adult immune function.

It is clear that prenatal stress and prenatal glucocorticoid levels can have sex-specific effects on the development of the brain and behavior in mammals (e.g., Weinstock, 2007; Viltart and Vanbesien-Mailliot, 2007). With respect to immune function, evidence for sex-specific effects is mixed, however. Although some studies have been able to demonstrate a sex-specific effect of prenatal or neonatal glucocorticoid exposure on adult immunity (Klein and Rager, 1995; Coe et al., 1996; Bakker et al., 1998; Couret et al., 2009) others have not (Tuchscherer et al., 2002; Otten et al., 2007). Given the well-documented sex-specific effects on brain development and behavior, sex-specific effects of glucocorticoids on immune function may well be general. Different results may be the consequence of different experimental procedures (see "Timing and dose of prenatal steroid exposure"). There are, for example, some indications that sex-specific effects depend on the timing of prenatal stress (Couret et al., 2009).

Sex-specific effects of prenatal steroid hormones are interesting from an evolutionary perspective because there may be the possibility that mothers could use steroid hormone transfer as a mechanism for differential sex allocation/secondary sex ratio adjustment (e.g., Groothuis et al., 2005b; Gil, 2008). Depending on the environmental conditions, maternal steroid hormone transfer to the offspring may favor one sex over the other, making them more or less disease-prone (however, this will also depend on which other offspring characteristics are simultaneously affected by maternal steroid hormones).

TIMING AND DOSE OF PRENATAL STEROID EXPOSURE

As indicated above, both the concentration and the timing of prenatal steroid exposure in the developing embryo are important. Timing is especially important in mammals in which the embryo develops inside the mother. Experiments designed to study the effects of prenatal/neonatal sex steroid or glucocorticoid exposure will produce different results when the steroids (or prenatal stress) are administered at different time points during gestation. For example, offspring of sows that received cortisol during the first and third period of gestation showed higher fever responses to LPS compared to offspring of sows that received cortisol during the second period of gestation or control offspring (de Groot et al., 2007). Administration of adrenocorticotropic hormone had an immunosuppressive effect on the lymphocyte proliferation of piglets 1 day after birth in response to the T-cell mitogen concanavalin A during late gestation but not during midgestation (Otten et al., 2007). This clearly shows that there are specific critical time windows when the offspring are more susceptible to maternal steroid hormone exposure. Hence, environmental effects that induce prenatal stress may also have very different effects on the offspring depending on when during gestation they occur.

The concentration of steroids that the developing embryo is exposed to is also important with respect to the development of the immune system. In birds, for example, exposure to high doses of yolk androgens produces immunosuppressive effects, whereas exposure to low doses results in no or even immunoenhancing effects (e.g., Norton and Wira, 1977; Navara et al., 2005; Cucco et al., 2008). Consequently, there may be no such thing as a general effect of yolk androgens on offspring immunity.

ADAPTIVE OR NON-ADAPTIVE TRANSFER OF MATERNAL STEROID HORMONES?

Whether or not mothers can actually regulate the amount of hormones transferred to the whole brood/litter or to particular offspring is an aspect that is still under debate (see, for example, Groothuis and Schwabl, 2008; Moore and Johnston, 2008, for a discussion in egg-laying vertebrates). This question is of interest, however, because the type of hormone transfer will affect the way selection can act on this process. If the transfer is "passive" (i.e., maternally derived hormone levels simply reflect the hormone levels in maternal circulation), then evolution of hormone transfer will be more tightly linked to selection on female characteristics. Evolution of optimal steroid hormone levels with respect to offspring immunity may be constrained, because changes in maternally transferred hormone levels may be negatively associated with changes in female traits such as immunity or reproduction (see also McGothlin and Ketterson, 2008). Hence, steroid hormones transferred may not necessarily be adaptive with respect to development of the offspring phenotype. However, if the transfer is "active" (i.e., females actually regulate, at least to some degree, the amount of hormones transferred to [individual] offspring), then the system is more flexible and patterns of maternal hormone transfer may indeed reflect strategic female allocation decisions, as suggested by the differential allocation hypothesis (e.g., Groothuis and Schwabl, 2008). If mothers are able to regulate the amount of hormones transferred to particular offspring then they may be able to prepare offspring (and the development of their immune system) for the prevailing environmental conditions after birth, and in such a case hormone patterns may indeed be adaptive. However, it may also give rise to a possible mother–offspring conflict in which some offspring may be favored over others (Müller et al., 2007; Groothuis and Schwabl, 2008). For birds (and other oviparous) species, there is some suggestive evidence that egg hormone levels do not simply reflect maternal circulating blood hormone levels at the time of egg formation (reviewed in Groothuis and Schwabl, 2008). However, it is unlikely that maternal circulating and egg hormone levels are completely independent, and therefore they advocate for a conservative use of language such as "differential allocation" or "maternal manipulation" (Groothuis and Schwabl, 2008). In mammals, this question has as yet received little attention. This is possibly because of an assumed tight link between maternal circulating hormones and hormones transferred to the embryo, at least during the prenatal period. In sum, the adaptive or non-adaptive role of maternal hormones for offspring immune function will have to be reassessed when the mechanisms of hormone transfer have been studied in more detail.

FUTURE DIRECTIONS

It is evident that mothers have ample potential to modulate offspring immunity through transfer of steroid hormones. The above overview, however, makes it clear that the effects of prenatal steroids on offspring immunity are complex. As noted by Viltart and Vanbesien-Mailliot (2007) with respect to the effects of prenatal stress in mammals: "data presented in the literature is [*sic*] often contradictory and results obtained so far greatly depend on the animal species, the nature of the stressor, the duration of stress (acute vs. chronic), the immune compartment investigated (blood vs. thymus, spleen or lymph nodes), and the age at which the offspring is examined (neonates vs. juveniles vs. adults)." From an evolutionary/ecological perspective it is, however, exactly this complexity that is interesting. There is a need for studies investigating prenatal steroid hormone effects in different contexts (e.g., different availability of food, different competitive environment in the brood/litter, different frequencies of diseases/parasites). Only if we understand how prenatal steroids act on the offspring's immune system in different contexts will it be possible to assess the fitness consequences of prenatal steroid exposure and thereby the mother's option to affect the phenotype of her offspring through epigenetic effects. Furthermore, it is important to use integrative approaches. Use of a single immune assay is likely not sufficient to exclude immunomodulatory effects of prenatal steroid hormones (see also Martin et al., 2008). Also, particularly with respect to mammals, studies on wild species in their natural environment are needed to validate the results from laboratory experiments. It is likely that the environment interacts with the mother in a way that influences the transfer of maternal steroid hormones (e.g., through factors that cause maternal stress).

MATERNAL TRANSFER OF ANTIOXIDANTS AND OFFSPRING IMMUNE FUNCTION

Free radicals are very reactive compounds or atoms since they contain at least one unpaired electron in the outer electron shell (Halliwell and Gutteridge, 2007). Most of these compounds or atoms contain oxygen and are collectively called reactive oxygen species (ROS), although similar compounds containing nitrogen (reactive nitrogen species [RNS]) occur together with ROS. For simplicity, we will use the nomenclature pertaining to the oxygen-based free radicals (ROS and oxidative stress), although it should be kept in mind that RNS is produced in the same manner as ROS (Surai, 2002; Poyton et al., 2009) and has similar effects on the body. The detrimental effects of the free radicals depend on their high reactivity. Having unpaired electrons makes the free radicals very prone to attract electrons from other atoms or molecules, which thereby will be converted to secondary free radicals, and a cascading chain reaction starts. These free radicals will eventually react with important biological molecules, causing damage to DNA, RNA, proteins, and lipids, a set of processes known as oxidative stress, resulting in cell injury or cell death (Halliwell and Gutteridge, 2007). Oxidative damage to lipids (i.e., lipid peroxidation) is of special interest as it negatively affects cellular metabolism through changes in the function of cell membranes (Surai, 2002; Hulbert et al., 2007). Thus, the action of free radicals has the potential to accumulate dysfunctions at the cellular level, and this accumulation, if not repaired, has been proposed as a possible mechanism behind senescence (e.g., Balaban et al., 2005, Muller et al., 2007). It should be noted, however that free radicals are not always detrimental but do also have beneficial effects in the body. Besides being used by the immune system to combat intruders in the body (see below), free radicals are also used in intracellular signaling pathways. A small proportion of free radicals in cells are produced for this signaling purpose and for the regulation of smooth muscles and blood flow (Dröge, 2002; Monaghan et al., 2009).

Free radicals can be produced by both exogenous processes (e.g., ultraviolet radiation and pollutants) and endogenous processes (e.g., as a byproduct of metabolic activity or by immune cells as a weapon to destroy invading pathogens) (von Schantz et al., 1999). During all aerobic respiration, a small proportion of oxygen is not completely reduced to water in the electron transport chain of the mitochondria. Such partially reduced oxygen results in the formation of ROS, mostly superoxide anions, that react with and cause oxidative damage to molecules both in the mitochondria and in other parts of the cell (Dowling and Simmons, 2009; Poyton et al., 2009). Thus, formation of ROS is an unwanted side effect of respiration with largely negative consequences to cells. A second source of endogenously formed ROS is the free radicals formed during a so-called respiratory burst in certain immune cells, for example phagocytic leukocytes such as macrophages and neutrophils (Surai, 2002; Costantini and Møller, 2009; Dowling and Simmons, 2009). Phagocytic cells, but also B and T lymphocytes, contain nicotinamide adenine dinucleotide phosphate (NADPH) oxidase, which is a prerequisite for the controlled formation of ROS. When an antigen is inducing an immune response, these immune cells increase oxygen uptake 10 to 20 times above resting levels (the respiratory burst). In this process the NADPH oxidase is oxidized and oxygen is reduced to form the superoxide free radical, which then induces the formation of various other ROS (Halliwell and Gutteridge, 2007; Costantini and Møller, 2009). The actual killing of intruding microorganisms is mediated by superoxide free radicals, giving rise to hydrogen peroxide

(H_2O_2) that readily crosses bacteria cell walls and directly kills the pathogens by itself or by forming very toxic hydroxyl radicals (Surai, 2002; Halliwell and Gutteridge, 2007).

The controlled production of ROS by immune cells is an important first-line defense against pathogens, as evidenced by reduced survival of infected knockout mice that lack the possibility to produce ROS (Mastroeni et al., 2000). However, it is also a dangerous weapon as ROS are unspecific in their actions. Thus, even though they are produced as a response to an immune challenge, they may damage not only the pathogen but also the host's own cells and tissues, causing immunopathology. This risk is particularly high during chronic inflammations, when ROS produced by phagocytic immune cells may cause extensive oxidative damage also to the host (Råberg et al., 1998; Sorci and Faivre, 2009). However, also experimental challenge of the immune system has been shown to result in increased levels of oxidative stress (Bertrand et al., 2006; Costantini and Dell'Omo, 2006; Hõrak et al., 2007).

However, just the production of ROS, by the immune system or through metabolism, does not by itself imply that the risk of oxidative stress is high (Costantini and Verhulst, 2009; Monaghan et al., 2009). Two lines of defense, enzymatic and non-enzymatic antioxidants, can neutralize (scavenge) ROS before they do any harm to important molecules. Thus, when the levels of the antioxidants manage to balance the amount of ROS produced, no oxidative stress will occur. However, oxidative damage will result whenever the production of ROS increases, without any concomitant increase in antioxidants, or when the levels of antioxidants decrease without a reduction in ROS production (Monaghan et al., 2009). The effectiveness of the antioxidant system is thus imperative for the use of oxygen for metabolism and of phagocytic cells for immune defense.

Antioxidants are of both endogenous and exogenous origin. Endogenously derived antioxidant enzymes include, for example, superoxide dismutase (SOD), uric acid, and glutathione peroxidases (GPx). GPx is important from a maternal-effect point of view as its antioxidant function is dependent on diet-derived selenium (Se) (Surai, 2002; Halliwell and Gutteridge, 2007). Antioxidants of exogenous origin form another important part of the antioxidant system. These cannot be synthesized by most vertebrates, and their action as free radical scavengers depends on ingestion from dietary sources. The most important antioxidants in this category include vitamin C (although most vertebrate groups can synthesize these antioxidants from glucose) and polyphenols, which are hydrophilic, and vitamin E and carotenoids, which are lipophilic (Catoni et al., 2008a). These antioxidants are part of a fine-balanced set of interactions among themselves (Surai et al., 1998, 2001a). Antioxidants with similar solubility may compete for absorption, leading to deficiencies in one antioxidant if another one is provided in excess. Furthermore, when vitamin E has been quenching free radicals and is oxidized, it can be recycled to its active reduced form by other antioxidants (e.g., vitamin C). It is even so that oxidized carotenoids become pro-oxidants and can thereby themselves cause damage to cells. However, when vitamin C or E is present, these antioxidants can reduce oxidized carotenoids and recycle them. Thus, carotenoids are effective free radical scavengers only when vitamin C or E is present (Catoni et al., 2008a). This balance between carotenoids and other antioxidants may well be disrupted if carotenoids are experimentally added to the diet, resulting in no or even a reduced effect on the total antioxidant capacity, and this may be the reason for the recently questioned antioxidant effect of carotenoids (Hartley and Kennedy, 2004; Costantini and Møller, 2008).

Thus, antioxidants have the potential to enhance an immune response by taking care of misdirected free radicals generated by immune cells as a weapon to inactivate antigens (von Schantz

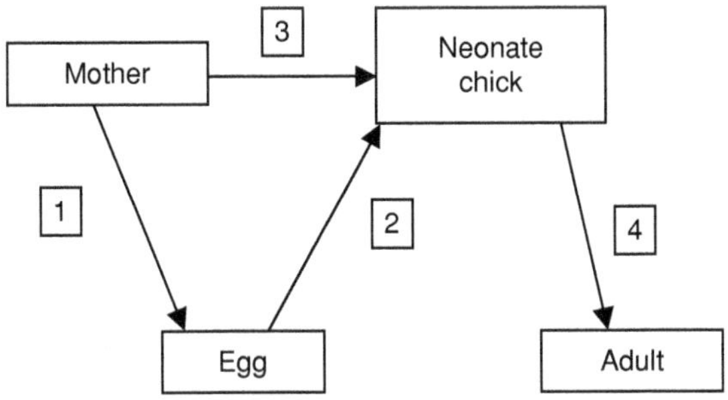

FIGURE 6.1: Schematic overview of possible paths (1–4) of antioxidant transfer between different states of an organism that may be important for the immune system

et al., 1999). This immuno-enhancing effect may be evident during different stages of vertebrate life (neonate vs. chick/immature vs. adult; Fig. 6.1). If this is true, then mothers have the potential to enhance the immune responses of their young (short-term maternal effect; paths 1–3 in Fig. 6.1) and even when they become adults (long-term maternal effect, paths 1–4 in Fig. 6.1) by maternal transfer of antioxidants. In oviparous vertebrates, such transgenerational maternal effects will act through maternally derived antioxidants being deposited in the egg (path 1 in Fig. 6.1), which may then be used by the hatched chick (path 2 in Fig. 6.1). In viviparous vertebrates, maternally derived antioxidants can be transferred both over the placenta and during lactation (path 3 in Fig. 6.1). We will start to review the evidence for immune-enhancing effects of antioxidants in adults and in neonates/chicks, respectively. Then we will evaluate the importance of transgenerational maternal effects on the immune system by reviewing the evidence for paths 1 to 4 in Figure 6.1.

THE EFFECT OF ANTIOXIDANTS ON THE IMMUNE SYSTEM—STUDIES OF ADULTS

In general, the most important immunostimulating effect of antioxidants is their free radical-scavenging properties. Antioxidants directly enhance the phagocytosis of macrophages and polymorphonuclear neutrophils (PMNs) by protecting host cells, including the phagocytic cells themselves, from ROS leaking away from phagocytic cells (e.g., Heinzerling et al., 1974; Bendich, 1989; Chew, 1996; Zhao et al., 1998; Surai, 2002; McGraw et al., 2006). Furthermore, antigen recognition of lymphocytes is dependent on receptors on their membranes, and because membranes of immune cells are rich in polyunsaturated fatty acids (PUFAs), which are very prone to being peroxidized by ROS (Hulbert et al., 2007), a deficit of antioxidants may harm several aspects of the immune system (Bendich, 1989; Klasing, 1998; Surai, 2002). This includes the extensive cell-to-cell communication imperative for an effective immune response, which is dependent on receptors on cell membranes (Murphy et al., 2008). Thus, the prevention of lipid peroxidation of such receptor structures may be an important part of the immunoenhancing properties of antioxidants (Bendich, 1989; Klasing, 1998; Surai, 2002). Another, quite different effect of an unbalanced production of ROS is the risk of inducing genomic changes in viruses, making them more virulent than before (Beck, 1999). To avoid these potentially adverse effects of an overproduction of ROS by phagocytic immune cells, the immune system may adaptively be downregulated when antioxidants are in low supply and correspondingly upregulated when extra antioxidants are provided (cf. von Schantz et al., 1999).

Investigations into the immunostimulating effects of antioxidants stem mainly from studies of poultry, farm, and laboratory animals and are based on adult individuals (reviews in Chew, 1996; Hughes, 2002; Surai, 2002). Most of these studies include the manipulation of diets, resulting in a deficit or an experimental surplus of antioxidants such as carotenoids, Se, and vitamins E and A. In the vast majority of these studies, supplementation of antioxidants results in improved immune responses, and diets with antioxidant deficits compromise the immune responses of chickens, mice, horses, pigs, cattle, sheep, cats, and dogs (Finch and Turner, 1996; Surai, 2002). Surai (2002) cites 46 studies presenting evidence for an immunostimulating effect of vitamin E and/or Se and a further 33 studies showing the same effect for carotenoids. Studies performed on vertebrates other than farm and laboratory animals are scarcer, in particular those performed on vertebrates in the wild, and the results from these studies are not as unequivocal as those from farm and laboratory animals. In these studies, the antioxidant manipulation is severely biased to carotenoids, as these compounds are at the heart of theories of honest competence signaling between potential mates and between parents and offspring (e.g., von Schantz et al., 1999; Møller et al., 2000). In birds, carotenoid supplementation has been found to enhance the bacteria-killing activity of blood (McGraw and Klasing, 2006; McGraw et al., 2006), cell-mediated immune responses (Blount et al., 2003; McGraw and Ardia, 2003; Cucco et al., 2007), and humoral immune responses (McGraw and Ardia, 2003). In one of the few experiments involving flavonids, blackcaps (*Sylvia atricapilla*) on a flavonid-enriched diet mounted a stronger humoral immune response to an immune challenge than did control birds (Catoni et al., 2008b). However, several studies have also failed to find an effect of carotenoid diet supplementation on especially cell-mediated immune responses, measured as an unspecific immune response to the plant-derived mitogen phytohemagglutinin (PHA) (Navara and Hill, 2003; McGraw and Ardia, 2004; Hõrak et al., 2006; McGraw et al., 2006; Smith et al., 2006). In carotenoid-fed female lesser black-backed gulls (*Larus fuscus*), the circulating plasma concentrations of immunoglobulins (Igs) was lower compared to control females (Blount et al., 2002b). Although this may seem an example of an immunosuppressive effect of carotenoids, it might well be an indication of health, because high circulating levels of Igs may be a sign of an ongoing infection. Indications of the immune-stimulating effect of carotenoids acting through free radical scavenging comes from immune-challenged male blackbirds (*Turdus merula*) that withdrew carotenoids from a secondary sexual trait, probably to scavenge free radicals produced during the immune response (Faivre et al., 2003), and from the removal of parasites from red grouses (*Lagopus lagopus scoticus*), which simultaneously enhanced carotenoid-based coloration and reduced the level of oxidative stress (Mougeot et al., 2010). Also, experimental supplementations of primarily carotenoids to different fish species (mostly salmonids) have resulted in enhanced cell-mediated and humoral immune responses (e.g., Amar et al., 2000) and in enhanced survival after a parasite challenge (Christiansen et al., 1995). However, also supplementation of vitamin E has a large immunoenhancing effect on fishes (Ortuño et al., 2000; Sahoo and Mukherjee, 2002).

Thus, especially among poultry and farm animals, there is strong evidence for an immunoenhancing effect of antioxidants (Surai, 2002). However, if mothers were able to confer advantages by providing their offspring with antioxidants, such antioxidants must also be transferred to offspring and have a positive effect on the young.

THE EFFECT OF ANTIOXIDANTS ON THE IMMUNE SYSTEM OF NEONATES/CHICKS

At birth and hatching, young are exposed to a sudden burst of free radicals as they start pulmonary respiration (Vleck and Bucher, 1998). Furthermore, the postnatal immune system relies heavily on innate immunity, as the adaptive immune system is still undeveloped (see above). The innate immune system is characterized by the macrophage-phagocyte system producing free radicals as the most important mean of dealing with antigens, thus further augmenting the production of free radicals (Klasing and Leshchinsky, 1999). Hence, early in life, young can be predicted to incur a high risk of oxidative damage. This is also an important period for the development of the immune system, with a rapid expansion of the leukocyte population and the start of growth of the lymphoid organs, ultimately affecting the emergence of unique clones of lymphocytes that are important for postnatal immunity (Klasing, 1998; 2004).

Some diet manipulations of vitamin E, Se, and carotenoids have been performed early in the postnatal life in mostly avian species. The growth of the primary lymphoid organs in birds, the bursa of Fabricius and the thymus, was impaired following a diet deficient in vitamin E and Se (Marsh et al., 1986). Furthermore, supplementation of antioxidants to young chicks has been found to increase cell-mediated immune responses in four non-passerine bird species (Swain et al., 2000; Fenoglio et al., 2002; Cucco et al., 2006; Selvaraj et al., 2006), whereas such supplementation had no effect in two passerine bird species (Biard et al., 2006; de Ayala et al., 2006). Furthermore, antibody titers to immune challenges have been found to depend on the level of antioxidants in the diet of young chicks (Nockels, 1979; Marsh et al., 1981; Swain et al., 2000). Other correlates of the immune system that have been found to be enhanced by the level of antioxidants in the diet include macrophage free radical production (Selvaraj et al., 2006) and survival after infection with a disease (Nockels, 1979; Bu et al., 1996).

Thus, there are ample examples of studies reporting positive influences of dietary antioxidants on the immune system also of post-embryonic young. This opens up the evolutionary possibility of maternal effects in the form of antioxidants being transferred from the mother to the young, either through the egg or through lactation. We will now evaluate the evidence for such transgenerational maternal effects by separating the effects on four different paths (Fig. 6.1).

CAN THE MOTHER DEPOSIT ANTIOXIDANTS IN THE EGG IN RELATION TO HER OWN INTAKE?

Experimental supplementation of lipophilic antioxidants (i.e., vitamin E, carotenoids, Se) to females before egg laying generally results in an concomitant increase of these antioxidants in the egg both in poultry (Haq et al., 1996; Hossain et al., 1998; Surai, 2000; 2002; Lin et al., 2005) and other avian species (Blount et al., 2002a, 2002b; Royle et al., 2003; Biard et al., 2005; Karadas et al., 2006; Remeš et al., 2007, but see Cucco et al., 2007). Carotenoids are prominent antioxidants not only in avian eggs, but also in the eggs of reptiles (Blount et al., 2000) and several species of fish. For example, supplementation of carotenoids to female guppies (*Poecilia reticulata*) increased the carotenoid content of eggs produced by these females (Grether et al., 2008), and carotenoid-based color in female two-spotted gobies (*Gobiusculus flavescens*) predicted the level of carotenoids in their eggs (Svensson et al., 2006).

Besides the potential advantages of providing hatchlings with protection against ROS attacks, maternal allocation of antioxidants to the eggs can also serve as protection against lipid peroxidation of the lipid-rich yolk and the developing embryo. Several studies also report that the eggs of females receiving antioxidant-enriched diets produce egg yolk less susceptible to lipid peroxidation (Surai and Speake, 1998; Blount et al., 2002a, 2002b). Of even more interest from an immunological point of view is that maternally deposited antioxidants may protect maternally deposited antibodies in the egg yolk from ROS attack (Haq et al., 1996; Blount et al., 2000).

The positive relationship between antioxidants in the diet of females and their eggs seems to indicate that availability of these compounds may constrain the possibility of a mother to transmit antioxidants to her eggs. Because antioxidants also may be important for the efficiency of the female's own immune system, a trade-off between providing the egg with such a defense and self-maintenance will determine her allocation of antioxidants compounds to the egg (Blount et al., 2002b; Biard et al., 2005; Hargitai et al., 2009). The most prominent source of carotenoids for egg-producing passerine females in the wild is caterpillars (Partali et al., 1987). As the density of this food resource increases steadily during early spring, the seasonal increase in mean egg carotenoids in a clutch found in collared flycatchers (Hargitai et al., 2009), for example, further emphasizes that availability constrains the ability of females to allocate these antioxidants to their eggs. A further indication of such a trade-off comes from the fact that females exposed to immune system challenge during egg laying allocated lower levels of carotenoids to their eggs than unchallenged females (Saino et al., 2002b). Furthermore, in a comparative study, Biard et al. (2009) found a positive relationship between the levels of antioxidants deposited in eggs and clutch size. If species with a large clutch size in general are biased towards current reproduction in the trade-off between current and future reproduction, we would predict that they also should allocate more antioxidants to their eggs.

Scarcity of dietary antioxidants, and the resulting trade-off between maternal effects on offspring and self-maintenance of the mother, may also constrain the maternal deposition of antioxidants within a clutch. The levels of carotenoids in eggs generally decrease over the laying sequence (Royle et al., 2001, 2003; Hõrak et al., 2002; Saino et al., 2002b; Groothuis et al., 2006; Safran et al., 2008; Saino et al., 2008). This has been proposed to be a consequence of an adaptive brood-reduction strategy (Hõrak et al., 2002) but could also be a consequence of constrained deposition of carotenoids late in the laying sequence due to female depletion (Groothuis et al., 2006). Indications for the latter interpretation also come from studies on two gull species in which maternal deposition of androgens also has been measured within the laying sequence (Royle et al., 2001; Groothuis et al., 2006). Although the bulk of the antioxidants incorporated in an egg are derived from the diet of the female shortly before laying a particular egg, she has the ability to store a limited amount of antioxidants before commencing egg laying, especially in the liver (Surai et al., 1998). The use of this storage in the beginning of the laying sequence, with an eventual depletion of these stores, may account for the observed antioxidant decline with laying order. A more direct indication of the limiting supply of antioxidants for egg-producing females is evident from studies on a wild population of the lesser black-backed gull (*Larus fuscus*). Females receiving experimental carotenoid supplementation were more prone to produce a replacement clutch following experimental removal of the first clutch than were females fed a control diet without carotenoid supplementation (Blount et al., 2004). Thus, the increased amount of carotenoids in the diet may have made it possible for the females to allocate more carotenoids to reproduction while at the same time using an optimal amount for self-maintenance.

DO ANTIOXIDANTS IN THE EGG AFFECT THE IMMUNE SYSTEM OF CHICKS?

The first days after hatching in avian species are characterized by rapid growth and a sudden change to pulmonary respiration (Vleck and Bucher, 1998). Both these processes are likely to increase the production of ROS and thus stress the importance of an efficient antioxidant defense early in life. Shortly before hatching, antioxidants are transported from the egg yolk to primarily the liver of the embryo (Surai, 2002). When the chick is 1 day old, the liver has much higher concentrations of, for instance, carotenoids (Surai et al., 1996; Surai and Speake, 1998; Surai et al., 2001b), vitamin E (Surai et al., 1996; Surai, 2000), and Se (Surai, 2000) than other tissues in the chick, and concomitant with this, the susceptibility of lipid peroxidation is greatly reduced in the liver (Surai et al., 1996; Surai and Speake, 1998; Surai, 2000). Maternally derived antioxidants stored in the liver generally decline with age (Hossain et al., 1998), although somewhat increased concentrations have been noted in one study of chicken for up to 4 weeks after hatching (Koutsos et al., 2003). Thus, as the antioxidants are used up during free radical scavenging in the period shortly after hatching, the store in the liver declines irrespective of dietary supplies to the chicks. This indicates a reduced ability to assimilate or accumulate antioxidants early in life (Hossain et al., 1998). Such dietary additions of antioxidants may nevertheless be of some importance because maternally derived antioxidants and those provided in the diet were found to have additive positive effects on macrophage numbers in chickens (Koutsos et al., 2007).

Besides an immunoenhancing effect (see below in this section), antioxidants from the egg may be advantageous due to their ability to scavenge ROS produced by the metabolic machinery, allowing a higher growth rate than is possible for chicks with smaller egg-derived antioxidant stores. Non-experimental levels of antioxidants in eggs are positively related to chick size and body mass without having any influence on the immune response to PHA (Rubolini et al., 2006c; Saino et al., 2008). Another advantage of maternal transfer of antioxidants is that increased concentrations of carotenoids in the yolk have been found to enhance the capacity of chicks to assimilate carotenoids later in life (Koutsos et al., 2003).

The effect of antioxidants in the egg on the immunity of hatched chicks can be manipulated by direct injection of antioxidants into the egg. When chicken eggs were artificially enriched with vitamin E, hatchlings produced a higher proportion of phagocytic macrophages (Gore and Qureshi, 1997). Furthermore, chicks from vitamin E-enriched eggs produced more antibodies in response to a challenge with sheep red blood cells (SRBCs) when they were 7 days old, and continued to do so in response to a secondary challenge when 21 days old (Gore and Qureshi, 1997). The same manipulation to eggs of the barn swallow (*Hirundo rustica*) resulted in an enhanced

T-cell–mediated immune response of 12-day-old chicks, although in this study artificial injection of a carotenoid did not result in an increased antibody response to an immunogen (Saino et al., 2003b). Also in yellow-legged gulls (*Larus michahellis*), carotenoid enrichment of the egg increased the T-cell–mediated immune response of 8-day-old chicks, although only in males (Romano et al., 2008). Thus, in these three studies, addition of antioxidants to the egg yolk enhanced different parts of the immune system, but the effects varied with respect to species and sex.

However, to further evaluate the effect of maternally transferred antioxidants on chick immunity, we need to manipulate antioxidant availability to the mother and look for consequences in the chicks, thus integrating over both paths 1 and 2 in Figure 6.1. These types of experiments in poultry have generally produced evidence for such a maternal effect, whereas studies on other avian groups are less consistent (Table 6.3). One-day-old chicks from hens fed a diet enriched with vitamin E and carotenoids had higher lymphocyte proliferation rates and higher antibody titers, both when 1 and 7 days old, than chicks from hens fed a control diet (Haq et al., 1996). Furthermore, chicks of hens with a surplus vitamin E diet responded with higher antibody titers to a challenge with NDV than did chicks originating from hens on a control diet (Hossain et al., 1998).

Thus, in these studies the humoral immune system was enhanced by maternally transferred antioxidants. One possible mechanism for this maternal effect may be the protection conferred by the antioxidants against lipid peroxidation of transferred matAbs (see above). Hens immunized with *Brucella abortus* antigen and fed diets enriched with vitamin E produced chicks with higher passive immunity as they had higher Ab titers against the antigen when 2 days old than chicks from hens on a control diet (Jackson et al., 1978; Nockels, 1979). At 7 days of age, the antibody titers

TABLE 6.3: Studies on the effect of diets supplemented with antioxidants given to females before egg laying on the immune response of their chicks

		Immune response			
Species	Antioxidant	Humoral	WWS	Innate	Reference
Chicken (*Gallus gallus*)	Vit. E; Car	+			Haq et al., 1996
Chicken (*Gallus gallus*)	Vit. E	+			Hossain et al., 1998
Chicken (*Gallus gallus*)	Vit. E	+			Nockels, 1979
Chicken (*Gallus gallus*)	Vit. E	+			Jackson et al., 1978
Chicken (*Gallus gallus*)	Car		+		Koutsos et al., 2007
Grey partridge (*Perdix perdix*)	Car		0		Cucco et al., 2006
Great tit (*Parus major*)	Car		0		Remeš et al., 2007
Great tit (*Parus major*)	Car		+		Berthouly et al., 2007
Great tit (*Parus major*)	Car		(+)		Berthouly et al., 2008
Blue tit (*Cyanistes caeruleus*)	Car		0		Biard et al., 2007
Blue tit (*Cyanistes caeruleus*)	Car			+	Biard et al., 2005

Vit. E = vitamin E; Car = carotenoids.
Immune responses measured as humoral (i.e., variation in antibody titers), wing web swelling (WWS) assays (i.e., the swelling resulting from injection of phytohaemagglutinin [PHA] or lipopolysaccharide [LPS] in the wing web), and innate (i.e., variation in leukocyte concentration). "+" and "(+)" denotes a positive and a tendency for a positive effect, respectively of female diet on chick immunity, and "0" denotes no effect.

had decreased as they were catabolized; however, chicks from hens supplemented with vitamin E still had higher antibody titers. Furthermore, when these chicks were themselves challenged with the same antigen at 14 days of age, chicks of mothers fed extra vitamin E also had higher antibody titers 1 and 2 weeks post-immunisation (Nockels, 1979). Moreover, the swelling of the wing web following injection with PHA can be affected by the diet of the mother; chicks from hens on a carotenoid-enriched diet responded stronger but later to the mitogen (Koutsos et al., 2007).

Whereas immunity in chickens mostly has been evaluated as differences in antibody titers, immunity in wild birds has usually been measured as the strength of wing web swelling following an injection with either PHA or LPS (Table 6.3). Another difference between these two groups of birds is that in chicken vitamin E is usually the antioxidant of interest, whereas in wild birds it is carotenoids (Table 6.3). Thus, the less consistent effects of female ingestion of antioxidants on chick immunity among wild birds may have different explanations besides the fact that one group is kept in captivity and the other is generally not. Furthermore, the majority of studies on wild birds have been performed on tits, which depend on caterpillars for egg laying and feeding of young. Caterpillars are known to be very rich in carotenoids (Partali et al., 1987); thus, limitations of this particular antioxidant may be less problematic in this group of birds. Also, in cases where some advantageous effects have been noted among these species, the results are not straightforward. In great tits (*Parus major*), the beneficial effect depends on the nature of stressors put on the nestlings (Berthouly et al., 2007, 2008). In blue tits (*Cyanistes caeruleus*), carotenoid supplementation to females was primarily manifested in a faster development of the immune system of their chicks (Biard et al., 2005, 2007).

Another way to investigate effects of maternally derived antioxidants on chick immunity is to compare growth rates and survival when actually being exposed to a parasite. The blood-sucking mite (*Ornithonyssus bursa*) affects the growth rate and condition of nestling hihi (*Notiomystis cincta*). However, these negative effects could be alleviated if mothers were supplemented with carotenoids before egg laying (Ewen et al., 2009). The experimental design precludes a specific test of the mechanism involved, which could be due to either immunoenhancement or increased begging intensities. Such offset of the negative effects of parasitism was not, however, found in the great tit-hen flea host–parasite system (Berthouly et al., 2007, 2008).

DO ANTIOXIDANTS TRANSFERRED DURING LACTATION AFFECT NEONATE IMMUNITY?

Investigations of this type of maternal effect are mostly confined to farm or laboratory animals, but the general mechanisms parallel those from avian studies. The primary route for transferring maternally derived antioxidants to young in mammals seem to be through colostrum and to some extent milk (Loudenslager et al., 1986; Babinszky et al., 1991; Debier et al., 2002), whereas placental transfer seems to be of less importance (Lauridsen et al., 2002). Experimental supplementation of vitamin E and selenium to the diet of mothers generally results in a concomitant increase in the concentration of these antioxidants in their colostrum and milk, and this is translated into higher concentrations in the plasma of their young (Loudenslager et al., 1986; Babinszky et al., 1991; Daniels et al., 2000; Capper et al., 2005). Furthermore, the phagocytic and microbicidal activities in the colostrum were impaired when sows were fed a diet deficient in vitamin E and Se (Wuryastuti et al., 1993).

Just as in the avian system, data on the potential advantages of maternal transfer of antioxidants are rather common, but direct tests of the effects of maternally derived antioxidants on the immune system of the young are much scarcer. Also in parallel with studies on birds, a potentially immunoenhancing mechanism of maternally derived antioxidants has been suggested to be protection of maternally derived antibodies. Thus, antioxidants may protect the membranes of immunoglobulins from lipid peroxidation when these are transferred from the mother to her offspring (Sikka et al., 2002). In line with this, antibodies in piglets and buffalo calves early after birth, before they start to produce own Ab, were more numerous if the females had been fed extra vitamin E (Babinszky et al., 1991; Sikka et al., 2002). Moreover, immunized neonates of supplemented females responded with higher antibody production (Babinszky et al., 1991). In sheep, ewes on a control or vitamin E-supplemented diet were immunized around a month before lambing date. However, in this case no increase in antibody titers was found in either the colostrum or the serum of the lambs (Daniels et al., 2000). Thus, although the maternally derived antioxidants seem to be immunoenhancing in some circumstances, this is not always the case.

DO MATERNALLY DERIVED ANTIOXIDANTS AFFECT THE IMMUNE SYSTEM ALSO IN MATURE OFFSPRING?

Virtually nothing is known about the long-term effects of maternal antioxidants on offspring intrinsic immune responses. We suggest two potentially important effects of transgenerational transfer of antioxidants that can be mediated also into adult life. First, carotenoid deficiencies during the embryonic or early post-hatching period have been suggested to reduce assimilation and accumulation of carotenoids at the adult stage (Koutsos et al., 2003; Biard et al., 2009). This may be of great importance for the immunocompetence of adults, because there are numerous indications of an immunoenhancing property of carotenoids in adults. Second, the protection of matAbs conferred by maternally derived antioxidants may be of great importance for future immunocompetence (e.g., Nockels, 1979; Sikka et al., 2002). Shortly after hatching or birth, neonates do not produce any Ab by themselves. During this period the immediate and organizational effects of matAbs will have short-term advantages but also possible long-term advantages on the mature immune system (see "Offspring uptake of matAb" above).

FUTURE PROSPECTS

The many investigations showing immunoenhancing effects of antioxidants in adults or young point to a great potential for important transgenerational maternal effects on the immune system. However, there are so far relatively few studies evaluating the whole sequence of mechanisms, from the specific acquisition of dietary antioxidants by mothers to their effect on the immune system of their progeny. Those studies performed are severely biased towards poultry, laboratory, and farm animals. Thus, we need many more studies on wild vertebrates in their natural environments to be able to evaluate the importance of this maternal effect in an evolutionary framework. A promising experimental design would be to immunize mothers before egg laying or gestation, as has been done to evaluate the degree and effect of maternally derived antibodies, and at the same time provide one group of mothers with supplemental dietary antioxidants. Then the importance of antioxidants can be evaluated with reference to the possible advantages of passively acquired immunity.

The antioxidant of interest differs between different research traditions. In poultry and laboratory and farm animals, the efficient antioxidants vitamin E and Se are generally manipulated and evaluated in relation to neonate immunity. In wild birds and fishes, carotenoids are much more common as the antioxidant of interest. This bias depends on an interesting aspect of carotenoids: as a potentially honest signal of condition or need (e.g., von Schantz et al., 1999; Møller et al., 2000). Although this is important for evaluating hypotheses about mate choice and parent–offspring interactions, vitamin E and Se may be more important antioxidants from a purely immunoenhancing perspective.

The fact that different antioxidants depend on each other for effective free radical scavenging (Surai et al., 1998; Catoni et al., 2008a) makes manipulations problematic. Especially dietary manipulations of carotenoids, which might turn into pro-oxidants if other antioxidants are in short supply (Catoni et al., 2008a), may result in non-optimal blends of antioxidants and even into situations where oxidative stress is increased. The varying results of such manipulations (Table 6.3) may to some degree be explained by such unintentional aberrations in the balance of the components of the antioxidant system. Thus, a blend of different antioxidants may be the best way to manipulate natural effects of maternally derived antioxidants on neonate immune responses.

TRADE-OFFS AND SYNERGIES BETWEEN DIFFERENT MATERNAL FACTORS AND IMMUNE FUNCTION

Because the various maternal effects potentially have advantageous (matAbs and antioxidants) or disadvantageous (androgens) effects on the immune system of offspring, relative combinations may be determined by trade-offs or adaptive synergies between these maternal compounds. Most studies on such blends of maternal effects have been performed on oviparous vertebrates because the relative deposition is predetermined already when the egg is formed and can be measured after the egg is laid.

In a number of studies (see "Maternal hormone transfer and offspring immunity in humans and other mammals"), high doses of prenatal androgens have been shown to suppress immune responses in neonates. Because adding carotenoids to the diet of androgen-boosted chicks has been found to restore immunity, it has been hypothesized that transfer of high androgen levels could be balanced by transfer of high carotenoid levels (Cucco et al., 2008). Such a positive relationship between the clutch averages of androgens and antioxidants (carotenoids, vitamin E) has been found in house finches (*Carpodacus mexicanus*; Navara et al., 2006a). Moreover, experimentally elevated yolk testosterone levels have been found to modulate the antioxidant defense in zebra finch nestlings in a sex-specific way (Tobler and Sandell, 2009). Male nestlings exposed to elevated yolk testosterone levels had lower total antioxidant levels than female nestlings; however, this difference disappeared at the fledgling stage. Apart from the amount of androgen exposure and offspring sex, other factors such as food availability are likely to influence androgen–antioxidant interactions. However, the interactions between carotenoids (and other antioxidants) and androgens may be complex. Small amounts of androgens have sometimes been found to be immunoenhancing (Cucco et al., 2008) and may be explained by a testosterone-dependent increase in carotenoid absorption from the diet (Blas et al., 2006; Peters, 2007).

Also for matAbs and antioxidants, we could predict a positive relationship because these maternal effects may potentially have a synergistic positive effect on neonate immunity. However, data from an experiment on lesser black-backed gulls instead suggest a trade-off between matAbs and maternal carotenoid transfer, because it was found that carotenoid-supplemented females transferred more carotenoids but less matAbs into their eggs than unsupplemented females (Blount et al., 2002b). It is possible that allocation of large amounts of carotenoids into the egg yolk may in some way interfere with the antibody transfer.

Relationships between maternally derived compounds and egg-laying order in birds could potentially reveal interactions between the maternal factors. Siblings from the same clutch may be exposed to different mixtures of androgens, matAbs, and antioxidants in the yolk, depending on their position in the clutch. Assuming there is some female regulation over the compounds transferred to the eggs, mothers may provide particular offspring with advantageous or disadvantageous combinations of these compounds. Synergistic effects may be present in magpies, where matAb and testosterone levels both increased with laying order (Pihlaja et al., 2006). If testosterone suppresses immune responses (see above), it would be adaptive for mothers to increase the transfer of matAbs alongside an increase in androgens to last-laid eggs, thereby augmenting the immune protection of these nestlings (Pihlaja et al., 2006), at least if they are pursuing a "brood-survival" strategy. However, in studies where more than one maternal effect has been measured in the same egg (all studies conducted on gull species), a commonly found relationship is a decrease within the laying sequence for matAbs and antioxidants and an increase for androgens (Royle et al., 2001; Blount et al., 2002b; Groothuis et al., 2006). This has led to the conclusion that deposition of matAbs and antioxidants is constrained by limited availability, whereas that of androgens is an adaptive maternal strategy (Royle et al., 2001; Groothuis et al., 2006). Thus, in these gull species, where last-hatched chicks usually are much smaller than their siblings, mothers are prioritizing a rapid growth rate by increasing the concentration of androgens in the eggs of these chicks instead of immunity.

Of particular interest is also that females may transfer androgens and antioxidants differently with respect to the same environmental stimulus. In zebra finches, for example, mothers transfer more androgens to their egg when mated to attractive males, but the decrease in yolk androgens over the laying order is similar, irrespective of mate attractiveness (Gil et al., 1999). The amount of maternally derived antioxidants, however, is associated with male attractiveness in another way. When paired with an attractive male, the decrease in yolk antioxidants over the laying order was weaker than when females were paired with unattractive males (Williamson et al., 2006). Eggs from mothers paired with unattractive males contained relatively more antioxidants than eggs from mothers paired with attractive males early in the laying order, whereas it was the opposite for eggs late in the laying order. Although the results on androgens and antioxidants come from two different studies, they could suggest that female zebra finches separately adjust androgen and antioxidant levels in their eggs.

In general, we will need a lot more studies investigating the interactions between different maternal effects. Manipulating the environment and the state of the female can help us unravel how maternal effects are combined with possible important influences on evolutionary strategies for the development of the immune system in offspring.

GENERAL CONCLUSIONS

It should be clear from this chapter that the study of maternal effects on offspring immunity has broadened our view on the development, epigenetics, and evolvability of the immune system. However, the study of maternal effects in general, and the relationship between maternal effects and offspring immune function in particular, has a rather short history in wild vertebrates. Although the interest has increased exponentially over the past 10 years, it is obvious from this review that patterns are still rather unsettled. The potential importance of maternal effects for offspring immune function is clear from studies of model organisms, but also begins to emerge based on a growing number of studies in wild vertebrates.

Several aspects within the framework of "maternal modulation of offspring immune function" deserve more attention. One important but difficult task for the future is to assess the relative importance of the different maternal effects on immunity in relation to each other, but also in relation to genetic effects. Via transfer of antibodies, hormones, and antioxidants, mothers modulate immune function in different ways, but few studies have tried to assess the combined effect of these molecules on offspring immunity. Hence, we know little about how different maternal effects interact with each other. They may, for example, cancel each other out (e.g., when elevated steroid hormone transfer is associated with elevated transfer of antioxidants/antibodies) or reinforce each other. Moreover, experimental studies that manipulate the amount of immune stimulators the offspring are exposed to rarely take into account genetic variation. Being able to factor out the genetic contribution to immune function, however, not only facilitates the detection of experimental effects (as it will reduce some of the "noise"—that is, variation that comes about through other effects than the experimental manipulation) but will also allow an assessment of the strength of maternal effects in relation to genetic effects. This is highly relevant if we are to understand the importance of maternal effects on immunity in an evolutionary context.

Moreover, there are several important aspects of maternal transfer of antibodies that deserve more study, both under closely controlled conditions in indoor experiments and also in field experiments. First, more knowledge is needed regarding transfer of Ab from mother to offspring in different types of vertebrates; is it Ab from the whole lifetime repertoire of antigens to which the mother has been exposed, or is it mainly Ab against antigens encountered just prior to pregnancy that is transferred across generations? Second, for how long is matAb present in the circulation of the offspring, and to what extent does it directly protect the neonate? Third, what are the long-term effects of matAb, in interaction with other maternal effects, on the mature offspring's own immune function and fitness?

In the future, more studies in the wild are needed. Only if we know the impact of maternal effects under ecologically relevant conditions can we predict how selection will act on these effects and their impact on evolutionary processes. Ideally, with advancement of new methods, the study of maternal effects in the wild will become more attractive and feasible. This would also facilitate studies of the last issue we would like to highlight for future studies. To obtain a comprehensive picture it is crucial to compare maternal effects in different species and contexts. It is striking that patterns differ between species, as well as within a species, depending on, for instance, population differences, time of season, egg sequence, offspring sex, experimental procedures, and so forth. It is now time for ecologists to explore these differences to unravel trade-offs and potentially adaptive strategies related to maternal effects and offspring immune function.

REFERENCES

Abdel-Moniem, AS and Abel-Gawad, MMA (2006). Genetic variations in maternal transfer and immune responsiveness to infectious bursal disease virus. Vet. Microbiol. 114:16–24.

Addison, B, Klasing, KC, Robinson, WD, Austin, SH, and Ricklefs, RE (2009). Ecological and life-history factors influencing the evolution of maternal antibody allocation: a phylogenetic comparison. Proc. R. Soc. B 276:3979–3987.

Agrawal, AA, Laforsch, C, and Tollrian, R (1999). Transgenerational induction of defences in animals and plants. Nature 401:60–63.

Albrecht, P, Ennis, FA, Saltzman, EJ, and Krugman, S (1977). Persistence of maternal antibody in infants beyond 12 months: mechanism of measles vaccine failure. J. Pediatr. 91:715–718.

Amar, EC, Kiron, V, Satoh, N, Okamoto, N, and Watanabe, T (2000). Effects of dietary β-carotene on the immune response of rainbow trout *Oncorhynchus mykiss*. Fish. Sci. 66:1068–1075.

Anderson, RW (1995). On the maternal transmission of immunity: A "molecular attention" hypothesis. BioSystems 34:87–105.

Andersson, S, Uller, T, Lohmus, M, and Sundström, F (2004). Effects of egg yolk testosterone on growth and immunity in a precocial bird. J. Evol. Biol. 17:501–505.

Apanius, V (1998). Ontogeny of immune function. In Starck, JM and Ricklefs, RE (eds.). Avian Growth and Development: Evolution Within the Altricial-Precocial Spectrum. Oxford University Press, Oxford, pp. 203–222.

Apanius, V and Nisbet, ICT (2003). Serum immunoglobulin G levels in very old common terns *Sterna hirundo*. Exp. Geront. 38:761–764.

Arnold, AP and Breedlove, SM (1985). Organizational and activational effects of sex steroids on brain and behaviour: A reanalysis. Horm. Behav. 19:469–498.

Babinszky, L, Langhout, DJ, Verstegen, MW, den Hartog, LA, Joling, P, and Nieuwland, M (1991). Effect of vitamin E and fat source in sows' diets on immune response of suckling and weaned piglets. J. Anim. Sci. 69:1833–1842.

Bakker, JM, van den Dobbelsten, GPJM, Kroes, H, Kavelaars, A, Heijnen, CJ, Tilders, FJH, and van Rees, EP (1998). Long-term gender-specific effects of manipualtion during pregnancy on immune and endocrine responsiveness in rat offspring. J. Neuroimmunol. 82:56–63.

Balaban, RS, Nemoto, S, and Finkel, T (2005). Mitochondria, oxidants, and aging. Cell 120:483–495.

Beck, MA (1999). Selenium and host defence towards viruses. Proc. Nutr. Soc. 58:707–711.

Becker, JB, Breedlove, SM, Crews, D, and McCarthy, MM (2002). Behavioral Endocrinology, 2nd ed. MIT Press, London.

Bendich, A (1989). Carotenoids and the immune response. J. Nutr. 119:112–115.

Berthouly, A, Cassier, A, and Richner, H (2008). Carotenoid-induced maternal effects interact with ectoparasite burden and brood size to shape the trade-off between growth and immunity in nestling great tits. Funct. Ecol. 22:854–863.

Berthouly, A, Helfenstein, F, and Richner, H (2007). Cellular immune response, stress resistance and competitiveness in nestling great tits in relation to maternally transmitted carotenoids. Funct. Ecol. 21:335–343.

Bertrand, S, Criscuolo, F, Faivre, B, and Sorci, G (2006). Immune activation increases susceptibility to oxidative tissue damage in zebra finches. Funct. Ecol. 20:1022–1027.

Biard, C, Gil, D, Karadas, F, Saino, N, Spottiswoode, CN, Surai, PF, and Møller, AP (2009). Maternal effects mediated by antioxidants and the evolution of carotenoid-based signals in birds. Am. Nat. 174:696–708.

Biard, C, Surai, PF, and Møller, AP (2005). Effects of carotenoid availability during laying on reproduction in the blue tit. Oecologia 144:32–44.

Biard, C, Surai, PF, and Møller, AP (2006). Carotenoid availability in diet and phenotype of blue and great tit nestlings. J. Exp. Biol. 209:1004–1015.

Biard, C, Surai, PF, and Møller, AP (2007). Analysis of pre- and post-hatching maternal effects mediated by carotenoids in the blue tit. J. Evol. Biol. 20:326–339.

Blas, J, Pérez-Rodríguez, L, Bortolotti, GR, Viñuela, J, and Marchant, TA (2006). Testosterone increases bioavailability of carotenoids: insights into the honesty of sexual signaling. Proc. Natl. Acad. Sci. USA 103:18633–18637.

Blount, JD, Houston, DC, and Møller, AP (2000). Why egg yolk is yellow. Trends Ecol. Evol. 15:47–49.

Blount, JD, Houston, DC, Surai, PF, and Møller, AP (2004). Egg-laying capacity is limited by carotenoid pigment availability in wild gulls *Larus fuscus*. Proc. R. Soc. B (Suppl.) 271:S79–S81.

Blount, JD, Metcalfe, NB, Birkhead, TR, and Surai, PF (2003). Carotenoid modulation of immune function and sexual attractiveness in zebra finches. Science 300:125–127.

Blount, JD, Surai, PF, Houston, DC, and Møller, AP (2002a). Patterns of yolk enrichment with dietary carotenoids in gulls: the roles of pigment acquisition and utilization. Funct. Ecol. 16:445–453.

Blount, JD, Surai, PF, Nager, RG, Houston, DC, Møller, AP, Trewby, ML, and Kennedy, MW (2002b). Carotenoids and egg quality in the lesser black-backed gull *Larus fuscus*: a supplemental feeding study of maternal effects. Proc. R. Soc. B 269:29–36.

Bly, JE, Grim, AS, and Morris, IG (1986). Transmission of passive immunity from mother to young in a teleost fish: haemagglutinating activity in serum and eggs of plaice, *Pleuronectes platessa* L. Comp. Biochem. Physiol. A 84:309–314.

Boulinier, T and Staszewski, V (2008). Maternal transfer of antibodies: raising immuno-ecology issues. Trends Ecol. Evol. 23:282–288.

Brugha, R, Ramsay, M, Forsey, T, and Brown, D (1996). A study of maternally derived measles antibody in infants born to naturally infected and vaccinated mothers. Epidemiol. Infect. 117:519–524.

Bu, Z, Huang, K, and Chen, W (1996). Study on the cell mediated immunity mechanism of selenium inhancing resistance of chicken to IBD. Chin. J. Vet. Sci. 16:273–276.

Buechler, K, Fitze, PS, Gottstein, B, Jacot, A, and Richner, H (2002). Parasite-induced maternal response in a natural bird population. J. Anim. Ecol. 71:247–252.

Campagne, G, Garba, A, Schuchat, A et al. (1998). Response to conjugate *Haemophilus influenzae B* vaccine among infants in Niamey Niger. Am. J. Trop. Med. Hyg. 59:837–842.

Capper, JL, Wilkinson, RG, Kasapidou, E, Pattinson, SE, Mackenzie, AM, and Sinclair, LA (2005). The effect of dietary vitamin E and fatty acid supplementation of pregnant and lactating ewes on placental and mammary transfer of vitamin E to the lamb. Brit. J. Nutr. 93:549–557.

Carlier, Y and Truyens, C (1995). Influence of maternal infection on offspring resistance towards parasites. Parasitol. Today 11:94–99.

Catoni, C, Peters, A, and Schaefer, M (2008a). Life history trade-offs are influenced by diversity, availability and interactions of dietary antioxidants. Anim. Behav. 76:1107–1119.

Catoni, C, Schaefer, HM, and Peters, A (2008b). Fruit for health: the effect of flavonoids on humoral immune response and food selection in a frugivorous bird. Funct. Ecol. 22:649–654.

Chandra, RK (1975). Antibody formation in first and second generation offspring of nutritionally deprived rats. Science 190:289–290.

Chew, BP (1996). Importance of antioxidant vitamins in immunity and health in animals. Anim. Feed Sci. Tech. 59:103–114.

Christiansen, R, Glette, J, Lie, Ø, Torrissen, OJ, and Waagbø, R (1995). Antioxidant status and immunity in Atlantic salmon, *Salmo salar* L., fed semi-purified diets with and without astaxanthin supplementation. J. Fish Dis. 18:317–328.

Coe, CL and Lubach, GR (2005). Prenatal origins of individual variation in behavior and immunity. Neurosci. Biobehav. Rev. 29:39–49.

Coe, CL, Lubach, GR, and Karazewski, JW (1996). Prenatal endocrine activation alters postnatal cellular immunity in infant monkeys. Brain, Behav. Immun. 10:221–234.

Collaer, ML and Hines, M (1995). Human behavioral sex differences: A role for gonadal hormones during early development? Psychol. Bull. 118:55–108.

Costantini, D and Dell'Omo, G (2006). Effects of T-cell-mediated immune response on avian oxidative stress. Comp. Biochem. Physiol. A 145:137–142.

Costantini, D and Møller, AP (2008). Carotenoids are minor antioxidants for birds. Funct. Ecol. 22:367–370.

Costantini, D and Møller, AP (2009). Does immune response cause oxidative stress in birds? A meta-analysis. Comp. Biochem. Physiol. A 153:339–344.

Costantini, D and Verhulst, S (2009). Does high antioxidant capacity indicate low oxidative stress? Funct. Ecol. 23:506–509.

Couret, D, Prunier, A, Mounier, A-M, Thomas, F, Oswald, IP, and Merlot, E (2009). Comparative effects of a prenatal stress occuring during early or late gestation on pig immune response. Physiol. Behav. 98: 498–504.

Cucco, M, Guasco, B, Malacarne, G, and Ottonelli, R (2006). Effects of β-carotene supplementation on chick growth, immune status and behaviour in the grey partridge, *Perdix perdix*. Behav. Proc. 73: 325–332.

Cucco, M, Guasco, B, Malacarne, G, and Ottonelli, R (2007). Effects of β-carotene on adult immune condition and antibacterial activity in the eggs of the grey partridge, *Perdix perdix*. Comp. Biochem. Physiol. A 147:1038–1046.

Cucco, M, Guasco, B, Malacarne, G, Ottonelli, R, and Tanvez, A (2008). Yolk testosterone levels and dieatary carotenoids influence growth and immunity of grey partridge chicks. Gen. Comp. Endocrinol. 156: 418–425.

Cunningham, EJA and Russell, AF (2000). Egg investment is influenced by male attractiveness in the mallard. Nature 404:74–77.

Daniels, JT, Hatfield, PG, Burgess, DE, Kott, RW, and Bowman, JGP (2000). Evaluation of ewe and lamb immune response when ewes were supplemented with vitamin E. J. Anim. Sci. 78:2731–2736.

Davison, F, Magor, KE, and Kaspers, B (2008). Structure and evolution of avian immunoglobulines. In: Davison, F, Kaspers, B, and Schat, KA (eds.). Avian Immunology. Academic Press, London, pp. 107–128.

de Ayala, RM, Martinelli, R, and Saino, N (2006). Vitamin E supplementation enhances growth and condition of nestling barn swallows (*Hirundo rustica*). Behav. Ecol. Sociobiol. 60:619–630.

de Groot, J, Kranendonk, G, Fillerup, M, Hopster, H, Boersma, W, Hodgson, D, van Reenen, K, and Taverene, M (2007). Response to LPS in female offspring from sows treated with cortisol during pregnancy. Physiol. Behav. 90:612–618.

Debier, C, Pomeroy, PP, Baret, PV, Mignolet, E, and Larondelle, Y (2002). Vitamin E status and the dynamics of its transfer between mother and pup during lactation in grey seals (*Halichoerus grypus*). Can. J. Zool. 80:727–737.

Dowling, DK and Simmons, LW (2009). Reactive oxygen species as universal constraints in life-history evolution. Proc. R. Soc. B 276:1737–1745.

Dröge, W (2002). Free radicals in the physiological control of cell function. Physiol. Rev. 82:47–95.

Eishi, Y, Hirokawa, K, and Hatakeyama, S (1983). Long-lasting impairment of immune and endocrine systems of offspring induced by injection of dexamethasone into pregnant mice. Clin. Immunol. Immunopath. 26:335–349.

Ewen, JG, Thorogood, R, Brekke, P, Cassey, P, Karadas, F, and Armstrong, DP (2009). Maternally invested carotenoids compensate costly ectoparasitism in the hihi. Proc. Natl. Acad. Sci. USA 106:12798–12802.

Faivre, B, Grégoire, A, Préault, M, Cézilly, F, and Sorci, G (2003). Immune activation rapidly mirrored in a secondary sexual trait. Science 300:103.

Fenoglio, S, Cucco, M, and Malacarne, G (2002). The effect of carotenoid-rich diet on immunocompetence and behavioural performances in moorhen chicks. Ethol. Ecol. Evol. 14:149–156.

Finch, JM and Turner, RJ (1996). Effects of selenium and vitamin E on the immune response of domestic animals. Res. Vet. Sci. 60:97–106.

Folstad, I and Karter, AK (1992). Parasites, bright males and the immunocompetence handicap. Am. Nat. 139:603–622.

Fonseca, ESM., Massoco, CO, and Palermo-Neto, J (2002). Effects of prenatal stress on stress-induced changes in behavior and macrophage activity of mice. Physiol. Behav. 77:205–215.

Gagliano, M and McCormick, MI (2009). Hormonally mediated maternal effects shape offspring survival potential in stressful environments. Oecologia 160:657–665.

Gallizzi, K, Guenon, B, and Richner, H (2008a). Maternally transmitted parasite defence can be beneficial in the absence of parasites. Oikos 117:223–230.

Gallizzi, K, Gern, L, and Richner, H (2008b). A flea-induced pre-hatching maternal effect modulates tick feeding behaviour on great tit netslings. Funct. Ecol. 22:94–99.

Gasparini, J, McCoy, KD, Haussy, C, Tveraa, T, and Boulinier, T (2001). Induced maternal response to the Lyme disease spriochaete *Borrelia burgdorferi* sensu lato in a colonial seabird, the kittiwake *Rissa tridactyla*. Proc. R. Soc. B 268:647–650.

Gasparini, J, McCoy, KD, Tveraa, T, and Boulinier, T (2002). Related concentrations of specific immunoglobulins against the Lyme disease agent *Borrelia burgdorferi* sensu lato in eggs, young and adults of the kittiwake (*Rissa tridactyla*). Ecol. Lett. 5:519–524.

Gasparini, J, McCoy, KD, Staszewski, V, Haussy, C, and Boulinier, T (2006). Dynamics of anti-Borrelia antibodies in black-legged kittiwake (*Rissa tridactyla*) chicks suggest a maternal educational effect. Can. J. Zool. 84:623–627.

Gasparini, J, Boulinier, T, Gill, VA, Gil, D, Hatch, SA, and Roulin, A (2007). Food availability affects the maternal transfer of androgens and antibodies into eggs of a colonial seabird. J. Evol. Biol. 20: 874–880.

Gasparini, J, Piault, R, Bize, P, and Roulin, A (2009). Pre-hatching maternal effect inhibit nestling humoral immune response in the tawny owl *Strix aluco*. J. Avian Biol. 40:271–278.

Gil, D (2008). Hormones in avian eggs: physiology, ecology and behavior. Adv. Study Behav. 38:337–398.

Gil, D, Graves, J, Hazon, N, and Wells, A (1999). Male attractiveness and differential testosterone investment in zebra finch eggs. Science 286:126–128.

Gill, JW, Hosking BJ, and Egan, AR (1998). Prenatal programming of mammalian growth – a review of the role of steroids. Livestock Prod. Sci. 54:251–267.

Glick, B (1961). The influence of dipping eggs in male hormone solutions on lymphatic tissue and antibody response of chickens. Endocrinol. 69:984.

Glick, B and Sadler, CR (1961). The elimination of the bursa of Fabricius and reduction of antibody production in birds from eggs dipped in hormone solutions. Poultry Sci. 40:185–189.

Gore, AB and Qureshi, MA (1997). Enhancement of humoral and cellular immunity by vitamin E after embryonic exposure. Poultry Sci. 76:984–991.

Gotz, AA, Wittlinger, S, and Stefanski, V (2007). Maternal social stress during pregnancy alters immune function and immune cell numbers in adult male Long-Evans rat offspring during stressful life-events. J. Neuroimmunol. 185:95–102.

Grether, GF, Kolluru, GR, Lin, K, Quiroz, MA, Robertson, G, and Snyder, AJ (2008). Maternal effects of carotenoid consumption in guppies (*Poecilia reticulata*). Funct. Ecol. 22:294–302.

Grindstaff, JL (2008). Maternal antibodies reduce costs of an immune response during development. J. Exp. Biol. 211:654–660.

Grindstaff, JL, Brodie, III ED, and Ketterson, ED (2003). Immune function across generations: integrating mechanism and evolutionary process in maternal antibody transfer. Proc. R. Soc. B 270:2309–2319.

Grindstaff, JL, Demas, GE, and Ketterson, ED (2005). Diet quality affects egg size and number but does not reduce maternal antibody transmission in Japanese quail *Coturnix japonica*. J. Anim. Ecol. 74:1051–1058.

Grindstaff, J, Hasselquist, D, Nilsson, J-Å, Sandell, M, Smith, HG, and Stjernman, M (2006). Transgenerational priming of immunity: maternal exposure to a bacterial antigen enhances offspring humoral immunity. Proc. R. Soc. B 273:2551–2557.

Groothuis, TGG, Eising, CM, Blount, JD, Surai, P, Apanius, V, Dijkstra, C, and Müller, W (2006). Multiple pathways of maternal effects in black-headed gull eggs: constraint and adaptive compensatory adjustment. J. Evol. Biol. 19:1304–1313.

Groothuis, TGG, Eising, C, Dijkstra, C, and Müller, W (2005a). Balancing between costs and benefits of maternal hormone deposition in avian eggs. Biol. Lett. 1:78–81.

Groothuis, TGG, Müller, W, von Engelhardt, N, Carere, C, and Eising, C (2005b). Maternal hormones as a tool to adjust offspring phenotype in avian species. Neurosci. Biobehav. Rev. 29:329–352.

Groothuis, TGG and Schwabl, H (2008). Hormone-mediated maternal effects in birds: mechanisms matter but what do we know of them? Phil. Trans. R. Soc. B 363:1647–1661.

Gustafsson, E, Mattsson, A, Holmdahl, R, and Mattsson, R (1994). Pregnancy in B-cell-deficient mice: postpartum transfer of immunoglobulins prevents neonatal runting and death. Biol. Reprod. 51: 1173–1180.

Guzman, C, Camacho-Arroyo, I, Leon-Nava, MA, and Morales-Montor, J (2009). Neonatal exposure to estradiol induces resistance to helminth infection and changes in the expression of sex steroid hormone receptors in the brain and spleen in adult mice of both sexes. Brain Behav. Immun. 23:709–715.

Halliwell, B and Gutteridge, JMC (2007). Free Radicals in Biology and Medicine, 4th ed. Oxford University Press, Oxford.

Hamal, KR, Burgess, SC, Pevzner, IY, and Erf, GF (2006) Maternal antibody transfer from dams to their egg yolks, egg whites, and chicks in meat line of chickens. Poultry Sci. 85:1364–1372.

Haq, A-U, Bailey, CA, and Chinnah, A (1996). Effect of β-carotene, canthaxanthin, lutein, and vitamin E on neonatal immunity of chicks when supplemented in the broiler breeder diets. Poultry Sci. 75:1092–1097.

Hargitai, R, Arnold, KE, Herényi, M, Prechl, J, and Török, J (2009). Egg composition in relation to social environment and maternal physiological condition in the collared flycatcher. Behav. Ecol. Sociobiol. 63:869–882.

Hargitai, R, Prechl, J, and Török, J (2006). Maternal immunoglobulin concentration in collared flycatcher (*Ficedula albicollis*) eggs in relation to parental quality and laying order. Funct. Ecol. 20:829–838.

Hartley, RC and Kennedy, MW (2004). Are carotenoids a red herring in sexual display? Trends Ecol. Evol. 19:353–354.

Hassan, JO and Curtiss, R (1996). Effect of vaccination of hens with an avirulent strain of *Salmonella typhimurium* on immunity of progeny challenged with wild-type *Salmonella strains*. Infect. Imm. 64:938–944.

Hasselquist, D and Kempenaers, B (2002). Parental care and adaptive brood sex ratio manipulation in birds. Phil. Trans. R. Soc. B 357:363–372.

Hasselquist, D (2007). Comparative immunoecology in birds: hypotheses and tests. J. Ornithol. 148 (Suppl. 2): S571–S582.

Hasselquist, D and Nilsson, J-Å (2009). Maternal transfer of antibodies in vertebrates: trans-generational effects on offspring immunity. Phil. Trans R. Soc. B 364:51–60.

Heeb, P, Werner, I, Kölliker, M, and Richner, H (1998). Benefits of induced host responses against an ectoparasite. Proc. R. Soc. B 265:51–56.

Heinzerling, RH, Tengerdy, RP, Wick, LL, and Lueker, DC (1974). Vitamin E protects mice against *Diplococcus pneumoniae* type I infection. Infect. Immunity 10:1292–1295.

Heller, ED, Leitner, G, Drabkin, N, and Melamed, D (1990). Passive immunization of chicks against *Escherichia coli*. Avian Pathol. 19:345–354.

Hõrak, P, Saks, L, Zilmer, M, Karu, U, and Zilmer, K (2007). Do dietary antioxidants alleviate the cost of immune activation? An experiment in greenfinches. Am. Nat. 170:625–635.

Hõrak, P, Surai, PF, and Møller, AP (2002). Fat-soluble antioxidants in the eggs of great tits *Parus major* in relation to breeding habitat and laying sequence. Avian Sci. 2:123–130.

Hõrak, P, Zilmer, M, Saks, L, Ots, I, Karu, U, and Zilmer, K (2006). Antioxidant protection, carotenoids and the cost of immune challenge in greenfinches. J. Exp. Biol. 209:4329–4338.

Hossain, SM, Barreto, SL, Bertechini, AG, Rios, AM, and Silva, CG (1998). Influence of dietary vitamin E level on egg production of broiler breeders, and on growth and immune response of progeny in comparison with the progeny from eggs injected with vitamin E. Anim. Feed Sci. Tech. 73:307–317.

Hu, F, Crespi, EJ, and Denver, RJ (2008). Programming neuroendocrine stress axis activity by exposure to glucocorticoids during postembryonic development in the frog, *Xenopus laevis*. Endocrinol. 149:5470–5481.

Huber, SK, Owen, JP, Koop, JAH, King, MO, Grant, PR, Grant, BR, and Clayton, DH (2010). Ecoimmunity in Darwin's finches: invasive parasites trigger acquired immunity in the medium ground finch (*Geospiza fortis*). PLoS One 5(1): e8605.

Hughes, DA (2002). Antioxidant vitamins and immune function. In: Calder, PC, Field, CJ, and Gill, HS (eds.). Nutrition and Immune Function. CABI Publishing, New York, pp. 171–191.

Hulbert, AJ, Pamplona, R., Buffenstein, R., and Buttemer, WA (2007). Life and death: metabolic rate, membrane composition, and life span of animals. Physiol. Rev. 87:1175–1213.

Ilmonen, P, Taarna, T, and Hasselquist, D (2000). Experimentally activated immune defence in female pied flycatchers results in reduced breeding success. Proc. R. Soc. B 267:665–670.

Jablonka, E and Lamb, MJ (1999). Epigenetic Inheritance and Evolution. Oxford, Oxford University Press.

Jackson, DW, Law, GRJ, and Nockels, CF (1978). Maternal vitamin E alters passively acquired immunity of chicks. Poultry Sci. 57:70–73.

Kallio, ER, Poikonen, A, Vaheri, A, Vapalahti, O, Henttonen, H, Koskela, E, and Mappes, T (2006). Maternal antibodies postpone hantavirus infection and enhance individual breeding success. Proc. R. Soc. B 273:2771–2776.

Kanitz, E, Otten, W, and Tuchscherer, M (2006). Changes in endocrine and neurochemical profiles in neonatal pigs prenatally exposed to increased maternal cortisol. J. Endocrinol. 191:207–220.

Kanitz, E, Tuchscherer, M, Puppe, B, Tuchscherer, A, and Stabenow, B (2004). Consequences of repeated early isolation in domestic pigs (*Sus scrofa*) on their behavioral, neuroendocrine and immunological responses. Brain Behav. Immun. 18:35–45.

Kapoor, A, Dunn, E, Kostaki, A, Andrews, MH, and Matthews, SG (2006). Fetal programming of hypothalamo-pituitary-adrenal function: prenatal stress and glucocorticoids. J. Physiol. 572:31–44.

Karadas, F, Surai, P, Grammenidis, E, Sparks, NHC, and Acamovic, T (2006). Supplementation of the maternal diet with tomato powder and marigold extract: effects on the antioxidant system of the developing quail. Br. Poultry Sci. 47:200–208.

Karell, P, Kontiainen, P, Pietiäinen, H, Siitari, H, and Brommer, JE (2008). Maternal effects on offspring Igs and egg size in relation to natural and experimentally improved food supply. Funct. Ecol. 22:682–690.

Kay, G, Tarcic, N, Poltyrev, T, and Weinstock, M (1998). Prenatal stress depresses immune function in rats. Physiol. Behav. 63:397–402.

Ketterson, ED and Nolan Jr, V (1992). Hormones and life histories: an integrative approach. Am. Nat. 140: S33–S62.

Kilpimaa, J, Alatalo, RV, and Siitari, H (2006). Prehatching maternal investment and offspring immunity in the pied flycatcher (*Ficedula hypoleuca*). J. Evol. Biol. 20:717–724.

King, MO, Owen, JP, and Schwabl, HG (2010). Are maternal antibodies really that important? Patterns in the immunological development of altricial passerine house sparrows (*Passer domesticus*). PLoS ONE 5:e9639.

Kissling, RE, Eidson, ME, and Stamm, DD (1954). Transfer of maternal neutralizing antibodies against eastern equine encephalomyelitis virus in birds. J. Infect. Dis. 95:179–181.

Klasing, KC (1998). Nutritional modulation of resistance to infectious diseases. Poultry Sci. 77:1119–1125.

Klasing, KC (2004). The cost of immunity. Acta Zool. Sin. 50:961–969.

Klasing, KC and Leshchinsky, TV (1999). Functions, costs, and benefits of the immune system during development and growth. In: Adams, NJ and Slotow, R (eds.). Proc. 22 Int. Ornithol. Congr. H. BirdLife South Africa, Johannesburg, pp. 2817–2835.

Klein, SL (2000). The effects of hormones on sex differences in infection: from genes to behavior. Neurosci. Biobehav. Rev. 24:627–638.

Klein, SL, Marson, AL, Scott, AL, Ketner, G, and Glass, GE (2002). Neonatal sex steroids affect responses to Seoul virus infection in male but not female Norway rats. Brain Behav. Immun. 16:736–746.

Klein, SL and Rager, DR (1995). Prenatal stress alters immune function in the offspring of rats. Dev. Psychobiol. 28:321–336.

Kondo, Y, Goto, C, and Abe, A (2004). Effects of estrogen treatment during the embryonic period on chick antibody production. J. Poultry Sci. 41:85–93.

Koutsos, EA, Clifford, AJ, Calvert, CC, and Klasing, KC (2003). Maternal carotenoid status modifies the incorporation of dietary carotenoids into immune tissues of growing chickens (*Gallus gallus domesticus*). J. Nutr. 133:1132–1138.

Koutsos, EA, López, JCG, and Klasing, KC (2007). Maternal and dietary carotenoids interactively affect cutaneous basophil responses in growing chickens (*Gallus gallus domesticus*). Comp. Biochem. Physiol. B 147:87–92.

Kowalczyk, K, Daiss, J, Halpern, J, and Roth, TF (1985). Quantitation of maternal-fetal IgG transport in the chicken. Immunology 54:755–762.

Kristan, DM (2002). Maternal and direct effects of the intestinal nematode *Heligmosomoides polygyrus* on offspring growth and susceptibility to infection. J. Exp. Biol. 205:3967–3977.

Kristan, DM (2004). Intestinal nematode infection affects host life history and offspring susceptibility to parasitism. J. Anim. Ecol. 73:227–238.

Kuby, J (1992). Immunology. Freeman, New York.

Landete-Castillejos, T, Garcia, A, Gomez, JA, Laborda, J, and Gallego, L (2002). Effects of nutritional stress during lactation on immunity costs and indices of future reproduction in the Iberian red deer (*Cervus elaphus hispanicus*). Biol. Repr. 67:1613-1620.

Lange, H, Kobarg, J, Yazynin, S, Solterbeck, M, Henningsen, M, Hansen, H, and Lemke, H (1999). Genetic analysis of the maternally induced affinity enhancement in the non-Ox1 idiotypic antibody repertoire of the primary immune response to 2-phenyloxazolone. Scand. J. Immunol. 49:55–66.

Lauridsen, C, Engel, H, Jensen, SK, Craig, AM, and Traber, MG (2002). Lactating sows and suckling piglets preferentially incorporate RRR- over All-rac-α-Tocopherol into milk, plasma and tissues. J. Nutr. 132:1258–1264.

Lawrence, EC, Arnaud-Battandier, F, Grayson, J, Koski, IR, Dooley, NJ, Muchmore, AV, and Blaese, RM (1981). Ontogeny of humoral immune function in normal chickens: a comparison of immunoglobulin-secreting cells in bone marrow, spleen, lungs and intestine. Clin. Exp. Immunol. 43:450–457.

Lemke, H, Coutinho, A, and Lange, H (2004). Lamarckian inheritance by somatically acquired maternal IgG phenotypes. Trends Immunol. 25:180–186.

Lemke, H and Lange, H (1999). Is there a maternally induced immunological imprinting phase á la Konrad Lorenz? Scand. J. Immunol. 50:348–354.

Lemke, H, Hansen, H, and Lange, H (2003). Non-genetic inheritable potential of maternal antibodies. Vaccine 21:3428–3431.

Lemke, H, Tanasa, RI, Trad, A, and Lange, H (2009). Benefits and burden of the maternally-mediated immunological imprinting. Autoimmun. Rev. 8:394–399.

Leposavic, G, Perisic, M, Kosec, D, Arsenovic-Ranin, N, Radojevic, K, Stojic-Vulkanic, Z, and Pilipovic, I (2009). Neonatal testosterone imprinting affects thymus development and leads to phenotypic rejuvenation and masculinization of the peripheral blood T-cell compartment in adult female rats. Brain Behav. Immun. 23:294–304.

Levitt, N, Lindsay RS, Holmes, MC, and Seckl, JR (1996). Dexamethasone in the last week of pregnancy attenuates hippocampal glucocorticoid receptor gene expression and elevates blood pressure in the rat. Neuroendocrinol. 64:412–418.

Lin, Y-F, Tsai, H-L, Lee, Y-C, and Chang, S-J (2005). Maternal vitamin E supplementation affects the antioxidant capability and oxidative status of hatching chicks. J. Nutr. 135:2457–2461.

Little, TJ, O'Connor, B, Colegrave, N, Watt, K, and Read, AF (2003). Maternal transfer of strain-specific immunity in an invertebrate. Current Biol. 13:489–492.

Loudenslager, MJ, Ku, PK, Whetter, PA, Ullrey, DE, Whiehair, CK, Stowe, HD, and Miller, ER (1986). Importance of diet of dam and colostrum to the biological antioxidant status and parenteral iron tolerance of the pig. J. Anim. Sci. 63:1905–1914.

Loyau, A, Jalme MS, Mauget, R, and Sorci G (2007). Male sexual attractiveness affects the investment of maternal resources into the eggs in peafowl (*Pavo cristatus*). Behav. Ecol. Sociobiol. 61:1043–1052.

Lozano, GA and Ydenberg, RC (2002). Transgenerational effects of maternal immune challenge in tree swallows (*Tachycineta bicolour*). Can. J. Zool. 80:918–925.

Lundin, BS, Dahlman-Höglund, A, Pettersson, I, Dahlgren, UIH, Hansson, LÅ, and Telemo, E (1999). Antibodies given orally in the neonatal period can affect the immune response for two generations: evidence for active maternal influence on the newborn's immune system. Scand. J. Immunol. 50:651–656.

Lunn, SF, Cowen, GM, and Fraser, HM (1997). Blockade of the neonatal increase in testosterone by a GnRH antagonist: the free androgen index, reproductive capacity and postmortem findings in the male marmoset monkey. J. Endocrinol. 154:125–131.

Marsh, JA, Combs Jr, GF, Whitacre, ME, and Dietert, RR (1986). Effect of selenium and vitamin E dietary deficiencies on chick lymphoid organ development. Proc. Soc. Exp. Biol. Med. 182:425–436.

Marsh, JA, Dietert, RR, and Combs Jr, GF (1981). Influence of dietary selenium and vitamin E on the humoral immune response of the chick. Proc. Soc. Exp. Biol. Med. 166:228–236.

Martin, JT (2000). Sexual dimorphism in immune function: the role of prenatal exposure to androgens and estrogens. Eur. J. Pharmacol. 405:251–261.

Martin, LB, Han, P, Lewittes, J, Kuhlman, JR, Klasing, KC, and Wikelski, M (2006). Phytohemagglutinin-induced skin swelling in birds: histological support for a classic immunoecological technique. Funct. Ecol. 20:290–299.

Martin, LB, Weil, ZM, & Nelson, RJ (2008). Seasonal trade-offs between reproduction and immune activity. Phil. Trans. R. Soc. B 363:321–339.

Mastroeni, P, Vazquez-Torres, A, Fang, FC, Xu, Y, Khan, S, Hormaeche, CE, and Dougan, G (2000). Antimicrobial actions of the NADPH phagocyte oxidase and inducible nitric oxide synthase in experimental salmonellosis. II. Effects on microbial proliferation and host survival in vivo. J. Exp. Med. 192:237–247.

McCormick, MI (1999). Experimental test of the effect of maternal hormones on larval quality of a coral reef fish. Oecologia 118:412–422.

McGlothlin, JW and Ketterson, ED (2008). Hormones and the continuum between adaptation and constraint. Phil. Trans. R. Soc. B 363:1611–1620.

McGraw, KJ and Ardia, DR (2003). Carotenoids, immunocompetence, and the information content of sexual colors: an experimental test. Am. Nat. 162:704–712.

McGraw, KJ and Ardia, DR (2004). Immunoregulatory activity of different dietary carotenoids in male zebra finches. Chemoecol. 14:25–29.

McGraw, KJ, Crino, OL, Medina-Jerez, W, and Nolan, PM (2006). Effect of dietary carotenoid supplementation on food intake and immune function in a songbird with no carotenoid coloration. Ethology 112: 1209–1216.

McGraw, KJ and Klasing, KC (2006). Carotenoids, immunity, and integumentary coloration in red junglefowl (*Gallus gallus*). Auk 123:1161–1171.

Merlot, E, Couret, D, and Otten, W (2008). Prenatal stress, fetal imprinting and immunity. Brain Behav. Immun. 22:42–51.

Michalek, SM, Rahman, AF, and McGhee, JR (1975). Rat immunoglobulins in serum and secretions: comparison of IgM, IgA and IgG in serum, colostrum, milk and saliva of protein malnourished and normal rats. Proc. Soc. Exp. Biol. Med. 148:1114-1118.

Monaghan, P, Metcalfe, NB, and Torres, R (2009). Oxidative stress as a mediator of life history trade-offs: mechanisms, measurements and interpretation. Ecol. Lett. 12:75–92.

Moore, MC and Johnston, GIH (2008). Toward a dynamic model of deposition and utilization of yolk steroids. Int. Comp. Biol. 48:411–418.

Mor, A and Avtalion, RR (1990). Transfer of antibody activity from immunized mother to embryos in tilapias. J. Fish Biol. 37:249–255.

Morales, J, Sanz, JJ, and Moreno, J (2006). Egg colour reflects the amount of yolk maternal antibodies and fledging success in a songbird. Biol. Letters 2:334–336.

Moreno, J, Lobato, E, Morales, J, Merino, S, Martinez-De La Puente, J, and Tomas, G (2008). Pre-laying nutrition mediates maternal effects on offspring immune capacity and growth in the pied flycatcher. Oecologia 156:727–735.

Moret, Y and Schmid-Hempel, P (2001). Immune defence in bumble-bee offspring. Nature 414:506.

Mougeot, F, Martínez-Padilla, J, Blount, JD, Pérez-Rodríguez, L, Webster, LMI, and Piertney, SB (2010). Oxidative stress and the effect of parasites on a carotenoid-based ornament. J. Exp. Biol. 213:400–407.

Mousseau, TA and Fox, CW (1998). Maternal Effects as Adaptations. Oxford University Press, Oxford.

Muller, FL, Lustgarten, MS, Jang, Y, Richardson, A, and van Remmen, H (2007). Trends in oxidative aging theories. Free Radic. Biol. Med. 43:477–503.

Müller, W, Boonen, S, Groothuis, TGG, and Eens, M (2010). Maternal yolk testosterone in canary eggs: toward a better understanding of mechanisms and function. Behav. Ecol. 21:493–500.

Müller, W, Deptuch, K, Lopez-Rull, I, and Gil, D (2007). Elevated yolk androgen levels benefit offspring development in a between-clutch context. Behav. Ecol. 18:929–936.

Müller, W, Groothius, TGG, Dijkstra, C, Siitari, H, and Alatalo, R (2004). Maternal antibody transmission and breeding densities in the black-headed gull *Larus ridibundus*. Funct. Ecol. 18:719–724.

Müller, W, Groothuis, TGG, Kasprzik, A, Dijkstra, C, Alatalo, RV, and Siitari, H (2005). Prenatal androgen exposure modulates cellular and humoral immune function of black-headed gull chicks. Proc. R. Soc. B 272:1971–1977.

Murphy, K, Travers, P, and Walport, M (2008). Janeway's Immunobiology, 7th ed. Garland Science, New York.

Møller, AP, Biard, C, Blount, JD, Houston, DC, Ninni, P, Saino, N, and Surai, PF (2000). Carotenoid-dependent signals: indicators of foraging efficiency, immunocompetence or detoxification ability. Avian Poultry Biol. Rev. 11:137–159.

Nager, RG, Ruegger, C, and van Noordwijk, AJ (1997). Nutrient or energy limitation on egg formation: a feeding experiment in great tits. J. Anim. Ecol. 66:495–507.

Navara, KJ, Badyaev, AV, Mendonça, MT, and Hill, GE (2006b). Yolk antioxidants vary with male attractiveness and female condition in the house finch (*Carpodacus mexicanus*). Physiol. Biochem. Zool. 79: 1098–1105.

Navara, KJ and Hill, GE (2003). Dietary carotenoid pigments and immune function in a songbird with extensive carotenoid-based plumage coloration. Behav. Ecol. 14:909–916.

Navara, KJ, Hill, GE, and Mendonça, MT (2005). Variable effects of yolk androgens on growth, survival, and immunity in Eastern bluebird nestlings. Physiol. Biochem. Zool. 78:570–578.

Navara, KJ, Hill, GE, and Mendonça, MT (2006a). Yolk testosterone stimulates growth and immunity in house finch chicks. Physiol. Biochem. Zool. 79:550–555.

Nemeth, NM, Oesterle, PT and Bowen, RA (2008). Passive immunity to West Nile virus provides limited protection in a common passerine species. Am. J. Trop. Med. Hyg. 79:283–290.

Ng, PLK and Higgins, DA (1998). Bile immunoglobulin of the duck (*Anas platyrhynchos*): I. Preliminary characterization and ontogeny. Immunology 58:323–327.

Nicoara, C, Zäch, K, Trachsel, D, Germann, D, and Matter, L (1999). Decay of passively acquired maternal antibodies against measles, mumps, and rubella viruses. Clin. Diag. Lab. Immunol. 6:868–871.

Nilsson, J-Å and Råberg, L (2001). The resting metabolic cost of egg laying and nestling feeding in great tits. Oecologia 128:187–192.

Nockels, CF (1979). Protective effects of supplemental vitamin E against infection. Fed. Proc. 38:2134–2138.

Norton, JM and Wira, CR (1977). Dose-related effects of the sex hormones and cortisol on the growth of the bursa of Fabricius in chicken embryos. J. Steroid Biochem. 8:985–987.

Okamoto, Y, Tsutsumi, H, Kumar, NS, and Ogra, PL (1989). Effect of breast feeding on the development of anti-idiotype antibody response to F glycoprotein of respiratory syncytial virus in infant mice after postpartum maternal immunization. J. Immunol. 142:2507–2512.

Ortuño, J, Esteban, MA, and Meseguer, J (2000). High dietary intake of α-tocopherol acetate enhances the nonspecific immune response of gilthead seabream (*Sparus aurata* L.). Fish Shellfish Immunol. 10:293–307.

Otten, W, Kanitz, E, Tuchscherer, M, Puppe, B, and Nuernberg, G (2007). Repeated administrations of adrenocorticotropic hormone during gestation in gilts: effects on growth, behavior and immune responses of their piglets. Livestock Sci. 106:261–270.

Owen-Ashley, NT, Hasselquist, D, and Wingfield, JC (2004). Androgens and the immunocompetence handicap hypothesis: unraveling direct and indirect pathways of immunosuppression in song sparrows. Am. Nat. 164:490–505.

Parmentier, HK, Nieuwland, MGB, Rijke, E, De Vries Reilingh, G, and Sharma, JW (1996). Divergent antibody responses to vaccines and divergent body weights of chicken lines selected for high and low humoral responsiveness to sheep red blood cells. Avian Dis. 40:634–644.

Partali, V, Liaaen-Jensen, S, Slagsvold, T, and Lifjeld, JT (1987). Carotenoids in food chain studies–II. The food chain of Parus spp. monitored by carotenoid analysis. Comp. Biochem. Physiol. 87B:885–888.

Patterson, R, Younger, JS, Weigle, WO, and Dixon, FJ (1962). The metabolism of serum proteins in the hen and chick, and secretion of serum proteins by the ovary of the hen. J. Gen. Physiol. 45:501–513.

Peters, A (2007). Testosterone and carotenoids: an integrated view of trade-offs between immunity and sexual signalling. BioEssays 29:427–430.

Pihlaja, M, Siitari, H, and Alatalo, RV (2006). Maternal antibodies in a wild altricial bird: effects on offspring immunity, growth and survival. J. Anim. Ecol. 75:1154–1164.

Pilz, KM, Quiroga, M, Schwabl, H, and Adkins-Regan, E (2004). European starling chicks benefit from high yolk testosterone levels during a drought year. Horm. Behav. 46:179–192.

Pitala, N, Ruuskanen, S, Laaksonen, T, Doligez, B, Tschirren, B, and Gustafsson, L (2009). The effects of experimentally manipulated yolk androgens on growth and immune function of male and female nestling collared flycatchers *Ficedula albicollis*. J. Avian Biol. 40:225–230.

Poulin, R and Thomas, F (2008). Epigenetic effects of infection on the phenotype of host offspring: parasites reaching across host generations. Oikos 117:331–335.

Poyton, RO, Ball, KA, and Castello, PR (2009). Mitochondrial generation of free radicals and hypoxic signalling. Trends Endocrinol. Metab. 20:332–340.

Råberg, L, Grahn, M, Hasselquist, D, and Svensson, E (1998). On the adaptive significance of stress-induced immunosuppression. Proc. R. Soc. B 265:1637–1641.

Råberg, L, Nilsson, J-Å, Ilmonen, P, Stjernman, M, and Hasselquist, D (2000). The cost of an immune response: vaccination reduces parental effort. Ecol. Letters 3:382–386.

Råberg, L, Vestberg, M, Hasselquist, D, Holmdahl, R, Svensson, E, and Nilsson, J-Å (2002). Basal metabolic rate and the evolution of the adaptive immune system. Proc. R. Soc. B 269:817–821.

Radder, RS and Shine, R (2007). Are the phenotypic traits of hatchling lizards affected by maternal allocation of steroid hormones to the egg? Gen. Comp. Endocrinol. 154:111–119.

Recabarren, SE, Rojas-Garcia, PP, Recabarren, MP, Norambuena, K, and Sir-Petermann, T (2009). Impact of prenatal exposure to testosterone in biometrics and endocrine parameters of newborn lambs. Arch. Medicina Vet. 41:43–51.

Reeves, WC, Sturgeon, JM, French, EM, and Brookman, B (1954). Transovarian transmission of neutralizing substances to western equine and St. Louis encephalitis viruses by avian hosts. J. Infect. Dis. 95: 168–178.

Reid, JM, Arcese, P, Keller, LF, and Hasselquist, D (2006). Long-term maternal effect on offspring immune response in song sparrows *Melospiza melodia*. Biol. Lett. 2:573–576.

Remeš, V, Krist, M, Bertacche, V, and Stradi, R (2007). Maternal carotenoid supplementation does not affect breeding performance in the great tit (*Parus major*). Funct. Ecol. 21:776–783.

Rettenbacher, S, Möstl, E, and Groothuis, TGG (2009). Gestagens and glucocorticoids in chicken eggs. Gen. Comp. Endocrinol. 164:125–129.

Reuman, PD, Paganini, CM, Ayoub, EM, and Small Jr, PA (1983). Maternal–infant transfer of influenza-specific immunity in the mouse. J. Immunol. 130:932–936.

Rhen, T and Crews, D (2002). Variation in reproductive behaviour within a sex:neural system and endocrine activation. J. Neuroendocrinol. 14:517–531.

Richner, H, Oppliger, A, and Christe, P (1993). Effects of an ectoparasite on reproduction in great tits. J. Anim. Ecol. 62:703–710.

Ricklefs, RE and Wikelski, M (2002). The physiology/life-history nexus. Trends Ecol. Evol. 17:462–468.

Romano, M, Caprioli, M, Ambrosini, R, Rubolini, D, Fasola, M, and Saino, N (2008). Maternal allocation strategies and differential effects of yolk carotenoids on the phenotype and viability of yellow-legged gull (*Larus michahellis*) chicks in relation to sex and laying order. J. Evol. Biol. 21:1626–1640.

Roszman, TL and Brooks, WH (1997). Interactive signaling pathways of the neuroendocrine-immune network. Chem. Immunol. 69:203–222.

Royle, NJ, Surai, PF, McCartney, RJ, and Speake, BK (1999). Parental investment and egg yolk lipid composition in gulls. Funct. Ecol. 13:298–306.

Royle, NJ, Surai, PF, and Hartley, IR (2001). Maternally derived androgens and antioxidants in bird eggs: complementary but opposing effects? Behav. Ecol. 12:381–385.

Royle, NJ, Surai, PF, and Hartley, IR (2003). The effect of variation in dietary intake on maternal deposition of antioxidants in zebra finch eggs. Funct. Ecol. 17:472–481.

Rubolini, D, Romano, M, Boncoraglio, G, Ferrari, RP, Martinelli, R, Galeotti, P, Fasola, M, and Saino, N (2005). Effects of elevated egg corticosterone levels on behavior, growth, and immunity of yellow-legged gull (*Larus michahellis*) chicks. Horm. Behav. 47:592–605.

Rubolini, D, Romano, M, Bonisoli Alquati, A, and Saino, N (2006c). Early maternal, genetic and environmental components of antioxidant protection, morphology and immunity of yellow-legged gull (*Larus michahellis*) chicks. J. Evol. Biol. 19:1571–1584.

Rubolini, D, Romano, M, Martinelli, R, Leoni, B, and Saino, N (2006a). Effects of prenatal yolk androgens on armaments and ornaments of the ring-necked pheasant. Behav. Ecol. Sociobiol. 59:549–560.

Rubolini, D, Romano, M, Martinelli, R, and Saino, N (2006b). Effects of elevated yolk testosterone on survival, growth and immunity of male and female yellow-legged gull chicks. Behav. Ecol. Sociobiol. 59:344–352.

Safran, RJ, Pilz, KM, McGraw, KJ, Correa, SM, and Schwabl, H (2008). Are yolk androgens and carotenoids in barn swallow eggs related to parental quality? Behav. Ecol. Sociobiol. 62:427–438.

Sahoo, PK and Mukherjee, SC (2002). Influence of high dietary α-tocopherol intakes on specific immune response, nonspecific resistance factors and disease resistance of healthy and aflatoxin B_1-induced immunocompromised Indian major carp, *Labeo rohita* (Hamilton). Aquaculture Nutr. 8:159–167.

Saino, N, Bertacche, V, Bonisoli-Alquati, A, Romano, M, and Rubolini, D (2008). Phenotypic correlates of yolk and plasma carotenoid concentration in yellow-legged gull chicks. Physiol. Biochem. Zool. 81:211–225.

Saino, N, Bertacche, V, Ferrari, RP, Martinelli, R, Møller, AP, and Stradi, R (2002b). Carotenoid concentration in barn swallow eggs is influenced by laying order, maternal infection and paternal ornamentation. Proc. R. Soc. B 269:1729–1733.

Saino, N, Ferrari, RP, Martinelli, R, Romano, M, Rubolini, D, and Møller, AP (2002a). Early maternal effects mediated by immunity depend on sexual ornamentation of the male partner. Proc. R. Soc. B 269:1005–1009.

Saino, N, Ferrari, R, Romano, M, Martinelli, R, and Møller, AP (2003b). Experimental manipulation of egg carotenoids affects immunity in barn swallow nestlings. Proc. R. Soc. B 270:2485–2489.

Saino, N, Romano, M, Ferrari, RP, Martinelli, R, and Møller, AP (2003a). Maternal antibodies but not carotenoids in barn swallow eggs covary with embryo sex. J. Evol. Biol. 16:516–522.

Sams, MG, Lochmiller, RL, Hellgren EC, Payton, ME, and Varner, LW (1995). Physiological responses of neonatal white-tailed deer reflective of maternal dietary protein intake. Can. J. Zool. 73:1928-1936.

Sandell, MI, Tobler, M, and Hasselquist, D (2009). Yolk androgens and the development of avian immunity: an experiment in jackdaws (*Corvus monedula*). J. Exp. Biol. 212:815–822.

Sato, H, Albrecht, P, Reynolds, DW, Stagno, S, and Ennis, FA (1979). Transfer of measles, mups, and rubella antibodies from mother to infant. Its effect on measles, mups, and rubella immunization. Am. J. Dis. Child. 133:1240–1243.

Schijns, VEJC, Sharma, J, and Tarpey, I (2008). Practical aspects of poultry vaccination. In: Davison, F, Kaspers, B, and Schat, KA (eds.). Avian Immunology. Academic Press, London, pp. 373–394.

Schlabritz-Loutsevitch, NE, Hodara, VL, Parodi, LM, Hubbard, GB, Jenkins, SL, Dudley, DJ, Nathanielsz, PW, and Giavedoni, LD (2006). Three weekly courses of betamethasone administered to pregnant baboons at 0.6, 0.65 and 0.7 of gestation alter fetal and maternal lymphocyte populations at 0.95 of gestation. J. Reprod. Immunol. 69:149–163.

Schumacher, IM, Rostal, DC, Yates, RA, Brown, DR, Jacobson, ER, and Klein, PA (1999). Persistence of maternal antibodies against *Mycoplasma agassizii* in desert tortoise hatchlings. Am. J. Vet Res. 60:826–831.

Schuurs, AHWM and Verheul, HAM (1990). Effects of gender and sex steroids on the immune response. J. Steroid Biochem. 35:157–172.

Schwabl, H (1993). Yolk is a source of maternal testosterone for developing birds. Proc. Natl. Acad. Sci. USA 90:11446–11450.

Schwabl, H (1996). Maternal testosterone in the avian egg enhances postnatal growth. Comp. Biochem. Physiol. 114:271–276.

Schwabl, H, Mock, DW, and Gieg, JA (1997). A hormonal mechanism for parental favouritism. Nature 386:231.

Schwanz, LE (2008a). Chronic parasitic infection alters reproductive output in deer mice. Behav. Ecol. Sociobiol. 62:1351–1358.

Schwanz, LE (2008b). Persistent effects of maternal parasitic infection on offspring fitness: implications for adaptive reproductive strategies when parasitized. Funct. Ecol. 22:691–698.

Seckl, JR (2004). Prenatal glucocorticoids and long-term programming. Eur. J. Endocrinol. 151:U49–U62.

Selvaraj, RK, Koutsos, EA, Calvert, CC, and Klasing, KC (2006). Dietary lutein and fat to modify macrophage properties in chicks hatched from carotenoid deplete or replete eggs. J. Anim. Physiol. Anim. Nutr. 90:70–80.

Shaheen, SO, Hines, M, Newson, RB, Wheeler, M, Herrick, DRM, Strachan, DP, Jones, RW, Burney, PGJ, and Henderson, AJ (2007). Maternal testosterone in pregnancy and atopic outcomes in childhood. Allergy 62:25–32.

Sheldon, BC (2000). Differential allocation: tests, mechanisms and implications. Trends Ecol. Evol. 15: 397–402.

Siegrist, C-A (2003). Mechanisms by which maternal antibodies influence infant vaccine responses: review of hypotheses and definition of main determinants. Vaccine 21:3406–3412.

Sikka, P, Lall, D, Arora, U, and Sethi, RK (2002). Growth and passive immunity in response to micronutrient supplementation in new-born calves of Murrah buffaloes given fat-soluble vitamins during late pregnancy. Livestock Prod. Sci. 75:301–311.

Smith, HG, Råberg, L, Ohlsson, T, Granbom, M, and Hasselquist, D (2006). Carotenoid and protein supplementation have differential effects on pheasant ornamentation and immunity. J. Evol. Biol. 20:310–319.

Smith, NC, Wallach, M, Petracca, M, Braun, R, and Eckert, J (1994). Maternal transfer of antibodies induced by infection with *Eimeria maxima* partially protects chickens against challenge with *Eimeria renella*. Parasitology 109:551–557.

Soler, JJ, De Neve, L, Perez-Contreras, T, Soler, M, and Sorci, G (2003). Trade-off between immunocompetence and growth in magpies: an experimental study. Proc. R. Soc. B 270:241–248.

Sorci, G and Clobert, J (1995). Effects of maternal parasite load on offspring life-history traits in the common lizard (*Lacerta vivipara*). J. Evol. Biol. 8:711–723.

Sorci, G and Faivre, B (2009). Inflammation and oxidative stress in vertebrate host-parasite systems. Phil Trans. R. Soc. B 364:71–83.

Staszewski, V, Gasparini, J, McCoy, KD, Tveraa, T, and Boulinier, T (2007a). Evidence of an interannual effect of maternal immunization on the immune response of juveniles in a long-lived colonial bird. J. Anim. Ecol. 76:1215–1223.

Surai, PF (2000). Effect of selenium and vitamin E content of the maternal diet on the antioxidant system of the yolk and the developing chick. Brit. Poultry Sci. 41:235–243.

Surai, PF (2002). Natural Antioxidants in Avian Nutrition and Reproduction. Nottingham University Press, Nottingham.

Surai, PF, Ionov, IA, Kuklenko, TV, Kostjuk, IA, MacPherson, A, Speake, BK, Noble, RC, and Sparks, NHC (1998). Effect of supplementing the hen's diet with vitamin A on the accumulation of vitamins A and E, ascorbic acid and carotenoids in the egg yolk and in the embryonic liver. Br. Poultry Sci. 39:257–263.

Surai, PF, Noble, RC, and Speake, BK (1996). Tissue-specific differences in antioxidant distribution and susceptibility to lipid peroxidation during development of the chick embryo. Biochim. Biophys. Acta 1304:1–10.

Surai, PF and Speake, BK (1998). Distribution of carotenoids from the yolk to the tissues of the chick embryo. J. Nutr. Biochem. 9:645–651.

Surai, PF, Speake, BK, and Sparks, NHC (2001a). Carotenoids in avian nutrition and embryonic development. 2. Antioxidant properties and discrimination in embryonic tissues. J. Poultry Sci. 38:117–145.

Surai, PF, Speake, BK, Wood, NAR, Blount, JD, Bortolotti, GR, and Sparks, NHC (2001b). Carotenoid discrimination by the avian embryo: a lesson from wild birds. Comp. Biochem. Physiol. B 128:743–750.

Svensson, PA, Pélabon, C, Blount, JD, Surai, PF, and Amundsen, T (2006). Does female nuptial coloration reflect egg carotenoids and clutch quality in the two-spotted goby (*Gobiusculus flavescens*, Gobiidae)? Funct. Ecol. 20:689–698.

Swain, BK, Johri, TS, and Majumdar, S (2000). Effect of supplementation of vitamin E, selenium and their different combinations on the performance and immune response of broilers. Br. Poultry Sci. 41:287–292.

Swain, P and Nayak, SK (2009). Role of maternally derived immunity in fish. Fish Shellfish Immunol. 27:89–99.

Takemori, T and Rajewsky, K (1984). Specificity, duration and mechanism of idiotype suppression induced by neonatal injectionof monoclonal anti-idiotype antibodies into mice. Eur. J. Immunol. 14:656–667.

Takemura, A and Takano, K (1997). Transfer of maternally-derived immunoglobulin (IgM) to larvae in tilapia, *Oreochromis mossambicus*. Fish Shellfish Immunol. 7:355–363.

Tanasa RI, Trad A, Lange H, Grötzinger J, and Lemke H (2010). Allergen IgE-isotype-specific suppression by maternally derived monoclonal anti-IgG-idiotype. Allergy 65:16–23.

Tobler, M, Hasselquist, D, Smith, HG, and Sandell, MI (2010). Short- and long-term consequences of prenatal testosterone for immune function: an experimental study in the zebra finch. Behav. Ecol. Sociobiol. 64:717–727.

Tobler, M and Sandell, M (2009). Sex-specific effects of prenatal testosterone on nestling plasma antioxidant capacity in the zebra finch. J. Exp. Biol. 212:89–94.

Trivers, RL (1974). Parent–offspring conflict. Am. Zool. 14:249–264.

Tschirren, B, Saladin, V, Fitze, PS, Schwabl, H, and Richner, H (2005). Maternal yolk testosterone does not modulate parasite susceptibility or immune function in great tit nestlings. J. Anim. Ecol. 74:675–682.

Tschirren, B, Siitari, H, Saladin, V, and Richner, H (2009). Transgenerational immunity in a bird-ectoparasite system: do maternally transferred antibodies affect parasite fecundity or the offspring's susceptibility to fleas? Ibis 151:160–170.

Tuchscherer, M, Kanitz, E, Otten, W, and Tuchscherer, A (2002). Effects of prenatal stress on cellular and humoral immune responses in neonatal pigs. Vet. Immunol. Immunopath. 86:195–203.

Uller, T, Astheimer, L, and Olsson, M (2007). Consequences of maternal yolk testosterone for offspring development and survival: experimental test in a lizard. Funct. Ecol. 21:544–551.

Uller, T, Hollander, J, Astheimer, L, and Olsson, M (2009). Sex-specific developmental plasticity in response to yolk corticosterone in an oviparous lizard. J. Exp. Biol. 212:1087–1091.

Uller, T, Isaksson, C, and Olsson, M (2006). Immune challenge reduces reproductive output and growth in a lizard. Funct. Ecol. 20:873–879.

Uller, T and Olsson, M (2003). Prenatal exposure to testosterone increases ectoparasite susceptibility in the common lizard (*Lacerta vivipara*). Proc. R. Soc. B 270:1867–1870.

Verheul, HAM, Tittes, EV, Kelder, J, and Schuurs, A (1986). Effects of steroids with different endocrine profiles on the development, morphology and function of the bursa of Fabricius in chickens. J. Steroid Biochem. Mol. Biol. 25:665–675.

Verhulst, S, Dieleman, SJ, and Parmentier, HK (1999). A tradeoff between immunocompetence and sexual ornamentation in domestic fowl. Proc. Natl. Acad. Sci. USA 96:4478–4481.

Viltart, O and Vanbesien-Mailliot, CCA (2007). Impact of prenatal stress on neuroendocrine programming. Scientific World J. 7:1493–1537.

Vleck, CM and Bucher, TL (1998). Energy metabolism, gas exchange, and ventilation. In: Starck, JM and Ricklefs, RE (eds.). Avian Growth and Development. Oxford University Press, Oxford, pp. 89–116.

von Schantz, T, Bensch, S, Grahn, M, Hasselquist, D, and Wittzell, H (1999). Good genes, oxidative stress and condition-dependent sexual signals. Proc. R. Soc. B 266:1–12.

Weinstock, M (2007). Gender differences in the effects of prenatal stress on brain development and behavior. Neurochem. Res. 32:1730–1740.

Welberg, LAM and Seckl, JR (2001). Prenatal stress, glucocorticoids and the programming of the brain. J. Neuroendocrinol. 13:113–128.

Welberg, LAM, Seckl, JR, and Holmes, MC (2001). Prenatal glucocorticoid programming of brain corticosteroid receptors and corticotrophin-releasing hormone: possible implications for behavior. Neuroscience 104:71–79.

West, AP, Herr, AB, and Bjorkman, PJ (2004). The chicken yolk sac IgY receptor, a functional equivalent of the mammalian MHC-related Fc-receptor, is a phospholipase A2 receptor homolog. Immunity 20:601–610.

Williams, TD (2005). Mechanisms underlying costs of egg production. Bioscience 55:39–48.

Williamson, KA, Surai, PF, and Graves, JA (2006). Yolk antioxidants and mate atrractiveness in the zebra finch. Funct. Ecol. 20:354–359.

Wuryastuti, H, Stowe, HD, Bull, RW, and Miller, ER (1993). Effects of vitamin E and selenium in immune responses of peripheral blood, colostrum, and milk leukocytes of sows. J. Anim. Sci. 71:2464–2472.

Zapata, A, Diez, B, Cejalvo, T, Frias, CG, and Cortes, A (2006). Ontogeny of the immune system in fish. Fish Shellfish Immunol. 20:126–136.

Zahwa, H, Yorty, JL, and Bonneau, RH (2008). Elevated maternal corticosterone during lactation hinders the neonatal adaptive immune response to herpes simplex virus (HSV) infection. Brain Behav. Immun. 22:339–353.

Zhao, W, Han, Y, Zhao, B, Hirota, S, Hou, J, and Xin, W (1998). Effects of carotenoids on the respiratory burst of rat peritoneal macrophages. Biochim. Biophys. Acta 1381:77–88.

CHAPTER 7

TRADE-OFFS LIMITING MHC HETEROZYGOSITY

Jason L. Kubinak, Adam C. Nelson, James S. Ruff, and Wayne K. Potts

INTRODUCTION

HISTORICAL PERSPECTIVE

In the late 1940s, J.B.S. Haldane posited that the striking serological diversity in vertebrates must be due to co-evolution with pathogens (Haldane, 1949). George Snell's pioneering work on tissue rejection and the discovery of major histocompatibility complex (MHC) genes led to the identification of MHC proteins as the major contributors to this diversity (Snell, 1948). Today, classical MHC genes are widely appreciated as the most polymorphic loci in vertebrates (Klein and Figueroa, 1986). For example, there are currently over 800 known allelic variants of the human MHC class II DRB gene circulating in the human population (Robinson et al., 2003). How this variation is maintained in the face of constant purifying selection has made MHC genes one of the most attractive models for understanding mechanisms facilitating adaptive evolution and the maintenance of genetic diversity. Elucidation of the central role MHC genes play in vertebrate immune recognition by Doherty and Zinkernagel in the 1970s, and the subsequent decades of research exploring the relationship between MHC polymorphism and pathogens, has now provided convincing support for Haldane's original hypothesis. But the developing story of MHC genes is much richer, and now extends well beyond the interface between host and pathogen. Indeed, the past three decades have led to the discovery that beyond their importance in immunity, MHC genes are also key mediators of various aspects of behavior. Recently, a neurobiological mechanism of MHC-mediated social signaling has been established (Leinders-Zufall et al., 2004), providing a functional link between MHC-mediated aspects of immunity and behavior.

Recent evidence suggests that there are fitness costs as well as benefits associated with MHC diversity (Wegner et al., 2003; Bonneaud et al., 2004), and these trade-offs make the picture much more complicated than the simple notion that MHC diversity is good for both individuals and populations. How costs associated with MHC genotype affect patterns of MHC variation observed in nature is an open area of research in part because there has been a bias in our thinking regarding MHC variation, that more is better. However, while extreme variation observed at the population level indicates that variation is favored, it does not necessarily mean that maximum levels of variation are beneficial for individuals. Rather, there are multiple ways in which too much MHC variation could be detrimental to individual fitness. Therefore, to truly understand the current significance of MHC variation in nature, and the contemporary forces of selection operating on it, we must begin to include the costs associated with this diversity in our analyses.

IMPETUS FOR A NEW APPROACH

An objective of ecological immunology is to understand the tradeoffs between immunity and other aspects of animal physiology (Sheldon and Verhulst, 1996), and in this light we have purposely strayed from the typical path of MHC reviews. Numerous reviews have been written on the nature of selection operating to maintain the remarkable levels of MHC polymorphism we observe in nature, with most models of selection focusing on the benefits of high MHC diversity (Apanius et al., 1997; Hughes and Yeager, 1998; Hedrick, 2002; Piertney and Oliver, 2006). However, immunity is but one of a multitude of physiological processes that operate in an individual to increase viability and fecundity. Therefore, the benefit of MHC-mediated immunity must be weighed against its associated physiological costs. To the extent that MHC variation controls the intensity and likelihood of an immune response, it is easy to imagine physiological trade-offs imposing selective constraints on the level of immune responsiveness within a given individual.

In this review, we first briefly outline the major models of selection operating to promote and maintain MHC variation at the population level in order to provide an overall appreciation of the adaptive significance of this diversity. Then we explore the question of why individuals do not harbor limitless MHC variation by describing potential costs (i.e. selective constraints) associated with increasing levels of individual MHC heterozygosity. Following this, we discuss the ecological outcomes of MHC genotype on individual fitness, with particular attention on pathogen resistance. Our hope is that by emphasizing this cost–benefit contrast we highlight important deficiencies in our understanding of the forces shaping patterns of MHC diversity and its' functionality.

Ecological patterns of MHC variation are also influenced by the role of MHC during social communication. Although many aspects of MHC-mediated social signaling can be related to enhancing the immunocompetence of offspring, others, such as individual and kin recognition, are unrelated to MHC's role in adaptive immunity. We address these non-immunological functions of MHC in conjunction with other important, but commonly overlooked aspects of MHC diversity. In the last major section we discuss potential practical applications of MHC ecological immunology in conservation genetics and population viability, as well as in modulating disease transmission. Throughout this chapter we guide the reader to what we believe are fruitful areas of future MHC research.

FUNCTIONAL SIGNIFICANCE

Due to the vital role MHC molecules play in both shaping the T-cell repertoire and initiating an adaptive immune response, some have referred to the MHC as the "center of the immunological universe" (Trowsdale, 1995). MHC molecules are trans-membrane proteins whose immunological function is to bind short peptides for presentation to circulating T lymphocytes. Variability within "classical" MHC genes is located within exons encoding the antigen-binding cleft of MHC proteins. By convention, the term "classical" refers to the highly polymorphic class I and class II MHC genes, which are the focus of this review. High rates of non-synonymous versus synonymous substitutions within exons of these loci are a characteristic feature of MHC genes, which also implies that strong diversifying selection operates to generate *de novo* genetic variation at these loci (Hughes and Yeager, 1998). On the surface of almost all cell types, class I MHC molecules present self and nonself peptides derived from the breakdown of intracellular proteins for presentation to CD8+ killer T cells (named for their cytotoxic effector functions). Class II MHC molecules, which are expressed only on the surface of so-called "professional" antigen-presenting cells (e.g., dendritic cells, macrophages, B lymphocytes), bind self and nonself peptides derived from the breakdown of extracellular proteins within endo-lysosomes and present them to CD4+ helper T cells (named for their critical role as enhancers of various aspects of both the innate and adaptive immune response). Regulatory T cells (Tregs) are a subclass of helper T cells whose primary function is to control the immune response and limit immunopathology (disease caused by an immune response). Figure 7.1 provides a schematic representation of MHC-mediated immunity.

The development of a highly specific cell-mediated immune response directed against nonself (i.e. foreign) peptides is critically dependent upon the relationship between MHC-peptide complexes and T cells during maturation of T cells in the thymus. The T-cell receptor (TCR) that directly interacts with MHC–peptide complexes also has a highly polymorphic region whose diversity is generated via somatic recombination (i.e., it is not germline encoded like the diversity within MHC genes). The region of high polymorphism within the TCR coincides with the residues associated with direct binding to peptide antigens bound within the antigen-binding cleft of MHC proteins. The interaction between the TCR and peptide–MHC complex, along with associated interactions between co-stimulatory proteins (i.e., CD8 or CD4 proteins) with non-polymorphic regions of MHC molecules, stimulates a signaling cascade leading to activation of that T cell.

Two MHC-mediated selection events during T-cell maturation determine the resulting T-cell repertoire (i.e., the extent of diversity found among TCRs). During the first stage of maturation T cells are positively selected that bind MHC–self-peptide complexes (failure to bind results in programmed cell death). Following this selection event, T cells having TCRs that bind to MHC-self-peptide complexes with high affinity are selected against (i.e., negative selection). These MHC-mediated selection events ensure that mature T cells are capable of responding to MHC–peptide complexes harboring foreign peptides, but not self-peptides. This reduces (but does not eliminate) the chance of autoimmunity (disease caused by immune response directed against self).

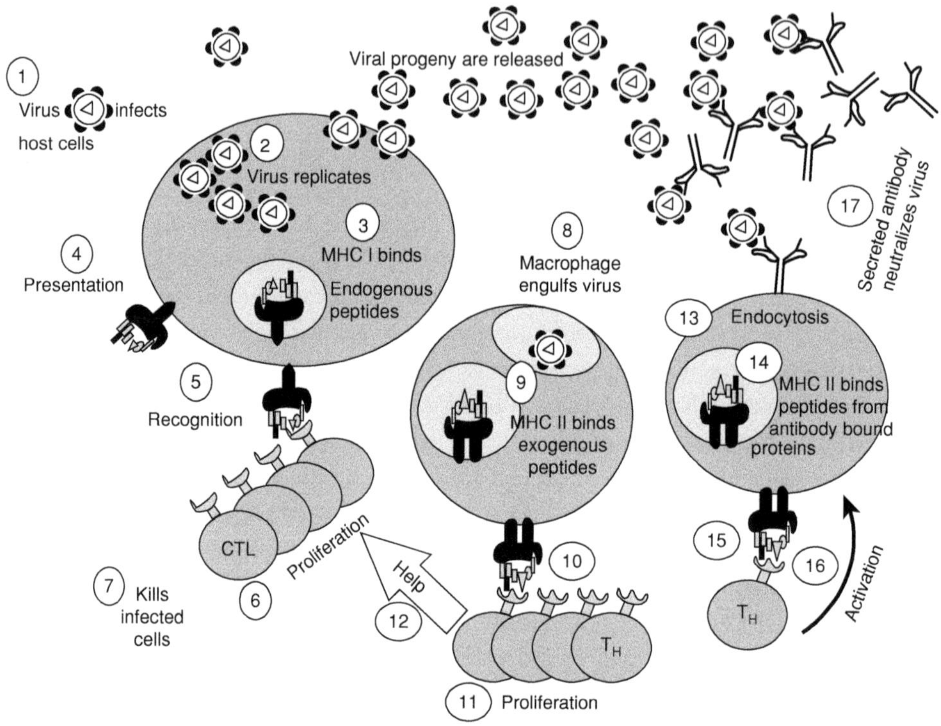

FIGURE 7.1: Simple overview of MHC-mediated adaptive immunity using an example of viral infection. A virus enters a host cell (1) and replicates (2). New infectious viral particles are released to infect more host cells. In infected cells, self-proteins and foreign proteins are degraded by antigen-processing pathway where classical MHC class I molecules bind self- and non-self-peptides (3). *Activation of cytotoxic T lymphocytes (CTL):* Self-peptides as well as foreign peptides are presented by classical MHC class I molecules on the surface of the infected cell for presentation (4) to circulating CTLs. A CTL with the appropriate receptor recognizes the foreign peptide–MHC complex (5), begins to clonally proliferate (6), and then goes on to kill infected cells (7). *Activation of helper T lymphocytes (T_H cells)* that enhance CTL activity: Macrophages at the site of infection phagocytose viral particles/proteins, break them down, and load the resulting peptides onto classical MHC class II molecules for presentation to circulating T_H cells (10). Upon stimulation, T_H cells proliferate (11) and then enhance nearby activated CTLs by secreting cytokines (12). *Activation of antibody-secreting B cells:* Membrane-bound antibody receptors bind and endocytose extracellular virus particles (13). Degraded viral proteins are loaded onto classical MHC class II molecules for presentation to circulating T_H cells (15). Upon activation, T_H cells secrete cytokines that stimulate the B cell to begin secreting antibodies (16) that bind to and neutralize extracellular virus particles (17). CTL killing of infected cells and antibody neutralization of extracellular virus particles continues until the infection is cleared. *Not pictured:* Regulatory T cells (a subset of T_H cells) control the intensity and longevity of an adaptive immune response by secreting inhibitory cytokines. Adapted from Edwards and Potts (1996).

SELECTION FOR MHC DIVERSITY

Three major models have emerged as the most likely evolutionary mechanisms favoring MHC diversity: heterozygote advantage, antagonistic co-evolution, and disassortative mating preferences (and related reproductive mechanisms). The relative importance of these three mechanisms in promoting the evolution and maintenance of MHC diversity remains controversial. There have been many excellent reviews addressing these models of selection (Apanius et al., 1997; Edwards and Hedrick, 1998; Meyer and Thomson, 2001; Bernatchez and Landry, 2003; Garrigan, 2003; Sommer, 2005; Piertney et al., 2006; Spurgin and Richardson, 2010). Here our goal is to briefly describe the models so we have a starting point from which to evaluate the trade-offs presented by the biological processes involved in each model.

HETEROZYGOTE ADVANTAGE

Heterozygote advantage occurs when heterozygous individuals have a fitness advantage over both respective homozygotes. This well-established form of balancing selection will maintain genetic diversity, and this mechanism has been applied to MHC genes (Doherty and Zinkernagel, 1975; Hughes and Nei, 1988; Hughes and Nei, 1989). This process is diversity-maintaining because rare MHC alleles gain an advantage because they are more often found in the heterozygous condition. Thus, rare alleles are rescued from extinction and they are favored until their frequencies increase to that of other alleles. Common MHC alleles suffer the opposite fate because they find themselves disproportionately in low-fitness homozygous individuals. This form of heterozygote advantage is also termed overdominance, which refers to the fact that MHC heterozygotes are more fit than either MHC homozygote individually. This should not be confused with dominance. When a beneficial trait is dominant, the heterozygote is identical to the superior homozygote, and thus in a polymorphic system the mean fitness of heterozygotes will be higher than the mean fitness of homozygotes, which we refer to as "broad-sense heterozygote advantage." This mechanism will not maintain genetic diversity because the best allele will go to fixation.

MHC heterozygotes could achieve high fitness via two different mechanisms. First, heterozygotes may be able to present more peptides to T lymphocytes, and this might allow them to more effectively recognize and defend against infectious agents. This is the mechanism most commonly given, but it is almost certainly false, because in 21 independent studies of infectious agents infecting MHC-congenic mice (*Mus musculus*), heterozygotes had an advantage in only 2 cases (McClelland et al., 2003b). MHC heterozygotes typically showed intermediate resistance compared to both homozygotes, but they were usually more like the most resistant homozygote. In hindsight this result is not too surprising, because most immune responses are driven by T-cell recognition of one or a few MHC-presented peptides (immunodominance) (Yewdell and Bennink, 1999) rather than by all or most of the presented peptides. Thus, even though a heterozygote may present more peptides, the immune response is designed to respond to only a limited number of them. This is likely because each peptide carries a potential cost of autoimmunity (Apanius et al., 1997).

The second mechanism that could confer an advantage on heterozygotes is based on the observation that all MHC alleles are resistant to some pathogens, but susceptible to others. If resistance to infectious agents is generally dominant, then a heterozygote will express the resistance profiles of both alleles and susceptibilities of each allele will be masked. In a survey of the literature, MHC resistance was observed to be (partially) dominant about 70% of the time (McClelland et al., 2003b). If resistance to infectious agents is generally dominant, then during co-infection by pathogens with opposite resistance/susceptibility profiles for two alleles at the same locus, a heterozygote may express the resistance profiles of both alleles and susceptibilities of each allele will be masked. McClelland and coworkers (2003b) experimentally tested this hypothesis by co-infecting MHC congenic mice with *Salmonella* and Theiler's virus. These two pathogens were chosen because they show opposite resistance/susceptibility profiles across a number of MHC genotypes—in other words, an MHC genotype that is resistant to *Salmonella* tends to be susceptible to Theiler's virus, and vice versa. MHC heterozygotes were never superior to the best homozygote during single pathogen infections, but during co-infections MHC heterozygotes had an overall (41%) lower combined pathogen load. These results are inconsistent with the more-peptides-presented mechanism, but strongly support the resistance-is-dominant mechanism.

In summary, heterozygote advantage remains a viable explanation for the evolution and maintenance of MHC diversity, although its relative importance is yet to be determined. The conventional explanation that heterozygotes have an advantage over homozygotes during single infections because they bind and present more antigenic peptides is not supported by multiple experimental infection studies in MHC-congenic mice (McClelland et al., 2003b). However, in the context of infections with multiple pathogens, evidence does suggest that heterozygotes gain a fitness advantage over homozygotes. Thus, the fitness benefits associated with heterozygote advantage may arise over the course of the many infections experienced during an individuals lifetime.

ANTAGONISTIC CO-EVOLUTION

Antagonistic co-evolution theory predicts that pathogens adapt to hosts by evading MHC-dependent immune recognition, which then favors counter-adaptations by hosts, such as favoring new MHC alleles. This would promote MHC diversity because common MHC alleles would be the primary target of pathogen adaptation, thereby lowering their fitness and providing an advantage to individuals carrying rare alleles. Rare alleles will on average have an advantage because they have not recently been a major target of pathogen adaptation. This process, which was anticipated by Haldane (1949), but first formally proposed by Bodmer (1972), establishes never-ending cycles of negative frequency-dependent selection that will maintain genetic diversity in both hosts and pathogens.

Serial passage (whereby a pathogen is manually transferred from one host to another) is a proven technique for inducing pathogen adaptation in experimental time (Ebert, 1998). Serial passage studies have been conducted in numerous host–pathogen systems, and a set of general patterns emerge: (1) increased virulence in 23 of 24 studies (96%), (2) increased pathogen fitness in 9 of 9 studies (100%), (3) increased pathogen growth rate in 15 of 15 studies (100%), and (4) attenuated pathogen virulence to a former host in 32 of 34 studies (94%). These results strongly support the idea that rapid pathogen evolution in nature is likely to be associated with substantial fitness costs for their hosts. However, fitness and virulence attenuates in former hosts, presumably due to fitness trade-offs associated with pathogen adaptation. These fitness trade-offs are critical for the diversity-maintaining properties of antagonistic co-evolution, because if pathogen adaptation permanently reduced an allele's fitness, then that allele would go extinct.

There is no doubt that pathogens can adapt to evade MHC-mediated immune recognition, as demonstrated in T-cell leukemia virus (Furukawa et al., 2001), hepatitis C virus (Erickson et al., 2001), and immunodeficiency viruses of humans (HIV) and simians (SIV) (reviewed in McMichael and Rowland-Jones, 2001). These viruses all establish chronic infections, which allow substantial time for pathogen adaptation to occur. Thus, there is some question whether this process operates with acutely infectious pathogens that are quickly cleared. However, serial passage of influenza in mice shows that rapid and substantial adaptation (five orders of magnitude of virulence increase) can occur in just 36 days of passage (Brown et al., 2001). This evidence suggests that pathogen adaptation can be a rapid process that can be directed towards escaping MHC recognition.

Experimental evolution studies designed to look for MHC-specific pathogen adaptation to MHC have been conducted in only two host–pathogen systems. First, during serial passages in mice of the fungal pathogen *Cryptococcus neoformans*, MHC was found to influence the trajectory of virulence evolution, but there was no evidence of MHC-specific pathogen adaptation (McClelland et al., 2004). Second, during the course of a single infection of SIV in macaques, mutations accumulate in cytotoxic T-cell epitopes (MHC-presented viral peptides) that result in partial viral escape from the immune response (Evans et al., 1999; Fernandez et al., 2005). These mutations are likely favored in the virus because they disturb either MHC presentation or cytotoxic T-cell recognition, thus reducing the effectiveness of the immune response. The escape mutations often revert (i.e., mutate back to wild-type) when passaged in an individual with a different MHC genotype (Friedrich et al., 2004; Fernandez et al., 2005). These reversions suggest the viral mutations carry costs that are selected against when not needed during infection of a host with a different MHC. As described above, this trade-off is requisite for the diversity-maintaining properties of antagonistic co-evolution.

In summary, antagonistic co-evolution between pathogens and hosts is another force of selection capable of favoring MHC diversity. Though this hypothesis is attractive for explaining patterns of MHC diversity, its relative importance is still uncertain due to the limited number of experiments testing its plausibility and the difficulty of detecting this process in natural populations. Serial passage experimentation is a powerful tool for testing fundamental assumptions regarding the importance of antagonistic co-evolution as a mechanism capable of influencing MHC evolution.

MHC-MEDIATED SEXUAL SELECTION

MHC-mediated disassortative mating preferences were discovered more than 30 years ago in mice (Yamazaki et al., 1976). Since then, experiments have demonstrated that MHC-associated odors can be detected in numerous species (including humans) (Yamazaki et al. 1999) and that MHC-mediated mating preferences occur in 19 species spanning most vertebrate taxa (Ruff et al., in press). These and other data suggest that MHC-mediated mating preferences are a general trait found in vertebrates (Ziegler et al., 2005), despite the failure to find such behavior in a few studies (Piertney and Oliver, 2006). MHC-disassortative mating preferences would contribute to the maintenance of MHC diversity because individuals expressing rare alleles would have a mating advantage. The strength of reported mating preferences is sufficient to account for existing levels of MHC diversity, and the resulting balancing selection is substantially stronger than reported selection coefficients for either of the two pathogen-mediated forms of selection described above (Potts et al., 1991).

MHC-disassortative mating preferences could function to (1) produce MHC-heterozygous offspring, (2) create a moving target against rapidly evolving pathogens, and/or (3) avoid inbreeding (Sommer, 2005; Piertney and Oliver, 2006). In all cases the result is offspring with reduced levels of either infectious or genetic disease (Slev et al., 2006). Both pathogen-driven mechanisms thought to diversify MHC genes (described above) would favor the evolution of MHC-disassortative mating preferences. In the heterozygote advantage model, MHC heterozygotes are thought to have a superior immune system, so reproductive mechanisms should evolve to preferentially produce high-fitness MHC heterozygous offspring. In the antagonistic co-evolution model, pathogens are constantly adapting to parental MHC genotypes, so mating with MHC-dissimilar individuals ensures that offspring will have a novel MHC genotype from both parents. In this scenario, MHC-dissassortative mating produces a moving target against rapidly evolving pathogens. Thus, MHC-disassortative mating preferences are consistent with all of the conventional immunological thinking about pathogen-driven diversifying selection operating on MHC.

The existence of MHC-mediated mate choice is a remarkable biological story about immune recognition genes that also influence complex behaviors in order to reduce disease in progeny. Why, then, is this story often missing in immunology textbooks? One reason is that the mechanism of how individuals detect MHC genotype in themselves and others had remained elusive. However, the recent discovery that specialized sensory neurons bind peptides in an MHC-like fashion has revealed the long-sought odorants used to recognize the MHC genotype/phenotype of other individuals (Leinders-Zufall et al., 2004; Spehr et al., 2006). The odorants are the same MHC peptides used during immune recognition, which seamlessly links selection acting on MHC-mediated behaviors with selection acting on the immune system; both processes affect peptide-binding properties of MHC molecules (Kelliher et al., 2005; Boehm and Zufall, 2006). This discovery provides the molecular logic underlying MHC-mediated mating preferences observed in humans and many vertebrates (Slev et al., 2006).

MHC-mediated sexual selection can also maintain diversity by post-copulatory mechanisms. In particular, selective fertilization (Wedekind et al., 1996; Rülicke et al., 1998) and selective abortion (Alberts and Ober, 1993) have been reported to preferentially produce MHC-heterozygous offspring. These cryptic behaviors have been interpreted as alternative mechanisms to achieve the same functions as disassortative mating preferences and contribute to maintaining MHC diversity in an analogous way—by providing an advantage to rare MHC alleles.

TRADE-OFFS IN IMMUNE FUNCTION AND ECOLOGICAL OUTCOMES OF MHC DIVERSITY

TRADE-OFFS ASSOCIATED WITH MHC VARIATION: THE MECHANISMS

If MHC diversity is good for resistance to infectious disease, then why do organisms not have dozens of polymorphic MHC loci? Typically species have between two and six MHC loci, even though given the roughly 540 million years since their origin (Kasahara et al., 2004), duplication

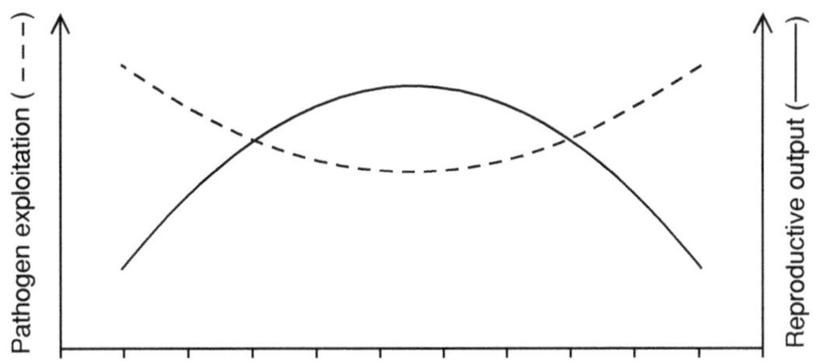

FIGURE 7.2: A predicted optimal relationship between pathogen exploitation and host fitness, mediated by MHC heterozygosity. The above model assumes a cost associated with higher levels of MHC heterozygosity, which can manifest in multiple ways (see text). The benefit of MHC heterozygosity in controlling pathogen exploitation is limited by costs associated with higher levels of individual variation.

events could have produced many more. A reasonable assumption is that there are "optimal" levels of variation beyond which individual fitness declines. A plausible model incorporating trade-offs associated with MHC heterozygosity (discussed below) predicts that an intermediate number of MHC alleles maximizes the benefits and minimizes the costs associated with individual MHC variation (Fig. 7.2). Although this model has theoretical (Woelfing et al., 2009) and empirical support (Wegner et al., 2003), it is not likely to apply to all species. In some situations maximal MHC variation may indeed be optimal. We discuss five classes of trade-offs associated with MHC variation that may define optimal levels of individual MHC heterozygosity: T-cell deletion, immunopathology, dominant susceptibility alleles, resource allocation constraints, and MHC cell-surface concentration influencing T-cell activation. This list is not exhaustive; we expect more trade-offs will be found as researchers increasingly appreciate the costs of MHC variation in nature.

The primary intent of this section is to highlight key mechanisms limiting individual MHC heterozygosity. Since immune system genes are on average the most polymorphic genes of any organ system, presumably due to similar forms of balancing selection caused by infectious agents (Klein, 1986; Wandstrat et al., 2004; Parham, 2005), many of the processes we discuss may also apply to other polymorphic immune defense genes.

T-CELL DELETION AND REDUCED IMMUNOCOMPETENCE

MHC-mediated restriction of individual T-cell repertoire diversity (i.e. the number of T cells carrying unique TCRs) is a well-established phenomenon (Nikolich-Zugich et al., 2004). During the maturation of T cells in the thymus, MHC proteins play a major role in determining which T-cell lineages will survive and become circulating lymphocytes and which will die via programmed cell death (apoptosis) (Goldrath and Bevan, 1999). It is thought that increasing numbers of MHC alleles within an individual will lead to increased deletion of T-cell lineages, resulting in reduced T-cell repertoire diversity. Therefore, MHC-mediated T-cell deletion could largely determine the effectiveness of an immune response by limiting the extent to which individuals can recognize and respond to foreign antigens (Vidovic and Matzinger, 1988; Vidovic, 1989) and consequently predicts an optimal degree of MHC heterozygosity (Nowak et al., 1992; De Boer and Perelson, 1993; but see Borghans et al., 2003). The cost of T-cell deletion associated with increasing individual MHC variation is the conventional explanation given for why maximal heterozygosity might not be optimal (Wegner et al., 2003). To our knowledge, this theoretically attractive hypothesis has little empirical support. The only study to accurately measure T-cell repertoire diversity in response to different levels of MHC variation was conducted in mice and showed no effect as MHC allele numbers went from 3 to 5 (Casrouge et al., 2000). However, the maximum number of alleles in a fully heterozygous mouse is 10, so this study may not have evaluated the T-cell repertoire diversity across the relevant range of MHC variation (e.g., see Fig, 7.2). There is now abundant evidence that reduced T-cell repertoire diversity substantially impairs individual immunocompetence (Nikolich-Zugich et al., 2004). Thus, T-cell deletion as a mechanism for creating trade-offs between MHC heterozygosity and fitness remains a viable hypothesis.

IMMUNOPATHOLOGY AND AUTOIMMUNE DISEASE

Immunopathology is disease caused by an immune response. Recent appreciation of its effect on individual health has led to the awareness that maximal immune response can often be very dangerous to the host (Graham et al., 2005) and that in some circumstances immunopathology can be the major cause of infection-triggered disease (Belkaid, 2007). Autoimmune disease on the other hand represents a situation where an immune response is misdirected at self. There are numerous links between specific MHC alleles and a variety of autoimmune diseases (Tiwari and Terasaki, 1985). For example, specific MHC alleles have been associated with predisposition to developing multiple sclerosis, lupus, diabetes, and Crohn's disease, as well as others (Fernando et al., 2008). We will discuss how an inability to control an immune response properly can promote both kinds of disease.

Regulatory T cells (Tregs) are important for determining which types of effector mechanisms are used during an immune response and also in modulating those responses (Graham et al., 2005; Beissert et al., 2006). Thus, they control the intensity of an immune response and consequently mitigate immunopathology. As discussed previously, MHC molecules play a large role in limiting T-cell repertoire diversity during lymphocyte maturation, and numerous studies have shown how this reduces individual immunocompetence (Nikolich-Zugich et al., 2004). However, T-cell repertoire restriction can affect an individual's ability to properly coordinate an immune response by limiting diversity within the Treg pool as well. For example, in a study using Rag2$^{-/-}$ C57BL/6 mice (mice incapable of producing functional T cells) that consequently suffer severe multi-organ inflammatory disease, it was observed that adoptive transfer of Tregs into these animals eliminated disease, but only when the introduced Treg repertoire was diverse rather than limited (Milner et al., 2007).

Deletion of self-reactive T cells during thymic education is not absolute, and Tregs are known to suppress the activity of these cells in peripheral tissues, thereby limiting autoimmune responses (Sakaguchi et al., 2001). For instance, in a study using a diabetic mouse model, it was observed that reduced Treg repertoire diversity promoted the onset of diabetes in individuals

(Ferreira et al., 2009). Moreover, this limited Treg repertoire was found to be partially due to the presence of a specific MHC class II allele found in this mouse strain, providing evidence that specific MHC alleles can contribute to susceptibility to autoimmune diseases by reducing T-cell repertoire diversity.

Finally, it has been shown that spontaneous abortion is largely due to an immune response directed against the developing embryo/fetus (see Kundert et al., 2008, and references therein) and that insults to normal Treg functioning precipitate this physiological cost (Saito et al., 2007). While the relationship between Treg repertoire diversity and spontaneous abortion has so far been unstudied, the link between perturbation to these modulators of immune response and rejection of developing young in the womb demonstrates another important cost associated with improper control of the immune response. This is distinct from post-copulatory mate choice because it represents an unintentional consequence of a faulty immune response.

DOMINANT MHC SUSCEPTIBILITY ALLELES TO INFECTIOUS AND AUTOIMMUNE DISEASES

Almost every carefully studied autoimmune and infectious disease has been associated with one or more MHC alleles that confer susceptibility to that disease (Tiwari and Terasaki, 1985). Moreover, it is likely that all MHC alleles can be classified as susceptibility alleles because if there were alleles resistant to all diseases they would out-compete alleles expressing susceptibilities. These susceptibility alleles are presumably maintained in populations because the cost of their susceptibility to some diseases is balanced by the resistance they provide to other diseases. The uniformity of MHC allelic frequencies supports this hypothesis (Klitz et al., 1986).

Most MHC susceptibility alleles to *autoimmune* diseases are dominant, meaning that they cause the disease phenotype despite being co-dominantly expressed at the cellular level (Tiwari and Terasaki, 1985), and about 30% of MHC susceptibility alleles to *infectious* diseases are dominant (McClelland et al., 2003b). The accumulation of dominant susceptibility alleles will increase costs incrementally as individual MHC diversity increases, while the benefits of dominant resistance will show a diminishing-returns curve as it asymptotes toward an effective immune response (Fig. 7.3). Such cost–benefit curves establish optimal amounts of individual MHC variation, and this could contribute to the observation that maximal MHC heterozygosity is not always optimal (Wegner et al., 2003; Bonneaud et al., 2004; Madsen and Ujvari, 2006; Woelfing et al., 2009).

Some authors have claimed that because autoimmunity tends to strike later in life it would not be an important selective factor favoring an optimal level of individual MHC variation (Woelfing et al., 2009). However, whereas debilitating symptoms tend to have a late onset, subclinical health degradation can occur much earlier. For example, about 90% of pancreatic islet cells are destroyed before clinical type 1 diabetic symptoms commence (Matthews et al., 2001). Also, it is becoming clear that major fitness declines associated with natural ecological stresses can largely disappear under the benign conditions of animal captivity or modern human societies (Meagher et al., 2000; Carroll et al., 2004; Ilmonen et al., 2008). Hence, subtle fitness consequences are largely unappreciated, but their cumulative effects may carry a substantial cost. Combining this fact with early-onset autoimmune diseases (such as diabetes) and the estimate that 30% of MHC susceptibility

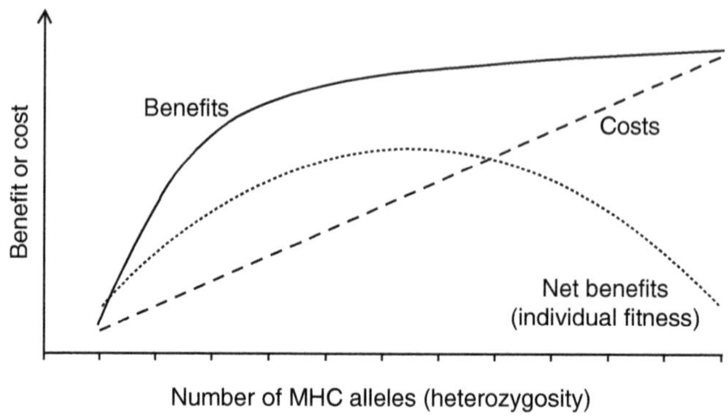

FIGURE 7.3: This model illustrates a diminishing-returns curve for the benefit of individual heterozygosity. The benefits of heterozygosity will asymptote. However, the accumulation of dominant susceptibility alleles will increase costs incrementally, resulting in an optimal number of MHC alleles.

alleles to infectious diseases are dominant indicates that dominant MHC susceptibility alleles could be an important force of selection favoring an optimal level of individual MHC variation.

RESOURCE ALLOCATION CONSTRAINTS

T-cell receptor binding to MHC–peptide complexes on the surface of a cell stimulates the cellular and humoral arms of the adaptive immune system, producing a potent immune response against specific pathogens. However, because energy and nutrients are usually limiting, organisms must balance the benefit of combating infectious agents via an immune response against the costs associated with diverting limited resources away from other critical physiological processes. For example, studies artificially stimulating an immune response (which removes the cost associated with pathogen replication) have demonstrated that immunity can be physiologically very costly to an organism in terms of growth (Klasing et al., 1987; Fair et al., 1999; Bonato et al., 2009), survival (Moller and Saino, 1994; Norris and Evans, 2000; Hanssen et al., 2004; Barribeau et al., 2008), and reproduction (Ilmonen et al., 2000; Bonneaud et al., 2003, 2004; Hanssen, 2006; Ilmonen et al., 2007; Cai et al., 2009; Bascunan-Garcia et al., 2010).

In the first study of its kind to link MHC heterozygosity with greater pathogen resistance, Doherty and Zinkernagel demonstrated that the intensity of virus-directed immune responses were substantially greater in MHC heterozygotes versus homozygotes (Doherty et al., 1975). This predicts that MHC-diverse individuals should suffer greater physiological costs than their less-diverse counterparts within the context of an immune response. In a population study on wild mice cross-bred with C57BL/10 mice (for the purpose of creating heterozygous and homozygous individuals of known MHC haplotype) it was observed that heterozygous females infected with *Salmonella enterica* had significantly reduced fitness (as measured by pup production) than their infected homozygous counterparts, yet both had similar bacterial loads (Ilmonen et al., 2007). These data suggest that heterozygous individuals pay greater physiological costs than homozygotes and enjoy no associated benefit in resistance associated with their high MHC diversity.

This could be due to quality, rather than quantity, being a stronger component of resistance. If a single allele drives the resistance profile of *Salmonella* infection in these mice, then both homozygotes and heterozygotes will have the same allele and therefore will be equally resistant. As a result, any extra intensity of immune response in heterozygotes would simply require more limited resources while providing no additional benefit. Future studies in this area should focus on elucidating links between the degree of immune response and its impact on physiology between MHC-heterozygous versus -homozygous individuals.

MHC CELL-SURFACE CONCENTRATION AND T-CELL ACTIVATION

MHC molecules bind and present self-peptides as well as foreign peptides to circulating T cells. Therefore, the likelihood of mounting a T-cell–mediated immune response might be contingent upon the rate at which a T cell with the appropriate TCR locates and binds to its cognate peptide–MHC complex. Because hundreds of thousands of self-peptide–MHC complexes are expressed on the surface of every cell, it has been proposed that individual MHC heterozygosity must be restricted in order for foreign peptide–MHC complexes to be sufficiently abundant for detection by T cells (van den Berg and Rand, 2003; Woelfing et al., 2009). In essence, too much MHC variation may negatively affect T-cell activation by decreasing the relative number of specific peptide–MHC complexes on the surface of a cell. This would decrease the likelihood of an appropriate TCR–peptide–MHC interaction from occurring and decrease the likelihood of eliciting an immune response. As it stands, there is little empirical support for this. However, the observation that surface expression of MHC molecules is upregulated during infections (Rodriguez et al., 2003; Peatman et al., 2008) suggests that the amount of specific peptide–MHC complexes on the surface of a cell is probably important for determining the effectiveness of an immune response.

TRADE-OFFS ASSOCIATED WITH MHC VARIATION: ECOLOGICAL OUTCOMES

There are multiple costs associated with higher levels of MHC diversity that may function in discrete or concerted fashion to reduce individual fitness. However, pathogens and the diseases they cause are potent environmental stressors that also affect individual survival and reproduction. From the perspective of pathogen resistance, then, some degree of individual MHC variation is clearly beneficial to host health, and decades of research corroborate this conclusion. Here, we summarize this evidence with respect to the primary mechanisms by which MHC genotype confers resistance (or susceptibility) to pathogens. Furthermore, because MHC also plays a pivotal role in social communication for many vertebrate groups, we emphasize that mate choice for MHC-encoded resistance to parasites also plays a role in selection for MHC heterozygosity. Finally, we review evidence for the prediction that in order to optimize the net benefits of MHC heterozygosity, viability and sexual selection will often favor individuals with an intermediate number of MHC alleles.

VIABILITY SELECTION

Pathogens are a major cause of mortality/morbidity in animals and are therefore a primary agent of selection. Given its role in detecting the presence of foreign invaders, MHC clearly mitigates the cost of parasitism. One intuitive prediction is that MHC heterozygosity will provide greater resistance to pathogens. There is now abundant evidence to conclude that some degree of MHC heterozygosity is beneficial to individuals due to increased resistance to pathogens. In humans, increased individual MHC heterozygosity has been associated with greater resistance to hepatitis B (Thursz et al., 1997), hepatitis C (Hraber et al., 2007), and HIV-1 (Carrington et al., 1999). Similarly, MHC class IIB-heterozygous common carp (*Cyprinus carpio L.*) and Atlantic salmon (*Salmo salar*) are more resistant to infection with a protozoan blood parasite (Rakus et al., 2009) and a bacterium (Turner et al., 2007), respectively.

Despite these links, heterozygotes are not always more resistant to single infectious agents than their homozygous counterparts. For instance, in a study comparing the resistance profiles of inbred MHC heterozygote mice against that of both MHC homozygotes to infection with a pathogenic fungus, it was observed that MHC heterozygotes were more resistant than equivalent homozygotes in one haplotypic combination, but less resistant than both homozygotes in another (McClelland et al., 2003a). Underdominant effects (i.e., heterozygotes being less resistant than both homozygotes) such as this strongly support the notion that MHC heterozygosity can have an immunological cost. Additionally, a recent study on brook charr (*Salvelinus fontinalis*) characterizing resistance/susceptibility profiles of specific MHC alleles to infection with a bacterium (*Aeromonas salmonicida*) demonstrated heterozygote disadvantage due to susceptibility alleles being dominant to resistance alleles (Croisetiere et al., 2008). Finally, as discussed earlier, MHC heterozygotes were no more resistant than homozygotes to *Salmonella* infection (Illmonen et al., 2007). These studies clearly demonstrate that MHC heterozygosity may not always be advantageous (also see Optimality section below).

Animals in nature are likely to be infected with more than one parasite at any given time, so a more realistic approach to studying heterozygote advantage is to test for resistance to multiple parasites. For instance, pathogens having largely opposite resistance/susceptibility patterns can offset the costs associated with MHC heterozygosity. As discussed, an experiment by McClelland and colleagues demonstrated that the cost of carrying susceptibility alleles was o ffset by the advantage they bestowed on heterozygotes during multi-pathogen infections (McClelland et al., 2003b). Therefore, it seems that the benefit of MHC heterozygosity will emerge when individuals are under pressure from multiple pathogens. Indeed, in humans, MHC variation across multiple populations was shown to be positively associated with a measure of pathogen diversity (i.e., virus richness) (Prugnolle et al., 2005). Similarly, an analysis of MHC diversity across populations of Atlantic salmon demonstrated that individuals within populations under pressure from a more diverse pathogen community also have higher levels of individual variation (Dionne et al., 2007). Finally, demonstration that MHC heterozygosity is beneficial during infection with more than one parasite has been shown in studies on long-tailed giant rats (*Leopoldamys sabanus*) (Lenz et al., 2009), striped mice (*Rhabdomys pumilio*) (Froeschke and Sommer, 2005), Arctic charr (*Salvelinus alpinus*) (Kekalainen et al., 2009), water voles (*Arvicola terrestris*) (Oliver et al., 2009), and brushtail possums (*Trichosurus cunnihami*) (Banks et al., 2010).

Taken together, these studies demonstrate the fitness benefits of MHC heterozygosity and suggest that selection should generally favor some degree of individual MHC variation. However, a pressing question in vertebrate biology is whether too much MHC variation is costly. Thus, although it can be argued that parasite diversity is likely to select for a wider variety of MHC genotypes, future research should bear in mind that there could be a hidden upper limit to this diversity.

SEXUAL SELECTION

Although animals in natural populations are often beset by pathogens and parasites, infection is usually not lethal, and in these cases viability selection may not be particularly strong. However, non-lethal infections can greatly diminish the mating success of parasitized individuals, and sexual selection can be an equally, if not a more powerful force than viability selection (Hoekstra et al., 2001). MHC-mediated sexual selection is pervasive across vertebrates and has a variety of functions (Ruff et al., in press). With respect to the immune response, MHC-mediated social signaling can yield direct and indirect benefits to the individual in several ways. First, MHC-mediated disease resistance can provide the opportunity for greater physiological investment in the expression of secondary sexual traits. Second, individuals can choose MHC-dissimilar mates such that they avoid horizontal transmission of disease, or so that their offspring have increased resistance to pathogens. Third, individuals can chose mates such that their offspring have an optimal level of MHC heterozygosity, such that the physiological costs of MHC variation are minimized and the benefits of parasite resistance are maximized. All three of these mechanisms promote MHC heterozygosity.

Because immunity is a physiologically costly trait and energy/nutrient resources are finite (Lochmiller and Deerenberg, 2000), there should exist trade-offs between resources allocated to immune function and other aspects of animal physiology. A direct prediction arising from such a tradeoff might be differential investment in the expression of secondary sexual characteristics important to mating success. Correlations between the immune response and the expression of sexual characteristics have been described in many taxa (Moller, 1996; Verhulst et al., 1999; Roulin et al., 2001; Uller et al., 2006; Aguilera and Amat, 2007; Amat et al., 2007; Garvin et al., 2008; Gasparini et al., 2009; Lopez et al., 2009), and disease resistance allows greater expression of sexual traits (Hill and Farmer, 2005; Dawson and Bortolotti, 2006; Ezenwa and Jolles, 2008). Consistent with this trade-off hypothesis, there are links between MHC genotype and sexual trait expression. In some cases, the presence of specific MHC alleles correlates with enhanced trait expression. For instance, von Schantz and colleagues (1996) were the first to report an association between MHC and sexual trait expression; they found that spur length in male pheasants (*Phasianus colchicus*) was correlated with fitness and was dependent on MHC genotype. Similarly, great snipe (*Gallinago media*) males that harbored specific MHC alleles were larger, and larger males are preferred by females (Ekblom et al., 2004). In other cases, MHC heterozygosity correlates with greater investment in trait expression. For example, the length of male peacock (*Pavo cristatus*) trains reflects levels of individual heterozygosity, with longer trains coming from more MHC-diverse individuals (Hale et al., 2009). In a very thorough study, Ditchkoff and colleagues (2001) found that MHC heterozygous male white-tailed deer (*Odocoileus virginianus*) had larger antlers and body size as well as lower pathogen loads, demonstrating a clear link between MHC heterozygosity, its impact on degree of parasitism, and the consequent expression of sexual characters.

An alternative way that MHC genotype can influence trade-offs between sexual trait expression and immune investment is if MHC-mediated social signals are themselves costly to produce. Three-spined stickleback females (*Gasterosteus aculeatus*) use excreted male MHC pheromones to guide mate choice decisions (Reusch et al., 2001; Milinski et al., 2005). Remarkably, only males in the reproductive state invest in the expression of this signal. The authors suggest that shedding MHC–peptide complexes will create localized deficiencies of this critical immunological component, and this therefore likely represents a trade-off between immune defense and MHC-mediated social signaling (Milinski et al., 2009). Although such a pattern is consistent with their hypothesis, a direct experiment is needed.

Social interactions (e.g., fighting or copulation) are an excellent opportunity for parasites to be horizontally transmitted between hosts, and as a result of this, hosts may develop mechanisms to limit such transmission events (Zuk, 1992; Moller et al., 1994; Penn and Potts, 1998; Kavaliers et al., 2004). MHC-disassortative mating could be a mechanism by which individuals gain a direct benefit of disease avoidance during mating. For example, two monogamous Malagasy rodent species were found to harbor much lower MHC diversities than their promiscuous congener (Sommer et al., 2002). Such evidence suggests that mechanisms promoting MHC diversity (like MHC-disassortative mate choice) are more important with higher rates of contact between individuals (where there is more opportunity for disease transmission). Also, evidence from humans supports the hypothesis that mate choice for MHC-dissimilar individuals reduces the opportunity for horizontal transmission of HIV. A study of a large cohort of Zambian couples observed a significantly decreased chance of horizontal transmission of HIV between MHC-dissimilar couples (Dorak et al., 2004). Obviously more studies are needed from a broader array of vertebrates and their associated pathogens before any conclusions can be drawn, but evidence thus far supports the hypothesis that MHC-mediated mate choice may serve to decrease transmission of infectious diseases.

MHC-mediated mating preferences can also have the indirect benefit of generating offspring with enhanced immunocompetence. One possible mechanism of gaining such indirect benefits is mating with individuals carrying specific MHC resistance alleles. A second possible mechanism is by mating with MHC-dissimilar individuals; this assumes individual MHC heterozygosity is positively associated with resistance to disease. Two studies demonstrated that MHC dissimilarity between mothers and their children significantly reduces the chance of their children becoming infected (MacDonald et al., 1998; Polycarpou et al., 2002), suggesting a relationship between MHC-mediated mate choice and disease resistance in offspring. More recently, a study of salmon empirically addressed the relationship between MHC-mediated mate choice and offspring immunocompetence by measuring correlations between the extent of MHC variation and parasite loads across two cohorts of animals: offspring that had been derived from artificial crosses where parents were unable to choose mates, and offspring that had been born to parents capable of freely choosing their mates (Consuegra and de Leaniz, 2008). Strikingly, when the salmon were allowed to choose mates they produced more heterozygous offspring than random artificial crosses, and this increased heterozygosity was associated with a four-fold reduction in the chance of offspring being infected by a nematode pathogen (*Anasakis spp.*). Therefore, in this study sexual selection favored MHC heterozygosity and this was associated with enhanced offspring immunocompetence.

OPTIMALITY SELECTION

If MHC heterozygosity enables parasite resistance but only at the cost of accumulating susceptibility alleles and increasing physiological costs of immunity, then we would predict an optimal degree of MHC variation for any given individual (Figs. 7.2 and 7.3). Even if these costs are by themselves minimally harmful, it seems likely that their cumulative effect on individual viability would be sufficient to constrain the evolution of maximally diverse individuals. In addition, if these costs are strong enough to restrict species to only between 2 and 6 MHC loci after more than 500 million years of MHC evolution, then it seems probable that natural selection would also lead to the evolution of mechanisms to limit these costs. Mate choice for optimally diverse MHC individuals represents one possible strategy to achieve that end.

Our model in Figure 7.2 predicts that intermediate levels of individual heterozygosity are favorable if maximal heterozygosity is under negative selection. This model is largely based on the one previously proposed (Wegner et al., 2003), which first provided experimental support for the notion that individuals with an intermediate number of alleles would have the greatest resistance to parasitism (see also Wegner et al., 2004). This result was later confirmed by observation data from a free-living population (Wegner et al., 2008). In another study on house sparrows (*Passer domesticus*), Bonneaud and colleagues reported an optimal MHC diversity beyond which female birds suffered reproductive fitness declines in the form of initial and secondary clutch size (Bonneaud et al., 2004). Immune stimulation exacerbated the effect, but did not alter the relationship between MHC diversity and physiological cost. What makes these data particularly revealing is that a costly signal of MHC diversity was revealed outside the context of infection. Additional studies in water pythons (*Liasis fuscus*) and bank voles (*Myodes glareolus*) have shown that intermediate levels of individual MHC variation are associated with the greatest resistance to infectious agents (Madsen and Ujvari, 2006; Kloch et al., 2010). Altogether, these studies provide evidence for the notion of fitness costs associated with maximal MHC heterozygosity.

The optimality hypothesis also predicts mate choice strategies that will produce an intermediate level of MHC heterozygosity in the offspring. Indeed, Aeschlimann and colleagues (2003) found that female sticklebacks chose mates that complemented their set of MHC alleles to produce an intermediate level of MHC heterozygosity in their offspring. To test the effect of such a mating strategy from the perspective of immunocompetence, Wegner and colleagues (2003) experimentally infected individuals expressing varying levels of MHC heterozygosity with multiple pathogens. Intermediate rather than maximal MHC heterozygosity conferred the greatest advantage to individuals during multi-pathogen infections (as determined by relative pathogen loads). These experiments were one of the first to provide direct evidence for MHC-mediated mating preferences yielding pathogen resistance in offspring. Since then, the existence of mate choice strategies that seek to optimize individual MHC heterozygosity has also been found in brown trout (*Salmo trutta*) (Forsberg et al., 2007) and in house sparrows (*Passer domesticus*) (Bonneaud et al., 2006), suggesting that other species employ similar strategies to optimize levels of offspring MHC variation. We argue that any of the costs (or their cumulative effects) outlined above could explain these patterns. For instance, female rejection of maximally divergent males may function to avoid disruption of co-adapted or locally adapted MHC genes (Bonneaud et al., 2006). Regardless, the realization that maximal immune response is not always optimal emphasizes a critical point: to understand the ecological and immunological significance of the MHC, it is necessary to analyze the trade-offs between heterozygosity and fitness. Finally, it must be emphasized that depending on the organism and its co-evolving pathogens, intermediate heterozygosity may not be the best strategy; in some cases, maximal heterozygosity can be optimal (Brown, 1997; Schwensow et al., 2008).

CAVEATS TO INTERPRETING PATTERNS OF MHC DIVERSITY IN NATURE

MHC-MEDIATED SOCIAL SIGNALING

When considering the tradeoffs of MHC heterozygosity, it is crucial to remember that there is more at play than simply the costs and benefits associated with immune recognition; indeed, individual MHC variation may have as many ecological implications mediated by its role during social communication as it has due to its role in adaptive immunity. Therefore, there are trade-offs associated with MHC variation not only in regards to immunocompetence, but also between its functionality in both immune recognition and social communication. The relative importance between these two functions is still not understood; however, the pervasive presence of MHC-mediated social signaling across vertebrates has led some to argue that the function of MHC in social communication is the ancestral condition and that it was co-opted for use in the immune system at a latter date (Brown, 1983; Boehm, 2006). Regardless of which function came first, when attempting to understand patterns of MHC diversity in nature it is imperative to consider both the immunological and signaling roles of MHC.

Social signaling meditated by the MHC was first discovered in regards to mate preferences in laboratory mice (Yamazaki et al., 1976) and has now been described in at least 19 species of vertebrates, including representatives of bony fish and all classes of tetrapods (for review see Ruff et al., in press). Major discoveries have revealed that MHC's role in social communication goes beyond the mating preferences first identified and includes many behaviors facilitated by both individual and kin recognition. The odorant signal was recently discovered to be peptides, which bind and activate sensory neurons in the olfactory epithelia with binding properties similar to peptides binding MHC molecules (Leinders-Zufall et al., 2004). Although many consequences of MHC-mediated social signaling are related to the production of offspring with superior immune systems (mentioned above as a possible force driving MHC diversity), many consequences of this communication are independent of immune function, and it is these that will be highlighted here.

MHC is able to act as a signal of individuality because it is highly polymorphic and individuals within a population are not likely to express identical genotypes, thus providing a specific MHC signature to most individuals. MHC has been suggested to promote individual recognition since 1975 (Thomas, 1975). Since then, MHC-mediated individual recognition has been invoked in many studies (e.g., Kwak et al., 2009). One of the most important examples of MHC-based individual recognition is pregnancy block, also known as the Bruce effect, which occurs when recently mated female mice are exposed to the odors of an unfamiliar male, resulting in pregnancy failure (Bruce, 1959). The signal responsible for pregnancy block is considered to be individual-specific because the unfamiliar male and the mate both express odors capable of inducing pregnancy block, requiring females to learn the identity of their mate. The ecological advantage of pregnancy block may be derived from the fact that male mice are highly infanticidal and that a new territorial male will often kill all pups that are not his own (Piertney and Oliver, 2006; Slev et al., 2006).

MHC polymorphism can also allow for the effective identification of kin (Grafen, 1990, 1992). Recognition of relatives can increase indirect fitness of individuals by promoting cooperation with kin (Hamilton, 1964) and can increase direct fitness by preventing inbreeding (Brown, 1983). MHC's role in kin recognition has been shown to mediate cooperation (Manning et al., 1992), parent–offspring identification (Yamazaki et al., 2000), and mating preferences that prevent inbreeding (Brown, 1983). The rules that individuals use to determine kin based on MHC signals vary between species. Whereas all species use a type of phenotype matching system, many use their own MHC genotype as a template for identification (e.g., African clawed frogs [*Xenopus laevis*] [Villinger and Waldman, 2008] and Atlantic salmon [Rajakaruna et al., 2006]), while others imprint upon the MHC-associated odors of nest mates (e.g., house mice [Manning et al., 1992] and zebrafish [*Brachydanio rerio*] [Gerlach et al., 2008]). Both of these phenotype matching systems are enhanced by increasing levels of MHC diversity between individuals and therefore favor heterozygosity. Inbreeding avoidance may be the most important MHC-mediated kin-recognition function as both sibling-level and cousin-level inbreeding have been shown to have devastating impacts on vertebrate (house mouse) fitness (Meagher et al., 2000; Ilmonen et al., 2008). These strong fitness effects suggest that the prevention of inbreeding could be a primary function of MHC-disassortative mating preferences.

A trait that is involved in multiple physiological functions may not be able to serve any single function maximally; as a consequence therefore, trade-offs exist. The dual functionality of MHC alleles in both immune recognition and social communication may for example lead to situations where alleles are maintained within a population due to their function in social communication despite them conferring susceptibility to infectious or autoimmune diseases. Interestingly, this dual role also creates fundamental differences in the nature of selection operating on levels of individual MHC variation. In regards to social communication, selection should favor maximal MHC variation because it produces more individual-specific signals as well as enhances recognition of kin. In regards to MHC's immunological function, evidence reviewed here suggests that maximal variation can be costly to individuals. Thus, selection should favor an optimal level of individual variation. Currently, the degree to which selection on MHC is influenced by MHCs role in social communication versus immune recognition has not been considered experimentally.

BEYOND MHC ALLELIC DIVERSITY

MHC copy number variation occurs when individuals of the same species have a variable number of MHC loci. Such variation is difficult to detect because it can easily be mistaken for allelic variation, and thus represents an important component of MHC diversity that is often overlooked. MHC gene copy number variation has been reported in advanced neoteleosts (such as cichlids, sticklebacks, and cod [Miller et al., 2002]), swordtail fish (*Xiphophorus* spp.) (Figueroa et al., 2001), Amazon mollys (*Poecilia formosa*) (Schaschl et al., 2008), macaques (*Macaca mulatta*) (Bonhomme et al., 2008), bank voles (Kloch et al., 2010), sea lions (*Zalophus californianus*)

(Bowen et al., 2004), swine (*Sus scrofa*) (Tanaka-Matsuda et al., 2009), mice (Kumanovics et al., 2002), rats (*Rattus norvegicus*) (Roos and Walter, 2005), Tasmanian devils (*Sarcophilus harrisii*) (Siddle et al., 2010), and cattle (*Bos taurus*) (Birch et al., 2006). This differential copy number acts in a similar fashion to that of allelic heterozygosity at any particular locus, with the end result being the expression of more types of MHC molecules on cellular surfaces. So, copy number variants are likely to be subject to the same costs of MHC variation discussed above. In fact, Klein and colleagues (1993) predicted that MHC copy number would expand and contract with shifting parasite-mediated selection. It is noteworthy, then, that of the few studies that have found optimal MHC diversity, two of them were organisms with known copy-number variation (Wegner et al., 2003; Kloch et al., 2010). Future studies seeking to understand the control of MHC diversity will need to resolve the impact of MHC gene copy number on levels of individual variation.

MHC supertypes are defined as groups of alleles that share peptide-binding properties, and this idea, proposed in the 1990s, has allowed for the grouping of MHC molecules based on their functionality and not simply by their nucleotide or peptide sequences (Sidney et al., 1996). Recent advances have allowed for both the characterization of these supertypes as well as assessment of supertype frequencies within populations (Schwensow et al., 2007). This ability to classify MHC diversity in a functional metric has recently led to newly appreciated attributes of individual MHC variation for both ecologists and immunologists. For instance, within the ecological literature there is now an example where MHC-based mating preferences seem to emerge at the supertype level (Schwensow et al., 2008). This singular finding raises the question of whether previous cases where MHC mating preferences were not observed were due to a failure to consider supertype-level information. Supertype-level analysis has also been shown to be of importance within the immunological literature, where MHC supertype identification can elucidate population-level patterns of autoimmune disease (Antal et al., 2008) and pathogen susceptibility (Lazaryan et al., 2006) within humans.

UNDERSTANDING MHC ECO-IMMUNOLOGY: APPLICATIONS

Genetic variation is the raw material on which natural selection operates and provides the adaptive flexibility needed by populations to persist in their dynamic environments. MHC polymorphism has become a prototypical example of such critical adaptive variation. There are multiple practical applications associated with MHC diversity from the perspective of mitigating disease, which we address below. One classic argument is that because MHC variability is so important for resistance to disease, captive breeding programs of endangered species should focus on retaining maximal MHC diversity across individuals. By extension, the same principle holds for domesticated livestock species at risk of low levels of genetic diversity. The most immediate benefit of breeding resistance in livestock might be a reduced need to use antibiotics to control infection, which may consequently reduce selection for antibiotic resistance. Similarly, maximizing MHC diversity within founder individuals intended for reintroduction to nature may increase the likelihood of them persisting long enough to become a viable population again. Finally, by understanding how

MHC diversity within populations affects the spread of infectious diseases, we may be able to use this knowledge to reduce the transmission of zoonotic pathogens.

CONSERVATION GENETICS AND POPULATION VIABILITY

While it is difficult to retrospectively ascribe infectious disease as the causal agent of an extinction event, the abundance of examples where pathogens have been implicated in population declines (O'Brien et al., 1985; Roelke et al., 1993; Sanjayan et al., 1996; Kretzmann et al., 1997; Osterhaus et al., 1997; Coltman et al., 1999; Acevedo-Whitehouse et al., 2003, 2005) supports the notion that infectious diseases are potent regulators of host population demographics (see also May, 1988; Wilcove et al., 1998; Daszak, 2000; Altizer et al., 2003; De Castro and Bolker, 2005). Loss of genetic diversity is an unavoidable consequence of abrupt contractions in effective population size. Nevertheless, evidence suggests that selection can maintain MHC diversity, even in the face of severe genetic bottlenecks (Richardson and Westerdahl, 2003; Aguilar et al., 2004; Jarvi et al., 2004). For example, kit foxes (*Urocyon littoralis*) on the island of San Nicolas off the coast of Southern California are genetically monomorphic at hypervariable neutral microsatellite loci, indicating a very recent and severe bottleneck event (Aguilar et al., 2004). In spite of this, very high levels of genetic diversity at MHC loci were observed with selection coefficients exceeding 0.5; the magnitude of this selection underscores the adaptive importance of diversity at MHC genes, at least for this species. Observations such as this, coupled with the numerous studies indicating that individual MHC variation is fitness-enhancing, suggest that long-term genetic conservation efforts should focus on maintaining MHC diversity.

Eighty to ninety percent of all reintroductions of captive-bred animals are unsuccessful (Frankham et al., 2002), and evidence from studies listed above suggests parasites play a large role in these failures. Hughes (1991) first suggested that breeding programs should maximize diversity at MHC loci, but opposing views were aired (e.g., Miller and Hedrick, 1991), and the issue remains controversial. While it is true that "targeted" breeding strategies by their very nature reduce genetic diversity and that current breeding strategies seeking to maximize mean heterozygosity among individuals also retain some degree of MHC diversity (Radwan et al., 2010), there are situations where a targeted breeding strategy might be warranted. For example, in a species whose demographics are largely being influenced by a density-dependent infectious agent, selective breeding programs designed to increase the proportion of resistant individuals within the population could have the effect of blocking routes of disease transmission and impeding the dissemination of disease. Additionally, founders for reintroduction could be chosen to represent the full MHC diversity in the captive breeding population. What percentage of the 80% to 90% of failures of reintroduced captive-bred animals is due to reduced MHC diversity will of course never be known. Despite this, it is hoped that the potential value of the MHC-targeted breeding programs mentioned above will expand the breadth of MHC research into the field of disease transmission and applied animal husbandry practices.

There are potential benefits of targeted MHC animal breeding on human health as well. Zoonotic diseases are one of the fastest-emerging groups of infectious diseases, due in large part to the increasing interface between humans and wild animals (Daszak, 2000). Any method that decreases transmission of disease through a wild population (e.g., flooding the contact network with resistant host genotypes) could theoretically drive a pathogen population to extinction (Swinton et al., 1998; De Castro and Bolker, 2005) and reduce the exposure rate of humans to these infectious agents. Additionally, we are currently facing an unprecedented emergence of antibiotic-resistant microbes in our environment. Evidence is accumulating that the prophylactic use of antibiotics on livestock is not only promoting the evolution of resistance (Silbergeld et al., 2008), but also facilitating the movement of antibiotic resistance genes via horizontal transfer into related and potentially pathogenic species of microbes living in the human gut (van den Bogaard and Stobberingh, 2000). Targeted breeding for resistance in livestock breeds could reduce the need to use antibiotics, which may reduce the selective pressure we are placing on pathogens to evolve resistance. Unfortunately, it is becoming increasingly clear that the war on microbes will not be won with antibiotics alone, so it is imperative that we explore all available options to remove or reduce the selective forces favoring the evolution of antibiotic resistance in microbes.

MHC AND TRANSMISSION OF DISEASE

MHC immune recognition triggers a cascade of events that leads to an immune response necessary to control pathogen replication and the dissemination of disease. In light of this, it is surprising that very few empirical studies have attempted to understand the importance of MHC diversity on the spread of infectious disease through a population.

The most convincing evidence that reduced MHC diversity facilitates disease transmission comes from a recent study on the transmission of devil facial tumor disease (DFTD) through Tasmanian devil populations. Three lines of evidence support the contention by Siddle and colleagues that reduced MHC diversity has facilitated the spread of DFTD through the devil population (Siddle et al., 2007). First, a general immune evasion tactic employed by most tumors and all transmissible tumors is downregulation of MHC expression so as to avoid being detected as "foreign." DFTD tumors do not do this, suggesting that they are recognized as "self" in all affected individuals. Second, mixed-lymphocyte assays failed to detect alloreactivity between individuals, suggesting very low MHC diversity within affected populations (which was subsequently confirmed by genotyping). Finally, the acceptance of allogeneic grafts (a highly unusual event under normal circumstances) has been observed in cheetahs (*Acinonyx jubatus*) and pocket gophers (*Thomomys bottae*), which both had low MHC diversity as well (Siddle et al., 2007, and references therein). Much more work will need to be done to establish a causal relationship between DFTD transmission and reduced MHC diversity, but evidence so far is compelling. Further evidence that MHC diversity is an impediment to disease transmission comes from the HIV-immunology literature, where studies have observed lower rates of horizontal HIV transmission between MHC-discordant Zambian couples (Dorak et al., 2004), as well as between MHC-discordant mothers and their children (MacDonald et al., 1998; Polycarpou et al., 2002). Though suggestive of a link between MHC diversity and pathogen transmission, these correlative studies suffer from either small sample sizes or an inability to control for important diversity elsewhere in the genome. Finally, the unique immunobiology of infection with HIV or transmissible tumors (both elicit MHC-mediated allograft-rejection responses) may exaggerate the link between MHC diversity and disease transmission, and therefore might not be representative of the general relationship.

These studies raise the possibility that management efforts involving MHC could be used to restrict the movement of zoonotic infectious agents through the environment. This could be achieved by maximizing the proportion of individuals carrying resistance alleles, which would consequently result in more transmission "dead ends" in the population contact network. Such ecological manipulations are well off in the future. However, with the emerging field of experimental evolution, the increasing availability of MHC-congenic lines of animals, and the use of semi-natural animal enclosures (where environment can be controlled), the empirical tools are now in hand to fill this gap in our knowledge of MHC-mediated disease ecology.

Future empirical work on MHC diversity should focus on addressing the following questions. What role does MHC diversity play in the transmission of infectious disease? How does the impact of MHC diversity on transmission of infectious disease in turn affect virulence evolution? Does reduced MHC diversity facilitate pathogen adaptation, enhance transmission, and result in significantly more stress (via pathogen exploitation) on host populations? Such questions will be difficult to address (especially under purely natural settings), but answers to them are of paramount importance for the development of maximally effective long-term conservation strategies.

CONCLUSIONS

Despite the clear advantages associated with certain levels of individual MHC variation, associations between specific MHC alleles and susceptibility to a variety of infectious and autoimmune diseases abound. Hence, there are trade-offs in health associated with this diversity, which might explain the paradox of why individuals express such a limited portion of the diversity present in their populations. We have outlined five non-mutually exclusive mechanisms that could account for this limitation on individual variation, but expect that more will be brought to light as the costs associated with individual MHC variation are uncovered. These costs will define the optimal amount of diversity that is beneficial to individual health. To more fully appreciate the ecological significance of MHC genes it is important to explore the costs as well as the benefits associated with this variation. Future ecological work on the MHC should consider the following questions: (1) Is optimality a general phenomenon dictating levels of individual MHC heterozygosity? (2) What are the trade-offs restricting evolution of maximal MHC heterozygosity? (3) To what extent do these trade-offs influence patterns of MHC variation in nature? Future work in the field should strive to appreciate the ecological benefit of MHC diversity within the context of its associated physiological and immunological costs to individual health and lifetime reproductive success.

Viability and sexual selection act on variation in the MHC and favor those genotypes that confer resistance to pathogens. There is now abundant evidence to conclude that compared to MHC homozygosity, MHC heterozygosity is a fitness-enhancing trait. However, maximal individual MHC heterozygosity is not always optimal, as recent studies have demonstrated that maximal diversity can be costly to individuals. Similarly, sexual selection patterns mirror these viability patterns as MHC-disassortative mating preferences promote MHC heterozygosity in offspring.

However, in some species mate-choice decisions produce offspring with intermediate rather than maximal MHC diversity, appears to provide indirect genetic benefits to an individual by producing more immunocompetent offspring. We imagine MHC research pertaining to "individual optimal MHC variation" blossoming over the next decade as those in the field become increasingly aware of the costs associated with individual MHC variation.

The importance of MHC in mediating multiple aspects of animal behavior will likely confound purely immunological interpretations of patterns of MHC variation we see in nature. Inbreeding avoidance and/or increasing cooperative behavior within kin-groups may be more important selective forces shaping MHC variation than pathogens in some situations; this has gone largely unappreciated. Finally, MHC variation now extends well beyond simple nucleotide diversity. The discovery that MHC allelic diversity can be condensed into functional groups (i.e., supertypes), as well as the appreciation that MHC gene copy number is variable within species, requires researchers in the field to broaden their survey of MHC heterozygosity to encompass these additional metrics of individual variation.

It has been suggested that MHC variation should be a focus of genetic conservation strategies because MHC variation is so important for disease resistance. Currently, there is very little direct evidence that reduced MHC variation has an impact on population viability. However, very little work has been done to directly address this hypothesis. Because infectious diseases contribute to population declines, it is prudent that we begin to explicitly address the relationship between immunologically important adaptive variation and population persistence. Information from such work may guide the design of more effective conservation strategies. Furthermore, the central role MHC plays in disease resistance makes it a prime candidate for studies that seek to understand how adaptive genetic variation in host populations affects the spread of infectious agents through populations. With the increasing interest in the potential application of MHC variation as a tool for mitigating the impacts of disease on individuals, we expect this area of research to expand in the near future.

Finally, for over 500 million years, natural selection has operated on MHC genes to produce the most polymorphic set of loci known in vertebrates. The importance of individual MHC variation in pathogen resistance, and the emerging appreciation of its role in social signaling, attests to the value of MHC variation in enhancing individual viability and reproductive success. Increasing evidence is now suggesting that there may be an optimal level of individual MHC variation beyond which individuals suffer fitness declines. This recent advance in the field has highlighted the importance of understanding the physiological costs associated with MHC-mediated immunity. In the spirit of ecological immunology, which seeks to understand the evolution and function of the immune system within the context of tradeoffs, no other genetic system lends itself to such analyses as the MHC.

ACKNOWLEDGMENTS

This manuscript was written while Jason L. Kubinak, Adam C. Nelson, and Wayne K. Potts were funded in part by NSF grants DEB 0910052, IOS 0909801, and DEB 0918969, respectively. We thank Shannon M. Gaukler for constructive comments on the manuscript.

REFERENCES

Acevedo-Whitehouse, K, Gulland, F, Greig, D, and Amos, W (2003). Inbreeding: Disease susceptibility in California sea lions. Nature 422:35.

Acevedo-Whitehouse, K, Vicente, J, Gortazar, C, Hofle, U, Fernandez-de-Mera, IG, and Amos, W (2005). Genetic resistance to bovine tuberculosis in the Iberian wild boar. Mol. Ecol. 14:3209–3217.

Aeschlimann, PB, Haberli, MA, Reusch, TBH, Boehm, T, and Milinski, M (2003). Female sticklebacks *Gasterosteus aculeatus* use self-reference to optimize MHC allele number during mate selection. Behav. Ecol. Sociobiol. 54:119–126.

Aguilar, A, Roemer, G, Debenham, S, Binns, M, Garcelon, D, and Wayne, RK (2004). High MHC diversity maintained by balancing selection in an otherwise genetically monomorphic mammal. Proc. Natl. Acad. Sci. U. S. A. 101:3490–3494.

Aguilera, E and Amat, JA (2007). Carotenoids, immune response and the expression of sexual ornaments in male greenfinches (*Carduelis chloris*). Naturwissenschaften 94:895–902.

Alberts, SC and Ober, C (1993). Genetic variability in the major histocompatibility complex: a review of non-pathogen-meditated selective mechanisms. Year. Phys. Anthrop. 36:71–89.

Altizer, S, Harvell, D and Friedle, E (2003). Rapid evolutionary dynamics and disease threats to biodiversity. Trends Ecol. Evol. 18:589–596.

Amat, JA, Aguilera, E, and Visser, GH (2007). Energetic and developmental costs of mounting an immune response in greenfinches (*Carduelis chloris*). Ecol. Res. 22:282–287.

Antal, Z, Jarchum, I, and DiLorenzo, TP (2008). HLA class I supertypes in type 1 diabetic children in an urban children's hospital. Ann. N. Y. Acad. Sci. 1150:86–89.

Apanius, V, Penn, D, Slev, P, Ruff, LR, and Potts, WK (1997). The nature of selection on the major histocompatibility complex. Crit. Rev. Immunol. 17:179–224.

Banks, SC, Dubach, J, Viggers, KL, and Lindenmayer, DB (2010). Adult survival and microsatellite diversity in possums: effects of major histocompatibility complex-linked microsatellite diversity but not multilocus inbreeding estimators. Oecologia 162:359–370.

Barribeau, SM, Villinger, J, and Waldman, B (2008). Major histocompatibility complex based resistance to a common bacterial pathogen of amphibians. PLoS One 3:e2692.

Bascunan-Garcia, AP, Lara, C, and Cordoba-Aguilar, A (2010). Immune investment impairs growth, female reproduction and survival in the house cricket, *Acheta domesticus*. J. Insect Physiol. 56: 204–211.

Beissert, S, Schwarz, A, and Schwarz, T (2006). Regulatory T-cells. J. Invest. Dermat. 126:15–24.

Belkaid, Y (2007). Regulatory T-cells and infection: a dangerous necessity. Nat. Rev. Immunol. 7: 875–888.

Bernatchez, L and Landry, C (2003). MHC studies in nonmodel vertebrates: what have we learned about natural selection in 15 years? J. Evol. Biol. 16:363–377.

Birch, J, Murphy, L, MacHugh, ND, and Ellis, SA (2006). Generation and maintenance of diversity in the cattle MHC class I region. Immunogenetics 58:670–679.

Bodmer, WF (1972). Evolutionary significance of the HL-A system. Nature 237:139–145.

Boehm, T (2006). Co-evolution of a primordial peptide-presentation system and cellular immunity. Nat. Rev. Immunol. 6:79–84.

Boehm, T and Zufall, F (2006). MHC peptides and the sensory evaluation of genotype. Trends Neurosci. 29:100–107.

Bonato, M, Evans, MR, Hasselquist, D, Cloete, SWP, and Cherry, MI (2009). Growth rate and hatching date in ostrich chicks reflect humoral but not T-cell-mediated immune function. Behav. Ecol. Sociobiol. 64:183–191.

Bonhomme, M, Doxiadis, GGM, Heijmans, CMC, Vervoort, V, Otting, N, Bontrop, RE, and Crouau-Roy, B (2008). Genomic plasticity of the immune-related MHC class I B region in macaque species. BMC Genomics 9:514.

Bonneaud, C, Chastel, O, Federici, P, Westerdahl, H, and Sorci, G (2006) Complex MHC-based mate choice in a wild passerine. Proc. R. Soc. London B. 273:1111-1116.

Bonneaud, C, Mazuc, J, Chastel, O, Westerdahl, H, and Sorci, G (2004). Terminal investment induced by immune challenge and fitness traits associated with major histocompatibility complex in the house sparrow. Evolution 58:2823–2830.

Bonneaud, C, Mazuc, J, Gonzalez, G, Haussy, C, Chastel, O, Faivre, B, and Sorci, G (2003). Assessing the cost of mounting an immune response. Amer. Nat. 161:367–379.

Borghans, JA, Noest, AJ, and De Boer, RJ (2003). Thymic selection does not limit the individual MHC diversity. Eur. J. Immunol. 33:3353–3358.

Bowen, L, Aldridge, BM, Gulland, F, Van Bonn, W, DeLong, R, Melin, S, Lowenstine, LJ, Stott, JL, and Johnson, ML (2004). Class II multiformity generated by variable MHC- DRB region configurations in the California sea lion (*Zalophus californianus*). Immunogenetics 56:12–27.

Brown, EG, Liu, H, Kit, LC, Baird, S, and Nesrallah, M (2001). Pattern of mutation in the genome of influenza A virus on adaptation to increased virulence in the mouse lung: identification of functional themes. Proc. Natl. Acad. Sci. U. S. A. 98:6883–6888.

Brown, JL (1983). Some paradoxical goals of cells and organisms: the role of the MHC. In: Pfaff, DW. Ethical Questions in Brain and Behavior. Springer Verlag, New York, pp. 111–124.

Brown, JL (1997). A theory of mate choice based on heterozygosity. Behav. Ecol. 8:60–65.

Bruce, HM (1959). An exteroceptive block to pregnancy in the mouse. Nature 164:105.

Cai, XQ, Yang, M, Zhong, WQ, and Wang, DH (2009). Humoral immune response suppresses reproductive physiology in male Brandt's voles (*Lasiopodomys brandtii*). Zoology 112:69–75.

Carrington, M, Nelson, GW, Martin, MP, Kissner, T, Vlahov, D, Goedert, JJ, Kaslow, R, Buchbinder, S, Hoots, K, and O'Brien, SJ (1999). HLA and HIV- 1:heterozygote advantage and B*35-Cw*04 disadvantage. Science 283:1748–1752.

Carroll, LS, Meagher, S, Morrison, L, Penn, DJ, and Potts, WK (2004). Fitness effects of a selfish gene are revealed in an ecological context. Evolution 58:1318–1328.

Casrouge, A, Beaudoing, E, Dalle, S, Pannetier, C, Kanellopoulos, J, and Kourilsky, P (2000). Size estimate of the alpha beta TCR repertoire of naive mouse splenocytes. J. Immunol. 164:5782–5787.

Coltman, DW, Pilkington, JG, Smith, JA, and Pemberton, JM (1999). Parasite-mediated selection against inbred Soay sheep in a free-living, island population. Evolution 53:1259–1267.

Consuegra, S and de Leaniz, CG (2008). MHC-mediated mate choice increases parasite resistance in salmon. Proc. R. Soc. London B. 275:1397–1403.

Croisetiere, S, Tarte, PD, Bernatchez, L, and Belhumeur, P (2008). Identification of MHC class II beta resistance/susceptibility alleles to *Aeromonas salmonicida* in brook charr (*Salvelinus fontinalis*). Mol. Immunol. 45:3107–3116.

Daszak, P (2000). Emerging infectious diseases of wildlife: Threats to biodiversity and human health. Science 287:1756–1756.

Dawson, RD and Bortolotti, GR (2006). Carotenoid-dependent coloration of male American kestrels predicts ability to reduce parasitic infections. Naturwissenschaften 93:597–602.

De Boer, R and Perelson, AS (1993). How diverse should the immune system be? Proc. R. Soc. London B. 252:171–175.

De Castro, F and Bolker, B (2005). Mechanisms of disease-induced extinction. Ecol. Lett. 8:117–126.

Dionne, M, Miller, KM, Dodson, JJ, Caron, F, and Bernatchez, L (2007). Clinal variation in MHC diversity with temperature: Evidence for the role of host-pathogen interaction on local adaptation in Atlantic salmon. Evolution 61:2154–2164.

Ditchkoff, SS, Lochmiller, RL, Masters, RE, Hoofer, SR, and Van Den Bussche, RA (2001). Major-histocompatibility-complex-associated variation in secondary sexual traits of white-tailed deer (*Odocoileus virginianus*):evidence for good-genes advertisement. Evolution 55:616–625.

Doherty, PC and Zinkernagel, RM (1975). Enhanced immunological surveillance in mice heterozygous at the H-2 gene complex. Nature 256:50–52.

Dorak, MT, Tang, FM, Penman-Aguilar, A, Westfall, AO, Zulu, I, Lobashevsky, ES, Kancheya, NG, Schaen, MM, Allen, SA, and Kaslow, RA (2004). Transmission of HIV-1 and HLA-B allele-sharing within serodiscordant heterosexual Zambian couples. Lancet 363:2137–2139.

Ebert, D (1998). Experimental evolution of parasites. Science 282:1432–1435.

Edwards, SV and Potts, WK (1996) Polymorphism of genes in the major histocompatibility complex (MHC): implications for conservation genetics of vertebrates. In: Smith, TB and Wayne, RK. Molecular Genetic Approaches in Conservation. Oxford University Press, New York, pp. 214–237.

Edwards, SV and Hedrick, PW (1998). Evolution and ecology of MHC molecules: from genomics to sexual selection. Trends Ecol. Evol. 13:305–311.

Ekblom, R, Saether, SA, Grahn, M, Fiske, P, Kalas, JA, and Hoglund, J (2004). Major histocompatibility complex variation and mate choice in a lekking bird, the great snipe (*Gallinago media*). Mol. Ecol. 13: 3821–3828.

Erickson, AL, Kimura, Y, Igarashi, S, Eichelberger, J, Houghton, M, Sidney, J, McKinney, D, Sette, A, Hughes, AL, and Walker, CM (2001). The outcome of hepatitis C virus infection is predicted by escape mutations in epitopes targeted by cytotoxic T lymphocytes. Immunity 15:883–895.

Evans, DT, O'Connor, DH, Jing, P, Dzuris, JL, Sidney, J, da Silva, J, Allen, TM, Horton, H, Venham, JE, Rudersdorf, RA, Vogel, T, Pauza, CD, Bontrop, RE, DeMars, R, Sette, A, Hughes, AL, and Watkins, DI (1999). Virus-specific cytotoxic T-lymphocyte responses select for amino-acid variation in simian immunodeficiency virus Env and Nef. Nat. Med. 5:1270–1276.

Ezenwa, VO and Jolles, AE (2008). Horns honestly advertise parasite infection in male and female African buffalo. Anim. Behav. 75:2013–2021.

Fair, JM, Hansen, ES and Ricklefs, RE (1999). Growth, developmental stability and immune response in juvenile Japanese quails (*Coturnix coturnix japonica*). Proc. R. Soc. London B. 266:1735–1742.

Fernandez, CS, Stratov, I, De Rose, R, Walsh, K, Dale, CJ, Smith, MZ, Agy, MB, Hu, SL, Krebs, K, Watkins, DI, O'Connor D, H, Davenport, MP, and Kent, SJ (2005). Rapid viral escape at an immunodominant simian-human immunodeficiency virus cytotoxic T-lymphocyte epitope exacts a dramatic fitness cost. J. Virol. 79:5721–5731.

Fernando, MMA, Stevens, CR, Walsh, EC, De Jager, PL, Goyette, P, Plenge, RM, Vyse, TJ, and Rioux, JD (2008). Defining the role of the MHC in autoimmunity: A review and pooled analysis. PLoS Genet. 4:e1000024.

Ferreira, C, Singh, Y, Furmanski, AL, Wong, FS, Garden, OA, and Dyson, J (2009). Non-obese diabetic mice select a low-diversity repertoire of natural regulatory T-cells. Proc. Natl. Acad. Sci. U. S. A. 106: 8320–8325.

Figueroa, F, Mayer, WE, Sato, A, Zaleska-Rutczynska, Z, Hess, B, Tichy, H, and Klein, J (2001). MHC class I genes of swordtail fishes, *Xiphophorus*: variation in the number of loci and existence of ancient gene families. Immunogenetics 53:695–708.

Forsberg, LA, Dannewitz, J, Petersson, E, and Grahn, M (2007). Influence of genetic dissimilarity in the reproductive success and mate choice of brown trout: females fishing for optimal MHC-dissimilarity. J. Evol. Biol. 20:1859–1869.

Frankham, R, Ballou, JD, and Briscoe, DA (2002). Conservation Genetics. Cambridge University Press, Cambridge.

Friedrich, TC, Dodds, EJ, Yant, LJ, Vojnov, L, Rudersdorf, R, Cullen, C, Evans, DT, Desrosiers, RC, Mothe, BR, Sidney, J, Sette, A, Kunstman, K, Wolinsky, S, Piatak, M, Lifson, J, Hughes, AL, Wilson, N, O'Connor, DH, and Watkins, DI (2004). Reversion of CTL escape-variant immunodeficiency viruses *in vivo*. Nat. Med. 10:275–281.

Froeschke, G and Sommer, S (2005). MHC class II DRB variability and parasite load in the striped mouse (*Rhabdomys pumilio*) in the southern Kalahari. Mol. Biol. Evol. 22:1254–1259.

Furukawa, Y, Kubota, R, Tara, M, Izumo, S, and Osame, M (2001). Existence of escape mutant in HTLV-I tax during the development of adult T-cell leukemia. Blood 97:987–993.

Garrigan D, HP (2003). Perspective: Detecting adaptive molecular polymorphism: Lessons from the MHC. Evolution 57:1707–1722.

Garvin, JC, Dunn, PO, Whittingham, LA, Steeber, DA, and Hasselquist, D (2008). Do male ornaments signal immunity in the common yellowthroat? Behav. Ecol. 19:54–60.

Gasparini, J, Bize, P, Piault, R, Wakamatsu, K, Blount, JD, Ducrest, AL, and Roulin, A (2009). Strength and cost of an induced immune response are associated with a heritable melanin-based colour trait in female tawny owls. J. Anim. Ecol. 78:608–616.

Gerlach, G, Hodgins-Davis, A, Avolio, C, and Schunter, C (2008). Kin recognition in zebrafish:a 24-hour window for olfactory imprinting. Proc. R. Soc. London B. 275:2165–2170.

Goldrath, AW and Bevan, MJ (1999). Selecting and maintaining a diverse T-cell repertoire. Nature 402: 255–262.

Grafen, A (1990). Do animals really recognize kin? Anim. Behav. 39:42–54.

Grafen, A (1992). Of mice and the MHC. Nature 360:530.

Graham, AL, Allen, JE, and Read, AF (2005). Evolutionary causes and consequences of immunopathology. Annu. Rev. Ecol. Evol. Syst. 36:373–397.

Haldane, JBS (1949). Disease and evolution. La Ric. Scient. 19:68–75.

Hale, ML, Verduijn, MH, Moller, AP, Wolff, K, and Petrie, M (2009). Is the peacock's train an honest signal of genetic quality at the major histocompatibility complex? J. Evol. Biol. 22:1284–1294.

Hamilton, WD (1964). The genetical evolution of social behaviour. I. II. J. Theor. Biol. 7:1–52.

Hanssen, SA (2006). Costs of an immune challenge and terminal investment in a long-lived bird. Ecology 87:2440–2446.

Hanssen, SA, Hasselquist, D, Folstad, I, and Erikstad, KE (2004). Costs of immunity: immune responsiveness reduces survival in a vertebrate. Proc. R. Soc. London B. 271:925–930.

Hedrick, PW (2002). Pathogen resistance and genetic variation at MHC loci. Evolution 56:1902–1908.

Hill, GE and Farmer, KL (2005). Carotenoid-based plumage coloration predicts resistance to a novel parasite in the house finch. Naturwissenschaften 92:30–34.

Hoekstra, HE, Hoekstra, JM, Berrigan, D, Vignieri, SN, Hoang, A, Hill, CE, Beerli, P, and Kingsolver, JG (2001). Strength and tempo of directional selection in the wild. Proc. Natl. Acad. Sci. U. S. A. 98: 9157–9160.

Hraber, P, Kuiken, C, and Yusim, K (2007). Evidence for human leukocyte antigen heterozygote advantage against hepatitis C virus infection. Hepatology 46:1713–1721.

Hughes, AL (1991). MHC polymorphism and the design of captive breeding programs. Conserv. Biol. 5: 249–251.

Hughes, AL and Nei, M (1988). Pattern of nucleotide substitution at major histocompatibity complex class I loci reveals overdominant selection. Nature 335:167–170.

Hughes, AL and Nei, M (1989). Nucleotide substitution at class II MHC loci: evidence for overdominant selection. Proc. Natl. Acad. Sci. USA 86:958–962.

Hughes, AL and Yeager, M (1998). Natural selection and the evolutionary history of major histocompatibility complex loci. Front. Biosci. 3:d509–516.

Ilmonen, P, Penn, DJ, Damjanovich, K, Clarke, J, Lamborn, D, Morrison, L, Ghotbi, L, and Potts, WK (2008). Experimental infection magnifies inbreeding depression in house mice. J. Evol. Biol. 21: 834–841.

Ilmonen, P, Penn, DJ, Damjanovich, K, Morrison, L, Ghotbi, L, and Potts, WK (2007). Major histocompatibility complex heterozygosity reduces fitness in experimentally infected mice. Genetics 176:2501–2508.

Ilmonen, P, Taarna, T, and Hasselquist, D (2000). Experimentally activated immune defence in female pied flycatchers results in reduced breeding success. Proc. R. Soc. London B. 267:665–670.

Jarvi, SI, Tarr, CL, McIntosh, CE, Atkinson, CT, and Fleischer, RC (2004). Natural selection of the major histocompatibility complex (MHC) in Hawaiian honeycreepers (Drepanidinae). Mol. Ecol. 13:2157–2168.

Kasahara, M, Suzuki, T, and Pasquier, LD (2004). On the origins of the adaptive immune system: novel insights from invertebrates and cold-blooded vertebrates. Trends Immunol. 25:105–111.

Kavaliers, M, Agmo, A, Choleris, E, Gustafsson, JA, Korach, KS, Muglia, LJ, Pfaff, DW, and Ogawa, S (2004). Oxytocin and estrogen receptor alpha and beta knockout mice provide. Genes Brain Behav. 3:189–195.

Kekalainen, J, Vallunen, JA, Primmer, CR, Rattya, J, and Taskinen, J (2009). Signals of major histocompatibility complex overdominance in a wild salmonid population. Proc. R. Soc. London B. 276:3133–3140.

Kelliher, KR, Spehr, M, Li, XH, Zufall, F, and Leinders-Zufall, T (2005). Relative roles of the main and accessory olfactory systems in behavioral responses to MHC class I peptides: Bruce effect. Chem. Senses 30:A18.

Klasing, KC, Laurin, DE, Peng, RK, and Fry, DM (1987). Immunologically mediated growth depression in chicks: influence of feed intake, corticosterone and interleukin-1. J. Nutr. 117:1629–1637.

Klein, J (1986). Natural History of the Histocompatibility Complex. Wiley, New York.

Klein, J and Figueroa, F (1986). Evolution of the major histocompatibility complex. Crit. Rev. Immunol. 6:295–386.

Klein, J, Ono, H, Klein, D, and O'huigin, C (1993). The Accordion Model of MHC Evolution. Springer-Verlag, Heidelberg.

Klitz, W, Thomson, G, and Baur, MP (1986). Contrasting evolutionary histories among tightly linked HLA loci. Am. J. Hum. Genet. 39:340–349.

Kloch, A, Babik, W, Bajer, A, Sinski, E, and Radwan, J (2010). Effects of an MHC-DRB genotype and allele number on the load of gut parasites in the bank vole *Myodes glareolus*. Mol. Ecol. 19 (Suppl. 1):255–265.

Kretzmann, MB, Gilmartin, WG, Meyer, A, Zegers, GP, Fain, SR, Taylor, BF, and Costa, DP (1997). Low genetic variability in the Hawaiian Monk seal. Conserv. Biol. 11:482–490.

Kumanovics, A, Madan, A, Qin, S, Rowen, L, Hood, L, and Fischer Lindahl, K (2002). QUOD ERAT FACIENDUM:sequence analysis of the H2-D and H2-Q regions of 129/SvJ mice. Immunogenetics 54:479–489.

Kundert, JA, Sealey, AL, Li, Y, Capecchi, MR, and Schmidt, EE (2008). Syngeneic immune-dependent abortions in mice suggest paternal alloantigen-independent mechanisms. Am. J. Reprod. Immunol. 60:290–297.

Kwak, J, Opiekun, MC, Matsumura, K, Preti, G, Yamazaki, K, and Beauchamp, GK (2009). Major histocompatibility complex-regulated odor types: peptide-free urinary volatile signals. Physiol. Behav. 96:184–188.

Lazaryan, A, Lobashevsky, E, Mulenga, J, Karita, E, Allen, S, Tang, JM, and Kaslow, RA (2006). Human leukocyte antigen B58 supertype and human immunodeficiency virus type 1 infection in native Africans. J. Virol. 80:6056–6060.

Leinders-Zufall, T, Brennan, P, Widmayer, P, et al. (2004). MHC class I peptides as chemosensory signals in the vomeronasal organ. Science 306:1033–1037.

Lenz, TL, Wells, K, Pfeiffer, M, and Sommer, S (2009). Diverse MHC IIB allele repertoire increases parasite resistance and body condition in the long-tailed giant rat (*Leopoldamys sabanus*). BMC Evol. Biol. 9:269.

Lochmiller, RL and Deerenberg, C (2000). Trade-offs in evolutionary immunology: just what is the cost of immunity? Oikos 88:87–98.

Lopez, P, Gabirot, M, and Martin, J (2009). Immune activation affects chemical sexual ornaments of male Iberian wall lizards. Naturwissenschaften 96:65–69.

MacDonald, KS, Embree, J, Njenga, S, Nagelkerke, NJ, Ngatia, I, Mohammed, Z, Barber, BH, Ndinya-Achola, J, Bwayo, J, and Plummer, FA (1998). Mother-child class I HLA concordance increases perinatal human immunodeficiency virus type 1 transmission. J. Infect. Dis. 177:551–556.

Madsen, T and Ujvari, B (2006). MHC class I variation associates with parasite resistance and longevity in tropical pythons. J. Evol. Biol. 19:1973–1978.

Manning, CJ, Wakeland, EK, and Potts, WK (1992). Communal nesting patterns in mice implicate MHC genes in kin recognition. Nature 360:581–583.

Matthews CE, GR, Savinov A, Serreze DV, and Leiter EH (2001). Unusual resistance of ALR/Lt mouse beta cells to autoimmune destruction: Role for beta cell-expressed resistance determinants. Proc. Natl. Acad. Sci. U. S. A. 98:235–240.

May, RM (1988). Conservation and disease. Conserv. Biol. 2:28–30.

McClelland, EE, Granger, DL, and Potts, WK (2003a). MHC-dependent susceptibility to *Cryptococcus neoformans* in mice. Infect. Immun. 71:4815–4817.

McClelland, EE, Penn, DJ, and Potts, WK (2003b). Major histocompatibility complex heterozygote superiority during coinfection. Infect. Immun. 71:2079–2086.

McClelland, EE, Adler, FR, Granger, DL, and Potts, WK (2004) Major histocompatibility complex controls the trajectory but not host-specific adaptation during virulence evolution of the pathogenic fungus *Cryptococcus neoformans*. Proc. R. Soc. London B. 271(1548):1557–1564.

McMichael, AJ and Rowland-Jones, SL (2001). Cellular immune responses to HIV. Nature 410:980–987.

Meagher, S, Penn, DJ, and Potts, WK (2000). Male-male competition magnifies inbreeding depression in wild house mice. Proc. Natl. Acad. Sci. U.S.A. 97:3324–3329.

Meyer, D and Thomson, G (2001). How selection shapes variation of the human major histocompatibility complex: a review. Ann. Hum. Genet. 65:1–26.

Milinski, M, Griffiths, S, Wegner, KM, Reusch, TB, Haas-Assenbaum, A, and Boehm, T (2005). Mate choice decisions of stickleback females predictably modified by MHC peptide ligands. Proc. Natl. Acad. Sci. U. S. A. 102:4414–4418.

Milinski, M, Griffiths, SW, Reusch, TB, and Boehm, T (2009). Costly major histocompatibility complex signals produced only by reproductively active males, but not females, must be validated by a "maleness signal" in three-spined sticklebacks. Proc Biol Sci. [Epub 2009 Oct. 21.]

Miller, KM, Kaukinen, KH, and Schulze, AD (2002). Expansion and contraction of major histocompatibility complex genes:a teleostean example. Immunogenetics 53:941–963.

Miller, PS and Hedrick, PW (1991). MHC polymorphism and the design of captive breeding programs:simple solutions are not the answer. Cons. Biol. 5:556–558.

Milner, J, Ward, J, Keane-Myers, A, Min, B, and Paul, WE (2007). Repertoire-dependent immunopathology. J. Autoimmun. 29:257–261.

Moller, AP (1996). Sexual ornamentation, condition, and immune defence in the house sparrow *Passer domesticus*. Behav. Ecol. Sociobiol. 39:317–322.

Moller, AP and Saino, N (1994). Parasites, immunology of hosts, and host sexual selection. J. Parasitol. 80: 850–858.

Nikolich-Zugich, J, Slifka, MK, and Messaoudi, I (2004). The many important facets of T-cell repertoire diversity. Nat. Rev. Immunol. 4:123–132.

Norris, K and Evans, MR (2000). Ecological immunology: life history trade-offs and immune defense in birds. Behav. Ecol. 11:19–26.

Nowak, MA, Tarczy-Hornoch, K, and Austyn, JM (1992). The optimal number of major histocompatibility complex molecules in an individual. Proc. Natl. Acad. Sci. U.S.A. 89:10896–10899.

O'Brien, SJ, Roelke, ME, Marker, L, Newman, A, Winkler, CA, Meltzer, D, Colly, L, Evermann, JF, Bush, M, and Wildt, DE (1985). Genetic basis for species vulnerability in the cheetah. Science 227:1428–1434.

Oliver, MK, Telfer, S, and Piertney, SB (2009). Major histocompatibility complex (MHC) heterozygote superiority to natural multi-parasite infections in the water vole (*Arvicola terrestris*). Proc. R. Soc. London B. 276:1119–1128.

Osterhaus, A, Groen, J, Niesters, H, van de Bildt, M, Martina, B, Vedder, L, Vos, J, van Egmond, H, Abou-Sidi, B, and Barham, ME (1997). Morbillivirus in monk seal mass mortality. Nature 388:838–839.

Parham, P (2005). MHC class I molecules and KIRs in human history, health and survival. Nat. Rev. Immunol. 5:201–214.

Peatman, E, Terhune, J, Baoprasertkul, P, Xu, P, Nandi, S, Wang, S, Somridhivej, B, Kucuktas, H, Li, P, Dunham, R, and Liu, Z (2008). Microarray analysis of gene expression in the blue catfish liver reveals early activation of the MHC class I pathway after infection with *Edwardsiella ictaluri*. Mol. Immunol. 45:553–566.

Penn, D and Potts, WK (1998). Chemical signals and parasite-mediated sexual selection. Trends Ecol. Evol. 13:391–396.

Piertney, SB and Oliver, MK (2006). The evolutionary ecology of the major histocompatibility complex. Heredity 96:7–21.

Polycarpou, A, Ntais, C, Korber, BT, Elrich, HA, Winchester, R, Krogstad, P, Wolinsky, S, Rostron, T, Rowland-Jones, SL, Ammann, AJ, and Ioannidis, JP (2002). Association between maternal and infant class I and II HLA alleles and of their concordance with the risk of perinatal HIV type 1 transmission. AIDS Res. Hum. Retroviruses 18:741–746.

Potts, WK, Manning, CJ, and Wakeland, EK (1991). Mating patterns in seminatural populations of mice influenced by MHC genotype. Nature 352:619–621.

Prugnolle, F, Manica, A, Charpentier, M, Guegan, JF, Guernier, V, and Balloux, F (2005). Pathogen-driven selection and worldwide HLA class I diversity. Curr. Biol. 15:1022–1027.

Radwan, J, Biedrzycka, A, and Babik, W (2010). Does reduced MHC diversity decrease viability of vertebrate populations? Biol. Conserv. 143:537–544.

Rajakaruna, RS, Brown, JA, Kaukinen, KH, and Miller, KM (2006). Major histocompatibility complex and kin discrimination in Atlantic salmon and brook trout. Mol. Ecol. 15:4569–4575.

Rakus, KL, Wiegertjes, GF, Jurecka, P, Walker, PD, Pilarczyk, A, and Irnazarow, I (2009). Major histocompatibility (MH) class II B gene polymorphism influences disease resistance of common carp (*Cyprinus carpio L.*). Aquaculture 288:44–50.

Reusch, T, Haeberli, MA, Aeschlimann, PB, and Milinski, M (2001). Female sticklebacks count alleles in a strategy of sexual selection explaining MHC polymorphism. Nature 414:300–302.

Richardson, DS and Westerdahl, H (2003). MHC diversity in two *Acrocephalus* species: the outbred great reed warbler and the inbred Seychelles warbler. Mol. Ecol. 12:3523–3529.

Robinson J, WM, Parham P, de Groot N, Bontrop R, Kennedy LJ, Stoehr P, and Marsh SGE (2003). IMGT/HLA and IMGT/MHC: sequence databases for the study of the major histocompatibility complex. Nucl. Acids Res. 31:311–314.

Rodriguez, M, Zoecklein, LJ, Howe, CL, Pavelko, KD, Gamez, JD, Nakane, S, and Papke, LM (2003). Gamma interferon is critical for neuronal viral clearance and protection in a susceptible mouse strain following early intracranial Theiler's murine encephalomyelitis virus infection. J. Virol. 77:12252–12265.

Roelke, ME, Forrester, DJ, Jacobson, ER, Kollias, GV, Scott, FW, Barr, MC, Evermann, JF, and Pirtle, EC (1993). Seroprevalence of infectious disease agents in free-ranging Florida panthers (*Felis concolor coryi*). J. Wildl. Dis. 29:36–49.

Roos, C and Walter, L (2005). Considerable haplotypic diversity in the RT1-CE class I gene region of the rat major histocompatibility complex. Immunogenetics 56:773–777.

Roulin, A, Riols, C, Dijkstra, C, and Ducrest, AL (2001). Female plumage spottiness signals parasite resistance in the barn owl (*Tyto alba*). Behav. Ecol. 12:103–110.

Rülicke, T, Chapuisat, M, Homberger, FR, Macas, E, and Wedekind, C (1998). MHC-genotype of progeny influenced by parental infection. Proc. R. Soc. London B. 265:711–716.

Ruff, JS, Nelson, AC, Kubinak, JL, and Potts, WK (in press). The major histocompatibility complex during social communication. In: Lopez-Larrea, C. Ancient Origin of Self Recognition Systems in Nature. Landes Bioscience, Austin, TX.

Saito, S, Shima, T, Nakashima, A, Shiozaki, A, Ito, M, and Sasaki, Y (2007). What is the role of regulatory T-cells in the success of implantation and early pregnancy? J. Assist. Reprod. Genet. 24:379–386.

Sakaguchi, S, Sakaguchi, N, Shimizu, J, Yamazaki, S, Sakihama, T, Itoh, M, Kuniyasu, Y, Nomura, T, Toda, M, and Takahashi, T (2001). Immunologic tolerance maintained by CD25+ CD4+ regulatory T-cells: their common role in controlling autoimmunity, tumor immunity, and transplantation tolerance. Immunol. Rev. 182:18–32.

Sanjayan, MA, Crooks, K, Zegers, G, and Foran, D (1996). Genetic variation and the immune response in natural populations of pocket gophers. Conserv. Biol. 10:1519–1527.

Schaschl, H, Tobler, M, Plath, M, Penn, DJ, and Schlupp, I (2008). Polymorphic MHC loci in an asexual fish, the amazon molly (*Poecilia formosa; Poeciliidae*). Mol. Ecol. 17:5220–5230.

Schwensow, N, Fietz, J, Dausmann, K, and Sommer, S (2008). MHC-associated mating strategies and the importance of overall genetic diversity in an obligate pair-living primate. Evol. Ecol. 22:617–636.

Schwensow, N, Fietz, J, Dausmann, KH, and Sommer, S (2007). Neutral versus adaptive genetic variation in parasite resistance: importance of major histocompatibility complex supertypes in a free-ranging primate. Heredity 99:265–277.

Sheldon, BC and Verhulst, S (1996). Ecological immunology: costly parasite defenses and trade-offs in evolutionary ecology. Trends Ecol. Evol. 11:317–321.

Siddle, HV, Kreiss, A, Eldridge, MDB, Noonan, E, Clarke, CJ, Pyecroft, S, Woods, GM, and Belov, K (2007). Transmission of a fatal clonal tumor by biting occurs due to depleted MHC diversity in a threatened carnivorous marsupial. Proc. Natl. Acad. Sci. U. S. A. 104:16221–16226.

Siddle, HV, Marzec, J, Cheng, Y, Jones, M, and Belov, K (2010). MHC gene copy number variation in Tasmanian devils: implications for the spread of a contagious cancer. Proc. R. Soc. London B. 277:2001–2006.

Sidney, J, Grey, HM, Kubo, RT, and Sette, A (1996). Practical, biochemical and evolutionary implications of the discovery of HLA class I supermotifs. Immunol. Today 17:261–266.

Silbergeld, EK, Graham, J, and Price, LB (2008). Industrial food animal production, antimicrobial resistance, and human health. Annu. Rev. Public Health 29:151–169.

Slev, PR, Nelson, AC, and Potts, WK (2006). Sensory neurons with MHC-like peptide binding properties: disease consequences. Curr. Opin. Immunol. 18:608–616.

Snell, GD (1948). Methods for the study of histocompatibility genes. J. Genet. 49:87–108.

Sommer, S (2005). The importance of immune gene variability (MHC) in evolutionary ecology and conservation. Front. Zool. 2:16.

Sommer, S, Schwab, D, and Ganzhorn, JU (2002). MHC diversity of endemic Malagasy rodents in relation to geographic range and social system. Behav. Ecol. Sociobiol. 51:214–221.

Spehr, M, Kelliher, KR, Li, XH, Boehm, T, Leinders-Zufall, T, and Zufall, F (2006). Essential role of the main olfactory system in social recognition of major histocompatibility complex peptide ligands. J. Neurosci. 26:1961–1970.

Spurgin, LG and Richardson, DS (2010). How pathogens drive genetic diversity: MHC, mechanisms and misunderstandings. Proc. R. Soc. London B. 277:979–988.

Swinton, J, Harwood, J, Grenfell, BT, and Gilligan, CA (1998). Persistence thresholds for phocine distemper virus infection in harbour seal *Phoca vitulina* metapopulations. J. Anim. Ecol. 67:54–68.

Tanaka-Matsuda, M, Ando, A, Rogel-Gaillard, C, Chardon, P, and Uenishi, H (2009). Difference in number of loci of swine leukocyte antigen classical class I genes among haplotypes. Genomics 93:261–273.

Thomas, L (1975). Symbiosis as an immunologic problem. The immune system and infectious diseases. In: Neter, E and Milgrom, F. Fourth International Convocation of Immunology. Karger, Basel, pp. 2–11.

Thursz, MR, Thomas, HC, Greenwood, BM, and Hill, AV (1997). Heterozygote advantage for HLA class-II type in hepatitis B virus infection. Nat. Genet. 17:11–12.

Tiwari, JL and Terasaki, PI (1985). HLA and Disease Associations. Springer-Verlag, New York.

Trowsdale, J (1995). "Both man & bird & beast": comparative organization of MHC genes. Immunogenetics 41:1–17.

Turner, SM, Faisal, M, and DeWoody, JA (2007). Zygosity at the major histocompatibility class IIB locus predicts susceptibility to *Renibacterium salmoninarum* in Atlantic salmon (*Salmo salar* L.). Anim. Genet. 38:517–519.

Uller, T, Andersson, S, and Eklof, J (2006). Juvenile cell-mediated immune response is negatively correlated with subsequent adult ornament size in quail. Evol. Ecol. 20:1–9.

van den Berg, HA and Rand, DA (2003). Antigen presentation on MHC molecules as a diversity filter that enhances immune efficacy. J. Theor. Biol. 224:249–267.

van den Bogaard, AE and Stobberingh, EE (2000). Epidemiology of resistance to antibiotics. Links between animals and humans. Int. J. Antimicrob. Agents 14:327–335.

Verhulst, S, Dieleman, SJ, and Parmentier, HK (1999). A tradeoff between immunocompetence and sexual ornamentation in domestic fowl. Proc. Natl. Acad. Sci. U. S. A. 96:4478–4481.

Vidovic, D (1989). Elimination of self-tolerogen turns nonresponder mice into responders. Immunogenetics 30:194–199.

Vidovic, D and Matzinger, P (1988). Unresponsiveness to a foreign antigen can be caused by self-tolerance. Nature 336:222–225.

Villinger, J and Waldman, B (2008). Self-referent MHC type matching in frog tadpoles. Proc. R. Soc. London B. 275:1225–1230.

von-Schantz, T, et al. (1996). MHC genotype and male ornamentation: genetic evidence for the Hamilton-Zuk model. Proc. R. Soc. London B. 263:265–271.

Wandstrat, AE, Nguyen, C, Limaye, N, Chan, AY, Subramanian, S, Tian, XH, Yim, YS, Pertsemlidis, A, Garner, HR, Jr., Morel, L, and Wakeland, EK (2004). Association of extensive polymorphisms in the SLAM/CD2 gene cluster with murine lupus. Immunity 21:769–780.

Wedekind, C, Chapuisat, M, Macas, E, and Rülicke, T (1996). Non-random fertilization in mice correlates with MHC and something else. Heredity 77:400–409.

Wegner, KM, Kalbe, M, Kurtz, J, Reusch, TB, and Milinski, M (2003). Parasite selection for immunogenetic optimality. Science 301:1343.

Wegner, KM, Kalbe, M, Milinski, M, and Reusch, TB (2008). Mortality selection during the 2003 European heat wave in three-spined sticklebacks: effects of parasites and MHC genotype. BMC Evol. Biol. 8:124.

Wegner, KM, Kalbe, M, Schaschl, H, and Reusch, TB (2004). Parasites and individual major histocompatibility complex diversity—an optimal choice? Microbes Infect. 6:1110–1116.

Wilcove, DS, Rothstein, D, Dubow, J, Phillips, A, and Losos, E (1998). Quantifying threats to endangered species in the United States. Bioscience 48:607–615.

Woelfing, B, Traulsen, A, Milinski, M, and Boehm, T (2009). Does intra-individual major histocompatibility complex diversity keep a golden mean? Philos. Trans. R. Soc. London B. 364:117–128.

Yamazaki, K, Beauchamp, GK, Curran, M, Bard, J, and Boyse, EA (2000). Parent-progeny recognition as a function of MHC odortype identity. Proc. Natl. Acad. Sci. U.S.A. 97:10500–10502.

Yamazaki, K, Beauchamp, GK, Singer, A, Bard, J, and Boyse, EA (1999). Odor types: Their origin and composition. Proc. Natl. Acad. Sci. U.S.A. 96:1522–1525.

Yamazaki, K, Boyse, EA, Mike, V, Thaler, HT, Mathieson, BJ, Abbott, J, Boyse, J, and Zayas, ZA (1976). Control of mating preferences in mice by genes in the major histocompatibility complex. J. Exp. Med. 144:1324–1335.

Yewdell, JW and Bennink, JR (1999). Immunodominance in major histocompatibility complex class I-restricted T lymphocyte responses. Annu. Rev. Immunol. 17:51–88.

Ziegler, A, Kentenich, H, and Uchanska-Ziegler, B (2005). Female choice and the MHC. Trends Immunol. 26:496–502.

Zuk, M (1992). The role of parasites in sexual selection: current evidence and future directions. Adv.Study Behav. 21:39–68.

CHAPTER 8

THE ENERGETICS OF IMMUNITY

Mechanisms Mediating Trade-offs in Ecoimmunology

Gregory Demas, Timothy Greives, Emily Chester, and Susannah French

"Nothing in biology makes sense except in the light of evolution."

Theodosius Dhabzhansky

INTRODUCTION

The oft-cited quote by Dhabzhansky above reflects the importance of applying an evolutionary approach to the study of biological questions. Although this almost certainly holds true for the study of ecoimmunology, an equally strong case can be made that nothing makes sense in biology without an understanding of the underlying *mechanisms* as well. As the concept of phenotypic plasticity has elaborated, not only does a particular response need to be adaptive to an organism to persist in a population, but the individual must possess the requisite physiological machinery to express a plastic response. In other words, no matter how adaptive a response may be, organisms cannot express it if they lack the requisite physiological machinery to do so. Similarly, when examining adaptive trade-offs within the field of ecoimmunology, it is important to understand not only the ecological relevance and evolutionary history of these responses, but also the underlying physiological mechanisms that mediate such trade-offs. The goal of the present review is to discuss recent advances on the topic of energetic trade-offs in ecoimmunology, and also to attempt to integrate both proximate and ultimate viewpoints into a common theoretical framework. It is our goal that such a framework will help elucidate some of the apparent discrepancies within the field and also guide future endeavors in this area of research.

As the old maxim "feed a cold, starve a fever" insinuates, there is an important biological link between energy balance and immune function and, thus, disease susceptibility and recovery.

Immunity, as with all other physiological processes, requires a continuous input of energy to sustain optimal functioning. Although this statement may appear obvious, the role of energy balance has only recently begun to be considered in the context of immune function and disease. Within the past decade or so, the concept of energetic trade-offs among competing physiological and behavioral systems has gained increasing popularity across several scientific disciplines, including comparative physiology, behavioral ecology, evolutionary biology, and physiological ecology. The underlying assumption of the trade-off concept is that animals require a relatively steady supply of energy to sustain biological functions. Energy, however, is not a limitless resource; finite energy reserves must serve all physiological processes, and thus energy must be allocated to a wide variety of often competing physiological functions. Furthermore, the total energy budget is not static; substantial fluctuations exist in both energy availability as well as energy expenditure across environmental contexts. For example, during times of reduced energy availability (e.g., winter), energy must be re-allocated from less important physiological functions (e.g., growth, reproduction) to those most important for immediate survival (e.g., thermoregulation).

The trade-off concept has provided a useful framework with which to consider interactions among physiological processes nested within the context of an animal's ecology. This approach has been successfully applied to trade-offs between immune function and other energetically demanding processes and, in fact, played an integral role in the establishment of the field of ecological immunology (reviewed in Sheldon and Verhulst, 1996; Lochmiller and Deerenberg, 2000). The goal of the present review is to provide an overview of energetic trade-offs with immunity, including a discussion of the recent empirical evidence in support of such trade-offs. While the primary focus will be on the proximate level of analysis, we also discuss the adaptive function of energetic trade-offs with immunity.

ENERGETIC COSTS OF IMMUNE FUNCTION

Is immunity actually costly in terms of energetic output? Traditional wisdom suggests that mounting an immune response requires resources that could otherwise be allocated to other biological functions (e.g., growth, reproduction). This idea was reinforced by an oft-cited seminal study early last century that suggested that for every 1°C increase in body temperature due to the induction of fever there is a 7% to 12% increase in oxygen consumption (Barr et al., 1922). More accurately, however, the study by Barr and colleagues assessed metabolic responses to typhoid vaccine via indirect calorimetry and demonstrated a ~125% increase in metabolic heat production following vaccine-induced fever. Extrapolating from van't Hoff's law, which suggests that chemical reactions increase two- to three-fold following a 10°C rise in temperature, the authors concluded that each degree of fever would result in a 7% to 12% increase in energy. Surprisingly, few subsequent studies provided further quantification of the energetic costs of a specific immune response since this initial report. It is only within the past 10 to 15 years that a more precise quantification of the energetic costs of mounting an immune response has been demonstrated (reviewed in Lochmiller and Deerenberg, 2000; Demas, 2004; French et al., 2009b). Although considerable debate exists as to the exact cost of maintaining immune defenses, as well as the adaptive significance of alterations in specific immune responses (Klasing, 1998; Svensson et al., 1998), mounting an immune response is unquestionably energetically costly (Nelson and Demas, 1996; Sheldon and Verhulst, 1996; Lochmiller and Deerenberg, 2000).

DIRECT METABOLIC COSTS

Preliminary research in the late 1990s examined the energetic costs of mounting an antibody response in house mice (*Mus musculus*). Specifically, mice were injected with the non-replicating antigen keyhole limpet hemocyanin (KLH) to induce an antibody response in the absence of fever or sickness. Mice immunized with KLH display ~25% increases in oxygen consumption and metabolic heat production compared with pre-immunization baseline values (Demas et al., 1997a) (Fig. 8.1).

Since this initial study, a growing number of studies have been published that have provided further evidence for the energetic costs of immunity using a variety of antigenic stimuli and across several vertebrate and invertebrate species (Moret and Schmid-Hempel, 2000; Bonneaud et al., 2003; Martin et al., 2003)(Table 8.1). For example, blue tits (*Parus caeruleus*) were immunized with diphtheria–tetanus toxin to induce an antibody responses, and basal metabolic rate (BMR), a measure of energy expenditure while at rest in a neutrally temperate environment and after a fasting period, was assessed via open circuit respirometry (Svensson et al., 1998). The energetic costs of mounting an antibody response were 8% to 13% of BMR. Further, birds subjected to increased energy turnover in the form of cold stress also displayed decreased humoral immunity. A similar study was conducted more recently in zebra finches (*Taeniopygia guttata*) using lipopolysaccharide (LPS) to induce immune activation (Burness et al., 2010). LPS is an endotoxin present on most gram-negative bacteria and its presence invokes a robust inflammatory response marked by the release of pro-inflammatory cytokines, fever, and sickness responses. Specifically, body temperature, food intake, and BMR were determined in birds undergoing an acute phase

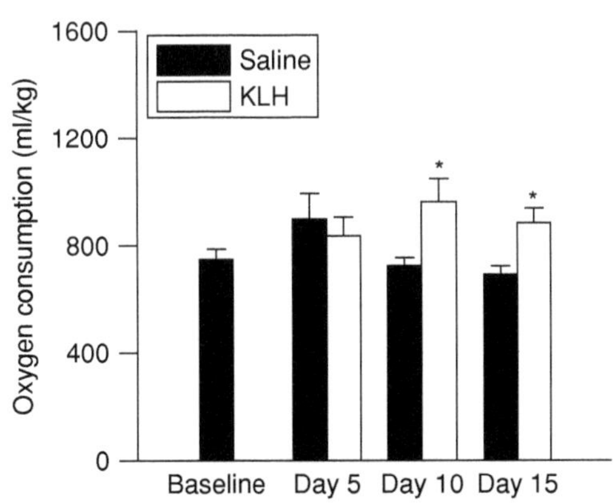

FIGURE 8.1: Energetic costs of immunity. Mean (±SEM) oxygen (O_2) consumption in house mice before immunization (baseline) and following immunization with either the novel antigen keyhole limpet hemocyanin (KLH) or control injections (saline). No differences in O_2 consumption are present early during antibody production (Day 5), but significant differences (*) between group means are present by Day 10 post-immunization and persist by Day 15 post-immunization.

TABLE 8.1: Energetic costs of immune responses in a range of non-human animal species

Species	Manipulation	Result	Reference
Collared flycatcher (*Ficedula albicollis*)	Increased brood size	Reduced antibody response against Newcastle virus	Nordling et al., 1998
Zebra finch (*Taeniopygia guttata*)	Increased brood size	Reduced antibody response against sheep red blood cells	Derrenberg et al., 1997
Tree swallow (*Tachycineta bicolor*)	Increased brood size	Reduced antibody response against sheep red blood cells	Ardia et al., 2003
Great tit (*Parus major*)	Increased brood size	Increased prevalence of Plasmodium	Richner et al., 1995
Bumblebee (*Bombus terrestris*)	Injected with lipopolysaccharide and latex beads	Reduced survival compared with controls	Moret and Schmid-Hempel, 2000
Mosquito (*Anopheles gambiae*)	Injected with lipopolysaccharide	Reduced egg production	Ahmed et al., 2002
House sparrow (*Passer domesticus*)	Injected with lipopolysaccharide	Reduced reproductive success	Bonneaud et al., 2003
Chicken (*Gallus domesticus*)	Injected with sheep red blood cells	Lowered fat deposition despite increased food consumption	Henken and Brandsma, 1982
Tree lizard (*Urosaurus ornatus*)	Injected with follicle-stimulating hormone to increase reproductive investment	Suppressed wound healing following punch biopsy	French et al., 2007a
Norway rat (*Rattus norvegicus*)	Infected with nematode *Nippostrongylus brasiliencsis*	Decreased body weight compared to controls	Ovington, 1985
Chicken (*Gallus domesticus*)	Infected with *Eimeria sp.*	Less weight gain compared to controls	Takhar and Farrekk, 1979
Chicken (*Gallus domesticus*)	Selected for resistance to Marek's disease	Lowered adult body weight compared to controls	Warner et al., 1987
Chicken (*Gallus domesticus*)	Selected for high antibody response against sheep red blood cells	Smaller comb size	Verhulst et al., 1999
Sheep (*Ovis aries*)	Selected for reduced intestinal helminth load	Lowered lamb growth rate compared to controls	Bisset et al., 2001

Pig (*Sus domesticus*)	Vaccinated against porcine respiratory and reproductive syndrome	21% decrease in body weight	Spurlock et al., 1997
West African dwarf goat (*Capra hircus*)	Infected with *Trypanosoma vivax*	28% increase in heat production	Zwart et al., 1991
House mouse (*Mus musculus*)	Injected with keyhole limpet hemocyanin	20–30% increase in oxygen consumption	Demas et al.a, 1997
Guinea pig (*Cavia porcellus*)	Infected with *Legionella pneumophila*	33% increase in oxygen consumption	Cooper et al., 1989
White cabbage butterfly pupa (*Pieris brassicae*)	Nylon implant	8% increase in metabolic rate	Freitak et al., 2003
Blue tit (*Parus caeruleus*)	Immunized with diphtheria–tetanus vaccine	8–13% increase in metabolic rate	Svensson et al., 1998
Great tit (*Parus major*)	Injected with sheep red blood cells	9% increase in metabolic rate	Ots et al., 2001
Collared dove (*Streptopelia decaocto*)	Injected with sheep red blood cells	8.5% increase in metabolic rate	Eraud et al., 2005
Common sparrow (*Passer domesticus*)	Injected with phytohemagglutinin	29% increase in metabolic rate	Martin et al., 2003
White-footed mouse (*Peromyscus leucopus*)	Injected with sheep red blood cells	17% increase in metabolic rate	Derting and Virk, 2005
Black rat (*Rattus rattus*)	Infected with *Fasciola hepatica*	56% increase in metabolic rate	Magnanou et al., 2006

Taken from Muehlenbein et al., 2010.

responses (APR), induced via injections of LPS, in birds under *ad libitum* feeding in the presence of relatively high (34°C) and low (15°C) ambient temperatures. The results demonstrated that LPS exposure was energetically costly, resulting in a 16% increase in total expenditure. However, cold-exposed, LPS-treated birds actually increased their resting energy expenditure but lost less body mass than birds maintained in their thermo-neutral zone (Burness et al., 2010). While the results of both of the studies described above do not support the idea of adaptive resource (energy) re-allocation or energetic trade-offs between immunity and other costly responses, they nevertheless provide additional support that immune function has a quantifiable energetic cost. It is important to note that both studies, similar to many lab-based studies in ecoimmunology, were conducted under conditions of *ad libitum* food availability; thus it is possible that potential energetic trade-offs were masked under conditions where resources were not limiting. Additional studies are required to determine whether the energetic costs of immunity have any functional significance under free-living conditions where resources are more limiting. Similar energetic costs have been associated with the cell-mediated branch of the immune system.

In house sparrows (*Passer domesticus*), activation of a cell-mediated immune response with injections of phytohemagglutinin (PHA) significantly elevates BMR by ~30%, an increase equivalent to the energy required of a female to produce half of an egg in this species (Martin et al., 2003). Furthermore, house sparrows that mount a humoral immune response to injections of LPS display significant decreases in body mass and food intake and reduced reproductive success (Bonneaud et al., 2003). Similar increases in resting metabolic rate have been observed in great tits (~5%) in response to PHA challenge (Nilsson et al., 2007) and sheep red blood cell (SRBC) challenge, which led to ~9% higher O_2 consumption and a decrease in body mass (Ots et al., 2001), suggesting that these responses are all highly energetically costly. Similarly, indirect evidence of behavioral fever in ectothermic animals demonstrates the significant energetic demands of mounting an immune response. Ectothermic vertebrates injected with LPS raise their metabolic rate to cope with the "infection" via behavioral thermoregulation (i.e., behavioral fever) (Sherman et al., 1991; Deen and Hutchison, 2001).

Conversely, inherent metabolic rate can influence an animal's ability to mount an immune response. More recently, the effects of metabolic rate on antibody production were assessed in mice selected for either high or low BMR (Ksiazek et al., 2003). Antibody production in response to immunization with SRBC was significantly reduced in mice with high BMR relative to low BMR. Animals in both lines, however, displayed comparable increases in both food intake and digestive efficiency in response to the immune challenge (Ksiazek et al., 2003). Furthermore, a similar immunization of wild-caught white-footed mice (*Peromyscus maniculatus*) led to decreases in dry masses of the small intestines and testes by 22% and 74%, respectively, suggesting that the costs of immune activation are met via re-allocation of energy away from other costly physiological systems (Derting and Compton, 2003). Lastly, activation of an immune response can alter growth rates in rodents. Specifically, immune activation via injections of LPS impaired growth and decreased both lean and fat mass in laboratory mice (Laugero and Moberg, 2000). Collectively, these and other studies suggest that immune function is energetically costly and that mounting an immune response can decrease energy investment into other physiological responses, including digestion and growth. Interestingly, within one species, tropical house wrens (*Troglodytes aedon*), a measure of constitutive (or innate) immunity, bacterial killing ability, is positively related with BMR when comparing individuals caring for young (e.g., the feeding of a day 12 nestling), suggesting that individuals in better condition are able to allocate more resources to this "first line of defense"(Tieleman et al., 2005). However, this same measure of innate immunity displays a negative correlation with BMR when compared across a range of tropical species (Tieleman et al., 2005).

Activation or maintenance of the immune system may be costly in terms of resources as well as the cost associated with oxidative damage. Indeed, zebra finches injected with LPS have a weakened ability to "resist" free radical attack, as measured by the time until 50% of red blood cells are lysed *in vitro* in the presence of AAPH, a water-soluble azo compound that generates free radicals (Bertrand et al., 2006). A recent meta-analysis revealed that an immune response induces oxidative stress in birds (Costantini and Møller, 2009), suggesting a mechanism by which immune function can be costly, leading to trade-offs. Immune activation appears to have an energetic cost in humans as well (Muehlenbein, 2010). For example, administration of the pro-inflammatory cytokine interleukin (IL)-6 in healthy volunteers increases metabolic rates by ~25% (Tsigos et al., 1997). Physiological insults, including sepsis, burns, and surgery, are associated with a 25% to 55% increase in resting metabolic rate (RMR) compared with healthy humans (reviewed in Muehlenbein, 2010). Furthermore, relatively mild infections can increase glucose utilization by 200% or more (Chioléro et al., 1997), whereas severe infections can increase metabolic rates by nearly 60%, decrease body mass by ~30%, and trigger significant wasting of lean tissues (Kreymann et al., 1993).

Collectively, the results of these studies support the assumption that immune responses, like all other biological processes, require energy. Increased energy dedicated to immune responses, although likely adaptive in the short term for increasing disease resistance, can ultimately come at the cost of decreased fitness, especially if immune activation is prolonged or excessive (Lochmiller and Deerenberg, 2000). The risk of infection and death is highest, however, when insufficient energy reserves are available to sustain optimal immunity (Nelson et al., 2002).

FOOD AVAILABILITY

Much of the evidence for the high energetic requirements of the immune system and resistance to disease that result in trade-offs is available through studies that manipulate resource availability, whereby limiting required resources suppresses immunity and increases disease susceptibility. For example, food restriction and reproductive investment reduce both fat stores and innate immunity in tree lizards (French et al., 2007a). The interactive effects of food availability and reproduction (testosterone) are important for innate immunity in the sagebrush lizard (*Sceloporus graciosus*) (Ruiz et al., 2010). Specifically, supplementation of food to testosterone-treated males increased bacteria killing ability.

Additionally, the effects of food restriction on seasonal changes in reproduction and immune function were recently assessed in Siberian hamsters (*Phodopus sungorus*) (D.A. Zysling and G.E. Demas, unpublished). Hamsters were housed in long or short days and were food restricted (30% reduction) or fed *ad libitum*. Immune responses were quantified by measuring specific antibody production and bacterial killing capacity. Food restriction decreased body and relative reproductive masses, but not immune function, in long-day animals. Antibody responses, but not bacterial killing ability, were enhanced in food-restricted, short-day animals as compared with *ad libitum*-fed controls. We also found differential effects of body fat on immune responses depending on the immune measure. Specifically, there was a significant positive relationship between body fat and antibody responses in *ad libitum*-fed but not restricted animals. In contrast, there was a significant positive correlation between body fat and bacterial killing ability in food-restricted but not *ad libitum*-fed animals. The effects of food restriction on immune function appear to vary based on the specific restriction regimen, the response measured, and the physiological state of the organism, including energy balance, metabolic rate, and reproductive status.

In addition to the effects of food intake on immunity, specific metabolic fuels have also been demonstrated to affect immune function in seasonally breeding rodents. For example, reductions in glucose availability via the metabolic inhibitor 2-deoxy-D-glucose (2-DG) inhibit splenic T lymphocyte in a dose-dependent fashion in rats (Lysle et al., 1988) and mice (Miller et al., 1994). Interestingly, 2-DG inhibits mitogen-induced splenocyte proliferation and anti-KLH antibody production in long- but not short-day deer mice (*Peromyscus maniculatus*) (Demas et al., 1997b) and Siberian hamsters. Long-day animals have reproductively functional gonads, whereas short-day deer mice display reduced investment in reproduction (i.e., gonadal regression). Short-day housed animals appear buffered against 2-DG–induced immune suppression (Demas et al., 1997b; Zysling and Demas, 2007). Because 2-DG is a glucose analog that inhibits cellular utilization of glucose and induces a state of glucoprivation, these results suggest that reductions in energy availability suppress immunity. Collectively, these findings suggest an important connection between energy availability and the regulation of immunity.

NUTRITION AND IMMUNITY

It is important to note that immunological resource requirements are precise; specific proteins and nutrients, and not merely calories, are required and can facilitate trade-offs with other physiological functions (for detailed reviews, see Klasing, 1998, 2007). Immune responses can utilize a large amount of certain nutrients, such as the nearly six-fold increase in lysine utilization during an acute phase response (Klasing, 2007). A host of nutrients have immunomodulatory effects; they enhance, regulate, and/or impair immune function depending on the amount present (Klasing, 1998, 2007). For example, in chickens a moderate amount of vitamin E enhances the function of T cells, whereas too much vitamin E weakens antibody responses to an antigen (Erf et al., 1998; Leshchinsky and Klasing, 2001). Increases in nutrients above a threshold of tolerance may result in toxicity or immunopathology (Klasing, 2007).

There are many types of dietary components (e.g., fats, carbohydrates, proteins, vitamins A, D, and E, biotin, iron, selenium) that affect the immune system. Of these, one type has been well studied as a mediator of trade-offs between physiological traits. Carotenoids are capable of modulating a whole host of immune parameters, including stimulating T-cell and antibody production, and acting as free radical scavengers to protect cells from reactive oxygen species (ROSs) generated during immune responses (reviewed in Chew and Park, 2004). We will briefly highlight a few examples for the evidence of these dietary compounds in altering immunity. In guppies (*Poecilia reticulata*), carotenoids significantly affect resistance to naturally occurring parasites (*Gyrodactylus turnbulli*) (Kolluru et al., 2006). In the laboratory, zebra finches that received dietary carotenoid supplementation displayed an enhanced wing-web swelling in response to an injection of PHA and greater antibody production to SRBCs (McGraw and Ardia, 2003). Additional support for a role of carotenoids in enhancing immune function comes from an investigation of house sparrows experimentally infected with coccidia. These birds displayed a decreased response to SRBC; however, supplementation with carotenoids partially rescued the coccidian-induced reductions in SRBC antibody response.

In addition to their role in enhancing immunity, these compounds also have been found to play an important role in the development of sexual signals (e.g., Peters, 2007). Because carotenoids must be obtained from the environment and these environmental resources are limited, and because these compounds are also used in the development of sexual signaling traits, it has been hypothesized that trade-offs between these two factors will have evolved, possibly allowing carotenoid-based signals to serve as an "honest signal" of quality. As expected, if a trade-off exists between sexual signaling and immune function, immune challenges or parasitic infections in birds have been shown to lead to a dulling/fading of carotenoid-based ornaments—for example, in European blackbirds (Faivre et al., 2003; Baeta et al., 2008), zebra finches (McGraw and Ardia, 2003), mallard ducks (Peters et al., 2004), and blue-footed boobies (Torres and Velando, 2007).

For example, blackbirds experimentally infected with parasites and provided supplemental carotenoids maintained brighter bill coloration and decreased parasitism compared with un-supplemented infected birds (Baeta et al., 2008). Further, the sex steroid hormone testosterone is known to increase the bioavailability of carotenoids in circulation through increases in their carrier proteins; thus, carotenoids may provide an avenue for presenting an honest signal (Peters, 2007).

Carotenoids have received a great deal of study in the context of trade-offs, yet other dietary compounds (e.g., vitamins, iodine and fluorine, specific fatty acids) are also critical for maintenance of a competent immune system, and it is unknown as to whether their role as immunomodulators comes at a cost to their other physiological functions. Also, the metabolic costs associated with immunomodulation by levels of nutrients have received very little attention.

BODY FAT AND IMMUNITY

Despite the apparent link between energy availability and immunity, relatively little is known regarding the physiological mechanisms by which energy regulates immune function. On one hand, a chronic positive imbalance between energy intake and expenditure leads to obesity and can impair immune function and increase disease susceptibility in both clinical populations and genetically obese animal models (Martí et al., 2001). On the other hand, marked reductions in energy availability without concomitant reductions in energy output can also lead to substantial suppression of immunity (Chandra, 1996; Nova et al., 2002). For most mammalian species, and small rodents in particular, white adipose tissue (WAT) deposits represent a substantial portion of the total energetic budget, and thus WAT likely plays an important role in maintaining energetically expensive physiological processes, including immune function. Consistent with this idea, reductions in total body fat are correlated with impaired immunity in a wide range of species, including humans (Norgan, 1997; Spurlock et al., 1997; Klasing, 1998; Lin and Shiau, 2003), and experimental reductions in body fat can impair antibody formation (Demas et al., 2003).

Furthermore, immunological disorders (e.g., AIDS) trigger marked changes in whole-body lipid metabolism, suggesting an important link between adipose tissue and immunity (Pond, 1996).

During times of increased energetic demands, WAT can be broken down to its constituent components, free fatty acids (FFAs) and glycerol, and subsequently converted to glucose to be utilized as energy. Several lines of research suggest an important role of FFAs in the regulation of immune function (reviewed in Pond, 1996). FFAs provide a major fuel source for lymphocytes and may be used preferentially over glucose (Ardawi and Newsholme, 1985). Furthermore, FFAs can either enhance or inhibit mitogen-induced proliferation of rodent and human lymphocytes *in vitro* (reviewed in Pond, 1996), and unsaturated FFAs are incorporated into proliferating lymphocytes (Calder et al., 1994). Alterations in FFAs can also affect responsiveness to infection and alter the severity of chronic infections (Erickson et al., 1992; Yaqoob et al., 1994).

In addition to its role as an energy deposit, adipose tissue also serves as an important endocrine organ (Ahima and Flier, 2000a). A wide variety of signaling molecules, including hormones (e.g., leptin, adiponectin) and cytokines (e.g., IL-6, tumor necrosis factor-α), are now known to be secreted by adipose tissue, and these factors may play an important role in autocrine or paracrine regulation of specific immune responses. For example, the cytokine tumor necrosis factor-α (TNFα), an important mediator of cytotoxic and inflammatory immune responses, is secreted by adipocytes and is believed to play a major role in changes in lipid metabolism. Furthermore, adipose tissue provides the necessary precursors for synthesis of lipid-based hormones (e.g., prostaglandins, leukotrienes, and thromboxanes), all of which have been implicated in specific immunological responses (reviewed in Pompéia et al., 2000; Rocca and FitzGerald, 2002).

Substantial, naturally occurring fluctuations in total body fat throughout the year can be found in seasonally breeding rodents (reviewed in Bartness, 1996). In addition, most rodent species studied to date display seasonal changes in immune function that are, in general, positively correlated with their energy stores. The exact seasonal response varies according to the species: some rodents gain significant amounts of body fat during short winter-like days, whereas other species undergo short-day decreases in body fat. As a general rule, animals that gain weight in the short days of winter (e.g., Syrian hamsters, *Mesocricetus auratus*) display increased immune function (Drazen et al., 2002), whereas animals that lose a significant portion of their total body fat in short days (e.g., Siberian hamsters, *Phodopus sungorus*; prairie voles, *Microtus ochrogaster*) display reduced immunity (Nelson et al., 1998; Demas, 2002). This relationship is particularly interesting given that all of the above-mentioned species display gonadal regression and thus decreases in circulating gonadal steroid hormones. Thus, seasonal fluctuations in gonadal steroid hormones cannot fully explain the differential changes in immune responses among these species.

Interestingly, some species of tropical rodents that do not undergo seasonal fluctuations in reproductive and body fat responses (e.g., Aztec mice, *Peromyscus aztecus*) also fail to undergo photoperiodic changes in immunity (Demas and Nelson, 2003). These data support the idea that seasonal changes in immune function correlate with seasonal fluctuations in total body fat. A previous study has demonstrated that experimental reductions in total body fat in long-day animals (that approximate short-day levels of body fat) suppress humoral immune responses in two seasonally breeding rodents, prairie voles and Siberian hamsters (Demas et al., 2003). Specifically, surgical removal of body fat (lipectomy; LIPx) was performed in which either epididymal WAT (EWAT) or inguinal WAT (IWAT) pads were removed bilaterally and animals immunized with the antigen KLH. Initially, anti-KLH antibody responses are reduced in LIPx hamsters and voles compared with control animals. After a prolonged period of time (i.e., 12 weeks), compensatory increases in IWAT are seen in EWAT-lipectomized hamsters, restoring total body fat to control levels. Interestingly, humoral immunity also returns to normal control levels in these animals. In IWAT-lipectomized hamsters, in contrast, no compensatory increase in EWAT occurs, and immune function remains impaired in these animals. Furthermore, no compensatory increases in WAT are seen in LIPx prairie voles and, as expected, humoral immunity remains impaired in these animals. These data suggest that humoral immune responses track seasonal fluctuations in body fat.

TRADE-OFFS AMONG COMPETING BIOLOGICAL PROCESSES

GROWTH AND DEVELOPMENT

Activation of the immune system not only may incur metabolic costs, but may also lead to deficits in growth and development due to a trade-off among competing energy investments (e.g., growth vs. immune response). For example, (*Hirundo rustica*) nestlings in nests heavily infected with ectoparasites display an increase in feather growth (presumably to enable earlier fledging from infected nests) but deceased rates of skeletal growth, suggesting a trade-off between active infection and actual measures of body (i.e., tarsus) growth. Similarly, testosterone-treated male lizards (*Lacerta vivipara*) had higher growth rates but showed a significant decrease in growth rate when infected with ectoparasites, suggesting an energetic trade-off between growth and immunity (Uller and Olsson, 2003). Experimental parasite infection (*Plasmodium*) of juvenile canaries negatively affects song complexity and neural development (Spencer et al., 2005). Conversely, pond-breeding amphibians showed increased development rates but weaker cellular immune system responses to PHA and lower total leukocyte numbers, suggesting a development–immune trade-off (Gervasi and Foufopoulos, 2008).

Induced investment into immune function in magpie nestlings using dietary supplementation of methionine, which is known to increase investment in T-cell immunity in chickens (Tsiagbe et al., 1987), leads to a more robust wing-web swelling response to PHA in treated magpie (*Pica pica*) chicks but a concurrent reduction of investment in relative growth (Soler et al., 2003); a similar result has been documented in blue tit (*Cyanistes caeruleus*) nestlings (Brommer, 2004) and juvenile dragon lizards (*Ctenophorus fordi*) during development (Uller et al., 2006).

Activation of the immune system may also alter or slow development, or trade-off with different arms of the immune system. Greenfinches challenged with SRBCs, inducing a humoral immune response, displayed a less robust swelling response when injected with PHA (Horak et al., 2006), possibly suggesting a trade-off between different arms of the immune system (e.g., humoral vs. cell-mediated responses). However, barn swallow nestlings from nests infected with parasites showed changes in growth, but no effect of experimental ectoparasite infestation was found on response to PHA (Saino et al., 1998). Infection and subsequent immune activation has been shown to affect growth rates in mammalian populations as well. For example, flea infestation affects both immune responses and growth rates in common voles (*Microtus arvalis*). Voles parasitized by rat fleas (*Nosopsyllus fasciatus*) showed reduced body masses and antibody production in response to injections of KLH (Devevey et al., 2008). Furthermore, KLH immunizations decreased body mass and growth compared with non-immunized voles (Devevey et al., 2008), suggesting that immune activation and antibody production directly affects growth in this species. Unlike voles, however, immunization of young guinea pigs with KLH did not affect growth or metabolic rate compared with non-immunized animals, despite relatively robust antibody responses (Pilorz et al., 2005). In humans, helminth infection, immunoglobulin E, and C-reactive protein (CRP) were examined in blood samples taken from a population of forager-horticulturalists from southeastern Ecuador (Blackwell et al., 2010). Levels of both IgE and CRP were negatively correlated with stature in young children, suggesting a trade-off between helminth immune responses and growth in this population.

REPRODUCTION

Reproduction is believed to be a costly life-history stage, and thus it would be predicted that strong investment in reproduction should trade off with resources being directed to immunity, and vice versa. The notion that "pace of life" (the energy budget for progressing through the major life-history stages) is tuned to the longevity of the specific organism leads to differences in the amount and timing of reproductive investment throughout an organism's life-time. This likely leads to trade-offs, including differences in immune investment (Ricklefs and Wikelski, 2002). Thus, the current chapter will not deal with these ideas at length. Briefly, we provide a recent example that examined pace of life in the field (as opposed to the lab): when injected with LPS, free-living Southern (CA) song sparrows displayed greater behavioral (i.e., territory defense and lethargy) and physiological (i.e., fever) deficits compared with individuals of the same species but from more northerly populations (WA and AK) (Adelman et al., 2010). These data support the idea that selection has acted differently between populations, leading to distinct and discernable patterns of investment in current reproductive events versus survival across these different populations.

Many studies report that reproductive investment, or even reproductive state, has profound effects on immune function (reviewed in Klein, 2000), suggesting competition for energy resources between the reproductive and immune systems. For example, Saad and colleagues observed pregnancy-induced involution of lymphatic tissue in a viviparous lizard, *Chalcides ocellatus* (1989). In a viviparous skink (*Egernia whitii*), elevated levels of corticosterone and progesterone and decreased circulating lymphocytes occur during gestation (Cartledge et al., 2005). Breeding-related immune changes in the cyprinid fish, *Rutilus rutilus,* including hematocrit levels and granulocyte activity, have been reported; however, phagocytosis activity, spleen size, and natural antibody concentration did not change with breeding stage (Kortet et al., 2003). Levels of ROSs, toxic metabolic byproducts, relate directly to reproductive investment in painted dragon lizards (*Ctenophorus pictus*) (Olsson et al., 2009). Further, spawning male Arctic charr (*Salvelinus alpines*) have a reduced spleen size and higher macroparasite infections during spawning, suggesting an energetic trade-off between reproduction and immune investment (Skarstein et al., 2001). Adult female dragon lizards (*Ctenophorus fordi*) decreased reproductive investment when undergoing an

LPS immune challenge (Uller et al., 2006). Similarly, in pied flycatchers and blue tits, injection of females with a diphtheria–tetanus vaccine leads to a significant decrease in feeding rate of nestlings (Raberg et al., 2000) and body mass of fledglings (Ilmonen et al., 2000). Studies that experimentally manipulate reproductive investment show analogous results. A recent meta-analysis reveals support for the hypothesis that increased reproductive effort significantly alters immune responses, and hence susceptibility to parasites (Knowles et al., 2009). For example, increasing reproductive effort via follicle-stimulating hormone (FSH) in the tree lizard significantly suppresses cutaneous wound healing (French et al., 2007a). Conversely, other studies have attempted to ascertain the effects of decreased investment into immunity or reproduction. Using the blue tit model system, it was revealed that treatment of individuals with drugs known to reduce levels of blood parasites leads to an increase in nestling feeding rates and a decrease in ectoparasites (Tomàs et al., 2006), and a significant increase in fledging success (Merino et al., 2000). Similarly, decreased reproductive effort via castration in striped plateau lizards (*Sceloporus virgatus*) decreases parasite burden (Cox and John-Alder, 2007). All of the above studies suggest an energetic antagonism between the two systems.

Not all studies, however, support the hypothesis of a trade-off between the immune and reproductive systems. One study injected starlings with SRBCs to induce investment in the immune system; these individuals generated an increased number of antibodies against SRBCs, while controls did not increase anti-SRBC antibodies. After induction of the immune response, all birds were forced to re-nest; individuals that invested in an antibody response were equally likely to be successful in rearing young compared with control injected birds (Williams et al., 1999). Furthermore, unlike male lizards described above, ovariectomized (OVx) female brown anole lizards (*Anolis sagrei*) were more likely to display higher parasitemia than control (sham) females (Cox et al., 2010). OVx females, however, did display increased growth, immune function (PHA-induced swelling), and survival than control lizards (Cox et al., 2010) (Fig. 8.2).

While most studies suggest an energetic struggle between the reproductive and immune systems, it is also likely that this competition shifts throughout an animal's lifetime.

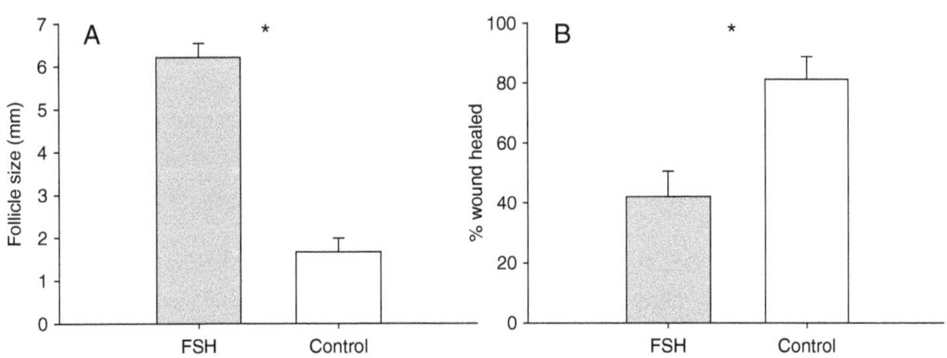

FIGURE 8.2: Trade-offs with reproduction. A. Mean (±SEM) follicle diameter of female ornate tree lizards (*Urosaurus ornatus*) treated either with follicle-stimulating hormone (FSH) or vehicle (control). B. Mean (±SEM) percent wound healed over period of follicular investment in FSH and control treated females. Experimentally elevated reproductive investment via FSH treatment resulted in significantly larger follicles being produced and suppressed wound-healing ability relative to control females.

TERMINAL INVESTMENT

An important prediction of life-history evolution suggests that as animals near reproductive senescence they are likely to invest more in their current reproductive attempts and presumably less in competing physiological functions such as immunity. (Williams, 1966; Pianka and Parker, 1975). When the value of current reproduction is high relative to the value of future reproduction and survival, the stress response is expected to be suppressed to ensure that critical resources are not diverted away from reproduction (Wingfield et al., 1995; Silverin et al., 1997; Meddle et al., 2003).

Here we provide two examples in support of this idea. In the long-lived blue-footed booby (*Sula nebouxii*), challenge of reproductively mature birds with LPS decreases reproductive output (i.e., number of fledglings) (Velando et al., 2006), a result that would be predicted if a trade-off between these two traits exists. Interestingly, closer examination of older males that normally fledge fewer young reveals increased reproductive output within this subgroup after LPS challenge. These data suggest that immune status in older males may signal high value of the current reproductive investment relative to future reproduction (Velando et al., 2006). An additional example was conducted on house sparrows that were injected with a *Nobivac-Paramyox,* a vaccine against avian paramyxovirus, to induce a humoral immune response; these individuals were more likely to lay a replacement clutch (following experimentally induced "failure" of the original clutch), suggesting greater "terminal investment" (Bonneaud et al., 2004).

MIGRATION

In addition to substantial energetic investment in reproduction, many birds experience another life-history stage, migration, that requires significant amounts of energy. If a trade-off among essential resources exists, then it would be predicted that, during the migratory phase, individuals would direct energy away from investment in immune function and towards physiological processes necessary for long-distance flight. The body mass gained by thrushes during a migratory stopover is positively correlated with the robustness of a response to a PHA injection (Owen and Moore, 2008). This suggests that energetic resources are a critical component governing a trade-off during this life-history stage. Further, in laboratory studies Swainson's thrushes (*Catharus ustulatus*) show reduced response to PHA when induced to display migratory restlessness, and in redwing thrushes (*Turdus iliacus*) the amount of the bacteria *Borrelia burgdorferi,* the causative agent of Lyme disease, is significantly greater in birds that were experimentally induced (using photoperiod) to exhibit migratory restlessness, compared with birds that did not display this behavior (Gylfe et al., 2000).

The long-distance migrant the red knot (*Calidris canutus*) does not display immunosuppression when exposed to long flight in wind tunnels (Hasselquist et al., 2007), suggesting that the reduction in immune function is likely anticipatory to migration and is not the direct result of increased energy consumption used for long-distance flight. Further supporting this hypothesis is the observation that an important immune tissue, the thymus, regresses prior to the onset of autumn migration in mallards, house sparrows, and robins (reviewed in Nelson and Demas, 1996). These data also suggest a downregulation in the amount of energy invested in immune function during periods of migration.

When considering immune function in migratory birds, they should seemingly invest more energy and resources into immune defenses to enable them to defend against the diversity of pathogens encountered on their breeding and wintering grounds. In an analysis of several species of migratory birds, the immune organs bursa of Fabricius and spleen are larger in migratory species compared with closely related non-migratory species (Møller and Erritzøe, 1998), suggesting that natural selection not only may act to shape differences in immune function within individuals across the year, but may also favor differential baseline investment in immune organs or function, depending on the inclusion of migration or not. Investigations of partial migrant species such as European blackbirds, where individuals within a population display either a migratory or non-migratory phenotype, may provide an ideal model to systematically probe the effects of migratory disposition on immune function without the confounds that arise in comparisons of several different species.

SURVIVAL

Survival to the next breeding season is a critical component governing individual lifetime reproductive success. Thus, selection not only may act on reproductive traits (as described above), but is also likely to be influenced by the need to survive. Indeed, immune responses sometimes may incur costs in terms of annual survival (as was suggested by the blue-footed booby discussion earlier); thus, optimizing the magnitude of the immune response is likely important. In common eiders (*Somateria mollissima*), immune responsiveness influences annual return rate; eiders that produced a strong antibody response to SRBCs had only a 27% return rate, compared with a 72% return rate for females that did not generate an antibody response. Response to multiple challenges further reduced return rate; birds responding to both SRBC and diphtheria toxin injections had significantly lower return rates than those responding to one or none of the challenges (Hanssen et al., 2004). Similarly, in female side-blotched lizards (*Uta stansburiana*), elevated antibody production is related to decreased survival and fitness metrics (Svensson et al., 2001). In blue tits, responsiveness to an immune challenge during winter resulted in stabilizing selection, supporting the hypothesis that an immune response may incur a fitness cost (Raberg et al., 2003) However, in other studies, no effect of immune response on survival was reported (Raberg et al., 2000; Cichon et al., 2001). This may suggest that selective forces are capable of optimizing immune responsiveness to enable survival to future breeding seasons, depending on an organism's pace of life.

MATERNAL TRADE-OFFS

Energetic trade-offs among immunity and other physiological and behavioral response are not limited to a single individual or generation within a given species. Production of offspring is energetically costly and can limit the allocation of resources to other simultaneous physiological functions, leading to trade-offs between the mother and her offspring. Metabolic expenditure is significantly elevated during pregnancy and lactation in mammals (Mauget et al., 1997) and vitellogenesis in birds (Chappell et al., 1995; Nilsson and Raberg, 2001; Martin et al., 2003) and reptiles (Angilletta and Sears, 2000). Thus, the reproductive period is a time of an extreme energetic bottleneck; in conjunction with changes in metabolism (Mauget et al., 1997), there are also changes in behavior, body fat, and specific hormone concentrations (Day et al., 2002), as well as immunity.

Specific immune components are suppressed during the reproductive stages, such as cell-mediated immunity in a variety of mammalian species (Weinberg, 1984). At the same time, females must also maintain their antibody production in sufficiently high quantities to ensure adequate transfer to the offspring. At present, it is unclear what physiological mechanisms mediate these energetic decisions. Mothers have the ability to adjust offspring immunity through the transfer of immune factors (e.g., antibodies [immunoglobulin G or Y], antioxidants, lysozyme) to offspring, which can affect the development of the immune system and ultimately survival. The decision of how much of these materials to invest in offspring can have profound effects on the offspring (Price, 1998). Such maternal effects are a source of phenotypic plasticity that potentially has great evolutionary and ecological significance. Passive immunity has been characterized in mammals (Ehrlich, 1892), birds (Klemperer, 1893; Grindstaff et al., 2003), reptiles (Schumacher et al., 1999), and fish (Bly et al., 1986), as well as several invertebrate species (Moret, 2006; Sadd and Schmid-Hempel, 2009). Given the broad range of taxa that display passive immunity, it stands to reason that it is an adaptive maternal effect; however, current understanding of the costs and the fitness implications of passive immunity is unclear.

Maternal provisioning of these factors is not without costs, which may limit the amount allocated to offspring (Addison et al., 2009). The direct cost of the transfer of immunoglobulin proteins is presumably low (Grindstaff et al., 2003; Addison et al., 2009); however, collared flycatcher eggs each contain 10% to 20% of the female's steady-state antibody levels (Hargitai et al., 2006); thus, maintaining maternal levels may be energetically demanding. Little is known about the physiological mechanisms that mediate the differential allocation of energy between mother and offspring. Recent evidence, described in more detail below, however, has implicated the adipose tissue hormone leptin as a biochemical signal communicating with the immune system (French et al., 2009a) (Fig. 8.3).

Immune systems in neonatal vertebrates are immature and naïve to pathogens (Brambell, 1958; Grindstaff et al., 2003), and fighting parasite and pathogen infections is costly (Klasing et al., 1987). Maternally derived antibodies provide humoral immune defense that represents the lifetime pathogen exposure of the mother (Grindstaff et al., 2003). Mothers and offspring are likely to be exposed to similar pathogen environments; thus, the specific antibodies from the mother will also benefit the offspring. Additionally, elevated specific antibodies from recent or recurring diseases may accurately reflect the risk of infection to offspring on a temporal or spatial scale (Hasselquist and Nilsson, 2009). For example, in areas where encephalitis viruses are present, infected females of various avian species provide antibodies to offspring, inducing resistance to the infection (Kissling et al., 1954; Reeves et al., 1954; Sooter et al., 1954). Likewise, in breeding areas where more ticks carrying *Borrelia burgdorferi* are found, the prevalence of *Borrelia*-specific antibodies in kittiwake eggs is higher (Gasparini et al., 2001).

Progeny that have experience with a particular pathogen through passive immunity have better survival rates and growth rates than naïve offspring following challenge with that pathogen. For example, Japanese quail (*Coturnix japonica*) possessing specific maternal antibodies to the

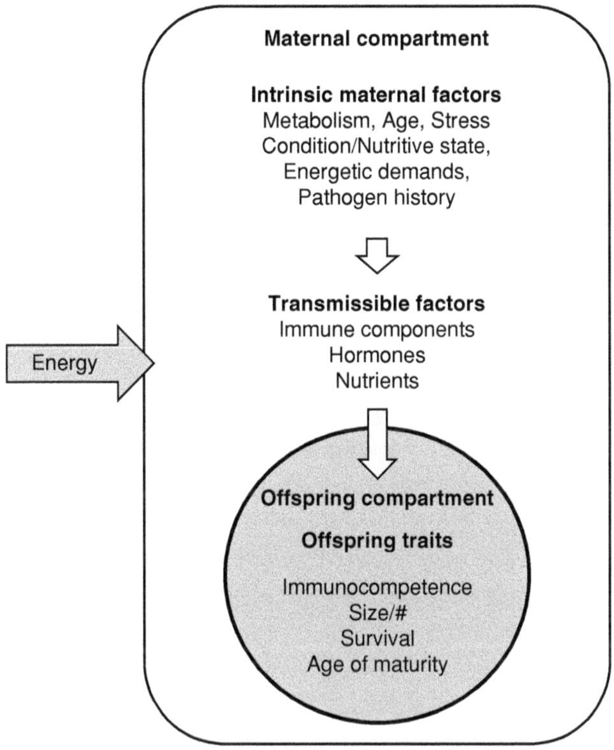

FIGURE 8.3: Maternal effects on immunity. Transfer of maternal factors that influence offspring immune system development and function depends on energy availability and other intrinsic states. For example, food intake, metabolic demands, and/or stressors can affect circulating hormone levels, including glucocorticoids, estrogens, or androgens. These hormones may pass to the embryos through the placenta or into the eggs, where they can alter immune development. Similarly, a mother's experience with the local disease environment may affect the levels of specific antibodies, complement proteins, and antioxidants that are transferred to the offspring. Passive transfer of immunity from mother to offspring can be protective while the neonatal immune system is immature. Among these factors, carotenoids have been implicated in altering the developmental trajectory of the offspring, effectively shortening the developmental time and lessening the vulnerability that comes with living in a nest.

same antigen they are challenged possessing had ameliorated growth suppression (body mass) relative to quail with elevated maternal antibodies for a different antigen (Grindstaff, 2008). Neonatal mice (*Mus musculus*) exposed to a lethal dose of influenza virus died if they lacked influenza antibodies. Mice with specific antibodies derived from mothers not only survived but also showed reduced virus titers and greater survival upon secondary influenza challenge in adulthood (Reuman et al., 1983). It appears that passive immunity reduces activation of the neonate immune system, which frees resources toward growth or maintenance.

Maternal diet can also influence the size and quantity of offspring. In birds, mothers deposit various substances into their eggs, including antioxidants such as carotenoids and vitamins A and E. Antioxidants prevent and restrict the formation of free radicals from metabolic processes and immune defense, and protect DNA, lipids, proteins, and carbohydrates from the damaging effects of free radicals (Royle et al., 2003; Tanvez et al., 2009). Animals are unable to synthesize carotenoids *de novo*, and increasing evidence supports the idea that carotenoids are a limiting resource due to their scarcity (Grether et al., 2001). Additionally, they are involved in the development and activation of the immune system (De Neve et al., 2008). Thus, allocation of antioxidants represents a trade-off between offspring and maternal self-maintenance (Blount et al., 2002). Zebra finches (*Taeniopygia guttata*) supplemented with a mixture of antioxidants transferred over 70% of the daily intake of the most potent compounds into egg yolk, with deposition declining with laying sequence (Royle et al., 2003). Interestingly, other studies have found the pattern of deposition across laying order to be reversed, with an increase in later eggs. This suggests that different allocation strategies exist in different species to maximize fitness.

In summary, much of the current work has determined that the maternal history of immune challenge is a determining factor for immune development through transfer of immune factors to offspring. Emerging evidence suggests that the regulation and means of passage for each of these factors may be different. More studies are needed to determine what dictates the amount of each substance that passes to offspring, the costs of transfer (both metabolic and the trade-offs with future reproduction), and the benefits received by mother and offspring. With this knowledge, it is possible to understand the potential adaptive value of passive immunity and how these maternal effects are shaped by selection.

PHYSIOLOGICAL MECHANISMS UNDERLYING ENERGETIC TRADE-OFFS

As discussed in the introduction of this chapter, it is important to understand the physiological mechanisms underlying energetic trade-offs with immunity. Recent research has implicated a number of neuroendocrine candidates that likely play an important role in regulating energetic trade-offs between immune function and other physiological and behavioral responses. While we will not attempt to cover all of these factors, some of which are described in other chapters in this volume (see Chapters 9 and 10), several of the more promising endocrine candidates are reviewed below.

LEPTIN AND INSULIN

An increasing number of studies, several of which are discussed above, suggest that body fat plays an important role in regulating immune function in many vertebrate species (Demas, 2004). However, the precise neuroendocrine mechanisms by which energy availability is translated into a physiological signal indicating current energy balance are not fully understood. In the past few years alone, however, several endocrine factors have been identified as potential candidates for providing biochemical signals of current energy availability (Woods et al., 2000). One obvious endocrine candidate linking available energy stores with immune function is leptin. Leptin, a peptide hormone first identified in 1994 (Zhang et al., 1994), is synthesized and secreted almost exclusively by adipose tissue. Moreover, circulating concentrations of leptin are directly proportional to adipose tissue mass such that high levels of the hormone indicate adequate energy stores, whereas low circulating levels of leptin are consistent with an energy deficit.

Initial studies on leptin suggested that the primary function of this hormone was that of a satiety factor, as treatment of mice with physiological levels of leptin triggered marked reductions in food intake and body fat (Zhang et al., 1994). Interestingly, decreases in body fat were still evident, even when food intake was kept constant, suggesting that leptin also exerted a direct effect on energy metabolism, independent of food intake (Scarpace et al., 2000; Elmquist, 2001; Rayner and Trayhurn, 2001). Since these initial findings, it has become increasingly clear that leptin is a pleiotropic molecule involved in a wide range of physiological functions, including reproduction, energy balance, and immune function (Clarke and Henry, 1999; Ahima and Flier, 2000b).

Many diverse actions within the immune system are influenced by leptin. For example, specific immune responses are disrupted in mice with impaired leptin signaling due to genetic defects (e.g., ob/ob mice, db/db mice). Ob/ob mice that are unable to synthesize leptin experience atrophy of specific lymphoid tissues (e.g., spleen, thymus), accompanied by decreases in the number of circulating lymphocytes and increases in the number of circulating monocytes (Lord et al., 1998). Leptin deficiency also results in reduced sensitivity to T-cell–activating stimuli and the cytotoxic response of splenocytes. Similarly, db/db mice that are unresponsive to leptin due to a mutation of the leptin receptor display similar immunological deficits. Db/db mice have reduced splenic and thymic masses and have a reduced capacity to reject skin grafts or to generate cytotoxic cells (Fernandes et al., 1978). More recently, a series of important studies using both *in vitro* and *in vivo* tests have provided strong support for a role of leptin in mediating immune responses (Lord et al., 1998). Leptin appears to differentially regulate the proliferation of naïve and memory T cells. Specifically, exogenous leptin added to T-cell cultures enhances proliferation in response to allogenic stimulator cells in both naïve and memory T-cell types, but the effect of leptin is significantly more pronounced in naïve compared with memory T cells (Lord et al., 1998).

Leptin also appears to mediate seasonal changes in immune function. For example, Siberian hamsters housed in short winter-like days display decreases in total body fat as well as reduced humoral immunity, and the degree of immune suppression is correlated with circulating leptin concentrations (Drazen et al., 2000). In addition, treatment of short-day hamsters with exogenous leptin to mimic long summer-like day levels of the hormone attenuates the short-day impairment in immunity (Drazen et al., 2001). Exogenous leptin has no effect on humoral immunity in long-day hamsters, however, suggesting the effect of leptin on humoral immunity is photoperiod-dependent (Drazen et al., 2001). Furthermore, exogenous leptin may attenuate metabolic stress-induced decreases in humoral immunity in Siberian hamsters (Drazen et al., 2001). Specifically, animals treated with the metabolic inhibitor 2-DG display reduced humoral immunity; exogenous leptin, however, attenuates 2-DG–induced immunosuppression. These results suggest that leptin acts as a neuroendocrine signal, communicating current energy availability, whether in the form of readily utilizable energy (i.e., glucose), or energy stores (i.e., WAT depots), to the immune system. Recent data from our laboratory have provided support for this hypothesis. Specifically, surgical removal of IWAT suppressed serum IgG concentrations in response to the antigen KLH in Siberian hamsters, as previously reported. Treatment of LIPx animals with exogenous leptin via osmotic minipump, however, attenuates LIPx-induced decreases in humoral immunity (Demas and Sakaria, 2005). As in previous studies, exogenous leptin does not alter humoral immunity in control animals.

Collectively, these studies provide compelling support for the hypothesis that leptin provides an endocrine signal from body fat to the immune system indicating current energy reserves. But how does leptin communicate with the immune system? As with most hormones, leptin can act directly by binding to receptors on lymphoid cells (e.g., circulating lymphocytes) to affect immune responses. Alternatively, leptin may also act indirectly, through the nervous or endocrine systems, to influence immunity. Unlike a wide variety of hormones (e.g., androgens, estrogens, glucocorticoids) in which specific receptors have been identified directly on peripheral immune cells (Grossman, 1984, 1985), high-affinity leptin receptors have yet to be identified on peripheral lymphoid tissues. Although it is plausible that such receptors exist and simply have not been identified to date, the putative lack of leptin receptors on immune tissues suggests that the effects of leptin on immune function may be indirect.

Do non-mammalian vertebrates have leptin? Although a gene sequence homologous to mouse leptin has been reported in avian species (Taouis et al., 1998), the likelihood that this gene represents an avian leptin gene is still debated (Sharp et al., 2008; Simon et al., 2009). Similarly in reptiles, although putative leptin has been identified in only a few lizard species, the physiological effects (e.g., temperature regulation, food intake) are robust, rendering it a likely mediator of energetic trade-offs (Niewiarowski et al., 2000; Sciarrillo et al., 2005; Putti et al., 2009). Interestingly, a report details that turkeys (Nicholas strain) provided with the protein generated by the leptin-like chicken gene had enhanced T-cell proliferation in response to concavalin A (Lohmus et al., 2004), and zebra finches receiving recombinant murine leptin had an enhanced wing-web swelling in response to a PHA injection (Alonso-Àlvarez et al., 2007). Leptin delays testicular regression in Italian wall lizards (*Podarcis sicula*), thus regulating reproduction (Putti et al., 2009). French and colleagues (2011) further showed that leptin is partially responsible for energetic trade-offs in tree lizards (*Urosaurus ornatus*), whereby leptin treatment reinstates immune function to normal levels in immunosuppressed reproductive females. These results would be predicted if the effects of leptin are similar in birds, reptiles, and mammals, and they suggest that even if birds and reptiles do not generate the leptin protein, they do have a functional receptor able to respond to a leptin-like ligand.

Leptin also appears to serve as an important biochemical signal mediating energetic trade-offs with immune function across generations (see Chapter 6). For example, pregnant Siberian hamsters implanted with minipumps containing exogenous leptin during late gestation gave birth to significantly larger litters and completely inhibited the post-gestational infanticide commonly displayed by rodents (French et al., 2009a). Interestingly, however, bacterial killing capacity, a measure of innate immunity, was impaired in leptin-treated relative to control pregnant hamsters. These findings suggest that female hamsters rely on leptin as a signal to modulate the degree of maternal investment into offspring. There is a limit to the signal, however, as increased energy investment in pups impaired maternal immunity despite elevated levels of leptin (French et al., 2009a). A similar finding has recently been reported in birds (Lõhmus and Björklund, 2009). Specifically, great tits (*Parus major*) were given injections of exogenous leptin or control injections following their first clutch. Leptin-treated birds were significantly more likely to lay a second clutch than control animals, suggesting that leptin or a leptin-like molecule is mediating energetic trade-offs in birds (Lõhmus and Björklund, 2009). Immune function was not measured in this study but presumably would be influenced by such trade-offs in this species.

More recently, another peripheral metabolic hormone, insulin, has been shown to act as a signal coordinating energetic trade-offs with immunity (Garcia et al., 2010). Insulin is a pancreatic hormone that plays a central role in regulating carbohydrate and fat metabolism in the body. Short-day housed hamsters, which have previously been shown to exhibit decreased antibody production, receiving exogenous treatment with a long-lasting form of insulin, protamine zinc insulin, showed increased antibody production compared with short-day controls, and production was stimulated to levels comparable to long-day animals (Garcia et al., 2010). These results suggest that insulin, similar to leptin, may serve as an important signaling molecule in the allocation of current energy reserves to a range of physiological and behavioral processes, including immune function, at least in mammals. No studies, to our knowledge, have investigated a role for insulin in immune function or disease in non-mammalian vertebrates; this is an area that should receive experimental attention.

GLUCOCORTICOIDS

Glucocorticoids (e.g., corticosterone, cortisol), in contrast with leptin and insulin discussed above, are important energy-mobilizing hormones that have definite effects on the immune system and have received much more experimental attention. The earliest endocrine-immune studies in any species linked seasonal changes in immune responses with glucocorticoids. For example, the first endocrine-immune studies in reptiles linked seasonal changes in immune responses in ocellated skinks (*Chalcides ocellatus*) and Caspian pond turtles (*Mauremys caspica*) with concurrent changes in circulating glucocorticoid levels, energy availability, and reproductive state (Saad and Elridi, 1984; Leceta and Zapata, 1986; Saad, 1988).

Glucocorticoids seem a likely mediator of energetic trade-offs; however, the majority of research in this area on nontraditional models has focused on stress-related associations between glucocorticoids and the immune system. Experimental studies involving treatment with adrenal steroids directly test the effects of specific hormones on the immune system. Hydrocortisone treatment altered lymphoid tissue masses in the ocellated skink (*Chalcides ocellatus*) (Saad et al., 1984b, 1984a). Similarly, corticosterone treatment decreased a myriad of circulating immune cells in juvenile alligators (*Alligator mississippiensis*) (Morici et al., 1997).

Similar results are present in studies examining immunity at the response level. For example, male chickens provided dietary corticosterone displayed a decreased antibody response compared with controls to a challenge of SRBCs (Van Niekerk et al., 1989). More recently, it was reported that chickens provided with corticosterone via their drinking water had an increase in the amount of West Nile virus in circulation and an increase in the amount of virus shed compared with controls. Interestingly, a separate set of birds exposed to corticosterone showed a reduced antibody response to SRBC but an increased antibody response to West Nile virus (Jankowski et al., 2010). Cortisol and aldosterone treatment inhibits lymphocyte proliferation in the South African clawed frog (*Xenopus laevis*) (Rollins-Smith and Blair, 1993), and the effects are reversed by treatment with the glucocorticoid receptor antagonist RU486 and the mineralocorticoid receptor antagonist RU26752 (Rollins-Smith et al., 1997), suggesting a direct effect of glucocorticoids on immune activation (see Chapter 5). In leopard frogs (*Rana pipiens*), the immunosuppressive effects of hydrocortisone cause a normally sub-lethal infection with *Mycobacterium marinum* to become acute and lethal (Ramakrishnan et al., 1997). Lastly, in mallard ducks, the glucocorticoid dexamethasone suppressed antibody response to SRBCs but increased the activity of natural killer cells, an important component of the innate immune system (Fowles et al., 1993), suggesting differential investment in different arms of the immune system.

In house sparrows, corticosterone implants significantly decrease responses to PHA in a temperate population during the non-breeding but not the breeding season; this corresponds to the effectiveness of the implants elevating plasma corticosterone, where corticosterone was elevated during the non-breeding season compared with controls, but no difference between controls and experimental animals were observed in the breeding season (Martin et al., 2005). Corticosterone responses to stress in marine iguanas (*Amblyrhynchus cristatus*) influence multiple immunological responses, including bacterial killing ability, cutaneous wound healing, and hemolytic complement activity, differently in breeding versus non-breeding marine iguanas, suggesting that corticosterone modulation of immunity is associated with energetic investment into reproduction (French et al., 2010). In a tropical population of sparrows, corticosterone implants significantly elevated plasma levels during the non-breeding season, but there was no effect of corticosterone treatment on response to PHA in either the breeding or non-breeding season (Martin et al., 2005).

Functional studies altering hypothalamo-pituitary-adrenal (HPA) activity via external stressors (i.e., restraint, predator stress) also demonstrate that elevations in HPA activity and downstream corticosterone levels can suppress immune parameters. A recent investigation of the effects of stress and glucocorticoids on immune function examined European starlings exposed to chronic psychological stress (which leads to a decrease in total baseline corticosterone). The birds receiving the chronic stress treatment displayed a similar response to PHA compared with unstressed controls (Cyr et al., 2007). Restraint stress increases circulating corticosterone and decreases wound healing in tree lizards (*Urosuarus ornatus*) (French et al., 2006). Although glucocorticoids are definitively involved in both energy and immune modulation, they are not inevitably immunosuppressive.

As in mammalian systems, the interactions between the HPA axis and the immune system are complex. Specifically, interactions are context-dependent: for instance, in tree lizards corticosterone treatment suppresses wound healing only when an individual is energetically compromised, such as during periods of heavy reproductive investment (French et al., 2007b). In certain cases the immunosuppressive effects of glucocorticoids may even be adaptive. For example, increases in corticosteroids during metamorphosis in many amphibian species helps to decrease lymphocyte numbers, presumably to help reorganize the immune system and prevent autoimmune responses against newly developing tissues (Rollins-Smith et al., 1997; Rollins-Smith, 1998). Glucocorticoids are clearly involved in both energy and immune modulation, and therefore are likely regulators of trade-offs with the immune system as well (Fig. 8.4).

GONADAL STEROIDS

We have already shown how investment of energetic resources in alternative physiological processes, especially reproduction, can limit or suppress immunity, providing indirect evidence for the high energetic demands of the immune system. Gonadal steroids, key regulators of

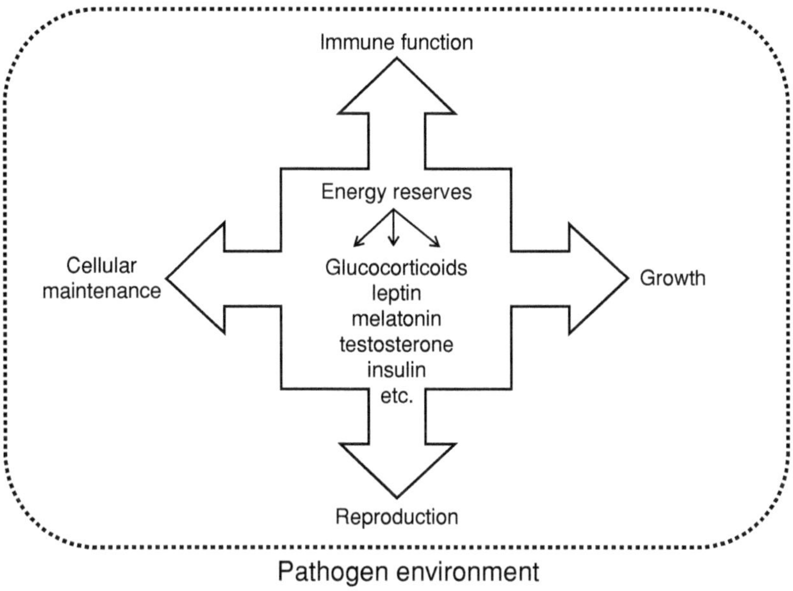

FIGURE 8.4: Physiological trade-offs. Most animals have a limited amount of energy available to allocate to a number of different physiological processes (outside of box). Differential investment of relatively finite resources to one process presumably limits investment in the other, competing processes. Such trade-offs are mediated, in part, by a variety of endocrine signals that interact with peripheral immune tissues (inside of box). The organism, in turn, is nested within a larger environmental context that includes environmentally relevant parasites and pathogens.

reproduction across vertebrates, have therefore been suggested to play a role in mediating trade-offs between reproduction and immune function (Folstad and Karter, 1992). For example, several studies of seasonally breeding animals suggest that decreases in immune tissue occur at the onset of gonadal recrudescence and beginning of breeding in both field and laboratory studies and in a variety of taxa (reviewed in Chapter 9).

To experimentally probe for a relationship between testosterone and immune function, numerous studies have experimentally elevated testosterone (T) levels using long-release implants and measured humoral or cell-mediated immune responses, with most studies finding that experimental T elevation leads to decreased immune function (e.g., Duffy et al., 2000; Peters, 2000; Casto et al., 2001). For example, elevating T in male sand lizards (*Lacerta agilis*) significantly increases parasite load (Olsson et al., 2000). In the turtle *Mauremys caspica,* injections of T induce lymphoid mass involution and lymphopenia (Saad et al., 1991). T implants suppressed lymphocyte numbers and increased parasite load in male lizards (*Psammodromus algirus*) (Veiga et al., 1998). During sexual maturation in fish, when gonadal steroids are elevated in both males and females (Maule et al., 1996; Harris and Bird 2000), many individuals experience immune deficiencies (Slater et al., 1995; Slater and Schreck, 1997) and increased parasitic infection (Pickering and Christie, 1980). Specifically, natural elevations in T that occur during sexual maturation directly suppress leukocytes and antibody-producing cells in Chinook salmon (*Oncorhynchus tshawytscha*) (Slater et al., 1995; Slater and Schreck, 1997). Further, receptors for reproductive steroids have been found on fish leukocytes, suggesting a direct link between the reproductive system and modulation of the immune system (Slater et al., 1995; Patino and Maule, 1997). Even pre-hatching exposure to T in lizards can increase offspring parasite burden (Uller and Olsson, 2003).

However, the relationship between endogenous levels of T and immunosuppression is conflicting (Saino et al., 1995; Hasselquist et al., 1999; Greenman et al., 2005). A recent meta-analysis provided limited support for the immunosuppressive effect of T in some species but concluded that the question is still open (Roberts et al., 2004). Few studies have investigated the relationship between natural variation in T between individuals and their constitutive immune defenses (e.g., complement activity). An investigation in dark-eyed juncos (*Junco hyemalis*) during the breeding season found a negative relationship between serum T levels and two measures of innate immunity, complement activity and levels of IgG (Greives et al., 2006). Seasonal changes in circulating T levels are inversely related to the seasonal immune profiles of the oscellated skink (*Chalcides ocellatus*) (Saad et al., 1990). A negative relationship between circulating T levels and immunoglobulin levels was also reported for breeding male king penguins (*Aptenodytes patagonicus*) (Bourgeon et al., 2007). Meanwhile, an investigation of the common yellowthroat (*Geothlypis trichas*) found no relationship between natural levels of T and wing-web swelling response to PHA, nor to levels of circulating IgG (Garvin et al., 2008).

Studies investigating the effects of T on females are limited and also provide conflicting evidence. In female dark-eyed juncos, T implants decreased wing-web swelling response to PHA (Zysling et al., 2006), whereas a recent study in female zebra finches found a slight enhancement in wing-web swelling in females with T implants (McGraw and Ardia, 2007). It is important to note that studies demonstrate that steroid hormones other than T, including estradiol, DHT, and corticosterone, also have significant effects on the immune system (Al-Afaleq and Homeida, 1998; Mondal and Rai, 1999; Owen-Ashley et al., 2004). For example, both estradiol and DHT suppressed macrophage phagocytic activity in the wall lizard (*Hemidactylus flaviviridis*) (Mondal and Rai, 1999).

The relationship is further complicated by the finding that manipulation of T by providing long-duration-release implants also often leads to an increase in circulating glucocorticoids (Ketterson et al., 1991; Schoech et al., 1999; Casto et al., 2001). Thus, it is unclear if the effects observed are due to a direct immunosuppressive effect of T, or if these results are responding to the T-induced increase in glucocorticoids. A study investigating the effects of T implants on house sparrow immunity found that birds with T implants had decreased antibody production to SRBC compared with castrated controls (Evans et al., 2000). Interestingly, when statistically controlling for the implant-induced upregulation of corticosterone, the authors found that T had a slight immunoenhancing effect (Evans et al., 2000). As shown, many of these hormones change concurrently, making direct links between them and their immune effects difficult. These same hormones also vary with important life-history processes and according to season, further complicating their effects.

MELATONIN

Recently, a seasonally variable hormone, melatonin, has received a lot of attention (see Chapter 9). As many temperate species alter their immune responses throughout the year depending on life-history stage (reviewed above), it is possible that melatonin, which varies in the amount and duration released during the night depending on the season, may act as a physiological mechanism driving downstream changes in immunity. In castrated short-day photostimulated starlings, melatonin was found to "rescue" the response of spleen tissue to a T-cell mitogen, Con A, whereas photostimulated castrated starlings receiving empty capsules had a significantly reduced response to Con A compared with melatonin-implanted birds, and photorefractory birds (Bentley et al., 1998). Further, administration of melatonin to quail housed in constant light, which abolished (or greatly reduced) endogenous melatonin release, induced more robust humoral and cellular immune responses compared to birds in constant light or long days receiving no melatonin (Moore and Siopes, 2000). However, a paper recently reported that in red knots, changes in principal component indices of immune function do not match the seasonal melatonin profile of birds from a separate group of red knots kept under identical housing conditions (Buehler et al., 2009). Future studies investigating natural variation in immune function and melatonin levels within the same individuals throughout the year are needed.

As mentioned above, migratory birds tend to show reduced immune function compared with birds not in migratory condition. Interestingly, migratory birds that fly at night also show reduced peak nocturnal melatonin levels (Fusani and Gwinner 2005), presumably to enable nighttime migration. It is unclear, however, whether melatonin acts as a physiological mechanism inducing a reduction in immune function, or whether exogenous melatonin would alleviate migratory-induced immunosuppression. A recent meta-analysis investigated whether parasite infection affects mortality in a wide variety of species (e.g., birds, fish, amphibians) and across latitudes. This analysis revealed a significant increase in mortality in hosts infected with parasites. It also suggests that organisms residing nearer the Equator compared with temperate organisms have greater odds of parasite-associated mortality, likely due to the density and diversity of parasites (Robar et al., 2010). There are numerous additional examples linking changes in melatonin with a variety of immune responses. We will not attempt to review them here, as they receive extensive coverage in Chapters 9 and 10 of this volume; readers wishing to learn more are encouraged to read these chapters.

CONCLUSIONS

The goal of this chapter was to review the relevant literature in support of the idea that immune function, like virtually all other physiological and behavioral processes, requires energy and as such has the potential to be modified by changes in energy availability. If energy were a limitless resource, then animals would be capable of investing more or less energy in any given biological function without direct consequences for other energetically expensive processes. Unfortunately, however, in most cases energy is not limitless but instead varies across a wide range of geographic, spatial, and temporal conditions and life-history stages. For example, energy input may be limited by decreases in available food and specific nutrients, resulting in fewer resources to be used by an individual. Alternately, even if energy input is unaffected, then energy output may increase in the form of an elevated need for thermogenesis, reproduction, or foraging. Either scenario, or even a combination of them, can result in an energetic bottleneck by which animals must reallocate energy into temporally critical responses and perhaps limit energy to less immediately relevant responses. Animals simply cannot survive when energy output is greater than energy input. Although we hope to have convinced the reader that mounting an immune response has measurable cost, there is no doubt that important questions remain. Does a decrease in immunity during times of high energy output really represent an adaptive reallocation of energy, or is it simply the byproduct of less energy? As the studies reviewed in this chapter suggest, this question has yet to be resolved, with some studies suggesting seemingly adaptive trade-offs between physiological responses (e.g., reproduction) and immune function. Other studies, however, have demonstrated that despite significant increases in energy following an immune response, there was little or no evidence to suggest that these responses were offset by decreases in other responses (e.g., growth), raising doubt as to the adaptive nature of such changes. What is clear, however, is that this area is still in its infancy, and considerably more research is necessary to address this and other questions.

The second goal of this chapter was to discuss several possible physiological mechanisms that regulate environmentally induced changes in immune function. While our review of these factors was by no means exhaustive, we discussed several of the more promising candidates for which we have reasonable data. Some of these factors (glucocorticoids, sex steroids) (see Chapters 2, 3, and 9) have a relatively large literature across a wide variety of species and taxa; others (e.g., leptin) have garnered more recent, and thus limited, attention (French et al., 2011). Here again, more research will be needed to address at least two (and likely many more) questions. First, what function(s) do relatively understudied metabolic hormones (e.g., leptin, insulin, and ghrelin) serve in coordinating potential trade-offs with immune function, and are these mechanisms universal across vertebrate taxa and environmental contexts? Second, in common with research in basic and clinical metabolism, a key goal will be to determine which among the many and growing list of factors play important or even core roles in regulating immune responses and which serve as secondary players. As we suggested in the beginning of this chapter, the most significant progress will be made in energetics of immunity, and in the field ecoimmunology more broadly, when both ultimate and proximate perspectives are combined. This approach will contribute to an integrative and comprehensive understanding of how the immune system is regulated by environmental factors and in turn what the adaptive significance of such changes is with respect to disease resistance and susceptibility.

REFERENCES

Addison, B., Klasing, K., Robinson, W., Austin, S. and Ricklefs, R. 2009. Ecological and life-history factors influencing the evolution of maternal antibody allocation: A phylogenetic comparison. *Proceedings of the Royal Society B: Biological Sciences* 276, 3979.

Adelman, J. S., Cordoba-Cordoba, S., Spoelstra, K., Wikelski, M. and Hau, M. 2010. Radiotelemetry reveals variation in fever and sickness behaviours with latitude in a free-living passerine. *Functional Ecology* 24, 813–823.

Ahima, R. S. and Flier, J. S. 2000a. Adipose tissue as an endocrine organ. *Trends in Endocrinology and Metabolism* 11, 327–332.

Ahima, R. S. and Flier, J. S. 2000b. Leptin. *Annual Review of Physiology* 62, 413–437.

Ahmed, A. M., Baggott, S. L., Maingon, R. and Hurd, H. 2002. The costs of mounting an immune response are reflected in the reproductive fitness of the mosquito *Anopheles gambiae*. *Oikos* 97, 371–377.

Al-Afaleq, A. I. and Homeida, A. M. 1998. Effects of low doses of oestradiol, testosterone and dihydrotestosterone on the immune response of broiler chicks. *Immunopharmacol. Immunotoxicol.* 20, 315–327.

Alonso-Àlvarez, C., Bertrand, S. and Sorci, G. 2007. Energetic reserves, leptin and testosterone: a refinement of the immunocompetence handicap hypothesis. *Biology Letters* 3, 271.

Angilletta, M. J. and Sears, M. W. 2000. The metabolic cost of reproduction in an oviparous lizard. *Functional Ecology* 14, 39–45.

Ardawi, M. S. and Newsholme, E. A. 1985. Metabolism in lymphocytes and its importance in the immune response. *Essays in Biochemistry* 21, 1–44.

Ardia, D. R., Schat, K. A. and Winkler, D. W. 2003. Reproductive effort reduces long-term immune function in breeding tree swallows (*Tachycineta bicolor*). *Proceedings of the Royal Society of London. Series B: Biological Sciences* 270, 1679–1683.

Baeta, R., Faivre, B., Motreuil, S., Gaillard, M. and Moreau, J. 2008. Carotenoid trade-off between parasitic resistance and sexual display: an experimental study in the blackbird (*Turdus merula*). *Proceedings of the Royal Society B: Biological Sciences* 275, 427.

Barr, D. P., Russell, M. D., Cecil, L. and Du Boise, E. F. 1922. Clinical calorimetry XXXII: Temperature regulation after the intravenous injections of protease and typhoid vaccine. *Archives of Internal Medicine* 29, 608–634.

Bartness, T. J. 1996. Photoperiod, sex, gonadal steroids, and housing density affect body fat in hamsters. *Physiology & Behavior* 60, 517–529.

Bentley, G., Demas, G., Nelson, R. and Ball, G. 1998. Melatonin, immunity and cost of reproductive state in male European starlings. *Proceedings of the Royal Society B: Biological Sciences* 265, 1191.

Bertrand, S., Criscuolo, F., Faivre, B. and Sorci, G. 2006. Immune activation increases susceptibility to oxidative tissue damage in zebra finches. *Functional Ecology* 20, 1022–1027.

Bisset, S. A., Morris, C. A., McEwan, J. C. and Vlassoff, A. 2001. Breeding sheep in New Zealand that are less reliant on anthelmintics to maintain health and productivity. *New Zealand Veterinary Journal* 49, 236–246.

Blackwell, A. D., Snodgrass, J. J., Madimenos, F. C. and Sugiyama, L.S. 2010. Life history, immune function, and intestinal helminths: Trade-offs among immunoglobulin E, C-reactive protein, and growth in an Amazonian population. *American Journal of Human Biology* 22, 836-48

Blount, J., Surai, P., Nager, R., Houston, D., M¨ller, A., Trewby, M. and Kennedy, M. 2002. Carotenoids and egg quality in the lesser black-backed gull *Larus fuscus*: a supplemental feeding study of maternal effects. *Proceedings of the Royal Society of London. Series B: Biological Sciences* 269, 29.

Bly, J., Grimm, A. and Morris, I. 1986. Transfer of passive immunity from mother to young in a teleost fish: haemagglutinating activity in the serum and eggs of plaice, *Pleuronectes platessa* L. *Comparative Biochemistry and Physiology Part A: Physiology* 84, 309–313.

Bonneaud, C., Mazuc, J., Chastel, O., Westerdahl, H. and Sorci, G. 2004. Terminal investment induced by immune challenge and fitness traits associated with major histocompatibility complex in the house sparrow. *Evolution* 58, 2823–2830.

Bonneaud, C., Mazuc, J., Gonzalez, G., Haussy, C., Chastel, O., Faivre, B. and Sorci, G. 2003. Assessing the cost of mounting an immune response. *American Naturalist* 161, 367–379.

Bourgeon, S., Viera, V., Raclot, T. and Groscolas, R. 2007. Hormones and immunoglobulin levels in king penguins during moulting and breeding fasts. *Ecoscience* 14, 519.

Brambell, F. W. R. 1958. The passive immunity of the young mammal. *Biological Reviews* 33, 488–531.

Brommer, J. 2004. Immunocompetence and its costs during development: an experimental study in blue tit nestlings. *Proceedings of the Royal Society of London. Series B: Biological Sciences* 271, S110.

Buehler, D., Koolhaas, A., Vanít Hof, T., Schwabl, I., Dekinga, A., Piersma, T. and Tieleman, B. 2009. No evidence for melatonin-linked immunoenhancement over the annual cycle of an avian species. *Journal of Comparative Physiology A: Neuroethology, Sensory, Neural, and Behavioral Physiology* 195, 445–451.

Burness, G., Armstrong, C., Fee, T. and Tilman-Schindel, E. 2010. Is there an energetic-based trade-off between thermoregulation and the acute phase response in zebra finches? *Journal of Experimental Biology* 213, 1386–1394.

Calder, P. C., Yaqoob, P., Harvey, D. J., Watts, A. and Newsholme, E. A. 1994. Incorporation of fatty acids by concanavalin A-stimulated lymphocytes and the effect on fatty acid composition and membrane fluidity. *Biochemical Journal* 300, 509–518.

Cartledge, V. A., Gartrell, B. and Jones, S. M. 2005. Adrenal and white cell count responses to chronic stress in gestating and postpartum females of the viviparous skink *Egernia whitii* (Scincidae). *Comparative Biochemistry and Physiology Part A: Molecular & Integrative Physiology* 141, 100–107.

Casto, J. M., Nolan, V. and Ketterson, E. D. 2001. Steroid hormones and immune function: Experimental studies in wild and captive dark-eyed juncos (*Junco hyemalis*). *American Naturalist* 157, 408–420.

Chandra, R. K. 1996. Nutrition, immunity and infection: From basic knowledge of dietary manipulation of immune responses to practical application of ameliorating suffering and improving survival. *Proceedings of the National Academy of Sciences of the United States of America* 93, 14304–14307.

Chappell, M. A., Zuk, M., Kwan, T. H. and Johnsen, T. S. 1995. Energy cost of an avian vocal display—crowing in red junglefowl. *Animal Behaviour* 49, 255–257.

Chew, B. and Park, J. 2004. Carotenoid action on the immune response. *Journal of Nutrition* 134, 257S.

Chioléro, R., Revelly, J.-P. and Tappy, L. 1997. Energy metabolism in sepsis and injury. *Nutrition* 13, 45–51.

Cichon, M., Dubiec, A. and Chadzinska, M. 2001. The effect of elevated reproductive effort on humoral immune function in collared flycatcher females. *Acta Oecologica—International Journal of Ecology* 22, 71–76.

Clarke, I. J. and Henry, B. A. 1999. Leptin and reproduction. *Reviews of Reproduction* 4, 48–55.

Cooper, A. L., Fitzgeorge, R. B., Baskerville, A., Little, R. A. and Rothwell, N. J. 1989. Bacterial infection (*Legionella pneumophila*) stimulates fever, metabolic rate and brown adipose tissue activity in the Guinea pig. *Life Sciences* 45, 843–847.

Costantini, D. and Møller, A. 2009. Does immune response cause oxidative stress in birds? A meta-analysis. *Comparative Biochemistry and Physiology Part A: Molecular & Integrative Physiology* 153, 339–344.

Cox, R. M. and John-Alder, H. B. 2007. Increased mite parasitism as a cost of testosterone in male striped plateau lizards *Sceloporus virgatus*. *Functional Ecology* 21, 327–334.

Cox, R. M., Parker, E. U., Cheney, D. M., Liebl, A. L., Martin, L. B. and Calsbeek, R. 2010. Experimental evidence for physiological costs underlying the trade-off between reproduction and survival. *Functional Ecology* 24, 1262–1269.

Cyr, N. E., Earle, K., Tam, C. and Romero, L. M. 2007. The effect of chronic psychological stress on corticosterone, plasma metabolites, and immune responsiveness in European starlings. *General and Comparative Endocrinology* 154, 59–66.

Day, D. E., Mintz, E. M. and Bartness, T. J. 2002. Diet choice exaggerates food hoarding, intake and pup survival across reproduction. *Physiology & Behavior* 75, 143–157.

De Neve, L., Fargallo, J., Vergara, P., Lemus, J., Jarén-Galán, M. and Luaces, I. 2008. Effects of maternal carotenoid availability in relation to sex, parasite infection and health status of nestling kestrels (*Falco tinnunculus*). *Journal of Experimental Biology* 211, 1414.

Deen, C. M. and Hutchison, V. H. 2001. Effects of lipopolysaccharide and acclimation temperature on induced behavioral fever in juvenile *Iguana iguana*. *Journal of Thermal Biology* 26, 55–63.

Deerenberg, C., Arpanius, V., Daan, S. and Bos, N. 1997. Reproductive effort decreases antibody responsiveness. *Proceedings of the Royal Society of London Series B Biological Sciences* 264, 1021–1029.

Demas, G. E. 2002. Splenic denervation blocks leptin-induced enhancement of humoral immunity in Siberian hamsters (*Phodopus sungorus*). *Neuroendocrinology* 76, 178–184.

Demas, G. E. 2004. The energetics of immunity: A neuroendocrine link between energy balance and immune function. *Hormones and Behavior* 45, 173–180.

Demas, G. E., Chefer, V., Talan, M. I. and Nelson, R. J. 1997a. Metabolic costs of mounting an antigen-stimulated immune response in adult and aged C57BL/6J mice. *American Journal of Physiology—Regulatory, Integrative and Comparative Physiology* 42, R1631–R1637.

Demas, G. E., DeVries, A. C. and Nelson, R. J. 1997b. Effects of photoperiod and 2-deoxy-D-glucose-induced metabolic stress on immune function in female deer mice. *American Journal of Physiology—Regulatory, Integrative and Comparative Physiology* 272, R1762–1767.

Demas, G. E., Drazen, D. L. and Nelson, R. J. 2003. Reductions in total body fat decrease humoral immunity. *Proceedings of the Royal Society of London Series B Biological Sciences* 270, 905–911.

Demas, G. E. and Nelson, R. J. 2003. Lack of immunological responsiveness to photoperiod in a tropical rodent, *Peromyscus aztecus hylocetes. Journal of Comparative Physiology B Biochemical Systemic and Environmental Physiology* 173, 171–176.

Demas, G. E. and Sakaria, S. 2005. Leptin regulates energetic tradeoffs between body fat and humoural immunity. *Proceedings of the Royal Society of London Series B Biological Sciences* 272, 1845–1850.

Derting, T. and Virk, M. 2005. Positive effects of testosterone and immunochallenge on energy allocation to reproductive organs. *Journal of Comparative Physiology B Biochemical, Systemic, and Environmental Physiology* 175, 543–556.

Derting, T. L. and Compton, S. 2003. Immune response, not immune maintenance, is energetically costly in wild white-footed mice (*Peromyscus leucopus*). *Physiological and Biochemical Zoology* 76, 744–752.

Devevey, G., Niculita-Hirzel, H., Biollaz, F., Yvon, C., Chapuisat, M. and Christe, P. 2008. Developmental, metabolic and immunological costs of flea infestation in the common vole. *Functional Ecology* 22, 1091–1098.

Drazen, D. L., Demas, G. E. and Nelson, R. J. 2001. Leptin effects on immune function and energy balance are photoperiod dependent in Siberian hamsters (*Phodopus sungorus*). *Endocrinology* 142, 2768–2775.

Drazen, D. L., Jasnow, A. M., Nelson, R. J. and Demas, G. E. 2002. Exposure to short days, but not short-term melatonin, enhances humoral immunity of male Syrian hamsters (*Mesocricetus auratus*). *Journal of Pineal Research* 33, 118–124.

Drazen, D. L., Kriegsfeld, L. J., Schneider, J. E. and Nelson, R. J. 2000. Leptin, but not immune function, is linked to reproductive responsiveness to photoperiod. *American Journal of Physiology—Regulatory, Integrative and Comparative Physiology* 278, R1401–R1407.

Duffy, D. L., Bentley, G. E., Drazen, D. L. and Ball, G. F. 2000. Effects of testosterone on cell-mediated and humoral immunity in non-breeding adult European starlings. *Behavioral Ecology* 11, 654–662.

Ehrlich, P. 1892. Ueber imrnunitat durch vererbung und saugung. *Infectkt. Krankh.* 2, 183–203.

Elmquist, J. K. 2001. Hypothalamic pathways underlying the endocrine, autonomic, and behavioral effects of leptin. *Physiology & Behavior* 74, 703–708.

Eraud, C., Duriez, O., Chastel, O. and Faivre, B. 2005. The energetic cost of humoral immunity in the collared dove, *Streptopelia decaocto*: is the magnitude sufficient to force energy-based trade-offs? *Functional Ecology* 19, 110–118.

Erf, G., Bottje, W., Bersi, T., Headrick, M. and Fritts, C. 1998. Effects of dietary vitamin E on the immune system in broilers: altered proportions of CD4 T cells in the thymus and spleen. *Poultry Science* 77, 529–537.

Erickson, K. L., Hubbard, N. E. and Somers, S. D. 1992. Dietary fat and immune function. In *Nutrition and Immunology* (ed. R. K. Chandra), pp. 81–104. St. John's, Newfoundland: ARTS Biomedical.

Evans, M. R., Goldsmith, A. R. and Norris, S. R. A. 2000. The effects of testosterone on antibody production and plumage coloration in male house sparrows (*Passer domesticus*). *Behavioral Ecology and Sociobiology* 47, 156–163.

Faivre, B., Grégoire, A., Préault, M., Cézilly, F. and Sorci, G. 2003. Immune activation rapidly mirrored in a secondary sexual trait. *Science* 300, 103.

Fernandes, G., Handwerger, B. S., Yunis, E. J. and Brown, D. M. 1978. Immune response in the mutant diabetic C57BL/Ks-dt+ mouse: Discrepancies between in vitro and in vivo immunological assays. *Journal of Clinical Investigation* 61, 243–250.

Folstad, I. and Karter, A. 1992. Parasites, bright males, and the immunocompetence handicap. *American Naturalist*, 139, 603–622.

Fowles, J. R., Fairbrother, A., Fix, M., Schiller, S. and Kerkvliet, N. I. 1993. Glucocorticoid effects on natural and humoral immunity in mallards. *Developmental & Comparative Immunology* 17, 165–177.

Freitak, D., Ots, I., Vanatoa, A. and Horak, P. 2003. Immune response is energetically costly in white cabbage butterfly pupae. *Proceedings of the Royal Society of London B* 270, S220–S222.

French, S.S., and Dearing, M.D., Demas, G.E. 2011 Leptin as a physiological mediator of energetic trade-offs in eco-immunology. Integrative and Comparative Biology, In press.

French, S. S., DeNardo, D. F., Greives, T. J., Strand, C. R. and Demas, G. E. 2010. Human disturbance alters endocrine and immune responses in the Galapagos marine iguana (*Amblyrhynchus cristatus*). *Hormones and Behavior* 58, 792–799.

French, S. S., DeNardo, D. F. and Moore, M. C. 2007a. Trade-offs between the reproductive and immune systems: Facultative responses to resources or obligate responses to reproduction? *American Naturalist* 170, 79–89.

French, S. S., Greives, T. J., Zysling, D. A., Chester, E. M. and Demas, G. E. 2009a. Leptin increases maternal investment. *Proceedings of the Royal Society B Biological Sciences* 276, 4003–4011.

French, S. S., Matt, K. S. and Moore, M. C. 2006. The effects of stress on wound healing in male tree lizards (*Urosaurus ornatus*). *General and Comparative Endocrinology* 145, 128–132.

French, S. S., McLemore, R., Vernon, B., Johnston, G. I. H. and Moore, M. C. 2007b. Corticosterone modulation of reproductive and immune systems trade-offs in female tree lizards: long-term corticosterone manipulations via injectable gelling material. *Journal of Experimental Biology* 210, 2859–2865.

French, S. S., Moore, M. C. and Demas, G. E. 2009b. Ecological immunology: The organism in context. *Integrative and Comparative Biology* 49, 246–253.

Fusani, L. and Gwinner, E. 2005. Melatonin and nocturnal migration. *Annals of the New York Academy of Sciences* 1046, 264–270.

Garcia, N. W., Greives, T. J., Zysling, D. A., French, S. S., Chester, E. M. and Demas, G. E. 2010. Exogenous insulin enhances humoural immune responses in short-day, but not long-day, Siberian hamsters (*Phodopus sungorus*). *Proceedings of the Royal Society B Biological Sciences* 277, 2211–2218.

Garvin, J., Dunn, P., Whittingham, L., Steeber, D. and Hasselquist, D. 2008. Do male ornaments signal immunity in the common yellowthroat? *Behavioral Ecology* 19, 54.

Gasparini, J., McCoy, K., Haussy, C., Tveraa, T. and Boulinier, T. 2001. Induced maternal response to the Lyme disease spirochaete *Borrelia burgdorferi sensu lato* in a colonial seabird, the kittiwake *Rissa tridactyla*. *Proceedings of the Royal Society of London Series B Biological Sciences* 268, 647.

Gervasi, S. S. and Foufopoulos, J. 2008. Costs of plasticity: Responses to desiccation decrease post-metamorphic immune function in a pond-breeding amphibian. *Functional Ecology* 22, 100–108.

Greenman, C. G., Martin, L. B., 2nd and Hau, M. 2005. Reproductive state, but not testosterone, reduces immune function in male house sparrows (*Passer domesticus*). *Physiological and Biochemical Zoology* 78, 60–68.

Greives, T. J., McGlothlin, J. W., Jawor, J. M., Demas, G. E. and Ketterson, E. D. 2006. Testosterone and innate immune function inversely covary in a wild population of breeding dark-eyed juncos (*Junco hyemalis*). *Functional Ecology* 20, 812–818.

Grether, G. F., Hudon, J. and Endler, J. A. 2001. Carotenoid scarcity, synthetic pteridine pigments and the evolution of sexual coloration in guppies (*Poecilia reticulata*). *Proceedings of the Royal Society of London Series B Biological Sciences* 268, 1245–1253.

Grindstaff, J. 2008. Maternal antibodies reduce costs of an immune response during development. *Journal of Experimental Biology* 211, 654.

Grindstaff, J. L., Brodie, E. D. and Ketterson, E. D. 2003. Immune function across generations: Integrating mechanism and evolutionary process in maternal antibody transmission. *Proceedings of the Royal Society of London Series B Biological Sciences* 270, 2309–2319.

Grossman, C. J. 1984. Regulation of the immune-system by sex steroids. *Endocrine Reviews* 5, 435–455.

Grossman, C. J. 1985. Interactions between the gonadal-steroids and the immune system. *Science* 227, 257–261.

Gylfe, Å., Bergström, S., Lundstrom, J. and Olsen, B. 2000. Epidemiology: reactivation of *Borrelia* infection in birds. *Nature* 403, 724–725.

Hanssen, S., Hasselquist, D., Folstad, I. and Erikstad, K. 2004. Costs of immunity: Immune responsiveness reduces survival in a vertebrate. *Proceedings: Biological Sciences*, 271, 925–930.

Hargitai, R., Prechl, J. and Torok, J. 2006. Maternal immunoglobulin concentration in collared flycatcher (*Ficedula albicollis*) eggs in relation to parental quality and laying order. *Functional Ecology* 20, 829–838.

Harris, J. and Bird, D. J. 2000. Modulation of the fish immune system by hormones. *Veterinary Immunology and Immunopathology* 77, 163–176.

Hasselquist, D., Lindstrom, A., Jenni-Eiermann, S., Koolhaas, A. and Piersma, T. 2007. Long flights do not influence immune responses of a long-distance migrant bird: A wind-tunnel experiment. *Journal of Experimental Biology* 210, 1123.

Hasselquist, D., Marsh, J. A., Sherman, P. W. and Wingfield, J. C. 1999. Is avian humoral immunocompetence suppressed by testosterone? *Behavioral Ecology and Sociobiology* 45, 167–175.

Hasselquist, D. and Nilsson, J. 2009. Maternal transfer of antibodies in vertebrates: trans-generational effects on offspring immunity. *Philosophical Transactions of the Royal Society B Biological Sciences* 364, 51.

Henken, A. M. and Brandsma, H. A. 1982. The effect of environmental temperature on immune response and metabolism of the young chicken. 2. Effect of the immune response to sheep red blood cells on energy metabolism. *Poultry Science* 61, 1667–1673.

Horak, P., Zilmer, M., Saks, L., Ots, I., Karu, U. and Zilmer, K. 2006. Antioxidant protection, carotenoids and the costs of immune challenge in greenfinches. *Journal of Experimental Biology* 209, 4329.

Ilmonen, P., Taarna, T. and Hasselquist, D. 2000. Experimentally activated immune defence in female pied flycatchers results in reduced breeding success. *Proceedings of the Royal Society of London Series B Biological Sciences* 267, 665.

Irene Tieleman, B., Williams, J., Ricklefs, R. and Klasing, K. 2005. Constitutive innate immunity is a component of the pace-of-life syndrome in tropical birds. *Proceedings of the Royal Society B Biological Sciences* 272, 1715.

Jankowski, M. D., Franson, J. C., Möstl, E., Porter, W. P. and Hofmeister, E. K. 2010. Testing independent and interactive effects of corticosterone and synergized resmethrin on the immune response to West Nile virus in chickens. *Toxicology* 269, 81–88.

Ketterson, E. D., Nolan, V., Jr., Wolf, L., Ziegenfus, C., Dufty, A. M., Jr., Ball, G. F. and Johnsen, T. S. 1991. Testosterone and avian life histories: The effect of experimentally elevated testosterone on corticosterone and body mass in dark-eyed juncos. *Hormones and Behavior* 25, 489–503.

Kissling, R. E., Eidson, M. E. and Stamm, D. D. 1954. Transfer of maternal neutralizing antibodies against Eastern equine encephalomyelitis virus in birds. *Journal of Infectious Diseases* 95, 179–181.

Klasing, K. 1998. Nutritional modulation of resistance to infectious diseases. *Poultry Science* 77, 1119–1125.

Klasing KC. 2007. Nutrition and the immune system. *British Poultry Science* 48:525–537.

Klasing, K., Laurin, D., Peng, R. and Fry, D. 1987. Immunologically mediated growth depression in chicks: influence of feed intake, corticosterone and interleukin-1. *Journal of Nutrition* 117, 1629.

Klein, S. L. 2000. Hormones and mating system affect sex and species differences in immune function among vertebrates. *Behavioural Processes* 51, 149–166.

Klemperer, F. 1893. Uber naturliche immunitat und ihre verwerthung fur die immunisierungstherapie. *Arch. Exp. Path. Pharmak.* 31, 356–382.

Knowles, S., Nakagawa, S. and Sheldon, B. 2009. Elevated reproductive effort increases blood parasitaemia and decreases immune function in birds: a meta-regression approach. *Functional Ecology* 23, 405–415.

Kolluru, G. R., Grether, G. F., South, S. H., Dunlop, E., Cardinali, A., Liu, L. and Carapiet, A. 2006. The effects of carotenoid and food availability on resistance to a naturally occurring parasite (*Gyrodactylus turnbulli*) in guppies (*Poecilia reticulata*). *Biological Journal of the Linnean Society* 89, 301–309.

Kortet, R., Taskinen, J., Sinisalo, T. and Jokinen, I. 2003. Breeding-related seasonal changes in immunocompetence, health state and condition of the cyprinid fish, *Rutilus rutilus, L. Biological Journal of the Linnean Society* 78, 117–127.

Kreymann, G., Grosser, S., Buggisch, P., Gottschall, C., Matthaei, S. and Greten, H. 1993. Oxygen consumption and resting metabolic rate in sepsis, sepsis syndrome, and septic shock. *Critical Care Medicine* 21, 1012–1019.

Ksiazek, A., Konarzewski, M., Chadzinska, M. and Cichon, M. 2003. Costs of immune response in cold-stressed laboratory mice selected for high and low basal metabolism rates. *Proceedings of Biological Sciences* 270, 2025–2031.

Laugero, K. D. and Moberg, G. P. 2000. Summation of behavioral and immunological stress: Metabolic consequences to the growing mouse. *American Journal of Physiology—Endocrinology and Metabolism* 279, E44–49.

Leceta, J. and Zapata, A. 1986. Seasonal variations in the immune response of the tortoise *Mauremys caspica. Immunology* 57, 483–487.

Leshchinsky, T. and Klasing, K. 2001. Relationship between the level of dietary vitamin E and the immune response of broiler chickens. *Poultry Science* 80, 1590–1599.

Lin, Y.-H. and Shiau, S.-Y. 2003. Dietary lipid requirement of grouper, *Epinephelus malabaricus*, and effects on immune responses. *Aquaculture* 225, 243–250.

Lochmiller, R. L. and Deerenberg, C. 2000. Trade-offs in evolutionary immunology: Just what is the cost of immunity? *Oikos* 88, 87–98.

Lõhmus, M. and Björklund, M. 2009. Leptin affects life history decisions in a passerine bird: A field experiment. *PLoS One* 4, e4602.

Lohmus, M., Olin, M., Sundström, L., Troedsson, M., Molitor, T. and El Halawani, M. 2004. Leptin increases T-cell immune response in birds. *General and Comparative Endocrinology* 139, 245–250.

Lord, G. M., Matarese, G., Howard, J. K., Baker, R. J., Bloom, S. R. and Lechler, R. I. 1998. Leptin modulates the T-cell immune response and reverses starvation-induced immunosuppression. *Nature* 394, 897–901.

Lysle, D. T., Cunnick, J. E., Wu, R., Caggiula, A. R., Wood, P. G. and Rabin, B. S. 1988. 2-Deoxy-glucose modulation of T-lymphocyte reactivity: Differential effects on lymphoid compartments. *Brain, Behavior, and Immunity* 2, 212–221.

Magnanou, E., Fons, R., Feliu, C. and Morand, S. 2006. Physiological responses of insular wild black rat (*Rattus rattus*) to natural infection by the digenean trematode *Fasciola hepatica. Parasitology Research* 99, 97–101.

Martí, A., Marcos, A. and Martínez, J. A. 2001. Obesity and immune function relationships. *Obesity Reviews* 2, 131–140.

Martin, I., Lynn, B., Gilliam, J., Han, P., Lee, K. and Wikelski, M. 2005. Corticosterone suppresses cutaneous immune function in temperate but not tropical house sparrows, *Passer domesticus. General and Comparative Endocrinology* 140, 126–135.

Martin, L. B., Scheuerlein, A. and Wikelski, M. 2003. Immune activity elevates energy expenditure of house sparrows: A link between direct and indirect costs? *Proceedings of the Royal Society of London Series B Biological Sciences* 270, 153–158.

Mauget, C., Mauget, R. and Sempere, A. 1997. Metabolic rate in female European roe deer (*Capreolus capreolus*): Incidence of reproduction. *Canadian Journal of Zoology/Revue Canadienne De Zoologie* 75, 731–739.

Maule, A. G., Schrock, R., Slater, C., Fitzpatrick, M. S. and Schreck, C. B. 1996. Immune and endocrine responses of adult Chinook salmon during freshwater immigration and sexual maturation. *Fish & Shellfish Immunology* 6, 221–233.

McGraw, K. and Ardia, D. 2007. Do carotenoids buffer testosterone-induced immunosuppression? An experimental test in a colourful songbird. *Biology Letters* 3, 375.

McGraw, K. J. and Ardia, D. R. 2003. Carotenoids, immunocompetence, and the information content of sexual colors: An experimental test. *American Naturalist* 162, 704–712.

Meddle, S. L., Owen-Ashley, N. T., Richardson, M. I. and Wingfield, J. C. 2003. Modulation of the hypothalamic-pituitary-adrenal axis of an Arctic-breeding polygynandrous songbird, the Smith's longspur, *Calcarius pictus*. *Proceedings of the Royal Society of London Series B Biological Sciences* 270, 1849–1856.

Merino, S., Moreno, J., Sanz, J. and Arriero, E. 2000. Are avian blood parasites pathogenic in the wild? A medication experiment in blue tits (*Parus caeruleus*). *Proceedings of the Royal Society B Biological Sciences* 267, 2507.

Miller, E. S., Klinger, J. C., Akin, C., Anne Koebel, D. and Sonnenfeld, G. 1994. Inhibition of murine splenic T lymphocyte proliferation by 2-deoxy-D-glucose-induced metabolic stress. *Journal of Neuroimmunology* 52, 165–173.

Møller, A. and Erritzøe, J. 1998. Host immune defence and migration in birds. *Evolutionary Ecology* 12, 945–953.

Mondal, S. and Rai, U. 1999. Sexual dimorphism in phagocytic activity of wall lizard's splenic macrophage and its control by sex steroids. *General and Comparative Endocrinology* 116, 291–298.

Moore, C. and Siopes, T. 2000. Effects of lighting conditions and melatonin supplementation on the cellular and humoral immune responses in Japanese quail *Coturnix coturnix japonica*. *General and Comparative Endocrinology* 119, 95–104.

Moret, Y. 2006. "Trans-generational immune priming": Specific enhancement of the antimicrobial immune response in the mealworm beetle, *Tenebrio molitor*. *Proceedings of the Royal Society B Biological Sciences* 273, 1399–1405.

Moret, Y. and Schmid-Hempel, P. 2000. Survival for immunity: The price of immune system activation for bumblebee workers. *Science* 290, 1166–1168.

Morici, L. A., Elsey, R. M. and Lance, V. A. 1997. Effects of long-term corticosterone implants on growth and immune function in juvenile alligators, *Alligator mississippiensis*. *Journal of Experimental Zoology* 279, 156–162.

Muehlenbein, M. P. 2010. Evolutionary medicine, immunity and infectious diseases. In *Human Evolutionary Biology* (ed. M. P. Muehlenbein), pp. 351–377. New York: Cambridge University Press.

Nelson, R. J. and Demas, G. E. 1996. Seasonal changes in immune function. *Quarterly Review of Biology* 71, 511–548.

Nelson, R. J., Demas, G. E. and Klein, S. L. 1998. Photoperiodic mediation of seasonal breeding and immune function in rodents: A multi-factorial approach. *American Zoologist* 38, 226–237.

Nelson, R. J., Demas, G. E., Klein, S. L. and Kriegsfeld, L. J. 2002. *Seasonal Patterns of Stress, Immune Function, and Disease*. New York: Cambridge University Press.

Niewiarowski, P. H., Balk, M. L. and Londraville, R. L. 2000. Phenotypic effects of leptin in an ectotherm: A new tool to study the evolution of life histories and endothermy? *Journal of Experimental Biology* 203, 295–300.

Nilsson, J., Granbom, M. and Raberg, L. 2007. Does the strength of an immune response reflect its energetic cost? *Journal of Avian Biology* 38, 488–494.

Nilsson, J. A. and Raberg, L. 2001. The resting metabolic cost of egg laying and nestling feeding in great tits. *Oecologia* 128, 187–192.

Nordling, D., Andersson, M., Zohari, S. and Gustafsson, L. 1998. Reproductive effort reduces specific immune response and parasite resistance. *Proceedings of the Royal Society of London Series B Biological Sciences* 265, 1291–1298.

Norgan, N. G. 1997. The beneficial effects of body fat and adipose tissue in humans. *International Journal of Obesity and Related Metabolic Disorders* 21, 738–746.

Nova, E., Samartin, S., Gomez, S., Morande, G. and Marcos, A. 2002. The adaptive response of the immune system to the particular malnutrition of eating disorders. *European Journal of Clinical Nutrition* 56. Suppl 3, S34–37.

Olsson, M., Wapstra, E., Madsen, T. and Silverin, B. 2000. Testosterone, ticks and travels: a test of the immunocompetence-handicap hypothesis in free-ranging male sand lizards. *Proceedings of the Royal Society of London Series B Biological Sciences* 267, 2339–2343.

Olsson, M., Wilson, M., Uller, T., Mott, B. and Isaksson, C. 2009. Variation in levels of reactive oxygen species is explained by maternal identity, sex and body-size-corrected clutch size in a lizard. *Naturwissenschaften* 96, 25–29.

Ots, I., Kerimov, A., Ivankina, E., Ilyina, T. and Hırak, P. 2001. Immune challenge affects basal metabolic activity in wintering great tits. *Proceedings of the Royal Society of London Series B Biological Sciences* 268, 1175.

Ovington, K. S. 1985. Dose-dependent relationships between *Nippostrongylus brasiliensis* populations and rat food intake. *Parasitology* 91, 157–167.

Owen-Ashley, N. T., Hasselquist, D. and Wingfield, J. C. 2004. Examining the indirect pathways of testosterone-induced immunosuppression in male songbirds. I. Effects of corticosterone and estradiol on acquired immune function. *Integrative and Comparative Biology* 44, 616–616.

Owen, J. and Moore, F. 2008. Relationship between energetic condition and indicators of immune function in thrushes during spring migration. *Canadian Journal of Zoology* 86, 638–647.

Patino, R. and Maule, A. G. 1997. Estrogen receptors in leukocytes from immature channel catfish. *Developmental and Comparative Immunology* 21, 123–123.

Peters, A. 2000. Testosterone treatment is immunosuppressive in superb fairy-wrens, yet free-living males with high testosterone are more immunocompetent. *Proceedings of the Royal Society of London Series B Biological Sciences* 267, 883–889.

Peters, A. 2007. Testosterone and carotenoids: An integrated view of trade-offs between immunity and sexual signalling. *Bioessays* 29, 427–430.

Peters, A., Delhey, K., Denk, Angelika G. and Kempenaers, B. 2004. Trade-offs between immune investment and sexual signaling in male mallards. *American Naturalist* 164, 51–59.

Pianka, E. R. and Parker, W. S. 1975. Age-specific reproductive tactics. *American Naturalist* 109, 453–464.

Pickering, A. D. and Christie, P. 1980. Sexual differences in the incidence and severity of ectoparasitic infestation of the brown trout, *Salmo trutta. Journal of Fish Biology* 16, 669–683.

Pilorz, V., Jäckel, M., Knudsen, K. and Trillmich, F. 2005. The cost of a specific immune response in young guinea pigs. *Physiology & Behavior* 85, 205–211.

Pompéia, C., Lopes, L. R., Miyasaka, C. K., Procópio, J., Sannomiya, P. and Curi, R. 2000. Effect of fatty acids on leukocyte function. *Brazilian Journal of Medical and Biological Research* 33, 1255–1268.

Pond, C. M. 1996. Interactions between adipose tissue and the immune system. *Proceedings of the Nutrition Society* 55, 111–126.

Price, T. D. 1998. Maternal and paternal effects in birds: Effects on offspring fitness. In *Maternal Effects as Adaptations* (ed. T. A. Mousseau and C. W. Fox), pp. 202–226. Oxford: Oxford University Press.

Putti, R., Varricchio, E., Gay, F., Elena, C. and Paolucci, M. 2009. Leptin effects on testis and epididymis in the lizard *Podarcis sicula*, during summer regression. *General and Comparative Endocrinology* 160, 168–175.

Raberg, L., Nilsson, J. A., Ilmonen, P., Stjernman, M. and Hasselquist, D. 2000. The cost of an immune response: vaccination reduces parental effort. *Ecology Letters* 3, 382–386.

Raberg, L., Stjernman, M. and Hasselquist, D. 2003. Immune responsiveness in adult blue tits: Heritability and effects of nutritional status during ontogeny. *Oecologia* 136, 360–364.

Ramakrishnan, L., Valdivia, R., McKerrow, J. and Falkow, S. 1997. *Mycobacterium marinum* causes both long-term subclinical infection and acute disease in the leopard frog (*Rana pipiens*). *Infection and Immunity* 65, 767–773.

Rayner, D. V. and Trayhurn, P. 2001. Regulation of leptin production: Sympathetic nervous system interactions. *Journal of Molecular Medicine* 79, 8–20.

Reeves, W. C., Sturgeon, J. M., French, E. M. and Brookman, B. 1954. Transovarian transmission of neutralizing substances to Western equine and St. Louis encephalitis viruses by avian hosts. *The Journal of Infectious Diseases* 95, 168–178.

Reuman, P., Paganini, C., Ayoub, E. and Small, P. 1983. Maternal-infant transfer of influenza-specific immunity in the mouse. *Journal of Immunology* 130, 932–936.

Richner, H., Christe, P. and Oppliger, A. 1995. Paternal investment affects prevalence of malaria. *Proceedings of the National Academy of Sciences of the United States of America* 92, 1192–1194.

Ricklefs, R. and Wikelski, M. 2002. The physiology/life-history nexus. *Trends in Ecology & Evolution* 17, 462–468.

Robar, N., Burness, G. and Murray, D. 2010. Tropics, trophics and taxonomy: The determinants of parasite-associated host mortality. *Oikos* 119, 1273–1280.

Roberts, M. L., Buchanan, K. L. and Evans, M. R. 2004. Testing the immunocompetence handicap hypothesis: a review of the evidence. *Animal Behaviour* 68, 227–239.

Rocca, B. and FitzGerald, G. A. 2002. Cyclooxygenases and prostaglandins: shaping up the immune response. *International Immunopharmacology* 2, 603–630.

Rollins-Smith, L. A. 1998. Metamorphosis and the amphibian immune system. *Immunological Reviews* 166, 221–230.

Rollins-Smith, L. A., Barker, K. S. and Davis, A. T. 1997. Involvement of glucocorticoids in the reorganization of the amphibian immune system at metamorphosis. *Developmental Immunology* 5, 145–152.

Rollins-Smith, L. A. and Blair, P. J. 1993. The effects of corticosteroid hormones and thyroid hormones on lymphocyte viability and proliferation during development and metamorphosis of *Xenopus laevis*. *Differentiation* 54, 155–160.

Royle, N., Surai, P. and Hartley, I. 2003. The effect of variation in dietary intake on maternal deposition of antioxidants in zebra finch eggs. *Functional Ecology* 17, 472–481.

Ruiz, M., French, S. S., Demas, G. E. and Martins, E. P. 2010. Food supplementation and testosterone interact to influence reproductive behavior and immune function in *Sceloporus graciosus*. *Hormones and Behavior* 57, 134–139.

Saad, A. H. 1988. Corticosteroids and immune systems of non-mammalian vertebrates: A review. *Developmental and Comparative Immunology* 12, 481–494.

Saad, A. H. 1989. Pregnancy-related involution of the thymus in the viviparous lizard, *Chalcides ocellatus*. *Thymus* 14, 223–232.

Saad, A. H. and Elridi, R. 1984. Corticosteroids and seasonal changes in reptilian immune responses. *Developmental and Comparative Immunology* 8, 945–945.

Saad, A. H., Elridi, R., Zada, S. and Badir, N. 1984a. Effect of hydrocortisone on immune system of the lizard, *Chalcides ocellatus*. 1. Response of lymphoid tissues and cells to in vivo and in vitro hydrocortisone. *Developmental and Comparative Immunology* 8, 121–130.

Saad, A. H., Elridi, R., Zada, S. and Badir, N. 1984b. Effect of hydrocortisone on immune system of the lizard, *Chalcides ocellatus*. 2. Differential action on lymphocytes-T and lymphocytes-B. *Developmental and Comparative Immunology* 8, 835–844.

Saad, A. H., Khalek, N. A. and el Ridi, R. 1990. Blood testosterone level: a season-dependent factor regulating immune reactivity in lizards. *Immunobiology* 180, 184–194.

Saad, A. H., Torroba, M., Varas, A. and Zapata, A. 1991. Testosterone induces lymphopenia in turtles. *Veterinary Immunology and Immunopathology* 28, 173–180.

Sadd, B. M. and Schmid-Hempel, P. 2009. A distinct infection cost associated with trans-generational priming of antibacterial immunity in bumble-bees. *Biology Letters* 5, 798–801.

Saino, N., Calza, S. and Muller, A. 1998. Effects of a dipteran ectoparasite on immune response and growth trade-offs in barn swallow, *Hirundo rustica*, nestlings. *Oikos* 81, 217–228.

Saino, N., Møller, A. P. and Bolzern, A. M. 1995. Testosterone effects on the immune system and parasite infestations in the barn swallow (*Hirundo rustica*): An experimental test of the immunocompetence hypothesis. *Behavioral Ecology* 6, 397–404.

Scarpace, P., Matheny, M., Moore, R. and Kumar, M. 2000. Modulation of uncoupling protein 2 and uncoupling protein 3: Regulation by denervation, leptin and retinoic acid treatment. *Journal of Endocrinology* 164, 331–337.

Schoech, S. J., Ketterson, E. D. and Nolan, V., Jr. 1999. Exogenous testosterone and the adrenocortical response in dark-eyed juncos. *The Auk* 116, 64–72.

Schumacher, I. M., Rostal, D. C., Yates, R. A., Brown, D. R., Jacobson, E. R. and Klein, P. A. 1999. Persistence of maternal antibodies against *Mycoplasma agassizii* in desert tortoise hatchlings. *American Journal of Veterinary Research* 60, 826–831.

Sciarrillo, R., Virgilio, F., De Falco, M., Laforgia, V., Varano, L. and Paolucci, M. 2005. Localization and role of leptin in the thyroid gland of the lizard *Podarcis sicula* (Reptilia, Lacertidae). *Journal of Experimental Zoology Part A Comparative Experimental Biology* 303A, 628–634.

Sharp, P., Dunn, I., Waddington, D. and Boswell, T. 2008. Chicken leptin. *General and Comparative Endocrinology* 158, 2–4.

Sheldon, B. C. and Verhulst, S. 1996. Ecological immunology: Costly parasite defences and trade-offs in evolutionary ecology. *Trends in Ecology & Evolution* 11, 317–321.

Sherman, E., Baldwin, L., Fernandez, G. and Deurell, E. 1991. Fever and thermal tolerance in the toad *Bufo marinus*. *Journal of Thermal Biology* 16, 297–301.

Silverin, B., Arvidsson, B. and Wingfield, J. 1997. The adrenocortical responses to stress in breeding Willow Warblers *Phylloscopus trochilus* in Sweden: effects of latitude and gender. *Functional Ecology* 11, 376–384.

Simon, J., Rideau, N. and Taouis, M. 2009. Reply to viewpoints by PJ Sharp, IC Dunn, D Waddington and T Boswell [Chicken Leptin]. *General and Comparative Endocrinology* 161, 159.

Skarstein, F., Folstad, I. and Liljedal, S. 2001. Whether to reproduce or not: Immune suppression and costs of parasites during reproduction in the Arctic charr. *Canadian Journal of Zoology* 79, 271–278.

Slater, C. H., Fitzpatrick, M. S. and Schreck, C. B. 1995. Androgens and immunocompetence in Salmonids: Specific binding in and reduced immunocompetence of Salmonid lymphocytes exposed to natural and synthetic androgens. *Aquaculture* 136, 363–370.

Slater, C. H. and Schreck, C. B. 1997. Physiological levels of testosterone kill Salmonid leukocytes in vitro. *General and Comparative Endocrinology* 106, 113–119.

Soler, J., Neve, L., Perez–Contreras, T., Soler, M. and Sorci, G. 2003. Trade-off between immunocompetence and growth in magpies: An experimental study. *Proceedings of the Royal Society of London Series B Biological Sciences* 270, 241.

Sooter, C. A., Schaeffer, M., Gorrie, R. and Cockburn, T. A. 1954. Transovarian passage of antibodies following naturally acquired encephalitis infection in birds. *Journal of Infectious Diseases* 95, 165–167.

Spencer, K., Buchanan, K., Leitner, S., Goldsmith, A. and Catchpole, C. 2005. Parasites affect song complexity and neural development in a songbird. *Proceedings of the Royal Society B Biological Sciences* 272, 2037.

Spurlock, M. E., Frank, G. R., Willis, G. M., Kuske, J. L. and Cornelius, S. G. 1997. Effect of dietary energy source and immunological challenge on growth performance and immunological variables in growing pigs. *Journal of Animal Science* 75, 720–726.

Svensson, E., Raberg, L., Koch, C. and Hasselquist, D. 1998. Energetic stress, immunosuppression and the costs of an antibody response. *Functional Ecology* 12, 912–919.

Svensson, E., Sinervo, B. and Comendant, T. 2001. Condition, genotype-by-environment interaction, and correlational selection in lizard life-history morphs. *Evolution* 55, 2053–2069.

Takhar, B. S. and Farrell, D. J. 1979. Energy and nitrogen metabolism of chickens infected with either *Eimeria acervulina* or *Eimeria tenella*. *British Poultry Science* 20, 197–211.

Tanvez, A., Amy, M., Chastel, O. and Leboucher, G. 2009. Maternal effects and beta-carotene assimilation in canary chicks. *Physiology and Behavior* 96, 389–393.

Taouis, M., Chen, J., Daviaud, C., Dupont, J., Derouet, M. and Simon, J. 1998. Cloning the chicken leptin gene. *Gene* 208, 239–242.

Tomàs, G., Merino, S., Moreno, J., Morales, J. and Martinez-De la Puente, J. 2006. Impact of blood parasites on immunoglobulin level and parental effort: A medication field experiment on a wild passerine. *Ecology* 21, 125–133.

Torres, R. and Velando, A. 2007. Male reproductive senescence: the price of immune-induced oxidative damage on sexual attractiveness in the blue-footed booby. *Journal of Animal Ecology* 76, 1161–1168.

Tsiagbe, V. K., Cook, M. E., Harper, A. E. and Sunde, M. L. 1987. Enhanced immune responses in broiler chicks fed methionine-supplemented diets. *Poultry Science* 66, 1147–1154.

Tsigos, C., Papanicolaou, D. A., Defensor, R., Mitsiadis, C. S., Kyrou, I. and Chrousos, G. P. 1997. Dose effects of recombinant human interleukin-6 on pituitary hormone secretion and energy expenditure. *Neuroendocrinology* 66, 54–62.

Uller, T., Isaksson, C. and Olsson, M. 2006. Immune challenge reduces reproductive output and growth in a lizard. *Functional Ecology* 20, 873–879.

Uller, T. and Olsson, M. 2003. Prenatal exposure to testosterone increases ectoparasite susceptibility in the common lizard (*Lacerta vivipara*). *Proceedings of the Royal Society of London Series B Biological Sciences* 270, 1867–1870.

Van Niekerk, T., Garber, T. K., Dunnington, E. A., Gross, W. B. and Siegel, P. B. 1989. Response of White Leghorn chicks fed ascorbic acid and challenged with *Escherichia coli* or with corticosterone. *Poultry Science* 68, 1631–1636.

Veiga, J. P., Salvador, A., Merino, S. and Puerta, M. 1998. Reproductive effort affects immune response and parasite infection in a lizard: A phenotypic manipulation using testosterone. *Oikos* 82, 313–318.

Velando, A., Drummond, H. and Torres, R. 2006. Senescent birds redouble reproductive effort when ill: confirmation of the terminal investment hypothesis. *Proceedings of the Royal Society B Biological Sciences* 273, 1443.

Verhulst, S., Dieleman, S. J. and Parmentier, H. K. 1999. A tradeoff between immunocompetence and sexual ornamentation in domestic fowl. *Proceedings of the National Academy of Sciences of the United States of America* 96, 4478–4481.

Warner, A. H. and Sonnenfeld, M. 1987. Evaluation of skeletal-muscle thiol protease inhibitor activity in dystrophic hamsters and chickens. *Federation Proceedings* 46, 2224–2224

Weinberg, E. D. 1984. Pregnancy-associated depression of cell-mediated immunity. *Reviews of Infectious Diseases* 6, 814–831.

Williams, G. C. 1966. Natural selection costs of reproduction and a refinement of Lack's principle. *American Naturalist* 100, 687–690.

Williams, T., Christians, J., Aiken, J. and Evanson, M. 1999. Enhanced immune function does not depress reproductive output. *Proceedings of the Royal Society of London Series B Biological Sciences* 266, 753.

Wingfield, J. C., O'Reilly, K. M. and Astheimer, L. B. 1995. Modulation of the adrenocortical responses to acute stress in arctic birds: A possible ecological basis. *American Zoologist* 35, 285–294.

Woods, S. C., Schwartz, M. W., Baskin, D. G. and Seeley, R. J. 2000. Food intake and the regulation of body weight. *Annual Review of Psychology* 51, 255–277.

Yaqoob, P., Newsholme, E. A. and Calder, P. C. 1994. The effect of dietary lipid manipulation on rat lymphocyte subsets and proliferation. *Immunology* 82, 603–610.

Zhang, Y., Proenca, R., Maffei, M., Barone, M., Leopold, L. and Friedman, J. M. 1994. Positional cloning of the mouse obese gene and its human homologue. *Nature* 372, 425–432.

Zwart, D., Brouwer, B. O., van der Hel, W., van den Akker, H. N. and Verstegen, M. W. 1991. Effect of *Trypanosoma vivax* infection on body temperature, feed intake, and metabolic rate of West African dwarf goats. *Journal of Animal Science* 69, 3780–3788.

Zysling, D. and Demas, G. 2007. Metabolic stress suppresses humoral immune function in long-day, but not short-day, Siberian hamsters (*Phodopus sungorus*). *Journal of Comparative Physiology B Biochemical, Systemic, and Environmental Physiology* 177, 339–347.

Zysling, D., Greives, T., Breuner, C., Casto, J., Demas, G. and Ketterson, E. 2006. Behavioral and physiological responses to experimentally elevated testosterone in female dark-eyed juncos (*Junco hyemalis carolinensis*). *Hormones and Behavior* 50, 200–207.

CHAPTER 9

NEUROENDOCRINE MECHANISMS OF SEASONAL CHANGES IN IMMUNE FUNCTION

Zachary M. Weil and Randy J. Nelson

SEASONALITY

For animals that live outside the equatorial regions, life is characterized by continual but relatively predictable temporal fluctuations in environmental conditions. Just as day follows night, the mild temperatures and relatively abundant food availability of the spring and summer are typically followed by harsh conditions of the winter, including reduced food availability and low temperatures. Thus, winter represents a severe energetic bottleneck for most animals wherein reduced food availability coincides with an annual peak in thermoregulatory demands (Bronson, 1989). The divergent environmental conditions that confront individuals during the winter and summer require that energetic budgets be invested very differently at different times of the year (Nelson et al., 2002). The energetic challenges of winter can be extreme, but they are also remarkably predictable, and as such mechanisms have evolved among virtually all non-tropical species to confine the most energetically demanding processes, such as reproduction, to the early spring and summer (Bronson, 1985). Throughout the year, there are dramatic shifts in physiological trade-offs that must be accomplished to maximize the odds of surviving to the next breeding season and successfully breeding (Bronson, 1985; Wingfield and Farner, 1993).

One of the most prominent annual physiological adjustments is the timing of breeding. Small mammals confine the energetically expensive activities of egg deposition, pregnancy, and lactation, as well as feeding of young, to the spring and early summer (Bronson, 1989; Wingfield and Farner, 1993). Larger mammals that have longer gestational periods tend to breed in the fall such that young are born to coincide with the spring thaw (Malpaux et al., 1989; Rosa and Bryant, 2003). During the non-reproductive months reproductive organs regress, and this presumably results in a considerable energetic savings. In addition to changes in the reproductive system, other adaptations have evolved that help animals to survive the challenges of winter. For instance, reduced food availability is met by changes in the regulation of energy balance, and changes in body composition, increased predation pressure, and low ambient temperatures are countered by molt to a dense, camouflaged winter coat and increased tolerance to close contact with conspecifics (Madison et al., 1984; Wade and Bartness, 1984; Duncan and Goldman, 1985; Domjan, 1987; Kauffman et al., 2001; Prendergast et al., 2002; Demas et al., 2010). Finally, to maintain host defense in the face of winter stressors, various aspects of the immune system undergo seasonal adjustments (Nelson, 2002, 2004).

Importantly, these adjustments require significant time to develop, and as such it is critical for animals to be able to predict the deteriorating environmental conditions before they begin. One cue that is used by virtually all vertebrate taxa to predict future environmental conditions is the duration of the daily illumination (Goldman and Nelson, 1993). The near universality of photoperiodism (i.e., the ability to decode time of year by attending to the annual cycle of changing day lengths) speaks to the unique qualities of this signal. First, although photoperiod likely has minimal direct fitness consequences, day length can be monitored with minimal noise. Further, with just two pieces of information—the length of the day and whether the day length is increasing or decreasing—animals can precisely determine the date (Goldman, 2001; Prendergast et al., 2002). Other environmental conditions have more direct consequences for wild animals, including ambient temperature, rainfall, and food availability (Baker, 1938). Food or water restriction can directly affect the immunological and reproductive systems; however, these factors are inherently less useful as predictors of changing conditions because they do not show smooth monotonic changes, but rather are characterized by variability throughout the year. Further, by the time food availability and temperature decline, it may well be too late to begin phenotypic alterations, which often take significant time to develop (Prendergast et al., 2002).

PHOTOPERIODIC EFFECTS ON THE IMMUNE SYSTEM/WINTER IMMUNOENHANCEMENT HYPOTHESIS

The threat from pathogens and parasites is a universal feature of life on earth. One could predict that animals would mount the strongest immune responses possible at all times of their lives. Stated differently, the only upper limit on the strength of immune responses would be the development of autoimmunity and bystander damage to host tissues. However, immune function varies markedly across the year in most vertebrates, with generally stronger immune responses during short days (Nelson et al., 2002). There are several overlapping potential explanations for these fluctuations, including: (1) disease prevalence and virulence could fluctuate across time, (2) the efficacy of immune responses themselves could change, (3) environmental conditions could affect the ability to mount an immune response, and (4) the immune system is energetically costly to maintain and utilize, and therefore when energy is limited trade-offs among competing physiological processes must occur (Ricklefs and Wikelski, 2002).

We have proposed a synthesis of these ideas as an explanation for seasonal variation in immune responses (Nelson and Demas, 1996; Nelson et al., 2002). Harsh winter conditions are sufficient to kill small vertebrates either directly, via starvation or hypothermia, or indirectly by rendering them more susceptible to infections. Therefore, energetic investment is biased towards immune responses during winter to counteract the suppressive effects of winter stressors and maintain immunological homeostasis (Lochmiller and Deerenberg, 2000). Taken together, a life-history–based theory emerges indicating that during the spring and summer energy is disproportionately invested in costly activities associated with reproduction among small animals. During winter, freed of the costs of maintaining and using the reproductive system, animals actively invest in processes that increase the chances of survival until the next breeding season, notably strong immune defenses. This active boosting of immune responses can be assessed physiologically and can also serve as a proxy for overall investment in survival mechanisms.

An alternative hypothesis also exists, that rather than an active boosting of immune responses during short days, the cost of maintaining the reproductive system during the summer simply renders animals unable to exhibit maximal immune responses. As we shall see below, this argument is weakened by laboratory studies in which reproductive and immunological responses to changes in day length can be dissociated. (For further discussion of this point see Martin et al., 2008, and Chapter 15.)

Finally, it is overly simplistic to argue that immune responses are universally enhanced by exposure to short days because several components of immune responses are increased in long days (Martin et al., 2006, 2008). The organization of the vertebrate immune system is such that, unlike the reproductive system, all types of immune responses do not covary and are not under the control of a single neuroendocrine axis. In the laboratory most immunological processes are enhanced by exposure to short days, although other processes are inhibited depending on the specific species, stimulus, and endpoint measured (Yellon et al., 1999; Drazen et al., 2001; Bilbo et al., 2002). It may be illustrative to think of photoperiodic changes in immune responses as a reorganization of immunological function to prioritize capabilities, rather than an overall bolstering.

PHOTOPERIODIC CHANGES IN IMMUNE RESPONSES

The vertebrate immune system is a complex, multiply redundant system with several different cell types, tissues, and modes of host defense distributed throughout the body. As already stated, the strength of various immunological responses does not move in lockstep, and they are capable of varying independently. Thus, we will specify the type of immune response and describe the type of stimulus used in this chapter.

PHOTOPERIODIC EFFECTS ON IMMUNE TISSUES

Most of the studies presented in this chapter are based on studies of small mammals (mostly rodents) and birds. When illustrative, data from larger animals will also be discussed.

Additionally, we are not limiting this discussion to either field- or laboratory-based studies; both types of experiments are useful, although both approaches also have important limitations.

Many of the original studies on seasonal variation in immune function focused on the size and cellularity of immune tissues and the abundance of circulating immune cells. These studies, particularly those conducted in the field, are often confounded by the immediate exposure to pathogens and parasites, as the variables tend to shift radically during active infection. We have previously argued that these types of studies are fairly poor predictors of specific immune defenses, but they do serve to provide some basic information about how the immune system processes environmental conditions (Nelson and Prendergast, 2002; Martin et al., 2006). This perspective probably stems from reproductive endocrinology, when larger gonadal tissues virtually always correlate with increased function.

Immunological tissue masses tend to vary across the seasons, but the direction of the change varies across species, tissue examined, and laboratory versus field studies. For instance, splenic masses are larger in short days in Norway rats (*Rattus norvegius*), deer mice (*Peromyscus maniculatus*), and Syrian hamsters (*Mesocricetus auratus*) (Demas et al., 1996). In birds, spleen and thymic masses are smallest during the spring recrudescence in pied flycatchers (*Ficedula hypoleuca*) and willow tits (*Parus montanus*) (Fange and Silverin, 1985; Silverin et al., 1999). However, other studies have reported reduced splenic mass during winter in rodents such as short-tailed voles (*Microtus agrestis*) and European ground squirrels (*Spermophilus citellus*) (Newson, 1962; Shivatcheva and Hadjioloff, 1987). The variability in these studies points to the need for studies that examine immunological function rather than simply lymphatic tissue size (Nelson and Prendergast, 2002; Martin et al., 2006).

INNATE IMMUNE FUNCTION

The innate immune system is a broadly effective system of protection against foreign cells, but it is costly to utilize both energetically and in damage to host tissue. Evidence exists for suppressed innate immune function during the winter. It seems likely that mounting full-blown innate immune responses during the winter is incompatible with the energetic limitations animals suffer during those months. One common strategy to assess innate immune activity is to treat animals with the endotoxin lipopolysaccharide (LPS) and examine the physiological and behavioral responses (Exton, 1997). LPS is a nonreplicating component of gram-negative bacterial cell walls that can be used to assess the physiological and behavioral responses to infection independent of a replicating pathogen. The behavioral responses to LPS include lethargy, anorexia, adipsia, anhedonia, and a reduction in social interactions (Hart, 1988). These responses, collectively termed "sickness behavior," along with the induction of fever, are thought to be part of a coordinated, adaptive effort to aid in recovery from infection (Hart, 1988; Kent et al., 1992). The primary mediators of the sickness response are the pro-inflammatory cytokines, interleukin-1β (IL-1), IL-6, and tumor necrosis factor-α (TNFα) (Kent et al., 1992; Dantzer, 2001).

Housing captive white-crowned sparrows (*Zonotrichia leucophrys gambelii*) in short days antagonized the suppressive effect of LPS on food intake and weight loss (Owen-Ashley et al., 2006, 2008). This photoperiodic effect was strongly linked to body condition, indicating that energetic limitations may preclude strong innate immune responses during the winter.

However, free-living northwestern song sparrows (*Melospiza melodia morphna*) showed enhanced behavioral responses to LPS during the spring breeding season compared with conspecifics tested in the fall. This may reflect the relatively mild climate experienced by these birds in the state of Washington and the pattern of increased body mass during the winter (Owen-Ashley et al., 2006).

In Siberian hamsters (*Phodopus sungorus*), oxidative burst (an index of cytotoxic mechanisms) and phagocytic activity of *ex vivo* cells were inhibited in the laboratory by short day lengths. This *in vitro* reduction in innate immune activity in short days is mirrored by reduced responses to LPS *in vivo.* Hamsters housed in short days exhibited attenuated febrile responses, lost less weight, and displayed smaller anhedonic responses compared to those housed in long days (Bilbo et al., 2002). This set of behavioral responses is mediated, at least in part, by reduced cytokine response to LPS, as short days attenuated the induction of the pro-inflammatory cytokines TNFα and IL-1β at both the mRNA and protein levels (Prendergast et al., 2003; Pyter et al., 2005). Additionally, the central induction of the prostaglandin-inducing enzyme cyclooxygenase-2 was inhibited by short day lengths (Bilbo et al., 2003). High doses of LPS produce a syndrome very similar to lethal bacterial sepsis. Housing animals in short day lengths significantly reduced the lethality of high doses of LPS, indicating that this attenuation of cytokine responses has important fitness consequences (Prendergast et al., 2003). Taken together, these data indicate that the immunological and behavioral responses to LPS are inhibited by short day lengths.

There are a number of potential molecular mediators of short-day–induced suppression of innate immune responses. *In vivo* peritoneal macrophages isolated from short-day hamsters attenuated cytokine responses to LPS treatment; however, no significant differences were detected in the expression of the Toll-like receptor proteins responsible for the detection and response to LPS (Navara et al., 2007). Administration of recombinant pro-inflammatory cytokines instead of LPS also produced an elevated sickness response in long-day Siberian hamsters (Wen and Prendergast, 2007). Additionally, the short-day attenuation of sickness behavior in Siberian hamsters is not isolated to gram-negative–derived LPS. Administration of immunogens from gram-positive bacteria and viruses (muramyl dipeptide and polyinosinepolycytidylic acid, respectively) initiated behavioral sickness responses that were of a smaller magnitude in short-day hamsters (Baillie and Prendergast, 2008). These data indicate that both reduced cytokine response to inflammatory mediators and reduced sensitivity to cytokines mediate photoperiodic differences in LPS responses. Several intracellular signaling pathways are common to both Toll-like receptors and cytokine receptors (e.g., the NFκB pathway) and should be examined as a mediator of photoperiodic effects on the production of and sensitivity to pro-inflammatory cytokines.

From a proximate perspective, attenuated sickness and cytokine response to simulated infection are likely the result of energetic constraints on full expression of febrile responses, and the reduced overall cytokine response (Klasing, 2004). Ultimately, however, it remains an open question whether: (1) reduction in the intensity of the behavioral and physiological responses to infection has serious fitness consequences, (2) other compensatory mechanisms have evolved, or (3) the type of pathogen pressures at different times of the year favors one antibacterial strategy over another. Although fever is an extremely energetically demanding process, other behavioral aspects of the sickness response (e.g., anhedonia) would seem to be relatively inexpensive tactics to help recover from infection. It is possible that the reduced cytokine response to infectious stimuli in short days evolved to minimize febrile responses, and the inhibition of other sickness behaviors is simply a byproduct. However, there are reports of increased hypothalamic cytokine gene expression and sickness behavior in the absence of altered fever (Weil et al., 2006). The behavioral responses to infection are not a simple byproduct of poor health and have been shown to aid in the recovery from infection (Hart, 1988, 1990). For instance, infection with the bacteria *Listeria monocytogenes* reduces spontaneous food intake, and force-feeding animals after infection dramatically worsens survival rate (Murray and Murray, 1979). Other adaptive immunological defenses may be overused during the winter to avoid using the expensive innate immune system.

ACQUIRED IMMUNE SYSTEM

Cell-mediated immune responses are significantly enhanced during the winter or short day lengths in many animals, including various rodent, passerine, and ungulate species (Demas and Nelson, 1998a, 1998b; Bilbo et al., 2002; Auchtung et al., 2004; Greenman et al., 2005; Weil et al., 2006). Cell-mediated immune responses can be assessed in several different ways. Among the assays are the delayed-type hypersensitivity (DTH) response, which is an integrative measure of several immunological processes, including antigen processing and presentation, immunological memory, immune cell trafficking, and T-cell–mediated inflammatory responses. *In vitro* cell proliferation responses to mitogens are also employed.

Free-living cotton rats (*Sigmodon hispidus*) showed enhanced leukocyte proliferation to pokeweed mitogen *in vitro* in February relative to other months (Lochmiller et al., 1994). In laboratory studies, the B-cell mitogen concanavalin A markedly enhanced splenocyte proliferation in deer mice maintained in short days (Demas and Nelson, 1998a, 1998b) and male European starlings (*Sturnus vulgaris*; Bentley et al., 1998). The DTH response is particularly responsive to modulation by photoperiod. Photoperiods that stimulate molt in house sparrows markedly reduced phytohemagglutinin-induced swelling responses (Greenman et al., 2005; Martin, 2005). Short days enhanced DTH responses in the laboratory in meadow voles (*Microtus pennsylvanicus*), collared lemmings (*Dicrostonyx groenlandicus*), white-footed mice (*Peromyscus leucopus*), and Siberian hamsters. Photoperiodic adjustment in DTH responses in Siberian hamsters are mediated, at least in part, by more circulating leukocytes and a stronger stress-induced trafficking of leukocytes to inflamed tissues (Bilbo et al., 2002). Short photoperiods enhance the induction, retention, and retrieval of DTH responses; housing hamsters in short days either prior to the initial sensitization, during the delay before challenge, or at the time of challenge all enhanced the final responses. However, subsequent or past exposure to long days mitigated the short-day–induced enhancement, indicating that photoperiod is acting on multiple aspects of this immune response (Prendergast et al., 2004).

In other non-rodent species there are also examples of boosted cell-mediated immune function during short days. Rhesus monkeys (*Macacca mulatta*) express altered ratios of Th1 and Th2 cytokines at different times of the year (Mann et al., 1993, 2000). Specifically, the Th1 cytokines interferon-gamma and IL-2 were expressed at a higher level in peripheral blood mononuclear cells during the summer, whereas the canonical Th2 cytokine IL-4 was elevated during the winter. Additionally, peripheral blood proliferation and mitogen-stimulated leukocyte proliferation were both enhanced during the winter (Mann et al., 1993, 2000).

HUMORAL RESPONSES

Antibody responses in animals are generally elicited by treating animals with a large innocuous protein or cellular component. Over the subsequent days and weeks, the time course and magnitude of the soluble antibody isotypes immunoglobulin (Ig) M and IgG in circulation are assessed. Humoral immune responses are typically more variable as to whether they are enhanced or inhibited by exposure to short days. Cattle (*Bos taurus*) living outdoors attenuated spontaneous antibodies to J substance in the winter compared to the late summer (Stone, 1956). On the other hand, common voles (*Microtus arvalis*) captured during the winter had higher levels of spontaneous antibodies compared with those captured during the summer. In the laboratory, Siberian hamsters housed in short day have attenuated IgG and IgM responses to the antigens keyhole limpet hemocyanin (KLH) and sheep red blood cells (SRBCs) (Yellon et al., 1999; Drazen et al., 2000a; Prendergast et al., 2004), although Syrian hamster showed enhanced responses in short days (Drazen et al., 2002). Collared lemmings display no photoperiod changes in antibody responses (Weil et al., 2006).

The immune system is very sensitive to the changing seasons. In general, immune function, and in particular cell-mediated immune function, is enhanced by exposure to short day lengths, although other components of immune function are inhibited by short days. In any case, the immune system is robustly responsive to changing seasons and, in common with the reproductive and metabolic systems, is capable of predicting, attending, and responding to environmental cues in order to maximize winter survival.

NEUROENDOCRINE CONTROL OF THE IMMUNE SYSTEM

MELATONIN

In mammals, day length is transduced from an environmental signal into a physiological one via the nighttime secretion of pineal melatonin (Bartness et al., 1993; Goldman and Nelson, 1993). The pineal synthesizes and secretes the indoleamine melatonin during the dark (Cardinali and Vacas, 1987; Pandi-Perumal et al., 2006). Photic input to the retina inhibits melatonin synthesis. Therefore, central and peripheral tissues can monitor photoperiodic conditions by attending to the duration of circulating melatonin (Bartness et al., 1993). In the laboratory, prolonged exposure to long-duration melatonin signals or short day lengths is sufficient to induce virtually all of the seasonal adaptations observed in the field. The role of melatonin signaling and the pineal gland is somewhat different in non-mammalian vertebrates but is also involved in transducing seasonal time.

In mammals, melatonin is the principal coordinator of seasonality in general and photoperiodic regulation of the immune system in particular. Melatonin is an indoleamine synthesized in a two-step reaction from serotonin primarily in the pineal gland, retina, and gut, although there are some reports of *de novo* melatonin production in other cell types (Jaworek et al., 2005). In the pineal gland, melatonin production is under the direct control of a multisynaptic circuit originating from non-rod/non-cone ganglion photoreceptors. These cells give rise to a fiber tract that projects, via relays in the hypothalamus and spinal cord, to the superior cervical ganglion, which in turn innervates the pineal gland (Cardinali and Vacas, 1987; Moore, 1996). This circuit is tonically inhibited by light and disinhibited during the dark, allowing the pineal gland to directly track photic inputs (Hoffman et al., 1981; Moore, 1996). Unlike most hormones the most salient feature of the melatonin signal is the duration of the elevation of blood concentrations and not the magnitude of the concentration (Hoffman et al., 1986). Virtually the entirety of the short-day phenotype can be blocked by pinealectomy, and the short-day phenotype can be induced in long days with appropriately timed melatonin infusions (Goldman, 2001).

Melatonin itself has the ability to scavenge free radicals, but most actions of melatonin require the presence of high-affinity receptors. Melatonin receptors are high-affinity and G-protein–coupled, and in mammals there are two known receptors (MT1 and MT2; although the MT2 receptor is not present in Siberian hamsters [Weaver et al., 1996]) that are found on a variety of CNS cell types and, importantly, a variety of immune tissues and circulating cells (Dubocovich, 1995; Pozo et al., 1997; Barjavel et al., 1998; Konakchieva et al., 1999). In birds, a third melatonin receptor (MT3) has also been identified (Reppert et al., 1995). Melatonin has strong immunomodulatory actions in both seasonal and non-seasonal contexts and across taxa (Bentley et al., 1998; Beskonakli et al., 2000; Dahl et al., 2002; Moore and Siopes, 2003). Importantly, most commonly studied strains of laboratory mice carry a genetic mutation in the N-acetyltransferase gene. N-acetyltransferase is the rate-limiting step in melatonin biosynthesis, and as such, mice display strongly attenuated or completely absent elevation of melatonin in the dark (Goto et al., 1989; Reppert and Weaver, 1995). Consequently, studies manipulating melatonin in domesticated strains of mice must be interpreted with caution.

In domestic and laboratory rodents melatonin tends to be immunoenhancing. Surgical or functional pinealectomy (exposure to constant light) reduced the size of thymic glands *in vivo* (Csaba and Barath, 1975; Mahmoud et al., 1994), an effect that appears to be mediated by reductions in circulating lymphocytes (Maestroni et al., 1987). In contrast, afternoon injections of melatonin increased thymic and splenic masses in rodents (Vaughan et al., 1976). Functionally, cell-mediated and humoral immune responses are inhibited by pinealectomy, an effect that can be blocked by exogenous melatonin (Vermeulen et al., 1993; Yellon et al., 1999). Melatonin appears to be sufficient to enhance diverse immunological processes, including lymphocyte proliferation, antigen presentation, cytokine production, and antibody-dependent cellular cytotoxicity effects that may be mediated in part via interactions with endogenous opioids (Maestroni et al., 1987; Maestroni et al., 1988; Vermeulen et al., 1993). *In vitro* melatonin can induce and fine-tune the cytokine response to various mitogenic and inflammatory signals (Liu et al., 2001). For instance, melatonin appears to push T-helper cells towards the Th1 pathway via upregulating IL-12 production (Pioli et al., 1993).

Melatonin has somewhat divergent effects on various disease states—sometimes suppressing inflammatory responses and at other times enhancing them. For instance, melatonin is a potent protective factor in a variety of injury and ischemia models, in part via suppression of runaway inflammatory responses (Ates et al., 2004; Munoz-Casares et al., 2006; Pandi-Perumal et al., 2006). However, the animal model of multiple sclerosis is attenuated by treatment with luzindole, a melatonin receptor antagonist, and collagen-induced arthritis is exacerbated by exogenous melatonin in mice (Hansson et al., 1992; Constantinescu et al., 1997).

The majority of photoperiodic adjustments can be blocked by pinealectomy and restored with exogenous melatonin. For instance, short-photoperiod–induced attenuation of sickness responses and cytokine production can be induced by chronic but not short-term lengthening of the melatonin signal (Bilbo et al., 2002; Bilbo and Nelson, 2002). Pinealectomy had no effect in long days but blocked the short-day reductions in behavioral responses to LPS (Wen et al., 2007). However, the attenuated febrile response persisted in short-day animals without a pineal gland (Wen et al., 2007). This is particularly interesting in light of data indicating that melatonin implanted directly into the suprachiasmatic nucleus is sufficient to attenuate the behavioral responses to inflammatory events, but not the fever, whereas peripheral melatonin implants induce the whole short-day phenotype (Freeman et al., 2007).

Melatonin does not have universally immunoenhancing activities in seasonal animals but tends to recapitulate the effects of short day lengths. Short days provoke a variety of immunomodulatory effects depending on the species and branch of the immune system considered. For example, deer mice implanted with melatonin capsules show increased lymphocyte proliferation (Demas and Nelson, 1998a, 1998b). However, Siberian hamsters display suppressed antibody production and sickness responses following exposure to short day lengths, and these effects can be recapitulated by exogenous melatonin (Yellon et al., 1999; Bilbo and Nelson, 2002; Drazen et al., 2002).

When initially considering the proximate mediators of photoperiodic control over the immune system, it was initially difficult to determine whether melatonin acted directly on immune processes or indirectly via alterations of other neuroendocrine mediators such as glucocorticoids or sex steroid hormones (Demas and Nelson, 1998a, 1998b). One method for teasing apart these two possibilities was to administer melatonin to immune cells *in vitro*. Melatonin added to lymphocyte cultures increased proliferative responses to the mitogen concanavalin A independent of day length or the reproductive responses to day length (Drazen et al., 2000a; Kriegsfeld et al., 2001). Short day lengths attenuate lymphocyte proliferation in Siberian hamsters. Addition of melatonin to these cultures reduced proliferative responses in long but not short days, suggesting that the melatonin response was already established in hamsters chronically exposed to long-duration melatonin rhythms (Prendergast et al., 2001). Importantly, *in vivo* induction of the short-day pattern of immune responses requires extended exposure to long-duration melatonin signals (Bilbo and Nelson, 2002; Drazen et al., 2002).

The melatonin signal also shares some temporal characteristics with photoperiod-induced changes in immune responses. After prolonged exposure to short day lengths, animals become refractory to the suppressive effects of short day lengths on the reproductive system (Stetson et al., 1977; Karsch et al., 1986). This feature allows animals to recrudesce their reproductive systems independent of photic inputs in preparation for the spring breeding season (Bartness et al., 1993). Prendergast and colleagues asked whether photorefractoriness, which is observed in the reproductive system, also occurs in the immune system. Twelve weeks of exposure to short day lengths increased spleen mass, total circulating lymphocytes, and mitogen-stimulated lymphocyte proliferation in deer mice compared to those animals housed in long days. However, if mice were maintained in short day lengths for 32 weeks, then gonadal recrudescence occurred and immune responses more closely resembled long-day mice than those housed in short day lengths for 12 weeks (Prendergast and Nelson, 2001). The spontaneous recrudescence is not caused by a loss of long-duration pineal signals, as spontaneous regression occurs even in hamsters treated with exogenous melatonin (Bittman, 1978). Rather, there is a loss of sensitivity to melatonin in the reproductive system (Bittman, 1978). In the immune systems of Siberian hamsters, photorefractoriness was also associated with insensitivity to exogenous melatonin (Prendergast et al., 2002). Finally, animals that respond to photoperiod with changes in their reproductive systems are also more sensitive to effects of melatonin on the immune system than are closely related species that are not reproductively responsive to photoperiod (Demas et al., 1996).

SEX STEROID HORMONES

The original link between seasonal reproductive cycles and immunological tissue mass can likely be attributed to the finding in birds that the thymus atrophies markedly at the onset of puberty (Aime, 1912; Riddle, 1924) and that castration can induce thymic hypertrophy (Hammar, 1929). These discoveries were made before it was understood that the thymus was an immunological organ and represent some of the earliest links between reproductive function and immunity. Sex steroid hormones are the likely cause of these phenomena and are potent immune modulators and downstream targets of melatonin signaling in a seasonal context. As a broad generalization, females have stronger immune responses than do males, and this sex difference is mediated by the generally immune-enhancing role for estrogens and the generally immunosuppressive role for androgens (Grossman, 1984; Folstad and Karter, 1992; Muehlenbein and Bribiescas, 2005). While this is certainly an oversimplification, there is strong evidence that females are generally more resistant to infection and parasitization than are males and are more susceptible to autoimmune diseases (Klein, 2008). In the early 1990s Folstad and Karter proposed a mechanistic extension of the Hamilton Zuk hypothesis indicating that maintenance of secondary sexual characteristics siphons energy away from the immune system and therefore acts as an honest signal of male quality (Zuk, 1990; Folstad and Karter, 1992). A full review of this topic is beyond the scope of this chapter, and we will consider these hormones only in the context of seasonality.

Androgens and estrogens, like all steroid hormones, signal primarily via nuclear receptors that are also ligand-activated transcription factors, but can also signal via membrane-bound receptors. Chronic exposure to short day lengths reduces sex steroids via the withdrawal of gonadotropin-releasing hormone release into the pituitary portal blood supply and the downstream decline in circulating gonadotropins. Androgen and estrogen receptors are also found on a variety of immune cells and tissues (Sullivan and Wira, 1979; Samy et al., 2000).

One strong possibility for the regulation of the immune system by melatonin is via reduction of circulating sex steroids that disinhibit immune responses. This hypothesis has fallen out of favor as it is not strongly supported by the data. First, if estrogens are immunoenhancing and androgens immunosuppressing, then regression of the reproductive tract should have opposite effects on the immune system of males and females. However, immunological adjustments are relatively similar in both males and females, although the magnitude of the plasticity tends to be greater in males (Demas and Nelson, 1998a, 1998b; Bilbo and Nelson, 2003; Weil et al., 2006). Secondly, if sex steroid hormones are responsible for photoperiodic changes in immune responses, then gonadectomy and hormone replacement should be sufficient to recapitulate the effects of day length. This prediction was also not supported by experimental findings. Neither gonadectomy nor hormone replacement modulated short-day enhancements of lymphocyte proliferation in either male or female deer mice (Demas and Nelson, 1998a, 1998b). Similarly, the enhancement of DTH responses in male Siberian hamsters was not affected by either castration or testosterone replacement (Prendergast et al., 2005). In long day lengths, both estrogens and androgens increase lymphocyte proliferation in female and male Siberian hamsters respectively (Bilbo and Nelson, 2001). Testosterone also failed to recapitulate the effects of photostimulation on DTH responses in house sparrows (Greenman et al., 2005), and photostimulation modulated immune responses in castrated European starlings (*Sturnus vulgaris*) (Bentley et al., 1998).

Thus, sex steroids are not the principal driver of photoperiodic adjustments in immune responses among rodent species thus far examined. However, there could be smaller modulatory effects of sex steroids that are masked by the larger effects of melatonin. In a recent analysis, some aspects of photoperiodic adjustments in the immune system were modulated strongly by castration, including the total number of lymphocytes and the weight loss and anorectic responses to LPS (Prendergast et al., 2008). This study supports a previous one in which melatonin and sex steroids interacted to regulate the total number of circulating lymphocytes in juvenile Siberian hamsters (Prendergast et al., 2003). Taken together, the role of sex steroids in seasonal changes in immune responses appears to modulate the effects of melatonin on immune responses and to occur in a character-specific manner.

GLUCOCORTICOIDS

Glucocorticoid hormones are the end-product, primary effectors, and principal negative regulators of the hypothalamo-pituitary-adrenal (HPA) axis. The HPA axis is bidirectionally intertwined with the immune system. In common with sex steroid hormones, glucocorticoids signal through two steroid receptors: (1) mineralocorticoid (MR) and (2) glucocorticoid receptors (GR), which bind glucocorticoids with high and low affinity, respectively. This dual receptor system therefore can detect both tonic and phasic elevation of glucocorticoid concentrations. Classically, glucocorticoids are potent anti-inflammatory molecules that can be used clinically for the treatment of dozens of conditions involving runaway inflammation. MR and GR are present on most types of immune cells (Munck et al., 1984; Armanini et al., 1988; McEwen et al., 1997).

Glucocorticoids, cortisol in primates and carnivores and corticosterone in most other animals, are often considered stress hormones as their concentrations rise in response to a variety of psychological and physiological stressors. However, glucocorticoids are also important metabolic hormones involved in energy mobilization and glucose metabolism (Thompson and Lippman, 1974).

The first fully described neuroendocrine-immune loop was the relationship between innate immune responses in the periphery and the HPA axis. Inflammatory events cause pro-inflammatory cytokine production and release, which in turn activates the HPA axis at all three anatomical levels. The resulting glucocorticoids then serve to downregulate cytokine production in a negative feedback fashion. Therefore, glucocorticoids can be conceptualized as brakes on the immune system that can serve to prevent runaway inflammation. The downside of this relationship is that prolonged or chronic elevation of blood glucocorticoids can suppress immune responses (Munck et al., 1984; McEwen et al., 1997; Sapolsky et al., 2000).

One prominent theory regarding the ultimate causes of short-day–induced immunoenhancement is that in the wild, seasonal stressors would cause glucocorticoid concentrations to rise and therefore suppress the immune system (Nelson and Demas, 1996; Demas and Nelson, 1998a, 1998b). Seasonal changes in glucocorticoid hormones and their potential role in mediating plasticity in the immune system have received extensive study. In the laboratory, aspects of HPA axis physiology, including MR and GR gene expression, circulating glucocorticoid concentrations, and negative feedback, are altered by day length (Ronchi et al., 1998; Pyter et al., 2007).

In the laboratory, the effects of photoperiod on HPA axis physiology are much smaller than those of reproductive axis, and the direction of the change in glucocorticoid concentrations differs across species (Ronchi et al., 1998; Bilbo et al., 2002; Pyter et al., 2005). Glucocortioids also do not appear to be sufficient to mediate all of the effects of day length on the immune system. Immune responses are enhanced in short day lengths in some species that also have higher circulating glucocorticoids in short days (Bilbo et al., 2002; Pyter et al., 2005; Weil et al., 2006). The data must be interpreted with caution, as this single-time-point sampling approach does not take into account the possibility of differential sensitivity to glucocorticoids, or alterations in the 24-hour pattern of glucocorticoid secretion.

Recently, there has been a critical advancement in the understanding of the relationship between glucocorticoids and immune responses. Although it is certainly true that chronically elevated glucocorticoid levels are immunosuppressive, a similar phenomenon does not occur when glucocorticoids are acutely elevated. This makes some logical sense: if stress-induced increases in circulating glucocorticoids are typically associated with wounding or other injuries, then it makes little sense to predict that immune responses would be inhibited following acute stressors. In fact, although immune cells leave the bloodstream during acute stressors, they typically do not die but are trafficked towards the front lines of immune defenses, such as the skin, lymph nodes, and gut (Dhabhar and McEwen, 1997). This effect is mediated by acute increases in glucocorticoids (and sympathetic catecholamines) and can be recapitulated with exogenous hormones (Dhabhar and McEwen, 1999). In Siberian hamsters housed in short day lengths, acute stressors enhanced circulating cortisol (in contrast to most rodents studied, Siberian hamsters use cortisol rather than corticosterone as their primary glucocorticoid), DTH responses, and the trafficking of lymphocytes out of the blood. Short day lengths therefore enhance both basal immune responses and the capacity for an immunological boost following an acute stressor (Bilbo et al., 2002). Overall, glucocorticoids appear to play a modulatory role in response to changing day length and may have a greater overall role when environmental conditions are more challenging than standard laboratory conditions.

PROLACTIN

Prolactin is a peptide released from the anterior pituitary and has dozens of reported functions, ranging from the regulation of somatic growth and development, water and electrolyte balance, and the control of the immune system (Goffin et al., 1999). Prolactin receptors are expressed on many leukocytes, and removal of the anterior pituitary induces thymic involution and deficits in cell-mediated and humoral immunity that can be blocked with exogenous prolactin (Smith, 1930; Reber, 1993). Exogenous prolactin enhances various aspects of immune responses, and prolactin receptor antagonists are sometimes used to counteract excessive inflammation associated with autoimmunity (Matera et al., 1992; Vera-Lastra et al., 2002). Prolactin also antagonizes the effects of glucocorticoids on lymphocyte apoptosis (Dorshkind and Horseman, 2001).

Short-day–induced reductions in circulating prolactin are observed in many species (Goldman and Nelson, 1993). Prolactin is one of the few immunoenhancing signals that is consistently elevated in long day lengths, and so its role in short-day immunoenhancement remains unspecified. Deer mice were treated with the chemical carcinogen dimethylbenzathracene (DMBA); long- but not short-day animals developed squamous cell carcinomas (Nelson and Blom, 1994). However, treatment with the prolactin release inhibitor bromocriptine significantly attenuated tumorigenesis. In steers, short photoperiods enhance lymphocyte proliferation neutrophils chemotaxis, and this effect was blocked with the administration of long-day levels of prolactin (Auchtung and Dahl, 2004). More research is needed, but it seems highly probable that prolactin is an important downstream modulator of photoperiodic plasticity in the immune system.

AUTONOMIC NERVOUS SYSTEM

Although not a pure neuroendocrine system, the sympathoadrenal system and the autonomic nervous system (ANS) are additional mechanisms for control of peripheral immune responses by the CNS. The ANS is organized into sympathetic and parasympathetic branches that have been classically described as the fight/flight and rest/digest arms of the system, respectively. Sympathetic inputs exist to most organs and release the catecholamines epinephrine and norepinephrine onto target tissues. The exception to this rule is in the adrenal gland, where activation of the sympathetic inputs causes epinephrine to be released into the general circulation. The parasympathetic nervous system has generally opposing actions on target tissues, and these effects are mediated by synaptic release of acetylcholine onto target tissues. Virtually all of the tissues in the body are innervated by the ANS, and receptors for both epinephrine and acetylcholine are found on most immune cells (Gordon et al., 1978; Kin and Sanders, 2006).

The ANS seems an attractive system for photoperiodic plasticity, as changing the tone of either branch can significantly alter a number of peripheral physiological systems simultaneously. The study of this system in a photoperiodic context has been somewhat complicated because the final synapse connecting the retinal inputs to the pineal gland is sympathetic in nature, thus making pharmacological manipulations of this system more complicated. In any case, the classical rest and digest activities of the parasympathetic nervous system seem consistent with the shift in priorities from reproduction towards cellular maintenance and survival during winter (McEwen and Wingfield, 2003; Yun et al., 2004). Indeed, it is known that the parasympathetic nervous system mediates the slowing of cardiovascular and metabolic activity in hibernating rodents (Morhardt, 1970; Zosky, 2002). In terms of the cardiovascular system, short day lengths increase the activity of both the sympathetic and parasympathetic inputs to the heart and enhance stress-induced withdrawal of both branches in male Siberian hamsters (Weil et al., 2009). Consistent with this observation, short day lengths increased splenic norepinephrine content (Demas et al., 2002). Short day lengths reduce visceral fat depots in Siberian hamsters, and this effect can be blocked by removing sympathetic inputs to the fat pad and removing the adrenal medulla bilaterally (Demas and Bartness, 2001). Multisynaptic tract tracing studies have identified cells that send sympathetic axons to fat pads in hypothalamic regions and also express melatonin receptors (Song and Bartness, 2001).

There is also a role for the ANS in mediating photoperiodic control over the immune system. Adrenal demedullation, which reduces circulating catecholamines but leaves direct sympathetic inputs to target tissues intact, reduced anti-KLH IgG production in Siberian hamsters housed in long but not short day lengths. In contrast, chemical removal of the sympathetic inputs to the spleen reduced antibody production in short but not long days. Apparently, the two noradrenergic systems differentially affect the antibody response across the year, with the sympathoadrenal system playing a primary role in long days and sympathetic inputs to the spleen having a greater effect in short days (Demas et al., 2002). Basal and mitogen-induced lymphoproliferative responses are reduced in short-day Siberian hamsters (Prendergast et al., 2001). *In vitro* norepinephrine, the principal neurotransmitter of the sympathetic nervous system, further suppresses proliferative responses in short but not long days; this is mediated by beta-adrenergic receptors, as beta- but not alpha-receptor antagonists can attenuate the effect. Finally, denervation of the spleen attenuates leptin-induced increases in antibody production in short but not long days, suggesting that the ANS is involved in transducing energy-related signals to lymphoid cells (Demas, 2002). Importantly, it appears to be both an increase in the activity of sympathetic inputs to immunological tissues *and* differential sensitivities to the catecholaminergic inputs that mediate photoperiod-autonomic effects on the immune system.

EVOLUTION OF PHOTOPERIODIC CONTROL OF THE REPRODUCTIVE AND IMMUNE SYSTEMS

The neuroendocrine system can be conceptualized as a mechanism by which animals can coordinate physiological and behavioral systems internally and to the changing environment (Adkins-Regan, 2005). In the reproductive system, there are hormones that coordinate steroidogenesis and gametogenesis, and presumably, over evolutionary time, there was selection pressure to coordinate sexual behavior with the maturation of the gametes. Therefore, the same hormones that control the physiological processes in the reproductive system also coordinate mating behavior. Further, this system has evolved sensitivity to other neural and neuroendocrine signals such that signals such as food restriction and inflammatory mediators can shut it down.

An analogy can be drawn between the evolution of neuroendocrine/behavior interactions relating to mating and the photoperiod/melatonin/immune system. The widespread and cross-taxa pattern of seasonal breeding generally and photoperiodism specifically suggests that natural selection has favored a strategy of accurately timing reproductive processes to the time of year when resources are plentiful. In mammals melatonin signaling mediates the link between photic cues and the reproductive axis. We have presented multiple examples of photoperiodic and melatonergic modulation of immune function that occurs across taxa and in both field and laboratory settings. Considered together, these data suggest that two distinct evolutionary processes occurred wherein the reproductive and immune system could both separately respond to photoperiodic cues.

There is also an alternative explanation that regression of the immune system frees up the energy needed for full utilization of immune defenses. Although there are likely small modulatory effects of the reproductive system on immune defenses, considerable evidence documents that photoperiodic effects on the immune and reproductive systems can be functionally dissociated. In contrast, both the immune and reproductive systems appear to respond in similar, but not identical, ways to the same signal. Next, we provide support that immune and reproductive responses to changing day length can be functionally dissociated and argue that changes in the reproductive system do not directly mediate photoperiodic changes in immune function.

INTERMEDIATE, EXTREME, AND PERINATAL PHOTOPERIODS

The reproductive system of photoperiodic rodents attends both to the absolute duration of the photoperiod and to whether the photoperiod is increasing or decreasing. This makes intuitive sense because every individual photoperiod occurs twice each year. For instance, at approximately 40° north latitude, the relatively long photoperiod 14 hours of light and 10 hours of dark occurs at both the beginning of the breeding season (May) and at the end (August). Small rodents born in the spring generally develop rapidly in an attempt to breed before the end of the season (Forger and Zucker, 1985; Shaw and Goldman, 1995). In contrast, animals born late in the summer delay reproductive development until the following spring. Because photoperiods of LD 14:10 may indicate either late spring or fall conditions, it is imperative that photoperiodic animals be able to distinguish between them. Young photoperiodic rodents compare the current ambient photoperiod to one that was encoded prenatally. If Siberian hamsters are born to mothers housed in unambiguously long photoperiods (LD 16:8) and then transferred to day-length conditions of LD 14:10 at birth, then they will delay reproductive development. In contrast, hamsters housed prenatally in short day lengths and then transferred to LD 14:10 interpret that signal as a lengthening photoperiod and undergo rapid reproductive development. Do decreasing photoperiods that suppress development of the reproductive system also induce the short-day pattern of immune responses (Prendergast et al., 2004)? Again, Siberian hamsters were bred in long days, then at birth transferred to either short day lengths or to an intermediate-length photoperiod (LD 14:10). Animals transferred to short day lengths or intermediate photoperiods delayed reproductive development, indicating that the reproductive system can attend to both the absolute duration and the direction of change of day length. In contrast, the immune system appeared to respond principally to the absolute day length. DTH responses and the numbers of circulating lymphocytes and leukocytes were higher in animals transferred to short day lengths than all other groups; hamsters transferred from long to intermediate photoperiods resembled short-day animals reproductively but long-day animals immunologically (Prendergast et al., 2004); see Figure 9.1.

In a related study, Siberian hamsters were bred in either long or short days, then either transferred to the opposite photoperiod or maintained in their current photoperiod at weaning. Again, hamsters housed in short day lengths until weaning had small undeveloped reproductive systems, but if they were transferred into long days, then they quickly developed their reproductive

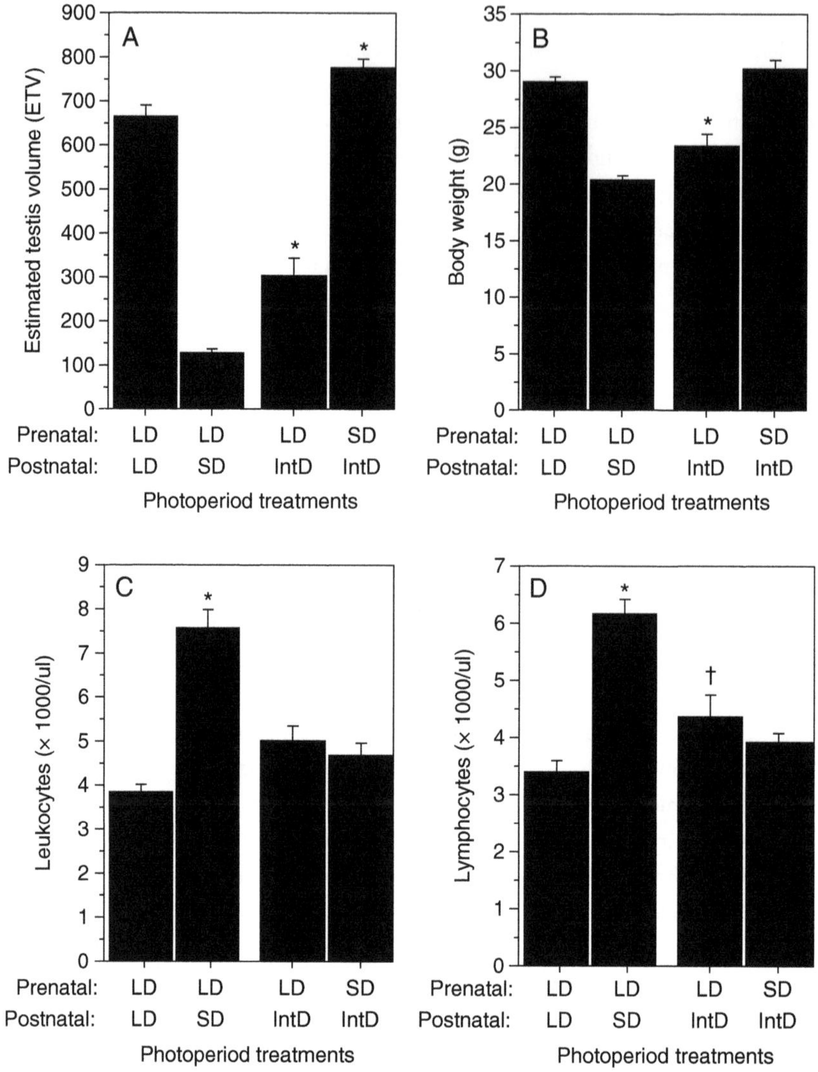

FIGURE 9.1: Mean (±SEM) estimated testis volumes (A) and body mass (B) on day 40 of male Siberian hamsters in Experiment 1. Hamsters were gestated in LD (16 h light/day) and either remained in the LD photoperiod postnatally (LD–LD) or were transferred on the day of birth (day 0) to either a SD (8 h light/day) photoperiod (LD–SD) or an IntD (14 h light/day) photoperiod (LD–IntD); SD–IntD hamsters were gestated in SD and transferred to IntD on the day of birth. *$p < 0.0083$ vs. all other groups. Mean (±SEM) blood leukocyte numbers (white blood cells [C], lymphocytes [D], monocytes [C]).

systems, and did so to the level of hamsters that had been housed in long days their entire lives. Similarly, hamsters transferred from long to short photoperiods at weaning halted reproductive development and maintained regressed gonads. Importantly, the immune system behaved differently: DTH responses were elevated in hamsters that had been born and maintained in short days throughout their lives but were markedly lower in hamsters that been in long days perinatally after weaning or during both developmental epochs (Weil et al., 2006). Thus, short day lengths at weaning were sufficient to inhibit reproductive development but did not similarly affect cell-mediated immune responses. This is consistent with data from another study that demonstrated that past photoperiod exposure can alter the immune system independently of the direction of the photoperiod change (Prendergast and Pyter, 2009).

Another way to address this question of differential photoperiodic effects on the reproductive and immune system is to investigate animals that live above the Arctic Circle and thus experience extreme annual changes in photoperiod ranging from nearly constant light to nearly constant dark. Collared lemmings inhabit parts of northern Canada and Greenland and inhabit the highest latitude home range of any small mammal (Millar, 2001). In a laboratory study, collared lemmings were housed in one of three photoperiods: LD 8:16, 16:8, or 22:2 (Weil et al., 2006); see Figure 9.2. The LD 22:2 photoperiod was chosen because it is reminiscent of constant light but provides a photic clue to which the lemmings can entrain.

These photoperiod treatments induced three distinct phenotypes. Short day lengths induced reproductive regression and a molt to a winter pelage, whereas large functional gonads were maintained in the two long photoperiods. However, DTH responses were elevated in both short and intermediate (LD 16:8) day lengths; swelling responses displayed a marked inhibition in the LD 22:2 condition. Lemmings housed in LD 16:8 maintained both a functional reproductive system and enhanced cell-mediated immune responses (Weil et al., 2006).

A final method to tease apart the photoperiodic effects on the reproductive and immune system is to investigate photoperiodic nonresponders. A subset of any photoperiod cohort (ranging from 10% to 100%) fails to regress their gonads in response to short days (Nelson, 1987). The proximate mediators of photoperiodic nonresponsiveness vary from species to species but typically involve either a failure to entrain properly to the new photoperiod or an insensitivity to the melatonin signal (Prendergast et al., 2001). Photoperiodic nonresponsiveness may serve as an alternative strategy for animals to take advantage of a mild winter with increased opportunities to mate (Nelson, 1987). This is highlighted by studies of hamsters selectively bred for reduced reproductive responsiveness

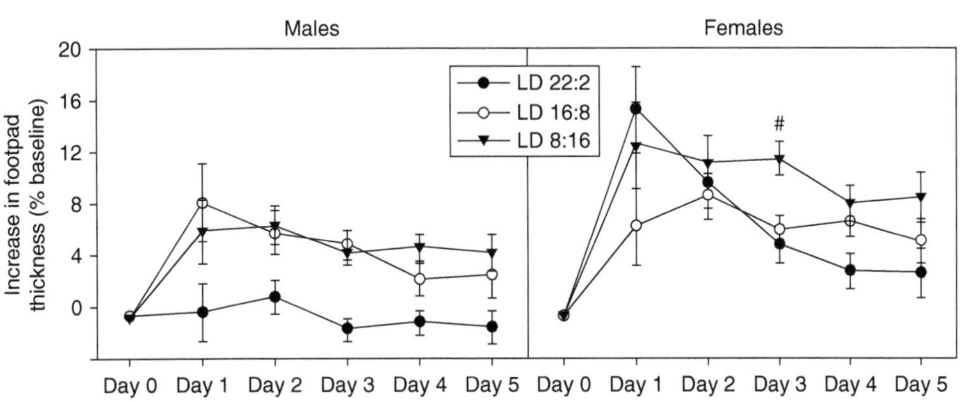

FIGURE 9.2: Photoperiod altered cell-mediated immune function. Delayed-type hypersensitivity (DTH) responses were higher in lemmings housed in LD 8:16 and 16:8 than those housed in LD 22:2. Females had stronger footpad DTH responses than males. Data are presented as mean percent of baseline footpad thickness ± SEM. *Significantly lower than all other photoperiod groups of the same sex. #Significantly higher than all other photoperiod groups of the same sex.

to short day lengths. When the nonresponsive line was transferred to short day lengths as adults, fewer animals regressed their gonads and the overall reduction in reproductive parameters was reduced. However, the photoperiodic phenotype could be partially "rescued" by transferring into short day lengths early in life, when presumably it is more important to attend to seasonal cues that guide reproductive development (Freeman and Goldman, 1997). In any case, photoperiod nonresponders by definition do not regress their gonads in short day lengths. This poses the obvious question: do the immune systems of photoperiod nonresponders resemble animals housed in long days or those housed in short days that did regress their gonads? Siberian hamsters were housed in long or short photoperiods and then injected with the antibody-generating protein KLH. Hamsters were then killed and their reproductive responses to photoperiod assessed. As in previous studies, short days induced two patterns of reproductive responses: (1) hamsters that regressed their gonads (responders; defined as those animals with a paired testes mass >4 standard errors lower than the long day mean) or (2) nonresponders that failed to meet this level of gonadal regression. The anti-KLH IgG response was inhibited by exposure to short day lengths, and this suppression occurred independently of reproductive responsiveness to day length (Drazen et al., 2000a). Photoperiodic nonresponders provide an additional example of experimental manipulations that indicate different critical cues for day-length–induced modification of the immune system.

NON-PHOTIC ENVIRONMENTAL FACTORS

Many of the laboratory studies described in this chapter have used photoperiodic signals to examine changes in the immune and reproductive systems. This is reasonable because in nature day length is the principal cue used to time seasonal changes in physiology and behavior. However, photoperiodism evolved to predict environmental conditions that have direct fitness consequences, such as low temperatures and food restriction, so it is critical that these environmental conditions be layered onto photoperiod manipulations to more fully understand how photoperiod alters fitness parameters. Additionally, these multifactorial studies can further our attempt to understand the evolution of seasonal changes in the immune and reproductive systems.

FOOD AND TEMPERATURE

Demas and Nelson (1996) housed deer mice in long or short days at either room temperature or in low temperatures (8°C). Short photoperiods induced reproductive regression and increased basal IgG and spleen sizes; cold temperatures reduced both of these immune measures. Importantly, however, the mice housed in cold short days showed reduced circulating IgG only to the level of the long-day mice housed in mild temperatures. Importantly, these animals were fed *ad libitum*, suggesting that they were able to compensate for the increased thermoregulatory expenditures by increasing food intake. In a follow-up study, the effects of food restriction was also examined (Demas and Nelson, 1998a, 1998b). Splenocyte proliferation was assayed in deer mice housed in either long days or short, at room temperature or 8°C, and either food-restricted or fed *ad libitum*. Low temperatures were sufficient to suppress immune responses, and this effect was inhibited by prior exposure to short day lengths. However, when mice living in low temperatures were food-restricted, short day lengths were no longer sufficient to block the suppressive effects on immune responses. Similarly, Bilbo and colleagues (2004) used a simple environmental manipulation, voluntary exercise, that enhances cell-mediated immune responses (Pedersen and Hoffman-Goetz, 2000). Siberian hamsters were housed in short or long days and then assigned to a cage that contained a running wheel or a standard cage. Finally, half of the hamsters in all conditions were food-restricted, whereas the remaining hamsters were fed *ad libitum*. Voluntary exercise increased DTH responses in all long-day hamsters, but only in short-day hamsters that had *ad libitum* food. Further, food restriction reduced DTH responses in short but not long days (Bilbo and Nelson, 2004).

SOCIAL INFLUENCES

The reproductive system is robustly responsive to social stimuli, as the presence of same- or opposite-sex conspecifics can have important modulatory effects on reproductive development and function (Vandenbergh, 1971). The immune system apparently can also attend to social cues and adapt in ways that depend on the season, sex, and reproductive status of conspecifics and whether the social interactions are agonistic or antagonistic. In a seasonal context, it seems rational that animals should attend to the reproductive status of their conspecifics. For instance, during mild winters, when abundant fat stores allow some females to maintain their reproductive systems and potentially bear a litter, it is advantageous for males to be able to overcome the suppressive effects of short day lengths and mate. Male Siberian hamsters, housed with intact females and transferred into short day lengths, do not regress their gonads and in fact are capable of bearing litters in the laboratory (Hegstrom and Breedlove, 1999). This leads to the question: if the anti-reproductive effects of short days are blocked by housing with opposite-sex conspecifics, then will the immune-enhancing effects also be antagonized? Male Siberian hamsters were housed with a female, a male littermate, or alone, then housed in long or short days for 10 weeks. At 10 weeks DTH responses to dinitrofluorobenzene were assessed. Short day lengths enhanced DTH responses in single-housed males and those housed with a same-sex conspecific but not those housed with an opposite-sex hamster (Weil et al., 2007); see Figure 9.3.

Reproductive parameters (circulating testosterone concentrations and testes mass) were enhanced by housing with a female in long days only. Unlike the previous study, housing with a female conspecific was not sufficient to inhibit reproductive regression in short day lengths (Weil et al., 2007). The different outcomes may reflect different endocrine states of the females; ovariectomized females were used in the study by Weil and colleagues, whereas gonadally intact females were used in the previous study. Importantly, these data provide three important clues about the evolution of the relationship between immune and reproductive changes across the year. First, the social housing-induced blockade of testicular regression apparently requires intact females, although whether it is ovarian steroids or some aspect of female behavior that is the salient cue requires further study. Second, short-day suppression of the immune system by reproductive cues occurred independently of both a fully functioning reproductive system and an ovary-containing female. Once again, these data indicate that similar and usually co-occurring stimuli mediate effects on the reproductive and immune systems, but that laboratory studies can reveal subtly different cues that each system attends to. Finally, the enhancement of the immune system by social

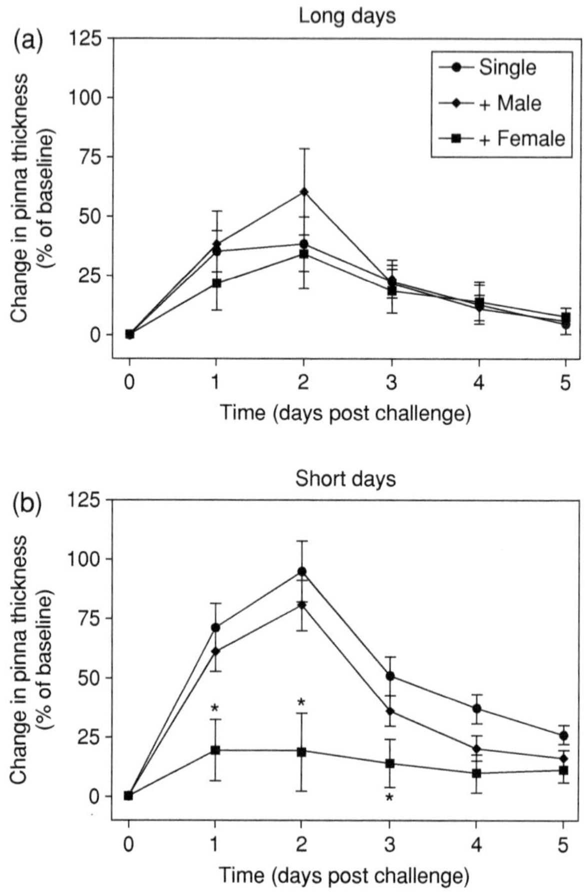

FIGURE 9.3: Photoperiod and social housing condition alter delayed-type hypersensitivity (DTH) responses. Short-day hamsters had enhanced DTH responses if housed singly or with a same-sex littermate. Data are presented as a percentage of baseline values (mean ± SEM) for long- (a) and short-day (b) hamsters. *Significantly lower than all other groups in the same photoperiod.

stimuli (and, by extension, short day lengths) cannot occur simply due to reproductive regression, because suppressed long-day–like immune responses were induced in the absence of a fully functional reproductive system. Similar results were reported for white-footed mice, in that housing with a female attenuated short-day–induced enhancements of DTH responses independent of activation of the reproductive system (Pyter et al., 2005).

Taken together, these studies are important because they demonstrate two important points. Under normal conditions, photoperiodic effects on the immune and reproductive systems are overlapping: (1) the specific properties of the melatonin/photoperiod cue are subtly different and (2) because reproductive and immunological changes can be dissociated, it is unlikely that the energetic costs of maintaining a functional reproductive system are the principal mediator of short-day enhancements of immune responses. The central thesis of this section was that understanding the different critical cues that regulate the photoperiod-induced plasticity in the reproductive and immunological systems could provide insight as to whether these two sets of adaptations coevolved or whether photoperiodic changes in one system lead to day-length sensitivity in the other. We favor the hypothesis that similar proximate factors (Baker, 1938) led to the evolution of seasonality in both systems, but that these evolutionary events did not necessarily happen together and that the argument that maintaining the reproductive system renders animals unable to mount large immune responses is not supported by the data.

CONCLUSION

The primary goal of this chapter was to review the extant data on seasonal and photoperiodic variation in immune responses and further to discuss the role of neuroendocrine signaling in mediating these adjustments. Although there is now a tremendous amount known about seasonal plasticity in the immune system, there are still a number of unanswered questions. For instance, the molecular link between melatonin receptor activation on the cell surface and changes in immunological parameters (cell numbers, cytokine production, phagocytosis, etc.) has not been fully elucidated. Further, it will be important to determine whether enhanced cell-mediated immune responses in most short-day rodents render them more susceptible to diseases with an inflammatory component, including autoimmune diseases. Additionally, it is not clear how immunological and reproductive tissues take the same melatonin signal and can respond in opposite ways, as occurs in the case of ambiguous photoperiods.

The study of ecoimmunology has been fundamentally interested in the relationship among environmental conditions, immunological defenses, and other physiological systems. Therefore, studying animals that have evolved in highly seasonal environments allows the use of a single ecologically relevant cue (photoperiod) to probe how organisms reapportion scarce resources among competing physiological priorities. Additionally, seasonal breeders allow insight into how physiological systems, such as the reproductive neuroendocrine and immune systems, act to parse out these competing priorities. A tremendous amount of research has been conducted on the evolution of breeding seasons, and this may have obscured the important seasonal adaptations that occur outside of the reproductive system. In fact, the immune system is in some ways as robustly plastic, in a seasonal context, as is the reproductive system. Further studying the reproductive and immune systems in isolation, as if they didn't interact, ignores the complex and reciprocal regulation that has evolved between the two systems. Further, understanding how the priorities within the immune system shift across the year within a single species is very likely to provide insights into the evolution of immunological systems across species. Finally, understanding the way in which organisms shift across the year is likely to provide insights into disease states as varied as seasonal affective disorder, obesity, and infectious disease.

REFERENCES

Adkins-Regan, E (2005). Hormones and Animal Social Behavior. Princeton University Press.

Aime, P (1912). Note sur le thymus chez les chelotniens. C. R. Biol. 889–890.

Armanini, D, Endres, S, Kuhnle, U and Weber, PC (1988). Parallel determination of mineralocorticoid and glucocorticoid receptors in T- and B-lymphocytes of human spleen. Acta Endocrinol. (Copenh.) 118:479–482.

Ates, B, Yilmaz, I, Geckil, H, Iraz, M, Birincioglu, M and Fiskin, K (2004). Protective role of melatonin given either before ischemia or prior to reperfusion on intestinal ischemia-reperfusion damage. J. Pineal Res. 37:149–152.

Auchtung, TL and Dahl, GE (2004). Prolactin mediates photoperiodic immune enhancement: effects of administration of exogenous prolactin on circulating concentrations, receptor expression and immune function in steers. Biol. Reprod. 71:1913–1918.

Auchtung, TL, Salak-Johnson, JL, Morin, DE, Mallard, CC and Dahl, GE (2004). Effects of photoperiod during the dry period on cellular immune function of dairy cows. J. Dairy Sci. 87:3683.

Baillie, SR and Prendergast, BJ (2008). Photoperiodic regulation of behavioral responses to bacterial and viral mimetics: a test of the winter immunoenhancement hypothesis. J. Biol. Rhythms 23:81–90.

Baker, J, (1938). The evolution of breeding seasons. In: deBeer, G (Ed.), Evolution: Essays on Aspects of Evolutionary Biology. Oxford University Press, London, pp. 161–177.

Barjavel, MJ, Mamdouh, Z, Raghbate, N and Bakouche, O (1998). Differential expression of the melatonin receptor in human monocytes 1. J. Immunol. 160:1191–1197.

Bartness, TJ, Powers, JB, Hastings, MH, Bittman, EL and Goldman, BD (1993). The timed infusion paradigm for melatonin delivery: what has it taught us about the melatonin signal, its reception and the photoperiodic control of seasonal responses? J. Pineal Res. 15:161–190.

Bentley, GE, Demas, GE, Nelson, RJ and Ball, GF (1998). Melatonin, immunity and cost of reproductive state in male European starlings. Proc. R. Soc. Lond. B Biol. Sci. 265:1191–1195.

Beskonakli, E, Palaoglu, S, Renda, N, Kulacoglu, S, Turhan, T and Taskin, Y (2000). The effect of pinealectomy on immune parameters in different age groups in rats: results of the weekly alteration of the zinc level and the effect of melatonin administration on wound healing. J. Clin. Neurosci. 7:320–324.

Bilbo, SD, Dhabhar, FS, Viswanathan, K, Saul, A, Yellon, SM and Nelson, RJ (2002). Short day lengths augment stress-induced leukocyte trafficking and stress-induced enhancement of skin immune function. Proc. Natl. Acad. Sci. U. S. A. 99:4067–4072.

Bilbo, SD, Drazen, DL, Quan, N, He, L and Nelson, RJ (2002). Short day lengths attenuate the symptoms of infection in Siberian hamsters. Proc. R. Soc. Lond. B Biol. Sci. 269:447–454.

Bilbo, SD and Nelson, RJ (2001). Sex steroid hormones enhance immune function in male and female Siberian hamsters. Am. J. Physiol. 280:R207–213.

Bilbo, SD and Nelson, RJ (2002). Melatonin regulates energy balance and attenuates fever in Siberian hamsters. Endocrinology. 143:2527.

Bilbo, SD and Nelson, RJ (2003). Sex differences in photoperiodic and stress-induced enhancement of immune function in Siberian hamsters. Brain. Behav. Immun. 17:462–472.

Bilbo, SD and Nelson, RJ (2004). Photoperiod influences the effects of exercise and food restriction on an antigen-specific immune response in Siberian hamsters. Endocrinology 145:556–564.

Bilbo, SD, Quan, N, Prendergast, BJ, Bowers, SL and Nelson, RJ (2003). Photoperiod alters the time course of brain cyclooxygenase-2 expression in Siberian hamsters. J. Neuroendocrinol. 15:958–964.

Bittman, EL (1978). Hamster refractoriness: the role of insensitivity of pineal target tissues. Science 202:648.

Bronson, FH (1985). Mammalian reproduction: an ecological perspective. Biol. Reprod. 32:1–26.

Bronson, FH (1989). Mammalian Reproductive Biology. University of Chicago Press, Chicago.

Cardinali, DP and Vacas, MI (1987). Cellular and molecular mechanisms controlling melatonin release by mammalian pineal glands. Cell. Mol. Neurobiol. 7:323–337.

Constantinescu, CS, Hilliard, B, Ventura, E and Rostami, A (1997). Luzindole, a melatonin receptor antagonist, suppresses experimental autoimmune encephalomyelitis. Pathobiology 65:190–194.

Csaba, G and Barath, P (1975). Morphological changes of thymus and the thyroid gland after postnatal extirpation of pineal body. Endocrinol. Exp. 9:59–67.

Dahl, GE, Auchtung, TL and Kendall, PE (2002). Photoperiodic effects on endocrine and immune function in cattle. Reprod. Suppl. 59:191–201.

Dantzer, R (2001). Cytokine-induced sickness behavior: where do we stand? Brain Behav. Immun. 15:7.

Demas, GE and Nelson, RJ (1998). Photoperiod, ambient temperature and food availability interact to affect reproductive and immune function in adult male deer mice. J. Biol. Rhythms 13:253.

Demas, GE (2002). Splenic denervation blocks leptin-induced enhancement of humoral immunity in Siberian hamsters (*Phodopus sungorus*). Neuroendocrinology 76:178–184.

Demas, GE and Bartness, TJ (2001). Direct innervation of white fat and adrenal medullary catecholamines mediate photoperiodic changes in body fat. Am. J. Physiol. 281:R1499–1505.

Demas, GE, Drazen, DL, Jasnow, AM, Bartness, TJ and Nelson, RJ (2002). Sympathoadrenal system differentially affects photoperiodic changes in humoral immunity of Siberian hamsters (*Phodopus sungorus*). J. Neuroendocrinol. 14:29–35.

Demas, GE, Klein, SL and Nelson, RJ (1996). Reproductive and immune responses to photoperiod and melatonin are linked in Peromyscus subspecies. J. Comp. Physiol. [A] 179:819–825.

Demas, GE and Nelson, RJ (1996). Photoperiod and temperature interact to affect immune parameters in adult male deer mice (*Peromyscus maniculatus*). J. Biol. Rhythms 11:94–102.

Demas, GE and Nelson, RJ (1998a). Exogenous melatonin enhances cell-mediated, but not humoral, immune function in adult male deer mice (*Peromyscus maniculatus*). J. Biol. Rhythms 13:245–252.

Demas, GE and Nelson, RJ (1998b). Short-day enhancement of immune function is independent of steroid hormones in deer mice (*Peromyscus maniculatus*). J. Comp. Physiol. [B] 168:419–426.

Demas, GE, Weil, ZM and Nelson, RJ, (2010). Photoperiodism in mammals: non-reproductive traits. In: Nelson RJ, Denlinger DL and Somers DE (Eds.), Photoperiodism: The Biological Calendar. Oxford University Press, New York, pp. 461–502.

Dhabhar, FS and McEwen, BS (1997). Acute stress enhances while chronic stress suppresses cell-mediated immunity in vivo: a potential role for leukocyte trafficking. Brain. Behav. Immun. 11:286–306.

Dhabhar, FS and McEwen, BS (1999). Enhancing versus suppressive effects of stress hormones on skin immune function. Proc. Natl. Acad. Sci. U. S. A. 96:1059–1064.

Domjan, M (1987). Photoperiodic and endocrine control of social proximity behavior in male Japanese quail (*Coturnix coturnix japonica*). Behav. Neurosci. 101:385–392.

Dorshkind, K and Horseman, ND (2001). Anterior pituitary hormones, stress, and immune system homeostasis. Bioessays 23:288–294.

Drazen, DL, Demas, GE and Nelson, RJ (2001). Leptin effects on immune function and energy balance are photoperiod dependent in Siberian hamsters (*Phodopus sungorus*). Endocrinology 142:2768–2775.

Drazen, DL, Jasnow, AM, Nelson, RJ and Demas, GE (2002). Exposure to short days, but not short-term melatonin, enhances humoral immunity of male Syrian hamsters (*Mesocricetus auratus*). J. Pineal Res. 33:118–124.

Drazen, DL, Klein, SL, Yellon, SM and Nelson, RJ (2000a). In vitro melatonin treatment enhances splenocyte proliferation in prairie voles. J. Pineal Res. 28:34–40.

Drazen, DL, Kriegsfeld, LJ, Schneider, JE and Nelson, RJ (2000b). Leptin, but not immune function, is linked to reproductive responsiveness to photoperiod. Am. J. Physiol. 278:R1401–1407.

Dubocovich, ML (1995). Melatonin receptors: are there multiple subtypes? Trends Pharmacol. Sci. 16: 50–56.

Duncan, MJ and Goldman, BD (1985). Physiological doses of prolactin stimulate pelage pigmentation in Djungarian hamster. Am. J. Physiol. 248:R664–667.

Exton, MS (1997). Infection-induced anorexia: active host defence strategy. Appetite 29:369–383.

Fange, R and Silverin, B (1985). Variation of lymphoid activity in the spleen of a migratory bird, the pied flycatcher (*Ficedula-Hypoleuca* - Aves, Passeriformes). J. Morphol. 184:33–40.

Folstad, I and Karter, AJ (1992). Parasites, bright males and the immunocompetence handicap hypothesis. Am. Nat. 139:603–622.

Forger, NG and Zucker, I (1985). Photoperiodic regulation of reproductive development in male white-footed mice (*Peromyscus leucopus*) born at different phases of the breeding season. J. Reprod. Fertil. 73:271–278.

Freeman, DA and Goldman, BD (1997). Photoperiod nonresponsive Siberian hamsters: effect of age on the probability of nonresponsiveness. J. Biol. Rhythms 12:110–121.

Freeman, DA, Kampf-Lassin, A, Galang, J, Wen, JC and Prendergast, BJ (2007). Melatonin acts at the suprachiasmatic nucleus to attenuate behavioral symptoms of infection. Behav. Neurosci. 121:689–697.

Goffin, V, Binart, N, Clement-Lacroix, P, Bouchard, B, Bole-Feysot, C, Edery, M, Lucas, BK, Touraine, P, Pezet, A, Maaskant, R, Pichard, C, Helloco, C, Baran, N, Favre, H, Bernichtein, S, Allamando, A, Ormandy, C and Kelly, PA (1999). From the molecular biology of prolactin and its receptor to the lessons learned from knockout mice models. Genet. Anal. 15:189–201.

Goldman, BD (2001). Mammalian photoperiodic system: formal properties and neuroendocrine mechanisms of photoperiodic time measurement. J. Biol. Rhythms 16:283–301.

Goldman, BD and Nelson, RJ, (1993). Melatonin and seasonality in mammals. In: Yu, HS and Reiter, RJ (Eds.), Melatonin: Biosynthesis, Physiological Effects and Clinical Applications. CRC, Boca Raton, pp. 225–253.

Gordon, MA, Cohen, JJ and Wilson, IB (1978). Muscarinic cholinergic receptors in murine lymphocytes: demonstration by direct binding. Proc. Natl. Acad. Sci. U. S. A. 75:2902.

Goto, M, Oshima, I, Tomita, T and Ebihara, S (1989). Melatonin content of the pineal gland in different mouse strains. J. Pineal Res. 7:195–204.

Greenman, CG, Martin, LB and Hau, M (2005). Reproductive state, but not testosterone, reduces immune function in male house sparrows (*Passer domesticus*). Physiol. Biochem. Zool. 78:60–68.

Grossman, CJ (1984). Regulation of the immune system by sex steroids. Endocr. Rev. 5:435.

Hammar, J (1929). Die Menschenthymus in Gesundheit und Krankheit: II. Das Organ unter anormalen Korperver-haltnissen. Z. Zelljorsch., Suppl 16.

Hansson, I, Holmdahl, R and Mattsson, R (1992). The pineal hormone melatonin exaggerates development of collagen-induced arthritis in mice. J. Neuroimmunol. 39:23–30.

Hart, BL (1988). Biological basis of the behavior of sick animals. Neurosci. Biobehav. Rev. 12:123.

Hart, BL (1990). Behavioral adaptations to pathogens and parasites: five strategies. Neurosci. Biobehav. Rev. 14:273–294.

Hegstrom, CD and Breedlove, SM (1999). Social cues attenuate photoresponsiveness of the male reproductive system in Siberian hamsters (*Phodopus sungorus*). J. Biol. Rhythms 14:54.

Hoffman, K, Illnerova, H and Vanecek, J (1981). Effect of photoperiod and of one minute light at night-time on the pineal rhythm on N-acetyltransferase activity in the Djungarian hamster *phodopus sungorus*. Biol. Reprod. 24:551–556.

Hoffman, K, Illnerova, H and Vanecek, J (1986). Change in duration of the nighttime melatonin peak may be a signal driving photoperiodic responses in the djungarian hamster. Neurosci. Lett. 67:68–72.

Jaworek, J, Brzozowski, T and Konturek, SJ (2005). Melatonin as an organoprotector in the stomach and the pancreas. J. Pineal Res. 38:73–83.

Karsch, FJ, Bittman, EL, Robinson, JE, Yellon, SM, Wayne, NL, Olster, DH and Kaynard, AH (1986). Melatonin and photorefractoriness: loss of response to the melatonin signal leads to seasonal reproductive transitions in the ewe. Biol. Reprod. 34:265–274.

Kauffman, AS, Cabrera, A and Zucker, I (2001). Energy intake and fur in summer- and winter-acclimated Siberian hamsters (*Phodopus sungorus*). Am. J. Physiol. 281:R519–527.

Kent, S, Bluthe, RM, Kelley, KW and Dantzer, R (1992). Sickness behavior as a new target for drug development. Trends Pharmacol. Sci. 13:24–28.

Kin, NW and Sanders, VM (2006). It takes nerve to tell T and B cells what to do. J. Leukoc. Biol. 79: 1093–1104.

Klasing, KC (2004). The costs of immunity. Acta Zoologica Sinica 50:961–969.

Klein, SL, (2008). Sex differences in infectious and autoimmune diseases. In: Becker, JB, Berkley, JK, Geary, N, Hampson, E, Herman, JP and Young, E (Eds.), Sex Differences in the Brain: From Genes to Behavior. Oxford University Press, USA, New York, pp. 329–353.

Konakchieva, R, Manchev, S, Pevet, P and Masson-Pevet, M (1999). Autoradiographic detection of 2-(125I)-iodomelatonin binding sites in immune tissue of rats. Adv. Exp. Med. Biol. 460:411–415.

Kriegsfeld, LJ, Drazen, DL and Nelson, RJ (2001). In vitro melatonin treatment enhances cell-mediated immune function in male prairie voles (*Microtus ochrogaster*). J. Pineal Res. 30:193–198.

Liu, F, Ng, TB and Fung, MC (2001). Pineal indoles stimulate the gene expression of immunomodulating cytokines. J. Neural Transm. 108:397–405.

Lochmiller, RL and Deerenberg, C (2000). Trade-offs in evolutionary immunology: just what is the cost of immunity? Oikos 88:87–98.

Lochmiller, RL, Vestey, MR and McMurry, ST (1994). Temporal variation in humoral and cell-mediated immune response in a *Sigmodon hispidus* population. Ecology 75:236–245.

Madison, D, Hill, J and Gleason, P (1984). Seasonality in the nesting behavior of *Peromyscus leucopus*. Am. Midl. Nat. 112:201–204.

Maestroni, GJ, Conti, A and Pierpaoli, W (1987). The pineal gland and the circadian, opiatergic, immunoregulatory role of melatonin. Ann. N. Y. Acad. Sci. 496:67–77.

Maestroni, GJ, Conti, A and Pierpaoli, W (1988). Role of the pineal gland in immunity. III. Melatonin antagonizes the immunosuppressive effect of acute stress via an opiatergic mechanism. Immunology 63: 465–469.

Mahmoud, I, Salman, SS and al-Khateeb, A (1994). Continuous darkness and continuous light induce structural changes in the rat thymus. J. Anat. 185 (Pt 1):143–149.

Malpaux, B, Robinson, JE, Wayne, NL and Karsch, FJ (1989). Regulation of the onset of the breeding season of the ewe: importance of long days and of an endogenous reproductive rhythm. J. Endocrinol. 122: 269–278.

Mann, D, Akinbami, M, Gould, K and Ansari, A (2000). Seasonal variations in cytokine expression and cell-mediated immunity in male rhesus monkeys. Cell. Immunol. 200:105–115.

Mann, D, Akinbami, M, Gould, K, Tanner, J and Wallen, K (1993). Neonatal treatment of male monkeys with a gonadotropin-releasing hormone agonist alters differentiation of central nervous system centers that regulate sexual and skeletal development. J. Clin. Endocrinol. Metab. 76:1319.

Martin, LB (2005). Trade-offs between molt and immune activity in two populations of house sparrows (*Passer domesticus*). Can. J. Zool. 83:780–787.

Martin, LB, Weil, ZM and Nelson, RJ (2006). Refining approaches and diversifying directions in ecoimmunology. Integr. Comp. Biol. 46:1030–1039.

Martin, LB, Weil, ZM and Nelson, RJ (2008). Seasonal changes in vertebrate immune activity: mediation by physiological trade-offs. Philos. Trans. R. Soc. Lond. B. Biol. Sci. 363:321–339.

Matera, L, Cesano, A, Bellone, G and Oberholtzer, E (1992). Modulatory effect of prolactin on the resting and mitogen-induced activity of T, B and NK lymphocytes. Brain. Behav. Immun. 6:409–417.

McEwen, BS, Biron, CA, Brunson, KW, Bulloch, K, Chambers, WH, Dhabhar, FS, Goldfarb, RH, Kitson, RP, Miller, AH, Spencer, RL and Weiss, JM (1997). The role of adrenocorticoids as modulators of immune function in health and disease: neural, endocrine and immune interactions. Brain Res. Brain Res. Rev. 23:79–133.

McEwen, BS and Wingfield, JC (2003). The concept of allostasis in biology and biomedicine. Horm. Behav. 43:2–15.

Millar, JS (2001). On reproduction in lemmings. Ecoscience 8:145–150.

Moore, CB and Siopes, TD (2003). Melatonin enhances cellular and humoral immune responses in the Japanese quail (*Coturnix coturnix japonica*) via an opiatergic mechanism. Gen. Comp. Endocrinol. 131:258–263.

Moore, RY (1996). Neural control of the pineal gland. Behav. Brain Res. 73:125–130.

Morhardt, JE (1970). Heart rates, breathing rates and the effects of atropine and acetylcholine on white-footed mice (*Peromyscus* sp.) during daily torpor. Comp. Biochem. Physiol. 33:441–457.

Muehlenbein, MP and Bribiescas, RG (2005). Testosterone-mediated immune functions and male life histories. Am. J. Hum. Biol. 17:527–558.

Munck, A, Guyre, PM and Holbrook, NJ (1984). Physiological functions of glucocorticoids in stress and their relation to pharmacological actions. Endocr. Rev. 5:25–44.

Munoz-Casares, FC, Padillo, FJ, Briceno, J, Collado, JA, Munoz-Castaneda, JR, Ortega, R, Cruz, A, Tunez, I, Montilla, P, Pera, C and Muntane, J (2006). Melatonin reduces apoptosis and necrosis induced by ischemia/reperfusion injury of the pancreas. J. Pineal Res. 40:195–203.

Murray, MJ and Murray, AB (1979). Anorexia of infection as a mechanism of host defense. Am. J. Clin. Nutr. 32:593–596.

Navara, KJ, Trainor, BC and Nelson, RJ (2007). Photoperiod alters macrophage responsiveness, but not expression of Toll-like receptors in Siberian hamsters. Comp. Biochem. Physiol. A 148:354–359.

Nelson, RJ (1987). Photoperiod-nonresponsive morphs: a possible variable in microtine population-density fluctuations. Am. Nat. 130:350–369.

Nelson, RJ (2004). Seasonal immune function and sickness responses. Trends Immunol. 25:187–192.

Nelson, RJ and Blom, JM (1994). Photoperiodic effects on tumor development and immune function. J. Biol. Rhythms 9:233–249.

Nelson, RJ and Demas, GE (1996). Seasonal changes in immune function. Q. Rev. Biol. 71:511.

Nelson, RJ, Demas, GE, Klein, SL and Kriegsfeld, L (2002). Seasonal Patterns of Stress, Immune Function and Disease. Cambridge University Press, New York.

Nelson, RJ and Prendergast, BJ (2002). Form over function? J. Biol. Rhythms 17:406–409.

Newson, J (1962). Seasonal differences in reticulocyte count, haemoglobin level and spleen weight in wild voles. Br. J. Haematol. 8:296.

Owen-Ashley, NT, Hasselquist, D, Råberg, L and Wingfield, JC (2008). Latitudinal variation of immune defense and sickness behavior in the white-crowned sparrow (*Zonotrichia leucophrys*). Brain. Behav. Immun. 22:614–625.

Owen-Ashley, NT, Turner, M, Hahn, TP and Wingfield, JC (2006). Hormonal, behavioral and thermoregulatory responses to bacterial lipopolysaccharide in captive and free-living white-crowned sparrows (*Zonotrichia leucophrys gambelii*). Horm. Behav. 49:15–29.

Pandi-Perumal, SR, Srinivasan, V, Maestroni, GJ, Cardinali, DP, Poeggeler, B and Hardeland, R (2006). Melatonin: Nature's most versatile biological signal? FEBS J. 273:2813–2838.

Pedersen, BK and Hoffman-Goetz, L (2000). Exercise and the immune system: regulation, integration and adaptation. Physiol. Rev. 80:1055.

Pioli, C, Caroleo, MC, Nistico, G and Doriac, G (1993). Melatonin increases antigen presentation and amplifies specific and nonspecific signals for T-cell proliferation. Int. J. Immunopharmacol. 15: 463–468.

Pozo, D, Delgado, M, Fernandez-Santos, JM, Calvo, JR, Gomariz, RP, Martin-Lacave, I, Ortiz, GG and Guerrero, JM (1997). Expression of the Mel1a-melatonin receptor mRNA in T and B subsets of lymphocytes from rat thymus and spleen. FASEB J. 11:466–473.

Prendergast, BJ and Pyter, LM (2009). Photoperiod history differentially impacts reproduction and immune function in adult Siberian hamsters. J. Biol. Rhythms 24:509.

Prendergast, BJ, Baillie, SR and Dhabhar, FS (2008). Gonadal hormone-dependent and -independent regulation of immune function by photoperiod in Siberian hamsters. Am. J. Physiol. 294:R384–392.

Prendergast, BJ, Bilbo, SD, Dhabhar, FS and Nelson, RJ (2004). Effects of photoperiod history on immune responses to intermediate day lengths in Siberian hamsters (*Phodopus sungorus*). J. Neuroimmunol. 149:31–39.

Prendergast, BJ, Bilbo, SD and Nelson, RJ (2004). Photoperiod controls the induction, retention and retrieval of antigen-specific immunological memory. Am. J. Physiol. 286:R54–60.

Prendergast, BJ, Bilbo, SD and Nelson, RJ (2005). Short day lengths enhance skin immune responses in gonadectomised Siberian hamsters. J. Neuroendocrinol. 17:18–21.

Prendergast, BJ, Hotchkiss, AK, Bilbo, SD, Kinsey, SG and Nelson, RJ (2003). Photoperiodic adjustments in immune function protect Siberian hamsters from lethal endotoxemia. J. Biol. Rhythms 18:51.

Prendergast, BJ, Hotchkiss, AK and Nelson, RJ (2003). Photoperiodic regulation of circulating leukocytes in juvenile Siberian hamsters: mediation by melatonin and testosterone. J. Biol. Rhythms 18:473–480.

Prendergast, BJ, Kriegsfeld, LJ and Nelson, RJ (2001). Photoperiodic polyphenisms in rodents: neuroendocrine mechanisms, costs and functions. Q. Rev. Biol. 76:293–325.

Prendergast, BJ and Nelson, RJ (2001). Spontaneous "regression" of enhanced immune function in a photoperiodic rodent *Peromyscus maniculatus*. Proc. R. Soc. Lond. B Biol. Sci. 268:2221–2228.

Prendergast, BJ, Nelson, RJ and Zucker, I, (2002). Mammalian seasonal rhythms: Behavior and neuroendrocrine substrates. In: Pfaff, DW (Ed.), Hormones, Brain and Behavior. Academic Press, San Diego, pp. 93–156.

Prendergast, BJ, Wynne-Edwards, KE, Yellon, SM and Nelson, RJ (2002). Photorefractoriness of immune function in male Siberian hamsters (*Phodopus sungorus*). J. Neuroendocrinol. 14:318–329.

Prendergast, BJ, Yellon, SM, Tran, LT and Nelson, RJ (2001). Photoperiod modulates the inhibitory effect of in vitro melatonin on lymphocyte proliferation in female Siberian hamsters. J. Biol. Rhythms 16:224–233.

Pyter, LM, Adelson, JD and Nelson, RJ (2007). Short days increase hypothalamic-pituitary-adrenal axis responsiveness. Endocrinology 148:3402–3409.

Pyter, LM, Neigh, GN and Nelson, RJ (2005). Social environment modulates photoperiodic immune and reproductive responses in adult male white-footed mice (*Peromyscus leucopus*). Am. J. Physiol. 288: R891–896.

Pyter, LM, Samuelsson, AR, Quan, N and Nelson, RJ (2005). Photoperiod alters hypothalamic cytokine gene expression and sickness responses following immune challenge in female Siberian hamsters (*Phodopus sungorus*). Neuroscience 131:779–784.

Pyter, LM, Weil, ZM and Nelson, RJ (2005). Latitude affects photoperiod-induced changes in immune response in meadow voles (*Microtus pennsylvanicus*). Can. J. Zool. Rev. Can. Zool. 83:1271–1278.

Reber, PM (1993). Prolactin and immunomodulation. Am. J. Med. 95:637–644.

Reppert, SM and Weaver, DR (1995). Melatonin madness. Cell 83:1059–1062.

Reppert, SM, Weaver, DR, Cassone, VM, Godson, C and Kolakowski, LF, Jr. (1995). Melatonin receptors are for the birds: molecular analysis of two receptor subtypes differentially expressed in chick brain. Neuron 15:1003–1015.

Ricklefs, RE and Wikelski, M (2002). The physiology/life-history nexus. Trends Ecol. Evol. 17:462–468.

Riddle, O (1924). Studies on the physiology of reproduction in birds. XIX. A hitherto unknown function of the thymus. Am. J. Physiol. 68:104–119.

Ronchi, E, Spencer, RL, Krey, LC and McEwen, BS (1998). Effects of photoperiod on brain corticosteroid receptors and the stress response in the golden hamster (*Mesocricetus auratus*). Brain Res. 780:348–351.

Rosa, HJD and Bryant, MJ (2003). Seasonality of reproduction in sheep. Small Ruminant Res. 48:155–171.

Samy, ATS, Schwacha, MG, Cioffi, WG, Bland, KI and Chaudry, IH (2000). Androgen and estrogen receptors in splenic T lymphocytes: effects of flutamide and trauma-hemorrhage. Shock 14:465.

Sapolsky, RM, Romero, LM and Munck, AU (2000). How do glucocorticoids influence stress responses? Integrating permissive, suppressive, stimulatory and preparative actions. Endocr. Rev. 21:55–89.

Shaw, D and Goldman, BD (1995). Influence of prenatal and postnatal photoperiods on postnatal testis development in the Siberian hamster (*Phodopus sungorus*). Biol. Reprod. 52:833–838.

Shivatcheva, TM and Hadjioloff, AI (1987). Seasonal involution of gut-associated lymphoid tissue of the European ground squirrel. Dev. Comp. Immunol. 11:791–799.

Silverin, B, Fange, R, Viebke, PA and Westin, J (1999). Seasonal changes in mass and histology of the spleen in willow tits *Parus montanus*. J. Avian Biol. 30:255–262.

Smith, PE (1930). The effect of hypophysectomy upon the involution of the thymus in the rat. Anat. Rec. 47:119–129.

Song, CK and Bartness, TJ (2001). CNS sympathetic outflow neurons to white fat that express MEL receptors may mediate seasonal adiposity. Am. J. Physiol. 281:R666–672.

Stetson, MH, Watson-Whitmyre, M and Matt, KS (1977). Termination of photorefractoriness in golden hamsters-photoperiodic requirements. J. Exp. Zool. 202:81–88.

Stone, W (1956). The J substance of cattle: III. Seasonal variation of the naturally occurring isoantibodies for the J substance. J. Immunol. 77:369.

Sullivan, D and Wira, C (1979). Sex hormone and glucocorticoid receptors in the bursa of Fabricius of immature chicks. J. Immunol. 122:2617.

Thompson, EB and Lippman, ME. (1974). Mechanism of action of glucocorticoids. Metab. 23:159–202.

Vandenbergh, J (1971). The influence of the social environment on sexual maturation in male mice. J. Reprod. Fertil. 24:383.

Vaughan, M, Vaughan, G, Blask, D and Reiter, R (1976). Influence of melatonin, constant light, or blinding on reproductive system of gerbils (*Meriones unguiculatus*). Cell. Mol. Life Sci. 32:1341–1342.

Vera-Lastra, O, Jara, L and Espinoza, LR (2002). Prolactin and autoimmunity. Autoimmun. Rev. 1:360–364.

Vermeulen, M, Palermo, M and Giordano, M (1993). Neonatal pinealectomy impairs murine antibody-dependent cellular cytotoxicity. J. Neuroimmunol. 43:97–101.

Wade, GN and Bartness, TJ (1984). Seasonal obesity in Syrian hamsters: effects of age, diet, photoperiod and melatonin. Am. J. Physiol. 247:R328–334.

Weaver, DR, Liu, C, and Reppert, SM. (1996). Nature's knockout: the Mel1b receptor is not necessary for reproductive and circadian responses to melatonin in Siberian hamsters. Mol. Endo. 10:1478–1487.

Weil, ZM, Bowers, SL, Pyter, LM and Nelson, RJ (2006). Social interactions alter proinflammatory cytokine gene expression and behavior following endotoxin administration. Brain. Behav. Immun. 20:72–79.

Weil, ZM, Martin, LB, 2nd and Nelson, RJ (2006). Photoperiod differentially affects immune function and reproduction in collared lemmings (*Dicrostonyx groenlandicus*). J. Biol. Rhythms 21:384–393.

Weil, ZM, Norman, GJ, DeVries, AC, Berntson, GG and Nelson, RJ (2009). Photoperiod alters autonomic regulation of the heart. Proc. Natl. Acad. Sci. U. S. A. 106:4525–4530.

Weil, ZM, Pyter, LM, Martin, LB, 2nd and Nelson, RJ (2006). Perinatal photoperiod organizes adult immune responses in Siberian hamsters (*Phodopus sungorus*). Am. J. Physiol. 290:R1714–1719.

Weil, ZM, Workman, JL and Nelson, RJ (2007). Housing condition alters immunological and reproductive responses to day length in Siberian hamsters (*Phodopus sungorus*). Horm. Behav. 52:261–266.

Wen, JC, Dhabhar, FS and Prendergast, BJ (2007). Pineal-dependent and -independent effects of photoperiod on immune function in Siberian hamsters (*Phodopus sungorus*). Horm. Behav. 51:31–39.

Wen, JC and Prendergast, BJ (2007). Photoperiodic regulation of behavioral responsiveness to proinflammatory cytokines. Physiol. Behav. 90:717–725.

Wingfield, JC and Farner, DS (1993). Endocrinology of reproduction in wild species. Avian Biol. 9:163–327.

Yellon, SM, Fagoaga, OR and Nehlsen-Cannarella, SL (1999). Influence of photoperiod on immune cell functions in the male Siberian hamster. Am. J. Physiol. 276:R97–R102.

Yellon, SM, Teasley, LA, Fagoaga, OR, Nguyen, HC, Truong, HN and Nehlsen-Cannarella, L (1999). Role of photoperiod and the pineal gland in T cell-dependent humoral immune reactivity in the Siberian hamster. J. Pineal Res. 27:243–248.

Yun, AJ, Lee, PY and Bazar, KA (2004). Temporal variation of autonomic balance and diseases during circadian, seasonal, reproductive and lifespan cycles. Med. Hypotheses 63:155–162.

Zosky, GR (2002). The parasympathetic nervous system: its role during torpor in the fat-tailed dunnart (*Sminthopsis crassicaudata*). J. Comp. Physiol. [B] 172:677–684.

Zuk, M (1990). Reproductive strategies and disease susceptibility: an evolutionary viewpoint. Parasitol. Today 6:231–233.

CHAPTER 10

PINEAL GLAND AND CIRCULATORY MELATONIN IN THE REGULATION OF IMMUNE STATUS OF SEASONALLY BREEDING MAMMALS

Chandana Haldar, Sameer Gupta, Seema Rai, Raise Ahmad, and Rajesh Yadav

INTRODUCTION

One of the forces that drove the evolution of the immune system was the day-to-day need for adjustment of the organism's health with environmental conditions (changing ecofactors, xenobiotics, etc.) for survival. This made immunity a primeval feature of living organisms and essential for reproduction and perpetuation of the species. Immunity and reproduction are the warp and weft of the continuum of life. Immunity and reproduction are closely interwoven mechanisms having a "trade-off" interrelationship. Successful reproduction is the result of a series of finely tuned neuroendocrine events where the pineal gland and its hormone melatonin play an important role along with the regulation of various events such as immunity, especially in seasonally breeding animals. Melatonin, the pineal hormone, probably maintains the general immune function by synchronizing and coordinating the organism's response to environmental demands. These demands/challenges may originate from seasonal variance of photoperiod or from other environmental factors (high and low temperature, restricted food availability, and social stress/pressure) against which individuals must defend with an appropriate response. Evidence has been gathered suggesting that the immune system is not a self-contained, closed system driven by cytokines alone, but rather is regulated by the neuroendocrine system. Taking all this into account, we were motivated to investigate the role of the pineal gland and melatonin in the regulation of general/natural immune functions in wild seasonal breeders.

Indeed, the function of the immune system is primarily to combat environmental threats when most environmental challenges are generally stressful. Whether melatonin could affect/ameliorate the response to stressful events, with particular regard to the effect of stress on the immune system, was a challenge to establish. For such a study wild seasonal breeders like hamsters, ferrets, and squirrels appeared to be the best mammalian models, since exposure to natural challenges by the seasonal breeders modulates various hematological and immune parameters (Haldar et al., 2001; Rai and Haldar, 2003). During challenging situations a high melatonin level was always noted. Anti-stress effects of melatonin suggest its use in restoring depressed immune function in adverse social and environmental condition (Maestroni et al., 1988; Haldar et al., 2004).

IMPACT OF SEASONAL VARIATION ON IMMUNITY

EFFECT OF LIGHT ON IMMUNE STATUS

INSOLATION AND IMMUNITY

The photoperiodic regulation of the melatonin rhythm makes the pattern of the hormone concentration an endocrine calendar for seasonally breeding animals (Attanasio et al., 1985; Reiter, 1993). In the tropical zone, the duration of the day in summer and winter is comparatively less than the temperate zone, but the annual cycle in rainfall, humidity, and amount of solar insolation plays a significant role in regulation of annual/seasonal physiology. Monsoon is a typical and exclusive climatic condition experienced in the Indian subcontinent in the months of July and August. During this period very dark clouds, mostly of cumulonimbus origin, overcast the sky (Gadgil, 2006; Haldar and Saxena, 1988a, 1988b) and although there is not much of a change in duration of photoperiod, there is a difference in intensity of insolation, direct radiation, and diffuse radiation reaching the earth's surface. Further, the ecological conditions during the monsoon are also harsh because of flooding that contributes to lack of food and shelter for wild animals.

One of the most important non-neural targets of melatonin is the immune system (Guerrero and Reiter, 2002; Poirel et al., 2003). Functional correlation between the pineal gland and immune system was implicated by chronobiological studies of immune cell function (Haus and Smolensky, 1999; Haldar et al., 2001). The existence of specific melatonin receptors on lymphoid organs provides evidence for a direct effect of melatonin on the immune system. Thus, melatonin (Nelson, 2004; Wen et al., 2006), via its receptor subtypes located on target tissues, is responsible for controlling the annual and daily events of immune function (Haldar et al., 2001; Carrillo-Vico et al., 2003a, 2005).

Increase in solar insolation by 2.09 $kWh/m^2/day$ increased expression of MT1 mRNA and protein in the Northern palm squirrel (*Funambulus pennanti*). During the months of May and June a small decrease was noted in the amount of insolation (0.05 $kWh/m^2/day$) and direct radiation due to approaching southwest monsoon, but because the amount of diffuse radiation remained high, this maintained the high level of MT1 mRNA as the clouds were not stagnant.

However, in the peak monsoon period (July and August) the amount of solar insolation was drastically reduced from 6.24 $kWh/m^2/day$ to 4.49 $kWh/m^2/day$ due to the dark cumulonimbus monsoon clouds (Gadgil, 2006). During this time, insolation was approximately equivalent to the amount of solar insolation experienced during winter (November–December), although photoperiod duration remained almost the same. Interestingly, this decrease induced a direct decrease in transcription and translation of the MT1 receptor gene. This level again rose slightly by 0.09 $kWh/m^2/day$ during September and October and finally decreased significantly by November–December to 4.105 $kWh/m^2/day$. The amount of solar insolation in terms of $kWh/m^2/day$ showed a direct relationship with the circulatory level of melatonin, splenic weight, MT1 receptor mRNA, and protein expression of splenic T-helper cells of spleen in *F. pennanti*, which could have high adaptive significance for survival of wild animals inhabiting the tropical regions and experiencing the monsoon type of climate (Lahiri and Haldar, 2009).

PHOTOPERIOD AND IMMUNITY

The differential effects of long (LD; 16L:8D), short (SD; 10L:14D), and natural day length (NDL; 12L:12D) during reproductively inactive (RIP) and active phases (RAP) were assessed in relation to immunity and reproductive function of the tropical rodent *F. pennanti*. These rodents presented high immunity and low testicular activity during RIP and vice versa during RAP. SD increased spleen and thymus weight, leukocyte and lymphocyte counts (TLC and LC), cell-mediated immunity (i.e., blastogenic response in terms of percent stimulation ratio [% SR] of splenocytes and thymocytes when challenged with concanavalin A), and delayed-type hypersensitivity (DTH) to oxazolone. SD during RIP increased the above-mentioned parameters and reduced testes weight when compared with 12L:12D groups. During RAP, LD reduced all the immunological parameters when compared with 12L:12D and SD groups in RAP and RIP phases. The LD group presented a marked decrease in the immunological parameters in comparison with RAP, suggesting that LD always had an inhibitory effect on immune status, being independent of the reproductive phases. The intensity of the stimulatory effects of SD and inhibitory effects of LD during both reproductive phases was significantly different. We exposed another set of squirrels to the above photoperiodic schedule for a prolonged period (30 weeks) during RAP. A clear testicular refractoriness followed by immunorefractoriness was observed in the group experiencing SD and LD for 30 weeks. The photorefractoriness presented by the testes was inversely related to the immunorefractoriness. The peripheral melatonin level of those squirrels reflected the photoperiodic signal perceived by squirrels for immunomodulation and gonadal function, suggesting that the immune system and gonadal function might have coevolved (Ahmad and Haldar, 2009).

We studied the change in expression of MT1 melatonin receptor (MT1) mRNA and protein located in spleen of *F. pennanti* under different photoperiodic conditions. Interestingly, melatonin receptor (MT1) mRNA and protein showed a significant decrease after continuous light exposure, indicating a possible role of light via melatonin in the modulation of immunity by regulating MT1 membrane receptor expression in *F. pennanti* (Lahiri, 2009). The present study comes in parallel with previous studies on Siberian and Syrian hamsters, where the regulation of MT1 mRNA and protein at the photoperiodic level directly depends on the RNA transcription level

(Maestroni, 1993; Shiu et al., 2000). In *Phodopus sungorus,* known as "nature's knockout" for MT2 receptors, the MT1 receptor mediates reproductive and circadian responses to melatonin. Melatonin binding and receptor mRNA levels are directly affected by light and plasma concentration of melatonin (Maestroni, 1993) in rodents.

TEMPERATURE AND IMMUNITY

Turner and Bagnara (1971) stated that "the exposure of warm-blooded animals to cold environments increases thyroid activity." The increased secretion of thyroid hormones during low ambient temperatures not only increases basic metabolic rate but also directly stimulates both primary and secondary lymphoid tissues. When an organism undergoes a thermal response, it reacts to the stimulus, and various physiological responses occur during physiological adaptation or acclimation that may exert effects on the immune system (Chaffee and Roberts, 1971; Kelley, 1980). During long-term hyperthermia, the functional activity of the lymphocytes was restored as animals became adapted to exposure to a high temperature, as noted by restoration of normal values of proliferative activity of spleen cells 40 days after the onset of hyperthermia in response to stimulation by mitogens and alloantigens. It is conceivable that during adaptation to thermal stimuli, physiological responses reach a level at which some of the immune processes may be affected, and then decline, and finally the effects disappear (Mount, 1979). The active stage of physiological response that occurs in the course of adaptation may persist much longer for the temperature shift to 36.5°C than for the shift to 8°C. Kikuchi (1988) proposed that such an active stage be referred to as the "effective period" of physiological adaptation. Meanwhile, during an immune response, a limited period in which immune reactions are sensitive to certain physiological adaptive processes appears to be present, and this period has been referred to as the "susceptible period." The immune response may be affected at a time when the "effective period" of physiological adaptation and the "susceptible period" of the immune process concur. Further, an ambient temperature shift also affects the immune processes, and the duration of the effect appears to depend on altered temperature levels. In the process of immune responses, there might be a susceptible period to temperature shift, especially at an early phase after immunization.

The interaction between photoperiod and temperature has been examined on immunoglobulin (IgG) levels and splenic mass in male deer mice (*Peromyscus maniculatus*) (Demas and Nelson, 1996). In another experiment, animals were maintained in LD 16L:8D or SD 8D:16L photoperiods and either in 20° or 8°C temperatures. Serum IgG levels were elevated in SD mice maintained at normal room temperature (i.e., 20°C) compared to LD animals housed at either 20° or 8°C. LD deer mice kept at 8°C had reduced IgG levels compared to LD mice maintained at 20°C, whereas SD mice exposed to low temperatures had IgG levels comparable to LD mice maintained at 20°C.

It is known that heat and cold stress change blood corticosteroid levels (Collins and Weiner, 1968; Chowers et al., 1970; Gisler, 1974) and corticosteroids affect immune function (Claman, 1972; Cupps and Fauci, 1982). Some *in vivo* and *in vitro* studies have demonstrated that endogenous or exogenously added corticosteroids exert some of their immunosuppressive effects on an early phase of immune responses, for example inhibition of monocyte antigen presentation, including expression of I region-associated (Ia) antigens and interleukin IL-1 production

(Snyder et al., 1982; Gerrard et al., 1984), and preventing IL-2 production of T lymphocytes (Gills et al., 1979; Larsson, 1980). However, further experiments are needed to conclude whether the temperature effect is independent of or dependent on photoperiod, as well as on adrenal secretion of corticosterone. But it is time to accept that ecofactors (temperature and humidity-induced infection) other than photoperiod interfere with immunity, depending upon the geographical distribution of fauna. The monsoon season of tropical zones has an inhibitory effect on the immune status of birds and mammals (rodents and ruminants). This is because high humidity, along with high temperature, in nature acts as an incubator for the growth and development of various bacteria and viruses that commonly infect the animals in general, and grazing animals in particular, as the bacteria and viruses settle mostly on plant leaves and grasses (Van Frankenhuyzen, 1994; Radostits et al., 2000).

PINEAL GLAND AND IMMUNE STATUS

The pineal gland is integral to the mediation between the environment and the endocrine system. The pineal gland, via the rhythmic synthesis and release of melatonin, influences the development and function of the immune system. Circulating messages produced by activated immune cells are reciprocally perceived by the pineal gland and provide feedback for the regulation of pineal function. The pineal gland and the immune system are, therefore, reciprocally linked by bidirectional communication. Melatonin controls fertility in seasonally breeding animals and affects the function of virtually every other endocrine organ (Reiter, 1973, 1983; Bishnupuri and Haldar, 2000). The finding that first Hypothesized a role for the pineal gland and melatonin in immunity was the capacity of allogeneic antigens to evoke immunity when injected in mice. Besides this, the concept about the general pineal function of optimizing the organism response to environmental challenges convinced us about the soundness of our hypothesis.

Reports related to the pineal gland are scanty and controversial, although the role of melatonin is being established as immunostimulator. In our review, our aim is to focus on the involvement of endogenous and exogenous melatonin in the regulation of immune function and its immunoregulatory mechanism(s). The first indication that the pineal gland may be involved in endocrine immunomodulation came from early reports on antitumor effects of pineal extracts in animals and humans (Hill et al., 1992; Lissoni et al., 1994). Daily afternoon injection of melatonin induced an increase in thymus weight in the gerbil (Vaughan et al., 1976) and spleen hypertrophy in the Syrian hamster (Vaughan et al., 1987). The pineal gland, *via* the rhythmic synthesis and release of melatonin, influences the development and function of the immune system. Circulating messages produced by activated immune cells are reciprocally perceived by the pineal gland and provide feedback for the regulation of pineal function (Maestroni, 1993). The regulatory function of melatonin on immune mechanisms is season-dependent (Figs. 10.1 and 10.2).

The bulk of the experimental evidence suggesting that melatonin can indeed be considered as an immunoregulatory hormone was elucidated in our laboratory by testing the natural immune parameters in a seasonal breeder *F. pennanti* (Figs. 10.3 and 10.4). Removal of the pineal gland decreased thymus and spleen weight (Csaba and Barath, 1975), and Champney and McMurray (1991) also found that pineal activation modifies spleen morphology and function. However, Milcu and Potop (1973) have reported that pineal gland has a stimulatory action on thymic

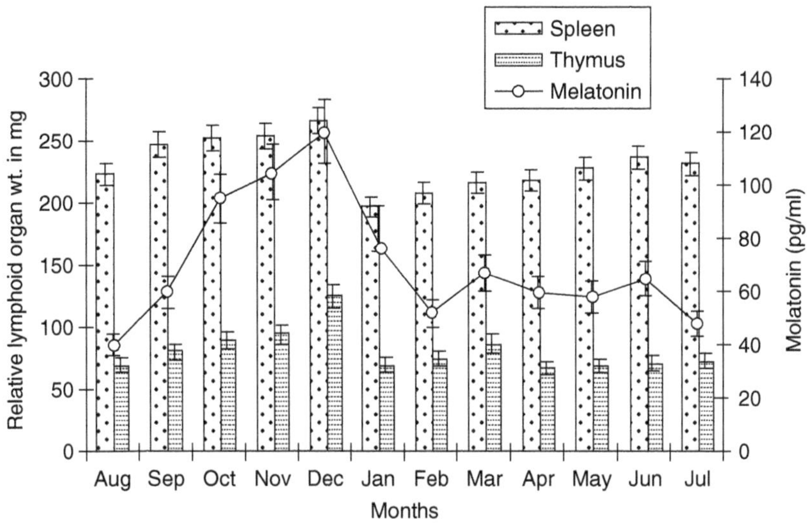

FIGURE 10.1: Annual variation in relative weight (mg/100 g body wt) of spleen and thymus in relation to circulatory melatonin level (pg/mL) in *F. pennanti*. Vertical bar represents mean ± SE; N = 7.

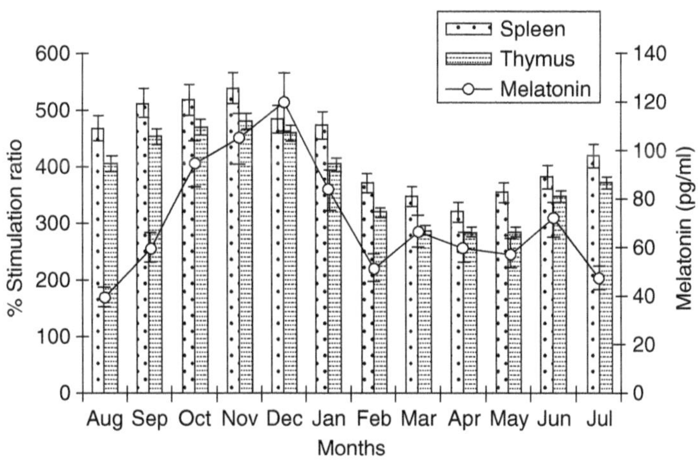

FIGURE 10.2: Blastogenic response of splenocytes and thymocytes in terms of percent stimulation ratio (%SR) (challenged with concanavalin A) and its relation with circulatory melatonin level (pg/mL) in *F. pennanti*. Vertical bar represents mean ± SE; N + 7.

growth, since the administration of pineal extract may determine the hyperplasia of the thymus gland.

These findings of Milcu and Potop (1973) and Iwata et al. (1981) supported one of our earliest findings, that exogenous melatonin treatment given to the pinealectomized squirrel prevented the decline in lymphoid organs following pinealectomy and improved all hematological parameters (Fig. 10.5). We found that melatonin maintained natural immune parameters in pinealectomized animals for a long duration, providing great diagnostic/clinical significance. Interestingly, only the evening injection could maintain such activity; the morning injection failed to do so.

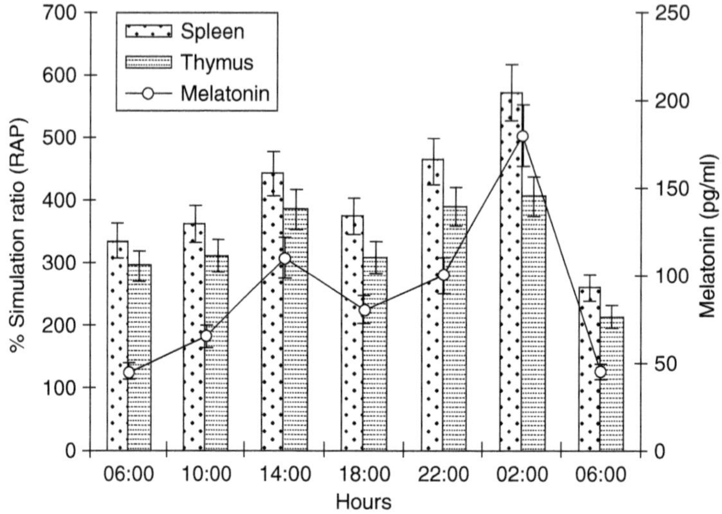

FIGURE 10.3: Daily variation in % stimulation ratio of spleen and thymus and its direct relation with circulatory melatonin level during reproductively active phase (RAP; May and June) of *F. pennanti.* Vertical bar represents mean ± SE; N = 7.

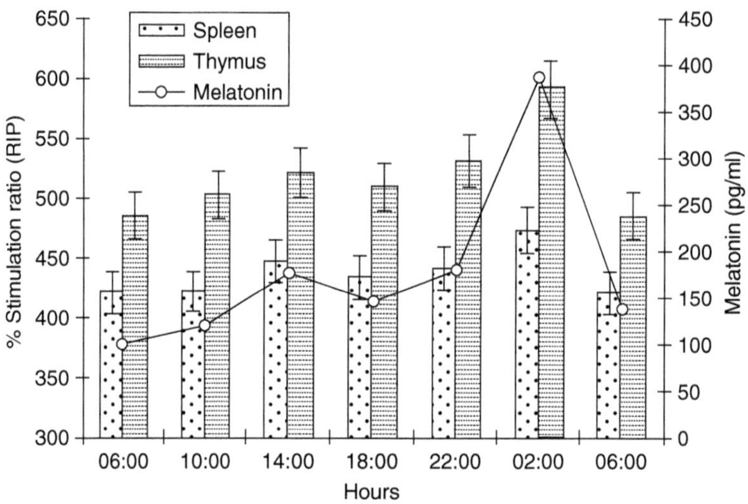

FIGURE 10.4: Daily variation in % stimulation ratio of spleen and thymus and its direct relation with circulatory melatonin level during reproductively inactive phase (RIP; November and December) of *F. pennanti.* Vertical bar represents mean ± SE; N = 7.

Melatonin possesses a stimulatory effect on hematopoiesis in mouse (Kuci et al., 1988) and rats colony forming unit-granulocyte macrophage (CFU-GM) (Haldar et al., 1992; Fig. 10.6). Various data suggest that the suppression of endogenous melatonin by pinealectomy, long day length, constant light (physiological pinealectomy), p-CPA (serotonergic blocker), or propranolol (β-adrenergic blocker) profoundly decreases the efficiency of the immune system, making melatonin a potential immunoenhancing agent.

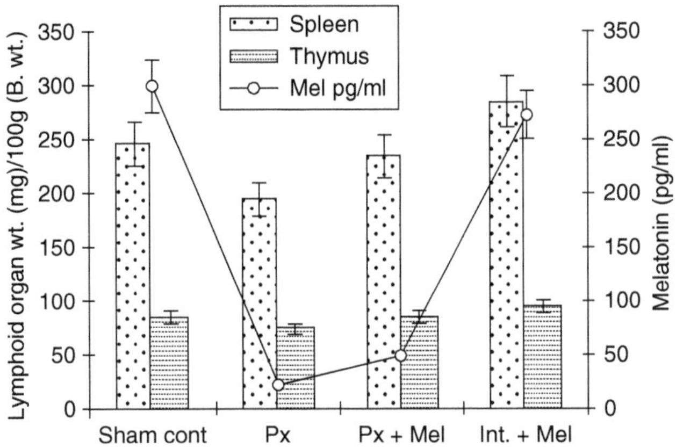

FIGURE 10.5: Effect of surgical pinealectomy and melatonin injection to pinealectomized and intact control squirrels on relative spleen and thymus weight (mg/100 g body wt) and circulatory melatonin level. Vertical bar represents mean ± SE; N = 7.

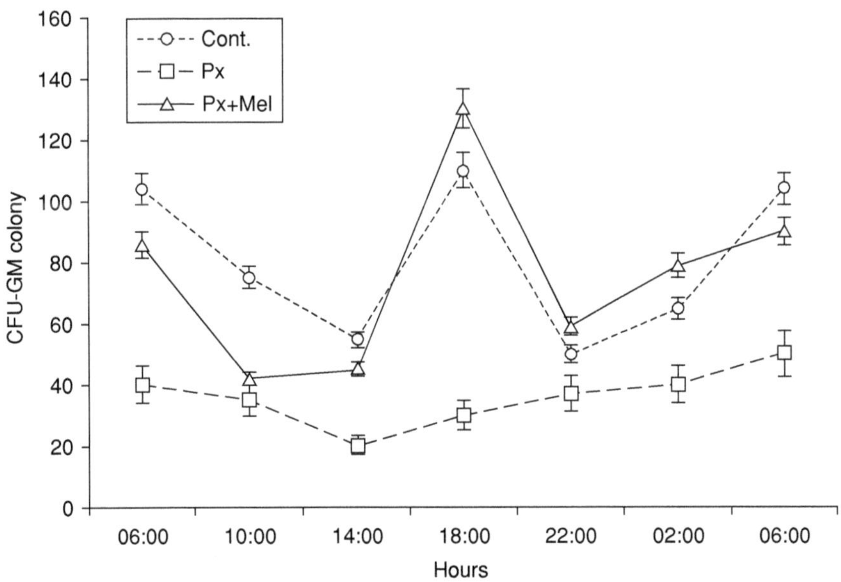

FIGURE 10.6: Circadian variation in colony-forming units for granulocytes–macrophages from bone marrow after surgical pinealectomy and melatonin injection to pinealectomized squirrels. Vertical bar represents mean ± SE; each point had ~100 colonies counted (N = 100 colonies).

The dietary intake of melatonin is known to influence pineal gland activity (Pevet and Haldar, 1982). Dietary intake of melatonin also stimulates thymus and spleen function by increasing its weight and thymocyte population. Circadian variations in hematological parameter red blood cells, packed cell volume and haemoglobin percentage (RBC, PCV, and Hb %) and natural immune parameters total leukocyte, lymphocyte and differential leukocyte count (TLC, % LC, and DLC) suggest a parallel relationship with the melatonin rhythm. This rhythm was independent of the circulatory steroid hormone (gonadal and adrenal steroids) levels during reproductive phases, suggesting that the rhythm of immunity is driven primarily by melatonin but not by steroids. Whenever examined, compromised immune function caused by pinealectomy could be ameliorated by melatonin replacement therapy (Maestroni, 1993). These data further support the role of melatonin as an anti-stress hormone proposed by Maestroni et al. (1998).

PINEALECTOMY AND SEASONAL IMMUNE STATUS

Most authors agree that pinealectomy and *in vivo* melatonin administration clearly show the immunoenhacing properties of this indoleamine. Many immune function tests by numerous investigators leave little doubt that, along with its many other functions, melatonin should also be considered an immunomodulatory compound. Removal of pineal gland in squirrels by an established method (Haldar, 1986) during the reproductive inactive phase (November and December) significantly reduces thymus and spleen weight, compared with a sham-pinealectomy control group. Pinealectomy also reduces TLC and LC when compared with a sham-surgery control group; levels are resumed upon melatonin treatment (Fig. 10.7). A significant decrease in lymphocytes and monocytes with increased neutrophil percentage was noted following pinealectomy, whereas eosinophils did not exhibit any marked changes. Further, melatonin administration in the evening hours to pinealectomized squirrels provokes a significant increase in thymus and spleen weight along with restoration of thymic cellular architecture when compared with pinealectomized animals (Haldar and Singh, 2001).

The strong inhibition of immune status (reduced spleen weight along with leukocyte and lymphocyte count with increased neutrophil percentage) was noted following thymectomy and pinealectomy; it was restored following melatonin administration (Fig. 10.8). Pinealectomy drastically decreases the plasma melatonin concentration when compared with sham-pinealectomized control animals. Exogenous melatonin treatment during dusk to pinealectomized animals

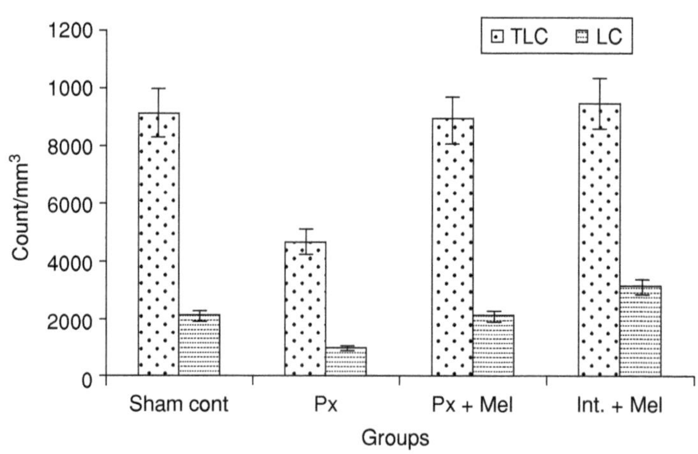

FIGURE 10.7: Variation in total leukocyte count (TLC) and lymphocyte count (LC) after surgical pinealectomy and melatonin injection to squirrels. Vertical bar represents mean ± SE; N = 7.

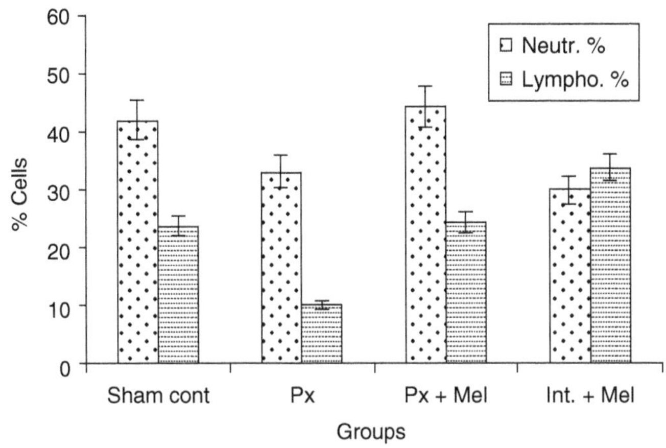

FIGURE 10.8: Variation in % neutrophils and % lymphocyte count after surgical pinealectomy and melatonin injection to squirrels. Vertical bar represents mean ± SE; N = 7.

significantly increased the plasma melatonin concentration compared with the pinealectomized group (Haldar and Singh, 2001). Pinealectomized squirrels, however, maintained full gonadal activity and plasma testosterone levels, while melatonin supplementation to pinealectomized squirrels caused a significant decrease in plasma testosterone levels. Administration of testosterone to pinealectomized squirrels showed further decrease in the weight of lymphoid organs, with severe depletion in their cellular architecture (Haldar and Vidhu, 1997; Rai and Haldar, 2003).

It has been suggested that surgical pinealectomy performed in C57BL/6 mice suppressed IL-2 production and natural killer (NK) activity in spleen cells. This suppression was reversed only by a single acute injection but not by repeated administration of melatonin. This feature and the fact that the effective single dose of melatonin was exceedingly high, together with the possible lack of melatonin in pineal gland of C57BL/6 mice, suggest that the immunological effects observed are not melatonin- mediated. However, a number of experiments carried out earlier on rats and mice do support the involvement of pineal gland in immune regulation (Maestroni and Conti, 1993).

MELATONIN AND SEASONAL IMMUNITY

Planetary movement is responsible for day length, which varies in a definite manner over eons, and is also responsible for seasonal changes in day length. This environmental factor (i.e., day length or photoperiod) is then responsible for the variation in seasonal and daily rhythms of melatonin responsible for regulating reproduction. Melatonin secreted from the pineal gland is a biological signal of day length and is a well-established modulator of immunity (Maestroni et al., 1986). The circulatory level of melatonin is controlled by the duration of photoperiod as well (Nelson, 2004). The long photoperiod of summer is associated with short melatonin pulse, and the short photoperiod of winter is associated with long melatonin pulse (Maestroni, 1993). The photoperiodic regulation of the melatonin rhythm thus makes the pattern of the hormone concentration an endocrine calendar for seasonally breeding animals (Haldar and Saxena, 1988a, 1988b).

The immune system exhibits prominent circadian and circannual rhythms of proliferation and function (Haus and Smolensky, 1999). It is yet to be studied whether these photoperiodic effects leave a long-term imprint on lymphocytes, predicting their future sensitivity to melatonin, or on other endogenous immune enhancers and mitogens. Recently, Drazen et al. (2000) have shown in prairie voles that splenocyte proliferation in response to the T-cell mitogen Con A is enhanced by the addition of 1 nM melatonin. Our knowledge regarding the particular role of melatonin in the immune response is increasing rapidly. The final effect of melatonin under conditions of "environmental stress" depends on a kind of "accelerated recovery" of the organism from seasonal low immunity during RAP rather than on interference with the organism's physiologic response to stress. This also suggests that the immuno-pharmacological effects observed might reflect the general physiologic adaptive function of melatonin (see Figs. 10.1 and 10.2).

IMMUNOLOGICAL EFFECTS OF MELATONIN

Melatonin is involved in the regulation of both cellular and humoral immunity. Melatonin not only stimulates the production of natural killer cells, monocytes, and leukocytes, but also alters the balance of T-helper (Th-1 and Th-2) cells, mainly towards Th-1 responses, and increases the production of relevant cytokines such as interleukin (IL)-2, IL-6, IL-12, and interferon (IFN)-γ. The nocturnal melatonin surge impairs the rolling and adherence of leukocytes to endothelial layers, limiting cell migration, and stimulates nocturnal production of IL-2 by T-helper lymphocytes, exerting an immunostimulatory effect. Otherwise, the release of TNF-α from activated macrophages suppresses the nocturnal melatonin surge, allowing full cell migration and inhibiting IL-2 production (Markus et al., 2007). In some cases, an inhibitory effect of melatonin on lymphocyte proliferation has been described as being coupled to inhibition of NK activity or IFN-γ and TNF-α production (Di Stefano and Paulesu, 1994). Contrary to this, Colombo et al. (1992) have shown that melatonin increases IFN-γ production by mouse splenocytes, with this stimulation being significantly higher in cells isolated at night than in cells isolated in the morning. Antigen presentation by splenic macrophages to T cells is also enhanced by melatonin; furthermore, this enhancement is coincident with MHC class II molecules as well as IL-1 and TNF-α (Pioli et al., 1993).

Melatonin is effective only when administered in the evening, not when the same dose is injected in the morning (Skwarlo-Sonta et al., 1991). This rhythmic mode of action has been described for most biological activities of melatonin. Further, melatonin is immunostimulatory only in antigen-primed condition (Maestroni and Conti, 1990). In a simple experiment, melatonin was injected in different doses for various periods of time in different species of rodents without finding any significant change of natural immune parameters. Therefore, the immune-enhancing action of melatonin in a normal rodent is Antigen dependent and is mainly exerted via humoral responses to T-cell-dependent antigens. Daily afternoon injection of melatonin induces an increase in thymus weight in the gerbil (Vaughan et al., 1976) and spleen hypertrophy in the Syrian hamster (Vaughan et al., 1987). Melatonin administration also increases the mitogenic response of mice splenocyte (Sze et al., 1993) to concanavalin A and lipopolysaccharide (LPS). Many cell-mediated immune parameters have been measured in mammals, including humans, demonstrating the immunostimulatory effect of melatonin. Afternoon injection of melatonin increases the production of IFN-γ in Syrian hamsters (Champney et al., 1998).

Melatonin was able to enhance the immune response only when administered during the primed condition of the immune system (Maestroni et al., 1988). An immunoenhancing effect of melatonin was most evident when the immune reactivity was depressed either by acute stress or by various pharmacological treatments. A consensus exists concerning a depressive influence of acute and/or chronic stress on various parameters of acquired and natural immunity (Reiter, 1983; Guerrero and Reiter, 2002). The observation that exogenous melatonin was able to restore the immune reactivity depressed by the inhibition of its endogenous production led to the investigation of its immuno-pharmacological properties. If the general function of circadian melatonin production is to optimize the ability of the organism to cope with environmental challenges, exogenous melatonin might well be able to enhance this ability, including the immune response.

INHIBITION OF MELATONIN SYNTHESIS AND IMMUNE REGULATION

Melatonin administration counteracted the effect of propranolol, a beta-adrenergic blocker, bringing spleen weight and blastogenesis to values close to those shown by animals kept under short photoperiod (Maestroni et al., 1986; Nelson et al., 1995; Demas and Nelson, 1998). A study on inbred mice, such as C57BL/6 and BALB/c and AKR/J, which do not produce melatonin, will constitute a beautiful model to investigate non–melatonin-mediated pineal functions. The pineal gland contains many known and unknown substrates such as 5-methoxyindoles other than melatonin and some protein peptides (Pevet et al., 1980; Haldar and Pevet, 1983), which may modulate some immune functions. Evening, but not morning, administration of propranolol abolished the melatonin peak and was associated with depressed primary responses to sheep red blood cells (SRBCs) and reduced autologous mixed lymphocyte reactions. A similar immunodepression was also obtained following treatment of the animals with para-chloro-phenyl-alanine, a blocker of serotonin synthesis (Rai et al., 2009). Most interesting is that evening administration of melatonin reversed the depression of the immune responses induced by these pharmacological interventions. These findings were then essentially confirmed in other labs, concluding that melatonin exerts a significant immunoregulatory role (Maestroni, 1993; Garcia-Maurino et al., 1997; Drazen et al., 2000).

MECHANISM OF ACTION OF MELATONIN

MELATONIN RECEPTOR AND IMMUNE SYSTEM

Melatonin displays a pleiotropic effect on immunity and immune system functions. The primary membrane melatonin receptors that have been cloned and characterized in mammals to date are MT1, MT2 (G-protein-coupled receptor; Reppert et al., 1994, 1995), and MT3, which was recently affinity-purified, belonging to the family of quinone reductases (Nosjean et al., 2000).

Melatonin via MT1 receptors ameliorates the inhibitory effect of prostaglandin E_2 on IL-2 production in human lymphocytes (Carrillo-Vico et al., 2003b, 2005) and via MT2 receptor mediates proliferation of splenocytes in laboratory mice (Drazen et al., 2001).

Melatonin can activate or inhibit signal transduction cascades independent of receptors or through receptors. The ability of melatonin to act independently from its receptors is attributed to its small and highly lipophilic nature and/or due to active uptake mechanisms (Reppert et al., 1995). Each of the G-protein-coupled melatonin receptor types, denoted MT1 and MT2, can couple to multiple signal transduction cascades, whereas the signal transduction cascades mediating MT3 responses are still unclear (Witt-Enderby et al., 2003). The nature of the mechanism involved in immunomodulatory activity of melatonin remains largely an unsettled question. Two high-affinity membrane-bound melatonin receptors, MT1 and MT2, have been localized and cloned; those belong to the super-family of G-protein-coupled receptors. Melatonin receptor activation includes a variety of responses that are mediated by both pertussis toxin-sensitive and -insensitive G protein (Mc Arthur et al., 1997). Nuclear binding receptors have also been identified in human lymphocytes and monocytes (Garcia Maurino et al., 2000). The expression of melatonin receptor genes within the immune system shows species and organ specificity. A ubiquitous expression of MT1 receptor mRNA and protein in T and B subsets of lymphocytes in spleen and thymus of rat and mouse, unlike MT2 receptors, has been reported (Pozo et al., 1997; Carrillo-Vico et al., 2003b). All these data suggest that more studies are needed before we generalize about the role of melatonin receptor activity on lymphoid organs and immune status of animals. A partial attempt by our lab suggests that there is a daily and seasonal rhythmicity in the receptor of melatonin MT1 activity in spleen, and this is inversely correlated with the circulatory level of melatonin (Fig. 10.9). This is because the autoregulatory activity of melatonin downregulates its own receptor density (Barrett et al., 1996).

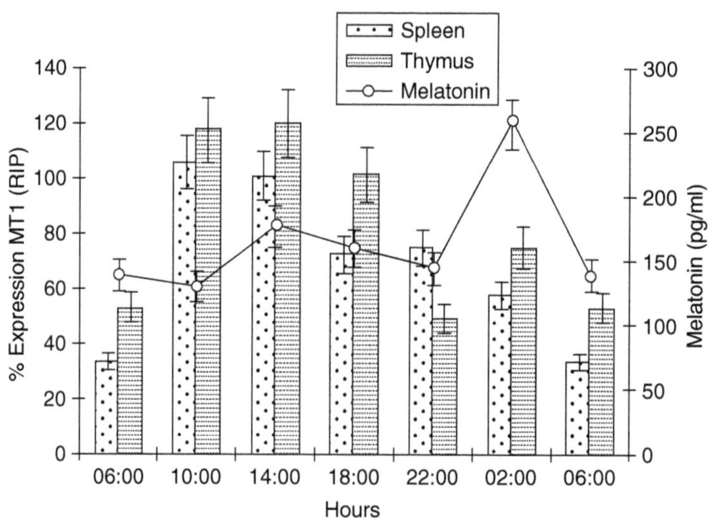

FIGURE 10.9: Variation in % expression of melatonin receptor expression MT1 in spleen and thymus following Western blot densitometric analysis. Note the inverse correlation with circulatory melatonin level (pg/mL) during reproductively inactive phase (RIP; November and December) of *F. pennanti.*

OPIOIDS AND MELATONIN INDUCED IMMUNE OPIOIDS

Melatonin is known to present multiple biological effects almost exclusively *in vivo*. However, its effect varies *in vitro,* which suggests that the immune enhancement of the anti-stress actions of melatonin is mediated by mechanisms other than those involving its direct action on immune function. In this case endogenous opioid systems were considered. According to Maestroni et al. (1988), it may be mediated by a family of opioid peptides, because the use of naltrexone, a specific opioid antagonist, prevents the immunoenhancing properties of melatonin. Moreover, beta-endorphin and dynorphin, which belong to the melatonin-induced opioid system, mimic the effect of melatonin (Maestroni and Conti, 1989). Further, it has been reported that at least thymic membranes bear specific opioid-binding sites (Raffi-El-Idrissi et al., 1998).

However, in surgically pinealectomized mice (i.e., no exposure to melatonin), neither the thymic cellularity nor the primary antibody response was influenced (Raffi-El-Idrissi et al., 1998). These results suggested that melatonin-induced immune opioids are indeed the physiological mediators of the immunoenhancing and anti-stress effects of melatonin observed in mice. Melatonin may influence the blood-forming system in mice via the induction or stimulation of T-helper cell-derived cytokines, including the melatonin-induced opioid system (Maestroni et al., 1994a, 1994b). These cytokines exert significant colony-stimulating activity; this hematopoietic effect of melatonin appears to be mediated by the type 1 κ-opioid receptor on bone marrow macrophages and by IL-1. In addition to melatonin-induced immune opioids, other lymphocyte products, such as IFN-γ and IL-2, are influenced by melatonin (Conti and Maestroni, 1994). Melatonin-induced immune opioids are also involved in the effect of melatonin on IFN-γ production. On the other hand, other reports showed that IFN-γ is able to modulate either directly, in pinealocytes, or indirectly, via neural mechanisms, the synthesis of melatonin in the pineal gland (Maestroni and Conti, 1990). Both lymphokines are, in fact, well known to stimulate NK activity and/or other natural immune parameters. It is thus possible that these lymphokines mediate the observed effects of melatonin on such natural immune parameters. These interesting findings suggest the existence of a bidirectional physiological connection between products of activated immunocompetent cells and melatonin.

Alterations of melatonin and endogenous opioid concentration and immune disturbance have been associated with disorders of reproductive functions. It has been proposed that the melatonin/immune/opioid network plays an important homeostatic role in preventing immune-based degenerative diseases. In fact, the proper functioning of such a network keeps the immune reactivity optimal despite the unavoidable stress and/or infectious events that occur during the life span of any organism.

FACTORS INTERFERING WITH MELATONIN AND IMMUNITY

STRESS AND IMMUNITY

Stress, a state of threatened homeostasis, may be induced by various physical or environmental factors (stressors). Stressful experiences may affect the well-being and immune functioning of humans and animals. Wild animals experience more stress in terms of ecofactors—that is, extreme conditions of temperature, search for food, and social pressure while avoiding being killed by predators. The molecular basis of these effects involves a network of multidirectional signaling and feedback regulations of neuroendocrine- and immunocyte-derived mediators. These are highly reproducible phenomena that illustrate a functional relationship between the central nervous system and the immune system. Fluctuations in environmental conditions do affect immunity—for example, cold temperature (Lifjeld et al., 2002; Hangalapura et al., 2003) and lack of food (Hangalapura et al., 2003; Bourgeon and Raclot, 2006) reduce immune activity in wild birds and domesticated fowl. Captivity too can reduce immune responses (Ewenson et al., 2003; Berzins et al., 2008), although the magnitude of these effects is species-specific (Matson et al., 2006).

The normal mammalian stress response involves a cascade of neuroendocrine events that mobilize energy reserves necessary to cope with immediate challenges, in part by inhibiting non-essential physiological processes and diverting resources from "maintenance" activities (Munck et al., 1984; Moberg, 2000; Sapolsky, 2002). A major component of this response involves the secretion of glucocorticoids (primarily cortisol and corticosterone) from the adrenal cortex (Nelson, 2000; Sapolsky, 2002). Interactions between glucocorticoids and immune function have been reported in relation to environmental stress and also in this article. Reports exist to demonstrate that environmental stresses elevate blood glucocorticoid levels and that high glucocorticoid levels suppress cellular and humoral immune function (Besedovesky and Del Ray, 1991; Ader and Cohen, 1993) and increase susceptibility to both infections and neoplastic processes (Hall and Goldstein, 1983). Changes in glucocorticoid concentrations within species have been linked to environmental factors, such as temperature (Weingrill et al., 2004), time of year (Perret and Predine, 1984), circadian rhythm (Coe and Levine, 1995; Smith and French, 1997), and food availability (Sapolsky, 1986), as well as to demographic factors, such as sex and age (Gesquiere et al., 2005; Mateo, 2006). Further, melatonin, a principal pineal neurohormone, has been reported to ameliorate the immunocompromising effects of glucocorticoids (Maestroni, 1993). Interestingly, exogenous melatonin treatment blocks the immunosuppressive effects of either catecholamine or clonidine administration in rats (Liebmann et al., 1996).

Therefore, the final results of stress-induced alteration of immune reactions are difficult to predict, and hypotheses to explain stress–immune interactions thus require refinement. Moreover, although there are large bodies of evidence demonstrating that stress can modify immune function, this does not necessarily imply that the immune system is compromised in maintaining self-tolerance. An important avenue of research needs to be pursued as how stressors affect immune processes during the period of acute and chronic stressors. This dynamic relationship between the nervous and immune systems is required to beselected to help in preventing neurodegenerative diseases. This modified perspective of stress and immunity should lead also to new insight into how animals endure challenging conditions in their environment.

OXIDATIVE STRESS

Melatonin is a potent free radical scavenger (Hardeland et al., 1993; Badr et al., 1999; Tan et al., 2002) and an indirect antioxidant (Mayo et al., 2002; Reiter and Tan, 2003; Rodriguez et al., 2004).

Melatonin is an efficient neutralizer of OH (Reiter et al., 1995) and is an electron-rich molecule interacting with free radicals to form metabolites that are also effective as free radical scavengers (Tan et al., 1999). Stable end products are excreted. Melatonin may stimulate several antioxidative enzymes (Pablos et al., 1995). *In vitro* melatonin is more effective than glutathione in scavenging the highly toxic (OH) radical and also more efficient than vitamin E in neutralizing the peroxyl radical. It also stimulates the main antioxidant enzyme of the brain, glutathione peroxidase. *In vivo* melatonin is a potent antioxidant (Reiter et al., 1995). It was shown that a diurnal variation of pineal SOD activity is dependent on the light–dark phase cycle and that probably the antioxidant power of the pineal gland is related to diurnal level of melatonin.

DIFFERENT KINDS OF RADIATION

The biological effects of radiation and ionizing radiation on cells are quite similar to those produced by endogenous stress. Indeed, ionizing radiation normally interacts with materials indirectly, via the formation of radical species. Much evidence has been gathered showing that enhanced free radical levels and impaired antioxidant pools are important factors underlying the pathophysiological mechanisms in a variety of diseases, including neurodegenerative diseases (Mates et al., 1999; Parthasarathy et al., 1999). Reports by Vijayalaxmi et al. (1995a, 1995b, 1998) demonstrate that melatonin-pretreated human peripheral blood lymphocytes, when exposed to gamma *radiation in vitro*, exhibited a significantly reduced (60% to 70%) incidence of chromosomal aberrations, as well as micronuclei and primary deoxyribonucleic acid (DNA) damage, compared with irradiated cells not pretreated with melatonin. Melatonin may act as a radical scavenger in peripheral blood cells during total body irradiation in rats (Koc et al., 2002). X-ray irradiation induced DNA damage and a marked decrease in antioxidants in the bone marrow of mice (Umegaki and Ishikava, 1994). A study by Sharma and Haldar (2006) demonstrates that X-ray irradiation in squirrels causes tissue damage in the spleen, as assessed by increased lipid peroxidation level and percent apoptosis, and in peripheral blood, as assessed by decreased total leukocyte count, as these cells are highly radiosensitive. Similarly, Koc et al. (2002) confirmed the marked effects of radiation on the stem cells of the leukocytes, lymphocytes, and platelets.

It has been suggested that survival after irradiation represents recovery of target tissues. In our study, we examined the effect of melatonin on recovery of peripheral blood cell counts in squirrels that received X-ray irradiation. During the RAP, exogenous melatonin supplementation can restore blood parameters up to control level in pinealectomized squirrels (Rai and Haldar, 2003). We considered that inhibition of free radical generation or intensified scavenging capacity, enhancement of DNA and membrane repair, replenishment of dead hematopoietic and other cells, and stimulation of immune cell activity by melatonin are important areas for radioprotection. A study was performed during the reproductively active phase of squirrels when peripheral melatonin levels are low (Haldar, 1996) so that the protective role of exogenous melatonin following X-ray irradiation becomes more evident. The data obtained suggest for the first time that in a seasonal breeder, while endogenous levels of melatonin are low, X-ray exposure caused oxidative damage, but exogenous pretreatment with melatonin significantly prevented the oxidative as well as hematopoietic damage, while post-treatment failed to do so (Sharma and Haldar, 2006).

The destructive action of UV irradiation results from both direct and indirect mechanisms involving endogenous sensitizers and the generation of active oxygen species (Foyer et al., 1994). UV radiation is known to cause inflammation due to the release of inflammatory cytokines, which in turn generate reactive oxygen species (ROS) that drive damage to lipids, proteins, and DNA (Peak et al., 1985; Vink et al., 1996). Exposure to UV radiation suppresses T-cell responses, such as the contact hypersensitivity and DTH responses, in both mice and humans (el-Ghorr and Norval, 1997; Narbutt et al., 2005). Our study on UV-C exposure showed significantly inhibited DTH responses to oxazolone-treated ear pinna (Sharma, 2008). It is known that UV radiation can induce a distinct T-regulatory type 1-like T-suppressor (Ts) population that may block the activation of T-helper (Th1) cell-mediated immune responses (Shreedhar et al., 1998). Such a block of Th1 cell-mediated immune response was evident in our DTH results of UV-C exposed squirrels. The immunomodulatory action of melatonin gets support from our data, as melatonin significantly protected against UV-C–induced DTH suppression.

Lymphocytes are the most studied and contain a variety of redox and free radical scavenging systems (Robinson et al., 1993). Hence, studies on lipid peroxidation and antioxidant enzymes in blood lymphocytes could be of immense significance in identifying intracellular oxidative damage in individuals who could be at risk for UV-C–induced oxidative damage. Our results have shown that UV-induced increase in Thio Barbituric Acid Reactive Substances TBARS level in splenocytes was accompanied by a significant decrease in its SOD activity, thus indicating the role of free radicals, as superoxide anion mediated damages following UV-C exposure. SOD enzyme is a superoxide anion scavenger. Thus, changes in the level of SOD activity might be due to the utilization of SOD in quenching superoxide anion generated due to UV-C effect (Sharma, 2008).

Our laboratory study suggests that UV-C radiation induces oxidative stress, resulting in apoptotic death via the caspase-3 pathway. Cysteine proteases (caspases) are central components of the apoptotic machinery, which is known to dismantle the cell by targeting a panoply of proteins for limited proteolysis (Earnshaw et al., 1999). Caspase-3 remains the most common and important effecter of apoptosis among the 14 different caspases. Cell death induced by a high dose of *in vivo* UV-C exposure involves the activation of caspase-3 in apoptotic pathway at 4 hours. Both the apoptotic index and caspase-3 level followed a similar pattern of rise after 4 hours of UV-C exposure compared to the control group, and a decline in the melatonin-pretreated group was observed compared to the UV-C irradiated group. Our results clearly indicate the involvement of caspase-3 as a major factor in inducing cell death after an acute dose of *in vivo* UV-C exposure (Sharma, 2008). However, the induction of caspase-3 was very low compared to the *in vitro* UV-C–induced apoptotic death, which might be due to the involvement of factors other than caspase-3 responsible for apoptosis.

Melatonin can reduce the UV-C–induced toxicity, which initiated an oxidative stress in the lymphatic tissue, as analyzed by reduced lipid damages, increased T-cell–mediated immune response, and reduced apoptotic cells (Sharma, 2008). The reduction in apoptotic index following melatonin treatment was because of the inhibition of caspase-3 activity, which indicated the involvement of melatonin in this pathway. Under *in vivo* conditions, melatonin is more potent in protecting tissues from radiation-induced cell death compared to its *in vitro* application (Sharma, 2008).

HORMONES INTERFERING WITH MELATONIN AND IMMUNITY

Practically all hormones and neurotransmitters have been shown to influence the immune response. Cytokines from activated immunocompetent cells have been reported to exert neuroendocrine and central nervous system effects (Black, 1994). Immunocompetent cells have been reported to bear hormone and neurotransmitter receptors capable of, respectively, translating specific endocrine signals and triggering the synthesis and release of the appropriate hormones (Maestroni, 1993; Black, 1994). The emerging picture is thus a complex and integrated network between the immune system and the neuroendocrine system that can be defined as the "immune-neuroendocrine network" (Maestroni, 1993).

ADRENAL HORMONES

Glucocorticoids

Development and activation of immune cells are prone to hormonal influences. The precise mechanism by which immune system is affected by the hypothalamo-hypophysial-adrenocortical (HPA) axis is unknown, but it probably involves increased cytokine release rates from activated immunological cells. The acute effects of these cytokines are mediated by stimulation of the release of corticotropin-releasing factor from the hypothalamus. Thus, not only are immunological cells influenced by glucocorticoids, but their products can also control the activity of the HPA axis (Besedovsky and del Rey, 1991). Substantial evidences links glucocorticoids with suppressed immune function (Caroleo et al., 1992; Maestroni, 1993). Short-term, temporary restraint stress and/or injection of a pharmacological dose of corticosteroid produce a dramatic involution of the lymphatic organs and a profound depression of the immune reactivity.

Evening administration of melatonin in mice enhanced the suppressed primary antibody production and thymus cellularity induced either by acute restraint stress or by corticosteroid treatment. Our study on lymphocyte proliferation in response to the mitogen concanavalin A (Con A) showed significant suppression of thymocytes, splenocytes, and lymph node cells after long-term (60 days) administration of dexamethasone. However, melatonin administration showed significant restoration in dexamethasone-suppressed proliferative response of lymphoid cells (Haldar et al., 2004). Melatonin treatment with dexamethasone antagonized the dexamethasone-induced lymphoid tissue involution, which supports previous reports (Maestroni, 1993). Another reliable immune parameter (i.e., DTH response to oxazolone) reflects the T-cell–mediated immune response, because DTH was significantly decreased following dexamethasone treatment. However, melatonin treatment, along with dexamethasone, showed antagonizing effects of the dexamethasone-induced DTH suppression (Haldar et al., 2004). Although necessary for the maintenance of overall metabolism and survival during periods of stress, one potentially undesirable effect of glucocorticoids is their inhibition of immune responses. Whereas this inhibition occurs by many mechanisms, one of the principal causes of glucocorticoid-induced immunosuppression is its induction of apoptosis in T-cell progenitor populations (Wyllie, 1980; Evans-Storms and Cidlowski, 1995). Melatonin appears to protect the organisms from this undesirable immunosuppression.

GLUCOCORTICOID EFFECTS ON IMMUNITY AND INSENSITIVITY

Corticosterone has been reported to suppress antibody production and nucleated spleen cells and causes thymic atrophy (Maestroni et al., 1986). Glucocorticoids can of course have downregulating effects on immune parameters (Sapolsky et al., 2000). Glucocorticoids can suppress maturation, differentiation, and proliferation of all immune cells (Sternberg, 2006) and trigger apoptosis in immature T and B cells and mature T cells (Sapolsky et al., 2000). Glucocorticoids also suppress IL-1 transcription, translation, and secretion and destabilize mRNA in leukocytes (Dhabhar and McEwen, 1997), rapidly reduce lymphocyte, monocyte, and granulocyte chemotaxis (Sapolsky et al., 2000), and reduce numbers of circulating lymphocytes (T cells > B cells and CD4+ > CD8+) as well as eosinophils, basophils, macrophages, and monocytes. The prime role of glucocorticoids in inflammation is to rein in defenses once infections are controlled. Glucocorticoid insensitivity has some parallels in free-living animals. Immune processes can become insensitive to glucocorticoids, but they can also be primed if glucocorticoid exposure happens early in life. Exposure of neonatal rats to endotoxin (a component of gram-negative bacteria) elevated basal glucocorticoids and increased the sensitivity of leukocytes to glucocorticoids in adulthood (Shanks et al., 2000). It is also important to address a recent (re)discovery that may have programming implications for glucocorticoid–immune interactions: production of glucocorticoids by immune organs during development (Schmidt and Soma, 2008). Glucocorticoids may often influence the development of immunity in growing animals, perhaps by driving expression of proteins responsible for lymphocyte receptor diversity (e.g., RAG I and II).

PROLACTIN

In vitro studies suggest that lymphocytes are an important target tissue for circulating prolactin. Prolactin stimulates ornithine decarboxylase (Hsu et al., 2006) and activates protein kinase C, which are pivotal enzymes in the differentiation, proliferation, and function of lymphocytes. Prolactin antibodies inhibit the proliferation of lymphocytes (Piccoletti et al., 1994). Prolactin induces IL-2 receptors on the surface of lymphocytes and acts as a progression factor during the proliferation of lymphocytes stimulated by IL-2, probably functioning in the nucleus without binding to its cell surface receptor (Gushchin et al., 1995). Some human B-lymphoblastoid cell lines produce prolactin constitutively (Di Mattia et al., 1988). A prolactin-like molecule is synthesized and secreted by Con A or phytohemagglutinin (PHA)-stimulated human peripheral blood mononuclear cells and functions in an autocrine manner as a growth factor for lymphoproliferation (Sabharwal et al., 1992). Prolactin enhances superoxide anion generation and hydrogen peroxide release from murine macrophages (Sakai et al., 1996). Growth hormone augments superoxide anion secretion of human neutrophils by binding to the prolactin receptor. Prolactin promotes the proliferation of Nb2 pre-T-cell lymphoma cells (Yu Lee, 1990). In these cells prolactin induces the biphasic expression of a transcription factor, IRF1 (interferon regulatory factor; one of the members of interferon response sequence (IRS) and may be involved in cell-cycle activation and S-phase progression (McAlexander and Yu Lee, 2001).

While examining the effects of glucocorticoids on the prolactin-dependent T-cell line Nb2, either pretreatment or concomitant treatment of Nb2 cells with prolactin significantly inhibited glucocorticoid-induced apoptosis in lymphatic tissue (Fletcher-Chiappini et al., 1993; Witorsch et al., 1993). Whereas treatment with glucocorticoid led to a 10-fold reduction in viable cell number, concomitant treatment of Nb2 cultures with both prolactin and glucocorticoid induced a relative stasis in the total number of viable cells. These studies also demonstrated that prolactin in defined culture could inhibit the dexamethasone-induced apoptosis of normal murine thymocytes. Thus, at the neuroimmune level, prolactin and glucocorticoids appear to antagonize their respective actions, resulting in the appropriate titration of immune response. Additional evidence supporting the interaction between melatonin and prolactin is that nighttime exposure to bright light, sufficient to induce a decrease in nocturnal melatonin secretion, results in a decreased prolactin secretion in women (Bispink et al., 1990).

Development and activation of immune cells are subject to hormonal influences, as illustrated by the roles of corticosteroids in thymus, pregnancy-related estrogens in B-cell development, or prolactin in T-cell generation and function (Morales et al., 1999). Lincoln and Clarke (2002) provide evidence that melatonin acts directly within the pituitary to regulate photoperiod-induced changes in prolactin secretion in seasonally breeding Soay sheep. Structural analysis of prolactin has revealed it to be related to members of the cytokine/hematopoietin family such as growth hormone, erythropoietin, granulocyte–macrophage colony-stimulating factor (GM-CSF), and the interleukins IL-2 to IL-7 (Bazan, 1990). The receptor for prolactin is present on T and B lymphocytes and macrophages (Pellegrini et al., 1992; Dardenne et al., 1994). Acting through its receptor, prolactin modulates immune system function by stimulating both cell proliferation and survival.

Because exposure to short day lengths suppresses circulating prolactin levels (Goldman and Darrow, 1983; Bittman, 1984), this hormone is a possible candidate for mediating some of the reported seasonal changes in immune function (Nelson et al., 1995). Hypophysectomy of rats compromised humoral and cell-mediated immunity; immune function can be resumed by prolactin replacement therapy (Reber, 1993). Prolactin is known to elevate respiratory burst and phagocytosis of peritoneal macrophages from both young and old mice (Chen and Johnson, 1993). Elevated levels of prolactin have been observed in patients during acute cardiac allograft rejection. It has been suggested that cyclosporin A (CsA), which was used to suppress graft rejection, can act as an antagonist to prolactin binding to the prolactin receptor on lymphocytes. In addition to triggering resting lymphocytes to cell division, the hormone can also control the magnitude of their response to polyclonal stimuli.

SEX STEROID HORMONES

Melatonin can also act indirectly on immune function via its effects on sex and adrenal steroid hormones, as well as prolactin. Sex steroid hormones and prolactin can affect immune function (Nelson and Demas, 1996). Both melatonin administration and maintenance in short days can result in significant changes in circulating levels of these hormones (Bartness et al., 1993). Both clinical and experimental evidence support the hypothesis that gonadal steroids regulate immune function. This conclusion is based on the following observations: (1) a sexual dimorphism exists in the immune response, (2) the immune response is altered by gonadectomy and sex steroid hormone replacement, (3) the immune response is altered during pregnancy, when the sex steroid hormone concentrations are increased, and (4) the organs responsible for the immune response contain specific receptors for gonadal steroids. Recent study from our lab (Ahmad and Haldar, 2010) revealed that melatonin and androgen show a trade-off relationship at their receptor level to maintain seasonal adjustment of immunity in tropical rodents.

STEROID RECEPTOR EXPRESSION IN PRIMARY LYMPHOID ORGANS

Early studies showed that in rats and mice, castration induced thymic enlargement, an effect reversed by androgen replacement, suggesting the presence of specific androgen-binding sites in thymic cells (Fitzpatrick et al., 1985; Olsen et al., 1991). Androgen receptors have been demonstrated in thymocytes by Western blot and flow cytometry in mice (Viselli et al., 1995; Ahmad and Haldar, 2010) and radioligand binding assays in humans (Kovacs and Olsen, 1987). Several studies have also demonstrated estrogen receptor (ER) expression in thymocytes and thymic epithelial cells in both mice and humans (Nilsson et al., 1986; Kawashima et al., 1992). Estrogen interacts with two distinct nuclear receptors, ER-α and ER-β. The human (Mosselman et al., 1996) and mouse (Tremblay et al., 1997) ER-β homologues have recently been cloned. In the male ER-α knockout mice, the role of ER-α in thymic development and estrogen-induced thymocyte phenotypic shift has been clearly demonstrated (Staples et al., 1999) and the requirement for ER-α in thymic development in both sexes during the postnatal period has been established (Yellayi et al., 2000). A study in ER-α– and/or ER-β–disrupted mice established that ER subtypes have different roles in females and males. ER-α was shown to be essential for thymic and splenic development in males, whereas expression of ER-β was required for estrogen-mediated thymic cortex atrophy and thymocyte phenotypic shift in females (Erlandsson et al., 2001). In mammals, B cells are produced in the bone marrow. Studies in mice have confirmed the effects of androgens on developing B cells mediated through androgen receptors expressed in bone marrow stromal cells (Bellido et al., 1995; Olsen and Kovacs, 2001). These cells also contain functional ER (Bellido et al., 1993). Recent studies have demonstrated that B-cell lymphopoiesis is normal in female ER-α–disrupted mice (Smithson et al., 1998), suggesting that ER-β might be responsible for regulation of B-cell formation in bone marrow (Kincade et al., 2000).

STEROID RECEPTOR EXPRESSION IN PERIPHERAL B AND T CELLS

As in primary lymphoid organs, ER transcripts were identified in mature peripheral B and T lymphocytes both in healthy subjects and in patients with system lupus erythematosus (SLE) (Suenaga et al., 1998). By contrast, androgen receptors have not been documented in mature peripheral T and B lymphocytes (Olsen and Kovacs, 2001), rendering genomic actions of androgens unlikely. Nevertheless, it was demonstrated that testosterone induces calcium influx, presumably via nongenomic surface receptors, in activated murine splenic mature T cells (Benten et al., 1997, 1999);

the functional repercussions of this have not been established. In summary, estrogen and androgen receptors are expressed in primary lymphoid organs and ERs are expressed in mature peripheral B and T cells, supporting the action of sex steroids in immune system development and modulation.

THYROID HORMONES

The thyroid gland and its hormones have been reported to influence reproduction and metabolism in a positive manner. However, research to date provides strong evidence for a reciprocal relationship between the immune system and hormones of the hypothalamus–pituitary–thyroid axis. Thyroxine (T_4) causes marked enlargement of the thymus and increased peripheral lymphocytes (Hassman et al., 1985), whereas thyroidectomy resulted in hypoplasia of the lymphoid organ (Haldar and Singh, 2001). Pinealectomy induced hypothyroidism, while melatonin injection to pinealectomized rodents restored it to control level (Shavali and Haldar, 1998).

It has been demonstrated that thymus growth is negatively influenced by removal of thyroid (Haldar and Singh, 2001). However, thyroid and its hormone T_4 are known to modulate the thymus-dependent immune function (Fabris et al., 1982; Ong et al., 1986) in mammals only. The thyroid hormone has receptors on lymphocytes and thymocytes in rats (Csaba et al., 1977). Further, melatonin receptors have also been detected on the circulating lymphocytes (Poon and Pang, 1992) as well as on thymocytes and splenocytes (Lopez-Gonzalez et al., 1993; Martin-Cacao et al., 1993; Rafii-El-Idrissi et al., 1995) of mammals and humans. Hence, a direct effect of thyroxine and melatonin on lymphoid organ could be conceived. It could be proposed that there may be a common site of action through which melatonin and L-thyroxine synergize the thyroxine-modulated immune function *in vivo* and vice versa.

Our data for *Funambulus pennanti* during its reproductively inactive phase, when peripheral T_4 is low, correlated with significant elevation of all immune parameters. However, combined treatment with L-thyroxine (35 mg/100 g body weight) and melatonin (25 μg/100 g body weight) had no additive effect. On the other hand, *in vitro* supplementation of T_4 (10^{-6} M), either alone or in combination with melatonin (0.5 μM), did not induce any significant change in thymocyte and splenocyte proliferation. Therefore, a trade-off effect of L-thyroxine and melatonin on the immune system (T- and B-cell differentiation) is suggested for this rodent (Rai et al., 2005).

PARACHLOROPHENYLALANINE, A SEROTONIN DEPLETER

Parachlorophenylalanine (PCPA), a β-adrenergic blocker and a known serotonin depleter, is used to perform a chemical pinealectomy (Px). PCPA has been proposed to be an indirect antagonist of melatonin as it inhibits the melatonin biosynthesis pathway by acting on tryptophan hydroxylase, the enzyme responsible for the synthesis of serotonin (serotonin is the substrate for melatonin). The role of PCPA, alone and in combination with melatonin was recorded in immunomodulation and the modulation of free radical load in lymphoid organs of a seasonal breeder, Indian palm squirrel (*F. pennanti*). We observed that PCPA suppressed immune status more in aged than in adult animals by reducing melatonin level by chemical Px. Melatonin level decreased in control aged squirrels, and so there was a decrease in immune parameters with a concomitant increase in free radical load of spleen. Decreased immune status can be restored following melatonin injection, which decreased free radical load of spleen, and this suggests that the immune organs of aged squirrels were sensitive to melatonin. Increased free radical load and decreased peripheral melatonin could be one of the reasons for immunosenescence (Rai et al., 2009).

PINEAL GLAND, MELATONIN, AND MATERNAL IMMUNITY

Pregnancy is associated with profound immunological changes that are characterized by a strong activation of certain components of the innate immune defense and a downregulation of adaptive immune function. This shift in the balance in immune system towards an innate dominance is thought to be important for the maintenance of a pregnancy. Based on our observations in the diurnal seasonal breeder *F. pennanti*, a seasonal breeder *R. norvegicus*, and nocturnal seasonal breeder *C. sphinx*, we for the first time suggest that melatonin injection to the pregnant female significantly increases lymphocyte proliferation in spleen and consequently increases the circulating level of lymphocyte in peripheral blood and percentage stimulation ratio of splenocyte, thereby increasing immune status during pregnancy. We reported earlier (Bishnupuri and Haldar, 2000; Bishnupuri and Haldar, 2001) that the melatonin level during pregnancy increases significantly, and now we suggest that this is for the increase of maternal immunity. This could be an adaptation during pregnancy to protect the mother from various external threats. This specific immune adaptation is conveyed to the fetus through placental transfer of melatonin, thereby controlling fetal immunity as well. The functional explanation for immune suppression during early pregnancy has been to protect the fetus from being attacked as foreign tissue by the maternal immune system. It was interesting to note that subsequent elevated melatonin levels in the mother also served to maintain immunity and pregnancy. After delivery, melatonin acts as an immunomodulator for both mother and fetus.

CONCLUSION

The immunoregulatory role of melatonin could be taken as part of the general physiological function adjusting the ability of the organism to tolerate various environmental challenges. The findings reported here document the existence of a bidirectional flow of information between the neuroendocrine pineal gland and immune system (as part of a complex system of checks and balances that also includes the immune system). The pineal gland might be considered part of a complex sensorial system (CNS) that also includes the immune system. Environmental challenges such as photoperiod variations, microbial pathogens, and other stressors are sensed by different parts of the same diffuse sensory system and the information is integrated and elaborated to give the appropriate response. Temperature may finally modulate the pineal melatonin response to a given photoperiod, which in turn will influence the specific immune response along with other melatonin-sensitive neuroendocrine mechanisms. The subsequent differential effects might thus be seen as indirect immunological consequences and immunosuppressive states as consequences. Such evidence brings forth new and interesting possibilities to improve our understanding and perhaps control a variety of serious diseases. The correlation of melatonin and hormonal balance highlights the significance of melatonin as an important hormone for gestation and maternal and fetal immunity during gestation. Melatonin thus has varied and profound roles in the modulation of immunity/immune system/immune status.

FUTURE PERSPECTIVES

Immunity, once thought to be a closed-circuit system, is now considered to be regulated by the neuroendocrine system and its hormones. The pineal gland as a photoperiodic transducer and melatonin as a chemical signal of dark period has come a long way to establish itself as a potent immunomodulator and may be a prospective immunotherapeutic agent under immunodepressive conditions. However, to date, most efforts have not advanced much beyond describing the changes that occur in immune function and disease prevalence with season and the potential involvement of photoperiod and melatonin in these processes. The next step is to identify the cellular and molecular mechanisms that mediate the effects of season and photoperiod on immune function. Despite the use of phototherapy in many cases, a more specific study needs to be done to assign a role to melatonin in the regulation of immunity and reproduction. Many areas, like the role of melatonin in implantation and maintenance of gestational physiology, need to be explored (where the embryo itself is a foreign body) to define the role of melatonin in regulation of maternal immunity. The non–receptor-mediated action of melatonin will be highly beneficial in reducing oxidative stress induced by various immune disorders (in lymphatic tissues), radiation-induced oxidative damage during therapy, microwaves and radiation induced by cell phone towers, electromagnetic waves of high-tension wires and towers, and so forth, which will be a major problem in coming decades. The interaction of melatonin in the modulation of various hormones like sex steroids (testosterone, estradiol), prolactin, and thyroxine needs to be explored further to use the knowledge in maintenance of immune homeostasis related to the disorders caused by anomalies of those hormones. The large number of studies that have already been carried out on neuroendocrine/melatonin in modulation of immunity indicates that the time has come when we must focus on some clinical trials to provide answers to the immunotherapeutic potentials of melatonin.

ACKNOWLEDGMENTS

This review is dedicated to one of the initial researchers on pineal gland and immunity, Prof. G.J.M. Maestroni, Center for Experimental Pathology, Intituto Cantonale di Patologia, 6604 Locarno, Switzerland. The authors thank all those colleagues whose work has been cited in the present review.

REFERENCES

Ader R, Cohen N. (1993). Pshychoneuroimmunology: conditioning and stress. *Ann Rev Psychol* 44: 53–85.

Ahmad R, Haldar C. (2009). Photoperiod-testicular-immune interaction in a seasonal breeder Indian palm squirrel *Funambulus pennanti* during the reproductively inactive and active phases. *J Neuroendocrinol* 21: 2–9.

Ahmad R, Haldar C. (2010). Melatonin and androgen receptor expression interplay modulates cell-mediated immunity in tropical rodent *Funambulus pennanti*: An *in-vivo* and *in-vitro* study. *Scand J Immunol* 71: 420–430.

Attanasio A, Borrell PP, Gupta D. (1985). Circadian rhythms in serum melatonin from infancy to adolescence. *J Clinical Endocrinol Metab* 61: 388–389.

Badr FM, El Habit OH, Harraz MM. (1999). Radioprotective effect of melatonin assessed by measuring chromosomal damage in mitotic and meiotic cells. *Mutat Res* 444: 367–372.

Barrett P, MacLean A, Davidson G, Morgan PJ. (1996). Regulation of the Mel 1a melatonin receptor mRNA and protein levels in the ovine pars tuberalis: evidence for a cyclic adenosine 3",5"-monophosphate-independent Mel 1a receptor coupling and an autoregulatory mechanism of expression. *Mol Endocrinol* 10: 892–902.

Bartness TJ, Powers JB, Hastings MH, Bittman EL, Goldman BD. (1993). The timed infusion paradigm for melatonin delivery: what has it taught us about the melatonin signal, its reception, and the photoperiodic control of seasonal responses? *J Pineal Res* 15: 161–190.

Bazan JF. (1990). Haematopoietic receptors and helical cytokines. *Immunol Today* 11: 350–354.

Bellido T, Girasole G, Passeri G, Yu XP, Mocharla H, Jilka RL, Notides A, Manolagas SC. (1993). Demonstration of estrogen and vitamin D receptors in bone marrow-derived stromal cells: up regulation of the estrogen receptor by 1, 25-dihydroxyvitamin-D3. *Endocrinol* 133: 553–562.

Bellido T, Jilka RL, Boyce BF, Girasole G, Broxmeyer H, Dalrymple SA, Murray R, Manolagas SC. (1995). Regulation of interleukin- 6, osteoclastogenesis, and bone mass by androgens. The role of the androgen receptor. *J Clin Invest* 95: 2886–2895.

Benten WP, Lieberherr M, Giese G, Wrehlke C, Stamm O, Sekeris CE, Mossmann H, Wunderlich F. (1999) Functional testosterone receptors in plasma membranes of T cells. *FASEB J* 13: 123–133.

Benten WP, Lieberherr M, Sekeris CE, Wunderlich F. (1997) Testosterone induces Ca21 influx via non-genomic surface receptors in activated T cells. *FEBS Lett* 407: 211–214.

Berzins LL, Tilman-Schindel E, Burness G. (2008). Sex-specific effects of handling time on an index of immune activity in zebra finches. *Physiol Biochem Zool* 81: 383–387.

Besedovsky HO, del Rey A. (1991). Feed-back interaction between immunological cells and the hypothalamo-pituitary-adrenal axis. *Netherland J Med.* 39: 274–280.

Bishnupuri KS, Haldar C. (2000). Impact of photoperiodic exposures during late gestation and lactation periods on the pineal and reproductive physiology of the Indian palm squirrel, *Funambulus pennanti*. *J Reprod Fertil* 118:295–301.

Bishnupuri KS, Haldar C. (2001). Maternal transfer of melatonin alters the growth and sexual maturation of young Indian palm squirrel *Funambulus pennanti*. *Biol Signals Recept* 10: 317.

Bispink G, Zimmermann R, Weise HC, Leidenberger F. (1990). Influence of melatonin on the sleep-independent component of prolactin secretion. *J Pineal Res* 8: 97–106.

Bittman EL. (1984). Melatonin and photoperiodic time measurement: Evidence from rodents and ruminants, 155–192. In: *The Pineal Gland* (Reiter RJ.), Raven Press, New York, pp. 155–192.

Black PH. (1994). Immune system-central nervous system interactions: effect and immunomodulatory consequences of immune system mediators on the brain. *Antimicrob Agents Chemother* 38: 7–12.

Bourgeon S, Raclot T. (2006). Corticosterone selectively decreases humoral immunity in female eiders during incubation. *J Exp Biol* 209: 4957–4965.

Caroleo MC, Frasca D, Nistico G, Doria G. (1992). Melatonin as immunomodulator in immune deficient mice. *Immunopharmacol* 23: 81–89.

Carrillo-Vico A, Garcia-Perganeda A, Naji L, Calvo JR, Romero MP, Guerrero JM. (2003a). Expression of membrane and nuclear melatonin receptor mRNA and protein in the mouse immune system. *Cell Mol. Life Sci.* 60: 2272–2278.

Carrillo-Vico A, Garcia-Maurino S, Calvo JR, Guerrero JM. (2003b). Melatonin counteracts the inhibitory effect of PGE2 on IL-2 production in human lymphocytes via its mt1 membrane receptor. *FASEB J* 17: 755–757.

Carrillo-Vico A, Guerrero JM, Lardone PJ, Reiter RJ. (2005). A Review of the multiple actions of melatonin on the immune system. *Endocrine* 27: 189–200.

Chaffee RRJ, Roberts JC. (1971). Temperature acclimatization of birds and mammals. *Ann Rev Physiol* 33: 115–205.

Champney TH, Allen GC, Zannelli M, Beausang L. (1998). Time-dependent effects of melatonin on immune measurements in male Syrian hamsters. *J Pineal Res* 25: 142–146.

Champney TH, Mc Murray DN. (1991). Spleen morphology and lymphoproliferative activity in short photoperiod exposed hamsters. In: *Role of Melatonin and Pineal Peptides in Neuroimmunomodulation* (Eds.), F Fraschini, R J Reiter)., Plenum Press, London, pp. 219–225.

Chen Y, Johnson AG. (1993). In vivo activation of macrophages by prolactin from young and aging mice. *Int J Immunopharmacol* 15: 39–43.

Chowers I, Conforti N, Feldman S. (1970). Effect of changing levels of glucocorticosteroids on body temperature on exposure to cold. *Am J Physiol* 218: 1563–1567.

Claman HN. (1972). Corticosteroids and lymphoid cells. *N Engl J Med* 287: 388.

Coe CL, Levine S. (1995) Diurnal and annual variation of adrenocortical activity in the squirrel-monkey. *Am J Primatol* 35: 283–292.

Collins KJ, Weiner JS. (1968). Endocrinological aspects of exposure to high environmental temperatures. *Physiol Rev* 48: 785–839.

Colombo MP, Modesti A, Parmiani G, Forni G. (1992). Local cytokine availability elicits tumor rejection and systemic immunity through granulocyte-T-lymphocyte cross-talk. *Cancer Res* 52: 4853–4857.

Conti A, Maestroni GJM. (1994). Melatonin induced immuno-opioids: role in lymphoproliferative and autoimmune diseases. *Adv Pineal Res* 7: 83–100.

Csaba C, Sudar F, Dobozy O. (1977). Triiodothyronine receptors in lymphocytes of new born and adult rats. *Horm Metab Res* 79: 499–501.

Csaba G, Barath, P. (1975). Morphological changes of thymus and the thyroid gland after postnatal extirpation of pineal body. *Endocrinol Exper* 9: 59–66.

Cupps TR, Fauci AS. (1982). Corticosteroid mediated immunoregulation in man. *Immunol Rev* 65: 133–155.

Dardenne M, de Morales MCL, Kelly PA, Gagnerault MC. (1994). Prolactin receptor expression in human hematopoietic tissues analyzed by flow cytofluorimetry. *Endocrinol* 134: 2108–2114.

Demas GE, Nelson RJ. (1996). Photoperiod and temperature interact to affect immune parameters in adult male deer mice (*Peromyscus maniculatus*). *J Biol Rhythms* 11: 94–102.

Demas GE, Nelson RJ. (1998). Exogenous melatonin enhances cell-mediated, but not humoral, immune function in adult male deer mice (*Peromyscus maniculatus*). *J Biol Rhythms* 13: 245–252.

Dhabhar F, McEwen B. (1997). Acute stress enhances while chronic stress suppresses cell-mediated immunity in vivo: a potential role for leukocyte trafficking. *Brain Behav Immun* 11: 286–306.

Di Mattia, GE, Gellersen B, Bohnet HG, Friesen HG. (1988). A human B-lymphoblastoid cell line produces prolactin. *Endocrinol* 122: 2508–2518.

Di Stefano A, Paulesu L. (1994). Inhibitory effect of melatonin on production of IFN- γ of TNF- α in peripheral blood mononuclear cells of some blood donors. *J Pineal Res* 17: 164–169.

Drazen DL, Bilu D, Bilbo SD, Nelson RJ. (2001). Melatonin enhancement of splenocyte proliferation is attenuated by luzindole, a melatonin receptor antagonist. *Am J Physiol Regul Integr Comp Physiol* 280: R1476–R1482.

Drazen DL, Klein SL, Yellon SM, Nelson RJ. (2000). In vitro melatonin treatment enhances splenocyte proliferation in prairie voles. *J Pineal Res* 28: 34–40.

Earnshaw WC, Martins LM, Kaufmann SH. (1999). Mammalian caspases: structure, activation, substrates, and functions during apoptosis. *Ann Rev Biochem* 68: 383–424.

el-Ghorr AA, Norval M. (1997). The role of interleukin-4 in ultraviolet B light-induced immunosuppression. *Immunol* 92: 26–32.

Erlandsson MC, Ohlsson C, Gustafsson JA. (2001). Role of oestrogen reseptors alpha and beta in immune organ development and in oestrogen-mediated effect on thymus. *Immunol* 103: 17–25.

Evans-Storms RB, Cidlowski JA. (1995). Regulation of apoptosis by steroid hormones. *J Steroid Biochem Mol Biol* 53: 1–8.

Ewenson E, Zann R, Flannery G. (2003). PHA immune response assay in captive zebra finches is modulated by activity prior to testing. *Anim Behav* 66: 797–800.

Fabris N, Muzzioli M, Mocchegiani E. (1982). Recovery of age- dependent immunological deterioration in Balb/C mice by short-term treatment with L-thyroxine. *Mech Ageing Dev* 18: 327–343.

Fitzpatrick FT, Kendall MD, Wheeler MJ, Adcock IM, Greenstein BD. (1985). Reappearance of thymus of ageing rats after orchidectomy. *J Endocrinol* 106: R17–R19.

Fletcher-Chiappini SE, Compton MM, LaVoie HA, Day EB, Witorsch RJ. (1993). Glucocorticoid-prolactin interactions in Nb2 lymphoma cells: antiproliferative versus anticytolytic effects. *Proc Soc Exp Biol Med* 202: 345–352.

Foyer CH, Lelandais M, Kunert KJ. (1994). Photooxidative stress in plants. *Physiol Plant* 92: 696–717.

Gadgil S. (2006). The Indian monsoon 2. How do we get rain? *Resonance.* Springer India. 11: 8–21.

Garcia-Maurino S, Gonzalez-Haba MG, Calvo JR, Rafii-El-Idrissi M, Sanchez-Margalet V, Goberna R, Guerrero JM. (1997). Melatonin enhances IL-2, IL-6, and IFN-gamma production by human circulating CD4+ cells: a possible nuclear receptor-mediated mechanism involving T helper type 1 lymphocytes and monocytes. *J Immunol* 159: 574–581.

Garcia-Maurino S, Pozo D, Calvo JR, Guerrero JM. (2000). Correlation between nuclear melatonin receptor expression and enhanced cytokine production in human lymphocytic and monocytic cell lines. *J Pineal Res* 29: 129–137.

Gerrard JW. (1984). The biological importance of Ig E. *Immunol. Allergy Pract* 25: 381–384.

Gesquiere LR, Altmann J, Khan MZ, Couret J, Yu JC, Endres CS, Lynch JW, Ogola P, Fox EA, Alberts SC, Wango EO. (2005). Coming of age: steroid hormones of wild immature baboons (*Papio cynocephalus*). *Am J Primatol* 67: 83–100.

Gillis S, Crabtree GR, Smith KA. (1979). Glucocorticoid induced inhibition of T-cell growth factor production. I. The effect of mitogen-induced lymphocyte proliferation. *J Immunol* 123:1624–1631.

Gisler RH. (1974). Stress and the hormonal regulation of the immune response in mice. *Psychother Psychosom* 23: 197–208.

Goldman BD, Darrow JM. (1983). The pineal gland and mammalian photoperiodism. *Neuroendocrinol* 37: 386.

Guerrero JM, Reiter RJ. (2002). Melatonin-immune system interrelationship. *Curr Top Med Chem* 2: 167–179.

Gushchin GV, Cheney C, Glaser R and Malarkey WB. (1995). Temporal relationships and IL-2 dependency of prolactin-induced lymphocyte proliferation. *J Neuroimmunol* 60: 93-9.

Haldar C, Haussler D, Gupta D. (1992). Effect of the pineal gland on circadian rhythmicity of colony forming units for granulocytes and macrophages (CFU-GM) from rat bone marrow cell culture. *J Pineal Res* 12: 79–83.

Haldar C, Rai S, Singh R. (2004). Melatonin blocks dexamethasone induced immunosuppression in a seasonally breeding rodent Indian palm squirrel, *Funambulus pennanti. Steroids* 69: 367–377.

Haldar C, Saxena N. (1988a). Pineal gland and humidity effects on testicular function of the Indian palm squirrel, *Funambulus pennanti. J Pineal Res* 5: 411–418.

Haldar C, Saxena N. (1988b). Tropical environmental impact on reproduction: role of pineal gland. In: *Threat Hab* V.K. Agrawal and R.P. Agrawal, Editors, Threat Habitate., pp. 307–320.

Haldar C, Singh R, Guchhait P. (2001). Relationship between annual rhythm in melatonin and immune system in the tropical palm squirrel *F. pennanti. Chronobiol Internat* 18: 61–69.

Haldar C, Singh R. (2001). Pineal modulation of thymus and immune function in a seasonally breeding rodent, *F. pennanti. J Exp Zool* 286: 90–98.

Haldar C, Vidhu S. (1997). Effect of pinealectomy and testosterone on gonadal regression and accessory sex organs in Indian palm squirrel, *Funambulus pennanti. Ind J Exp Biol* 35: 594–596

Haldar C. (1986). Method of pinealectomy in some vertebrates. *Ind J Exp Biol* 24: 319–322.

Haldar C. (1996). Patterns of plasma melatonin in Indian tropical palm squirrel, *F. pennanti.* In: *Role of Pineal Gland: Its Molecular Signals* (Ed. Haldar C). HPCL, New Delhi, pp. 95–117.

Haldar-Misra C, Pevet P. (1983). Influence of melatonin on the process of protein and/or peptide secretion in the pineal gland of the rat and hamster. An *in vitro* study. *Cell Tissue Res* 231: 73–82.

Hall NR, Goldstein AL. (1983). Role of thymosin and the neuroendocrine system in the regulation of immunity. In: Immunoregulation (Eds. Fabris N, Garaci E, Hadden J, Mitchison NA, (eds). Immunoregulation, New York: Plenum Press,; pp. 141.

Hangalapura BN, Nieuwland MGB, Reilingh GD, Heetkamp MJW, van den Brand H, Kemp B, Parmentier HK. (2003). Effects of cold stress on immune responses and body weight of chicken lines divergently selected for antibody responses to sheep red blood cells. *Poultry Sci* 82: 1692–1700.

Hardeland R, Reiter RJ, Poeggeler B, Tan DX. (1993). The significance of the metabolism of the neurohormone melatonin: Antioxidative protection and formation of bioactive substances. *Neurosci Biobehav Rev* 17: pp. 347–357.

Hassman R, Weetman AP, Gunn C, Stringer BMJ, Wynford Thomas D, Hall R, McGregor AM. (1985). The effects of hypothyroidism on auto immune thyroiditis in rat. *Endocrinol* 166: 1253.

Haus E, Smolensky MH. (1999). Biologic rhythms in the immune system. *Chronobiol Int* 16: 581–622.

Hill S, Spriggs LL, Simon MA, Muraoka H, Blask DE. (1992). The growth inhibitory action of melatonin on human breast cancer cells is linked to the estrogen response system. *Cancer Lett* 64: 249.

Hsu PC, Hour TC, Liao YF, Hung YC, Liu CC, Chang WH, Kao MC, Tsay GJ, Hung HC, Liu GY. (2006). Increasing ornithine decarboxylase activity is another way of prolactin preventing methotrexate-induced apoptosis: crosstalk between ODC and BCL-2. Apoptosis 11: 389–399.

Iwata T, Incefy GS, Cunningham–Rrundles S, Cunningham–Rrundles C, smithwick E, Geller N, Reilly RO, Good RA. (1981). Circulating thymic hormones activity in patients with primary and secondary immunedeficiency diseases. *Am J Med* 71: 385–393.

Kawashima I, Seiki K, Sakabe K, Ihara S, Akatsuka A, Katsumata Y. (1992). Localization of estrogen receptors and estrogen receptor mRNA in female mouse thymus. *Thymus* 20: 115–121.

Kelley KW. (1980). Stress and immune function. A bibliographic review. *Ann Rech Vet* 11: 445.

Kikuchi M. (1988). Effective period and susceptible period during the influence of ambient temperature change on immune response. *Int J Biometeorol* 26 (Suppl): 1101–1115.

Kincade PW, Medina KL, Payne KJ, Rossi MI, Tudor KS, Yamashita Y, Kouro T. (2000). Early B-lymphocyte precursors and their regulation by sex steroids. *Immunol Rev.* 175: 128–137.

Koc M, Buyukokuroglu ME, Taysi S. (2002). The effect of melatonin on peripheral blood cells during total body irradiation in rats. *Biol Pharm Bull* 25: 656–657.

Kovacs WJ, Olsen NJ. (1987). Androgen receptors in human thymocytes. *J Immunol* 139: 490–493.

Kuci S, Becker J, Veit G, Handgretinger GR, Attanasio A, Bruchett G, Treuner J, Niethammerd, and Gupta D. (1988). Circadian variations in the immunomodulatory role of pineal gland. *Neuroendocr Lett* 10: 65–80.

Lahiri S, Haldar C. (2009). Response of melatonin receptor MT1 in spleen of a tropical Indian rodent, *Funambulus pennanti* to natural solar insolation and different photoperiodic conditions. *Chronobiol Int* 26: 1559–1574.

Lahiri S. (2009). Expression of membrane melatonin receptor and receptor mRNA on lymphoid tissue of a tropical seasonal breeder, *Funambulus pennanti* , under different experimental conditions. Ph. D. Thesis, submitted to Department of Zoology, BHU, India.

Larsson EL. (1980). Cyclosporin A and dexamethasone suppress T cell responses by selectively acting at distinct sites of the triggering process. *J Immunol* 124: 2828–2833.

Liebmann PM, Hofer D, Felsner P, Wölfler A, Schauenstein K. (1996). β-blockade enhances adrenergic immunosuppression in rats via inhibition of melatonin release. *J Neuroimmunol* 67:137–42.

Lifjeld JT, Dunn PO, Whittingham LA. (2002). Short-term fluctuations in cellular immunity of tree swallows feeding nestlings. *Oecologia* 130: 185–190.

Lincoln GA, Clarke IJ. (2002). Noradrenaline and dopamine regulation of prolactin secretion in sheep: role in prolactin homeostasis but not photoperiodism. *J Neuroendocrinol* 14: 36–44.

Lissoni P, Barni S, Tancini G, Ardizzoia A, Ricci G, Aldeghi R, Brivio F, Tisi E, Rovelli F, Rescaldani R. (1994). A randomised study with subcutaneous low-dose interleukin 2 alone vs interleukin 2 plus the pineal neurohormone melatonin in advanced solid neoplasms other than renal cancer and melanoma. *Br J Cancer* 69: 196–199.

Lopez–Gonzalez MA, Calvo JR, Osuna C, Sequra JJ, Gurrero JM. (1993). Characterization of melatonin binding sites in human peripheral blood neutrophils. *Biotechnol Ther* 4: 253–262.

Maestroni GJM, Conti A, Lissoni A. (1994a). Hematopoietic rescue via T- cell dependent endogenous GM–CSF by the pineal neurohormone melatonin in tumor bearing mice. *Cancer Res* 54: 2429–2432.

Maestroni GJM, Conti A, Lissoni P. (1994b). Colony stimulating activity and hematopoietic rescue from cancer chemotherapy compounds are induced by melatonin via endogenous interleukin-4. *Cancer Res* 54: 4740–4743.

Maestroni GJM, Conti A, Pierpaoli W. (1986). Role of the pineal gland in immunity. Circadian synthesis and release of melatonin modulates the antibody response and antagonizes the immunosuppressive effect of corticosterone. *J Neuroimmunol* 13: 19–30.

Maestroni GJM, Conti A, Pierpaoli W. (1988). Role of the pineal gland in immunity. III. Melatonin antagonizes the immunosuppressive effect of acute stress via an opiatergic mechanism. *Immunol* 63: 465–469.

Maestroni GJM, Conti A, Pierpaoli W. (1998). Melatonin stress and the immune system. In: *Pineal Research Review,* (Ed.), RJ Reiter. Liss, New York, pp. 203–223.

Maestroni GJM, Conti A. (1989). Beta–endorphin and dynorphin mimic the circadian immunoenhancing and anti stress effect of melatonin. *Int J Immunopharmacol* 11: 333–340.

Maestroni GJM, Conti A. (1990). The pineal neurohormone melatonin stimulates activated CD4+, Thy-1+ cells to release opioid agonist(s) with immunoenhancing and anti-stress properties. *J Neuroimmunol* 28: 167–176.

Maestroni GJM, Conti A. (1993). Melatonin in relation to the immune system. In: *Melatonin Biosysthesis, Physiological effects and clinical applications.* (Eds.), HS Yu, RJ Reiter, CRC, Boca Raton, CRC Press. pp. 289–310.

Maestroni GJM. (1993). The immunoneuroendocrine role of melatonin. *J Pineal Res* 14: 1–10.

Markus RP, Ferreira ZS, Fernandes PACM, Cecon E. (2007). The immune-pineal axis: a shuttle between endocrine and paracrine melatonin sources. *Neuroimmunomodulation* 14: 126–133.

Martin-Cacao A, Lopez-Gonzalez NA, Reiter RJ, Calvo JR, Guerrero JM. (1993). Binding of 2[125I] melatonin by rat thymus membranes during postnatal development. *Immunol Lett* 36: 59–64.

Mateo JM. (2006). Developmental and geographic variation in stress hormones in wild Belding's ground squirrels (*Spermophilus beldingi*). *Horm Behav* 50: 718–725.

Mates JM, Perez-Gomez C, De Castro IN. (1999). Antioxidant enzymes and human diseases. *Clin Biochem* 32: 595–603.

Matson KD, Tieleman BI, Klasing KC. (2006). Capture stress and the bactericidal competence of blood and plasma in five species of tropical birds. *Physiol Biochem Zool* 79: 556–564.

Mayo JC, Sainz RM, Antolin I, Herrera F, Martin V, Rodriquez C. (2002). Melatonin regulation of antioxidant enzyme gene expression. *Cell Mol Life Sci* 59: 1706–1713.

McAlexander MB, Yu-Lee LY. (2001) Prolactin activation of IRF-1 transcription involves changes in histone acetylation. *FEBS Lett* 12: 91–4.

McArthur AJ, Hunt AE, Gillette MU. (1997). Melatonin action and signal transduction in the rat suprachiasmatic circadian clock: activation of protein kinase C at dusk and dawn. *Endocrinol* 138: 627–634.

Milcu SM, Potop I. (1973). Biological activity of thymic protein extracts. In: *Thymic hormones*. (Eds.), Luckey TD. N Urban and J C Schawarzenberg, Munich, pp. 97–134.

Moberg GP. (2000). Biological response to stress: implications for animal welfare. In: *The Biology of Animal Stress*. CABI Publishing, New York, pp. 1–21.

Morales J, Homey B, Vicari AP, Hudak S, Oldham E, Hedrick J, Orozco R, Copeland NG, Jenkins NA, McEvoy LM, Zlotnik A. (1999). CTACK, a skin-associated chemokine that preferentially attracts skin-homing memory T cells. *Proc Natl Acad Sci USA* USA 96: 14470–14475.

Mosselman S, Polman J, Dijkema R. (1996). ER- b: Identification and characterization of a novel human estrogen receptor. *FEBS Lett* 392: 49–53.

Mount LE. (ed.) (1979). Adaptation to the thermal environment. In: *Adaptation to Thermal Environment: Man and his Productive Animals*. Edward Arnold, London, pp. 116–144.

Munck A, Guyre PM, Holbrook NJ. (1984). Physiological functions of glucocorticoids in stress and their relation to pharmacological actions. *Endocr Rev* 5: 25–44.

Narbutt J, Lesiak A, Skibinska M, Wozniacka A, van Loveren H, Sysa-Jedrzejowska A, Lewy-Trenda I, Omulecka A, Norval M. (2005). Suppression of contact hypersensitivity after repeated exposures of humans to low doses of solar simulated radiation. *Photochem Photobiol Sci* 4: 517–522.

Nelson RJ, Demas GE, Klein SL, Kriegsfeld LJ. (1995). The influence of season, photoperiod, and pineal melatonin on immune function. *J Pineal Res* 19: 149–165.

Nelson RJ, Demas GE. (1996). Seasonal changes in immune function. *Q Rev Biol* 71: 511–548.

Nelson RJ, Demas GE. (1996). Seasonal changes in immune function. *Rev Biol* 71: 511–548.

Nelson RJ. (2000). *An introduction to behavioral endocrinology*. Sinauer, Sunderland, MA.

Nelson RJ. (2004). Seasonal immune function and sickness responses. *Trends Immunol* 25: 187–191.

Nilsson B, Bergqvist A, Lindblom D, Ljungberg OO, Sodergard R, von Schoultz B. (1986). Characterization and localization of specific oestrogen binding in the human thymus. *Gynecol Obstet Invest* 21:150–157.

Nosjean O, Ferro M, Coge F, Beauverger P, Henlin JM, Lefoulon F, Fauchere JL, Delagrange P, Canet E, Boutin JA. (2000). Identification of the melatonin melatonin- binding site MT3 as the quinone reductase 2. *J Biol Chem* 275: 31311–31317.

Olsen NJ, Kovacs WJ. (2001). Effects of androgens on T and B lymphocyte development. *Immunol Res* 23: 281–288.

Olsen NJ, Watson MB, Henderson GS, Kovacs WJ. (1991). Androgen deprivation induces phenotypic and functional changes in the thymus of adult male mice. *Endocrinol* 129: 2471–2476.

Ong ML, Malbin DG, Malkin A. (1986). Alteration of lymphocytes reactivities by thyroid hormones. *Int J Immunopharmacol* 8: 755–762.

Pablos MI, Agapito MT, Gutierrez R, Recio JM, Reiter RJ, Barlow-Walden L, Acuna-Castroviejo D, Menendez-Pelaez A. (1995). Melatonin stimulates the activity of the detoxifying enzyme glutathione peroxidase in several tissues of chickens. *J Pineal Res* 19: 111–115.

Parthasarathy S, Santanam N, Ramachandran S, Meilhac O. (1999). Oxidants and antioxidants in atherogenesis; an appraisal. *J Lipid Res* 40: 2143–2157.

Peak MJ, Peak JG, Jones CA. (1985). Different (direct and indirect) mechanisms for the induction of DNA-protein crosslinks in human cells by far- and near-ultraviolet radiation (290 and 405 nm). *Photochem Photobiol* 42: 141–146.

Pellegrini I, Lebrun JJ, Ali S, Kelly PA. (1992). Expression of prolactin and its receptor in human lymphoid cells. *Mol Endocrinol* 6: 1023–1031.

Perret M, Predine J. (1984). Effects of long-term grouping on serum cortisol levels in *Microcebus murinus* (Prosimii). *Horm Behav* 18: 346–358.

Pevet P, Ebels I, Swaab DF, Mud MT, Arimura A. (1980). Presence of AVT-, alpha-MSH-, LHRH- and somatostatin-like compounds in the rat pineal gland and their relationship with the UMO5R pineal fraction. An immunocytochemical study. *Cell Tissue Res* 206: 341–353.

Pevet P, Haldar-Misra C. (1982). Morning injections of large doses of melatonin, but not of 5-methoxytryptamine, prevent in the hamster the antigonadotropic effect of 5- methoxytryptamine administered late in the afternoon. *J. Neural Transm* 55: 85–93.

Piccoletti R, Maroni P, Bendinelli P, Bernelli-Zazzera, A. (1994). Rapid stimulation of mitogen-activated protein kinase of rat liver by prolactin. *Biochem J* 303: 429–433.

Pioli C, Caroleo C, Nistico G, Doria G. (1993). Melatonin increases antigen presentation and amplifies specific and non-specific signals for T-cell proliferation. *Int J Immunopharmacol* 15: 463–468.

Poirel VJ, Cailotto C, Streicher D, Pevet P, Masson-Pevet M, Gauer F. (2003). MT1 melatonin Receptor mRNA tissular localization by PCR amplification. *Neuroendocrinol Lett* 24: 33–38.

Poon AMS, Pang SF. (1992). 2[125I] iodomelatonin binding sites in spleens of guinea pigs. *Life Sci* 50: 1719–1726.

Pozo D, Delgado M, Fernandez-Santos JM, Calvo JR, Gomariz RP, Martin-Lacave I, Ortiz GG, Guerrero JM. (1997). Expression of the Mel1a-melatonin receptor mRNA in T and B subsets of lymphocytes from rat thymus and spleen. *FASEB J* 11: 466–473.

Radostits OM, Arundel JH, Gay. CC (2000). *Veterinary Medicine: A Textbook of the Diseases of Cattle, Sheep, Pigs, Goats and Horses.* Elsevier Health Sciences, USA.

Rafii-El-Idrissi M, Calvo JR, Harmouch A, Garcia-Mourino S, and Guerrero JM. (1998). Specific binding of melatonin by purified cell nuclei from spleen and thymus of the rat. *J Neuroimmunol* 86: 190–197.

Rafii-el- Idrissi M, Calvo JR, Pozo D, Harmouch A, Guerrero JM. (1995). Specific binding of 2[125I] iodomelatonin by rat splenocytes: Characterization and its role on regulation of cyclic AMP production. *J Neuroimmunol* 57: 171–178.

Rai S, Haldar C, Singh R. (2009). Modulation of immunity in young-adult and aged sSquirrel, *Funambulus pennanti* by melatonin and p-chlorophenylalanine. *Immun Aging* 6: 5.

Rai S, Haldar C, Singh SS. (2005). Trade off between L-thyroxin and melatonin in immune regulation of Indian palm squirrel, *Funambulus pennanti* during reproductively inactive phase. *J Neuroendoicrinol* 182: 103–110.

Rai S, Haldar C. (2003). Pineal control of immune status and hematological changes in blood and bone marrow of male squirrels (*Funambulus pennanti*) during their reproductively active phase. *Comp Biochem Physiol (C)* 136: 319–328.

Reber PM. (1993). Prolactin and immunomodulation. *Am J Med* 95: 637–644.

Reiter RJ, Melchiorri D, Sewerynek E, Poggeler B, Barlow–Walden L, Chuang JI, Ortij GG, Acuna-Castroviejo D. (1995). A review of the evidence supporting melatonin's role as an antioxident. *J Pineal Res* 18: 1–11.

Reiter RJ, Tan DX. (2003). Melatonin: A novel protective agent against oxidative injury of the ischemic/reperfused heart. *Cardiovasc Res* 58: 10–19.

Reiter RJ. (1973). Pineal control of a seasonal reproductive rhythm in male golden hamsters exposed to natural day length and temperature. *Endocrinol* 92: 423–430.

Reiter RJ. (1983). The pineal gland: An intermediary between the environment and the endocrine system. *Psychoneuroendocrinol* 8: 31.

Reiter RJ. (1993). The melatonin rhythm: both a clock and a calendar. *Experientia* 49: 654–664.

Reppert SM, Godson C, Mahle CD, Weaver DR, Slaugenhaupt SA, Gusella JF. (1995). Molecular characterization of a second melatonin receptor expressed in human retina and brain: the Mel1b melatonin receptor. *Proc Natl Acad Sci USA* 92: 8734–8738.

Reppert SM, Weaver DR, Ebisawa T. (1994). Cloning and characterization of a mammalian melatonin receptor that mediates reproductive and circadian responses. *Neuron* 13: 1177–1185.

Robinson MK, Rodrick MC, Jacobs DO, Rounds JD, Collins KH, Saporoschetz IB, Mannick JA, Wilmore DW. (1993). Glutathione depletion in rats impairs T cell and macrophage immune functions. *Arch of Surg* 126: 29–35.

Rodriguez C, Mayo JC, Sainz RM, Antolin I, Herrera F, Martin V, Reiter RJ. (2004). Regulation of antioxidant enzymes: A significant role for melatonin. *J Pineal Res* 36: 1–9.

Sabharwal P, Glaser R, Lafuse W, Varma S, Liu Q, Arkins S, Kooijman R, Kutz L, Kelley KW, Malarkey WB. (1992). Prolactin synthesized and secreted by human peripheral blood mononuclear cells: an autocrine growth factor for lymphoproliferation. *Proc Natl Acad Sci USA* 89: 7713–6.

Sakai M, Kobayashi M, Kawauchi H. (1996). In vitro activation of fish phagocytic cells by GH, prolactin and somatolactin. *J Endocrinol* 151: 113–118.

Sapolsky RM, Romero LM, Munck AU. (2000). How Do Glucocorticoids Influence Stress Responses? Integrating Permissive, Suppressive, Stimulatory and Preparative Actions *Endocr Rev* 21: 55–89.

Sapolsky RM. (1986). Endocrine and behavioral correlates of drought in wild olive baboons (*Papio anubis*). *Am J Primatol* 11: 217–227.

Sapolsky RM. (2002). Endocrinology of the stress-response. In: *Behavioral Endocrinology,* Becker JB, Breedlove SM, Crews D, McCarthy MM, (eds), *Behavioral endocrinology* (2nd ed.), MIT Press, Cambridge, MA, pp. 409–450.

Schmidt KL, Soma, KK. (2008). Cortisol and corticosterone in the songbird immune and nervous system: local vs. systemic levels during development. *Am J Physiol Regulat Integ Comp Physiol* 295: R103–R110.

Shanks N, Windle RJ, Perks PA, Harbuz MS, Jessop DS, Ingram CD, Lightman SL. (2000). Early-life exposure to endotoxin alters hypothalamic–pituitary–adrenal function and predisposition to inflammation. *Proc Natl Acad Sci USA* 97: 5645–5650.

Sharma S, Haldar C. (2006). Melatonin prevents X-ray irradiation induced oxidative damage in peripheral blood and spleen of the seasonally breeding rodent, *Funambulus pennanti* during reproductively active phase. *Int J Radiat Biol* 82: 411–419.

Sharma S. (2008). Effect of melatonin on radiation-induced immune damages in *Funambulus pennanti.* PhD thesis, submitted to Department of Zoology, BHU, India.

Shavali SS, Haldar C. (1998). Effects of continuous light, continuous darkness and pPinealectomy on pineal thyroid-gonadal axis of the female Indian palm squirrel, *F. pennanti. J Neural Transm* 105: 407–413.

Shiu SY, Li L, Siu SW, Xi SC, Fong SW, Pang SF. (2000). Biological basis and possible physiological implication of melatonin recepotor mediated signaling in rat epididymis. *Biol Signal Recept* 2: 172–187.

Shreedhar VK, Pride MW, Sun Y, Kripke ML, Strickland FM. (1998). Origin and characteristics of ultraviolet-B radiation induced suppressor T-lymphocytes. *J Immunol* 161: 1327–1335.

Skwarlo-Sońta K, Thaela MJ, Głuchowska B, Stepien D, Jagura M. (1991). Effect of dose and time of melatonin injections on the diurnal rhythm of immunity in chicken. *J Pineal Res* 10: 30–35.

Smith TF, French JA. (1997). Social and reproductive conditions modulate urinary cortisol excretion in black tufted-ear marmosets (*Callithrix kuhli*). *Am J Primatol* 42: 253–267.

Smithson G, Couse JF, Lubahn DB, Korach KS, Kincade PW. (1998). The role of estrogen receptors and androgen receptors in sex steroid regulation of B lymphopoiesis. *J Immunol* 161: 27–34.

Snyder DS, Beller DI, Unanue ER. (1982). Prostaglandins modulate macrophage Ia expression. *Nature* 299: 163–165.

Staples JE, Gasiewicz TA, Fiore NC, Lubahn DB, Korach KS, Silverstone AE. (1999). Estrogen receptor alpha is necessary in thymic development and estradiol-induced thymic alterations. *J Immunol* 163: 4168.

Sternberg EM. (2006). Neural regulation of innate immunity: a coordinated nonspecific host response to pathogens. *Nat Rev Immunol* 6: 318–328.

Suenaga R, Evans MJ, Mitamura K, Rider V, Abdou NI. (1998). Peripheral blood T-cells and monocytes and B-cell lines derived from patients with lupus express estrogen receptor transcripts similar to those of normal cells. *J Rheumatol* 25: 1305–1312.

Sze SF, Liu WK, Ng TB. (1993). Stimulation of murine splenocytes by melatonin and methoxytryptamine. *J Neural Transm* 94: 115–126.

Tan DX, Manchester LC, Reiter RJ, Qi WB, Zhang M, Weintraub ST, Cabrera J, Sainz, RM, Mayo JC. (1999). Identification of highly elevated level of melatonin in bone marrow: its origin and significance. *Biochim Biophys Acta* 1472: 206–214.

Tan DX, Reiter RJ, Manchester LC, Yan MT, El-Sawi M, Sainz RM, Mayo JC, Kohen R, Allegra M, Hardeland R. (2002). Chemical and physical properties and potential mechanisms: Melatonin as a broad-spectrum antioxidant and free radical scavenger. *Curr Top Med Chem* 2: 181–198.

Tremblay GB, Tremblay A, Copeland NG, Gilbert DJ, Jenkins NA, Labrie F, Giguere V.(1997). Cloning, chromosomal localization and functional analysis of the murine estrogen receptor. *Mol Endocrinol* 11: 353–365.

Turner CD, Bagnara JT. (1971). *General Endocrinology*. WB Saunders Company, Philadelphia, pp. 205–208.

Umegaki K, Ishikava T. (1994). Decrease in vitamin E levels in the bone marrow of mice receiving whole body X-ray irradiation. *Free Radic Biol Med* 17: 439–444.

Van Frankenhuyzen K. (1994). Effect of temperature on the pathogenesis of *Bacillus thuringienses* Berliner in larvae of the spruce budworm, *Choristoneura fumiferana* Clem. (Lepidoptera: Tortricidae). *Can Entomol* 126: 1061–1065.

Vaughan MK, Hubbard GB, Champney GM, Vaughan GM, Little JC, Reiter RJ. (1987). Splenic hypertrophy and extramedullary hematopoiesis induced in male Syrian hamsters by short photoperiod or melatonin injection and reversed by melatonin pillete or pinealectomy. *Am J Anat* 179: 131–136.

Vaughan MK, Vaughan GM, Reiter RJ. (1976). Inhibition of human chorionic gonadotrpophin-induced hypertrophy of the ovaries and uterus in immature mice by some pineal indoles, 6-hydroxymelatonin and arginine vasotocin. *J Endocrinol* 68: 397–400.

Viselli SM, Olsen NJ, Shults K, Steizer G, Kovacs WJ. (1995). Immunochemical and flow cytometric analysis of androgen receptor expression in thymocytes. *Mol Cell Endocrinol* 109:19–26.

Vijayalaxmi, Reiter RJ, Herman TS, Meltz ML. (1998). Melatonin reduces gamma radiation-induced primary DNA damage in human blood lymphocytes. *Mutat Res* 397: 203–208.

Vijayalaxmi, Reiter RJ, Meltz ML. (1995a). Melatonin protects human blood lymphocytes from radiation-induced chromosome damage. *Mutat Res* 346: 23–31.

Vijayalaxmi, Reiter RJ, Sewerynek E, Poeggeler B, Leal BZ, Meltz ML. (1995b). Marked reduction of radiation-induced micronuclei in human blood lymphocytes pretreated with melatonin. *Radiat Res* 143: 102–106.

Vink AA, Strickland FM, Bucana C, Cox PA, Roza L, Yarosh DB, Kripke ML. (1996). Localization of DNA damage and its role in altered antigen-resenting cell function in ultraviolet irradiated mice. *J Exp Med* 183: 1491–1500.

Weingrill, T, Gray DA, Barrett L, Henzi SP. (2004). Fecal cortisol levels in free-ranging female chacma baboons: relationship to dominance, reproductive state and environmental factors. *Horm Behav* 45: 259–269.

Wen JC, Dhabar FS, Prendergast BJ. (2006). Pineal dependent and independent effect of photoperiod on immune function in Siberian hamsters (*Phodopus sungorus*). *Horm Behav* 51: 31–39.

Witorsch RJ, Day EB, LaVoie HA, Hashemi N, Taylor JK. (1993). Comparison of glucocorticoid-induced effects in prolactin-dependent and autonomous rat Nb2 lymphoma cells. *Proc Soc Exp Biol Med* 203: 454–460.

Witt-Enderby PA, Bennett J, Jarzynka MJ, Firestine S, Melan MA. (2003). Melatonin receptors and their regulation: biochemical and structural mechanisms. *Life Sci* 72: 2183–2198.

Wyllie AH. (1980). Glucocorticoid-induced thymocyte apoptosis is associated with endogenous endonuclease activation. *Nature* 284: 555–556.

Yellayi S, Teuscher C, Woods JA, Welsh TH, Jr Tung KS, Nakai M, Rosenfeld CS, Lubahn DB, Cooke PS. (2000). Normal development of thymus in male and female mice requires estrogenestrogen receptor-alpha signaling pathway. *Endocrine* 12: 207–213.

Yu-Lee LY. (1990). Prolactin stimulates transcription of growth-related genes in Nb2 T lymphoma cells. *Mol Cell Endocrinol* 68: 121–28.

CHAPTER 11

ENVIRONMENTAL CHALLENGES AND THE NEUROENDOCRINE MECHANISMS OF STRESS-INDUCED MODULATION OF HOST RESISTANCE TO MICROBIAL INFECTION

Jacqueline W. Mays, Nicole D. Powell, Michael T. Bailey, and John F. Sheridan

INTRODUCTION

The immune response to microbial challenge is deeply integrated with host physiology. Neural circuitry and endocrine feedback loops compose layers of gross regulation and fine-tuning mechanisms that are built into the immune system in both vertebrates and invertebrates. The integration of the immune and neuroendocrine systems and the complexity of each provide multiple pathways for regulation, counter-regulation, and trade-offs among resources, such as energy, throughout the natural history of an organism. The nervous and immune systems share molecular mediators of communication, including cytokines and neuropeptides, use physically similar connections, such as synapses, and cells of both systems display targeted migratory behavior during development and throughout life (Kerschensteiner et al., 2009). The similar *modus operandi* between the two systems has led to the question of a common ancestral cell that could recognize and interpret the environment, communicate with other cells, and exhibit plasticity—a cell that later diverged into the nervous and immune systems (Kioussis and Pachnis, 2009). Although this idea will not be the focus of this chapter, it does point to the importance of dissecting the interplay between the nervous system and the immune system. A better understanding of the current state of integration and co-regulation of the mammalian nervous and immune systems, often by way of the endocrine system, is essential to the refinement and discussion of such evolutionary theories.

Social interactions have shaped mammalian development and survival since the dawn of time. This development includes the evolution of a complex immune and nervous system to respond to environmental threats. In turn, the environment and social situation control many aspects of neuroendocrine regulation, particularly with respect to the immune system. The stress response activates the "fight or flight" response, as well as hormone release, and can modulate immune responses through neuroendocrine stimulation in a plethora of situation-dependant ways. Stressors may be acute or chronic, and include psychological, physical, and infectious insults to organism homeostasis. Each type of stressor may differentially affect the immune system. These differences depend on a number of variables, including the nature and duration of the stressor, as well as the individual experiencing the stressor. An organism's place within the social hierarchy, baseline stress reactivity, and immune status all determine the impact of a stressor on the immune response.

Measures of the quiescent immune system give us little functional information regarding the actual potential performance of the immune system in response to microbial challenge. One of the most effective ways to study the immune response is to stimulate the immune system through mitogenic stimulation or exposure to a non-replicating antigen, or via an infection with a parasitic, viral, or bacterial challenge. The nature of the challenge plays a significant role in determining the immune response. Factors including where the pathogen replicates, what type of surface receptors it displays, and the mechanisms it has evolved to evade the host's immune response are critical in determining the shape of the normal immune response.

In this chapter, we will introduce the general mammalian stress response and the major features of the antimicrobial immune response, as well as the important interactions among the nervous, endocrine, and immune systems during microbial challenge. These discussions will center around work done in an experimental mouse model system of psychosocial stress, repeated social defeat, with particular focus on how this work changed the innate immune response to a bacterial challenge and the adaptive immune response to influenza viral challenge. Finally, we will consider how neuroendocrine regulation of the immune system may ultimately enhance an organism's chance of survival following an infectious challenge.

SOCIAL STRUCTURE AND THE STRESS RESPONSE

In group-living animals, an individual's position within a social hierarchy heavily influences access to limited resources, with dominant animals obtaining first choice of mates, food, and territory. As a result, the animal's rank within the social hierarchy has the potential to significantly increase life quality (Sapolsky, 2005). The establishment and maintenance of rank within a hierarchy can induce a strong stress response. And, once established, both dominance and subordination can be potent stressors depending upon the animal species, the social form of the hierarchy, and the stability of the hierarchy (Sapolsky, 2005). In despotic (i.e., top-down) hierarchies, dominant animals frequently fight with and attack subordinate animals to maintain their dominance (Sapolsky, 2005). This fighting can be a potent stressor for the subordinates but tends to be a more potent stressor for the dominants (Creel et al., 1996; Cavigelli, 1999; Sapolsky, 2005). In species where reasserting dominance can be accomplished by posturing, it is the subordinates, rather than the dominants, that tend to have higher stress responses (Sapolsky, 2005). These generalizations, however, can differ depending upon the species, population, and conditions of study.

Common house mice (*Mus musculus*) tend to develop social/breeding groups that are composed of the parental generation and the prepubertal offspring (Miczek et al., 2001). As the male offspring reach puberty, they leave their parental group in search of new groups. Dominant males of stable breeding groups, however, patrol and defend their territories from these intruder males (Mackintosh, 1970). Although these dominant males might let the younger subordinates live as part of the group, these subordinates do not breed and are the subject of frequent attacks (Miczek et al., 2001).

These natural social groups are rarely developed in the laboratory, and more often, groups of male mice are forced to cohabitate. In general, cohabitating laboratory mice form a despotic dominance hierarchy with one dominant mouse and several subordinates (Poole and Morgan, 1976; Brain and Poole, 1974; Miczek et al., 2001). Frequent attacks and infighting do occur, such as after changing cage bedding, but in general, laboratory mice can cohabitate stably for long periods of time, providing the hierarchy itself remains stable. Disrupting the social hierarchy is often used in the laboratory to study the stress response, and has been very useful for studying the impact of the stress response on the immune system.

THE HYPOTHALAMIC-PITUITARY-ADRENAL AXIS

Physiological pathways become activated or suppressed upon exposure to environmental, physical, or social challenges. Two of the most widely studied physiological pathways are the hypothalamo-pituitary-adrenal axis (HPA) and the sympathetic nervous system (SNS). Activation of the HPA axis occurs through the release of corticotrophin releasing hormone (CRH) from neurosecretory cells found in the paraventricular nucleus of the hypothalamus. CRH travels a short distance from the hypothalamus in the hypothalamic portal blood supply to the anterior pituitary gland, where the CRH stimulates the release of adrenocorticotrophic hormone (ACTH). The ACTH is released into the systemic circulation, where it reaches cells in the zona fasciculata in the cortex of the adrenal glands, causing these cells to in turn produce steroid hormones. Most important to actions on the immune system is the adrenocortical production of glucocorticoids. The predominant glucocorticoid produced is species-specific. Humans and other primates primarily produce cortisol (as do some smaller species, including guinea pigs and hamsters), while most rodent species produce corticosterone. However, some avian species, including zebra finches, combine use of glucocorticoids and express corticosterone as the predominant serum glucocorticoid and cortisol within immune organs, such as the bursa of Fabricius (Schmidt and Soma, 2008).

The HPA axis is regulated in part through negative feedback inhibition. Thus, after acute stimulation of the HPA axis, the increased circulatory level of glucocorticoid ultimately causes an inhibition of the release of CRH and ACTH from the hypothalamus and pituitary glands, respectively. Stressor-induced elevations of glucocorticoid are important for the generation of glucose via gluconeogenesis, which primarily occurs in the liver. Glucose can also be liberated from the liver via the breakdown of glycogen through glycogenolysis, primarily through the effects of SNS-induced elevations of epinephrine. The half-life of glucocorticoids, however, is longer than the half-life of catecholamine hormones: norepinephrine and epinephrine have a half life of ~3 minutes in plasma, while blood corticosterone has a half-life of ~11 to 30 minutes (Whitby et al., 1961; Grota, 1976; Weise and Kopin, 1976). As a result, HPA axis-induced elevations of glucocorticoids tend to lead to a more persistent adaptation of an organism to new demands. The majority of glucocorticoid found in circulation is not found in a free form, but is rather bound to a specific carrier protein called corticosteroid binding globulin (CBG). The primary function of CBG is to regulate the bioavailability and metabolic clearance of glucocorticoids, since only the unbound hormone is biologically active. Thus, environmental, physical, and social stimuli that affect CBG have an indirect influence of the effects of HPA axis activation (Fleshner et al., 1995; Spencer et al., 1996).

SYMPATHETIC NERVOUS SYSTEM

Activation of the SNS originates in the brain stem from different nuclei, including the locus coeruleus, pons, and medulla. The SNS sends projections to nearly all organs of the body through a two-neurotransmitter system. Preganglionic neurons travel within the spinal column, and upon exiting the spinal columns they synapse in prevertebral ganglia, using acetylcholine as their neurotransmitter. The preganglionic neurons stimulate postganglionic neurons that travel from the prevertebral ganglia to their target organs, where they release norepinephrine as the terminal neurotransmitter. Because the SNS innervates the adrenal medulla, SNS activation can also cause an increase in circulatory levels of catecholamines, primarily epinephrine, with smaller amounts of norepinephrine.

The SNS is one arm of the autonomic nervous system that works to balance the parasympathetic nervous system (PNS). Activation of the SNS occurs very rapidly and is largely responsible for the well-known "fight-or-flight" stress response. SNS activation readies the body for physical activity through actions on the heart and lung (e.g., increased heart rate and respiration), blood vessels (e.g., increased vasodilation in skeletal muscle), and internal organs (e.g., increased glycogenolysis in the liver and reduced activity of non-digestive functions). Upon stressor termination, the PNS becomes activated to restore homeostasis and induce the "rest-and-digest" response. The PNS also uses preganglionic and postganglionic neurons, but in this case both the pre- and postganglionic neurons use acetylcholine as the neurotransmitter (Radek, 2010).

GLUCOCORTICOID AND ADRENERGIC RECEPTORS ON LEUKOCYTES

Receptors for glucocorticoid and catecholamine hormones are widespread throughout the immune system. For example, cytoplasmic glucocorticoid receptors are found in all types of leukocytes, but the receptor subtype and receptor density differ depending upon the type of leukocyte. For example, the lower-affinity type II glucocorticoid receptor is found in high levels in monocytes, but neutrophils have lower levels of this receptor subtype. Lymphocytes of the adaptive immune system contain both the high-affinity type I glucocorticoid receptor and the lower-affinity type II glucocorticoid receptor, but the receptor density of the type II glucocorticoid receptor tends to be lower than that found in monocytes (Lupien and McEwen, 1997).

Catecholamines (both norepinephrine and epinephrine) bind to adrenergic receptors that are also found throughout the immune system in a subtype-, and cellular subset-, specific manner. In general, adrenergic receptors can be broken down into two different G-protein–coupled receptor classes, α-adrenergic and β-adrenergic receptors. These receptors are further divided into subtypes: α_1, α_2, β_1, β_2, and β_3. The β-adrenergic receptors have been detected on the cell surface of all subpopulations of lymphocytes, macrophages, and neutrophils (Fuchs et al., 1988; Landmann, 1992), but their expression quantitatively differs with tissue origin, phenotype, and activation state (Fuchs et al., 1988). The presence of α-adrenergic receptors is less certain, but α-adrenergic receptor mRNA has been found in lymphocytes and the presence of α-adrenergic agonists influences the activity of a variety of different types of immune cells. In general, however, more is known about how β-adrenergic receptors influence immune activity than is known about how α-adrenergic receptors can influence immune activity.

Both the glucocorticoid receptors and the adrenergic receptors have the capacity to influence the immune response through their effects on transcription factor expression and activity. For example, one of the best-known effects of glucocorticoids is their ability to suppress the activity of major transcription factors NF-κB and AP-1 (Scheinman et al., 1995; Ashwell et al., 2000). These transcription factors are largely responsible for regulating inflammatory cytokine gene expression. Because glucocorticoid-mediated inhibition of these transcription factors in turn inhibits the production of inflammatory cytokines, stressor-induced elevations of glucocorticoid hormones are thought to be immunosuppressive. However, it should be noted that some stressors induce a state of glucocorticoid insensitivity (Avitsur et al., 2001; Stark et al., 2001), whereas in other paradigms glucocorticoids can be immunostimulatory (Frank et al., 2010; Sorrells and Sapolsky, 2010). Interestingly, the majority of the known glucocorticoid effects involve altering the production of cytokines, and cytokines are known to signal to the brain and further affect the activity of the HPA axis.

BIDIRECTIONAL RELATIONSHIP BETWEEN THE BRAIN AND THE IMMUNE SYSTEM

There exists bidirectional communication between the brain and the immune system through which the nervous system influences inflammatory and immune responses, mainly through the activation of the SNS and HPA, and in turn, the immune system can influence the nervous system through the release of cytokines and other mediators. This immune-to-nervous-system signaling is most clearly exemplified by illness behavior (see Chapter 12). Mammals develop a characteristic set of adaptive behaviors to infection and immunity that are collectively termed illness behaviors. These behaviors include excessive sleepiness and lethargy, anorexia, hypophagia, cachexia, cognitive impairment, decreased motor activity, and changes in thermal regulation (Kluger, 1991; Kent et al., 1992; Krueger and Majde, 1994; Plata-Salaman, 1998). Most of these behaviors can be replicated by the injection of pro-inflammatory cytokines, with interleukin (IL)-1 having the most notable effects. For example, mice injected with escalating doses of IL-1β increase the amount of time spent in REM sleep, decrease food intake, and decrease exploratory behavior (Kent et al., 1994; Krueger and Majde, 1995; Bluthe et al., 2000). Furthermore, IL-1β affects cognition, memory, and fear conditioning, demonstrating the unique, pluripotent biological activity that inflammation can have on central nervous system function (Pugh et al., 1999).

REPEATED SOCIAL DEFEAT: SOCIAL DISRUPTION STRESS

The literature is replete with paradigms that have been used to induce a stress response in rodents. However, paradigms that affect the social hierarchy in group-living mice are ethologically and ecologically relevant methods to study the brain, behavior, and immunity. One such paradigm that has been developed in our laboratory is called social disruption (SDR) and involves disrupting the social hierarchy by introducing an aggressive conspecific into a cage of resident mice (Avitsur et al., 2001; Stark et al., 2001). This aggressive intruder is preselected for aggressiveness towards its cage-mates, and thus will repeatedly attack and defeat the resident mice within the cage. This occurs during a 2-hour period at the transition from the light:dark period and is repeated for 6 consecutive days. In the period between cycles of SDR, the resident mice are left undisturbed. The repeated social defeat incurred during SDR results in a stress response that includes activation of the HPA axis and the SNS (Avitsur et al., 2001). Additionally, the brain regions associated with fear and anxiety are activated, as evidenced by enhanced expression of the transcription factor c-fos, and anxiety-like behavior is increased (Mongeau et al., 2003; Kinsey et al., 2007; Liu et al., 2009).

SOCIAL DISRUPTION INDUCES ANXIETY-LIKE BEHAVIOR

A battery of behavioral tests for anxiety and depressive-like behavior has shown that exposure to SDR induces an increase in anxiety-like behavior (Kinsey et al., 2007; Bailey et al., 2009). Tests of anxiety-like behavior in mice often take advantage of the conflict between the mouse's innate fear of open spaces and its strong exploratory drive (Crawley, 1981; Ohl, 2005). In two such tests of anxiety-like behavior, the open field test and the light:dark preference test, mice tend to avoid the open areas of the testing equipment. However, due to their strong exploratory drive, over time they venture into and explore the open areas. Mice are said to display anxiety-like behavior when they venture into the center of field less often, and spend more time near the relative safety of walls, corners, or enclosed areas of the testing apparatus (Ohl, 2005). In studies of SDR, mice were found to enter the center of the open field less often than non-stressed home cage control mice (Kinsey et al., 2007; Bailey et al., 2009). Moreover, when the SDR-exposed mice did enter the center of the open field, they spent less time there compared to controls (Kinsey et al., 2007). This finding was evident after a single 2-hour session of SDR and lasted for over 14 days after exposure to six cycles of SDR (Kinsey et al., 2007).

The conclusion that mice exposed to SDR develop anxiety-like behavior was confirmed in the light:dark preference test, where SDR-exposed mice spent more time in the relative safety of the dark/enclosed portion of the testing apparatus rather than in the open/exposed portion of the apparatus (Kinsey et al., 2007). Moreover, in a test of neophobia, mice exposed to SDR took longer to investigate a novel object compared to non-stressed home cage control mice (Bailey et al., 2007), further demonstrating the anxiety-like behavior of the SDR-exposed mice.

In contrast to the development of anxiety-like behavior, there is no evidence that SDR exposure leads to the development of depressive-like behavior. Murine tests of depressive-like behavior are based on learned helplessness, where mice are placed into an adverse situation from which they must try to escape. Over time, the mice learn that they cannot escape and thus stop trying. Mice are said to have increased depressive-like behavior when they spend less time struggling to escape the adverse situation than controls. We used two tests, the forced swim test and the tail suspension test, to determine whether SDR would induce depressive-like behavior. In both tests, mice exposed to SDR tended to struggle longer than the non-stressed home cage controls (Kinsey et al., 2007). Thus, there was no evidence of depressive-like behavior, and the occurrence of more struggling was consistent with the finding that the stressor induces anxiety-like rather than depressive-like behavior.

SUBORDINATE SOCIAL STATUS LEADS TO SOCIAL DISRUPTION-INDUCED IMMUNOMODULATION

While the development of anxiety-like behavior is associated with the SDR-induced alterations of immunological parameters, it is also evident that status within the social hierarchy is associated with the immune alterations. Tests of dominance in pair-housed mice were conducted before and after mice were exposed to an aggressive intruder mouse. Interestingly, even after the aggressive intruder was removed, the dominant resident mouse remained dominant to subordinate cagemates the majority of the time (Avitsur et al., 2007). Moreover, the subordinate, rather than the dominant, mouse developed the significant SDR-induced alteration in splenic immune cell activity. Increased splenic immune cell activity did not appear to be due to an increased number of bite wounds or attacks on the subordinate animal; the number of attacks by the aggressive intruder on the subordinate and dominate resident mice were equal (Avitsur et al., 2007). These data suggest that individual differences play an important role in modulation of the immune response by repeated social defeat, even among groups of genetically identical mice.

ANTIMICROBIAL IMMUNITY

The innate and adaptive immune responses evolved to protect a host from microbial challenges. Thus, the immune system is a unique physiological system designed to protect the host against invasion of its tissues by foreign agents living in its environment. Most of the time, however, a host lives in balance with its microbial flora in a relationship that has evolved over time and does not require activation of immune defenses. These commensal relationships generally benefit both host and microbe. However, microbial infections caused by viruses, bacteria, fungi, or parasites represent a unique set of circumstances in which the relationship between host and microbe changes. Such change might occur as a result of seasonal changes (temperature or length of day), reduced nutritional status (seasonal or due to competition for resources), exposure to chronic stressors (behavioral interactions, repeated defeat, low social status), or the development of acquired immunodeficiency (aging, certain viral infections), to name but a few; circumstances such as these likely will result in increased susceptibility of the host to infection. Infection also may occur as a result of exposure to a new pathogenic microbe that has emerged through mutation, selection, or recombination. Infection by a pathogenic microbe upsets homeostasis and may threaten survival of the host.

Unlike other physiological systems, the immune system is unique in that it is diffusely distributed throughout the body of the host and mobilizes in response to a pathogenic threat. Although the bone marrow and thymus are the primary organs of the immune system, lymphocytes and other immune cells constantly circulate throughout the body in blood and lymphatic vessels surveying for foreign antigens. Secondary lymphoid tissues, such as the spleen and lymph nodes, provide special microenvironments in which foreign antigens are presented to immune helper and effector cells. Antigen processing and presentation generates a pattern of gene and cellular activation designed to terminate replication of a specific pathogen. To reduce destruction of the host's tissue through immunopathological processes, activation of immune responses occurs in a highly regulated and restricted fashion. Regulation occurs on a number of levels from within and from outside the immune system. Regulation from within is accomplished through the elaboration of soluble factors (cytokines and chemokines), and cell interactions that are restricted by the expression of cell surface receptors (major histocompatibility complex [MHC], co-stimulatory molecules, receptors for cytokines and chemokines). Regulation from outside the immune system is accomplished by factors produced by the nervous and endocrine systems (e.g., hormones, neurotransmitters, neuropeptides).

The immune response to a microbial infection is dictated by the nature of the pathogen. To elicit an immune response, a molecule must first be recognized as non-self. Cells of the immune system are programmed during development to discriminate between native (self) or foreign (non-self) molecules based on surface receptor expression, molecular pattern, and peptide profile. Peptides are the processed form of proteins and result from cleavage of the whole protein into short chains of amino acids (peptides) by antigen-processing cells. As is well recognized in the field of ecoimmunology, innate immunity is an antigen-nonspecific defense mechanism that is not restricted by elements of the MHC. Innate responses can be mobilized within minutes to hours after pathogen exposure. Generally, this innate immune response represents a genetically inherited pattern of resistance and serves as the initial response by the host to terminate microbial replication and limit tissue damage. Pattern recognition receptors (PRRs) play a critical role in the detection of microbial invaders. These types of germline-encoded receptors are widely conserved across the spectrum of living organisms, and many were first characterized in insects (e.g., the Toll-like receptors [TLRs] in *Drosophila*) and then later detected in the mammalian immune system (Cooper and Herrin, 2010). PRRs recognize molecules with significant structural differences from self molecules, also called pathogen-associated molecular patterns (PAMPs), and thus represent some of the earliest triggers through which to activate the innate immune response. Currently, four distinct classes of PRRs have been identified. These include the best-known transmembrane proteins, including TLRs and C-type lectin receptors (CLRs), and the cytoplasmic proteins, which include both the retinoic acid-inducible gene (RIG)-I-like receptors (RLRs) and NOD-like receptors (NLRs) (Takeuchi and Akira, 2010). Immune cells that typically have PRRs include macrophages, dendritic cells, endothelial cells, mucosal epithelial cells, and lymphocytes. Viruses are recognized by TLRs -3, -7, -8, or -9 and cytoplasmic RLRs that sense PAMPs on the virus themselves, or that recognize replication intermediates as the virus begins to multiply within an infected cell. Virus recognition triggers the synthesis of cytokines called type I interferons (IFNα and ß) that block viral replication within infected host cells and turn on antiviral defenses in neighboring cells. Bacteria are recognized by additional TLRs, typically TLR-2, -4, -5, or -9; however, the sensing of bacterial PAMPs by PRRs also upregulates the transcription of genes involved in inflammation. Whether triggered by bacterial or viral PAMPS, detection leads to the expression of pro-inflammatory cytokines, chemokines, type I IFNs, antimicrobial proteins, and other less well-characterized molecules. These mediators activate innate immune cells in the area and recruit other immune cells to join the response against the invading pathogen through inflammation and chemokine expression. Key pro-inflammatory cytokines in this early response include tumor necrosis factor (TNF), IL-1, and IL-6. These pleiotropic proteins (each one of these cytokine has multiple biological activities) regulate cell death in inflamed tissues, modify vascular endothelial permeability to allow egress of immune cells from the blood, and aid in immune cell recruitment to the site of infection. Type I IFNs also activate natural killer (NK) cell cytotoxicity and IFNγ cytokine production, which serve both to attack pathogen-infected cells, critical during the earliest stages of viral infection, and to recruit additional cellular help. Dendritic cells (DCs) and macrophages, which express various PRRs, are important phagocytes that constantly sample the external environment, pick up available proteins, process them into short-amino-acid sequences, and present these peptides to adaptive immune cells. Activation of the humoral (B-cell, antibody-mediated) and cell-mediated adaptive immune response often takes place in lymphatic tissues away from the site of infection. Two basic types of T cells with different functions participate in this response and are named for expression of either CD4 or CD8 protein on their cell surface. $CD4^{+}$T cells are activated by antigen displayed in the context of MHC II on the surface of antigen-presenting cells (APCs). These APCs include macrophages, DCs, and B lymphocytes and are distinguished by (1) their ability to present antigen in the context of surface MHC, and (2) their ability to deliver the co-stimulatory signal necessary for full activation of the antigen-specific lymphocyte. $CD8^{+}$T cells are similarly activated by peptides presented in the context of surface MHC I. The process and extent of $CD8^{+}$T cell clonal expansion are affected by the cytokine milieu and strength of interaction with the APC during priming. B cells also expand and undergo a maturation process that leads to clones of antibody-producing plasma cells. After virus clearance, the immune system retains memory cells for the pathogen encountered. Virus-specific memory $CD8^{+}$

and $CD4^+$T cells are maintained both in the periphery and centrally, and memory B cells ensure continued production of antiviral antibody. Upon re-exposure to the same or a similar pathogen, the adaptive memory response will be days faster, and the presence of anti-pathogen antibodies can immediately inhibit pathogen entry. Consequently, immune memory is an adaptation of the immune system that increases the odds for successfully combating re-exposure to a similar pathogen.

All viral infections are not the same, and the immune response to any single virus may differ from this general scheme in the kinetics of the cellular response, the kinetics of viral replication and detection by the immune system, the cytokine profile expressed, and the amount of pathology induced by replicating virus and the antiviral immune response. Latent viral infections, such as those caused by herpes simplex virus type-1 (HSV-1), induce an appropriate primary immune response, which terminates viral replication and limits tissue damage. However, resolution of the primary infection is not sufficient, as HSV-1 achieves a latent state that results in periodic reactivation of the latent viral genome. These recrudescent episodes are generally associated with the experience of stress or immunosuppression. Persistent viral infections, such as infection by human immunodeficiency virus (HIV) and lymphocytic choriomeningitis virus (LCMV), may exhaust the adaptive immune response and result in the generation of defective T cells and an ineffective host immune response (Kaufmann and Walker, 2009; Mueller and Ahmed, 2009). Though an imperfect system, the individual parts of the complex mammalian immune system work in concert to combat pathogens and enhance the probability of survival.

SDR-INDUCED MODULATION OF IMMUNITY

Social stressors have been widely implicated in immune modulation to viruses among humans and other mammals. A body of experimental literature supports depression of the immune system in chronically stressed individuals. Spousal caregivers of patients with dementia respond less robustly to influenza vaccination than do their age- and gender-matched counterparts (Kiecolt-Glaser et al., 1996). A plethora of other examples exist, as chronic stress has been repeatedly demonstrated to depress most antimicrobial immune responses (Hunzeker et al., 2004; Ashcraft et al., 2008; Webster Marketon and Glaser, 2008; Elftman et al., 2010). However, when social stress alone is considered, the frame shifts and a wider variety is seen in both modes and mechanisms of immune system modulation. Critical components of the social stress paradigm include the social organization and the dominance hierarchy. In rodents, humans, and most social mammals in between, individuals organize themselves into groups and, within those groups, assume different ranks. As discussed earlier, male mice are one such species that develops a strong dominance hierarchy within a stable group, with one animal dominant over those below it and submissive to those above it in the hierarchy. Disruption of this hierarchy activates the mammalian stress response systems, including the HPA axis and the SNS, resulting in the release of corticosterone and catecholamines, respectively. The key concept for this discussion is that immune cells in mice and humans contain both glucocorticoid receptors and ß2 adrenergic receptors, which allow neuroendocrine signals to directly affect the immune response to an infectious challenge.

Seminal work by Barnard et al. examined the effect of social behavior and changes in social status on the immune response to the parasite *Babesia microti* (Barnard et al., 1993). This malaria-like parasite is transmitted by blood-feeding ticks and naturally infects both mice and humans. The normal disease course involves a transient parasitemia, followed by recovery and resistance to rechallenge in normal mice. Clearance of *B. microti* is T cell dependent, as congenitally hypothymic mice (nude, *nu/nu*) do not clear the parasite and maintain a chronic infection (Clark and Allison, 1974). In a series of studies, Barnard et al. examined the effects of aggressive behavior, changes in social status, and stability of the social hierarchy on parameters of *B. microti* infection in a random-bred strain of mice. In a study that investigated the connection between stability of the social hierarchy (grouping duration) and aggressive behavior prior to parasitic infection, they showed that the most aggressive males within a group had higher serum levels of corticosterone, reached a peak parasitemia earlier than less-aggressive subordinates, and were slower to clear the *B. microti* infection. This suggests that increased susceptibility to disease may be a cost to male mice of maintaining high social status, particularly in changing social environments (Barnard et al., 1993, 1994). In mice and many other species, a simple cause-and-effect relationship between hormone secretion, behavior, and the immune system does not exist, but rather complex interactions and interrelationships are the norm (Barnard et al., 1995). In these studies, serum corticosterone, testosterone, IgG antibody, and parasite clearance were the biological parameters assessed. Whole serum IgG was measured before and after grouping unfamiliar mice, and the ratio of the two samples was analyzed. Unfortunately, no assessment of the adaptive immune status of the animal could be made from this antibody measure, as no work was done to determine how well the natural antibody profile of the mice protected against *B. microti* challenge. It would have been more telling to measure *B. microti*-specific antibody production during or after the infectious challenge to measure basic changes in the B-cell response. Nevertheless, these carefully conducted studies laid the groundwork for studying the impact of social stressors and social hierarchy on the murine immune response to a pathogenic challenge.

Studies using the SDR model in male mice have allowed further unraveling of the complexities of the integration among environment, behavior, and immunity. Exposure to the SDR paradigm results in a unique phenotype, detailed in Table 11.1, that includes alterations in the location and function of multiple immune cell populations. In this model, the behavioral interaction is structured in the same manner, and only the defeated mice are studied. Aggressors are introduced from outside the stable social group. For most of the studies, mice were subjected to six 2-hour cycles of SDR and then given either a bacterial (Bailey et al., 2007) or viral challenge (Mays et al., 2010). Both the immune response and clearance of the pathogen were monitored. The innate immune system is the first line of defense to microbial infection, and modulation of the innate antimicrobial response will be the focus of the first part of the discussion of the SDR influence on the immune system. Innate immune cells originate in the bone marrow from common precursor cells and then circulate and reside in peripheral and central organs. These cells, including macrophages and DCs, display a variety of surface and intracellular receptors, including cytokine receptors, TLRs, and hormone receptors, which allow for constant sampling of the local microenvironment. The microenvironmental cues picked up by the macrophage or DC influence the ability of these cells to detect and eliminate infection through phagocytic processes as well as antigen processing and presentation to lymphocytes. The microenvironment resulting from six consecutive days of repeated social defeat has significant immune consequences, particularly for innate immune cells. Studies conducted by Bailey describe the impact that repeated social defeat has on $CD11b^+$

TABLE 11.1: Social Distruption Stress Phenotype

Behaviorial Changes	Circulation	Bone Marrow	Spleen	Lung
Increased anxiety-like behavior	Accumulation of neutrophils and monocytes	Increased myelopoiesis	Accumulation of neutrophils and monocytes	Monocyte accumulation
No depressive-like behavior	Reduction of B cells	Reduction of B cells	Accumulation of B cells	Neutrophil accumulation and activation
Increased response to SDR in subordinate mice	Reduction of T cells	Reduction of T cells	Reduction of T cells	Upregulation of chemokines: KC/CXCL1, MIP-2/CXCL2, and MCP-1/CCL2
Physical contact is required during SDR for development of complete SDR phenotype	Higher levels of IL-1 beta and IL-6	Higher levels of granulocyte-macrophage colony-stimulating factor (GM-CSF)	Splenomegaly	Increased integrin expression: P-selectin, E-selectin, and ICAM-1

Exposure to SDR results in specific changes in behavior and physiology. Observations were made following 3–6 cycles of SDR in male mice. Strains included inbred C57 and Balb/c mice and outbred CD-1 mice.

Data were compiled from currently published literature (Avitsur et al., 2001; Stark et al., 2002; Bailey et al., 2004; Engler et al., 2005; Avitsur et al., 2007; Kinsey et al., 2007; Curry et al., 2010).

monocytes (Bailey et al., 2007). The primary effector function of monocytes is the phagocytosis of extracellular pathogens and debris. Phagocytosis is mediated by the recognition of pathogenic particles by TLRs. Mice that experienced repeated social defeat and were infected intravenously with a gram-negative bacteria cleared bacteria from circulation more rapidly than control, non-stressed mice. *In vitro* studies with CD11b$^+$ monocytes showed that they had enhanced antimicrobial activity in response to infection with *E. coli.* This stress-specific enhancement of bacterial clearance was shown to be mediated by upregulation of TLR4, and importantly, stress-enhanced killing was not evident in mice lacking functional TLR4 receptors (Bailey et al., 2007).

Cytokine secretion in response to Toll-specific stimuli was also enhanced in DC populations in mice exposed to repeated social defeat. Inflammatory cytokine protein levels were higher in the supernatants of DCs from stressed compared to non-stressed mice that were stimulated with synthetic TLR ligands (Powell et al., 2009). Taken together, social stress-induced increases in expression of TLRs leads to enhanced detection of pathogens, more efficient pathogenic clearance, and enhanced cytokine secretion.

DCs serve as important sentinels of the body, with resident DCs posted in various tissues and sites of interface with the external environment, such as skin (Langerhans cells) and the respiratory tract (resident lung DCs). In an extension of important phagocytic functions and cytokine secretion, DCs are professional APCs that serve as a bridge between the innate and adaptive immune responses. For antigen presentation to result in the activation of naïve T cells, two important signals must be present during DC/T-cell interactions. Signal one is achieved when antigen in the context of MHC is recognized in a specific manner by the T-cell receptor. Signal two consists of co-stimulation of T cells by members of the B7 family of receptors, namely B7-1 (CD80) and B7-2 (CD86) on the APC. B7-1 and B7-2 are critical accessory molecules during antigen presentation and provide the necessary second signal during the activation of naïve T cells when bound to CD28 on the T-cell surface. If a DC presents an immunogenic peptide and provides the appropriate co-stimulation to a specific T cell, clonal expansion and T-cell differentiation will result in an adaptive immune response. Not surprisingly, DCs are particularly sensitive to their surrounding microenvironment due to an abundance of receptors over their large surface area. Consequently, small shifts in the local microenvironment can have a major impact on DCs' immunogenic or regulatory function. Powell further examined the impact of SDR on the activation state of DCs in the spleen (Powell et al., 2009). Interestingly, co-stimulatory molecules, including CD80, are significantly increased on DCs after repeated social defeat, and DCs from SDR mice also show increases in MHC I and CD44. Increased expression of these molecules on the cell surface, coupled with enhanced cytokine secretion by DCs in response to Toll-specific stimuli, indicates that the DCs are primed for enhanced antigen presentation during a pathogenic insult.

SDR results in tissue-specific effects that suggest an impact beyond organs of the immune system or those organs, such as the heart, commonly associated with stress-induced modulation. Other organs of the body, including the lungs, contain rich sympathetic innervation and abundant vasculature. This allows ample exposure of both the tissue and the cells contained therein to neuroendocrine factors at a systemic level via blood flow and locally through the release of catecholamines by the SNS. Curry et al. examined the tissue-specific effect of SDR on lung inflammation (Curry et al., 2010). Following six cycles of SDR, there was increased inflammation in lung tissues compared to control mice. Using H&E staining, a marked increase in cellularity was noted in lung tissue after SDR, and this corresponded to significant increases in the pulmonary neutrophil and monocyte cell populations. This was not due to leaky vasculature in the lung tissue, but rather was associated with upregulation of factors that recruit and retain cells into tissue. These included the production of pro-inflammatory cytokines IL-1beta in the lung and IL-6 in the sera. Chemokines KC/CXCL1 and MIP-2/CXCL2 have been implicated in the recruitment of neutrophils to sites of inflammation, while MCP-1/CCL2 induces monocyte chemotaxis. Protein levels of all three chemokines were increased in lung tissue after two cycles of SDR. Adhesion molecules are expressed by tissue to stop recruited cells at the target site following tissue infiltration. Gene and cell-surface expression of these markers were measured after SDR, and data showed an increase in adhesion molecule expression in lung tissue. This would aid in cellular recruitment to lung tissue and could be responsible for enlarged monocyte and neutrophil populations found in the lung after SDR.

STRESS AND ANTIVIRAL IMMUNITY

When an experimental stressor is added to a viral infection, an additional layer of regulation and modulation is added to challenge the immune system. Specific stress-reactive pathways, including the HPA axis and the SNS, facilitate intersystem communication via release of soluble mediators, including glucocorticoids, catecholamines, and cytokines (Sanders and Kohm, 2002; Webster and Sternberg, 2004). These neuroendocrine mediators enable environmental interactions to directly affect the course of a viral infection (Hermann et al., 1995; Truckenmiller et al., 2005).

Viruses may also, in turn, activate neuroendocrine circuitry. Dunn et al. first observed in 1989 that an influenza viral infection initiated stress-like neuroendocrine changes, and that such an infection alone could cause immunosuppression (Dunn et al., 1989). This work was recently confirmed and extended to demonstrate an important role for virus-induced HPA axis activation in controlling the careful balance between immunopathology and immune defense when an organism is faced with a secondary bacterial infection (Jamieson et al., 2010).

In addition to mammals, a similar phenomenon has been observed in invertebrates. The cricket, *Gryllus textensis,* secretes the neurohormonal mediator of its stress response, octopamine, in response to challenge with heat-inactivated bacteria, which most likely activates PRRs to initiate an innate immune response (Adamo, 2010) (see Chapter 13). In rhesus macaques, an extension of virus-induced modulation of the HPA axis has been observed. Here, chronic infection with simian immunodeficiency virus (SIV) was shown to alter the structure of lymph node innervation, thereby regulating immune responses in part through direct denervation of lymphoid organs (Sloan et al., 2006, 2008). The investigators hypothesized that cells of the immune system have evolved a coordinated gene-regulation program that acted to alter local neural architecture during chronic viral infection (Sloan et al., 2008). Consequently, interplay between a psychosocial stressor and a viral infection may have unexpected effects on the ensuing antiviral immune response.

An ocular model of herpes simplex virus 1 (HSV-1) has been used to study the effect of SDR and other stressors on the antiviral immune response. HSV-1 is a neurotropic virus that first causes an acute infection and primary immune response and then establishes latency within nerve ganglia, effectively going into hiding from the immune system. During the latent phases of HSV infection, the virus can reactivate and cause herpetic lesions. In the ocular mouse model of HSV-1 infection, virus is applied to the cornea, and the initial immune response takes place within the cornea and the trigeminal ganglia during the primary infection. When mice were infected with HSV-1 McKrae virus after six cycles of SDR, they mounted an enhanced innate immune response to a primary HSV-1 infection in the cornea and trigeminal ganglia. This included increased gene expression of antiviral cytokines in the cornea and trigeminal ganglia (Dong et al., 2010). When the cells in the trigeminal ganglia were examined using flow cytometry, more CD11 b^{+} monocytes/macrophages were found in the trigeminal ganglia of SDR mice than in non-stressed controls. These changes correlated with reduced gene expression of viral proteins in the trigeminal ganglia, suggesting that exposure to repeated defeat prior to HSV-1 infection induced a more effective innate immune response to the virus.

The mouse model of influenza virus infection is a well-characterized experimental system, particularly in the context of C57BL/6 mice. The transient nature of an influenza infection, as well as its compartmentalization to lung tissue, make it an excellent system in which to ask questions about the impact of nervous and endocrine mediators on viral infection. Furthermore, activation or deletion of parts of the neuroendocrine network has been demonstrated to strongly affect the outcome of an influenza infection (Hermann et al., 1994; Hermann et al., 1995; Hunzeker et al., 2004; Grebe et al., 2010; Mays et al., 2010).

Influenza is an acute infection with tropism for the airway epithelial cells. In the earliest stages of infection, influenza and other viruses are recognized by the immune system through TLR activation within minutes of viral uncoating. In the mouse model of influenza infection, the virus is administered intranasally, and then the immune response is typically measured in spleen, lung, and lymph node tissue. In light of the tissue-specific changes seen in lung tissue following SDR, it was of particular interest to us to examine the antiviral immune response to a pulmonary infection in this system. Work in mouse models of chronic stress has demonstrated broad suppression of anti-influenza immune responses. When an influenza viral infection was administered to mice after one cycle of restraint stress, and then the stressor was continued for four more cycles, the number and cytotoxicity of NK cells were reduced, virus replication was less well controlled, and cytokine expression was altered (Hunzeker et al., 2004; Tseng et al., 2005). This type of chronic stressor, however, alters the circadian rhythm of mice, and increases and maintains high levels of corticosterone. When the regulation of anti-influenza innate immunity was examined outside the context of experimental stressors, it was found that chemical sympathectomy of mice with the norepinephrine analogue 6-hydroxydopamine (6-OHDA) prior to infection with influenza A virus reduced morbidity and mortality after infection (Grebe et al., 2010). This reduction in mortality was associated both with a smaller lung influx of monocytes, neutrophils, and NK cells early (day 5 after infection) in the anti-influenza immune response, and with lower levels of proinflammatory cytokines at the site of infection (Grebe et al., 2010). These data suggest that both glucocorticoids and the SNS work together to fine-tune the innate immune response to influenza virus and limit immunopathology and death.

The adaptive immune response is critical for full clearance of an influenza viral infection and the development of protective immune memory. Adaptive immune cells, including $CD4^+$ and $CD8^+$ T cells and B cells, are activated specifically by innate immune cells carrying "keys," or specific immune determinants with which to activate cells specific for the invading pathogen. In the case of a primary influenza virus infection, the $CD8^+$T-cell response is crucial for virus clearance, and peaks around day 7 to 9 after infection. After invading $CD8^+$T cells attack and clear the virus-infected cells, this cell population is no longer needed, and ~90% of the cells apoptose. The remaining cells, which include both $CD4^+$ and $CD8^+$ T cells, maintain protective memory for the organism in the tissue and at central immune sites such as the spleen. Long-lasting memory B cells continue to produce antiviral antibodies. In light of changes seen in lung tissue after SDR, it was of interest to investigate the effect of social stress on the development, maintenance, and function of immunological memory to the influenza virus. After six cycles of SDR, mice were infected with influenza A virus, and then allowed to recover from the viral infection. At 6 weeks after infection, the memory T-cell populations were assessed. Surprisingly, there was a significant increase in the number of virus-specific memory T cells within the lung and spleen tissue in SDR-memory (SDR-MEM) mice (Mays et al., 2010). The enhanced memory cell populations measured in SDR mice were supported by an *in vivo* measure of delayed-type hypersensitivity (DTH) that actively recalls memory cells to a site of subcutaneous antigen injection. SDR-MEM mice had significantly enlarged footpad DTH responses to influenza antigen. When memory mice were rechallenged with influenza virus, a significant increase was again noted in the cellularity of the influenza-specific $CD8^+$T-cell response in the lung, which resulted in earlier cessation of viral gene replication. SDR was not universally beneficial to immune memory. Curiously, mice with significantly attenuated circulating levels of anti-influenza IgG after exposure to SDR and a primary influenza infection were able to clear virus earlier than mice with higher IgG levels following a second influenza infection (Mays et al., 2010). Further, lungs tissue showed more immunopathological damage in SDR-MEM mice when examined histologically at 6 weeks after primary infection, suggesting a less effective recovery from virus-induced or immune-response–induced pathology in SDR-MEM mice. Redundancy is clearly built into the immune system at every level such that reduction of one parameter may not accurately predict reduced efficacy of the immune response. Exposure to SDR, prior to an infectious challenge, resulted in a unique functional immunophenotype, detailed in Table 11.2. The experience of social stress, in these experiments, prior to a primary influenza viral infection resulted in long-lasting functional enhancement of immune memory, pointing to the intimate connection between environment, behavior, and immunity.

TABLE 11.2: Functional Immunophenotype Following Social Disruption

Functional Glucocorticoid Resistance	LPS	Bacteria	HSV	Influenza
Cells are less sensitive to glucocorticoid-induced apoptosis in culture with LPS stimulation	Increased cytokine production after *ex vivo* stimulation: splenic CD11b monocytes: IL-1β, IL-6, TNFα	Translocation of host bacteria to lymph nodes and liver	Increased infiltration of CD11b+ monocytes to site of infection (trigeminal ganglia [TG])	Increased number and function of virus-specific memory CD8+ and CD4+T cells
Conventional CD11c+ DCs				Decreased anti-influenza IgGAb
CD11b+ monocytes	Decreased resistance to endotoxic shock when injected with LPS	Increased ability to kill bacteria *in vitro* and *ex vivo*	Reduced gene expression of most viral components in the TG	Earlier viral clearance during an infectious recall challenge
Expression of glucocorticoid resistance requires TLR4 ligation (second signal)	Increased cytokine production after *ex vivo* stimulation: splenic CD11c DCs: TNFα, IL-6, and IL-10	Upregulation of gene expression for nitric oxide synthase and NADPH oxidase complex subunits in splenic macrophages	Increased gene expression of TNFα and IFNα in the cornea and TG, decreased expression of IFNβ	Increased gene expression of pro-inflammatory cytokines including IFNγ, TNFα, and IL-2 in lung tissue after resolution of a primary infection

Exposure to SDR prior to pathogen exposure reveals a distinct "immunophenotype." In these experiments, mice were exposed to 6 cycles of SDR and then tissues were harvested for analysis or mice were challenged with LPS, bacteria, or virus. The subsequent immune response was carefully monitored and analyzed.

Data were compiled from currently published literature (Avitsur et al., 2001, 2002a, 2002b; Quan et al., 2001; Stark et al., 2001; Bailey et al., 2006, 2007; Powell et al., 2009; Dong-Newsom et al., 2010; Mays et al., 2010).

ENERGY AND THE IMMUNE RESPONSE

A common thread that runs through the ecoimmunology literature, particularly when considered in the context of neuroimmunology, is the evolution of the immune system in a manner that maximizes survival of the species while spending minimal energy on the immune response, as potentiated by a system of trade-offs among the nervous, endocrine, and immune systems (Lee, 2006; Ottaviani et al., 2008). It has been suggested that the function of the neuroendocrine immune system in an eco-immunological context is to fine-tune the redistribution of energy within an organism according to environmental requirements using nervous and endocrine mediators that act on the immune system (Ottaviani et al., 2008). Additionally, it has been noted that infectious processes induce the expression of glucocorticoids, which not only function as stress hormones but also serve to liberate glucose from cellular stores. It has been alternatively, but not mutually exclusively, suggested that the role of the stress response and its associated neuroendocrine regulation is to modulate the release of glucocorticoids to make best use of available glucose during an immune response (Maier, 2003; Adamo, 2010). Unfortunately, there are scant data in the literature to support or refute this hypothesis. Surprisingly little is known about the energy requirements of an immune response to any pathogen, be it viral, bacterial, or parasitic. To our knowledge, no one has defined how much glucose is required for clonal expansion of a $CD8^+$T-cell subset, nor is it known how much ATP a B cell requires for affinity maturation, class switching, and antibody production. We know that this energy comes mostly from extracellular stores, but we don't know the temporal requirements of energy attainment throughout an immune response (Dziurla et al., 2010). T-cell activation through the T-cell receptor (TCR) is coupled to complex redox and metabolic pathways that provide ATP for cellular processes. Cross-linking of the TCRs mobilizes calcium and a secondary cascade of kinases, which in turn activate transcription factors initiating cell proliferation and cytokine production (Perl et al., 2002). Proliferation and protein production require energy, and much of this is made available by cellular mitochondria via the electron transport chain and oxidative phosphorylation. There are multiple redox and metabolic checkpoints that control activation and survival of T lymphocytes (Perl et al., 2002). Each checkpoint and receptor involved provides opportunity for modulation of the energy pathways of the cell. Presumably a massive energy requirement is entailed for the cellular replication and protein production involved with an active immune response. Such a requirement would shift energy from other resources to the immune system (Cham et al., 2008). However, with scant data, arguments about the shifting of resources are difficult to make.

Research has been performed using a systems approach to ask what the energy costs are for mounting an immune response, and how the immune response is altered by nutrition status (see Chapter 8). Numerous studies have used oxygen consumption to calculate metabolic rates during an immune response to a non-replicating antigen to indirectly measure the energy used to mount an immune response. Demas et al. found that oxygen consumption increased in an inbred strain of male mice during challenge with keyhole limpet hemocyanin (KLH), suggesting that the immune challenge increased the metabolic requirement of the mice (Demas et al., 1997). In the rodent species *Ctenomys talarum* (tuco-tuco), oxygen consumption and resting metabolic rate were similarly increased when animals were challenged with sheep red blood cells (Cutrera et al., 2010). Interestingly, human $CD4^+$ T cells have been shown to be very bioenergetically adaptive and to maintain function in the face of tissue hypoxia (Tripmacher et al., 2008; Dziurla et al., 2010). Greater sensitivity has been demonstrated in the function of both $CD4^+$ and $CD8^+$ T cells to low local glucose levels (Cham et al., 2008; Tripmacher et al., 2008). In *Drosophila,* antimicrobial peptides (AMPs) are an important class of molecules that fight pathogen infections. In normal energy states, the transcription factor FOXO serves as a key regulator of stress resistance, metabolism, and aging. In states of starvation, the production of AMPs was recently demonstrated to be independently activated by FOXO. This work made clear that in invertebrates, modulation of the immune system can be induced in response to oscillating energy status of cells and tissues (Becker et al., 2010). These studies and many others have clearly illustrated that the energy status of immune cells and lymphoid tissues changes during gene and cell activation. It would be instructive to measure how exposure to a psychosocial stressor alters the energy available for an immune response. The fight-or-flight response, controlled by the sympathetic nervous system, is designed to direct the body's energy resources where they are most needed during times of flight. It is well understood that within the autonomic nervous system, the SNS is responsible for "fight/flight" and the parasympathetic nervous system switches to "rest or digest." SDR could serve as a potential model for energy redistribution with respect to immune response through well-characterized activation of both the SNS and the HPA axis within a behavioral model. More work is clearly needed in this area, not only in measuring whole organism metabolism under various conditions, but also in very carefully defining the minimum amount of energy necessary for a protective immune response.

FUTURE DIRECTIONS IN NEUROIMMUNOLOGY AND ECO-IMMUNOLOGY

The study of neuroendocrine regulation of the immune response has great potential to unlock new information about cooperation between three major systems of the body. Understanding the coordination between the immune, endocrine, and nervous systems may lead to recognition of more parallel mechanisms and similar cellular machinery between cells, such as a neuron and a T cell, that seem unrelated at first glance. Theories such as that of Kioussis and Pachnis, in which a common evolutionary precursor is proposed for nervous and immune cells, will be greatly informed by co-investigations of neuroimmunology and ecoimmunology (Kioussis and Pachnis, 2009). The stress response, part of the daily life of most vertebrates, plays an important role in driving animal behavior, physiology, and immunity. The SNS and HPA axis are two important circuits that regulate the immune system in response to stressors in mammals. Evidence exists for variations of this circuitry in evolutionarily older organisms, suggesting that conservation and evolution of neuroendocrine regulation of the immune system may have conferred a survival advantage. It is not clear if this regulation aids in antiviral immunity due to more efficient use of resources (i.e., energy redistribution), or if this is due to specific neuroendocrine mediators such as cytokines, chemokines, and neuropeptides.

It is important to bear in mind that any single metric of immune system activation (cell count, Ab level, DTH response, etc.) does not predict the immune response to every challenge, particularly in the case of replicating pathogens. In a recent study of SDR and the development of immunological memory in inbred mice, lower circulating levels of anti-influenza IgG didn't correlate with less protection (Mays et al., 2010). In this case, the enlarged population of memory T cells likely compensated for the attenuated antibody response. The nature of the infection also determines how important a given facet of the immune response is for organism survival. Certain pathogens may require action by specific immune cells or cytokines to inhibit replication and stop spread, while others may be more susceptible to a general attack by the innate immune system. Further, depending on the site of infection for a given microbe, very different outcomes may result from immune activation. When a respiratory viral infection was applied within a model of repeated social reorganization (SRO), lung hypercellularity was observed during an influenza infection in male mice (Sheridan et al., 2000). Despite the massive cellular influx, infected SRO mice were more likely to die from lung consolidation than controls, which suggests that the lung tissue didn't function properly after an infiltrating bolus of cells. This issue is less common with non-respiratory viruses and tissues, such as the spleen, that can maintain function in the face of significant cellular influx. These data illustrate that critical parameters of the immune response to enhance organism survival and restore homeostasis may vary depending on the characteristics of an individual pathogen.

The immune system has evolved to include multiple overlapping pathways that regulate the innate and adaptive antimicrobial immune responses, including neuroendocrine pathways that maintain the delicate balance between protective immunity and immunopathology. This enables the environmental experience to alter the immune response, even in genetically identical animals, and can affect survival and shape the evolution of the immune system. Further investigation of the connections between neuroimmunology and evolution of the immune system is relevant to ecoimmunology and will inform both fields.

REFERENCES

Adamo, S. A. (2010) Why should an immune response activate the stress response? Insights from the insects (the cricket *Gryllus texensis*). Brain Behav Immun 24, 194–200.

Ashcraft, K. A., Hunzeker, J., and Bonneau, R. H. (2008) Psychological stress impairs the local CD8+ T cell response to mucosal HSV-1 infection and allows for increased pathogenicity via a glucocorticoid receptor-mediated mechanism. Psychoneuroendocrinology 33, 951–963.

Ashwell, J. D., Lu, F. W. M., and Vacchio, M. S. (2000) Glucocorticoids in T cell development and function. Ann Rev Immunol 18, 309–345.

Avitsur, R., Kinsey, S. G., Bidor, K., Bailey, M. T., Padgett, D. A., and Sheridan, J. F. (2007) Subordinate social status modulates the vulnerability to the immunological effects of social stress. Psychoneuroendocrinology 32, 1097–1105.

Avitsur R, Stark JL, Dhabhar FS, Padgett DA, Sheridan JF. (2002a) Social disruption-induced glucocorticoid resistance: kinetics and site specificity. J Neuroimmunol 124(1–2), 54–61.

Avitsur R, Stark JL, Dhabhar FS, Sheridan JF. (2002b) Social stress alters splenocyte phenotype and function. J Neuroimmunol. Nov;132(1–2), 66–71.

Avitsur, R., Stark, J. L., and Sheridan, J. F. (2001) Social stress induces glucocorticoid resistance in subordinate animals. Horm Behav 39, 247–257.

Bailey, M. T., Kinsey, S. G., Padgett, D. A., Sheridan, J. F., and Leblebicioglu, B. (2009) Social stress enhances IL-1beta and TNF-alpha production by *Porphyromonas gingivalis* lipopolysaccharide-stimulated CD11b+ cells. Physiol Behav 98, 351–358.

Bailey, M. T., Engler, H., Powell, N. D., Padgett, D. A., and Sheridan, J. F. (2007) Repeated social defeat increases the bactericidal activity of splenic macrophages through a Toll-like receptor-dependent pathway. Am J Physiol Regul Integr Comp Physiol 293, R1180–R1190.

Bailey, M. T., Engler, H., Sheridan, J. F. (2006) Stress induces the translocation of cutaneous and gastrointestinal microflora to secondary lymphoid organs of C57BL/6 mice. J Neuroimmunol 171, 29–37.

Bailey, M. T., Avitsur, R., Engler, H., Padgett, D. A., Sheridan, J. F. (2004) Physical defeat reduces the sensitivity of murine splenocytes to the suppressive effects of corticosterone. Brain Behav Immun 18, 416–24.

Barnard, C., Behnke, J., and Sewell, J. (1995) Social status and resistance to disease in house mice (*Mus musculus*): Status-related modulation of hormonal responses in relation to immunity costs in different social and physical environments. Ethology 102, 63–84.

Barnard, C. J., Behnke, J. M., and Sewell, J. (1993) Social behaviour, stress and susceptibility to infection in house mice (*Mus musculus*): effects of duration of grouping and aggressive behaviour prior to infection on susceptibility to *Babesia microti*. Parasitology 107 (Pt 2), 183–192.

Barnard, C. J., Behnke, J. M., and Sewell, J. (1994) Social behaviour and susceptibility to infection in house mice (*Mus musculus*): effects of group size, aggressive behaviour and status-related hormonal responses prior to infection on resistance to *Babesia microti*. Parasitology 108 (Pt 5), 487–496.

Becker, T., Loch, G., Beyer, M., Zinke, I., Aschenbrenner, A. C., Carrera, P., Inhester, T., Schultze, J. L., and Hoch, M. (2010) FOXO-dependent regulation of innate immune homeostasis. Nature 463, 369–373.

Bluthe, R. M., Laye, S., Michaud, B., Combe, C., Dantzer, R., and Parnet, P. (2000) Role of interleukin-1 beta and tumour necrosis factor-alpha in lipopolysaccharide-induced sickness behaviour: a study with interleukin-1 type I receptor-deficient mice. Eur J Neurosci 12, 4447–4456.

Brain, P. F. and Poole, A. E. (1974) Role of endocrines in isolation-induced intermale fighting in albino laboratory mice. 1. Pituitary-adrenocortical influences. Aggressive Behavior 1, 39–69.

Cavigelli, S. A. (1999) Behavioural patterns associated with faecal cortisol levels in free-ranging female ring-tailed lemurs, *Lemur catta*. Anim Behav 57, 935–944.

Cham, C. M., Driessens, G., O'Keefe, J. P., and Gajewski, T. F. (2008) Glucose deprivation inhibits multiple key gene expression events and effector functions in CD8+ T cells. Eur J Immunol 38, 2438–2450.

Clark, I. A. and Allison, A. C. (1974) *Babesia microti* and *Plasmodium berghei yoelii* infections in nude mice. Nature 252, 328–329.

Cooper, M. D. and Herrin, B. R. (2010) How did our complex immune system evolve? Nature Rev Immunol 10, 2–U11.

Crawley, J. N. (1981) Neuropharmacologic specificity of a simple animal model for the behavioral actions of benzodiazepines. Pharmacol Biochem Behav 15, 695–699.

Creel, S., Creel, N. M., and Monfort, S. L. (1996) Social stress and dominance. Nature 379, 212.

Curry, J. M., Hanke, M. L., Piper, M. G., Bailey, M. T., Bringardner, B. D., Sheridan, J. F., and Marsh, C. B. (2010) Social disruption induces lung inflammation. Brain Behav Immunity 24, 394–402.

Cutrera, A. P., Zenuto, R. R., Luna, F., and Antenucci, C. D. (2010) Mounting a specific immune response increases energy expenditure of the subterranean rodent *Ctenomys talarum* (tuco-tuco): implications for intraspecific and interspecific variation in immunological traits. J Exp Biol 213, 715–724.

Demas, G. E., Chefer, V., Talan, M. I., and Nelson, R. J. (1997) Metabolic costs of mounting an antigen-stimulated immune response in adult and aged C57BL/6J mice. Am J Physiol 273, R1631–R1637.

Dong-Newsom P, Powell ND, Bailey MT, Padgett DA, Sheridan JF. (2010) Repeated social stress enhances the innate immune response to a primary HSV-1 infection in the cornea and trigeminal ganglia of Balb/c mice. Brain Behav Immun 24, 273–80.

Dunn, A. J., Powell, M. L., Meitin, C., and Small, P. A., Jr. (1989) Virus infection as a stressor: influenza virus elevates plasma concentrations of corticosterone, and brain concentrations of MHPG and tryptophan. Physiol Behav 45, 591–594.

Dziurla, R., Gaber, T., Fangradt, M., Hahne, M., Tripmacher, R., Kolar, P., Spies, C. M., Burmester, G. R., and Buttgereit, F. (2010) Effects of hypoxia and/or lack of glucose on cellular energy metabolism and cytokine production in stimulated human CD4(+) T lymphocytes. Immunol Lett 13, 907–105.

Elftman, M. D., Hunzeker, J. T., Mellinger, J. C., Bonneau, R. H., Norbury, C. C., and Truckenmiller, M. E. (2010) Stress-induced glucocorticoids at the earliest stages of herpes simplex virus-1 infection suppress subsequent antiviral immunity, implicating impaired dendritic cell function. J Immunol 184, 1867–1875.

Engler, H., Engler, A., Bailey, M. T., Sheridan, J. F. (2005) Tissue-specific alterations in the glucocorticoid sensitivity of immune cells following repeated social defeat in mice. J Neuroimmunol 163, 110–9.

Fleshner, M., Deak, T., Spencer, R. L., Laudenslager, M. L., Watkins, L. R., and Maier, S. F. (1995) A long-term increase in basal levels of corticosterone and a decrease in corticosteroid-binding globulin after acute stressor exposure. Endocrinology 136, 5336–5342.

Frank, M. G., Barrientos, R. M., Hein, A. M., Biedenkapp, J. C., Watkins, L. R., and Maier, S. F. (2010) IL-1RA blocks *E. coli*-induced suppression of Arc and long- term memory in aged F344 x BN F1 rats. Brain Behav Immunity 24, 254–262.

Fuchs, B. A., Albright, J. W., and Albright, J. F. (1988) Beta-adrenergic receptors on murine lymphocytes: density varies with cell maturity and lymphocyte subtype and is decreased after antigen administration. Cell Immunol 114, 231–245.

Grebe, K. M., Takeda, K., Hickman, H. D., Bailey, A. M., Embryo, A. C., Bennink, J. R., and Yewdell, J. W. (2010) Cutting edge: sympathetic nervous system increases proinflammatory cytokines and exacerbates influenza A virus pathogenesis. J Immunol 184, 540–544.

Grota, L. J. (1976) Effects of early experience on the metabolism and production of corticosterone in rats. Dev Psychobiol 9, 211–215.

Hermann, G., Beck, F. M., and Sheridan, J. F. (1995) Stress-induced glucocorticoid response modulates mononuclear cell trafficking during an experimental influenza viral infection. J Neuroimmunol 56, 179–186.

Hermann, G., Beck, F. M., Tovar, C. A., Malarkey, W. B., Allen, C., and Sheridan, J. F. (1994) Stress-induced changes attributable to the sympathetic nervous system during experimental influenza viral infection in DBA/2 inbred mouse strain. J Neuroimmunol 53, 173–180.

Hunzeker, J., Padgett, D. A., Sheridan, P. A., Dhabhar, F. S., and Sheridan, J. F. (2004) Modulation of natural killer cell activity by restraint stress during an influenza A/PR8 infection in mice. Brain Behav Immun 18, 526–535.

Jamieson, A. M., Yu, S., Annicelli, C. H., and Medzhitov, R. (2010) Influenza virus- induced glucocorticoids compromise innate host defense against a secondary bacterial infection. Cell Host Microbe 7, 103–114.

Kaufmann, D. E., and Walker, B. D. (2009) PD-1 and CTLA-4 inhibitory cosignaling pathways in HIV infection and the potential for therapeutic intervention. J Immunol 182, 5891–5897.

Kent, S., Bluthe, R. M., Kelley, K. W., and Dantzer, R. (1992) Role of interleukin-1 in the behavioral effects of lipopolysaccharide in the rat. J Physiol London 452, 124.

Kent, S., Rodriguez, F., Kelley, K. W., and Dantzer, R. (1994) Reduction in food and water intake induced by microinjection of interleukin-1-beta in the ventromedial hypothalamus of the rat. Physiol Behav 56, 1031–1036.

Kerschensteiner, M., Meinl, E., and Hohlfeld, R. (2009) Neuro-immune crosstalk in CNS diseases. Neuroscience 158, 1122–1132.

Kiecolt-Glaser, J. K., Glaser, R., Gravenstein, S., Malarkey, W. B., and Sheridan, J. (1996) Chronic stress alters the immune response to influenza virus vaccine in older adults. Proc Natl Acad Sci U S A 93, 3043–3047.

Kinsey, S. G., Bailey, M. T., Sheridan, J. F., Padgett, D. A., and Avitsur, R. (2007) Repeated social defeat causes increased anxiety-like behavior and alters splenocyte function in C57BL/6 and CD-1 mice. Brain Behav Immun 21, 458–466.

Kioussis, D., and Pachnis, V. (2009) Immune and nervous systems: more than just a superficial similarity? Immunity 31, 705–710.

Kluger, M. J. (1991) Fever: role of pyrogens and cryogens. Physiol Rev 71, 93–127.

Krueger, J. M., and Majde, J. A. (1994) Microbial products and cytokines in sleep and fever regulation. Crit Rev Immunol 14, 355–379.

Krueger, J. M., and Majde, J. A. (1995) Cytokines and sleep. Int Arch Allergy Immunol 106, 97–100.

Landmann, R. (1992) Beta-adrenergic receptors in human-leukocyte subpopulations. Eur J Clin Invest 22, 30–36.

Lee, K. A. (2006) Linking immune defenses and life history at the levels of the individual and the species. Integrative Comp Biol 46, 1000–1015.

Liu, X., Tang, X., and Sanford, L. D. (2009) Stressor controllability and Fos expression in stress regulatory regions in mice. Physiol Behav 97, 321–326.

Lupien, S. J., and Mcewen, B. S. (1997) The acute effects of corticosteroids on cognition: Integration of animal and human model studies. Brain Res Rev 24, 1–27.

Mackintosh, J. (1970) Territory formation by laboratory mice. Animal Behav 18, 177.

Maier, S. F. (2003) Bi-directional immune-brain communication: Implications for understanding stress, pain, and cognition. Brain Behav Immun 17, 69–85.

Mays, J. W., Bailey, M. T., Hunzeker, J. T., Powell, N. D., Papenfuss, T., Karlsson, E. A., Padgett, D. A., and Sheridan, J. F. (2010) Influenza virus-specific immunological memory is enhanced by repeated social defeat. J Immunol 184, 2014–2025.

Miczek, K. A., Maxson, S. C., Fish, E. W., and Faccidomo, S. (2001) Aggressive behavioral phenotypes in mice. Behav Brain Res 125, 167–181.

Mongeau, R., Miller, G. A., Chiang, E., and Anderson, D. J. (2003) Neural correlates of competing fear behaviors evoked by an innately aversive stimulus. J Neurosci 23, 3855–3868.

Mueller, S. N. and Ahmed, R. (2009) High antigen levels are the cause of T cell exhaustion during chronic viral infection. Proc Natl Acad Sci U S A 106, 8623–8628.

Ohl, F. (2005) Animal models of anxiety. In: Handbook of Psychological Pharmacology, Eds. Holsboer F., Strohle A., Springer Verlag, Berlin-Heidelberg, pp. 35–69.

Ottaviani, E., Malagoli, D., Capri, M., and Franceschi, C. (2008) Ecoimmunology: is there any room for the neuroendocrine system? BioEssays 30, 868–874.

Perl, A., Gergely, P., Jr., Puskas, F., and Banki, K. (2002) Metabolic switches of T-cell activation and apoptosis. Antioxid Redox Signal 4, 427–443.

Plata-Salaman, C. R. (1998) Cytokines and feeding. News in Physiological Sciences 13, 298–304.

Poole, T. B. and Morgan, H. D. R. (1976) Social and territorial behavior of laboratory mice (*Mus musculus* L.) in small complex areas. Animal Behaviour 24, 476.

Powell, N. D., Bailey, M. T., Mays, J. W., Stiner-Jones, L. M., Hanke, M. L., Padgett, D. A., and Sheridan, J. F. (2009) Repeated social defeat activates dendritic cells and enhances Toll-like receptor dependent cytokine secretion. Brain Behav Immun 23, 225–231.

Pugh, C. R., Nguyen, K. T., Gonyea, J. L., Fleshner, M., Watkins, L. R., Maier, S. F., and Rudy, J. W. (1999) Role of interleukin-1 beta in impairment of contextual fear conditioning caused by social isolation. Behav Brain Res 106, 109–118.

Quan, N., Avitsur, R., Stark, J. L, He, L., Shah, M., Caligiuri M, Padgett DA, Marucha PT, Sheridan JF. (2001) Social stress increases the susceptibility to endotoxic shock. J Neuroimmunol 2,115, 36–45.

Radek, K. A. (2010) Antimicrobial anxiety: the impact of stress on antimicrobial immunity. J Leukoc Biol 88, 263–277.

Sanders, V. M., and Kohm, A. P. (2002) Sympathetic nervous system interaction with the immune system. Int Rev Neurobiol 52, 17–41.

Sapolsky, R. M. (2005) The influence of social hierarchy on primate health. Science 308, 648–652.

Scheinman, R. I., Cogswell, P. C., Lofquist, A. K., and Baldwin, A. S. (1995) Role of transcriptional activation of I-kappa-B-alpha in mediation of immunosuppression by glucocorticoids. Science 270, 283–286.

Schmidt, K. L. and Soma, K. K. (2008) Cortisol and corticosterone in the songbird immune and nervous systems: local vs. systemic levels during development. Am J Physiol Regul Integr Comp Physiol 295, R103–R110.

Sheridan, J. F., Stark, J. L., Avitsur, R., and Padgett, D. A. (2000) Social disruption, immunity, and susceptibility to viral infection. Role of glucocorticoid insensitivity and NGF. Ann N Y Acad Sci 917, 894–905.

Sloan, E. K., Nguyen, C. T., Cox, B. F., Tarara, R. P., Capitanio, J. P., and Cole, S. W. (2008) SIV infection decreases sympathetic innervation of primate lymph nodes: the role of neurotrophins. Brain Behav Immun 22, 185–194.

Sloan, E. K., Tarara, R. P., Capitanio, J. P., and Cole, S. W. (2006) Enhanced replication of simian immunodeficiency virus adjacent to catecholaminergic varicosities in primate lymph nodes. J Virol 80, 4326–4335.

Sorrells, S. F., and Sapolsky, R. M. (2010) Glucocorticoids can arm macrophages for innate immune battle. Brain Behav Immun 24, 17–18.

Spencer, R. L., Miller, A. H., Moday, H., Mcewen, B. S., Blanchard, R. J., Blanchard, D. C., and Sakai, R. R. (1996) Chronic social stress produces reductions in available splenic type II corticosteroid receptor binding and plasma corticosteroid binding globulin levels. Psychoneuroendocrinology 21, 95–109.

Stark, J. L., Avitsur, R., Padgett, D. A., Campbell, K. A., Beck, F. M., and Sheridan, J. F. (2001) Social stress induces glucocorticoid resistance in macrophages. Am J Physiol Regul Integr Comp Physiol 280, R1799–R1805.

Stark, J. L, Avitsur, R., Hunzeker, J., Padgett, D. A., Sheridan, J. F. (2002) Interleukin-6 and the development of social disruption-induced glucocorticoid resistance. J Neuroimmunol 124, 9–15.

Takeuchi, O., and Akira, S. (2010) Pattern recognition receptors and inflammation. Cell 140, 805–820.

Tripmacher, R., Gaber, T., Dziurla, R., Haupl, T., Erekul, K., Grutzkau, A., Tschirschmann, M., Scheffold, A., Radbruch, A., Burmester, G. R., and Buttgereit, F. (2008) Human CD4(+) T cells maintain specific functions even under conditions of extremely restricted ATP production. Eur J Immunol 38, 1631–1642.

Truckenmiller, M. E., Princiotta, M. F., Norbury, C. C., and Bonneau, R. H. (2005) Corticosterone impairs MHC class I antigen presentation by dendritic cells via reduction of peptide generation. J Neuroimmunol 160, 48–60.

Tseng, R. J., Padgett, D. A., Dhabhar, F. S., Engler, H., and Sheridan, J. F. (2005) Stress-induced modulation of NK activity during influenza viral infection: role of glucocorticoids and opioids. Brain Behav Immun 19, 153–164.

Webster Marketon, J. I., and Glaser, R. (2008) Stress hormones and immune function. Cell Immunol 252, 16–26.

Webster, J. I., and Sternberg, E. M. (2004) Role of the hypothalamic-pituitary-adrenal axis, glucocorticoids and glucocorticoid receptors in toxic sequelae of exposure to bacterial and viral products. J Endocrinol 181, 207–221.

Weise, V. K., and Kopin, I. J. (1976) Assay of cathecholamines in human plasma: studies of a single isotope radioenzymatic procedure. Life Sci 19, 1673–1685.

Whitby, L. G., Axelrod, J., and Weil-Malherbe, H. (1961) The fate of H3-norepinephrine in animals. J Pharmacol Exp Ther 132, 193–201.

CHAPTER 12

INFLAMMATION AND BEHAVIOR

Keith W. Kelley, Arnaud Aubert, and Robert Dantzer

INTRODUCTION

Setting a house on fire causes flames, a word derived from Latin (*flammula*). When the human body is figuratively on fire, it is inflamed (derived from the Latin word *inflammmatio*, to set on fire). Four Latin words originally defined inflammation: *tumor* (swelling), *calor* (heat), *rubor* (redness), and *dolor* (pain). Loss of organ function was subsequently included in this definition (*functio laesa*). Inflammation can be either acute, lasting only a few days, or chronic, lasting for months to years (Fig. 12.1). Originally, inflammation was considered to be only a local event caused by something like a splinter, burn, wound, or infection. It is now recognized that inflammation can occur throughout the body, which is known as systemic inflammation. When inflammation occurs in the CNS, the umbrella term is "neuroinflammation." Severe symptoms of neuroinflammation occur in multiple sclerosis and stroke. However, milder forms of neuroinflammation that do not involve the massive recruitment of blood-borne cells into the CNS cause well-known physiological symptoms because of the activation of cytokine signaling in the brain. Bodily signs that accompany systemic and neuroinflammation are well known to all of us: sleepiness, hypophagia, listlessness, fever, reduced interactions with others, aching joints, and general lack of motivation.

These symptoms of systemic inflammation are collectively known as sickness behavior, and they are characteristic of neuroinflammation (Dantzer and Kelley, 1989, 2007). Inflammation that continues unabated can even lead to development of mental health disorders, including clinical depression and anxiety (Dantzer et al., 2008; Miller et al., 2008). Systemic inflammation has now been strongly linked to a wide variety of disorders, such as classic autoimmune (e.g., rheumatoid arthritis, multiple sclerosis) and chronic infectious (e.g., tuberculosis) diseases, as well as major noninfectious pathologies that include cardiovascular diseases, type 2 diabetes, obesity, and cancer. Inflammation is no longer simply a tool of the pathologist; it has permeated nearly every aspect of modern medical research. A new forefront for understanding the role of the immune system in animal and human health and well-being is to better understand the many facets of inflammatory responses.

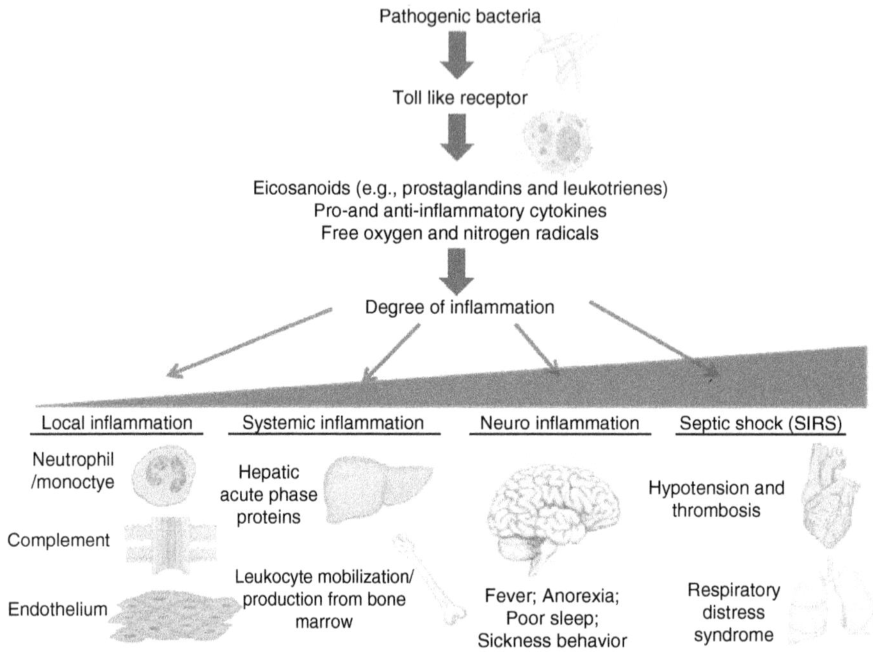

FIGURE 12.1: An overview of inflammation. When a pathogen overcomes the physical barriers of the body, such as the skin or gastrointestinal tract, the pathogen invades the body and sets in motion a series of events that lead to the induction of inflammatory events near the site of invasion. In the case of pathogenic bacteria, pathogen-associated molecular pattern molecules that are displayed by the pathogen bind to distinct types of Toll-like receptors that are expressed by local macrophages and dendritic cells in draining lymph nodes. These cells synthesize prostaglandins and leukotrienes, as well as pro-inflammatory (e.g., IL-1, TNF, IL-6) and anti-inflammatory (e.g., IL-10, IL-1 receptor antagonist) cytokines that act both locally and systemically. Complement may be activated (C3a, C5a), which promotes bacterial killing. Chemotactic factors are secreted that lead to activation of endothelial cells, causing the movement of other myeloid cells such as monocytes and neutrophils, as well as T and B lymphocytes, into the site of infection/injury. These and other events cause the cardinal signs of local inflammation: heat (*calor*), redness (*rubor*), swelling (*tumor/tugor*), pain (*dolor*), and ultimately loss of function (*functio laesa*). During systemic inflammation, the liver synthesizes acute phase proteins, such as C-reactive protein and serum amyloid A, leukocytes are mobilized from bone marrow, and hematopoiesis may be accelerated. Inflammation in the periphery is mirrored by inflammation in the brain. In this case, neuroinflammation is less severe than in classic pathological conditions such as stroke and multiple sclerosis that result in the recruitment of blood-borne cells into the CNS. However, cytokine signaling in the brain is enhanced, resulting in fever and headaches. Another result is poorer sleep quality, as assessed by more slow-wave sleep, greater sleep latency, less rapid-eye-movement sleep, and less stage 1 sleep. A variety of sickness behaviors are manifested, as determined by reduced motivation to accomplish normal daily activities. If local and systemic inflammation and neuro-inflammation are not controlled, symptoms of the systemic inflammatory response syndrome (SIRS) appear. Reduced cardiac perfusion, combined with nitric oxide-induced peripheral vasodilatation, leads to hypotension and blood coagulation. SIRS then progresses to septic shock and the respiratory distress syndrome develops, characterized by an increase in the rate of respiration, pulmonary edema, and hypoxia. Multiple organ failure and death soon follow.

FUNDAMENTAL CONCEPTS OF INFLAMMATION

LOCAL INFLAMMATION

Insults to the skin result in a localized inflammatory response. Resident macrophages, mast cells, and damaged fibroblasts release chemical mediators such as nitric oxide, vasoactive amines (e.g., histamine and serotonin), and pro-inflammatory cytokines (e.g., interleukin-1 [IL-1] and tumor necrosis factor [TNF]). These substances act in concert to activate the endothelium of postcapillary venules. This induces the expression of adhesion molecules on the endothelium and increases capillary size and permeability, allowing fluid to accumulate at the inflamed site. Histamine increases expression of various E (endothelial)-selectin adhesion proteins on the vascular endothelium and L (leukocyte)-selectin adhesion molecules that are expressed on nearly all populations of polymorphonuclear and mononuclear myeloid and lymphoid cells. These selectin proteins slow the flow of leukocytes through the blood vessel, which allows the cells to begin rolling. Firmer attachment to the vascular endothelium is mediated by binding of members of the β-integrin family expressed on leukocytes, such as leukocyte function-associated antigen-1 (LFA-1), to intercellular adhesion molecules (ICAM) on the inflamed endothelium. These events mobilize leukocytes near the site of injury. Chemokines such as IL-8 create a gradient to cause leukocytes to transmigrate across the vascular endothelium and basement membrane of the capillary towards the site of injury. Cadherins are adhesion proteins that mediate tight junctions between endothelial cells, and these endothelial cell junctions are disrupted as leukocytes extravasate across the basement membrane. A number of other local inflammatory events occur at the traumatized site, such as activation of the coagulation, fibrinolysis, and complement systems. These are important physiological events that are the major characteristics of local inflammation, but they are too peripheral to discuss extensively in this short primer. Once inflammatory cells reach the site of injury, they not only attempt to remove the inflammatory stimulus (e.g., splinter, bacteria) but also initiate the repair process that leads to healing. Removal of the initiating stimulus and repair of the tissue signals the end of a successful acute, local inflammatory response.

SYSTEMIC INFLAMMATION

The local inflammatory response, particularly when amplified in regional lymph nodes near the site of entry of pathogens into the body, often spill over into the blood. Cytokines are released into the blood, leading to the synthesis of acute phase proteins by the liver. The concentrations of some of these hepatic acute phase proteins, such as C-reactive protein (CRP), complement, α2-macroglobulin, serum amyloid A, ceruloplasmin, and ferritin, increase in the blood and are known as positive acute phase proteins. The functions of these proteins are diverse, including opsonization (CRP), inhibition of pro-inflammatory cytokine synthesis (CRP), binding and making iron unavailable to microbes (ferritin and ceruloplasmin), and inhibiting coagulation (CRP, α2-macroglobulin). Negative acute phase proteins are those whose concentration declines in blood, and they include such molecules as transcortin, albumin, and transferrin. Pro-inflammatory cytokines can activate afferent branches of the vagus nerve, such as those found in the liver, which sends signals to the nucleus of the solitary tract in the brain. Pro-inflammatory cytokines can also act in a variety of ways to stimulate the hypothalamic-pituitary-adrenal (HPA) axis. Ultimately, inflammation that is initiated systemically is represented as a mirror image of the same cytokines in the brain, and these pro-inflammatory cytokines induce all the symptoms that characterize the sickness response.

Like the local inflammatory response, systemic inflammation can be acute or chronic. A severe form of acute inflammation is known as the systemic inflammatory response syndrome (SIRS). If SIRS is caused by bacteria in the blood (septicemia), then this response is known as sepsis, which is life-threatening because it can lead to multiple organ failure. Other acute events, such as a coronary infarction, blunt force injury, severe burns, or extreme loss of blood, can induce SIRS. Stimuli that stimulate a chronic or remitting/relapsing systemic inflammatory response are autoimmune diseases such as rheumatoid arthritis and multiple sclerosis. Other disorders, such as asthma, inflammatory bowel disease, myocarditis, tuberculosis, and vasculitis, can chronically activate the systemic inflammatory response. Indeed, the distinction between local and systemic inflammation is now blurred. For example, inflammatory bowel disease and rheumatoid arthritis evolve from a local to a systemic event. This is similar to other disorders that begin in a specific organ and then can evolve into systemic inflammation, such as can occur in type 2 diabetes, obesity, atherosclerosis, and chronic heart disease. Both chemotherapy and radiotherapy can lead to systemic inflammation. New data indicate that systemic inflammation can lead to another form of inflammation known as neuroinflammation that affects both human and animal behaviors (Dantzer et al., 2008). And, of course, disorders such as cerebral stroke, Alzheimer disease, and Huntington disease can lead to neuroinflammation.

NEUROINFLAMMATION

It is now recognized that the CNS is not as immune privileged as was once thought. Although it is true that the blood–brain barrier constrains the unrestricted flow of large proteins into the brain, it is now recognized that cells of innate system not only exist but play a major role in both physiological and pathological responses that involve both the brain and the body. In a series of papers published in the *Journal of Immunology* and *Science* between 1981 and 1987, Hugo Besedovsky and colleagues discovered that the pro-inflammatory cytokine IL-1 acts in the hypothalamus to cause the release of corticotrophin releasing factor (Berkenbosch et al., 1987; reviewed in Besedovsky and Rey, 2007). Shortly thereafter, new molecular biology techniques permitted the identification of both mRNA and protein for IL-1 in the brain. Numerous studies since that time demonstrated that peripheral cytokines induce the synthesis of cytokines in the brain. For example, an intraperitoneal injection of LPS induces expression of IL-1, TNF, IL-6, and the IL-1 receptor antagonist in the brain. In this way, brain cytokines form a kind of mirror image of the major cytokines in the periphery. The main cellular sources of these cytokines were shown to be microglial cells and perivascular and meningeal macrophages.

Microglia are heterogeneous and serve many functions as macrophages that reside in the periphery. Similar to macrophages, both microglia and perivascular macrophages are derived from a common myeloid progenitor. These cells are crucial to inflammation in the brain, and the beneficial and detrimental roles of microglia continue to be the subject of hot debate (Soulet and Rivest, 2008). Animal behaviors that are induced by an "inflamed brain" can be critical for the ultimate survival or demise of animals and humans. For instance, pro-inflammatory cytokines can cause fever, and huddling and shivering behaviors help to reduce heat loss and promote heat production, respectively. Similarly, a classic symptom of neuroinflammation is a reduction in food consumption, which can prevent animals from searching for food and becoming a target for predators. The role of systemic inflammation and neuroinflammation in regulating a plethora of animal behaviors is more thoroughly discussed later in this chapter.

CYCLOOXYGENASE ENZYMES AND INFLAMMATION

Nonsteroidal anti-inflammatory drugs (NSAIDs) are very widely used to reduce the fever and pain (e.g., aching joints) that are caused by inflammation (Strauss, 2008). Most of the NSAIDs non-selectively inhibit the activity of cyclooxygenases (COX) 1, 2, and 3 because these are the rate-limiting enzymes for synthesis of eicosanoids. The polyunsaturated fatty acid arachidonic acid is liberated from glycerol in the cell membrane by phospholipase C or A_2, where it serves as a precursor for synthesis of the eicosanoids: prostaglandins, thromboxanes, prostacyclins, and leukotrienes. Prostaglandin E_2 (PGE_2) causes the vasodilation and fever that accompany acute inflammation, and the leukotrienes affect several physiological events that are too numerous to discuss in this short primer. COX-1 is constitutively expressed by most cells as a housekeeping gene, whereas COX-2 is induced during inflammation. Pro-inflammatory cytokines, such as IL-1α and IL-1β, TNFα, TNFβ, and interferon (IFN)-γ, are potent inducers of COX-2 in macrophages and microglia.

FREE RADICALS, CELLULAR REDOX, AND INFLAMMATION

Macrophages and neutrophils generate significant amounts of both free oxygen and nitrogen radicals during inflammatory insults. These free radicals are necessary to help kill microbial pathogens. Indeed, free oxygen radicals are necessary for efficient killing of many bacterial pathogens. For example, patients with chronic granulomatous disease (GCD) display a genetic defect in expression of the reduced form of nicotinamide adenine dinucleotide phosphate (NADPH). The disease is named because of the chronic expression of granulomas that do not resolve following infection. Most GCD patients present with numerous fungal and bacterial infections within the first 5 years of life and exhibit an earlier-than-expected mortality at less than 30 years of age. The NADPH oxidase is required for generation of the free radical superoxide anion, which by definition contains an unpaired electron in its outer shell. Transition metal ions, particularly iron, greatly increase the production of superoxide anion from hydrogen peroxide in a process known as the Fenton reaction. Bactericidal oxidants are subsequently generated from superoxide anion, such as hydrogen peroxide (which is not a free radical), hydroxyl radicals, and peroxynitrite anions. The latter is formed by a reaction between superoxide anion and nitric oxide.

The same pro-inflammatory cytokines that increase COX-2 activity also augment production of free oxygen and nitrogen radicals. However, cytokines such as IFNγ generally act in synergy with bacterial products such as lipopolysaccharide (LPS) to augment superoxide production through processes known as priming and triggering, respectively. Although production of free radicals by phagocytes is necessary for protection against microbial pathogens, they are damaging to mammalian cells when produced chronically and in excess because they readily oxidize proteins (particularly phosphatases), lipids, and DNA. ROS can activate NFκβ, as well as JNK and p38.

The reduction/oxidation state (redox) is defined by the concentration of electrons throughout the entire cell and is regulated by the gain (reduction) or loss (oxidation) of electrons. The redox state of a cell is maintained at near-constant levels by a variety of buffer systems. The major enzymatic intracellular constituents that protect against free radical damage are superoxide dismutase, catalase, and glutathione peroxidase. There are a number of endogenous antioxidants such as the thiols glutathione, thioredoxins, and lipoic acid. Selenium, carotenoids, vitamins C and E, and polyphenol compounds such as flavonoids are all compounds that can act as antioxidants and can be obtained in the diet.

CELL SIGNALING EVENTS DURING INFLAMMATION

Enormous advances have been made during the past decade in understanding how all types of cells respond to inflammatory stimuli. Two major discoveries have unequivocally advanced this field (see Doyle and O'Neill, 2006, for an excellent review): (a) the 20-year-old discovery of nuclear factor kappa B (NFκβ), which is the *sine qua non* of inflammatory signaling, and (b) pathogen-associated molecular patterns (PAMPs) that bind to Toll-like receptors (TLR) that are expressed on a variety of cells. PAMPs are expressed on a variety of microbial pathogens, such as gram-negative bacteria, viral RNAs, fungi, protozoa, and nucleotide motifs that are expressed on viruses and bacteria. Although not as well understood, it is also important to note that endogenous substances released by damaged cells can activate TLRs. These endogenous substances, such as heat shock proteins, pro-inflammatory cytokines, ATP, DNA, and reactive oxygen and nitrogen intermediates, have been referred to collectively as "danger signals" and are hypothesized to serve as endogenous messengers that inform the body of the presence of a pathogen (Matzinger, 1994). Although Charlie Janeway argued that the true danger signals are endogenous co-stimulator molecules (Janeway et al., 1996), the idea that the body has the tools to detect "danger" is solid.

PAMPs bind to one of the 10 TLRs that have been identified in humans and are expressed on macrophages, T, B, mast, dendritic, endothelial, and epithelial cells as well as cells in the parenchyma of most organs (13 TLRs have been identified in mice). Following binding, all TLRs cause activation of NFκβ, and all but TLR3 utilize a critical adaptor signaling protein known as MyD88 to activate it (Carpentier et al., 2008). Regulatory events of the five major Rel/NFκβ proteins—RelA (also known as p65), RelB, c-Rel, NF-B1 (p50/p105), and NF-β2 (p52/p100)—that control the canonical NFκβ pathway are fairly well understood (Wietek and O'Neill, 2007). Latent forms of NFκβ proteins are found in the cytoplasm, where they are bound to inhibitory proteins such as IκBα. Stimuli such as oxygen-free radicals, sphingomyelinases, pro-inflammatory cytokines, and ligands of the TLRs rapidly lead to serine phosphorylation of IκBα by IκB kinases (IKK), leading to IκBα degradation by ubiquination. The classical NFκβ heterodimer, p50/p65, is then able to translocate to the nucleus, undergo further phosphorylation events, and interact with other nuclear signaling proteins to induce transcription of a very wide variety of genes, including most of the major proteins involved in the inflammatory process (e.g., cyclooxygenase [COX]-2, inducible nitric oxide [iNOS], pro-inflammatory cytokines such as IL-1, TNF, IL-6, adhesion molecules, and chemokines).

NFκβ and the major members of the mitogen-activated protein kinase (MAPK) family of transcription factors (extracellular signal-regulated kinases [ERK] 1 and 2, c-Jun N-terminal kinase [JNK] and p38) generally synergize with one another to enhance transcription during an inflammatory event (Grivennikov et al., 2006). However, NFκβ and MAPKs can also antagonize each other. For example, activation of the type 1 TNF receptor activates Iκβ kinase (IKK), which leads to translocation of NFκβ to the nucleus, where it can stimulate the transcription of anti-apoptotic genes and promote cell survival. At the same time, the stress kinases JNK and p38 phosphorylate proteins that can lead to synthesis of death-inducing apoptotic proteins. The stress kinases also promote the synthesis of pro-inflammatory cytokines, ICAMs, and chemokines. Many of the major effects of JNK and p38 are mediated through phosphorylation of the transcription factor, activator protein (AP)-1 (De Bosscher et al., 2003). First recognized as a downstream product of protooncogenes, AP-1 is formed by hetero- or homodimers of four major families of immediate early gene transcription factors: fos, jun, activating transcription factor (ATF), and musculoaponeurotic fibrosarcoma (Maf).

Crosstalk that occurs in downstream intracellular pathways following activation of the same (e.g., IL-1) or different (e.g., cytokines and hormones) receptors affects important functional activities of cells that express both types of receptors. For example, activation of AP-1 and NFκβ occurs in all inflammatory cells in nearly all autoimmune diseases, and suppressing these two transcription factors (and the upstream JNK and p38 kinases) with synthetic glucocorticoid hormone agonists appears to be key to understanding both the anti-inflammatory properties of steroid treatment as well as the clinically important side effects of these steroids (De Bosscher et al., 2008). Another important concept in intracellular signaling concepts is that hormone and cytokine signaling is often reciprocal. Stated in another way, glucocorticoid hormones reduce synthesis of pro-inflammatory cytokines, but these same pro-inflammatory cytokines impair the biological actions of glucocorticoids. This concept of cytokine-induced hormone resistance has been documented not only for glucocorticoids (Pace et al., 2007) but also for major metabolic hormones that have major effects on growth and development of humans and animals, including insulin-like growth factor-I (IGF-I), growth hormone, and insulin (Kelley et al., 2007).

INFLAMMATION AND BEHAVIOR OF MAMMALS

BRAIN–IMMUNE COMMUNICATION

Physiology is derived from the Greek prefix *physis*, which means "nature or origin" and the suffix *logia*, or "the study of." All accepted definitions of physiology include the word "function."

Academic courses in systemic physiology focus on classic organ systems, such as the brain, heart, lungs, kidney, and endocrine glands. A major goal of these courses is to "connect the dots" that lead to interactive regulation of multiple organ systems. For example, events that occur in the heart affect the lungs, the kidneys affect the cardiovascular system by regulating plasma volume, and the neuroendocrine system regulates reproductive organs by hormone secretion. The central theme of systemic physiology is to convey a sound understanding of the numerous routes of communication among different organ systems. The idea that immunology is really just another physiologic system that interacts with numerous physiological systems, including the brain, was emphasized years ago (Kelley, 1988, 2004; Kelley et al., 1994)

Although it is well accepted that the brain responds to afferent information from the periphery to relay efferent signals to multiple organs, the possibility of reciprocal systems of communication existing between the brain and immune system was not seriously considered until late in the 19th century. Although fundamental physiological systems had been deciphered, the basic underpinnings of the immune system were unknown. Ilya Mechnikov and Paul Ehrlich shared the 1909 Nobel Prize for their work on immunity to bacterial pathogens. In 1960, Sir Frank Macfarlane Burnet and Peter Medawar were awarded the Nobel Prize in Medicine for their discovery of acquired immunological tolerance, and it was not until 1972 that Gerald Edelman and Rodney Porter received the Nobel Prize for determining the structure of antibodies. In 1978 the first cytokine was purified, interferon-α. Cloning of the interferon-α gene and expression of the recombinant protein soon followed in the early 1980s. The first two pro-inflammatory cytokines, IL-1 and IL-2, were also purified, cloned, and expressed in the 1980s, and it was often considered at that time that these were the only known proteins that were produced by leukocytes. The finding that lymphocytes and myeloid cells could synthesize and secrete soluble proteins opened new vistas of opportunity for biological scientists, in part because endocrinologists had provided the conceptual framework that information sharing among several organ systems could occur not only via nerves, but also by humoral molecules via the blood. Although sensitive assays were not yet available to detect cytokines in blood, attempts to use systemic injections of recombinant cytokines revealed gross physiological disturbances, many of which involved the CNS. These early clinical studies offered a clue about potential functions of endogenous cytokines that are produced during disease (Dantzer and Kelley, 1989). Emerging data clearly suggested that cytokines from the immune system communicated with the brain to change behaviors, and that hormones produced by the neuroendocrine system regulated the activity of leukocytes. New data on the physiology of immunology began to appear in leading journals. A syncretic form of immunology, psychology, and neuroscience was formed, and in 1987 it found a home in a new journal devoted specifically to this topic, *Brain, Behavior, and Immunity*.

It was not until the late 20th century that conventional physiological wisdom began to "connect the dots" regarding the possibility that cells of the immune system might somehow share reciprocal systems of communication with the brain. In 1984, a pioneering scientist by the name of Ed Blalock described the immune system as our sixth sense (Blalock, 1984). In other words, the classical sensory systems of touch, sight, smell, taste, and sound are unable to recognize that a foreign pathogen has broken though the first-line barriers and gained entrance to the body proper. The innate immune system, acting through the production of cytokines and other pro-inflammatory mediators, acts to inform the brain that an infection has occurred in the periphery. Although a heretical proposal when it first appeared, the tremendous amount of research progress that has been made during the past 20 years on systemic inflammation supports the original concept that the immune system can be viewed as our "sixth sense."

COMMUNICATION BETWEEN THE IMMUNE SYSTEM AND BRAIN

A century ago, it was discovered that peripheral injection of the dye aniline purple stained all organs of the body except the brain. In contrast, injection of the same dye into the cerebrospinal fluid caused staining in the brain but not other body tissues. These experiments clearly indicated that the brain was "compartmentalized," leading to the concept of the blood–brain barrier. This barrier was considered to impair the entry of immune cells and proteins into the brain. Although tight junctions in brain endothelial cells and astrocytic feet are now clearly accepted to form the major structural components of the blood–brain barrier, the traditional view of the brain as an immune-privileged site has undergone substantial revision during the past 20 years.

The early, pioneering experiments by Dwight Nance and colleagues offered the first evidence for an afferent neuronal communication pathway from the immune system to the brain. They demonstrated that an intraperitoneal injection of LPS induced a marker of early neuronal activation, *c-fos*, in the nucleus tractus solitarius of the brain stem that corresponds to the primary projection area of the vagus nerves, and in the paraventricular nucleus of the hypothalamus and brain limbic areas that correspond to secondary projections of the same nerves (Wan et al., 1993). They subsequently demonstrated that indomethacin, an NMDA antagonist, and a subdiaphragmatic vagotomy all acted to inhibit induction of *c-fos* in these brain regions following injection of LPS (Wan et al., 1994). These pioneering experiments offered prostaglandin synthesis, glutamate release as the major neurotransmitter in the nucleus tractus solitaries, and visceral afferents to this site as potential links to specific regions of the brain following activation of the innate immune system. These data were confirmed by a number of laboratories (Nance and Sanders, 2007) and extended to inflammation-induced behavioral changes in the form of sickness behavior (Watkins et al., 1994; Bluthe et al., 1996; Nance and Sanders, 2007). Other neural routes of afferent communication between the immune system and brain are being discovered. For example, the meninges are innervated by trigeminal afferent sensory neurons, and application to the dura of a mixture of inflammatory mediators (histamine, serotonin, bradykinin, and PGE_2) increased *c-fos* expression in hypothalamic neurons that are known to reduce feeding behavior (Malick et al., 2001).

A neuronal rather than a humoral route of communication from the immune system to the brain seemed to be most likely because cytokines are large, ~16 to 18 kDa lipophobic proteins that are excluded by the blood–brain barrier. However, several reports established that subdiaphragmatic vagotomy did not abolish *c-fos* induction or sickness behaviors when LPS was administered either subcutaneously or intravenously (reviewed by Dantzer and Kelley, 2007; Nance and Sanders, 2007). Two lines of evidence favor the existence of a blood-borne route as well. First, early after the discovery that cytokines could be found in the blood following an insult to the body, William Banks and colleagues discovered that several of these large, 16 to 18 kDa lipophobic cytokines can be actively transported from the blood into brain (Quan and Banks, 2007). The precise molecular nature of these transporters remains unspecified. Some areas of the brain lack tight endothelial junctions, and these specific structures are known as circumventricular organs

(Rivest, 2003). They include the median eminence, subfornical organ, area postrema, and the organum vasculosum of the lamina terminalis. Circumventricular organs may be one system by which blood-borne substances communicate with the brain. However, weak expression of E-series prostaglandin receptor type 3 (EP3), which is required for induction of LPS-induced fever, in the organum vasculosum of the lamina terminalis indicates that other communication systems besides circumventricular organs are likely to be employed.

Existing data offer a compelling argument for both neuronal and humoral afferent forms of communication from the immune system to the brain. Indeed, the fast-acting neuronal pathway may sensitize the slower-acting humoral pathway to the behavioral effects of inflammation on the brain (Vitkovic et al., 2000). These pathways were originally identified by the increased activation of *c-fos* in specific regions of the brain. Subsequently, the technique of reverse-transcriptase polymerase chain reaction (RT-PCR) replaced Northern blot and solution hybridization ribonuclease protection assays for more sensitive detection of steady-state transcripts. Just as the development of ELISA assays increased sensitivity of detection from the nanogram to the picogram blood levels of cytokine proteins, RT-PCR permitted the identification of mRNA in the brains of animals infected with pathogens. A plethora of experimental data has now clearly established that systemic inflammation leads to neuroinflammation, at least as evidenced by the expression of mRNA and protein for cytokines in the brain. These cytokines are expressed throughout most regions of the brain, and the current view is that the behavioral effects of these cytokines are determined by modulation of the relevant neuronal circuits that regulate these behaviors.

SICKNESS BEHAVIOR

Animals and humans have been getting sick since the beginning of time. Until 20 years ago, symptoms of sickness were considered to result from the debilitating consequences of the disease. Body aches and pains, nausea, headaches, fever (and associated shivering and chills), loss of appetite, unrelenting fatigue, sleepiness, and a lack of motivation to accomplish normal daily tasks are typical. If the pathogen is not life-threatening, then these behavioral sequelae have been considered to be banal consequences of the disease process. Acetylsalicylic acid (aspirin) and similar over-the-counter anti-inflammatory compounds have been taken as palliative remedies by adult humans for more than 100 years to relieve some of these symptoms, particularly fever and dull throbbing pain. NSAIDs such as indomethacin are effective in animals as well (Johnson et al., 1993). Sir John Robert Vane won the 1982 Nobel Prize in Physiology or Medicine for his discovery in the early 1970s that acetylsalicylic acid impairs the synthesis of prostaglandins. It does so by inhibiting the action of two enzymes needed for the synthesis of prostaglandins, COX-1 and COX-2. COX-2 is required for fever induced by LPS (Li et al., 1999). Medical textbooks have long highlighted this early discovery that NSAIDs such as acetylsalicyclic acid inhibit COX enzymes. However, the underlying mechanisms that would explain the immunophysiological principles of communication between the body and the brain involved in sickness behavior remained unknown until the discovery of cytokines.

During the early 1980s, the growing availability of recombinant cytokines greatly facilitated the search for host-derived, fever-producing molecules known as endogenous pyrogens. Shortly thereafter, IL-1, TNF, and IL-6 were identified as major endogenous pyrogens that were synthesized and released during autoimmune diseases and infections. It was also established that LPS could also initiate fevers and sickness behaviors in mice in which these cytokines or their receptors were blocked. These collective results suggest that PAMPs, such as those found in LPS and acting through specific TLRs (TLR-4 in the case of LPS), can induce fever in the absence of specific cytokines. It could also be that when one cytokine is blocked, another cytokine exerts its pyrogenic properties, which is an interpretation consistent with the very redundant properties of many cytokines. Both pyrogenic cytokines and most TLRs share common signaling components in the form of activation of the ubiquitous NFκβ transcription factor. The current view is that pyrogenic cytokines, as well as activation of TLRs, initiate the production of PGE_2 and that its binding to EP3 in the median preoptic nucleus of the hypothalamus is ultimately responsible for increasing the hypothalamic set-point thermostat that leads to fever following systemic LPS injection (Lazarus et al., 2007). Some of the most recent data on this issue recently appeared in *Neuron* from the laboratory of Paul Sawchenko (Serrats et al., 2010). These workers used a different end-point for the action of cytokines in the brain, which was activation of the HPA axis. These workers confirmed that both perivascular macrophages and brain endothelial cells can express COX-2, but in a distinct fashion. Perivascular macrophages express abundant COX-2 following intravenous injection of a low dose of IL-1, whereas a low dose of LPS induces abundant COX-2 expression only in the absence of perivascular macrophages. These data were interpreted to indicate that LPS probably binds to brain endothelial cells and subsequently activates perivascular macrophages. These cells would secrete a prostanoid that would deactivate endothelial cells, thus acting in a negative-feedback fashion. The IL-1 and other inflammatory mediators produced by activated endothelial cells would also act on nearby perivascular macrophages and further activate the production of PGE_2 by these cells.

Because fever and sickness behavior often occur simultaneously following infections or injections of LPS, it was important to learn whether fever alone is responsible for the behavioral changes that occur during systemic inflammation. To address that question, we administered small amounts of IL-1 both peripherally and into the ventricles of the brain of adult rats and measured metabolic rate, core body temperature, and two types of behavior: operant responding for food and investigation of a novel juvenile (Kent et al., 1992a, 1992b). As expected, IL-1 injected via either route increased metabolic rate, caused a 1.5°C increase in body temperature, and significantly reduced both behaviors. Rats were also pretreated with the IL-1 receptor antagonist, which inhibited the reduction in both food and social investigation when administered at the same location as IL-1. In contrast, this antagonist blocked the IL-1–induced increase in both metabolic rate and fever when it was administered along with IL-1 in the peritoneum. Even though rats given IL-1 and pretreated with the antagonist in the ventricles of the brain maintained a fever and elevated metabolic rate, their behavioral response for food and investigation of a novel juvenile were significantly improved. These data were interpreted to indicate that the mechanisms controlling fever and sickness behavior are different.

Two additional results of these experiments were very interesting. The first was that IL-1 injected centrally caused the same behavioral response as when injected peripherally at a 100- to 200-fold lower dose. This observation is consistent with the idea that very low amounts of IL-1 can induce dramatic changes in behavior. At that time, cancer patients were being experimentally treated with systemic injections of recombinant cytokines such as interferon-α and IL-2 (Dantzer and Kelley, 1989). These patients experienced numerous physiological and psychological symptoms following cytokine injection. Many scientists argued that the behavioral changes caused by recombinant cytokines were simply due to the toxicity of the high pharmacological doses that were tested. We now know that even a natural immune response to a pathogen leads to cytokine release and inflammation at the local, systemic, and brain levels, with accompanying changes in behavior. The second major result of the experiment described above clearly proved that cytokines in the periphery communicate with the brain. This finding was demonstrated by administering a blocking dose of the IL-1 receptor antagonist into the ventricles of the brain 15 minutes before IL-1 was given in the peritoneum. This experimental arrangement showed that the antagonist significantly reversed the reduction in both operant responding for food and social investigation that was caused by an intraperitoneal injection of IL-1. The specific route by which IL-1 in the peritoneum communicated with the brain was subsequently shown to occur via the afferent vagus nerve, as described above. Based upon these data, it was proposed that cytokine-induced sickness behavior, as measured in this and other experiments, be used as a new behavioral tool to assess the biological effects of new drugs in the pharmaceutical industry (Kent et al., 1992b).

DEPRESSIVE-LIKE BEHAVIOR

Hymie Anisman recently summarized data supporting the notion that the brain responds to afferent signals from the immune system in a manner similar to a stressor (Anisman, 2009). As early as 1998, the group of Maier and Watkins reported that inescapable shock increased IL-1 in the brain of adrenalectomized rats (Nguyen et al., 1998). Indeed, one of the more provocative findings in recent years has been the discovery that psychological stress, such as public speaking and mental arithmetic, not only activates the HPA axis but also increases the concentration of circulating cytokines such as IL-6, IL-1, and IFNγ (reviewed by Miller et al., 2009). These changes have been linked to alterations in mood. In animal models, chronic stressors such as social defeat (Krishnan et al., 2007), chronic mild stress (Goshen et al., 2008), and inescapable shock (Bilbo et al., 2008) induce a number of depressive-like behaviors in rodents, including loss of appetite, weight loss, anhedonia, a loss of social engagement, and behavioral despair as measured by an increase in immobility in inescapable tests such as forced swim and tail suspension. One of these groups made the insightful and novel observation that the chronic stress-induced increase in IL-1 in the hippocampus, as well as activation of the HPA axis, did not occur in IL-1 receptor type I knockout mice or in transgenic mice that overexpressed the IL-1RA only in the brain. The same was true for the reduction in hippocampal neurogenesis, as well as the behavioral reduction in both sucrose preference and social exploration, that occurred only in wild-type mice. These convincing findings demonstrated that IL-1 in the brain is necessary and sufficient for inducing symptoms of depressive-like behavior caused by chronic mild stress. A recent article offered support for the intriguing speculation that the association between chronic stress and depressive-like behaviors is caused by epigenetic changes (Autry and Monteggia, 2009). For example, brain-derived neurotrophic factor (BDNF) has been repeatedly implicated in the association between depressive-like behavior and neurogenesis. Eric Nestler's group has reported that an intruder paradigm associated with chronic defeat in mice leads to methylation of histones in the promoter of two of the five BDNF variant mRNAs in the hippocampus and a concomitant reduction in BDNF mRNA (Tsankova et al., 2006). The application of epigenetic principles to the field of brain, behavior, and immunity is ripe for much more intensive investigations.

Depressive-like behaviors in rodents can be induced by infectious agents, including *Escherichia coli* (and LPS, the outer membrane of gram-negative bacteria) (Yirmiya, 1996; Bilbo et al., 2008) and an attenuated form of *Mycobacterium bovis* known as bacille Calmette-Guérin (BCG) (Moreau et al., 2008; O'Connor et al., 2009b). Various components of sickness behavior precede development of depressive-like behavior. Sickness behavior induced by either LPS or BCG is preceded by heightened expression of pro-inflammatory transcripts for IL-1, TNF, and IL-6 in the brain. The expression of these transcripts, as well as development of both sickness and depressive-like behaviors, can be blocked by administration of the tetracycline antibiotic minocycline (Henry et al., 2008; O'Connor et al., 2009c). However, depressive-like behaviors can be specifically targeted by inhibiting an enzyme known as indoleamine 2,3 dioxygenase (IDO). This enzyme causes tryptophan to be degraded along the kynurenine rather than the serotonin pathway. When IDO induction is blocked either genetically or pharmacologically following infection with BCG, depressive-like behaviors do not appear as they do in control mice (O'Connor et al., 2009b). Furthermore, BCG-induced expression of both IDO and depressive-like behavior does not develop in mice that lack the gene for the IFNγ receptor, which demonstrates that IFNγ is required for induction of IDO in this model of chronic infection. TNF also synergizes with IFNγ to induce depressive-like behavior (O'Connor et al., 2009a). IDO-induced metabolites of kynurenine have been recently reported in the blood and cerebrospinal fluid of hepatitis C patients after a 3-month treatment regimen with IFNα (Raison et al., 2010).

A recent meta-analysis concluded that the inflammatory biomarkers of high-sensitivity CRP, IL-6, IL-1, and IL-1RA are all reliably and positively associated with symptoms of depression in humans (Howren et al., 2009). However, as noted previously by a number of authors (e.g., Irwin and Miller, 2007), the direction of this relationship is not understood: Does inflammation promote depression, or does depression increase inflammation? Similarly, the underlying neurobiological pathways and mechanisms that lead to depressive-like behavior in chronic stress models and inflammation models may well be different (Miller et al., 2009), although they are likely to synergize. A thorough discussion and review of the relationship between sickness behavior and inflammation, with a particular emphasis on IDO as a mediator of tryptophan metabolism in depression, has recently been published (Fig. 12.2; Dantzer et al., 2008).

INFLAMMATION AND BEHAVIOR OF INVERTEBRATES AND "LOWER" VERTEBRATES

EVOLUTIONARY ROOTS OF SICKNESS BEHAVIOR: EVIDENCE FROM COMPARATIVE PSYCHONEUROIMMUNOLOGY

Plants, insects, and vertebrates share many of the same fundamental molecular mechanisms of host defense against pathogens These shared mechanisms include the different forms of receptors

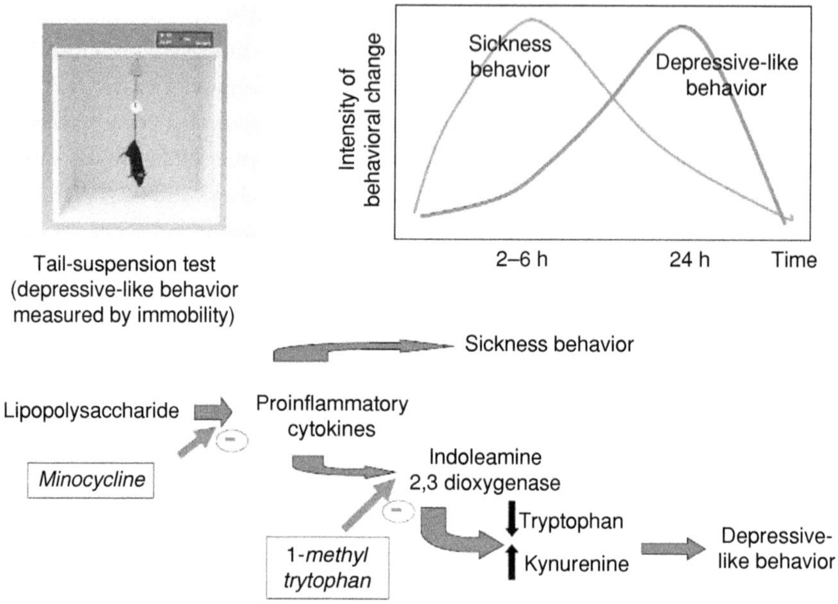

FIGURE 12.2: Inflammation and behavior of mammals. A common measure of depressive-like behavior in mice is the tail suspension set, in which immobility is considered as a measure of despair or resignation. This summary of LPS-induced inflammation depicts sickness behaviors as peaking within 6 hours of an intraperitoneal injection. If the systemic inflammatory response is not controlled and develops over time, then depressive-like behavior can appear on this background of sickness behavior. In the case of bacterial infections such as BCG, this depressive-like behavior can be manifested for several weeks. The lower portion of this graph indicates that LPS causes the induction of pro-inflammatory cytokines in both the periphery and the brain, leading to development of sickness behaviors. Heightened expression of cytokines in the brain, as well as development of both sickness and depressive-like behaviors, is blocked by the anti-inflammatory antibiotic minocycline. In contrast, neither expression of pro-inflammatory cytokines in the brain nor sickness behavior is affected by the specific competitive inhibitor of IDO, 1-methyl tryptophan (1-MT). However, pharmacological inhibition of IDO with 1-MT or deletion of the IDO gene prevents the increase in the ratio of tryptophan/kynurenine in blood and brain and inhibits development of depressive-like behavior. (Figure modified from Dantzer et al., 2008.)

(e.g., the Toll-like family receptors) that recognize pathogen-associated molecular patterns, the intracellular signaling pathways (e.g., NF-κβ activation) that eventually activate transcription of genes associated with the host's anti-pathogenic response, and finally the release of cationic peptides (i.e., cathelicidins and defensins) that play a role in the elimination of microbial agents (Borregaard et al., 2000). The discovery of such shared immune processes across phyla has established insects, and especially *Drosophila melanogaster*, as a valuable model for studying innate immunity (Shirazu-Hiza and Schneider, 2007). For example, injection of LPS into flies induces an immune response mediated by the activation of Relish, a homologue of NF-κβ (Imler et al., 2000). Invertebrates could therefore constitute appealing tools for a comparative approach to advance the continuing development of brain, behavior, and immunity (Adamo, 2006; Aubert, 2007).

Despite these similarities between the innate immune system of invertebrates and mammals (Vilmos and Kurucz, 1998), to date little effort has been invested to explore immune-related behavioral strategies in insects. So far, attention has been mainly focused on defensive strategies against parasites and parasitoids (Schmid-Hempel, 1998). However, recent studies have revealed bidirectional neuroimmune communications in invertebrates, homologous to those that have been described in mammals. For instance, the invertebrate immune system shares signaling features with the nervous system, as illustrated by the immunosuppressive action of opioid-like peptides. Moreover, it is now clearly established that parasites take advantage of these shared processes to influence the immune system to impair the host's defenses, as well as the nervous system to influence the host's behavior in a way that permits the parasitic cycle to progress (Salzet, 2000).

A growing set of studies shows that invertebrates display changes in behavior during an immune challenge similar to those originally observed in vertebrates. For example, LPS impairs associative learning abilities in *Apis mellifera* honeybees (Mallon et al., 2003). Similarly, administration of LPS to *Bombus terrestris* bumblebee workers maintained under a protein-depleted diet reduces performance in the proboscis extension reaction test, a standard learning task in insects (Riddell and Mallon, 2006). This test is a form of classical conditioning in which a sugar solution is used as unconditioned stimulus, a specific odor as the conditioned stimulus, and the extension of the proboscis (i.e., the elongated oral appendage allowing bumblebees to feed) as the response (unconditioned, then conditioned). The reason for using starved workers instead of bumblebees fed *ad libitum* is to avoid the synthesis of new proteins, especially octopamine (OA), a prominent stress-associated molecule. These results are in line with the idea that the octopaminergic system plays a key role in the link between immune and nervous systems of insects (Riddell and Mallon, 2006) (Fig. 12.3). Octopamine in invertebrates is analogous to vertebrate norepinephrine, and it is known to modulate immunity, as well as behavior (Roeder, 2005). Immune activation in response to LPS decreases the availability of OA that is necessary for the nervous system to sustain learning. Therefore, in the study by Riddell and Mallon, learning in bumblebees is impaired not only because of the LPS-induced depletion in OA, but also because OA cannot be re-synthesized due to the low-protein diet. Immune-mediated changes in OA can affect behavior independently of the direction of the changes in OA levels.

The host of the parasitic wasp *Cotesia congregata* reduced feeding behavior during the apex of the immune response (i.e., when the wasp's larvae were about to emerge) (Adamo, 2005). OA concentrations in the hemolymph were the highest at the same period, revealing a positive relationship between immune activation, OA concentrations, and diminished feeding behavior. Based on this and other evidence, Adamo proposed the possible involvement of the octopaminergic system in "immune-activated anorexia."

As noted, octopamine is a key hormone in the "stress" response of invertebrates. It is essential for the expression of defensive responses in insects, such as occurs by liberating lipids from fat bodies that provide the necessary energy for the fight/flight response (Orchard et al., 1993; Roeder, 2005). As in mammals, insects can become more susceptible to infection after an acute stress (Adamo and Parsons, 2006). Moreover, immune challenges can stimulate the release of OA (Dunphy and Downer, 1994). In addition, OA can decrease lysozyme-like activity in the cricket *Gryllus texensis* in the absence of infection, whereas OA increases this enzymatic activity in the presence of bacteria (Adamo, 2010). On the basis of these results, Adamo proposed that the OA "stress" system in insects could have evolved as a process that modulates or fine-tunes the immune systems for an optimal response.

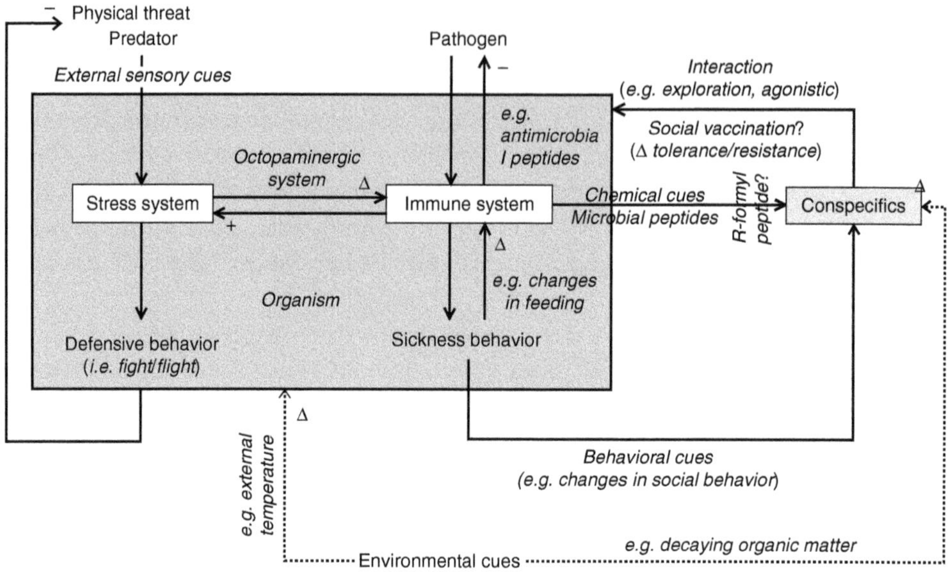

FIGURE 12.3: Comparative aspects of brain, behavior, and immunity and the evolutionary roots of stress/immune interactions. This figure summarizes current findings in invertebrates regarding relationships between the two main defensive systems. The immune and stress systems support and organize specific defenses against pathogens and predators or competitors, respectively. Octopamine (OA) is a chemical analogue of norepinephrine that plays a key role in the interaction between these two main physiological systems in invertebrates. Octopamine induces the liberation of lipids from fat bodies to support the fight/flight response. On the other hand, OA modulates the immune response depending on the biological context. For example, it decreases lysozyme-like activity in standard conditions but increases it in the presence of pathogens. Moreover, octopaminergic transmission is also enhanced during immune responses. Besides this inter-system communication, inter-individual communication has been demonstrated on the basis of immune-related changes in social behaviors of healthy conspecifics. The relevant cues, such as specific chemical signals or microbial peptides, and the adaptive functions of such changes, such as social vaccination, have yet to be fully elucidated. However, the selective response of R-formyl peptide in vomeronasal neurons of rodents to inflammatory mediators offers new leads for understanding potential ligands and their bioactivity in different phyla.

Another example of shared relationships between immunity and behavior can be found in the field of sleep research. In mammals, sleep deprivation affects immunity, and conversely immune stimulation increases slow-wave sleep via cytokine actions in the brain (Majde and Krueger, 2005). Similar relationships between sleep-like rest patterns (i.e., torpor) and activation of Relish have been described in *Drosophila* (Williams et al., 2007). Although sleep-like rest patterns in insects are not identical to sleep in mammals, these original results offer new leads to better understand not only sleep–immunity relationships, but also the evolution and general functions of sleep.

Collectively, these studies show that insects share at least three types of behavioral changes with mammals when facing an immune challenge: reduced eating, impaired learning, and increased torpor. Moreover, as the organization and responses of innate immune system share the same fundamental features across phyla, understanding the evolutionary history of immune-related behaviors in vertebrates could benefit from studies with insects (Adamo, 2006, 2010; Aubert, 2007). Furthermore, as a growing interest emerges in invertebrates, especially insects, to address the behavioral consequences of immune activation, a broader comparative perspective in neuroimmunology could also take advantage of "lower" vertebrate models. For example, fishes offer a glimpse of the evolution of the adaptive immunity, from the appearance of lymphocyte-like cells in jawless species to the development of a thymus and more specialized humoral factors in jawed species (Lieschke and Trede, 2009). The evolutionary roots of such "new" defense mechanisms likely coincide with the development of still unexplored "new" regulatory processes in the cytokine-to-brain communication pathways.

MOTIVATIONAL ASPECTS OF INFLAMMATION-ASSOCIATED SICKNESS

If we accept the notion that sickness-related changes in behavior are the outward signs of an organized strategy promoting survival, then it follows that sick animals should maintain the ability to adjust their actual behavioral repertoire. This should be true in relation not only to their endogenous inflammatory processes, but also to "relevant" exogenous signals. These concepts should apply to selectivity and flexibility in the behavioral responses of sick animals, which are the two core characteristics of what is called "motivation."

Although motivation is a key concept in behavioral sciences as it points to proximal causes of behavior, the exact definition and boundaries of this concept are quite variable. Depending on the scope and theoretical grounds (e.g., psychological, behaviorist, ethological, neurophysiological), motivation can serve as an umbrella for processes as different as instincts, drives, needs, incentives, arousal, impulses, or combinations of any of these. In the present context we will use the term "motivation" to refer to a "central state that organizes both perception and action," as previously proposed (Bolles and Fanselow, 1980). This definition clearly indicates that motivation is a dynamical dispositional state that defines behavioral goals and priorities, modulates sensory processes to focus attention toward these goals, and mobilizes resources to achieve these goals. The value of motivation as a hypothetical psychological construct and explanatory concept in behavioral studies is two-fold: (a) it avoids the necessity of simple mechanistic links between perception and action (i.e., between stimulus and response), and (b) it accounts for the mechanisms that allow an animal to select the most appropriate strategy in relation to the eliciting situation, depending on its internal environment and external context.

Because all motivational states ultimately result in the activation of goal-directed behaviors, it follows that behavioral output is the outcome of competition between different motivational systems. For example, feeding and mating are competitive motivational states because they imply competitive (even if not totally exclusive) behavioral repertoires. The actual behavioral expression is therefore supported by a hierarchical structure of motivational states, which is continuously updated according to external and internal inputs. During an inflammatory episode, a sick individual may be confronted with a lethal threat, which requires that its physiology and behavior be adapted to overcome infection. However, sickness can be a relatively long-term process that, if necessary, needs to be adjusted to adapt to more urgent needs or even more immediate threats. In motivational terms, this means that in the context of a physical danger, fear (i.e., motivation triggered by an immediate threat) would compete with sickness (i.e., motivation triggered by long-term inflammation), and fear-motivated behaviors would temporarily overcome sickness-motivated behaviors. Such dynamics ensures the necessary behavioral flexibility to adapt to environmental changes.

Instances of behavioral flexibility have been demonstrated in sick animals. A typical example can be found in a study of the effects of LPS on maternal behavior in mice (i.e., pup retrieval and nest building). Neonatal rodents are unable to thermoregulate, so the survival of pups strongly depends upon maternal behavior. In standard housing conditions, lactating mice that were made sick by the injection of an appropriate dose of LPS (400 μg/kg i.p.) remained able to retrieve pups that had been previously moved away from the nest, although they did not engage in nest building (i.e., no response of mice to the presence of scattered nest material). However, if the ambient temperature (22°C) was lowered to 6°C, then LPS-treated dams not only retrieved their whole litter as efficiently as before, but they also engaged in the building of a fully enclosed nest (Aubert et al., 1997a). These results have been interpreted as an example of the competition between sickness motivation that makes the animal tend to lie down and rest, and maternal motivation that makes the animal respond to relevant cues (i.e., vocalizations of pups and nest material). If maternal motivation triggered by pups always takes precedence over motivation to rest, then the motivational outcome concerning the competition between sickness and nesting depends on other parameters. Indeed, in standard thermal conditions, the sickness motivational system overrides nest material-elicited motivation, leading to the absence of nest building. When the environmental temperature falls substantially below the thermoneutral zone, nesting motivation rises in the immediate survival hierarchy and overcomes sickness motivation, causing the sick mother to engage in nest building in a manner similar to the non-LPS-injected control mother.

To show that sickness interferes not only with immediate adaptive responses but also with long-term behavioral strategies related to potential needs for resources, the effects of LPS (250 μg/kg i.p.) on food hoarding and immediate food consumption were compared in rats (Aubert et al., 1997b). This study showed that when rats were trained to rely on their behavior to access food resources (during a daily 30-minute session), LPS did not alter food hoarding, although it decreased immediate food consumption. These results indicate that the anticipatory response to future needs is stronger in the tested context than the sickness-induced motivation to rest. This differential effect of LPS on hoarding and feeding has been interpreted in relation to the concept of "dissociated motivations" (Whishaw and Kornelsen, 1993). According to this concept, hoarding behavior is an anticipatory behavior and is therefore mostly influenced by secondary incentives (i.e., the motivation aroused by the environmental cues that are associated with food), whereas food consumption is rather subjected to primary food cues. Therefore, it was concluded that LPS alters primary more drastically than secondary feeding motivation.

Convergent data were found regarding the dynamic competition between feeding and sickness motivation. For example, it has been reported that administration of IL-1β to mice (100 ng i.p.) decreased more extensively the consumption of sweetened milk in *ad libitum*-fed mice than in mice kept under a restricted diet. This effect occurred regardless of whether animals were tested in a free-feeding session (i.e., no effort to provide access to food) or in operant task conditioning (i.e., fixed amount of effort to provide access to food) (Larson et al., 2002). Moreover, using another reinforcement schedule, the effects of LPS on feeding motivation in mice depended on the amount of effort necessary to access food. These findings provide further support to the motivational perspective of sickness, because the behavior of cytokine-treated mice remains shaped by the ever-changing trade-off between diet-induced hunger and the effort required to access a food reward.

Further insights into how sickness as a motivational system could interact with others can be gained from a model proposed 30 years ago (Bolles and Fanselow, 1980). Their ideas were originally designed to account for the relationships between fear- and pain-related motivation. Influenced by the theories of behavioral ecology, they applied what they called a functional behaviorism perspective to the study of aversively motivated behaviors. The objective was to view pain not only as a simple sensory process, but also as the consequence of a motivational state. Based on several studies, they proposed the *Perceptual-Defensive-Recuperative* model, in which pain and fear are two antagonist motivational systems. In this model, fear-induced endogenous analgesia is the counter-regulatory response to a nociceptive stimulus that is dependent on activation of opioid-dependent or -independent endogenous processes. Fear-induced endogenous analgesia supports anti-predatory defensive behaviors (by preventing a disruptive influence of painful sensations on engagement of defensive behaviors such as fight or flight), whereas nociceptive stimuli normally promote a recuperative, resting profile. Interestingly, in the original description of the theoretical model, the role of pain-associated sickness in the expression of recuperative behaviors was emphasized (Bolles and Fanselow, 1980).

In summary, converging data demonstrate that the behavioral changes associated with sickness are not signs of temporary debilitation or incapacitation. Such a state might have been deduced from the general decrease of activity that occurs when sick individuals are observed in the relative comfort of conditions offered by standard laboratory housing. Sick individuals actually maintain their ability to interrupt inflammatory-associated recuperative behaviors in order to adapt to more salient or immediate concerns. The selectivity and flexibility of behavioral responsiveness of sick animals observed in various conflict situations is the result of a specific motivational state triggered by inflammatory processes and mediated by immune-to-brain communication pathways. Thus, as with any other motivational state, sickness is the result of new priorities taking over other drives and incentives. As part of this motivational dynamics, "sickness" participates in the regulation of individual behaviors.

BEHAVIOR, IMMUNITY, AND THE EVOLUTION OF SOCIALITY

Sickness appears to be conducive to social isolation, a conclusion that is based on the results of numerous studies in which the stimulation of rodents' innate immune system caused a reduction in social exploration of a juvenile conspecific (Kelley et al., 1997, 2003; Dantzer, 2004). As supported by various anecdotal observations in different species, social withdrawal could represent an adaptive pro-social (i.e., "altruistic") strategy to prevent spread of infection within the social group (Hart, 1988). One might argue that such pro-social behaviors would have evolved only in species with higher genetic relatedness, such as haplo-diploid social insects. However, this restriction can be ruled out as it is based on the assumption that "altruistic" behavior would have evolved under the pressure of kin selection (i.e., genetic relatedness), as it was formerly hypothesized. This early perspective was supported by leading scientists such as William D. Hamilton (1964) and

Edward O. Wilson (1975). However, many field data (especially in social insects) do not fit with this hypothesis. For example, termites are well-known eusocial insects emitting extensive "altruistic" behaviors. However, termites are diplo-diploids, and weak mean relatedness is common between social partners. Another example is provided by Argentine ants living in super-colonies involving many reproductive individuals. Hence, despite their hapo-diploidy, mean genetic relatedness is rather weak between workers, thus ruling out kin-selection hypothesis to account for altruistic/pro-social behaviors. The accumulation of such controversial field data led to the revision of this former perspective. Kin selection is generally no longer considered as the "key to altruism," even by authors previously involved in kin-selection hypothesis (for example, Nowak et al., 2010). It is now considered that pro-social behaviors were selected through general natural selection processes, whereas relatedness-based kin selection would rather be a possible supporting (i.e., accelerating) mechanism of natural selection processes. However, it has not been yet established whether immune-related changes in social interactions characterize a stable adaptive strategy evolved from selective pressures related to disease transmission within social groups (Aubert, 1999; Aubert and Renault, 2008). Some support for this hypothesis is provided by a recent study in ants, *Temnothorax unifasciatus*. When subjected to various lethal conditions (i.e., fungal infection, exposure to 95% CO_2, or spontaneous ending), ants isolate themselves from their nest-mates to die in isolation (Heinze and Bartosz, 2010). This study illustrates the relevance of invertebrate models, and especially insects, for understanding the evolutionary roots of immune-related behaviors. Because of their stable physical parameters (e.g., humidity, temperature), high population densities, and frequent interactions between individuals, social insect colonies (i.e., ants, honeybees, or termites) offer optimal conditions to study behaviors related to pathogen transmission (Sherman et al., 1998; Naug and Camazine, 2002; Pie et al., 2004).

Social insect colonies have often been described as super-organisms in which the colony represents the whole organism and the individuals represent unitary cells (Wilson, 1971; Hölldobler and Wilson, 1990). Beyond this pure metaphor, such a perspective predicts that the colony will protect itself from pathogen dissemination by eliminating infected insects, similar to an organism that would eliminate infected cells. This would be done either by direct elimination of the infected insect or by self-destruction. Such a mechanism (i.e., the "programmed" death of an organism) has been called "phenoptosis" and refers to individual cellular mechanisms that induce degeneration and death (Skulachev, 1999). However, as discussed above, a possible alternative is the active withdrawal of the infected individual from the colony.

In accordance with the phenoptosis hypothesis, field studies show that parasitized bumblebee workers remain outside overnight and actively seek colder temperatures rather than returning to their nest (Müller and Schmid-Hempel, 1993). This behavioral pattern certainly delays development of the parasite, although it also increases lethal predatory risks for the infected individual, both of which are beneficial to the colony (i.e., lower infection risk). Another example is the increased social interactions (e.g., allo-grooming), which reduce mortality from fungal infection in grouped ants or termites compared to isolated insects (Hughes et al., 2002; Calleri II et al., 2006; Ugelvig and Cremer, 2007). Aubert and Richard examined behavioral responses of healthy *Formica polyctena* ants to the presence of a "sick" nest-mate (Aubert and Richard, 2008). In this study, the innate immune system of individual insects was activated by a non-lethal dose of LPS to avoid any bias due to the presence of a living pathogen. LPS stimulates immune responses in insects through the stress kinase pathway, including JNK (Sluss et al., 1996; Boutros et al., 2002).

Contrary to the phenoptosis hypothesis, which would have predicted elimination or isolation of LPS-treated ants, there was an increase in social exploration (i.e., antennal contacts and body licking). This was accompanied by no increase in agonistic behaviors and a reduction in direct oral fluid exchanges (trophallaxis) (Aubert and Richard, 2008). Although LPS did induce immune activation, as indicated by increased expression of *Defensin 2* genes in fat bodies, it did not reduce motor or self-grooming capabilities. The reduction in trophallaxis, as well as the limited locomotion of treated ants (due to the number of surrounding healthy groomers), could be considered as limiting pathogen transmission through the colony (Hart, 1990). Increased allo-grooming can serve to physically remove external pathogens (Hughes et al., 2002; Calleri II et al., 2006), but it can also enable contact with a "behaviorally controlled" amount of microorganisms, immuno-active molecules (e.g., pathogen fragments), and/or antimicrobial peptides that are synthesized and released by the infected individual. Therefore, allo-grooming could act as a mechanism for "social vaccination" (i.e., the social "transfer" of immunity) of healthy nest-mates (Traniello et al., 2002). In support of this interpretation, naïve termites were found to be more resistant to the entomopathogenic fungus *Metarhizium anisopliae* after they were raised with termites previously infected with the same fungal pathogen (Traniello et al., 2002). Thus, the regulation of social contacts by immune-to-brain communication pathways would ultimately serve a "controlled exposure" strategy (Aubert and Richard, 2008) that promotes establishment of immune responses to subsequent exposure to the same pathogen, thereby promoting disease resistance (Hart, 1990).

Little is still known about the possible transcuticular transfer of immune-active molecules, and further research needs to be conducted to specifically address this important question. However, a growing knowledge of the physiology of the insect cuticle offers stimulating leads. Briefly, fat bodies and hematocytes release antimicrobial peptides and enzymes into the hemolymph following immune stimulation (Kim and Kim, 2005). Moreover, because hemolymph and cuticle are dynamically associated (Schal et al., 1998), various molecules from the hemolymph are constantly transported to the outer layer of the cuticle (Fan et al., 2003). Such a mechanism could constitute the basis of a possible transfer of immune-related compounds to the cuticle (Siva-Jothy et al., 2005).

The cuticle is not only a potential zone of transfer for immune-active molecules but is also a key organ involved in the regulation of social interactions and recognition of nest-mates (Wilson, 1971; Breed, 1983). Specifically, hydrocarbon profiles found on the outer epicuticular surface are dependent on genotype and are dynamically influenced by environmental factors, physiological state, and age (Howard and Blomquist, 2005). Thus, it could be hypothesized that changes in social behavior that occur in healthy insects when exposed to an infected conspecific could be due to changes in hydrocarbon profiles. Such an hypothesis has been tested in *A. mellifera* honeybees (Richard et al., 2008). After an LPS challenge, treated bees showed significant differences in cuticular hydrocarbon profiles (i.e., the chemical signature of the subjects' identity), which supports previous findings in bees infected by an ectoparasitic mite, *Varroa destructor* (Salvy et al., 2001) and confirms that chemical profiles can be modified by immune processes. Finally, experimentally changed chemical profiles induced by coating healthy bees with cuticular extracts obtained from LPS-treated bees caused significant changes in the social behaviors of naïve nest-mates (Richard et al., 2008). Therefore, immune-induced modulation of hydrocarbon profiles could play an important role in the colony response to pathogenic threats (Richard et al., 2008).

The possible existence of analogous behavioral processes in vertebrate models has been experimentally investigated. In a preliminary study, mice maintained in stable social groups of three individuals were observed in semi-natural climes over a 24-hour period after one of the three mice was treated with LPS. Surprisingly, no significant changes in social, agonistic, or spatial behaviors (including inter-individual distances) were found in healthy mice (Aubert et al., unpublished data). This absence of behavioral responses in healthy mice was quite puzzling because it did not fit with any of the main theoretical propositions concerning defensive strategies against pathogens (Hart, 1990). Moreover, the absence of any significant changes does not necessarily mean that healthy mice were unaware of the new condition of their LPS-treated social mate. It cannot be ruled out that environmental test conditions did not provide relevant functional cues to permit the occurrence of observable changes. In a subsequent study, the context in which social encounters between healthy and LPS-treated mice took place was controlled (Renault et al., 2008). This was done using 1,5-diaminopentane, also known as cadaverine. This molecule is a biogenic amine that results from the decomposition of animal tissues and induces hygienist behaviors in rats, such as non-defensive burying of marked objects or bodies (Montoya et al., 1981; De Boer and Koolhaas, 2003). Renault and colleagues failed to find any evidence of agonistic behaviors in healthy mice during interactions with LPS-treated conspecifics (Renault et al., 2008), thus excluding the hypothesis of coercive control of pathogen exposure However, under the influence of the cadaverine odor, healthy mice spent less time in contact with LPS-treated animals and remained further away from them. Moreover, social interactions were also qualitatively different: healthy mice expressed a higher proportion of muzzle sniffing toward LPS-treated animals and less anogenital contacts (Renault et al., 2008). Social exploration of the anogenital part of the body provides relevant information about social rank, gender, and also individual olfactory signatures as controlled by the expression of histocompatibility genes (Brown and MacDonald, 1985; Brown et al., 1987; Singh et al., 1987; Chapter 7 in this volume). In contrast, sniffing of the muzzle area informs about recent activities of the subject and associated locations, and specifically recently sampled or ingested food (Galef et al., 1988; Galef, 2002). Thus, the increase in the proportion of muzzle sniffing could be interpreted as a way for healthy subjects to acquire more information about recently absorbed food from sick individuals.

A key basic question is whether animals can discriminate a sick from a healthy individual. A critical role of the Toll-1 signaling pathway has been found in the expression of pathogen avoidance in the nematode worm *Caenorhabditis elegans* (Pujol et al., 2001). Such a pathway could be involved in "sickness" recognition because Toll signaling pathways are widely conserved across phyla and are major components of the vertebrate and invertebrate innate immune response (Kopp and Medzhitov, 1999). The oxytocin gene plays a crucial role in social interactions of female mice exposed to parasitized males (Kavaliers et al., 2005, 2006). Oxytocin-knockout female mice were impaired in their ability to develop an aversive response (as measured by analgesia and corticosterone plasma levels) to males subclinically infected with the gastrointestinal nematode parasite, *Heligimosomoides polygyrus*. These results are in accordance with the known role of oxytocin and vasopressin in the expression of social behaviors (Carter, 1998; Ferguson et al., 2001). Interestingly, central or systemic injection of IL-1β stimulates the release of vasopressin and oxytocin in rats (Landgraf et al., 1995; Brunton et al., 2006). These results call for further investigation into neuroendocrine regulation of immune-related changes in social behaviors.

In terms of social responses to pathogens, a stimulating lead was recently offered by the discovery of the responsiveness of formyl peptide receptors to disease/inflammation-related ligands on murine vomeronasal neurons (Rivière et al., 2009). Using *in vitro* and *in situ* approaches, a selective response of these receptors to well-known inflammatory mediators was reported (e.g., lipoxin A4 or formyl-methionyl-leucyl-phenylalanine [fMLF]) (Rivière et al., 2009). These findings indicate that formyl peptide receptors are likely to serve an olfactory function associated with the presence of pathogens or diseased conspecifics. If these findings can be replicated, it might be possible to decipher the sensory implications of disease-related ligands. Moreover, a comparative study of these receptors in vertebrates and invertebrates could provide a better understanding of the evolution of defensive behaviors to physical and pathogenic threats in animals.

Studies aimed at understanding immune changes in social behavior across species and phyla are likely to offer new insights to the resistance-versus-tolerance duality in defensive mechanisms (Råberg et al., 2007; Schneider and Ayres, 2008). The concept of resistance refers to the processes that restrain pathogen growth (e.g., immune responses to pathogen infection). The idea of tolerance refers to mechanisms that compensate for the damage of infection. Such a dichotomy fits well with an eco-immunological perspective. Indeed, behavioral processes preventing or limiting threats and infection belong to this second category of defense and should be taken into account to better understand terms of the evolutionary trade-off between these two lines of defense. As resistance and tolerance are commonly considered as inversely correlated (Råberg et al., 2007), one could argue that the weaker the ability to limit pathogen's growth, the stronger should be the perceptive and behavioral responses, including changes in social repertoire.

FUTURE DIRECTIONS

Growing evidence has permitted a better understanding of the adaptive features of immune-related behaviors and their neuroimmune bases, including the relevant brain areas, mediators, receptors, and signaling pathways. Like all good research, this new knowledge has led to more questions concerning the evolution and the mechanisms of individual and collective defenses against pathogens. At the individual level, there remains a need for identifying the specific exact neuronal circuits that are the target of inflammatory mediators. Much more also needs to be learned about the cellular mechanisms that are involved in immune-to-brain communication pathways as well as the modalities of interaction between the brain cytokine signaling pathways and neuronal networks of motivated behaviors. At the social and contextual level, multiple promising lines of research still need to be actively pursued to fully characterize behaviors of healthy group-mates in response to sick conspecifics, as well as the molecular signals and pathways that mediate avoidance or "sharing" strategies (e.g., formyl peptide receptors). Last but not the least, much remains to be learned in order to better exploit the extended homologies of innate immune responses in vertebrates and invertebrates, such as receptor mechanisms (e.g., Toll-like receptors), transduction pathways (e.g. NF-κβ-like factors), and cationic antimicrobial peptides (e.g., defensin, attacin, cecropin). Comparative analysis of behavioral strategies of organisms in response to infection will help clarify the evolutionary roots of individual behavioral defenses but also social management of pathogens and avoidance-versus-resistance strategies.

ACKNOWLEDGMENT

Partially supported by NIH grants to K.W.K. (R01 AG 029573 and R01 MH 51569) and R.D. (R01 MH 71349 and R01 MH 079829).

REFERENCES

Adamo, S.A., 2005. Parasitic suppression of feeding in the tobacco hornworm, *Manduca sexta*: parallels with feeding depression after an immune challenge. Archives of Insect Biochemistry and Physiology 60, 185–197.

Adamo, S.A., 2006. Comparative psychoneuroimmunology: evidence from the insects. Behavioral and Cognitive Neuroscience Reviews 5, 128–140.

Adamo, S.A., 2010. Why should an immune response activate the stress response? Insights from the insects (the cricket *Cryllus texensis*). Brain, Behavior and Immunity 24, 194–200.

Adamo, S.A., Parsons, N.M., 2006. The emergency life-history stage and immunity in the cricket, *Gryllus texensis*. Animal Behaviour 72, 235–244.

Anisman, H., 2009. Cascading effects of stressors and inflammatory immune system activation: implications for major depressive disorder. Journal of Psychiatry Neuroscience 34, 4–20.

Aubert, A., 1999. Sickness and behaviour in animals: a motivational perspective. Neuroscience and Biobehavioral Reviews 23, 1029–1036.

Aubert, A., 2007. Invertebrate studies and the evolution of comparative psychoneuroimmunology. Brain, Behavior and Immunity 21, 290–291.

Aubert, A., Goodall, G., Dantzer, R., Gheusi, G., 1997a. Differential effects of lipopolysaccharide on pup retrieving and nest building in lactating mice. Brain, Behavior and Immunity 11, 107–118.

Aubert, A., Kelley, K.W., Dantzer, R., 1997b. Differential effect of lipopolysaccharide on food hoarding behavior and food consumption in rats. Brain, Behavior and Immunity 11, 229–238.

Aubert, A., Renault, J., 2008. Cytokines and immune-related behaviors. In: Korneva, E., Phelps, C. (Eds.), Neuroimmune Biology. Cytokines and the Brain, Volume 6. Elsevier Science, Oxford.

Aubert, A., Richard, F.J., 2008. Social management of LPS-induced inflammation in *Formica polyctena* ants. Brain, Behavior and Immunity 22, 833–837.

Autry, A.E., Monteggia, L.M., 2009. Epigenetics in suicide and depression. Biological Psychiatry 66, 812–813.

Berkenbosch, F., van Oers, J., del Rey, A., Tilders, F., Besedovsky, H., 1987. Corticotropin-releasing factor-producing neurons in the rat activated by interleukin-1. Science 238, 524–526.

Besedovsky, H.O., Rey, A.D., 2007. Physiology of psychoneuroimmunology: a personal view. Brain, Behavior and Immunity 21, 34–44.

Bilbo, S.D., Yirmiya, R., Amat, J., Paul, E.D., Watkins, L.R., Maier, S.F., 2008. Bacterial infection early in life protects against stressor-induced depressive-like symptoms in adult rats. Psychoneuroendocrinology 33, 261–269.

Blalock, J.E., 1984. The immune system as a sensory organ. Journal of Immunology 132, 1067–1070.

Bluthe, R.M., Michaud, B., Kelley, K.W., Dantzer, R., 1996. Vagotomy blocks behavioural effects of interleukin-1 injected via the intraperitoneal route but not via other systemic routes. Neuroreport 7, 2823–2827.

Bolles, R.C., Fanselow, M.S., 1980. A perceptual-defensive recuperative model of fear and pain. Behavioral and Brain Sciences 3, 291–301.

Borregaard, N., Elsbach, P., Ganz, T., Garred, P., Svejgaard, A., 2000. Innate immunity: from plants to man. Immunology Today 21, 191.

Boutros, M., Agaisse, H., Perrimon, N., 2002. Sequential activation of signaling pathways during innate immune responses in Drosophila. Developmental Cell 3, 711–722.

Breed, M.D., 1983. Nestmate recognition in honey bees. Animal Behaviour 31, 86–91.

Brown, R.E., MacDonald, D.W., 1985. Social Odours in Mammals. Oxford University Press, Oxford.

Brown, R.E., Singh, P.B., Roser, B., 1987. The major histocompatibility complex and the chemosensory recognition of individuality in rats. Physiology and Behavior 40, 65–73.

Brunton, P.J., Sabatier, N., Leng, G., Russell, J.A., 2006. Suppressed oxytocin neuron responses to immune challenge in late pregnant rats: a role for endogenous opioids. European Journal of Neuroscience 23, 1241–1247.

Calleri II, D.V., McGrail Reid, E., Rosengaus, R.B., Vargo, E.L., Traniello, J.F.A., 2006. Inbreeding and disease resistance in a social insect: effects of heterozygosity on immunocompetence in the termite *Zootermopsis angusticollis*. Proceedings of the Royal Society B 273, 2633–2640.

Carpentier, P.A., Duncan, D.S., Miller, S.D., 2008. Glial Toll-like receptor signaling in central nervous system infection and autoimmunity. Brain, Behavior, and Immunity 22, 140–147.

Carter, S.C., 1998. Neuroendocrine perspectives on social attachment and love. Psychoneuroendocrinology 23, 779–818.

Dantzer, R., 2004. Innate immunity at the forefront of psychoneuroimmunology. Brain, Behavior and Immunity 18, 1–6.

Dantzer, R., Kelley, K.W., 1989. Stress and immunity: an integrated view of relationships between the brain and the immune system. Life Science 44, 1995–2008.

Dantzer, R., Kelley, K.W., 2007. Twenty years of research on cytokine-induced sickness behavior. Brain, Behavior, and Immunity 21, 153–160.

Dantzer, R., O'Connor, J.C., Freund, G.G., Johnson, R.W., Kelley, K.W., 2008. From inflammation to sickness and depression: when the immune system subjugates the brain. Nature Reviews Neuroscience 9, 46–56.

De Boer, S.F., Koolhaas, J.M., 2003. Defensive burying in rodents: ethology, neurobiology and psychopharmacology. European Journal of Pharmacology 463, 145–161.

De Bosscher, K., Van Craenenbroeck, K., Meijer, O.C., Haegeman, G., 2008. Selective transrepression versus transactivation mechanisms by glucocorticoid receptor modulators in stress and immune systems. European Journal of Pharmacology 583, 290–302.

De Bosscher, K., Vanden Berghe, W., Haegeman, G., 2003. The interplay between the glucocorticoid receptor and nuclear factor-kappaB or activator protein-1: molecular mechanisms for gene repression. Endocrine Reviews 24, 488–522.

Doyle, S.L., O'Neill, L.A., 2006. Toll-like receptors: from the discovery of NFkappaB to new insights into transcriptional regulations in innate immunity. Biochemical Pharmacology 72, 1102–1113.

Dunphy, G.B., Downer, R.G.H., 1994. Octopamine, a modulator of the haemocytic nodulation response of non-immune *Galleria mellonella*. Journal of Insect Physiology 40, 267–272.

Fan, Y., Zurek, L., Dykstra, M.J., Schal, C., 2003. Hydrocarbon synthesis by enzymatically dissociated oenocytes of the abdominal integument of the German cockroach, *Blattella germanica*. Naturwissenshaften 90, 121–126.

Ferguson, J.N., Aldag, J.M., Insel, T.R., Young, L.J., 2001. Oxytocin in the medial amygdala is essential for social recognition in the mouse. Journal of Neuroscience 21, 8278–8285.

Galef, B.G., Mason, J.R., Preti, G., Bean, N.J., 1988. Carbon disulfide: A semiochemical mediating socially-induced diet choice in rats. Physiology and Behavior 42, 119–124.

Galef, B.G.J., 2002. Social learning of food preferences in rodents: Rapid appetitive learning. In: Crawley, C.R.G., Rogawski, M.A., Sibley, D.R., Skolnick, P., Wray, S. (Eds.), Current Protocols in Neuroscience. Wiley, New York.

Goshen, I., Kreisel, T., Ben-Menachem-Zidon, O., Licht, T., Weidenfeld, J., Ben-Hur, T., Yirmiya, R., 2008. Brain interleukin-1 mediates chronic stress-induced depression in mice via adrenocortical activation and hippocampal neurogenesis suppression. Molecular Psychiatry 13, 717–728.

Grivennikov, S.I., Kuprash, D.V., Liu, Z.G., Nedospasov, S.A., 2006. Intracellular signals and events activated by cytokines of the tumor necrosis factor superfamily: From simple paradigms to complex mechanisms. International Review of Cytology 252, 129–161.

Hamilton, W.D., 1964. The genetical evolution of social behavior. Journal of Theoretical Biology 7, 1–16.

Hart, B.L., 1988. Biological basis of the behavior of sick animals. Neuroscience and Biobehavioral Reviews 12, 123–137.

Hart, B.L., 1990. Behavioral adaptations to pathogens and parasites: Five strategies. Neuroscience and Biobehavioral Reviews 14, 273–294.

Heinze, J., Bartosz, W., 2010. Moribund ants leave their nests to die in social isolation. Current Biology 20, 249–252.

Henry, C.J., Huang, Y., Wynne, A., Hanke, M., Himler, J., Bailey, M.T., Sheridan, J.F., Godbout, J.P., 2008. Minocycline attenuates lipopolysaccharide (LPS)-induced neuroinflammation, sickness behavior, and anhedonia. Journal of Neuroinflammation 5, 15.

Hölldobler, B., Wilson, E.O., 1990. The Ants. Belknap Press, Cambridge, MA.

Howard, W., Blomquist, G.J., 2005. Ecological, behavioral, and biochemical aspects of insect hydrocarbons. Annual Review of Entomology 50, 371–393.

Howren, M.B., Lamkin, D.M., Suls, J., 2009. Associations of depression with C-reactive protein, IL- 1, and IL-6: a meta-analysis. Psychosomatic Medicine 71, 171–186.

Hughes, W.O.H., Eilenberg, J., Boomsma, J.J., 2002. Trade-offs in group living: transmission and disease resistance in leaf-cutting ants. Proceedings of the Royal Society B 269, 1811–1819.

Imler, J.L., Tauszig, S., Jouanguy, E., Forestier, C., Hoffmann, J.A., 2000. LPS-induced immune response in Drosophila. Journal of Endotoxin Research 6, 459–462.

Irwin, M.R., Miller, A.H., 2007. Depressive disorders and immunity: 20 years of progress and discovery. Brain, Behavior, and Immunity 21, 374–383.

Janeway, C.A., Jr., Goodnow, C.C., Medzhitov, R., 1996. Danger—pathogen on the premises! Immunological tolerance. Curr Biol 6, 519–522.

Johnson, R.W., Curtis, S.E., Dantzer, R., Kelley, K.W., 1993. Central and peripheral prostaglandins are involved in sickness behavior in birds. Physiology & Behavior 53, 127–131.

Kavaliers, M., Choleris, E., Agmo, A., Braun, W.J., Colwell, D.D., Muglia, L.J., Ogawa, S., Pfaff, D.W., 2006. Inadvertent social information and the avoidance of parasitized male mice: A role for oxytocin. Proceedings of the National Academy of Science, USA 103, 4293–4298.

Kavaliers, M., Choleris, E., Pfaff, D.W., 2005. Recognition and avoidance of the odors of parasitized conspecifics and predators: Differential genomic correlates. Neuroscience and Biobehavioral Reviews 29, 1347–1359.

Kelley, K.W., 1988. Cross-talk between the immune and endocrine systems. Journal of Animal Science 66, 2095–2108.

Kelley, K.W., 2004. From hormones to immunity: the physiology of immunology. Brain, Behavior, and Immunity 18, 95–113.

Kelley, K.W., Bluthe, R.M., Dantzer, R., Zhou, J.H., Shen, W.H., Johnson, R.W., Broussard, S.R., 2003. Cytokine-induced sickness behavior. Brain, Behavior, and Immunity 17 Suppl 1, S112–118.

Kelley, K.W., Hutchison, K., French, R., Bluthe, R.M., Parnet, P., Johnson, R.W., Dantzer, R., 1997. Central interleukin-1 receptors as mediators of sickness. Annals of the New York Academy of Sciences 823, 234–246.

Kelley, K.W., Johnson, R.W., Dantzer, R., 1994. Immunology discovers physiology. Veterinary Immunology & Immunopathology 43, 157–165.

Kelley, K.W., Weigent, D.A., Kooijman, R., 2007. Protein hormones and immunity. Brain, Behavior, and Immunity 21, 384–392.

Kent, S., Bluthe, R.M., Dantzer, R., Hardwick, A.J., Kelley, K.W., Rothwell, N.J., Vannice, J.L., 1992a. Different receptor mechanisms mediate the pyrogenic and behavioral effects of interleukin 1. Proc Natl Acad Sci U S A 89, 9117–9120.

Kent, S., Bluthe, R.M., Kelley, K.W., Dantzer, R., 1992b. Sickness behavior as a new target for drug development. Trends Pharmacol Sci 13, 24–28.

Kim, T., Kim, Y.J., 2005. Overview of innate immunity in Drosophila. Journal of Biochemistry and Molecular Biology 38, 121–127.

Kopp, E.B., Medzhitov, R., 1999. The Toll-receptor family and control of innate immunity. Current Opinion in Immunology 11, 13–18.

Krishnan, V., Han, M.H., Graham, D.L., Berton, O., Renthal, W., Russo, S.J., Laplant, Q., Graham, A., Lutter, M., Lagace, D.C., Ghose, S., Reister, R., Tannous, P., Green, T.A., Neve, R.L., Chakravarty, S., Kumar, A., Eisch, A.J., Self, D.W., Lee, F.S., Tamminga, C.A., Cooper, D.C., Gershenfeld, H.K., Nestler, E.J., 2007. Molecular adaptations underlying susceptibility and resistance to social defeat in brain reward regions. Cell 131, 391–404.

Landgraf, R., Neumann, I., Holsboer, F., Pittman, Q.J., 1995. Interleukin-1-beta stimulates both central and peripheral release of vasopressin and oxytocin in the rat. European Journal of Neuroscience 7, 592–598.

Larson, S.J., Romanoff, R.L., Dunn, A.J., Glowa, J.R., 2002. Effects of interleukin-1beta on food-maintained behavior in the mouse. Brain, Behavior, and Immunity 16, 398–410.

Lazarus, M., Yoshida, K., Coppari, R., Bass, C.E., Mochizuki, T., Lowell, B.B., Saper, C.B., 2007. EP3 prostaglandin receptors in the median preoptic nucleus are critical for fever responses. Nature Neuroscience 10, 1131–1133.

Li, S., Wang, Y., Matsumura, K., Ballou, L.R., Morham, S.G., Blatteis, C.M., 1999. The febrile response to lipopolysaccharide is blocked in cyclooxygenase-2(-/-), but not in cyclooxygenase-1(-/-) mice. Brain Research 825, 86–94.

Lieschke, G.J., Trede, N.S., 2009. Fish immunology. Current Biology 19, R678–R682.

Majde, J.A., Krueger, J.M., 2005. Links between the innate immune system and sleep. Journal of Allergy & Clinical Immunology 116, 1188–1198.

Malick, A., Jakubowski, M., Elmquist, J.K., Saper, C.B., Burstein, R., 2001. A neurohistochemical blueprint for pain-induced loss of appetite. Proceedings of the National Academy of Sciences of the United States of America 98, 9930–9935.

Mallon, E.B., Brockmann, A., Schmid-Hempel, P., 2003. Immune response inhibits associative learning in insects. Proceedings of the Royal Society of London Series B-Biological Sciences 270, 2471–2473.

Matzinger, P., 1994. Tolerance, danger, and the extended family. Annual Review of Immunology 12, 991–1045.

Miller, A.H., Ancoli-Israel, S., Bower, J.E., Capuron, L., Irwin, M.R., 2008. Neuroendocrine-immune mechanisms of behavioral comorbidities in patients with cancer. Journal of Clinical Oncology 26, 971–982.

Miller, A.H., Maletic, V., Raison, C.L., 2009. Inflammation and its discontents: the role of cytokines in the pathophysiology of major depression. Biological Psychiatry 65, 732–741.

Montoya, C.P., Whishaw, I.Q., Sutherland, R.J., 1981. Cadaverine and burying in the laboratory rat. Bulletin of the Psychological Society 18, 118–120.

Moreau, M., Andre, C., O'Connor, J.C., Dumich, S.A., Woods, J.A., Kelley, K.W., Dantzer, R., Lestage, J., Castanon, N., 2008. Inoculation of Bacillus Calmette-Guerin to mice induces an acute episode of sickness behavior followed by chronic depressive-like behavior. Brain, Behavior, and Immunity 22, 1087–1095.

Müller, C.B., Schmid-Hempel, P., 1993. Exploitation of cold temperature as defence against parasitoids in bumblebees. Nature 363, 65–67.

Nance, D.M., Sanders, V.M., 2007. Autonomic innervation and regulation of the immune system (1987–2007). Brain, Behavior, and Immunity 21, 736–745.

Naug, D., Camazine, S., 2002. The role of colony organization on pathogen transmission in social insects. Journal of Theoretical Biology 215, 427–439.

Nguyen, K.T., Deak, T., Owens, S.M., Kohno, T., Fleshner, M., Watkins, L.R., Maier, S.F., 1998. Exposure to acute stress induces brain interleukin-1beta protein in the rat. Journal of Neuroscience 18, 2239–2246.

Nowak, M.A., Tarnita, C.E., Wilson, E.O., 2010. The evolution of eusociality. Nature 466, 1057–1062.

O'Connor, J.C., Andre, C., Wang, Y., Lawson, M.A., Szegedi, S.S., Lestage, J., Castanon, N., Kelley, K.W., Dantzer, R., 2009a. Interferon-gamma and tumor necrosis factor-alpha mediate the upregulation of indoleamine 2,3-dioxygenase and the induction of depressive-like behavior in mice in response to bacillus Calmette-Guerin. Journal of Neuroscience 29, 4200–4209.

O'Connor, J.C., Lawson, M.A., Andre, C., Briley, E.M., Szegedi, S.S., Lestage, J., Castanon, N., Herkenham, M., Dantzer, R., Kelley, K.W., 2009b. Induction of IDO by bacille Calmette-Guerin is responsible for development of murine depressive- like behavior. Journal of Immunology 182, 3202–3212.

O'Connor, J.C., Lawson, M.A., Andre, C., Moreau, M., Lestage, J., Castanon, N., Kelley, K.W., Dantzer, R., 2009c. Lipopolysaccharide-induced depressive-like behavior is mediated by indoleamine 2,3-dioxygenase activation in mice. Molecular Psychiatry 14, 511–522.

Orchard, I., Ramirez, J.M., Lange, A.B., 1993. A multifunctional role for octopamine in locust flight. Annual Review of Entomology 38, 227–249.

Pace, T.W., Hu, F., Miller, A.H., 2007. Cytokine-effects on glucocorticoid receptor function: relevance to glucocorticoid resistance and the pathophysiology and treatment of major depression. Brain, Behavior, and Immunity 21, 9–19.

Pie, M.R., Rosengaus, R.B., Traniello, J.F.A., 2004. Nest architecture, activity pattern, worker density and the dynamics of disease transmission in social insects. Journal of Theoretical Biology 226, 45–51.

Pujol, N., Link, E.M., Liu, L.X., Kurz, C.L., Alloing, G., Tan, M.W., Ray, K.P., Solari, R., Johnson, C.D., Ewbank, J.J., 2001. A reverse genetic analysis of components of the Toll signaling pathway in *Caenorhabditis elegans*. Current Biology 11, 809–821.

Quan, N., Banks, W.A., 2007. Brain-immune communication pathways. Brain, Behavior, and Immunity 21, 727–735.

Raberg, L., Sim, D., Read, A.F., 2007. Disentangling genetic variation for resistance and tolerance to infectious disease in animals. Science 318, 812–814.

Raison, C.L., Dantzer, R., Kelley, K.W., Lawson, M.A., Woolwine, B.J., Vogt, G., Spivey, J.R., Saito, K., Miller, A.H., 2010. CSF concentrations of brain tryptophan and kynurenines during immune stimulation with IFN-alpha: relationship to CNS immune responses and depression. Molecular Psychiatry 15, 393–403.

Renault, J., Gheusi, G., Aubert, A., 2008. Changes in social exploration of a lipopolysaccharides-treated conspecific in mice: Role of environmental cues. Brain, Behavior, and Immunity 22, 1201–1207.

Richard, F.J., Aubert, A., Grozinger, C.M., 2008. Modulation of social interactions by immune stimulation in honey bee, *Apis mellifera*, workers. BMC Biology 6, 50.

Riddell, C.E., Mallon, E.B., 2006. Insect psychoneuroimmunology: Immune response reduces learning in protein starved bumblebees (*Bombus terrestris*). Brain, Behavior, and Immunity 20, 135–138.

Rivest, S., 2003. Molecular insights on the cerebral innate immune system. Brain, Behavior, and Immunity 17, 13–19.

Rivière, S., Challet, L., Fluegge, D., Spehr, M., I., R., 2009. Formyl peptide receptor-like proteins are a novel family of vomeronasal chemosensors. Nature 459, 574–577.

Roeder, T., 2005. Tyramine and octopamine: Ruling behavior and metabolism. Annual Review of Entomology 50, 447–477.

Salvy, M., Martin, C., Bagnères, A.G., Provost, E., Roux, M., Le Conte, Y., Clement, J.L., 2001. Modifications of the cuticular hydrocarbon profile of *Apis mellifera* worker bees in the presence of the ectoparasitic mite *Varroa jacobsoni* in brood cells. Parasitology 122, 145–159.

Salzet, M., 2000. Invertebrate molecular neuroimmune processes. Brain Research Reviews 34, 69–79.

Schal, C., Sevala, V., Carde, R.T., 1998. Novel and highly specific transport of a volatile sex pheromone by hemolymph lipophorin in moths. Naturwissenschaften 85, 339–342.

Schmid-Hempel, P., 1998. Parasites in Social Insects. Princeton University Press, Princeton.

Schneider, D.S., Ayres, J.S., 2008. Two ways to survive infection: what resistance and tolerance can teach us about treating infectious diseases. Nature Reviews Immunology 8, 889–895.

Serrats, J., Schiltz, J.C., García-Bueno, B., Rooijen, N.v., Reyes, T.M., Sawchenko, P.E., 2010. Dual roles for perivascular macrophages in immune-to-brain signaling. Neuron 65, 94–106.

Sherman, P.W., Seeley, T., Reeve, H.K., 1998. Parasites, pathogens, and polyandry in honey bees. American Naturalist 151, 392–396.

Shirazu-Hiza, M.M., Schneider, D.S., 2007. Confronting physiology: how do infected flies die? Cell Microbiology 9, 2775–2783.

Singh, P.B., Brown, R.E., Roser, B., 1987. MHC antigens in urine as olfactory recognition cues. Nature 327, 161–164.

Siva-Jothy, M.T., Moret, Y., Rolff, J., 2005. Insect immunity: An evolutionary ecology perspective. Advances in Insect Physiology 32, 1–48.

Skulachev, V.P., 1999. Phenoptosis: Programmed death of an organism. Biochemistry Moscow 64, 1418–1426.

Sluss, H.K., Han, Z., Barrett, T., Davis, R.J., Ip, Y.T., 1996. A JNK signal transduction pathway that mediates morphogenesis and an immune response in Drosophila. Genes and Development 10, 2745–2758.

Soulet, D., Rivest, S., 2008. Bone-marrow-derived microglia: myth or reality? Current Opinion in Pharmacology 8, 508–518.

Strauss, K.I., 2008. Antiinflammatory and neuroprotective actions of COX2 inhibitors in the injured brain. Brain, Behavior, and Immunity 22, 285–298.

Traniello, J.F.A., Rosengaus, R.B., Savoie, K., 2002. The development of immunity in a social insect: Evidence for the group facilitation of disease resistance. Proceedings of the National Academy of Sciences of the United States of America 99, 6838–6842.

Tsankova, N.M., Berton, O., Renthal, W., Kumar, A., Neve, R.L., Nestler, E.J., 2006. Sustained hippocampal chromatin regulation in a mouse model of depression and antidepressant action. Nature Neuroscience 9, 519–525.

Ugelvig, L.V., Cremer, S., 2007. Social prophylaxis: group interaction promotes collective immunity in ant colonies. Current Biology 17, 1967–1971.

Vilmos, P., Kurucz, E., 1998. Insect immunity: evolutionary roots of the mammalian innate immune system. Immunology Letters 62, 59–66.

Vitkovic, L., Konsman, J.P., Bockaert, J., Dantzer, R., Homburger, V., Jacque, C., 2000. Cytokine signals propagate through the brain. Molecular Psychiatry 5, 604–615.

Wan, W., Janz, L., Vriend, C.Y., Sorensen, C.M., Greenberg, A.H., Nance, D.M., 1993. Differential induction of c-Fos immunoreactivity in hypothalamus and brain stem nuclei following central and peripheral administration of endotoxin. Brain Research Bulletin 32, 581–587.

Wan, W., Wetmore, L., Sorensen, C.M., Greenberg, A.H., Nance, D.M., 1994. Neural and biochemical mediators of endotoxin and stress-induced c-fos expression in the rat brain. Brain Research Bulletin 34, 7–14.

Watkins, L.R., Wiertelak, E.P., Goehler, L.E., Mooney-Heiberger, K., Martinez, J., Furness, L., Smith, K.P., Maier, S.F., 1994. Neurocircuitry of illness-induced hyperalgesia. Brain Research 639, 283–299.

Whishaw, I.Q., Kornelsen, R.A., 1993. 2 types of motivation revealed by ibotenic acid nucleus accumbens lesions. Dissociation of food carrying and hoarding and the role of primary and incentive motivation. Behavioural Brain Research 55, 283–295.

Wietek, C., O'Neill, L.A., 2007. Diversity and regulation in the NF-kappaB system. Trends in Biochemical Sciences 32, 311–319.

Williams, J.A., Sathyanarayanan, S., Hendricks, J.C., Sehgal, A., 2007. Interaction between sleep and the immune response in Drosophila: a role for the NFkappaB relish. Sleep 30, 389–400.

Wilson, E.O., 1971. The Insect Societies. Belknap Press, Cambridge, MA.

Wilson, E.O., 1975. Sociobiology: The New Synthesis. Belknap Press, Cambridge, MA.

Yirmiya, R., 1996. Endotoxin produces a depressive-like episode in rats. Brain Research 711, 163–174.

CHAPTER 13

THE IMPORTANCE OF PHYSIOLOGY FOR ECOIMMUNOLOGY

Lessons from the Insects

Shelley A. Adamo

INTRODUCTION: INSECT ECOIMMUNOLOGY: THE POWER OF A SIMPLER MODEL SYSTEM

Insects are attacked by the same classes of pathogens that attack other animals (e.g., viruses, bacteria, fungi, and parasites) (Fuxa and Tanada, 1987). Pathogens are a major cause of reduced fitness in insects, and some epizootics (i.e., plagues) can decimate entire populations (Tanada and Fuxa, 1987). However, insects survive despite being surrounded by pathogens by producing physical, chemical, and behavioral defenses. Therefore, insects face many of the same eco-immunological issues facing vertebrates. Insects must invest in internal and external defense systems for survival, but their maintenance and deployment result in costs that also limit fitness (Siva-Jothy et al., 2005). Insect ecological immunologists have studied how different insect species partition their energy budget between pathogen defense and other functions such as reproduction (Lawniczak et al., 2007).

Ecological immunologists working with insects have several advantages over their colleagues studying similar questions in vertebrates. Insects have short generation times, allowing researchers to study the impact of pathogen defense on fitness more directly. Furthermore, insect internal defense systems, although not simple, are simpler than those found in vertebrates. This relative simplicity makes measuring disease resistance and immune function more straightforward in insects than in vertebrates. It also makes it easier to tease apart some of the complex interactions between pathogen defense and other physiological systems that result in immunological trade-offs.

Terminology can be confusing in this field. In this chapter, the immune system refers to those physiological systems typically considered part of the insect immune system (e.g., Gillespie et al., 1997). The internal defense system is the larger collection of physiological mechanisms that contribute to host survival, some of which have not traditionally been considered part of the insect immune system (e.g., heat shock protein production). Disease resistance is the ability to survive an infection.

BRIEF OVERVIEW OF THE INSECT IMMUNE SYSTEM

Like vertebrate skin, the insect exoskeleton provides a barrier to infection (Boucias and Pendland, 1998). The exoskeleton is studded with an array of antimicrobial defenses (Boucias and Pendland, 1998). It is a formidable barrier, and many pathogens eschew this route (Andreadis, 1987). Most pathogens enter the insect through the gut (Andreadis, 1987). The digestive system is vulnerable because the tissues that absorb water and nutrients cannot be lined with a thick protective layer. Nevertheless, even these sensitive areas have their defenses. For example, the midgut is protected by a peritrophic membrane in most insects (Boucias and Pendland, 1998). The peritrophic membrane is an extracellular mesh that prevents large particles (i.e., particles the size of many viruses or larger) from abrading or attaching to the midgut (Boucias and Pendland, 1998).

Once pathogens manage to overcome these physical defenses, they must then face the internal immune defenses of the insect. Although these defenses are known in outline (see Beckage, 2008), there are still important aspects of insect immune function that remain poorly understood. For example, it remains controversial whether, and to what extent, immunological specificity and memory exist in invertebrates (e.g., see Hauton and Smith, 2007, and Pham and Schneider, 2008). However, we do know that much of the insect immune system is analogous to the innate immune system of vertebrates (Nation, 2008). That is, insects have proteins, lipoproteins, and lipid particles in the hemolymph (blood) capable of recognizing a range of pathogen-associated molecular patterns (PAMPs) such as peptidoglycan and ß-glucans (Schmidt et al., 2008; Marmaras and Lampropoulou, 2009). Hemocytes, insect immune cells residing in the hemolymph, also have receptors for a variety of compounds, both biotic and abiotic (Marmaras and Lompropoulou, 2009). Once receptors in the hemolymph or on hemocytes bind to a substance, both humoral (Kanost and Gorman, 2008) and cellular (Strand, 2008) immune responses are activated. Cellular immune responses include encapsulation, nodulation, and phagocytosis (Strand, 2008). During nodulation and encapsulation, invading particles are walled off by hemocytes from the rest of the body. During phagocytosis, hemocytes engulf and digest foreign particles. Humoral immune responses include the activation of proteins and other compounds that are found constitutively in the hemolymph (Nation, 2008), such as the cleavage of prophenoloxidase to form the active enzyme phenoloxidase (Kanost and Gorman, 2008). Once activated, these compounds participate in the destruction of pathogens. Another class of humoral compounds, antimicrobial peptides, is synthesized by the fat body in response to infection (Nation, 2008). They are released into the hemolymph and can destroy bacteria and other pathogens (Nation, 2008). The intracellular signaling pathways leading to the synthesis of these compounds are known in outline (Nation, 2008) and are a continuing area of research. Insect hemocytes also release cytokines (Strand, 2008). As in vertebrates, these compounds can activate or inhibit different immune processes (Strand, 2008). Their exact role in insect immunity is still being investigated.

BRIEF OVERVIEW OF NEURAL/ENDOCRINE SYSTEMS IN INSECTS

Like the insect immune system, the insect central nervous system (CNS) and endocrine system differ from those in vertebrates. Although there is a general "insect" plan, insect nervous systems and endocrine systems differ in their anatomy and function across life stages (e.g., larval or adult) and insect orders (Nijhout, 1994). This variability is greater than is found in these systems across the vertebrates. Therefore, although the description given below fits many insects, a large number of exceptions exist. This brief overview focuses on features that have relevance for eco-immunologists.

The CNS is distributed over several ganglia (Burrows, 1996). The head contains the brain (supraesophageal ganglion), attached to a series of other ganglia via paired connectives. The ganglia of the CNS are heavily interconnected (Burrows, 1996). The brain is thought to be the major center of information processing and influences activity in the other ganglia (Burrows, 1996). The brain contains structures such as the mushroom bodies, an area important for higher-order information processing (Burrows, 1996).

The CNS is protected from ionic concentration fluctuations and other biochemical disturbances by a robust blood–brain barrier (Nation, 2008). It is a selectively permeable barrier consisting of an acellular outer layer and a cellular inner layer (Nation, 2008). However, the barrier does contain an array of transporters (e.g., Mayer et al., 2009), and by this and other mechanisms (Nation, 2008), some substances can cross from the hemolymph into the CNS.

The insect CNS contains many of the same biogenic amine and peptidergic neurotransmitters as vertebrates (Burrows, 1996; Nation, 2008). However, it also contains some compounds unique to invertebrates. For example, the biogenic amine octopamine is a major neurotransmitter, neuromodulator, and neurohormone in insects (Roeder, 1999; 2005). Although octopamine is chemically similar to norepinephrine (Fig. 13.1), octopamine plays little, if any, role in the brain of vertebrates (Roeder, 1999). Conversely, insects do not use norepinephrine as a signaling molecule (Roeder, 1999). Octopamine and norepinephrine are thought to have arisen from a single ancestral system because of the similarities in the molecular sequences of their receptor and transporter molecules (Caveney et al., 2006).

Interestingly, octopamine and norepinephrine are both involved in coordinating the stress response, suggesting an ancient conserved function for these tyrosine-derived compounds (Roeder, 1999). Although octopaminergic neurons are not especially numerous within the insect CNS, they are pervasive and form connections with neurons throughout the CNS (Roeder, 1999).

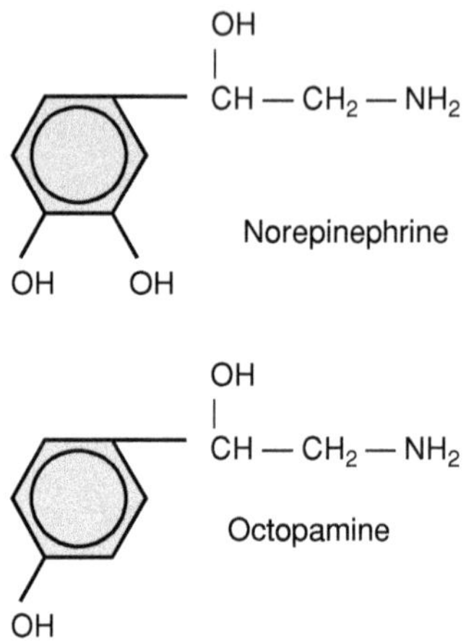

FIGURE 13.1: Chemical structures for norepinephrine and octopamine. Norepinephrine connects the nervous system and immune system in vertebrates and mollusks; octopamine performs a similar function in insects (Adamo, 2008a). Figure adapted from Cooper et al. (2003).

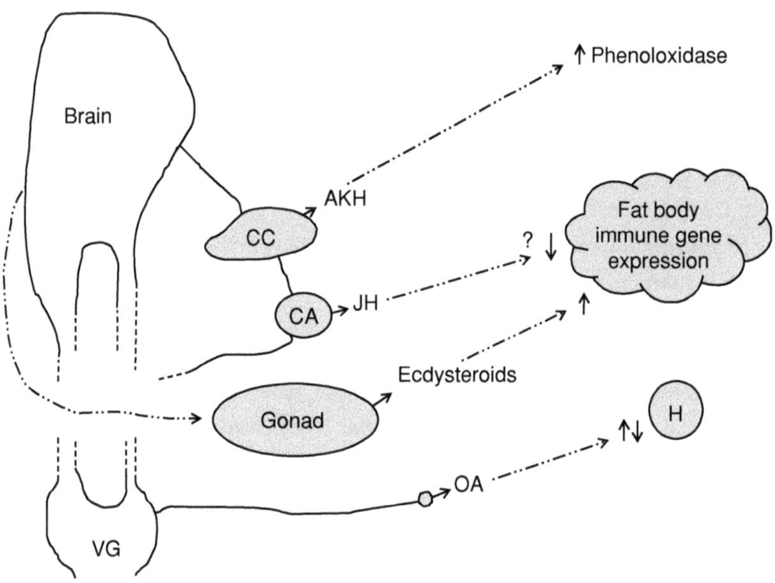

FIGURE 13.2: Neuroendocrine/endocrine connections to the immune system in insects. The alternating dashed and dotted lines represent movement through the hemolymph. Figure adapted from Loher and Zaretsky (1989). Abbreviations: corpora cardiacum (CC), corpora allatum (CA), ventral ganglion (VG), hemocyte (H), adipokinetic hormone (AKH), juvenile hormone (JH), octopamine (OA). References in text.

Some octopaminergic neurons also extend processes into the peripheral nerves and release octopamine into the hemolymph as a neurohormone (Fig. 13.2; Burrows, 1996; Roeder, 1999).

The brain also contains neurons that send their axons into the corpora cardiaca, paired insect neuroendocrine/endocrine glands (Nijhout, 1994). The corpora cardiaca and the brain region innervating it are loosely analogous to the pituitary gland and hypothalamus in vertebrates

(Kodrik, 2008). The corpora cardiaca release hormones such as the peptide adipokinetic hormone, a hormone released during physically intense behaviors such as flight (Fig. 13.2; Nijhout, 1994). Adipokinetic hormone is primarily involved with the mobilization of energy-rich compounds such as lipids (Orchard et al., 1993).

Attached to the corpora cardiaca are endocrine glands called the corpora allata (Fig. 13.2; Nijhout, 1994). These glands also receive innervation from the brain (Nijhout, 1994). The corpora allata release juvenile hormone, a hormone that is critical for normal insect development (Nijhout, 1994). In the adult insect, juvenile hormone also regulates reproduction in some, but not all, insects (Nijhout, 1994).

Ecdysteroids are another class of hormones that are important for both insect development and reproduction (Nijhout, 1994). In adult insects, ecdysteroids are secreted primarily by the ovarian follicular cells in females, and the sheath of the testes in males (Nijhout, 1994). The gonads are stimulated to release ecdysteroids by a brain-derived peptide (Fig. 13.2; e.g., Vafopoulou and Steel, 2005). In some insects (e.g.. mosquitoes), ecdysteroids induce the fat body to produce vitellogenins (egg proteins) in females and stimulate spermatogenesis in males (Nation, 2008). Unlike vertebrates, there is no evidence that testes and ovaries secrete different types of steroids (Lafont et al., 2005; Vafopoulou and Steel, 2005). One sex difference that does appear to be robust is that testes secrete less ecdysteroid than ovaries (Lafont et al., 2005).

IMMUNE/NEURAL/ENDOCRINE CONNECTIONS IN INSECTS

Immune/neural/endocrine connections exist in animals across phyla (Ottaviani and Franceschi, 1996). Intriguingly, both vertebrates and invertebrates use the same, or very similar, molecules to make some of these connections, suggesting that these connections may predate the split between vertebrates and invertebrates (Ottaviani and Franceschi, 1996). In insects there is evidence for both neural–immune connections and immune–neural connections (Adamo, 2006, 2008b).

One of the earliest indications that the immune system and nervous system were connected in insects was the observation that an injection of lipopolysaccharide (a component of gram-negative bacterial cell wall) is able to induce behavioral fever in cockroaches (Bronstein and Conner, 1984). In behavioral fever, insects change their temperature preference and migrate to warmer areas. In a variety of insects, migrating to a warmer temperature increases resistance to different types of pathogens, meaning that this putative immune–neural connection could provide the animal with an adaptive advantage (Moore, 2002). How behavioral fever actually decreases mortality is not entirely clear (Thomas and Blanford, 2003). It can increase some immunological functions and can also directly lower pathogen vitality (e.g., Ouedraogo et al, 2003).

The details of how immune systems induce behavioral change remain under investigation in insects. Nevertheless, the literature suggests some possible mechanisms. For example, eicosanoids, compounds synthesized by immune cells (e.g., hemocytes and fat body) (Stanley, 2000), appear to induce behavioral fever (Bundey et al., 2003). Eicosanoids may be able to activate neural circuits (e.g., for oviposition, Stanley, 2000), and therefore could link immune function with the nervous system.

Growth-blocking peptide (GBP) is a cytokine found in both immune cells (e.g., fat body, Hayakawa et al., 1998) and glial cells of the CNS (Hayakawa et al., 2000). In the periphery, GBP is immunomodulatory (Strand et al., 2000). GBP binds with a GBP receptor on plasmatocytes (Watanabe et al., 2006), leading to a change in hemocyte behavior (Strand et al., 2000). GBP also induces increased production of the neuromodulator dopamine within the CNS (Noguchi et al., 2003) and regulates dopamine release from neurons (Hayakawa et al., 2000), suggesting that neurons have receptors for GBP. Moreover, injections of GBP into larval lepidopterans can cause transient paralysis and other behavioral effects, suggesting that it does alter neural function (Hayakawa, 1995, 2006). Therefore, GBP appears to be a mediator of an immune–neural connection either directly or indirectly by its effect on dopamine concentrations within the CNS. Another insect cytokine, Spätzle, is released by hemocytes and activates antimicrobial peptide production in the fat body (Shia et al., 2009). Whether Spätzle can directly or indirectly influence neural activity remains to be tested. Spätzle is implicated in neural development (i.e., it may have neurotrophin-like qualities; Zhu et al. 2008), suggesting that it could influence both systems.

Octopamine appears to connect the nervous system to the immune system (Fig. 13.2; Adamo, 2008a). Octopamine is released by the nervous system and is also an immune modulator (see Roeder, 2005; Adamo, 2006). Octopamine can, for example, enhance phagocytosis (Baines et al., 1992). Hemocytes have receptors for octopamine (Orr et al. 1985; Easy and Adamo, in review), which suggests that changing octopamine titres will alter hemocyte function. Octopamine is also contained in hemocytes and increases in concentration in the hemolymph during an immune challenge (Adamo, 2010). Octopamine from the hemocytes and/or nervous system may also be involved in linking immune activity and neural activity, resulting in illness-induced anorexia (Adamo, 2005; Adamo et al., 2007).

There may also be direct neural–immune connections. The fat body, a major organ of the immune system (Gillespie et al., 1997), is innervated in insects (Hazarika and Gupta, 1987). Fat body innervation supports the possibility of direct neural influence on immune function, although such an effect remains to be demonstrated.

Although hormonal regulation of immune function has been established for decades in vertebrates (Sapolsky, 1992), its existence in insects has only recently been recognized. As late as 1993, there was little evidence that hormones had a major impact on insect immunity (Vinson, 1993). However, recent research demonstrates that hormones do regulate immune function (Fig. 13.2). Ecdysteroids can alter the gene expression of proteins known to be important for insect immunity in a range of insect groups (e.g., Dipterans: Ahmed et al., 1999; Lepidotoperans: Roxstrom-Lindquist et al., 2005). Juvenile hormone also influences immune function (Flatt et al., 2008) and can modify the effects of cytokines. For example, the cytokine plasmatocyte-spreading peptide typically induces plasmatocytes (a type of immune blood cell) to bind and spread on foreign surfaces in a larval lepidopteran (caterpillar) (Clark et al., 1997). Juvenile hormone reduces the effect of this cytokine on hemocytes, while ecdysteroids potentiate the response (Clark et al., 2005). Juvenile hormone may activate a $Na+/K+$ pump via a protein kinase C pathway leading to a reduction in cell volume, thus preventing plasmatocyte spreading (Kim et al., 2008). Therefore, the sensitivity of immune cells to cytokines can vary across development, depending on the hormonal milieu (Fig. 13.2). However, the functional and adaptive significance of endocrine effects on insect immunity remain unclear.

IMMUNE/NEURAL/ENDOCRINE CONNECTIONS AND ECOMMUNOLOGY

JUVENILE HORMONE, ECDYSTEROIDS, REPRODUCTION, AND IMMUNE FUNCTION

Immune activity and reproduction are often negatively correlated in insects (Siva-Jothy et al., 2005; Lawniczak et al., 2007). In some instances, the evidence suggests that immune function declines after reproductive behavior begins because of a depletion of resources needed for an immune response once reproduction is under way (Siva-Jothy et al., 2005). This depletion may not necessarily be energetic; it could be caused by competition for compounds needed by both functions (Siva-Jothy et al., 1998). There is good evidence that insects are resource-limited and that this limitation affects both immune and reproductive function. For example, Jacot et al. (2004) found that supplemental feeding of crickets (*Gryllus campestris*) in the field allowed them to call longer. Calling longer increases the ability of males to attract females, increasing male reproductive success (Jacot et al., 2004). These results suggest that crickets do not typically have access to sufficient food to maximize their potential reproductive output. Therefore, energetic investment in immune function may come at the cost of reproduction, and vice versa. As predicted, crickets forced to respond to an immune challenge showed a decreased investment in reproduction (i.e., they decreased the time spent calling; Jacot et al., 2004). These results support the hypothesis that reducing energy investment in immune function (e.g., by reducing the number of hemocytes) could enhance reproductive success in some species.

The mechanisms underlying hypothesized shifts in resources between immune function and reproduction remain clear. However, hormones are thought to be key mediators of life-history trade-offs (Harshman and Zera, 2007). In vertebrates, some hormones (e.g., stress hormones) are known to inhibit immune function (Sapolsky, 1992). Potentially, immunosuppressive hormones could mediate a physiological trade-off between immune defense and reproduction. There is evidence that vertebrates do, in fact, hormonally suppress immune function during reproduction (e.g., Mills et al., 2009). The interactions between hormones such as leptin and the immune and reproductive systems (e.g., French et al., 2009) support the concept that vertebrates decrease energetic investment in immunity in order to fuel reproduction (see Chapter 8).

Is there evidence that insects do the same? Selection should favor mechanisms that allow trade-offs between immune defense and reproduction even more strongly in insects than in vertebrates. Most insects lack parental care and have a single, short, breeding season. Therefore, physiological trade-offs that reduce life span would have less impact on reproductive success in many insects than it would in a vertebrate with obligate parental care and multiple breeding seasons. One potential mediator of immune/reproductive trade-offs in insects is juvenile hormone because it modulates both functions (Flatt et al., 2005). Juvenile hormone titers decline to undetectable levels during the last larval instar, providing one of the triggers for metamorphosis (Nijhout, 1994). However, juvenile hormone titers increase again in the adult insect (Nijhout, 1994). Juvenile hormone is important for egg production in females because it stimulates the synthesis of vitellogenins (Nijhout, 1994). Vitellogenins are proteins synthesized by the fat body, transported to the ovary, and used in the production of yolk proteins (Chapman, 1998). Removal of the corpora allata, the glands that secrete juvenile hormone, results in decreased or completely inhibited egg production in many females (e.g., crickets; Loher and Zaretsky, 1989). The production of vitellogenin is one of the main energetic costs of egg production (Carlson and Harshman, 1999). Antimicrobial peptides and proteins are also made in the fat body (Chapman, 1998). Therefore, juvenile hormone could potentially switch fat body resources from making immune factors to making proteins for the egg. There is some evidence that juvenile hormone can mediate this trade-off (see references in Flatt et al., 2008). In the next section, we will examine this question in two divergent insects, crickets (Orthoptera) and flies (*Drosophila melanogaster* [Diptera]).

ORTHOPTERANS (CRICKETS)

In crickets, mating induces an increase in juvenile hormone concentration in females, leading to a large increase in egg production (e.g., *Teleogryllus commodus*) (Loher and Zaretsky, 1989). However, mating also results in increased resistance to bacterial infection in females in a related cricket, *Gryllus texensis* (Shoemaker et al., 2006). Moreover, females that mate multiply have both increased egg production and increased life span compared with females that mate only once (*Gryllus lineaticeps*, Wagner et al., 2001). These results suggest that immune function, life span, egg production and juvenile hormone are all positively correlated in some crickets. Therefore, the data on female crickets do not support the hypothesis that juvenile hormone shifts resources from immune function to reproduction. However, these data do not directly address the juvenile hormone hypothesis because the direct effect of juvenile hormone on immune function was not assessed.

DIPTERANS (DROSOPHILA)

In *D. melanogaster*, juvenile hormone titers fall to undetectable levels in the pupae (Bownes and Rembold, 1987). Juvenile hormone titers then increase dramatically within 4 hours of eclosion in the adult (Bownes and Rembold, 1987). However, unlike the situation in female crickets, juvenile hormone titers decline steeply over the next two days and then remain low (Bownes and Rembold, 1987). Once adult, females become sexually mature in 8 to 10 hours, and 80% of females have mated by the fourth day (Markow and O'Grady, 2006). Female *D. melanogaster* live 20 or 30 days as adults and lay eggs at a sustained high rate for at least 10 days once adult (e.g., Tu and Tatar, 2003). mRNA expression of vitellogenin, an indication of the level of egg production, remains high in the fat body for at least the first 4 days of adulthood (Carlson and Harshman, 1999). Egg production within the ovary remains maximal for at least 12 days (Good and Tatar, 2001). In other words, egg production remains maximal even as juvenile hormone levels decline.

Juvenile hormone suppresses the expression of most immune-related genes in female *Drosophila* even in the absence of infection (Flatt et al., 2008). Juvenile hormone also reduces immune-related gene expression after infection (Flatt et al., 2008). Ecdysteroids can potentiate immune function in *D. melanogaster*, but this effect is lost in the presence of juvenile hormone (Flatt et al., 2008). These results fit the hypothesis that juvenile hormone may suppress immune function in order to shuttle resources to egg production. However, if the immunosuppressive effect of juvenile hormone is dose-dependent, then the hormone titers found in the adult (Bownes and Rembold, 1987) suggest that female immune function is suppressed during very early adult life, and much less so as the female fly ages. In a short-lived animal, immune investment should wane as the likelihood of future reproduction declines. If juvenile hormone is mediating an immune/reproductive trade-off, then juvenile hormone titers should increase with age in *D. melanogaster* females, as they do in cricket females (Loher and Zaretsky, 1989).

Moreover, according to the juvenile hormone hypothesis above, reproductive diapause should result in decreased juvenile hormone titers, leading to suppressed reproduction and a shift of resources back towards immune function. This shift would help ensure that females survive poor conditions so that they are alive to reproduce when conditions improve. However, there appears to be a controversy over whether the juvenile hormone titer increases or decreases during reproductive diapause. Gruntenko and Rauschenbach (2008) argue that juvenile hormone levels increase in mature females during stressful conditions leading to reproductive diapause. This conclusion is based on data from two different *Drosophila* species (*D. virilis* and *D. melangaster*; Gruntenko and Rauschenbach, 2008; Gruntenko et al., 2000) in which juvenile hormone degradation (via juvenile hormone hydrolysis) was found to decrease in heat-stressed females. The authors concluded that if degradation was reduced, then the juvenile hormone titer must be increased. However, other researchers (e.g., Saunders et al., 1990) found that juvenile hormone production by the corpora allata is decreased *in vitro* during reproductive diapause induced by changing photoperiod. Therefore, some researchers (e.g., Flatt et al., 2005) conclude that juvenile hormone titers must be low during diapause, given that hormonal production is reduced. Unfortunately, no one has measured actual juvenile hormone titers in sexually mature flies during reproductive diapause, so it is impossible to tell whether there is any net change given that both synthesis and degradation appear to be depressed.

In *Drosophila, in vitro* juvenile hormone synthesis does not appear to accurately estimate juvenile hormone titer. For example, Bownes and Rembold (1987) found that juvenile hormone titers fall within 24 hours of females becoming adult, even though Tu and Tatar (2003) found that juvenile hormone *in vitro* synthesis remained high at this time. This discrepancy suggests that juvenile hormone synthesis *in vitro* is a poor proxy for the actual hormonal titer. It is known in other insect species that *in vitro* juvenile hormone synthesis assays do not always correlate with the true hormone concentration (McNeil et al., 2004).

Although the levels of juvenile hormone decline in sexually mature females (Bownes and Rembold, 1987), they may still be high enough to suppress immune function. Mature adult females have higher juvenile hormone titers than do pupae and some larvae (depending on the developmental stage) (Bownes and Rembold, 1987). It is also possible that the spike in juvenile hormone levels observed during early adulthood has long-lasting effects on immune function. Many effects of juvenile hormone outlast the increase in hormonal concentration (Nijhout, 1994). However, such effects need to be demonstrated.

Therefore, whether juvenile hormone mediates a physiological trade-off between immune function and reproduction in *Drosophila* remains unclear. Nevertheless, it is clear that juvenile hormone has effects on life span, reproduction, and immune function in insects (e.g., Flatt et al., 2005, 2008). Unfortunately the difficulty in measuring juvenile hormone titers and the lack of well-characterized receptors (Nation, 2008) make studying eco-immunological phenomena involving juvenile hormone difficult. Moreover, the interactions among juvenile hormone, reproduction, and immune function will certainly be species-dependent because the effects of juvenile hormone on reproduction is species-dependent. In some species, juvenile hormone does not play a role in female reproduction (e.g., parasitic wasp [*Pteromalus puparum*]; Dong et al., 2009), while in others its role is indirect (e.g., burying beetles [*Nicrophorus orbicollis*]; Panaitof and Scott, 2006). These examples emphasize the importance of knowing the underlying physiology in order to test putative trade-offs (Harshman and Zera, 2007).

THE STRESS RESPONSE AND IMMUNE FUNCTION

In vertebrates, the effects of various types of stress on immune function form a large, and sometimes confused, literature. However, there is consensus that various types of stressors do alter immune function, although these effects vary depending on the species (Martin, 2009), the type of the stressor (Bowers et al., 2008), and its duration (Dhabhar, 2009). There is also consensus as to the major routes by which stressful stimuli are able to influence immune function in vertebrates. Stressful stimuli activate the hypothalamus-pituitary-adrenal gland (HPA) axis and the sympathetic nervous system (Sapolsky, 1992). This activation leads to the release of catecholamines such as epinephrine and norepinephrine as well as glucocorticoids (Sapolsky, 1992). Various immune cells in vertebrates have receptors for these compounds (Webster et al., 2002), and they induce a range of effects on immune function depending on variables such as the concentration of the compound (e.g., Nance and Sanders, 2007). The functional significance of these hormone-induced changes remains somewhat controversial (e.g., whether short-term flight-or-fight stress is immunosuppressive; see Glaser and Kiecolt-Glaser, 2005, Box 2; Dhabhar, 2009 vs. Walburn et al., 2009) and may depend on a variety of factors.

Understanding why the stress and immune responses are intimately connected requires an understanding of the physiological mechanisms linking the two and the functional consequences of these connections. Insects, with their simpler immune, nervous, and endocrine systems, could provide model organisms in which to study this question. Stress, defined as any external or internal event that threatens the animal's homeostasis, affects immune function in insects, just as it does in vertebrates (Adamo, 2008a). In insects, various forms of stress, such as pesticide poisoning, starvation, and extreme heat, depress immune function and disease resistance (Brey, 1994). However, as these types of stressors cause a number of pathological changes in multiple physiological systems, it is not surprising that they have negative effects on immunity. However, non-damaging stressful stimuli (e.g., flying) also influence immune function in insects (Adamo and Parsons, 2006). Moreover, there is evidence that some of the molecules linking the stress response and the immune system are similar in both insects and vertebrates, suggesting the possibility of an ancient conserved function (Adamo, 2008a). Therefore, examining this question in insects could have broad relevance.

As described earlier, insects, like vertebrates, have a physiological stress response. During the production of energetically demanding behaviors such as fighting or flying, many insects (e.g., orthopterans such as crickets, grasshoppers, and locusts) release the neurohormone octopamine (Orchard et al., 1993; Adamo et al., 1995). After a delay of about 20 minutes, a second hormone, adipokinetic hormone, is released from the corpora cardiaca (Orchard et al., 1993). Both adipokinetic hormone and octopamine help prime the animal for intense physical activity (Roeder, 1999, 2005).

In some insects (e.g., *Drosophila*) stressors such as heat stress also induce increases in octopamine, and this increase in octopamine leads to a change in two other hormones: juvenile hormone and ecdysteroids (Gruntenko and Rauschenbach, 2008). These latter hormonal changes inhibit reproduction in adult female flies, preventing them from laying eggs under adverse conditions (Gruntenko and Rauschenbach, 2008).

The insect stress response has not been as extensively studied as the stress response in vertebrates, so it is unclear whether this response is found in all insects (it has been measured in only a handful of species). Moreover, the range of stressors that can induce the stress response in different species is not known. In this section we will focus on the interactions between the stress response and the immune response in the context of short-term flight-or-fight behaviors (Fig. 13.3), also called "acute stress" in vertebrates.

The production of flight-or-fight behaviors (e.g., Fig. 13.3), and hence the activation of the stress response, results in a transient decline in disease resistance (Adamo and Parsons, 2006). Injections of octopamine (Adamo and Parsons, 2006) or adipokinetic hormone (Goldsworthy et. al., also 2005) reduce the ability of an individual to survive infection, suggesting that hormones may be involved in this decline. Insect hemocytes contain receptors for the stress neurohormone octopamine (Orr et al., 1985; Easy and Adamo, in review), creating a plausible mechanism by which stressful stimuli could influence immune function. Octopamine has a range of effects on hemocytes both *in vivo* and *in vitro* (see Adamo, 2008a; Kim et al., 2009; Adamo, 2010). Octopamine tends to upregulate individual immune responses such as phagocytosis (Baines et al., 1992), although not all of its effects enhance immune function (Adamo, 2010). Adipokinetic hormone has also been shown to increase immune function by enhancing phenoloxidase activity after an immune challenge (Goldsworthy et al., 2002). Given the mix of positive and negative effects, the functional significance of stress-induced changes in immune function remains unclear (e.g., Roeder, 2005).

In insects, the relationship between the stress response and immune response may be explained, at least in part, by a conflict over resources between immune function and other systems (i.e., a resource crunch) (Fig. 13.4). In crickets, both immune surveillance and lipid transport rely on the same protein, apolipophorin III (Weers and Ryan, 2006; Adamo et al., 2008). Unfortunately this protein appears to be unable to perform both functions simultaneously (Leon et al., 2006).

FIGURE 13.3: Fighting crickets (*Gryllus texensis*). Crickets are displaying the aggressive display "mandible flare."

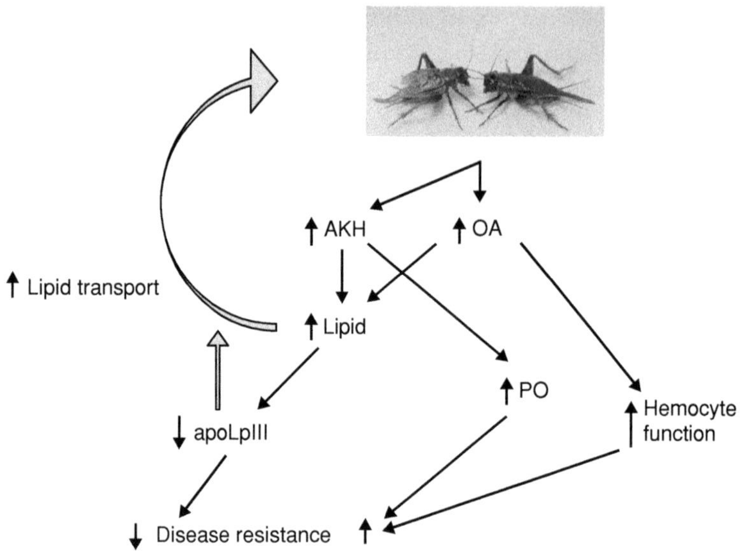

FIGURE 13.4: Hypothetical scheme demonstrating how octopamine (OA) and adipokinetic hormone (AKH) could maintain immune function during flight-or-fight behavior in crickets. High-intensity behaviors lead to increased OA and AKH hemolymph titers. These increases lead to an increase in lipid release from the fat body, forcing a concomitant shift in apolipophorin III (apoLpIII) from immune surveillance to lipid transport. The loss of immune surveillance leads to a decline in bacterial resistance. At the same time, OA upregulates hemocyte function and AKH enhances phenoloxidase activity (PO) to compensate for the loss of apoLpIII. See text for references.

When octopamine and adipokinetic hormone induce the mobilization of lipid from the fat body in order to fuel flight-or-fight behaviors (Orchard et al., 1993), the protein apolipophorin III (apoLpIII) changes its confirmation and combines with high-density lipophorin (HDLp) to form low-density lipophorin (LDLp), which has an increased lipid-carrying capacity (reviewed in Weers and Ryan, 2006). However, apoLpIII now loses its ability to act as an immune surveillance molecule, resulting in a decline in immune surveillance (Adamo et al., 2008). The decline in immune surveillance probably explains the increase in disease susceptibility after flying and fighting (Adamo et al., 2008). Therefore, in crickets, intense activity leads to transient immunosuppression because apoLpIII is co-opted into lipid transport and becomes unavailable as an immune surveillance molecule (Fig. 13.4). This hypothesis explains why crickets become immunosuppressed during flight-or-fight behaviors even if they have abundant energy stores (Adamo et al., 2008).

The ability of octopamine and adipokinetic hormone to mobilize lipid (Orchard et al., 1993) explains why both compounds can produce immunosuppression when injected into crickets. But why do both octopamine and adipokinetic hormone also have immunoenhancing effects? I hypothesize that both hormones also help to maintain immune system function as some of the components of the immune system are being siphoned off into lipid transport (Fig. 13.4). In other words, octopamine and adipokinetic hormone liberate lipid stores (needed to fuel flight-or-fight behavior) while simultaneously reconfiguring the immune system to maintain maximal function under the new physiological conditions. I predict that without the effects of octopamine and adipokinetic hormone on immune function, disease resistance would decline even more precipitously during flying or fighting in crickets. This hypothesis explains why octopamine and adipokinetic hormone can have both immunosuppressive and immunoenhancing effects (Fig. 13.4; Adamo et al., 2008; Adamo, 2010). This hypothesis is similar to an idea that is gaining popularity among vertebrate psychoneuroimmunologists. There is a growing consensus in the vertebrate literature that stress hormones are not globally immunoenhancing or immunosuppressing, but that they help to reconfigure the immune system to optimize its performance given the animal's present physiological state (e.g., Dhabhar, 2009). This convergence of ideas presents insect eco-immunologists with the opportunity to make important contributions to this field.

Why do crickets not make enough apoLpIII to support both immune surveillance and increased lipid transport? First, it would be energetically expensive to do so. ApoLpIII is already a very abundant protein in the hemolymph of many adult insects (Weers and Ryan, 2006). To produce more of this protein would decrease the energy available for reproduction and other activities. Furthermore, as apoLpIII concentrations increase, apoLpIII may begin to bind to the animal's own molecules, initiating an inappropriate immune response. Such autoimmunity could be costly. Therefore, shuttling apoLpIII between immune surveillance and lipid transport may be the most adaptive response, even though it results in transient immunosuppression during flying or fighting.

This "resource crunch" hypothesis should not be confused with the "energy crisis" hypothesis that postulates that acute stress is immunosuppressive in order to allow animals to channel more "energy" into flight-or-fight behavior (e.g., Råberg et al., 1998; Segerstrom, 2007). Others have pointed out difficulties with this hypothesis (Sapolsky, 1992; Dhabhar, 2002). For example, if energy is literally being removed from the immune system, a plausible mechanism needs to account for the following: (1) how this "robbery" occurs (apoptosis?), (2) what is taken (ATP? glucose?), (3) how it is transported to other tissues, and (4) how the "energy" can be reallocated within the time frame of a flight-or-fight episode that typically lasts less than an hour. If the argument is that the immune system is downregulated during acute stress to allow more energy to be shunted to the muscles, then this hypothesis needs to demonstrate that the factor that limits the ability of animals to run from a predator, fight an opponent, or run down prey is the lack of energy compounds such as glucose. In humans, the factors limiting intense physical activity are thought to be intrinsic to the muscle, such as the buildup of lactate due to low tissue oxygen levels and a decrease in glycogen levels (McArdle et al., 2006). It is unclear how suppressing immune function would alter either lactate production in the muscle or muscle glycogen levels. Proponents of this hypothesis must demonstrate how suppressing immune function will increase sprint speed and endurance in the short term.

In flying insects such as locusts, the first few minutes of flight are fueled by carbohydrates, but the stress response induces a suite of coordinated metabolic changes leading to a switch of the main fuel source to lipids (Chapman, 1998). During flying, the amount of physiological fuel available (e.g., lipid) can remain stable for extended periods of time (e.g., more than an hour) (Chapman, 1998). These observations suggest that flight speed in locusts is not limited by the availability of metabolic fuel. The other two requirements for maximal power output from insect muscle are oxygen availability and body temperature (Chapman, 1998). Immunosuppression would not increase either of these. Therefore, there is little support for the "energy crisis" hypothesis from the insect literature.

Understanding the ultimate function of interactions between the immune and stress responses also requires an understanding of why the immune response activates the stress response. In both vertebrates (Sternberg, 2006) and invertebrates (Adamo, 2010), the immune response activates components of the stress response, including the release of at least some stress hormones/neurohormones. Maier (2003) argues that the stress response includes the release of stress hormones because during evolution the stress response co-opted this mechanism from the evolutionarily older immune system in order to release the necessary energy compounds (Maier, 2003). However, in insects, the immune response induces the release of octopamine (Dunphy and Downer, 1994; Adamo, 2010) but not adipokinetic hormone, the more important energy-liberating stress hormone (Mullen et al., 2004), only partially supporting Maier's (2003) hypothesis.

Both the immune and stress responses occur when an animal is faced with a crisis that threatens its survival (e.g., by pathogens or predators). As Maier (2003) notes, both responses include an increased energetic demand and the increased potential for cellular damage. As in vertebrates, in insects various stressors activate intracellular defense mechanisms such as expression of heat shock proteins (Feder and Hofmann, 1999). The immune response also activates intracellular protective mechanisms (e.g., heat shock protein production; Altincicek et al., 2008). Octopamine activates intracellular defense mechanisms (Armstrong and Robertson, 2006) and therefore could be useful during both responses. Perhaps it is not surprising that both immune and stress responses evoke similar hormonal mechanisms to perform the same functions—liberate energy and activate cellular defense mechanisms. The similar requirements of both the immune and stress responses may partially explain their interconnectedness.

Immunosuppression after acute stress is transient, and can result in a rebound effect in insects. Twenty-four hours after experiencing a non-damaging stressor, caterpillars (*Galleria mellonella*) show enhanced resistance to pathogen exposure (Mowlds et al., 2008). This result suggests that immune function may rebound after a temporary reduction. A similar rebound effect also occurs in mollusks (Lacoste et al., 2002). Therefore, knowing the time course of the effects of stress on immune function is also important when conducting these studies. Choosing the wrong time point could lead to incorrect conclusions about the adaptive significance of stress-induced effects on immune function.

IMMUNOLOGICAL COMPLEXITIES AND THEIR IMPLICATIONS FOR ECOIMMUNOLOGY

Unfortunately for ecological immunologists, immunology is a complex and rapidly evolving field. Instead of being built on the "bedrock" of immunology, ecological immunology rests on the "shifting sands" of immunology. For example, as discussed above, there is an evolving consensus that stress hormones can be better understood as refocusing immune function towards a different maxima, as opposed to having globally suppressing or enhancing functions (e.g., Elenkov and Chrousos, 1999, 2006; Sternberg, 2006; Dhabhar, 2009). This shift in understanding will have repercussions for ecological immunologists, as described below. As a further example, the vertebrate innate immune response was previously viewed as a discrete set of signaling pathways activated by a pathogen binding to a receptor. Now, immunologists envision it as a complex network of interconnected pathways whose activities are dependent on many factors (Gardy et al., 2009). These pathways often exhibit cross-talk, complex feedback and feed-forward loops, and diverse mechanisms of regulation (Gardy et al., 2009). Insect immune systems show signs of similar complexity (e.g., Broderick et al., 2009). This complexity creates problems for ecological immunologists.

Using the previous view of immune systems, it seemed plausible that (a) higher values for assays of a particular immune function imply an increased immune ability, (b) immune assay values will correlate within an individual—that is, individuals that score high on one assay will score higher than average on other assays, and (c) the ability to resist most pathogens will be correlated with the score on an immune assay. These assumptions are necessary to allow eco-immunologists to use immune assays to study issues that are key to the field, such as determining whether females prefer the most "immunocompetent" male or whether reproduction suppresses immune function. Although these assumptions were known to have their limitations (Adamo, 2004a), recent work on insect immune systems demonstrates how these assumptions can lead to the misinterpretation of data. A better understanding of the immune system and its regulation will help ecological immunologists know when these assumption are likely to hold, and when they may not be valid. Below are examples demonstrating when these assumptions may be invalid.

First, some immune molecules participate in other physiological functions (e.g., apolipophorin III). These multifunctional molecules can lead to physiological trade-offs between the immune system and other physiological systems. These trade-offs can eliminate correlations among immune assays as well as correlations between some immune assays and disease resistance. For example, flight-or-fight behavior results in an increased hemocyte count in the cricket *G. texensis* (Adamo, 2010) but a decreased ability to survive a bacterial challenge (Adamo and Parsons, 2006). An increased hemocyte count is thought to indicate an increased immune capacity in crickets (Ryder and Siva-Jothy, 2000). The increased hemocyte count may reflect an attempt by the animal to enhance its circulating defense mechanisms in order to prepare for possible pathogen exposure (e.g. due to an increased risk of injury during flight-or-fight behaviors). However, the loss of apolipophorin III to lipid transport appears to increase disease susceptibility (Adamo et al., 2008) more than the increase in the hemocyte count increases disease resistance. Therefore, under conditions of acute stress, the hemocyte count may no longer correlate with relative disease resistance when compared with controls.

Other changes in behavioral state also affect immune function. For example, after mating, *Drosophila* females upregulate immune gene expression, suggesting an enhancement of immune function and, therefore, increased disease resistance (Fedorka et al., 2007). However, this upregulation may be compensating for other immunosuppressive changes, because the net result of mating is a decline in the ability of *Drosophila* females to survive at least some pathogen challenges (Fedorka et al., 2007; Short and Lazzaro, 2010). Therefore, after mating, increased immune gene expression does not necessarily mean increased immune ability in female *Drosophila*.

Second, under some conditions, hormones/neurohormones will alter some, but not all, immune functions (Adamo, 2010), potentially eliminating correlations among immune assays and between immune assays and disease resistance (Adamo, 2009). Therefore, immune assays taken from animals under different hormonal conditions (e.g., breeding vs. non-breeding) must be compared cautiously. For example, octopamine upregulates some immune functions, but it also leads to an overall decline in the ability of crickets to survive a bacterial challenge (Adamo, 2008a). Therefore, octopamine reduces the correlation between these immune assays and the ability to survive an infection when compared with controls.

Third, because immune systems are composed of multiple physiological pathways, there may be multiple ways to configure the immune system. For example, animals require mechanisms that destroy pathogens (called "resistance") and mechanisms that allow them to survive the effects of the pathogen ("tolerance") (Schneider and Ayres, 2008). A "tolerant" organism may have relatively high levels of enzymes designed to detoxify bacterial toxins, allowing them to withstand higher bacterial loads before dying. . An individual emphasizing "resistance" (the ability to destroy pathogens) may show enhanced production of antimicrobial compounds. Both individuals may still have approximately the same ability to survive most pathogens (Schneider and Ayres, 2008). Ayres and Schneider (2008) argue that typical measures of immune function assess only half of the defensive capacity of the host because mechanisms of "tolerance" are neglected. In fact, many physiological systems contribute to the ability of animals to survive the onslaught of both pathogens and the by-products of their own immune response. This ability is sometimes considered a component of an animal's general condition. For example, if two animals have the same ability to make antimicrobial peptides (an element of "resistance") but one has compromised malpighian tubule function, it will have reduced ability to remove toxins and will be less likely to survive the infection. Therefore, comparison of the results of specific immune assays across individuals may not reflect the relative ability of an animal to survive an infection. Moreover, immune defense may shift among different pathways to provide optimal defense in different situations (Schmid-Hempel and Ebert, 2003). Such shifts will make it difficult to provide a ranking of "immune ability" among individuals based on specific immune components. Animals that score higher on a particular immune assay may not have greater immune capacity, but may have differently configured immune pathways.

Fourth, the ability of different individuals to survive an infection may depend on the pathogen. The relative importance of different immune pathways for survival may differ depending on the pathogen's arsenal of abilities that allow it to overcome the immune system and extract nutrition from the host. Pathogens are not inert objects, and each has its own unique pattern of evading, suppressing, and manipulating host physiology (Boucias and Pendland, 1988). For example, some pathogens are able to enter a host without eliciting a measurable host response (Lemaitre et al., 1997). In this case, the ability of a host to survive an attack from this pathogen will require different mechanisms than those needed to survive an attack against a pathogen that activates a robust immune response. In fact, antimicrobial proteins secreted by *Drosophila* are not same for every pathogen, but depend on pathogen identity (Lemaitre et al., 1997). Therefore, ranking individuals in terms of a single immune ability may provide information about the ability to survive some pathogens, but not others (also see Ayres and Schneider, 2008). An ecologically valid ranking would need to emphasize those immune mechanisms that are most important in the battle against the major pathogens attacking the species (Adamo and Spiteri, 2005). Unfortunately, for most insects, the identity of the major pathogens attacking them in the field is unknown (Adamo and Spiteri, 2005).

Fifth, it may be impossible to maximize all physiological pathways involved with immune function simultaneously. Animals must then "choose" a particular immune system configuration. The two pathways may have different costs and benefits, but it may not be obvious which one is the lesser "immune investment" or the least "immunocompetent"—that is, results in the least protection from infection and death. For example, the phenoloxidase system is a major immunological pathway and is under heavy inhibition (Kanost and Gorman, 2008). Over-activation of this system can kill the host (Kanost and Gorman, 2008), while under-activation can also lead to host death by unopposed pathogens (Beck and Strand, 2007). If the relative amount of inhibitors versus activators changes, it could increase or decrease the ease with which the phenoloxidase cascade is activated. Bias towards activation may increase the insect's ability to neutralize pathogens, but it might also lead to greater immunopathology (e.g., a greater risk of its unnecessary deployment). The "optimal" balance between activators and inhibitors may vary depending a number of factors, including the identity of the pathogen, and is likely to be context-dependent.

A complete description of the internal defense system of an insect would include intracellular defenses (e.g., heat shock proteins), cells and molecules involved in pathogen attack, and all allied physiological systems. It would also require making separate assessments of these components in different compartments of the insect (e.g., local immunity in the gut and reproductive tract as well as systemic immunity). Local immunity and systemic immunity are connected in insects (Broderick et al., 2009), and this can sometimes lead to trade-offs among compartments (Mallon et al., 2003). Moreover, the immune response is not simply a function of the host; the pathogen itself plays a role and is itself modulated by the host's response (Gardy et al., 2009). Given the complexity of these interactions, the difficulty for eco-immunologists is obvious. How should immune function be assessed? How can we determine the effects of other life-history traits on the immune system? One way forward is to take a systems approach (lsley et al., 2009). Such an approach allows the combination of different layers of data (e.g., physiological and molecular; Geschwind and Konopka, 2009) to predict an outcome using mathematical models. In other words, the ability to survive a pathogen attack is the emergent property of a system that consists of all contributing pathways. It is probably impossible to explain or predict immune system function solely by examining its individual components (Gardy et al., 2009). Internal defense systems are not a product of discrete linear pathways; rather, they are composed of a complex set of integrated responses arising from a dynamic network of thousands of molecules subject to multiple influences (Gardy et al., 2009). Therefore, it may be impossible to assemble these pathways in such a way as to make clear predictions using present inspection methods.

Using a systems approach in vertebrates, Pulendran et al. (2009) found that different vaccines induced different gene expression patterns, and predicted that these "meta-signatures" could allow accurate testing of the "immune capacity" of an individual (i.e., its ability to resist a particular pathogen). Unfortunately, at present there is no systems-level model of an insect immune system of the type that exists for vertebrates (e.g., Gardy et al., 2009). Once one is developed, it may help eco-immunologists determine the most appropriate immune components to assess, given the scientific question. Such an approach may also help explain why assessing multiple immune components during a study often produces discordant results (i.e., some immune components are upregulated, while others are unaffected or downregulated). In some cases discordant results may be evidence of trade-offs within the immune system, but they may also be part of a consistent shift (either enhanced or suppressed immune function) that is masked by the system's complexity. A parallel example comes from neural circuits. Neurons interact in complex ways, and these interactions produce coordinated behavior in the same way that different physiological pathways work together to allow an animal to survive an infection. Behavioral output (i.e., the strength and direction of a motor movement), however, cannot always be predicted from the activity of individual neurons (Lockery et al., 1989), and I suggest that the same is true when attempting to determine survival after infection based on assays of individual immune components. In the nervous system, some neural connections are counter-intuitive compared with what would be expected to generate the movement under study. For example, in the leech, activation of pressure-sensitive mechanosensory cells (P-cells) causes the leech to bend (Lockery et al., 1989). Electrophysiological studies reveal that some of the connections between interneurons and motor neurons were "inappropriate" for the direction of the bend that was produced by P-cell stimulation (Lockery et al., 1989). However, mathematical models of the system (i.e., back-propagation) predicted the existence of such non-intuitive units. The non-intuitive units, when summed with the other interneurons, produced the observed behavior (Lockery et al., 1989). Similar unsupervised learning algorithms have been used to study vertebrate immunological networks (see Gardy et al., 2009). Individual immune component assay results may not always make intuitive sense, but they may still contribute to the emergent property of enhanced survival against pathogens given the physiological context.

HOW TO MEASURE AND INTERPRET IMMUNE FUNCTION

Does our incomplete understanding of internal defense systems mean that ecoimmunology is impossible? No, although it does require a cautious interpretation of data (see below). Fortunately, despite the many physiological pathways that make up an insect's internal defense system, not all of these pathways are of equal importance in determining disease resistance (Adamo, 2004b). In mammals, assaying only a few strategically chosen immune components can accurately predict which individuals will survive a specific pathogen attack (Luster et al., 1993). Unfortunately, we do not know the most important components of the insect internal defense system for different pathogens. Without this knowledge, the leech example exemplifies the danger of using one, or even a few, immune assays to predict the "emergent property" (i.e., disease resistance) of the system. The immune components chosen may not be linearly correlated with the emergent property of disease resistance. Nevertheless, eco-immunologists have devised partial solutions for this problem.

The consensus in ecoimmunology has been to move towards using multiple functional assays that attempt to estimate the relative strength of several immune components combined (e.g., Martin et al., 2008). For example, antimicrobial diffusion assays (Hultmark, 1998) combine a large number of separate immune-related events (e.g., recognition of foreign material, constitutive levels of several antimicrobial substances; Gillespie et al., 1997) into one assay. Assaying increasingly smaller units of the internal defense system will invariably increase the complexity, but may not add any extra explanatory power (e.g., Luster et al., 1993). Using functional assays that incorporate a large number of individual immune components is likely to give a better estimate of the emergent property of the system (disease resistance) than individual immune components (also see Zuk and Stoehr, 2002). Some general schemes have been proposed (e.g., Schmid-Hempel and Ebert, 2003, and Martin et al., 2008) to help ecological immunologists chose assays that cover a broad range of defense mechanisms. The difficulty in knowing which tests to perform reflects the fact that in insects it is unclear which immune functions/"tolerance" mechanisms/aspects of condition are most important for survival, and how these may differ for different pathogens. Therefore, insect ecological immunologists are forced to accept that their estimates of immune function and/or ability to survive an infection may vary in accuracy depending on the pathogen. Therefore the results may not translate well to actual "immunocompetence" in the field, depending on the identity of the major pathogens in the animal's environment.

Can combining immune assays with host resistance tests help increase our confidence in conclusions based on immune assays?

HOST RESISTANCE TESTS REVISITED

Host resistance tests are used as the gold standard for assessing changes in immune function in immunotoxicology (Descotes, 1999). Typically these tests consist of injecting a known dose of bacteria or other pathogen into an animal. The number of animals that die in control versus experimental groups is recorded (Robertson et al., 2007). These tests assess the animal's ability to survive a real infection from at least one pathogen. Therefore it reflects the sum of an even larger range of immune components than immune assays by including "tolerance" mechanisms and other assorted physiological parameters that contribute to survival after infection. However, Martin et al. (2008) point out that most host resistance tests inject a pathogen, as opposed to allowing a more natural route. Therefore, the injection bypasses local immunity in the cuticle and other defenses (e.g., behavioral) the animal may have to avoid infection. A host resistance test that would encompass all aspects of defense against pathogens could be difficult to set up in the lab. Standard host resistance tests might miss the ability of some animals to compensate for a reduced internal defense system with an enhanced external defense system. As discussed above, such an omission may be more important for some pathogens than others. Researchers then have to weigh the relative importance of incomplete assessment of an animal's defense against pathogens versus the difficulty of making a complete assessment reliably and accurately in the lab. Unfortunately the researcher must make this decision without knowing how important the missed aspects of defense are for survival against the typical pathogens attacking the organism Another problem for host resistance tests is that animals can be used only once. Therefore, it is not possible to test survivability to several pathogens in the same individual, in the same way it is possible to assess a variety of immune functions in a single individual. Nevertheless, changes in host resistance tests suggest, even more strongly than an immune assay test, that there has been a shift with functional consequences in the ability to resist disease. This can be very helpful information, especially when combined with different immune assay results. Unfortunately negative results on a host resistance test may merely imply that the pathogen chosen for the test is not affected by declines the compromised immune components.

Eco-immunologists should also consider that measures other than immune function can sometimes estimate disease resistance. In some cases, finding males in the best condition may correlate with disease resistance better than the results of some immune assays (see discussion in Adamo and Spiteri, 2005; 2009). Empirical data also suggest that the male in the best condition is often the most disease resistant (e.g., Milinski, 2006).

CONCLUSIONS

Behavior, evolutionary theory, neuroscience, and immunology all collide in ecological immunology. Each of these topics is itself complex, and the combination of all of them is no less so. This complexity can lead to spurious correlations and other data-interpretation problems. The studies discussed in this chapter demonstrate the importance of making detailed physiological measures (e.g., hormone titers) when examining putative physiological trade-offs. Moreover, knowing the physiological details (e.g., DiAngelo et al., 2009) can lead to new hypotheses about which traits may be most affected by immune activation. Although understanding the immune system is only one aspect of ecoimmunology (Schmid-Hempel, 2005), using incorrect inferences to correlate with complex ecological theories will not move the field forward nor allow it to increase our understanding of why immune systems are configured as they are.

Ecoimmunology is also difficult because immunology is a rapidly changing science. Because most theories in ecoimmunology make assumptions about the underlying physiology of a phenomenon (e.g., physiological trade-offs), shifts in our understanding of how immune systems work are likely to change our understanding of these phenomena. Recent work on insects, highlighted in this chapter, demonstrates that the ability to survive an infection requires a large network of physiological pathways, some of which have not traditionally been considered part of the immune system. The immune system itself is highly dynamic, and the strength of different immune pathways varies depending on the physiological context.

Given these physiological realities, immune function assays that assess the aggregate sum of a number of different immune components are likely to provide a better estimate of immune ability or disease resistance than assays that measure only one component. However, the ability to choose and interpret measures of immune function requires an in-depth understanding of an animal's immune system, and a systems approach may be helpful in gaining this understanding. Such knowledge will be especially needed if immune ability is being compared across different physiological states. Eco-immunologists should include in their papers an explanation for their choice of immune assays as well as a careful discussion justifying the interpretation of the results of these assays.

REFERENCES

Adamo SA (2004a) How should behavioural ecologists interpret measurements of immunity? Animal Behaviour 68:1443–1449.

Adamo SA (2004b) Estimating disease resistance in insects: phenoloxidase and lysozyme-like activity and disease resistance in the cricket *Gryllus texensis*. Journal of Insect Physiology 50:209–216.

Adamo SA (2005) Parasitic suppression of feeding in the tobacco hornworm, *Manduca sexta*: parallels with feeding depression after an immune challenge. Archives of Insect Biochemistry and Physiology 60: 185–197.

Adamo SA (2006) Comparative psychoneuroimmunology: Evidence from the insects. Behavioral and Cognitive Neuroscience Reviews 5:128–140.

Adamo SA (2008a) Norepinephrine and octopamine: linking stress and immune function across phyla. Invertebrate Survival Journal 5:12–19.

Adamo SA (2008b) Bidirectional connections between the immune system and the nervous system in insects. In: Insect Immunology (Beckage NE, ed), pp 129–149. San Diego: Academic Press.

Adamo SA (2009) The impact of physiological state on immune function in insects. In: Insect Infection and Immunity (Rolff J, Reynolds SE, eds), pp 173–186. Oxford: Oxford University Press.

Adamo SA (2010) Why should an immune response activate the stress response? Insights from the insects (the cricket *Gryllus texensis*). Brain, Behavior and Immunity 24:194–200.

Adamo SA, Fidler TL, Forestell CA (2007) Illness-induced anorexia and its possible function in the caterpillar, *Manduca sexta*. Brain, Behavior and Immunity 21:293–300.

Adamo SA, Linn CE, Hoy RR (1995) The role of neurohormonal octopamine during "fight or flight" behaviour in the field cricket *Gryllus bimaculatus*. Journal of Experimental Biology 198:1691–1700.

Adamo SA, Parsons NM (2006) The emergency life-history stage and immunity in the cricket, *Gryllus texensis*. Animal Behaviour 72:235–244.

Adamo SA, Roberts JL, Easy RH, Ross NW (2008) Competition between immune function and lipid transport for the protein apolipophorin III leads to stress-induced immunosuppression in crickets. Journal of Experimental Biology 211:531–538.

Adamo SA, Spiteri RJ (2005) Female choice for male immunocompetence: when is it worth it? Behavioral Ecology 16:871–879.

Adamo SA, Spiteri RJ (2009) He's healthy, but will he survive the plague? Possible constraints on mate choice for disease resistance. Animal Behaviour 77:67–78.

Ahmed A, Martin D, Manetti AGO, Hans SJ, Lee WJ, Mathiopoulos KD, Muller HM, Kafatos FC, Raikhel A, Brey PT (1999) Genomic structure and ecdysone regulation of the prophenoloxidase 1 gene in the malaria vector *Anopheles gambiae*. Proceedings of the National Academy of Sciences of the United States of America 96:14795–14800.

Altincicek B, Knorr E, Vilcinskas A (2008) Beetle immunity: identification of immune-inducible genes from the model insect *Tribolium castaneum*. Developmental and Comparative Immunology 32:585–595.

Andreadis TG (1987) Transmission. In: Epizootiology of insect diseases (Fuxa JR, Tanada Y, eds), pp 159–176. New York: Wiley.

Armstrong GAB, Robertson RM (2006) A role for octopamine in coordinating thermoprotection of an insect nervous system. Journal of Thermal Biology 31:149–158.

Ayres JS, Schneider DS (2008) A signaling protease required for melanization in *Drosophila* affects resistance and tolerance of infections. PloS Biology 6:2764–2773.

Baines D, DeSantis T, Downer R (1992) Octopamine and 5-hydroxytryptamine enhance the phagocytic and nodule formation activities of cockroach (*Periplaneta americana*) haemocytes. Journal of Insect Physiology 38:905–914.

Beck MH, Strand MR (2007) A novel polydnavirus protein inhibits the insect prophenoloxidase activation pathway. Proceedings of the National Academy of Sciences of the United States of America 104: 19267–19272.

Beckage NE, ed (2008) Insect Immunology. San Diego: Academic Press.

Boucias DG, Pendland JC (1998) Principles of Insect Pathology. Boston: Kluwer.

Bowers SL, Bilbo SD, Dhabhar FS, Nelson RJ (2008) Stressor-specific alterations in corticosterone and immune responses in mice. Brain, Behavior and Immunity 22:105–113.

Bownes M, Rembold H (1987) The titer of juvenile hormone during the pupal and adult stage of the life cycle of *Drosophila melanogaster*. European Journal of Biochemistry 164:709–712.

Brey PT (1994) The impact of stress on insect immunity. Bulletin of the Institut Pasteur 92:101–118.

Broderick NA, Welchman DP, Lemaitre B (2009) Recognition and response to microbial infection in *Drosophila*. In: Insect Infection and Immunity (Rolff J, Reynolds SE, eds), pp 13–33. Oxford: Oxford University Press.

Bronstein SM, Conner WE (1984) Endotoxin-induced behavioral fever in the Madagascar cockroach, *Gromphadorhina portentosa*. Journal of Insect Physiology 30:327–330.

Bundey S, Raymond S, Dean P, Roberts SK, Dillon RJ, Charnley AK (2003) Eicosanoid involvement in the regulation of behavioral fever in the desert locust, *Schistocerca gregoria*. Archives of Insect Biochemistry and Physiology 52:183–192.

Burrows M (1996) The Neurobiology of an Insect Brain. Oxford: Oxford University Press.

Carlson KA, Harshman LG (1999) Extended longevity lines of *Drosophila melanogaster*: abundance of yolk protein gene mRNA in fat body and ovary. Experimental Gerontology 34:173–184.

Caveney S, Cladman W, Verellen L, Donly C (2006) Ancestry of neuronal monoamine transporters in the Metazoa. Journal of Experimental Biology 209:4858–4868.

Chapman RF (1998) The Insects: Structure and Function, 4th ed. Cambridge: Cambridge University Press.

Clark KD, KIm Y, Strand MR (2005) Plasmatocyte sensitivity to plasmatocyte spreading peptide (PSP) fluctuates with the larval molting cycle. Journal of Insect Physiology 51:587–596.

Clark KD, Pech LL, Strand MR (1997) Isolation and identification of a plasmatocyte-spreading peptide from the hemolymph of the lepidopteran insect *Pseudoplusia includens*. Journal of Biological Chemistry 272:23440–23447.

Cooper JR, Bloom FE, Roth RR (2003) The Biochemical Basis of Neuropharmacology, 8th ed. New York: Oxford University Press.

Descotes J (1999) An Introduction to Immunotoxicology. London: Taylor and Francis.

Dhabhar F (2002) Stress-induced augmentation of immune function: the role of stress hormones, leukocyte trafficking, and cytokines. Brain, Behavior and Immunity 16:785–798.

Dhabhar FS (2009) Enhancing versus suppressive effects of stress on immune function: Implications for immunoprotection and immunopathology. Neuroimmunomodulation 16:300–317.

DiAngelo JR, Bland ML, Bambina S, Cherry S, Birnbaum MJ (2009) The immune response attenuates growth and nutrient storage in *Drosophila* by reducing insulin signaling. Proceedings of the National Academy of Sciences of the United States of America 106:20853–20858.

Dong SZ, Ye GY, Guo JY, Hu C (2009) Roles of ecdysteroid and juvenile hormone in vitellogenesis in an endoparasitic wasp, *Pteromalus puparum* (Hymenoptera: Pteromalidae). General and Comparative Endocrinology 160:102–108.

Dunphy GB, Downer RGH (1994) Octopamine, a modulator of the haemocytic nodulation response of non-immune *Galleria mellonella*. Journal of Insect Physiology 40:267–272.

Elenkov IJ, Chrousos GP (1999) Stress hormones, Th1/Th2 patterns Pro/Anti-inflammatory cytokines and susceptibility to disease. Trends in Endocrinology and Metabolism 10:359–368.

Elenkov IJ, Chrousos GP (2006) Stress system: Organization, physiology and immunoregulation. Neuroimmunodulation 13:257–267.

Feder ME, Hofmann GE (1999) Heat-shock proteins, molecular chaperones and the stress response: evolutionary and ecological physiology. Annual Review of Physiology 61:243–282.

Fedorka KM, Linder JE, Winterhalter W, Promislow D (2007) Post-mating disparity between potential and realized immune response in *Drosophila melanogaster*. Proceedings of the Royal Society of London B 274:1211–1217.

Flatt T, Heyland A, Rus F, Porpiglia E, Sherlock C, Yamamoto R, Garbuzov A, Palli SR, Tatar M, Silverman N (2008) Hormonal regulation of the humoral innate immune response in *Drosophila melanogaster*. Journal of Experimental Biology 211:2712–2724.

Flatt T, Meng-Ping T, Tatar M (2005) Hormonal pleiotropy and the juvenile hormone regulation of *Drosophila* development in life history. BioEssays 27:999–1010.

French SS, Greives TJ, Zysling DA, Chester EM, Demas GE (2009) Leptin increases maternal investment. Proceedings of the Royal Society B 276:4003–4011.

Fuxa JR, Tanada Y (1987) Epizootiology of Insect Diseases. New York: Wiley.

Gardy JL, Lynn DJ, Brinkman FSL, Hancock REW (2009) Enabling a systems biology approach to immunology: focus on innate immunity. Trends in Immunology 30:249–262.

Geschwind DH, Konopka G (2009) Neuroscience in the era of functional genomics and systems biology. Nature 461:908–915.

Gillespie JP, Kanost MR, Trenczek T (1997) Biological mediators of insect immunity. Annual Review of Entomology 42:611–643.

Glaser R, Kiecolt-Glaser JK (2005) Stress-induced immune dysfunction: implications for health. Nature Reviews Immunology 5:243–251.

Goldsworthy GJ, Opoku-ware K, Mullen L (2002) Adipokinetic hormone enhances laminarin and bacterial lipopolysaccharide-induced activation of the prophenoloxidase cascade in the African migratory locust, *Locusta migratoria*. Journal of Insect Physiology 48:601–608.

Goldsworthy GJ, Opoku-ware K, Mullen LM (2005) Adipokinetic hormone and the immune responses of locusts to infection. Annals of the New York Academy of Science 1040.

Good TP, Tatar M (2001) Age-specific mortality and reproduction respond to adult diet restriction in *Drosophila melanogaster*. Journal of Insect Physiology 47:1467–1473.

Gruntenko NE, Rauschenbach IY (2008) Interplay of JH, 20E and biogenic amines under normal and stress conditions and its effect on reproduction. Journal of Insect Physiology 54:902–908.

Gruntenko NE, Wilson TG, Monastirioti M, Rauschenbach IY (2000) Stress-reactivity and juvenile hormone degradation in *Drosophila melanogaster* strains have stress-related mutations. Insect Biochemistry and Molecular Biology 30:775–783.

Harshman LG, Zera AJ (2007) The cost of reproduction: the devil is in the details. Trends in Ecology and Evolution 22:80–86.

Hart BL (1988) Biological basis of the behavior of sick animals. Neuroscience and Biobehavioral Reviews 12:123–137.

Hauton C, Smith VJ (2007) Adaptive immunity in invertebrates: a straw house without a mechanistic foundation. BioEssays 29:1138–1146.

Hayakawa Y (1995) Growth-blocking peptide: an insect biogenic peptide that prevents the onset of metamorphosis. Journal of Insect Physiology 41:1–6.

Hayakawa Y (2006) Insect cytokine growth-blocking peptide (GBP) regulates insect development. Applied Entomology and Zoology 41:545–554.

Hayakawa Y, Ohnishi A, Endo Y (1998) Mechanism of parasitism-induced elevation of haemolymph growth-blocking peptide levels in host insect larvae (*Pseudaletia separata*). Journal of Insect Physiology 44: 859–866.

Hayakawa Y, Ohnishi A, Mizoguchi A, Yamashika C (2000) Distribution of growth-blocking peptide in the insect central nervous tissue. Cell and Tissue Research 300:459–464.

Hazarika HL, Gupta AP (1987) Anatomy of the retrocerebral complex of *Blattella germanica* L. (Dictyoptera Blattellidae). Zoologischer-Anzeiger 219:257–264.

Hultmark D (1998) Quantification of antimicrobial activity, using the inhibition zone assay. In: Techniques in Insect Immunology (Weisner A, Dunphy GB, Marmaras VJ, Morishima I, Sugumaran M, Yamakawa M, eds), pp 103–107. Fair Haven: SOS Publications.

Ilsley GR, Luscombe NM, Apweiler R (2009) Know your limits: assumptions, constraints and interpretation in systems biology. Biochimica et Biophysica Acta 1794:1280–1287.

Jacot A, Scheuber H, Brinkhof MWG (2004) Costs of an induced immune response on sexual display and longevity in field crickets. Evolution 58:2280–2286.

Kanost MR, Gorman MJ (2008) Phenoloxidases and insect immunity. In: Insect Immunology (Beckage NE, ed), pp 69–96. San Diego: Academic Press.

Kim GS, Nalini M, Kim Y, Lee DW (2009) Octopamine and 5-hydroxytryptamine mediate hemocytic phagocytosis and nodulate formation via eicosanoids in the beet armyworm, *Spodoptera exigua*. Archives of Insect Biochemistry and Physiology 70:162–176.

KIm Y, Jung S, Madanagopal N (2008) Antagonistic effect of juvenile hormone on hemocyte-spreading behavior of *Spodoptera exigua* in response to an insect cytokine and its putative membrane action. Journal of Insect Physiology 54:909–915.

Kodrik D (2008) Adipokinetic hormone functions that are not associated with insect flight. Physiological Entomology 33:171–180.

Lacoste A, Malham SK, Gelebart F, Cueff A, Poulet SA (2002) Stress-induced changes in the oyster *Crassostrea gigas*. Developmental and Comparative Immunology 26:1–9.

Lafont R, Dauphin-Villemant C, Warren HT, Rees HH (2005) Ecdysteroid chemistry and biochemistry. In: Comprehensive Molecular Insect Science (Gilbert LI, Gill S, Iatrou K, eds), pp 125–195. Amsterdam: Elsevier.

Lawniczak MKN, Barnes AI, Linklater JR, Boone JM, Wigby S, Chapman T (2007) Mating and immunity in invertebrates. Trends in Ecology and Evolution 22:48–55.

Lemaitre B, Reichhart JM, Hoffmann JA (1997) *Drosophila* host defense: differential induction of antimicrobial peptide genes after infection by various classes of microorganisms. Proceedings of the National Academy of Sciences of the United States of America 94:14614–14619.

Leon LJ, Idangodage H, Wan CL, Weers PMM (2006) Apolipophorin III: Lipopolysaccharide binding requires helix bundle opening. Biochemical and Biophysical Research Communications 348:1328–1333.

Lockery SR, Wittenberg G, Kristan WB, Cottrell GW (1989) Function of identified interneurons in the leech elucidated using neural networks trained by back-propagation. Nature 340:468–471.

Loher W, Zaretsky MD (1989) Structure and function of the endocrine system. In: Cricket Behavior and Neurobiology (Huber F, Moore TE, Loher W, eds), pp 114–146. Ithaca: Cornell University Press.

Luster M, Portier C, Pait D, Rosenthal G, Germolec D, Corsini E, Blaylock B, Pollock P, Kouchi Y, Craig W, White K, Munson A, Comment C (1993) Risk assessment in immunotoxicology: II. Relationships between immune and host resistance tests. Fundamental and Applied Toxicology 21:71–82.

Maier SF (2003) Bi-directional immune-brain communication: Implications for understanding stress, pain and cognition. Brain, Behavior and Immunity 17:69–85.

Mallon EB, Loosli R, Schmid-Hempel P (2003) Specific versus nonspecific immune defense in the bumblebee, *Bombus terrestris* L. Evolution 57:1444–1447.

Markow TA, O'Grady PM (2006) Drosophila. A Guide to Species Identification and Use. San Diego: Academic Press.

Marmaras VJ, Lampropoulou M (2009) Regulators and signalling in insect haemocyte immunity. Cellular Signalling 21:186–195.

Martin LB (2009) Stress and immunity in wild vertebrates: Timing is everything. General and Comparative Endocrinology 163:70–76.

Martin LB, Weil ZM, Nelson RJ (2008) Seasonal changes in vertebrate immune activity: mediation by physiological trade-offs. Philosophical Transactions of the Royal Society B 363:321–339.

Mayer F, Mayer N, Chinn L, Pinsonneault RL, Kroetz D, Bainton RJ (2009) Evolutionary conservation of vertebrate blood-brain barrier chemoprotective mechanisms in *Drosophila*. Journal of Neuroscience 29:3538–3550.

McArdle WD, Katch FI, Katch VL (2006) Essentials of Exercise Physiology, 3rd ed. Baltimore: Lippincott Williams and Wilkins.

McNeil JN, Maury M, Bernier-Cardou M, Cusson M (2004) *Manduca sexta* allatotropin and the in vitro biosynthesis of juvenile hormone by moth corpora allata: a comparison of *Pseudaletia unipuncta* females from two natural population and two selected lines. Journal of Insect Physiology 51:55–60.

Milinski M (2006) The major histocompatibility complex, sexual selection and mate choice. Annual Review of Ecology, Evolution and Systematics 37:159–186.

Mills SC, Grapputo A, Jokinen I, Koskela E, Mappes T, Oksanen TA, Poikonen T (2009) Testosterone-mediated effects on fitness-related phenotypic traits and fitness. American Naturalist 173:475–487.

Moore J (2002) Parasites and the Behavior of Animals. New York: Oxford University Press.

Mowlds P, Barron A, Kavanagh K (2008) Physical stress primes the immune response of *Galleria mellonella* lavae to infection by *Candida albicans*. Microbes and Infection 10:628–634.

Mullen LM, Lightfoot ME, Goldsworthy GJ (2004) Induced hyperlipaemia and immune challenge in locusts. Journal of Insect Physiology 50:409–417.

Nance DM, Sanders VM (2007) Autonomic innervation and regulation of the immune system (1987–2007). Brain, Behavior and Immunity 21:736–745.

Nation JL (2008) Insect Physiology and Biochemistry, 2nd ed. Boca Raton: CRC Press.

Nijhout HF (1994) Insect Hormones. Princeton, NJ: Princeton University Press.

Noguchi H, Tsuzuki S, Tanaka K, Matsumoto H, Hiruma K, Hayakawa Y (2003) Isolation and characterization of a dopa decarboxylase cDNA and the induction of its expression by an insect cytokine, growth blocking peptide in *Pseudaletia separata*. Insect Biochemistry and Molecular Biology 33:209–217.

Orchard I, Ramirez JM, Lange AB (1993) A multifunctional role for octopamine in locust flight. Annual Review of Entomology 38:227–249.

Orr GL, Gole JWD, Downer RGH (1985) Characterization of an octopamine-sensitive adenylate cyclase in hemocyte membrane fragments of the American cockroach *Periplaneta americana* L. Insect Biochemistry 15:695–701.

Ottaviani E, Franceschi C (1996) The neuroimmunology of stress from invertebrates to man. Progress in Neurobiology 48:421–440.

Ouedraogo RM, Cusson M, Goettel MS, Brodeur J (2003) Inhibition of fungal growth in thermoregulating locusts, *Locusta migratoria*, infected by the fungus *Metarhizium anisopliae* var *acridum*. Journal of Invertebrate Pathology 82:103–109.

Panaitof SC, Scott MP (2006) Effect of juvenile hormone on vitellogenin gene expression in the fat body of burying beetles, *Nicrophorus orbicollis*. Archives of Insect Biochemistry and Physiology 63: 82–91.

Pham LN, Schneider DS (2008) Evidence for specificity and memory in the insect innate immune response. In: Insect Immunology (Beckage NE, ed), pp 97–127. San Diego: Academic Press.

Pulendran B (2009) Learning immunology from the yellow fever vaccine: innate immunity to systems vaccinology. Nature Reviews Immunology 9:741–747.

Råberg L, Grahn M, Hasselquist D, Svensson E (1998) On the adaptive significance of stress-induced immunosuppression. Proceedings of the Royal Society of London B 265:1637–1641.

Robertson JL, Russell RM, Preisler HK, Savin NE (2007) Bioassays with Arthropods, 2nd ed. Boca Raton: CRC Press.

Roeder T (1999) Octopamine in invertebrates. Progress in Neurobiology 59:1–31.

Roeder T (2005) Tyramine and octopamine: ruling behavior and metabolism. Annual Review of Entomology 50:447–477.

Roxström-Lindquist K, Assefaw-Redda Y, Rosinska K, Faye I (2005) 20-Hydroxyecdysone indirectly regulates *Hemolin* gene expression in *Hyalophora cecropia*. Insect Molecular Biology 14:645–652.

Ryder JJ, Siva-Jothy MT (2000) Male calling song provides a reliable signal of immune function in a cricket. Proceedings of the Royal Society of London B 267:1171–1175.

Sapolsky RM (1992) Neuroendocrinology of the stress-response. In: Behavioral Endocrinology (Becker JB, Breedlove SM, Crews D, eds), pp 287–324. Cambridge, MA: MIT Press.

Saunders DS, Richard DS, Applebaum SW, Ma M, Gilbert LI (1990) Photoperiodic diapause in *Drosophila melanogaster* involves a block to the juvenile hormone regulation of ovarian maturation. General and Comparative Endocrinology 79:174–184.

Schmid-Hempel P (2005) Evolutionary ecology of insect immune defenses. Annual Review of Entomology 50:529–551.

Schmid-Hempel P, Ebert D (2003) On the evolutionary ecology of specific immune defense. Trends in Ecology and Evolution 18:27–32.

Schmidt O, Theopold U, Beckage NE (2008) Insect and vertebrate immunity: key similarities versus differences. In: Insect Immunology (Beckage NE, ed), pp 1–23. San Diego: Academic Press.

Schneider DS, Ayres JS (2008) Two ways to survive infection: what resistance and tolerance can teach us about treating infectious diseases. Nature Reviews Immunology 8:889–895.

Segerstrom SC (2007) Stress, energy and immunity: an ecological view. Current Directions in Psychological Science 16:326–330.

Shia AKH, Glittenberg M, Thompson G, Weber AN, Reichhart JM, LIgoxygakis P (2009) Toll-dependent antimicrobial responses in *Drosophila* laval fat body require Spatzle secreted by hemocytes. Journal of Cell Science 122:4505–4515.

Shoemaker KL, Parsons NM, Adamo SA (2006) Mating enhances parasite resistance in the cricket *Gryllus texensis*. Animal Behaviour 71:371–380.

Short SM and Lazarro BP (2010) Female and male genetic contributions to post-mating immune defence in female Drosophila melanogaster. Proceedings of the Royal Society B. 277: 3649–3657.

Siva-Jothy MT, Moret Y, Rolff J (2005) Insect immunity: an evolutionary ecology perspective. Advances in Insect Physiology 32:1–48.

Siva-Jothy MT, Tsubaki Y, Hooper RE (1998) Decreased immune response as a proximate cost of copulation and oviposition in a damselfly. Physiological Entomology 23:274–277.

Stanley DW (2000) Eicosanoids in Invertebrate Signal Transduction Systems. Princeton, NJ: Princeton University Press.

Sternberg EM (2006) Neural regulation of innate immunity: a coordinated nonspecific response to pathogens. Nature Reviews Immunology 6:318–328.

Strand MR (2008) Insect hemocytes and their role in immunity. In: Insect Immunology (Beckage NE, ed), pp 25–47. Boston: Elsevier.

Strand MR, Hayakawa Y, Clark KD (2000) Plasmatocyte spreading peptide (PSP1) and growth blocking peptide (GBP) are multifunctional homologs. Journal of Insect Physiology 46:817–824.

Tanada Y, Fuxa JR (1987) The pathogen population. In: Epizootiology of Insect Diseases (Fuxa JR, Tanada Y, eds), pp 113–157. New York: Wiley.

Thomas MB, Blanford S (2003) Thermal biology in insect-parasite interactions. Trends in Ecology and Evolution 18:344–350.

Tu MP, Tatar M (2003) Juvenile diet restriction and the aging and reproduction of adult *Drosophila melanogaster*. Aging Cell 2:327–333.

Vafopoulou X, Steel CGH (2005) Testis ecdysiotropic peptides in *Rhodnius prolixus*: biological activity and distribution in the nervous system and testis. Journal of Insect Physiology 51:1227–1239.

Vinson SB (1993) Interactions between the insect endocrine system and the immune system. In: Insect Immunity (Pathak JPN, ed), pp 103–112. Dordrecht: Kluwer Academic.

Wagner WE, Kelley RJ, Tucker KR, Harper CJ (2001) Females receive a life-span benefit from male ejaculates in a field cricket. Evolution 55:994–1001.

Walburn J, Vedhara K, Hankins M, Rixon L, Weinman J (2009) Psychological stress and wound healing in humans: A systematic review and meta-analysis. Journal of Psychosomatic Research 67:253–271.

Watanabe S, Tada M, Aizawa T, Yoshida M, Sugaya T, Taguchi M, Kouno T, Nakamura T, Mizuguchi M, Demura M, Hayakawa Y, Kawano K (2006) N-terminal mutational analysis of the interaction between growth-blocking peptide (GBP) and receptor of insect immune cells. Protein and Peptide Letters 13:815–822.

Webster JI, Tonelli L, Sternberg EM (2002) Neuroendocrine regulation of immunity. Annual Review of Immunology 20:125–163.

Weers PMM, Ryan RO (2006) Apolipophorin III: Role model apolipoprotein. Insect Biochemistry and Molecular Biology 36:231–240.

Zhu B, Pennack JA, McQuilton P, Forero MG, Mizuguchi K, Sutcliffe B, Gu CJ, Fenton JC, Hidalgo A (2008) *Drosophila* neurotrophins reveal a common mechanism for nervous system formation. PloS Biology 6:2476–2495.

Zuk M, Stoehr AM (2002) Immune defense and host life history. American Naturalist 160: S9–S22.

CHAPTER 14

INTERACTIONS BETWEEN HOST SOCIAL BEHAVIOR, PHYSIOLOGY, AND DISEASE SUSCEPTIBILITY

The Role of Dominance Status and Social Context

Bonnie Fairbanks and Dana M. Hawley

INTRODUCTION

Parasites and pathogens are often considered the bane of social animals (Alexander, 1974; Moller et al., 1993; Loehle, 1995; Altizer et al., 2003; Nunn and Altizer, 2006). Both within and among species, group sizes positively predict parasite and pathogen prevalence for a suite of vertebrates (e.g., Brown and Brown, 1986; Moore et al., 1988; Davies et al., 1991; Cote and Poulin, 1995; Nunn et al., 2003; Ezenwa, 2004) (but see Godfrey et al., 2006; Chapman et al., 2009; Woodroffe et al., 2009). These correlative patterns indicate that group living indeed carries costs in terms of increased transmission of infectious organisms, particularly those with a narrow host range (Loehle, 1995; Ezenwa, 2004). Less well understood are the mechanisms that contribute to the striking variation in infection outcomes *within* social groups (Hausfater and Watson, 1976; Hawley et al., 2007). Variation in infection outcome within groups is likely the rule; in some cases, only 10% to 40% of closely interacting group members become infected with a directly transmitted parasite or pathogen (Sydenstricker et al., 2006). This heterogeneity likely relates directly to the costs and benefits of group living, which distribute factors important to both exposure and susceptibility (i.e., resources, stress, aggression) disproportionately across group members. Surprisingly few studies have examined the sources of variation in exposure and resistance within social groups, despite the fact that parasites and pathogens represent one of the key costs to group living, and therefore are an important ecological factor mediating the evolution of sociality and the costs and benefits of behavioral strategies within groups (Moller et al., 1993).

Social behavior can lead to variation in pathogen susceptibility and exposure among individuals in a group via two broad proximate pathways (Fig. 14.1). First, social behavior may alter individual susceptibility to infectious organisms via changes in resource access that mediate individual condition and energetics (resource-mediated pathway; Lochmiller, 1996; left-hand arrows, Fig. 14.1). In addition to influencing immunity, variable resource access directly influences exposure to ingested or environmentally transmitted pathogens. Second, aggressive behavioral interactions may alter immunity via neuroendocrine changes such as circulating hormone levels, a mechanism demonstrated in numerous studies of laboratory vertebrates (e.g., Gross and Siegel, 1973; Barnard et al., 1998; de Groot et al., 1999). Aggressive interactions are also likely to have direct impacts on contact rate and therefore influence exposure to directly transmitted pathogens.

Resource- and aggression-mediated pathways rarely act independently to alter individual infection outcomes because condition and aggression are likely to covary both among and within groups—that is, when resources are sparse, competitive behavioral interactions are more frequent. Furthermore, subordinate animals in a group may be both highly susceptible to infection due to poor condition or high levels of received aggression, and have the highest rates of exposure to infectious organisms because they are forced to feed in non-optimal, contaminated habitats. The nature and extent of these types of patterns for group-living vertebrates—whereby certain group members are more susceptible and/or more exposed—remain largely unknown, and a number of patterns are possible. For example, subordinate group members may have higher rates of exposure to environmentally transmitted organisms if dominant individuals choose resources based on parasite load, or subordinates may have lower exposure to these types of pathogens if these pathogens are present at higher, infectious doses in the most valued resources. Additionally, dominant group members may in some cases have higher exposure to directly transmitted pathogens if they have more social contacts than subordinates; however, if subordinates receive wounds from

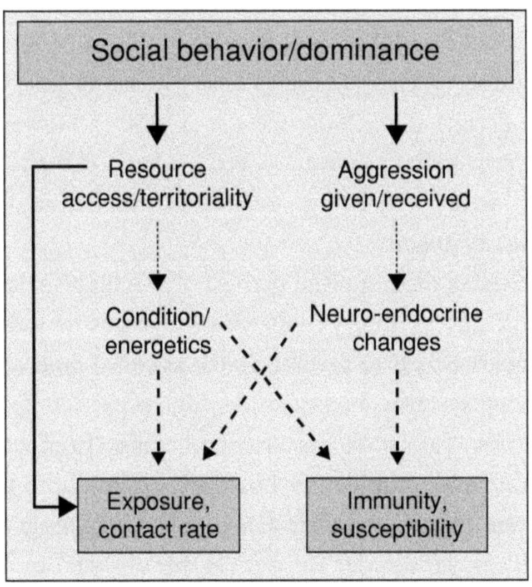

FIGURE 14.1: Potential behavioral (solid lines) and physiological (dotted lines) correlates of dominance status in vertebrates, and their dual consequences for disease outcomes (dashed lines). The majority of these relationships are likely bidirectional and more complex than depicted here. However, given the scarcity of data on these relationships and their cause and effect, we consider them as correlations.

aggressive encounters with dominants, this may be a pathway for exposure to directly transmitted organisms. Because these patterns and their physiological and behavioral mediators are likely to vary across host social contexts and transmission mode of the infectious organism, they are ripe for exploration in group-living vertebrates.

SCALING UP TO POPULATION-LEVEL PATHOGEN DYNAMICS: DOES INDIVIDUAL BEHAVIOR MATTER?

Historically, variation among individuals has been largely ignored by population-level disease studies, which generally assume that all individuals in a population have equivalent susceptibility and exposure to pathogens. However, empirical disease studies suggest that individual variation is the rule rather than the exception. In a U.S. gonorrhea outbreak, 60% of infections were caused by only 2% of individuals (Bansal et al., 2007), and in wild mouse populations, 80% of macroparasitic worms are transmitted by less than 20% of individuals (Ferrari et al., 2004). These key hosts are referred to as "superspreaders," or individuals who contribute disproportionately to pathogen transmission (Lloyd-Smith et al., 2005). The most infamous superspreader was Typhoid Mary, an asymptomatic cook who infected more than 54 individuals with *Salmonella enterica*. Although rarely that extreme, key hosts represent a broader phenomenon whereby only a subset of individuals in a population contribute most significantly to pathogen epidemics. The population-level consequences of key hosts are striking: models reveal that individual variation results in more severe pathogen outbreaks (Lloyd-Smith et al., 2005; Bansal et al., 2007), and furthermore, social hierarchies alone can affect the prevalence and persistence of disease within a group (Davidson et al., 2008). Because key hosts are defined by their contributions to secondary pathogen transmission, or R_0 (a pathogen's basic reproduction number), two variables largely characterize these hosts: (1) higher or longer *infectiousness* due to differences in susceptibility or immune strategies, and (2) more frequent *contacts* with susceptible conspecifics (Temime et al., 2009). For example, a suite of recent studies identified vertebrate males as key hosts for infectious parasites and pathogens (Perkins et al., 2003; Ferrari et al., 2004; Clay et al., 2009). However, all of the studies were performed on territorial, non-social mammal species. Given the physiological and behavioral suites of changes associated with group living, social factors such as rank may be strong predictors of key hosts in group-living organisms.

In this review, we synthesize the potential pathways by which social status and/or aggressive behavior may influence both susceptibility to disease and exposure to infectious agents. It is particularly difficult to tease apart effects of variables such as social rank on individual susceptibility versus exposure in free-living systems. For example, higher parasite loads are often observed in large, sexually mature (Ferrari et al., 2004) or territorial males (Ezenwa, 2004), which may result from higher exposure, higher susceptibility, or both. We discuss both types of mechanisms, and emphasize the need for more experimental studies that begin to tease apart the role of exposure versus susceptibility. Finally, we emphasize the critical role of social and ecological context on the nature and extent of relationships between dominance status and disease outcomes. Although studies to date suggest that correlations between dominance status and disease outcomes abound, these relationships are likely to vary both within and among species. We propose that the effect of dominance status on disease outcomes, much like previously described patterns of dominance and stress hormones, is largely context-dependent. In particular, effects of dominance on disease outcome may be a function of intraspecific aggression experienced and the stability of the social group. However, significantly more studies are needed to test these predictions across a suite of vertebrate taxa.

We divide our review into three sections. First, we discuss the causes and effects of dominance status in vertebrates, dividing the literature synthesis largely into behavioral versus physiological effects of dominance. Second, we synthesize and discuss the few studies that have examined the implications of behavioral and physiological correlates of dominance on health outcomes. Third, we discuss exciting topics and challenges for future research.

CAUSES AND EFFECTS OF DOMINANCE STATUS

WHY CREATE A SOCIAL HIERARCHY?

Animals that live in groups can benefit from the presence of their group-mates in a multitude of ways, including but not limited to increased predator detection and/or avoidance (Pulliam, 1973), increased success in foraging or hunting, and increased access to mates. However, group living also has costs, such as increased competition for food or mates and increased exposure to parasites and pathogens (Alexander, 1974). Social structures such as dominance relationships or ranks may produce disparities across individuals in the way that the costs and benefits of group living are distributed. For example, dominant individuals may have more access to food or mates, but may also have increased exposure to disease because of increased contact rates with conspecifics. Dominance relationships are hypothesized to be useful to both the dominant and subordinate individuals, because it saves both types of individuals the time, energy, and risk of injury associated with repeated conflicts (Creel, 2001). Overall, dominance may be a high-cost, high-benefit strategy, while subordinancy may be a low-cost, low-benefit strategy (Sands and Creel, 2004), allowing all individuals to retain some benefits of group living.

DEFINITIONS OF DOMINANCE

The study of dominance relationships, though fruitful, has some recurring problems. There are varied definitions of dominance, and researchers often fail to explicitly define the term. Because others have made extensive and important arguments for the need for definitions (Bernstein, 1981; Hand, 1986; Drews, 1993), here we briefly make our own definition and encourage investigators to read the papers cited above to understand the importance of properly defining dominance and rank, and the reasons behind the definitions given below.

Most simply, dominance is a relationship between two individuals. Although it may depend on the attributes of each individual, dominance is not an attribute of the individuals themselves. In other words, it is incorrect to call a single individual either dominant or subordinate; these terms must be used in relation to another individual. As a result, selection can act on attributes that affect dominance, such as aggression, but selection cannot act on dominance itself, as this is a relationship that exists between individuals. Although several have been offered, there is one definition of dominance that we find particularly compelling: *the dominance relationship is a relationship between two individuals in which there is a consistent winner and loser of agonistic encounters when this winning is based on the outcome(s) of previous encounter(s)* (Bernstein, 1981; Hand, 1986; Drews, 1993).

The presence of dominance relationships often leads to the establishment of a stable social hierarchy in animal groups, whereby animals are observed and assigned ranks based on relative wins and losses. Bernstein (1981) points out that, in many cases, there may be little biological significance to rank. An animal may know to whom it is dominant and to whom it is subordinate, but being 6th versus 8th in a group of 20 may not have meaningful biological significance. It is telling that in some studies a significant correlation between a given variable and rank is not found, but when animals are divided into dominant and subordinate groups rather than ranked, relationships between dominance and the given variable begin to appear (e.g., De Luca and Ginsberg, 2001). This distinction suggests that there are likely to be differences between an animal that is dominant in most of its relationships versus one that is subordinate in most of its relationships, but one must use caution when drawing inference between animals of close rank.

GENETIC CAUSES AND EFFECTS OF DOMINANCE

There are many behavioral, physiological, and genetic attributes that may contribute to an individual's ability to become socially dominant over others. Although genetic correlates of dominance have been characterized in only a single system to date (Tiira et al., 2006), aggressive and social behaviors more broadly have been linked with levels of genetic variation in free-living female rhesus macaques (Charpentier et al., 2008) and levels of inbreeding in mice, *Mus musculus* (Eklund, 1996; Meagher et al., 2000). Genetic effects can also act to modify the effects of social status; in female rhesus monkeys, Jarrell and coworkers (2008) demonstrated that genetic polymorphism at the serotonin uptake transporter gene interacts with social status to influence physiological changes such as serum leptin levels under high-stress social contexts. These genotype-by-environment interactions that result from social status underscore the potential complexities in dissecting the causes and consequences of dominance. As next-generation sequencing continues to increase the ease of genetic characterization of non-model organisms, we expect a suite of exciting studies at the interface of vertebrate genetics, behavior, and physiology that will begin to tease these complexities apart. Although the current chapter focuses on the effects rather than the causes of dominance, genetic components of rank are important to keep in mind as they may result in genotype-by-environment effects on the rank-related consequences of interest here.

The effects of dominance status on physiology, behavior, and disease is an enormous topic. Therefore, we limit this chapter to a review and synthesis of the literature that correlate behavior and/or physiology to social status, with an emphasis of potential mechanisms by which dominance may affect health via exposure or susceptibility. A suite of studies has looked at correlations between behavior and/or physiology and dominance, but fewer have made strong links between dominance, its health consequences, and the behavior or physiology that mediates this link through exposure to pathogens and/or susceptibility to disease.

AGGRESSION

Dominance status often correlates with the amount of aggression both given and received. For example, subordinates had more aggression directed toward them in captive male cynomolgus monkeys (Cohen et al., 1997) and male dwarf mongooses during the mating season (Creel et al., 1992). Several studies show dominant individuals initiating more aggression than subordinates (woodland caribou, *Rangifer tarandus caribou*: Barrette and Vandal, 1986; African wild dogs during the mating season: Creel et al., 1997; dwarf mongoose: Creel and Sands, 2003; captive female mountain gorillas, *Gorilla gorilla beringei*: Scott and Lockard, 2006). Bite wounds from aggression increase with rank in spotted hyenas (East et al., 2001), indicating that individuals of higher rank are presumably involved in more costly fights. In some cases, relationships between dominance and aggression are sex-specific: in Nubian ibexes (*Capra nubiana*), the dominant female initiated and was involved in more aggressive encounters than subordinate females (Shargal et al., 2008). In contrast, the dominant male, which is more solitary in this species, was involved in fewer encounters than subordinate males, although the dominant male initiated most of the aggressive interactions that he was involved in. Finally, in some group-living species such as wolves (*Canis lupus*), the number of aggressive encounters is unrelated to dominance rank (Sands and Creel, 2004). The above studies represent a small subset of the literature that relates aggression to dominance, but illustrate the widespread but not universal trend whereby dominant individuals typically initiate more aggressive interactions while subordinate individuals have more aggression directed toward them. Several features of aggression, both given and received, have important effects on hormone levels and immunity that will be discussed later in the chapter.

ACCESS TO RESOURCES

Access to resources is often highly skewed in dominance hierarchies (e.g., Barrette and Vandal, 1986; Scott and Lockard, 2006; Hogstad and Pedersen, 2007; Atwood and Gese, 2008). In fact, the near ubiquity of differential access to resources is the reason that dominance status is sometimes defined and/or measured by it (e.g., Metcalfe et al., 1995) Dominant animals may obtain the best food (Scott and Lockard, 2006), preferred territories or locations in the group (Rubenstein and Hohmann, 1989), or mating privileges (Zine and Krausman, 2000). Their differential resource access means that dominant individuals may in some systems be more or less likely to forage in infectious or contaminated habitats. The role of social status in foraging in parasite-rich versus -poor areas has not been examined to date, but represents a key area of future study given that selective foraging in parasite-free areas has been documented for several vertebrates (Ezenwa, 2004; Hutchings et al., 2007). Overall, the effects of differential access to resources on health may accrue through nutritional effects on immunity, or in some cases, through variable exposure via habitat use.

In summary, both aggression and access to resources are generally correlated with dominance rank. Interestingly, both of these traits can have dual effects on individual disease susceptibility by altering both exposure and immunity. While high aggression levels may help to maintain dominance status, they can also lead to higher exposure to parasites or pathogens through wounds and/or contact with infected individuals. Similarly, higher resource access for dominants can lead to stronger immunity and resistance due to nutritional benefits, but may also result in higher exposure to pathogens and parasites through preferred food sources or locations. The outcome of these dual effects will depend both upon the transmission biology of the pathogen/parasite of interest and the social context of the host under study, both of which are discussed below.

PHYSIOLOGICAL CORRELATES OF DOMINANCE

Extensive work has been done on physiological correlates of dominance (Table 14.1). We divide these into three categories: hormones, energetics, and immunocompetence.

STRESS HORMONES

Glucocorticoids are the most commonly measured hormonal stress response. When an animal experiences an acute stressor, the hypothalamic-pituitary-adrenal (HPA) axis responds with a cascade of secretions, ultimately producing glucocorticoids. Glucocorticoids are hormones that signal the body to stop functions unnecessary for immediate survival (i.e., growth, reproduction, digestion) and mobilize energy to areas used to avoid or eliminate the stressor (Sapolsky, 2005). This response is adaptive for short-term stressors, such as fleeing a predator or chasing prey, but long-term activation can be damaging. A healthy stress response includes low basal glucorticoid levels and a fast, high pulse of glucorticoids in the presence of an immediate stressor. In contrast, long-term activation of the stress response causes high basal glucorticoid levels and a sluggish glucorticoid response to an acute stressor (Sapolsky, 2005).

A cursory scan of the literature reveals apparent inconsistency among studies of stress hormones and dominance. Social rank and glucorticoids can have positive, negative, or no correlation (Table 14.1), depending on the species, the season, and/or whether the animal is in captivity or the wild. However, patterns have begun to emerge that explain some of this variation. Many early studies were conducted in captivity, and most captive studies show that subordinates have higher circulating basal glucorticoid levels. This has been credited to the fact that subordinates in captivity have limited ability to avoid dominant individuals and therefore experience more harassment and fighting than animals would in the wild (Creel, 2001; Sands and Creel, 2004). However, this does not explain all of the variation in glucorticoids and rank.

Recent reviews and a meta-analysis reveal that social context is key to understanding the direction of relationships between glucocorticoids and dominance status. Sapolsky (2005) reviews several types of primate societies and concludes that when dominants must repeatedly physically assert their rank, experience times of rank instability, or are cooperative breeders, glucorticoid levels tend to be higher in dominants than subordinates. Alternatively, subordinates generally show higher basal glucorticoid levels in societies where there is non-physical rank maintenance within a stable hierarchy, and subordinates are exposed to frequent social stress with little social support. A meta-analysis by Abbott and coworkers (2003) of several primates reveals similar results, concluding that subordinates will have higher basal glucorticoid levels when exposed to

TABLE 14.1: Behavioral and Physiological Correlates of Dominance.

Variable	Correlation w/Dominance	Possible Interactions	Potential Health Effects of Dominance	References
Aggression given	Positive	Testosterone	Increased exposure to disease through intraspecific contact and wounds.	Barrette and Vandal, 1986; Creel et al., 1997; East et al., 2001; Creel and Sands, 2003; Scott & Lockard, 2006; Shargal et al., 2008
Aggression received	Negative	Stress hormones	Increased exposure to disease through intraspecific contact and wounds. Increased susceptibility through increased stress response.	Creel et al., 1992; Cohen et al., 1997
Access to resources	Positive	Body condition	Decreased susceptibility due to lack of nutritional stress. Increased or decreased exposure through access to preferred food/sites/mates (depending on parasite/pathogen distribution).	Barrette & Vandal, 1986; Rubenstein & Hohmann, 1989; Zine & Krausman, 2000; Scott & Lockard, 2006; Hogstad & Pedersen, 2007; Atwood & Gese 2008
Stress hormones	Various, depending on social context	Immunocompetence	See reviews by Sapolsky (2004, 2005)	Creel, 2001; Abbott et al., 2003; Creel & Sands, 2003; Sands & Creel, 2004; Sapolsky, 2005
Testosterone	Positive or none	Aggression; immunocompetence	Increased susceptibility due to decreased immunity. Increased exposure through aggression.	*Positive correlation*: Creel et al., 1997; Beehner et al., 2006; Aguirre et al., 2007; Aubin-Horth et al. 2007; Drea, 2007; Hogstad & Pedersen, 2007; Johnston et al., 2007; Malaivijitnond et al., 2007; Barja et al., 2008; Shargal et al., 2008. *No correlation*: Hynes et al., 2005; Gauly et al., 2007; Taillon & Cote, 2008; Teichroeb & Sicotte, 2008
Female reproductive hormones	Positive	Immunocompetence	Increased susceptibility through nutritional stress and increased immunity during pregnancy	Creel et al., 1992; Creel et al., 1997; Barja et al., 2008
Metabolic rate	Positive	Body condition	Increased exposure through nutritional stress	Roskaft et al., 1986; Hogstad, 1987; Bryant & Newton, 1994; Cristol, 1995; Metcalfe et al., 1995; Senar et al., 2000; Vezina & Thomas, 2000
Body condition	Positive	Metabolic rate; immunocompetence	Decreased susceptibility to disease due to energy storage	Roskaft et al., 1986; Wiley, 1990; Cohen et al., 1997; Piper and Senar et al., 2000; Hynes et al., 2005
Immunocompetence	Positive	Sex hormones; stress hormones; body condition	Decreased susceptibility to disease due to increased defenses	Cohen et al., 1997; Zuk & Johnsen 2000; Lindstrom, 2004; Hawley, 2006; Hawley et al., 2006, 2007
Sickness behavior	Positive	Conspecific sickness behavior	Decreased susceptibility due to induction of fever and behavioral changes associated with innate immunity	Cohn and Sa-Rocha, 2006, 2009

Information for each correlate includes the direction of its relationship with dominance (e.g., positive = the correlate increases as dominance increases; causation is not inferred here), possible interactions with other correlates of dominance, possible health effects of the correlate, and references.

higher stressor rates than dominants and have low levels of social support. Finally, reviews of birds and mammals by Creel (2001) and of cooperative breeding mammals by Creel and Sands (2003) further confirm that social context is crucial to understanding trends in glucorticoids. In cooperative breeders, in which only dominants are guaranteed reproduction and subordinates help to raise young, dominants generally have a higher level of basal glucorticoids. Furthermore, it is rare in permanent groups for subordinates to be more stressed than dominants. Together, the above studies indicate that the effects of social rank on stress hormones are strongly context-dependent, with important implications for rank-related immunity and disease susceptibility.

SEX HORMONES

Because dominants typically have higher aggression rates, many researchers have considered whether the sex steroid testosterone correlates with social dominance. Some studies reveal a positive correlation between testosterone and rank in males (cichlids: Aubin-Horth et al., 2007; Iberian wolves: Barja et al., 2008; wild African wild dogs: Creel et al., 1997; willow tits: Hogstad and Pedersen, 2007; captive African wild dogs: Johnston et al., 2007; Nubian ibexes: Shargal et al., 2008). Interestingly, most of these are cooperative breeders, except Nubian ibexes and willow tits, although dwarf mongooses are cooperative breeders and do not show a positive relationship between testosterone and rank (Creel et al., 1992). Another study showed a correlation between dominance and testosterone only seasonally in animals that are not cooperative breeders (tropical hair rams: Aguirre et al., 2007), and still other studies show no correlation between testosterone levels and dominance (chickens: Gauly et al., 2007; captive tammar wallabies: Hynes et al., 2005; young white-tailed deer: Taillon and Cote, 2008; Ursine colobus monkeys: Teichroeb and Sicotte, 2008). Beehner and coworkers (2006) show a particularly interesting trend in male chacma baboons: testosterone was not significantly correlated with current rank, but predicted rank several months into the future, as well as future reproduction. Taken together, these studies suggest that in some social contexts, males exhibit elevated levels of testosterone during times of high aggression, competition, or reproduction, but testosterone does not universally correlate with current dominance rank or aggression.

Less research regarding sex hormones and rank has been done with females, and research on testosterone and dominance in females is particularly rare. However, in species where females tend to be dominant to males and/or "masculinized," testosterone or its precursors are often high in females (Aubin-Horth et al., 2007; Drea, 2007; Malaivijitnond et al., 2007). Although most work on testosterone in females has been done in species with masculinized females, differences in testosterone according to rank also occur in species without female masculinization. For example, dominant Nubian ibex females have higher testosterone levels than subordinate females and equal testosterone levels to subordinate males (Shargal et al., 2008).

In females, there are some cases where sex hormones other than testosterone or its precursors vary according to rank. In female dwarf mongooses, the dominant female has higher baseline and peak urinary estrogen conjugates during the mating season (Creel et al., 1992). Similarly, in Iberian wolves, the dominant female has higher levels of fecal progesterone and estradiol compared to subordinates (Barja et al., 2008). In African wild dogs, the dominant female in the group is the only female with a hormone cycle that facilitates reproduction (Creel et al., 1997). The above studies were all conducted on cooperative breeders, whereby the dominant female is often or always the only female in a group that reproduces. Overall, it seems that when a female holds a position of dominance that confers reproductive rights, she has a different hormone profile than subordinates in her group. More studies are needed to see if this holds only across similar social contexts, or if it is a common trend for female hormones to correlate with rank in all vertebrate societies.

ENERGETICS: METABOLIC RATE

Most studies of energetics and dominance have been performed on birds, making this area of research taxonomically narrow. One exception is a study in Atlantic salmon by Metcalfe and coworkers (1995), which showed that metabolic rate was established before hatching, and that those fish with higher metabolic rates were dominant (i.e., claim better feeding positions and obtain desirable food resources), thus suggesting that metabolic rate is one of the causes of dominance in this species. However, in several studies of birds where metabolic rate was correlated with dominance, it appears that metabolic rate is a consequence rather than cause of dominance. For example, in willow tits, metabolic rate positively correlated with dominance, and removing the alpha male caused the beta male to rise to alpha position, and his metabolic rate increased to a rate close to the former alpha (Hogstad, 1987). On the other hand, in dark-eyed juncos, although there was no correlation between rank and metabolic rate before treatment, a dominant male that was moved into a new group and subsequently decreased in rank showed an increase in metabolic rate (Cristol, 1995). These results suggest that a change in dominance is stressful, resulting in increased metabolic rate, but glucocorticoids were not measured in the former study, and did not change significantly in the latter. Alternatively, Hogstad (1987) suggests that the higher metabolic rate of alpha-male willow tits may result from higher activity levels.

In other studies of metabolic rate, dominance was not manipulated, but metabolic rate was sometimes positively correlated with rank (dippers, basal metabolic rate: Bryant and Newton, 1994; pied flycatchers and great tits, resting metabolic rate: Roskaft et al., 1986). However, not all studies measured the same kind of metabolic rate (e.g., resting metabolic rate, basal metabolic rate). The importance of this was highlighted by Senar and coworkers (2000), who measured both resting metabolic rate (oxygen consumption in the dark for a few hours) and active metabolic rate (oxygen consumption in the light). They found a negative correlation between active metabolic rate and rank, but no correlation between resting metabolic rate and rank. They suggest that subordinates are more susceptible to stress, causing the increase in metabolic rate only when subordinate birds could see their unusual surroundings during the measurement. Another study of dark-eyed juncos found no correlation between resting metabolic rate and rank (Vezina and Thomas, 2000). No study has been able to measure metabolic rate via oxygen consumption during normal activity, because subjects must be in a box or wearing a mask to measure this variable. Thus far, the paucity of studies and inconsistent measurements hinder synthesis regarding metabolic rate and dominance. Future studies should measure both resting/basal metabolic rate according to rank (indicating a physiological difference in ranks) and active metabolic rate during normal social activity (indicating an activity difference in ranks or stress levels due to social encounters). It is now possible to attach small heart-rate and/or temperature monitors to animals to collect data from which metabolic rate can be inferred (Cooke et al., 2004). In addition, a variety of taxa need to be studied, as bird flocks and pairs of salmon represent a small subset of the variety of social contexts that exist.

ENERGETICS: BODY CONDITION

Measures of body condition are often positively correlated with rank (captive male cynomolgus monkeys, weight: Cohen et al., 1997; captive male tammar wallabies, weight: Hynes et al., 2005; white-throated sparrows, fat storage: Piper and Wiley, 1990; great tits, heart weight: Roskaft et al., 1986; siskins, body mass: Senar et al., 2000). This is interesting in light of the fact that metabolic rate is often also positively correlated with dominance. Since energy that is acquired will either be used (metabolic rate) or stored (body condition), we might expect to see a trade-off wherein dominants with a high metabolic rate do not store as much energy, and vice versa. However, because both increase with rank (although not all studies compared both factors), we can conclude that dominant individuals must have a high energy intake to maintain both a high metabolic rate and a high body condition. Thus, in general, dominance does seem to have a positive influence on food resource acquisition, which in turn affects health.

IMMUNOCOMPETENCE: IMMUNE PARAMETERS

Of the few experiments that have related immunocompetence to social rank, most show a positive correlation between dominance and immunity. Cohen and coworkers (1997) created groups of male cynomolgus monkeys and kept half stable for 15 months, while the other half were rearranged every month, creating instability. All individuals were exposed to a common-cold–like virus, and regardless of social stability, dominants had significantly lower infection rates than subordinates. Interestingly, this was independent of glucocorticoid levels. In Lindström's (2004) study on greenfinches (*Carduelis chloris*), animals were grouped by similar plumage color, body mass, and age. Dominance was determined and then all were infected with Sindbis virus. The dominant animals had higher blood virus titers in the initial days of infection but cleared the virus more quickly than subordinate individuals. Together, these two studies suggest that dominants have lower susceptibility to and faster recovery from virus infection. Perhaps more interestingly, Lindström's (2004) results suggest that, in some systems, dominant and subordinate animals may have distinct resistance strategies, with dominant animals investing early in pathogen clearance. The role of social status on infection clearance (i.e., reduction of pathogen load) versus tolerance (i.e., the maintenance of individual fitness in the presence of a given pathogen load; Read et al., 2009) would be particularly interesting to examine, as suggested by this study. To date, no studies have considered how social context and social rank might influence immune strategies within and among groups.

Three other studies investigated specific immune responses to challenges based on dominance. In one of the few studies to manipulate rank and test for an effect, Hawley (2006) found that in house finches (*Carpodacus mexicanus*), forcing a decrease in rank for initially dominant birds caused a proportional decrease in antibody response to sheep red blood cells. This effect was likely mediated through a change in the extent of social defeat. No change in antibody response was seen in initially subordinate birds that changed rank. Additionally, Hawley and coworkers (2007) found a positive correlation between dominance and cell-mediated immunity, but a negative correlation between dominance and antibody response, suggesting that rank may mediate important trade-offs within the immune system. Dominant males also showed less severe symptoms and a shorter time symptomatic when inoculated with *Mycoplasma gallisepticum* (a naturally occurring bacterial pathogen in house finches) than subordinates, but dominant females did not show this difference. Finally, Zuk and Johnsen's (2000) study in male red jungle fowl (*Gallus gallus*) measured immune parameters before and after combining males into groups composed of two males and three females. They found that dominant males generally had higher overall immunocompetence both before and after flock formation based on four measures of immunocompetence, although the difference was not significant before flock formation (Fig. 14.2). This result intriguingly suggests that social context can exacerbate preexisting, potentially minimal, differences in immunity, thereby magnifying phenotypic variation in disease susceptibility within social groups.

IMMUNOCOMPETENCE: SICKNESS BEHAVIORS

A number of recent studies have examined the influence of social status on the expression of sickness behaviors, a constellation of behavioral changes that occur early in infection as part of the inflammatory response. These behaviors are thought to be adaptive by directing energy away from temporally unnecessary social behaviors and toward recovery and fever in order to clear infection (Hart, 1988). As such, the relative expression of sickness behaviors is predicted to vary with the ability of an individual to direct energy away from key social behaviors. For example, Cohn and Sa-Rocha (2006) found that subordinate mice injected with lipopolysaccharide (LPS), a component of the bacterial cell wall that stimulates sickness behaviors, did not express any signs of

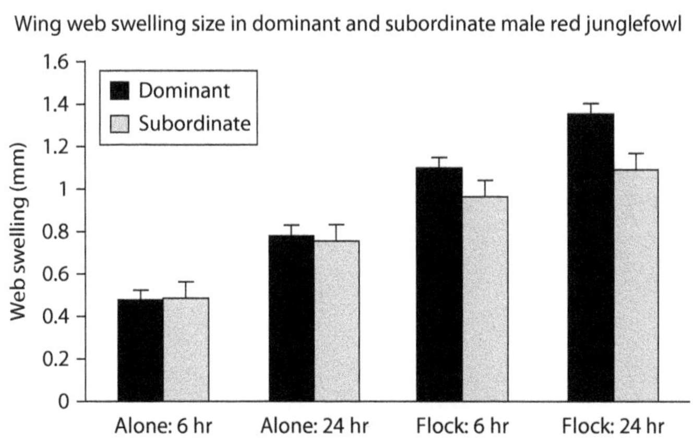

FIGURE 14.2: An individual's social environment can exacerbate preexisting differences in immunity. Here, the swelling response of male red junglefowl (*Gallus gallus*) in response to mitogen injection was measured when males were housed alone, and then in the presence of male and female group members (Zuk and Johnsen, 2000). Dominant males had higher mean swelling responses prior to group formation, but these differences were not significant until individuals were placed in a social, group-living context.

sickness behavior and were indistinguishable from control animals. Dominant mice, on the other hand, expressed sickness behavior within the social context. Cohn and Sa-Rocha (2006) hypothesized that subordinate mice were more likely to prioritize defense behaviors over recovery in a social context. Again, these results suggest that dominant and subordinate animals may exert distinct resistance strategies in response to a parasite or pathogen challenge: dominant animals may be more likely to devote available resources toward rapid clearance, while subordinate animals may be more likely to exhibit tolerance strategies in order to prioritize social behaviors in the presence of aggressive conspecifics.

Cohn and Sa-Rocha (2009) took their prior results a step further by testing how dominant versus subordinate mice treated with LPS responded to an intruder in their home cage. In each treatment, either the dominant or subordinate of a pair was injected with LPS; in the control treatment, both dominant and subordinate were injected with saline. They hypothesized that responses of the dominant and subordinate to an intruder would not only differ by rank but would respond to the expression of conspecific sickness behaviors. Consistent with their prior results, dominant animals spent more time expressing sickness behaviors in the social context. Intriguingly, subordinate animals increased their social behaviors when the co-housed dominant was expressing sickness behavior. This suggests that the social context not only influences sickness behavior across rank, but can influence the behavior of non-"infected" individuals in the group.

Overall, results to date, though limited to a few systems, suggest that dominant individuals generally have stronger immune responses, and this may be both a cause and effect of dominance. Individuals in higher overall condition as a result of genetic and non-genetic factors are likely to have a stronger immune system, and are also more likely to become dominant upon meeting a rival. Interpreting the results of studies that show an effect of changing dominance status or social context on immunocompetence is more difficult. Why does a drop in dominance rank result in decreased immune response in male house finches, while an increase in rank has no effect (Hawley, 2006)? In this study, it appeared that losing fights rather than resource access was the mechanism causing the degree of the decrease in immune response: the number of defeats experienced by dominant house finch males in their new social environment directly predicted the extent of immunosuppression. Similarly, in the Zuk and Johnsen (2000) study, differences in resource access or mass did not explain the socially induced differences in immunocompetence. Instead, social interactions, perhaps via changes in hormone levels, appear to mediate rank-related immune competence in a social context. Taken together, these studies suggest that neuroendocrine changes and social interactions, particularly aggression, likely underlie the patterns detected to date in terms of dominance and immune competence.

HEALTH CONSEQUENCES OF DOMINANCE

Physiological and behavioral correlates of dominance have diverse and complex interactions with one another. For example, dominance and aggression may affect testosterone levels, which interact bidirectionally with glucocorticoid levels (Sapolsky, 1985; Ketterson et al., 1991), and all of these can in turn affect immunity and responses to parasites and pathogens. This section will explore some of the many connections between dominance behavior and physiology, susceptibility, and exposure to pathogens.

LEVELS OF AGGRESSION

Higher rates of aggression, whether associated with testosterone or not, are often correlated with dominance rank (see *Behavioral Correlates of Dominance* section), and increased aggression can have major health implications. Aggression is particularly interesting and important because it can have dual effects on both exposure and susceptibility. First, aggression may increase contact with conspecifics and cause wounds, both of which lead to higher rates of infection through exposure to pathogens. Second, given or received aggression causes rapid neuroendocrine responses and heightened energy expenditure, both of which can decrease the strength of the immune response and, consequently, the ability to resist a parasite or pathogen. An example of the first mechanism occurs for Serengeti spotted hyenas, where, as previously mentioned, dominants experience more bite wounds than subordinates (East et al., 2001). Interestingly, East and coworkers found a correlation between dominance rank and rabies virus exposure, which they hypothesized was spread by the higher incidence of bite wounds in dominants and the submissive act of subordinates licking the open mouths of dominants. Although the rabies virus is asymptomatic in this species, making the fitness consequences difficult to interpret, this study provides a rare example from a free-living system of the potential health costs of aggression for dominant individuals.

The physiological effects of aggression on immunity have been well documented in numerous laboratory and domesticated vertebrates (Gross and Siegel, 1973; Barnard et al., 1998) and can vary with social status (Stefanski, 1998). Interestingly, in some cases a single social defeat can cause a constellation of physiological changes and alterations to the HPA axis that ultimately affect immunity (de Groot et al., 1999; Goncalves Carobrez et al., 2002), a mechanism explored in detail by Sheridan and colleagues in Chapter 11. Consistent with these studies, Hawley (2006) directly linked immunosuppression in male house finches with the extent of social defeat experienced in a new social context. These data suggest a strong role for neuroendocrine changes associated with defeat on the immune system. The higher aggression often received by subordinates and their lower average immunocompetence in the studies reviewed herein are consistent with a strong role for aggression in mediating immune competence and pathogen clearance in vertebrates. Overall, the studies to date highlight the potential health implications of dominance status, and spur exciting questions about physiological and behavioral mechanisms.

EXPOSURE VIA RESOURCE ACCESS OR TERRITORY DEFENSE

One of the less-studied consequences of dominance status on disease outcome is the cost of access to preferred resources, which may inadvertently increase the exposure of dominant individuals to infectious agents. Halvorsen (1986) showed that calves of dominant female reindeer had better access to favored food as well as a higher rate of infection with a nematode parasite, most likely because the gastropod host of the nematode used the reindeer's favored food source. Hausfater and Watson (1976) found higher parasite ova emission in dominant male baboons, and speculated that it could be due to food access, though they did not test this. Hogstad and Pedersen (2007) found that dominant willow tits had greater food access as well as higher cadmium levels in the liver. Plowright and coworkers (2008) found that Hendra virus infection in little red flying foxes increased with body size. They speculated this trend occurred because body size correlates with dominance, and dominant males have large harems and defend food sources, both of which cause high conspecific contact rates. Finally, Ezenwa (2004) showed that territorial male gazelle have higher intensities of gastrointestinal parasites than bachelor males, juvenile males, or females. The examples above are case-specific, and there may be times where the less-favored resources that subordinates are forced to use expose them to more pathogens, a possibility that has not yet been empirically demonstrated. Additionally, when dominants do have higher exposure to pathogens through resource access, it may be outweighed by the nutritional or other benefits gained (e.g., higher weight in dominants: Halvorsen, 1986; higher feather regrowth in dominants: Hogstad and Pedersen, 2007).

The higher prevalence of parasites found in dominant animals in several of the studies summarized above (Hausfater and Watson, 1976) appear to contradict the results of captive experimental inoculation studies, whereby dominant individuals show decreased susceptibility to disease and/or faster clearance of infectious organisms (Cohen et al., 1997; Lindstrom, 2004; Hawley et al., 2007). These conflicting patterns in captivity, whereby exposure was held constant, versus the wild indicate that heightened immunity of dominants, if present, may be insufficient to overcome greater exposure to pathogens through increased conspecific contact, increased food demand, and increased access to resources that may contain pathogens. On the other hand, dominant animals may, in some cases, show higher parasite loads or prevalence because they are practicing immune strategies such as pathogen tolerance by investing energy into territory defense or reproduction, rather than pathogen clearance, in order to maintain fitness. Finally, the costs and benefits of higher resource access may be context-dependent, whereby dominant animals are more exposed in some social systems or contexts than others.

BEHAVIORAL AVOIDANCE/PREFERENCE OF CONSPECIFICS

Selection should favor individuals who can recognize and avoid infected conspecifics in order to reduce their own risk of infection, particularly for directly transmitted pathogens with a limited host species range (Loehle, 1995). For many of these types of pathogens, evidence is accumulating that healthy animals actively avoid infected conspecifics (Kiesecker et al., 1999; Behringer et al., 2006; Kavaliers et al., 2006; but see Bouwman and Hawley, 2010), presumably to reduce their own risk of infection. The role of social status in mediating avoidance of infected conspecifics remains unknown, but partner choice is affected by social hierarchy for a suite of vertebrates (Cheney et al., 1986; Dugatkin and Sih, 1995; Schino, 2001; Smith et al., 2007). If dominants tend to be less infected or infectious, social partner choice may represent a critical mechanism by which dominant individuals, via selective access to "desirable" conspecifics, alter their probability of exposure. However, Rubenstein and Hohmann (1989) found that in island feral horses, dominant females prefer to be near the most dominant animal, which is the stallion. This preference increases the parasite load of the dominant females because the stallion attracts more flies than females, and the flies he attracts also bite the dominant females close to him. Thus, dominant animals may make partner choices based on factors other than infectiousness and thereby increase their exposure.

HORMONE-MEDIATED PATHWAYS

The health consequences of glucocorticoid levels and stress more broadly are now widely accepted (Sapolsky, 2004, 2005) but can result in immunostimulation or suppression depending on the stressor and the timing (Sapolsky 2005). Similarly controversial are the immunosuppressive effects of testosterone. Mixed support has been found that testosterone may suppress the immune system (reviewed for birds in Owen-Ashley et al., 2004) and increases the rate of ectoparasites, at least in some species (Grossman, 1985; Roberts et al., 2004). Testosterone is also often associated with increased aggression (Balthazart, 1983; Baum, 1992), although many other factors may influence the expression of aggressive behavior, especially in primates (see Booth et al., 2006, for the effects of testosterone in humans). Dominant animals with increased testosterone levels may also have higher contact rates with conspecifics, not only through aggressive interactions but also through courtship and mating. In an experimental testosterone-implant study with wild male white-footed mice, Grear and coworkers (2009) documented a significant increase in contact rates for testosterone-treated individuals and predicted a two-fold increase in parasite transmission on plots where testosterone levels were elevated. Interestingly, no wildlife study to date has directly examined how dominance rank or testosterone levels correlate with conspecific contact rate, particularly between infectious and susceptible individuals, the key variable driving the transmission of directly transmitted pathogens.

METABOLIC RATE

If an individual's rank is associated with a high metabolic rate for an extended period of time, its nutritional intake or energy reserves must be able to supply that metabolic demand. Since dominants usually have greater access to resources, it's likely that they can sustain a higher metabolic rate with less risk to health, as shown by their ability to also sustain higher body condition. However, the higher nutritional demand is not without possible costs; Hogstad and Pedersen (2007) found that dominant male willow tits had higher testosterone levels, more food access, and higher cadmium levels than subordinates. Earlier work (Hogstad, 1987) showed that dominants had higher metabolic rates, so it is possible that the higher metabolic rate led to higher food demand, which they were able to obtain, but which exposed dominants to greater amounts of environmental pollutants. The role of rank-related differences in metabolic rate on infection outcomes has not been examined but represents an important possible mechanism by which individuals within a group may show diverse infection outcomes.

CONTEXT-DEPENDENT PHYSIOLOGICAL COSTS OF DOMINANCE

Creel and Sand's (2003) suggestion that dominance may be a high-risk, high-yield strategy and subordinancy a low-risk, low-yield strategy seems borne out by the research summarized above; in general, dominant animals appear to have higher immune competence (Hawley, 2006) and resistance to experimental infection (Lindstrom, 2004; Hawley et al., 2007) in stable social groups. However, when group membership changes (Cristol, 1995; Hawley, 2006), social dominance may carry physiological costs such as elevated metabolic rate and immune suppression. The effect of social stability on rank-related immune competence within or among species has not been directly examined, but Hawley and coworkers (2006) manipulated the extent of group-level aggression in captive house finches by providing 4 versus 16 feeder-access points for flocks of 8 finches. The results indicate that the extent of aggression in the social group can mediate the relative costs and benefits of dominance: dominant house finches had higher immune competence in low-competition flocks (16 feeder-access points), and aggression rates decreased over time in these groups as membership stabilized. However, in high-competition groups (4 feeder-access points), high levels of aggressive interaction were maintained throughout the experiment. In this treatment, dominant individuals did not have higher immune competence. Interestingly, several of the high-ranking individuals in each flock did not mount an immune response when aggressive interactions were high. The discrepancy at high versus low social aggression corroborates results obtained in the previously discussed meta-analyses of social rank and glucocorticoid levels, whereby social stability is a strong determinant of rank-related costs and benefits (Creel, 2001; Creel and Sands, 2003; Sapolsky, 2005).

Although it is difficult to make generalizations with the limited number of studies available, we hypothesize that relationships between dominance and pathogen susceptibility will reflect the extent of stability and aggression in a social group. During times of the year where membership changes occur and/or aggression levels are enhanced, the physiological costs of dominance in terms of increased pathogen susceptibility will likely outweigh the benefits detected in previous studies. These situations may be particularly important for population-level spread of directly transmitted diseases, particularly if dominant animals have higher exposure to these infectious organisms during periods of intense intraspecific aggression and contact. Context-dependent costs of dominance under particular social contexts may result in positive covariation between exposure and susceptibility within dominant individuals, leading to key hosts and a high potential for epidemics (Fig. 14.3; Hawley et al., 2011). The role of social context in driving phenotypic covariation in pathogen-relevant traits is particularly important to understand because it is non-heritable, resulting in variation in disease dynamics with no potential for directional selection for pathogen resistance.

CHALLENGES AND SUGGESTIONS FOR FUTURE RESEARCH

Considering our current understanding of how social rank mediates pathogen exposure and susceptibility, how should future research be directed? One of the most pressing needs in the field of

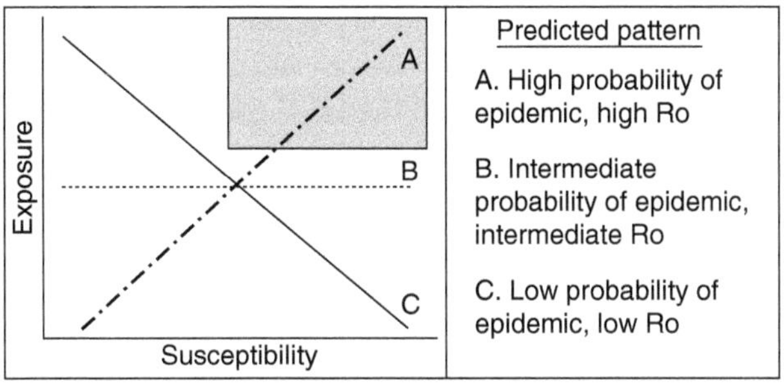

FIGURE 14.3: The presence and direction of phenotypic covariation between susceptibility and exposure among individuals will have important implications for population-level disease spread. Social status, because it has dual effects on susceptibility and exposure, may influence the nature and extent of these potential relationships. (A) Positive covariation between susceptibility and exposure, leading to high potential for population-level disease epidemics. (B) No detectable covariation between susceptibility and exposure, leading to intermediate epidemic potential. (C) Negative covariation between susceptibility and exposure leads to the lowest probability of population-level epidemics.

dominance more generally is to design experiments that reveal cause and effect. Correlations abound, but a clear understanding of how dominance results in behavioral or physiological traits will remain elusive until manipulative studies are performed. Below we discuss a few of the challenges for future research and potential exciting directions for elucidating how social context influences disease outcomes.

DISSECTING EFFECTS OF DOMINANCE ON SUSCEPTIBILITY VERSUS EXPOSURE

One of the key challenges in understanding how social status influences health is to dissect the relative influence of dominance interactions on exposure to infectious agents versus responses to infection once exposed. Correlative patterns, whereby dominant animals may have lower or higher pathogen prevalence, are the outcome of potentially complex interactions whereby dominant animals may be more frequently or less frequently exposed to an infectious organism, but may also clear that organism more quickly or slowly. Knowledge of pathogen exposure is therefore critical to ask whether variation in infection outcomes results from underlying immune variation, behavioral processes that control transmission, or both. Based on the results discussed above, we predict that dominant animals will generally have higher exposure to directly transmitted organisms due to higher contact rates with conspecifics, but the type of conspecific with which dominant animals primarily interact will be critical (see the section on infection-induced changes). Dominant animals with larger home ranges should also have higher exposure to pathogens that are environmentally transmitted, as suggested by the research above. On the other hand, subordinate animals may be more likely to be exposed to vector-transmitted parasites and pathogens due to less-preferred positions with the group, but the sole study to examine ectoparasite prevalence by rank found the opposite result, whereby the dominant stallions were most heavily infested (Rubenstein and Hohmann, 1989).

Teasing apart the relative role of susceptibility versus exposure on rank-related infection dynamics is particularly challenging because resistance and exposure may covary within individuals. In one of the few studies to date to explicitly examine covariation between host susceptibility and exposure, sheep that were genetically resistant to gastrointestinal helminths also avoided foraging in parasite-rich areas of habitat more effectively (Hutchings et al., 2007). Whether this type of covariation is present for a particular species or under particular social contexts may determine the extent to which individuals of a given rank serve as key hosts for secondary pathogen transmission. Although genetic covariation remains challenging to study in non-model organisms, understanding whether pathogen exposure and resistance phenotypically covary across social rank, and if so, whether that covariation is negative or positive, is both feasible and particularly important to consider given the population-level impacts of this covariation on infection dynamics (Hawley et al., 2011).

Only a handful of studies have attempted to tease apart the role of variation in susceptibility versus exposure on rank-related health. The simplest approach is to use novel immune antigen injections (as per Zuk and Johnsen, 2000), to which no individuals have been previously exposed, in order to determine how rank affects a measure of susceptibility (i.e., immune competence) without confounding effects of exposure or secondary transmission among group members. A second option is to experimentally eliminate variation in the timing and dose of exposure to a pathogen across social rank by simultaneously inoculating naïve individuals with identical doses of an ecologically relevant pathogen (as per Lindstrom 2004; Hawley et al., 2007). This approach allows the investigator to isolate the influence of social rank on recovery *per se* while ignoring the confounding effects of rank-related variation in exposure. In some cases, independent experimental manipulation of behavioral and/or immune phenotypes may be possible. Mougeot and coworkers (2005) directly measured the influence of testosterone on host immunity versus behavior by implanting male red grouse (*Lagopus scoticus scoticus*) with testosterone while simultaneously blocking androgen receptors, thus eliminating behavioral responses. Enhanced susceptibility of testosterone-implanted males to parasites was still detected, suggesting an immunological cause for sex differences in infection. These types of manipulations have not been performed across individuals of varying social rank, but would be particularly useful in elucidating behavioral versus physiological mechanisms.

The role of social rank in driving variation in exposure to pathogens is unusually challenging to study because measurements of exposure in free-living populations are tightly linked with immune responses. For example, seroconversion may indicate recent exposure to a pathogen to which specific antibodies are mounted; however, the use of seroconversion is limited to pathogens that cause measurable immunity, and measures of seroconversion can have high measurement errors, rendering interpretation difficult at the level of individuals. More refined techniques for measuring individual variation in exposure are needed to document how dominance influences exposure across a suite of social contexts (i.e., high vs. low aggression or stability). Social networks are a useful tool for quantifying variation in conspecific contact rate and types of interactions among group members, and network measures such as individual connectivity predict parasite prevalence for a suite of systems to date (e.g., Bansal et al., 2007; Perkins et al., 2009). Although social networks are increasingly being applied to understand disease transmission (Corner et al., 2003; Bansal et al., 2007; Godfrey et al., 2009; Perkins et al., 2009), they have not yet been used to examine how rank-related network characteristics influence parasite and pathogen exposure in social groups.

Creative solutions to the experimental quantification of transmission are beginning to be developed. Naug (2008) used non-pathogenic, intestinal microbeads to map how social networks in honeybees influence the spread of a pseudo-intestinal pathogen throughout the social group. Kulkarni and Heeb (2007) used a non-pathogenic soil bacterium, *Bacillus licheniformes*, as a model for bacterial transmission in zebra finches. They inoculated the feathers of paired male zebra finches with this "bacterial tracer" and found very high rates of allo-infection (~80%) in the gut of the co-housed individuals only 24 hours after treatment (Kulkarni and Heeb, 2007). They hypothesized that transmission resulted from allo-preening, but the exact mechanism was unknown. Similar studies could be conducted on paired males of known social rank for a variety of vertebrate species. Depending on the transmission mode of the pathogen of interest (a.k.a. environmental vs. direct, and if direct, sexual vs. non-sexual), experimental models of transmission may be a productive and informative way of examining how social context and dominance status influence the likelihood of exposure to parasites and pathogens of varying life-histories and transmission modes.

INFECTION-INDUCED BEHAVIORAL CHANGES AND SOCIAL RANK

For directly transmitted pathogens, the behavior of individuals *once infected* is the key determinant of a resulting pathogen epidemic (Lloyd-Smith et al., 2004). Therefore, while understanding how individual behavior might influence the probability of infection is useful for understanding the costs and benefits of behavioral strategies and group living, behaviors prior to infection are unlikely to be relevant to resulting population-level spread. There is a need to understand how aggressive behavior and social rank influence infection-induced behavioral changes such as sickness behavior in order to ultimately understand how social structure might affect broad-scale disease dynamics (Lloyd-Smith et al., 2004; Funk et al., 2009).

Evidence to date suggests that infection-induced changes in behavior can directly influence the spread of directly transmitted pathogens. In laboratory rats, healthy individuals actively avoid bedding of conspecifics that were injected with the bacterial mimic LPS (Arakawa et al., 2010), suggesting that olfactory cues are sufficient to alter contact rates between healthy and susceptible individuals in this species, and potentially in other vertebrates that utilize olfactory signals for conspecific communication. If rats, like mice (Cohn and Sa-Rocha, 2006), show status-related differences in sickness behavior expression, contact rates with infected dominants may be lower than with infected subordinates. The relative infectiousness of individuals across status would then determine resulting effects on transmission. In birds, which are thought to rely less heavily on olfactory cues than mammals, behavioral changes involved in the sickness response may alter contact rates. In male house finches, sickness behaviors resulting from infection with the naturally occurring bacterium *Mycoplasma gallisepticum* significantly attract healthy conspecifics seeking to avoid behavioral aggression at feeders (Bouwman and Hawley, 2010). These changes will likely increase contact rates between healthy finches and infected males, and hence may facilitate the transmission of *M. gallisepticum* in wild house finch populations.

The proximate mechanisms that underlie infection-induced behavioral changes comprise a network of neuroendocrine changes, which include glucocorticoid and testosterone levels as well as the expression of pro-inflammatory cytokines in the brain (Besedovsky and del Ray, 2001; Dantzer, 2004). Interestingly, these neuroendocrine pathways are known to respond to social context. Weil and coworkers (2006) challenged adult male rats with LPS under three different social contexts: housed alone, with a single juvenile male, or with an adult female. The presence of the juvenile male, but especially the adult female, significantly increased gene expression of two pro-inflammatory cytokines, TNFα and IL-1β, although behavioral differences were not noted. In all three treatments, LPS-injected males showed strong decreases in testosterone, a characteristic of the sickness response that appears to occur after antigen injection for a suite of vertebrates (Boonekamp et al., 2008). *In vitro* experiments indicate that the pro-inflammatory cytokines IL-1α (e.g., Hales 1992) and IL-1β (e.g., Lister and van der Kraak, 2002), released during an inflammation response in vertebrates, suppress testicular steroid production. However, no study to date has examined whether male vertebrates alter their testosterone levels when infected with an ecologically relevant pathogen, and furthermore, how decreases in testosterone vary with social status and ultimately alter contact rates with susceptible individuals, the most important component for emergent population-level transmission (Swinton et al., 2002).

The study of conserved vertebrate signaling molecules such as hormones or cytokines, both of which can have dual effects on host exposure and susceptibility, have the potential to help tease apart the mechanisms underlying rank-related variation in disease outcomes (Hawley and Altizer 2011). As described above, the steroid hormone testosterone can increase transmission-relevant behaviors such as contact rate (Grear et al., 2009) and has also been implicated in immunosuppression or immunomodulation across a range of taxa. Similarly, pro-inflammatory cytokines released during the acute phase response influence both within-host immune processes and, for the limited species of mammals and birds studied to date, behaviors relevant to transmission (e.g., sickness behavior and attraction or aversion of healthy conspecifics). If immunological and behavioral effects of rank positively co-occur during vertebrate infection (Fig. 14.3A; Hawley et al., 2011), key hormones such as testosterone may serve as common drivers of key hosts, broadly linking within-host (e.g., infectiousness) and among-host (e.g., contact rate) processes across individuals of varying social status. Their functional conservation across diverse vertebrate taxa makes these signaling molecules even more compelling candidates for considering mechanistic associations between dominance behavior, susceptibility, and transmission.

Overall, more mechanistic and integrative studies are needed on infection-induced behavioral changes in group-living organisms, as well as the influence of dominance status and aggression on these changes. Ultimately, these data will reveal whether common traits (i.e., dominance or high aggression) characterize the individuals that largely contribute to transmission in a group.

GOING BEYOND IMMUNE COMPETENCE AND PATHOGEN RESISTANCE

The studies to date concerning how dominance rank influences immunity and disease have almost universally assumed that higher responses are "better" (i.e., higher antibody titers reflect stronger resistance and higher fitness). There are multiple problems with this assumption that may limit our conclusions about the costs of benefits of dominance status. First, strong immune responses can have substantial fitness costs, including immunopathology (Graham, 2002). Second, traditional measures of immune competence are often not very predictive of outcomes to particular infections (Adamo, 2004). Finally, a "competent" immune system may employ tolerance strategies (e.g., the maintenance of host fitness in the presence of a pathogen load; Read et al., 2009), which is difficult to assess via infection outcome or antibody production alone.

The fitness consequences of immune investment and pathogen burden provide a critical context for understanding observed outcomes but are often overlooked (Baucom and DeRoode, 2011; Graham et al., 2011). For example, dominant individuals that invest in costly defenses in the absence of exposure to disease-causing agents could be making a flawed investment that will ultimately lead to lower fitness, but such hosts might readily be deemed "immune competent" by traditional ecoimmunology assays (e.g., immune cell recruitment, parasite agglutination and/or lysis, and antibody production). Though challenging, we suggest that future studies incorporate fitness outcomes to the extent possible in order to provide context to results obtained (Graham et al., 2011). It may be that dominant and subordinate individuals dynamically employ different strategies of clearance versus tolerance across varied social contexts, depending on the context-dependent fitness outcomes that result. The use of varying immune strategies (e.g., tolerance vs. clearance) by dominant individuals across social contexts is an area of exciting and unexplored territory.

SUMMARY

The dynamic role of social context on immunity has long been of interest for closely housed domesticated species (Gross and Siegel, 1973; Morrow-Tesch et al., 1994; Tuchscherer et al., 1998; De Groot et al., 2001). In contrast, ecoimmunology investigators are just beginning to appreciate the way in which social context and individual dominance rank mediate both immunity and behaviors relevant to exposure for group-living vertebrates. The limited studies to date on dominance status and disease susceptibility indicate that social status has contrasting and context-dependent effects on health; in some cases, dominant animals have higher exposure due to greater resource access or wounds, while in other species, dominant individuals appear to have higher immune competence and lower disease susceptibility. Further work is sorely needed to determine whether characteristics of social groups and/or the transmission mode of parasites and pathogens might help to parse out this variability in relationships between social status and disease susceptibility. The complexity of interactions between behavior, hormones, and immunity provide a significant challenge to understanding relationships between social rank and disease. However, the extensive work done on how social rank mediates stress hormones provides a testable framework for how social rank may also modulate immune responses and pathogen susceptibility. Variation in susceptibility via host immunity must then be incorporated with variation in exposure across rank to provide a complete understanding of how parasites and pathogens influence the costs and benefits of dominance and the evolution of sociality. Dissecting the mechanisms that contribute to the substantial heterogeneity in disease within and among social groups is critical to understand in order to predict the potential for disease outbreaks and the appropriateness of disease management strategies for socially structured vertebrate populations.

REFERENCES

Abbott, D, Keverne, E, Bercovitch, F, Shively, C, Mendoza, S, Saltzman, W, Snowdon, C, Ziegler, T, Banjevic, M, Garland, TJ and Sapolsky, RM. 2003. Are subordinates always stressed? A comparative analysis of rank differences in cortisol levels among primates. *Hormones and Behavior*, 43, 67–82.

Adamo, S. 2004. How should behavioural ecologists interpret measurements of immunity? *Animal Behaviour*, 68, 1443–1449.

Aguirre, V, Orihuela, A and Vazquez, R. 2007. Seasonal variations in sexual behavior, testosterone, testicular size and semen characteristics, as affected by social dominance, of tropical hair rams (*Ovis aries*). *Animal Science Journal*, 78, 417–423.

Alexander, R. 1974. The evolution of social behavior. *Annual Review of Ecology and Systematics*, 5, 324–383.

Altizer, S, Nunn, C, Thrall, P, Gittleman, J, Antonovics, J, Cunningham, A, Dobson, A, Ezenwa, V, Jones, K, Pedersen, A, Poss, M and Jones, K. 2003. Social organization and parasite risk in mammals: integrating theory and empirical studies. *Annual Review of Ecology and Systematics*, 34, 517–547.

Arakawa, H, Arakawa, K and Deak, T. 2010. Sickness-related odor communication signals as determinants of social behavior in rat: a role for inflammatory processes. *Hormones and Behavior*, 57, 330–341.

Atwood, TC and Gese, EM. 2008. Coyotes and recolonizing wolves: social rank mediates risk-conditional behaviour at ungulate carcasses. *Animal Behaviour*, 75, 753–762.

Aubin-Horth, N, Desjardins, JK, Martei, YM, Balshine, S and Hofmann, HA. 2007. Masculinized dominant females in a cooperatively breeding species. *Molecular Ecology*, 16, 1349–1358.

Balthazart, J. 1983. Hormonal correlates of behavior. In: *Avian Biology* (Ed. by Farner, D. S., King, J. and Parkes, K.), pp. 221–365. New York: Academic Press.

Bansal, S, Grenfell, BT and Meyers, LA. 2007. When individual behaviour matters: homogeneous and network models in epidemiology. *Journal of the Royal Society Interface*, 4, 879–891.

Barja, I, Silvan, G and Illera, JC. 2008. Relationships between sex and stress hormone levels in feces and marking behavior in a wild population of Iberian wolves (*Canis lupus signatus*). *Journal of Chemical Ecology*, 34, 697–701.

Barnard, C, Behnke, J, Gage, A, Brown, H and Smithurst, P. 1998. The role of parasite-induced immunodepression, rank, and social environment in the modulation of behaviour and hormone concentrations in male laboratory mice (*Mus musculus*). *Proceedings of the Royal Society, B*, 265, 693–701.

Barrette, C and Vandal, D. 1986. Social rank, dominance, antler size, and access to food in snow-bound wild woodland caribou. *Behaviour*, 97, 118–145.

Baucom, R and DeRoode, J. 2011. Ecological immunology and tolerance in plants and animals. *Functional Ecology*, 25, 18–28.

Baum, MJ. 1992. Neuroendocrinology of sexual behavior in the male. In: *Behavioral Endocrinology* (Ed. by Becker, J. B., Breedlove, S. M. and Crews, D.), pp. 97–130. Cambridge: MIT Press.

Beehner, J, Bergman, T, Cheney, D, Seyfarth, R and Whitten, P. 2006. Testosterone predicts future dominance rank and mating activity among male chacma baboons. *Behavioral Ecology and Sociobiology*, 59, 469–479.

Behringer, D, Butler, M and Shields, J. 2006. Avoidance of disease by social lobsters. *Nature*, 441, 421–421.

Bernstein, I, S. 1981. Dominance: The baby and the bathwater. *Behavioral and Brain Sciences*, 4, 419–457.

Besedovsky, H and del Ray, A. 2001. Cytokines as mediators of central and peripheral immune-neuroendocrine interactions. In: *Psychoneuroimmunology* (Ed. by Ader, R., Felten, D. and Cohen, N.), pp. 1–17. San Diego: Academic Press.

Boonekamp, J, Ros, A and Verhulst, S. 2008. Immune activation suppresses plasma testosterone level: a meta-analysis. *Biology Letters*, 4, 741–744.

Booth, A, Granger, DA, Mazur, A and Kivlighan, KT. 2006. Testosterone and social behavior. *Social Forces*, 85, 167–191.

Bouwman, K and Hawley, D. 2010. Sickness behaviour acting as an evolutionary trap? Male house finches preferentially feed near diseased conspecifics. *Biology Letters*, 6, 464–465.

Brown, C and Brown, M. 1986. Ectoparasitism as a cost of coloniality in cliff swallows (*Hirundo pyrrhonota*). *Ecology*, 67, 1206–1218.

Bryant, D and Newton, A. 1994. Metabolic costs of dominance in dippers, *Cinclus cinclus*. *Animal Behaviour*, 48, 447–455.

Chapman, C, Rothman, J and Hodder, S. 2009. Can parasite infections be a selective force influencing primate group size? A test with red colobus. In: *Primate Parasite Ecology: The Dynamics and Study of Host–Parasite Relationships* (Ed. by Huffman, M. and Chapman, C.), pp. 423–440. Cambridge: Cambridge University Press.

Charpentier, M, Prugnolle, F, Gimenez, O and Widdig, A. 2008. Genetic heterozygosity and sociality in a primate species. *Behavior Genetics*, 38, 151–158.

Cheney, D, Seyfarth, R and Smuts, B. 1986. Social relationships and social cognition in non-human primates. *Science*, 234, 1261–1466.

Clay, C, Lehmer, E, Previtali, A, St. Jeor, S and Dearing, M. 2009. Contact heterogeneity in deer mice: implications for Sin Nombre virus transmission. *Proceedings of the Royal Society, B*, 276, 1305–1312.

Cohen, S, Line, S, Manuck, SB, Rabin, BS, Heise, ER and Kaplan, JR. 1997. Chronic social stress, social status, and susceptibility to upper respiratory infections in nonhuman primates. *Psychosomatic Medicine*, 59, 213–221.

Cohn, D and Sa-Rocha, L. 2006. Differential effects of lipopolysaccharide in the social behavior of dominant and submissive mice. *Physiology and Behavior*, 87, 932–937.

Cohn, D and Sa-Rocha, L. 2009. Sickness and aggressive behavior in dominant and subordinate mice. *Ethology*, 115, 112–121.

Cooke, SJ, Hinch, SG, Wikelski, M, Andrews, RD, Kuchel, LJ, Wolcott, TG and Butler, PJ. 2004. Biotelemetry: a mechanistic approach to ecology. *Trends in Ecology and Evolution*, 19, 334–343.

Corner, L, Pfeiffer, D and Morris, R. 2003. Social-network analysis of *Mycobacterium bovis* transmission among captive brushtail possums (*Trichosurus vulpecula*). *Preventive Veterinary Medicine*, 59, 147–167.

Cote, I and Poulin, R. 1995. Parasitism and group-size in social mammals: a meta-analysis. *Behavioral Ecology*, 6, 159–165.

Creel, S, Creel, N, Wildt, DE and Monfort, SL. 1992. Behavioural and endocrine mechanisms of reproductive suppression in Serengeti dwarf mongooses. *Animal Behaviour*, 43, 231–245.

Creel, S, Creel, NM, Mills, MG and Monfort, SL. 1997. Rank and reproduction in cooperatively breeding African wild dogs: behavioral and endocrine correlates. *Behavioral Ecology*, 8, 298–306.

Creel, S. 2001. Social dominance and stress hormones. *Trends in Ecology and Evolution*, 16, 491–497.

Creel, S and Sands, J. 2003. Is social stress a consequence of subordination or a cost of dominance? In: *Animal Social Complexity* (Ed. by de Waal, F. B. and Tyack, P. L.), pp. 153–169. Cambridge, MA: Harvard University Press.

Cristol, DA. 1995. Costs of switching social groups for dominant and subordiante dark-eyed juncos (*Junco hyemalis*). *Behavioral Ecology and Sociobiology*, 37, 93–101.

Dantzer, R. 2004. Cytokine-induced sickness behaviour: a neuroimmune response to activation of innate immunity. *European Journal of Pharmacology*, 500, 399–411.

Davidson, RS, Marion, G and Hutchings, MR. 2008. Effects of host social hierarchy on disease persistence. *Journal of Theoretical Biology*, 253, 424–433.

Davies, C, Ayres, J, Dye, C and Deane, L. 1991. Malaria infection rate of Amazonian primates increases with body weight and group size. *Functional Ecology*, 5, 655–662.

de Groot, J, van Milligen, FJ, Moonen-Leusen, BW, Thomas, G and Koolhaus, JM. 1999. A single social defeat transiently suppresses the anti-viral immune response in mice. *Journal of Neuroimmunology*, 95, 143–151.

De Groot, J, Ruis, M, Scholten, J, Koolhaas, J and Boersma, W. 2001. Long-term effects of social stress on antiviral immunity in pigs. *Physiology and Behavior*, 73, 145–158

De Luca, D and Ginsberg, J. 2001. Dominance, reproduction, and survival in banded mongooses: towards an egalitarian social system? *Animal Behaviour*, 61, 17–30.

Drea, CM. 2007. Sex and seasonal differences in aggression and steroid secretion in *Lemur catta*: Are socially dominant females hormonally "masculinized"? *Hormones and Behavior*, 51, 555–567.

Drews, C. 1993. The concept and definition of dominance in animal behaviour. *Behaviour*, 124, 283–313.

Dugatkin, L and Sih, A. 1995. Behavioral ecology and the study of partner choice. *Ethology*, 99, 265–277.

East, ML, Hofer, H, Cox, JH, Wulle, U, Wiik, H and Pitra, C. 2001. Regular exposure to rabies virus and lack of symptomatic disease in Serengeti spotted hyenas. *Proceedings of the National Academy of Sciences*, 98, 15026–15031.

Eklund, A. 1996. The effects of inbreeding on aggression in wild male house mice (*Mus domesticus*). *Behaviour*, 133, 883–901.

Ezenwa, V. 2004. Host social behavior and parasitic infection: a multifactorial approach. *Behavioral Ecology*, 15, 446–454.

Ferrari, N, Cattadori, I, Nesereira, J, Rizzoli, A and Hudson, P. 2004. The role of host sex in parasite dynamics: field experiments on the yellow-necked mouse *Apodemus flavicollis*. *Ecology Letters*, 7, 88–94.

Funk, S, Gilad, E, Watkins, C and Jansen, V. 2009. The spread of awareness and its impact on epidemic outbreaks. *Proceedings of the National Academy of Sciences*, 106, 6872–6877.

Gauly, M, Duss, C and Erhardt, G. 2007. Influence of *Ascaridia galli* infections and anthelmintic treatments on the behaviour and social ranks of laying hens (*Gallus gallus domesticus*). *Veterinary Parasitology*, 146, 271–270.

Godfrey, S, Bull, C, Murray, K and Gardner, M. 2006. Transmission mode and distribution of parasites among groups of the social lizard *Egernia stokesii*. *Parasitology Research*, 99, 223–230.

Godfrey, S, Bull, C, James, R and Murray, K. 2009. Network structure and parasite transmission in a group living lizard, the gidgee skink, *Egernia stokesii*. *Behavioral Ecology and Sociobiology*, 63, 1045–1056.

Goncalves Carobrez, S, Gasparotto, O, Buwalda, B and Bohus, B. 2002. Long-term consequences of social stress on corticosterone and IL-1β levels in endotoxin-challenged rats *Physiology and Behavior*, 76, 99–105.

Graham, A. 2002. When T-helper cells don't help: immunopathology during concomitant infection. *Quarterly Review of Biology*, 77, 409–434.

Graham, A, Shuker, D and Little, T. 2011. Fitness consequences of immune responses: strengthening the empirical framework for ecoimmunological studies. *Functional Ecology*, 25, 5–17.

Grear, D, Perkins, S and Hudson, P. 2009. Does elevated testosterone result in increased exposure and transmission of parasites? *Ecology Letters*, 12, 528–537.

Gross, W and Siegel, H. 1973. Effects of social stress and steroids on antibody production. *Avian Diseases*, 17, 807–815.

Grossman, CJ. 1985. Interactions between the gonadal steroids and the immune system. *Science*, 227, 257–261.

Hales, D. 1992. Interleukin-1 inhibits Leydig-cell steroidogenesis primarily by decreasing 17-alpha-hydroxylase C17–20 lyase cytochrome P450 expression. *Endocrinology*, 131, 2165–2172.

Halvorsen, O. 1986. On the relationship between social status of host and risk of parasitic infection. *Oikos*, 47, 71–74.

Hand, JL. 1986. Resolution of social conflicts: dominance, egalitarianism, spheres of dominance, and game theory. *Quarterly Review of Biology*, 61, 201–220.

Hart, B. 1988. Biological basis of the behavior of sick animals. *Neuroscience Biobehavioral Review*, 12, 123–137.

Hausfater, G and Watson, D. 1976. Social and reproductive correlates of parasite ova emissions by baboons. *Nature*, 262, 688–689.

Hawley, DM. 2006. Asymmetric effects of experimental manipulations of social status on individual immune response. *Animal Behaviour*, 71, 1431–1438.

Hawley, DM, Lindström, KM and Wikelski, M. 2006. Experimentally elevated social competition compromises humoral immunocompetence in house finches. *Hormones and Behavior*, 49, 417–414.

Hawley, DM, Jennelle, C, Sydenstricker, K and Dhondt, A. 2007. Pathogen resistance and immunocompetence covary with social status in house finches (*Carpodacus mexicanus*). *Functional Ecology*, 21, 520–527.

Hawley, DM, Etienne, RS, Ezenwa, VO and Jolles, AE. 2011. Does animal behavior underlie covariation between hosts' exposure to infectious agents and susceptibility to infection? Implications for disease dynamics. In press: *Integrative and Comparative Biology*.

Hogstad, O. 1987. It is expensive to be dominant. *Auk*, 104, 333–336.

Hogstad, O and Pedersen, HC. 2007. Heavy metal load and dominance hierarchy in juvenile willow tits during winter. *Environmental Pollution*, 147, 245–247.

Hutchings, M, Knowler, K, McAnulty, R and McEwan, J. 2007. Genetically resistant sheep avoid parasites to a greater extent than do susceptible sheep. *Proceedings in Biological Sciences*, 274, 1839–1944.

Hynes, EF, Rudd, CD, Temple-Smith, PD, Sofronidis, G, Paris, D, Shaw, G and Renfree, MB. 2005. Mating sequence, dominance and paternity success in captive male tammar wallabies. *Reproduction*, 130, 123–130.

Jarrell, H, Hoffman, J, Kaplan, J, Berga, S, Kinkead, B and Wilson, M. 2008. Polymorphisms in the serotonin reuptake transporter gene modify the consequences of social status on metabolic health in female rhesus monkeys. *Physiology and Behavior*, 83, 807–819.

Johnston, S, Ward, D, Lemon, J, Gunn, I, MacCallum, C, Keeley, T and Blyde, D. 2007. Studies of male reproduction in captive African wild dogs. *Animal Reproductive Science*, 100, 338–355.

Kavaliers, M, Choleris, E, Agmo, A, Braun, J, Colwell, D, Muglia, L, Ogawa, S and Pfaff, D. 2006. Inadvertent social information and the avoidance of parasitized male mice: A role for oxytocin. *Proceedings of the National Academy of Sciences*, 103, 4293–4298.

Ketterson, ED, Nolan, V, Wolf, L, Ziegenfus, C, Duffy, AM, Ball, GF and Johnsen, TS. 1991. Testosterone and avian life histories: the effect of experimentally elevated testosterone on corticosterone and body mass in dark-eyed juncos. *Hormones and Behavior*, 25, 489–503.

Kiesecker, J, Skelly, D, Bear, K and Preisser, E. 1999. Behavioral reduction of infection risk. *Proceedings of the National Academy of Sciences*, 96, 9165–9168.

Kulkarni, S and Heeb, P. 2007. Social and sexual behaviours aid transmission of bacteria in birds. *Behavioral Processes*, 74, 88–92.

Lindstrom, KM. 2004. Social status in relation to Sindbis virus infection clearance in greenfinches. *Behavioral Ecology and Sociobiology*, 55, 236–241.

Lister, A and van der Kraak, G. 2002. Modulation of goldfish testicular testosterone production in vitro by tumor necrosis factor alpha, interleukin-1 beta, and macrophage conditioned media. *Journal of Experimental Zoology*, 292, 477–486.

Lloyd-Smith, J, Getz, W and Westerhoff, H. 2004. Frequency-dependent incidence in models of sexually transmitted diseases: portrayal of pair-based transmission and effects of illness on contact behaviour. *Proceedings of the Royal Society, B*, 271, 625–634.

Lloyd-Smith, J, Schreiber, S, Kopp, P and Getz, W. 2005. Superspreading and the effect of individual variation on disease emergence. *Nature*, 438, 355–359.

Lochmiller, R. 1996. Social barriers to pathogen transmission in wild animal populations. *Ecology*, 76, 326–335.

Loehle, C. 1995. Social barriers to pathogen transmission in wild animal populations. *Ecology*, 76, 326–335.

Malaivijitnond, S, Hamada, Y, Suryobroto, B and Takenaka, O. 2007. Female long-tailed macaques with scrotum-like structure. *American Journal of Primatology*, 69, 721–735.

Meagher, S, Penn, D and Potts, W. 2000. Male–male competition magnifies inbreeding depression in wild house mice. *Proceedings of the National Academy of Sciences*, 97, 3324–3329.

Metcalfe, NB, Taylor, AC and Thorpe, JE. 1995. Metabolic rate, social status and life-history strategies in Atlantic salmon. *Animal Behaviour*, 49, 431–436.

Moller, A, Dufva, R and Allender, K. 1993. Parasites and the evolution of host social behavior. *Advances in the Study of Behavior*, 22, 65–102.

Moore, J, Simberloff, D and Freehling, M. 1988. Relationships between bobwhite quail social-group size and intestinal helminth parasitism. *American Naturalist*, 131, 22–32.

Morrow-Tesch, J, McGlone, J and Salak-Johnson, J. 1994. Heat and social stress effects on pig immune measures. *Journal of Animal Science*, 72, 2599–2609.

Mougeot, F, Redpath, S, Piertney, S and Hudson, P. 2005. Separating behavioral and physiological mechanisms in testosterone-mediated trade-offs. *American Naturalist*, 166, 158–168.

Naug, D. 2008. Structure of the social network and its influence on transmission dynamics in a honeybee colony. *Behavioral Ecology and Sociobiology*, 62, 1719–1725.

Nunn, C, Altizer, S, Jones, K and Sechrest, W. 2003. Comparative tests of parasite species richness in primates. *American Naturalist*, 162, 597–614.

Nunn, C and Altizer, S. 2006. *Infectious Diseases in Primates: Behavior, Ecology and Evolution*. New York: Oxford University Press.

Owen-Ashley, N, Hasselquist, D and Wingfield, JC. 2004. Androgens and the immunocompetence handicap hypothesis: unraveling direct and indirect pathways of immunosuppression in song sparrows. *American Naturalist*, 164, 490–505.

Perkins, S, Cattadori, I, Tagliapietra, V, Rizzoli, A and Hudson, P. 2003. Empirical evidence for key hosts in persistence of a tick-borne disease. *International Journal of Parasitology*, 33, 909–917.

Perkins, S, Cagnacci, F, Stradiotto, A, Arnoldi, D and Hudson, P. 2009. Comparison of social networks derived from ecological data: implications for inferring infectious disease dynamics. *Journal of Animal Ecology*, 78, 1015–1022.

Piper, W and Wiley, R. 1990. The relationship between social dominance, subcutaneous fat, and annual survival in wintering white-throated sparrows (*Zonotrichia albicollis*). *Behavioral Ecology and Sociobiology*, 26, 201–208.

Plowright, RK, Field, HE, Smith, C, Divljan, A, Palmer, C, Tabor, G, Daszak, P and Foley, JE. 2008. Reproduction and nutritional stress are risk factors for Hendra virus infection in little red flying foxes (*Pteropus scapulatus*). *Proceedings of the Royal Society, B*, 275, 861–869.

Pulliam, H. 1973. On the advantages of flocking. *Journal of Theoretical Biology*, 38, 419–422.

Read, A, Graham, A and Raberg, L. 2009. Animal defenses against infectious agents: is damage control more important than pathogen control? *PLOS Biology*, 6, 2638–2641.

Roberts, M, Buchanan, K and Evans, M. 2004. Testing the immunocompetence handicap hypothesis: a review of the evidence. *Animal Behaviour*, 68, 227–239.

Roskaft, E, Jarvi, T, Bakken, M, Bech, C and Reinertsen, R. 1986. The relationship between social status and resting metabolic rate in great tits (*Parus major*) and pied flycatchers (*Ficedula hypoleuca*). *Animal Behaviour*, 34, 838–842.

Rubenstein, DI and Hohmann, ME. 1989. Parasites and social behavior of island feral horses. *Oikos*, 55, 312–320.

Sands, J and Creel, S. 2004. Social dominance, aggression and faecal glucocorticoid levels in a wild population of wolves, *Canis lupus*. *Animal Behaviour*, 67, 387–396.

Sapolsky, RM. 1985. Stress-induced suppression of testicular function in the wild baboon: role of glucocorticoids. *Endocrinology*, 116, 2273–2278.

Sapolsky, RM. 2004. Social status and health in humans and other animals. *Annual Review of Anthropology*, 33, 393–418.

Sapolsky, RM. 2005. The influence of social hierarchy on primate health. *Science*, 308, 648–652.

Schino, G. 2001. Grooming, competition and social rank among female primates: a meta-analysis. *Animal Behaviour*, 62, 265–271.

Scott, J and Lockard, JS. 2006. Captive female gorilla agonistic relationships with clumped defendable food resources. *Primates*, 47, 199–209.

Senar, JC, Polo, V, Uribe, F and Camerino, M. 2000. Status signalling, metabolic rate and body mass in the siskin: the cost of being a subordinate. *Animal Behaviour*, 59, 103–110.

Shargal, D, Shore, L, Roteri, N, Terkel, A, Zorovsky, Y, Shemesh, M and Steinberger, Y. 2008. Fecal testosterone is elevated in high ranking female ibexes (*Capra nubiana*) and associated with increased aggression and a preponderance of male offspring. *Theriogenology*, 69, 673–680.

Smith, J, Memenis, S and Holekamp, K. 2007. Rank-related partner choice in the fission-fusion society of the spotted hyena (*Crocuta crocuta*). *Behavioral Ecology and Sociobiology*, 61, 753–765.

Stefanski, V. 1998. Social stress in loser rats: Opposite immunological effects in submissive and subdominant males. *Physiology and Behavior*, 63, 605–613.

Swinton, J, Woolhouse, M, Begon, M, Dobson, A, Ferroglio, E and Grenfell, B. 2002. Microparasite transmission and persistence. In: *The Ecology of Wildlife Diseases* (Ed. by Hudson, P., Rizzoli, A., Grenfell, B., Heesterbeek, H. and Dobson, A.), pp. 83–101. New York: Oxford University Press.

Sydenstricker, K, Dhondt, A, Hawley, DM, Jennelle, C and Kollias, G. 2006. Characterization of experimental *Mycoplasma gallisepticum* infection in captive house finch flocks. *Avian Diseases*, 50, 39–44.

Taillon, J and Cote, SD. 2008. Are faecal hormone levels linked to winter progression, diet quality and social rank in young ungulates? An experiment with white-tailed deer (*Odocoilus virginianus*) fawns. *Behavioral Ecology and Sociobiology*, 62, 1591–1600.

Teichroeb, JA and Sicotte, P. 2008. Social correlates of fecal testosterone in male ursine colobus monkeys (*Colobus vellerosus*): The effect of male reproductive competition in aseasonal breeders. *Hormones and Behavior*, 54, 417–423.

Temime, L, Opatowski, L, Pannet, Y, Brun-Buisson, C, Yves Boelle, P and Guillemot, D. 2009. Peripatetic health-care workers as potential superspreaders. *Proceedings of the National Academy of Sciences*, 106, 19420–18425.

Tiira, K, Laurila, A, Enber, K, Piironen, J, Aikio, S, Ranta, E and Primmer, C. 2006. Do dominants have higher heterozygosity? Social status and genetic variation in brown trout, *Salmo trutta*. *Behavioral Ecology and Sociobiology*, 59, 657–665.

Tuchscherer, M, Puppe, B, Tuchscherer, A and Kantiz, E. 1998. Effects of social status after mixing on immune, metabolic, and endocrine responses in pigs. *Physiology and Behavior*, 64, 353–360.

Vezina, F and Thomas, DW. 2000. Social status does not affect resting metabolic rate in wintering dark-eyed junco (*Junco hyemalis*). *Physiological and Biochemical Zoology*, 73, 231–236.

Weil, Z, Bowers, S, Pyter, L and Nelson, R. 2006. Social interactions alter proinflammatory cytokine gene expression and behavior following endotoxin administration. *Brain, Behavior, and Immunity*, 20, 72–79.

Woodroffe, R, Donnelly, C, Wei, G, Cox, D, Bourne, F, Burke, T, Buttin, R, Cheeseman, C, Gettinby, G, Gilks, P, Hedges, S, Jenkins, H, Johnston, W, Mclnerney, J, Morrison, W and Pope, L. 2009. Social group size affects *Mycobacterium bovis* infection in European badgers (*Meles meles*). *Journal of Animal Ecology*, 78, 818–827.

Zine, MJ and Krausman, PR. 2000. Behavior of captive mountain sheep in a Mojave Desert environment. *Southwestern Naturalist*, 45, 184–195.

Zuk, M and Johnsen, TS. 2000. Social environment and immunity in male red jungle fowl. *Behavioral Ecology*, 11, 146–153.

CHAPTER 15

SEXUAL SELECTION AND PARASITES

Do Mechanisms Matter?

Anne C. Jacobs and Marlene Zuk

INTRODUCTION

Parasites have long been appreciated as a challenge that most, if not all, organisms must face, one that may affect many aspects of host biology and fitness. In particular, the effects of parasites on host mating decisions and mating behavior have become an area of active study in recent years. The potential importance of parasites in mate choice was first pointed out by Hamilton and Zuk (1982). The Hamilton-Zuk hypothesis postulates that exaggerated ornaments and sexual signals arose in part because they can signal parasite resistance, and that females should prefer flashier mates because the genes for resistance would be passed on to their offspring (Hamilton and Zuk, 1982).

The Hamilton-Zuk hypothesis is not alone in suggesting that females may benefit by choosing parasite-free mates. Other ideas, such as the transmission avoidance hypothesis and the resource provisioning model (also sometimes called the "good parent" hypothesis), also make this prediction (Clayton, 1991; Møller, 1994). The transmission avoidance hypothesis states that females should avoid parasitized mates to avoid becoming infected themselves, while the resource provisioning model assumes that parasite-free mates should be better at providing resources to females and offspring (Clayton, 1991). However, a key difference between these models and the Hamilton-Zuk hypothesis is that the Hamilton-Zuk hypothesis specifically states that parasite resistance is heritable and preference for healthy individuals is favored because females obtain genes for resistance for their offspring. This idea that females select for resistance genes is an appealing one, as co-evolution between hosts and pathogens may explain why genes for fitness do not go to fixation, even when subjected to strong selection pressure from female preference. The Hamilton-Zuk hypothesis has generated both intra- and inter-specific predictions about the connection between parasite levels and sexual signals. Within a species, one expects preferred mates to have flashier sexual signals and fewer parasites. When comparing multiple species, one would expect selection pressure on sexual ornaments to be strongest in species infected with more parasites, as females would benefit more from choosing resistant mates in such species. Thus one would predict that the species with the flashiest sexual signals would be infected by the greatest number of parasites.

After the publication of the Hamilton-Zuk hypothesis, many scientists set out to test the relationship between parasites and sexual signals (reviewed in Clayton, 1991; Zuk, 1992; Møller et al., 1999). Most of the studies focused on the intraspecific effects of parasites on mate choice. Researchers often picked a single parasite species to work with, studying its effects on host sexual signaling and mating preference. For instance, Møller (1990) found negative effects of an ectoparasitic mite on tail length in barn swallows, a species in which the length of the tail is an important sexual signal. Zuk et al. (1990) found that nematode infection decreased sexual signal quality while having few or no effects on non-sexually selected traits in jungle fowl, and that females selecting mates preferred unparasitized males with better sexual signals.

However, while the Hamilton-Zuk hypothesis set very clear and testable predictions about the relationship between parasites and sexual signals, the authors also cautioned that not every parasite species could be expected to influence the evolution of sexual signals in the same way. Parasites that killed hosts too quickly would leave only resistant individuals in the population, thus ensuring that every individual encountered would be resistant and removing the need for choosiness. On the other hand, benign parasites would not cause enough harm for choosy females to benefit by selecting mates with genes for parasite resistance. Thus, the hypothesis suggested that the parasites likely to be most important to the evolution of sexual signals were those that caused chronic, debilitating diseases (Hamilton and Zuk, 1982). This creates a potential problem for researchers testing the Hamilton-Zuk hypothesis, as picking the "wrong" parasite to work with might result in there being no relationship observed between parasites and sexual signaling.

While many studies have examined the effects of a parasite that causes a chronic infection on sexual signals, one may question how broadly the results of such studies can be interpreted. Examining the effects of a single parasite on sexual signals can provide useful information, but it may not tell us how an animal may respond to other parasites and pathogens. As animals are exposed to many species of parasites over the course of their lifetimes, it is difficult to evaluate the importance of any one in particular. Are females interested in males capable of resisting that one species, or do they prefer males that can respond to parasites in general and avoid or resist infection by a large number of pathogens? Answering the latter question when examining parasites alone is difficult, as it is often unfeasible to measure the total parasite burden of an animal. Thus, many scientists have instead chosen to look at immune response. The immune system has evolved to deal with most, if not all, of the parasites an animal may encounter, and thus should provide an acceptable idea of an animal's parasite resistance (Møller et al., 1999). This new direction brought scientists interested in sexual selection into the field of ecological immunology (Sheldon and Verhulst, 1996; Møller et al., 1999; Lawniczak et al., 2007).

Ecological immunology (ecoimmunology) is based on the principle that immune response is a necessary but costly trait in terms of energy and resources, and that these costs make immune defense a candidate for trade-offs with other life-history traits, such as reproductive effort (Zuk, 1990; Sheldon and Verhulst, 1996; Zuk and Stoehr, 2002; French et al., 2009). The issue of costs of immunity is an important one from the perspective of sexual signaling because it helps explain how signal honesty is maintained (Kotiaho, 2001). For signals to remain honest indicators of quality, some cost must be associated with both signals and immune response, creating a negative relationship between immunity and signal quality within an individual. These costs impose a handicap that can maintain signal reliability, as low-quality males will not be able to pay the costs associated with high signal quality (Zahavi, 1975; Nur and Hasson, 1984; Getty, 2002).

While the idea that immune response may be linked to sexual signals started as a development of the Hamilton-Zuk hypothesis, the Hamilton-Zuk hypothesis is not quite the same as many of the ideas put forward in ecoimmunology. Both seek to put immunology in a more evolutionary context by pointing out that parasites and pathogens represent an important selection pressure to their hosts and that defenses to fight off these pathogens should be under strong selection (Zuk, 1994; Sheldon and Verhulst, 1996; Zuk and Stoehr, 2002). However, the Hamilton-Zuk hypothesis emphasizes genetic-based parasite resistance as the factor a female should care most about in a mate. Only a resistant male would have low enough parasite burdens to produce a high-quality sexual signal. Also, the Hamilton-Zuk hypothesis relies on co-evolutionary cycles between hosts and their pathogens to generate a constant source of genetic variability in quality (Hamilton and Zuk, 1982). In contrast, ecological immunologists tend to focus more on the ways in which energy allocation to immunity versus signaling can influence sexual signal quality. This explanation does not include parasites directly, although parasites are certainly important in that they impose a cost on individuals who allocate too many resources away from the immune system. Instead of focusing on parasites and how they interact with their hosts directly, ecoimmunology relies more on the costs of immunity as the all-important explanation behind signal honesty. This causes ecological immunologists to emphasize the importance of an animal's general immune response, rather than resistance to a given pathogen (Westneat and Birkhead, 1998).

While fundamental differences exist between the Hamilton-Zuk hypothesis and ecoimmunology's emphasis on energetic trade-offs, both predict that sexual signaling and immunity should be tied together. Much of the work examining sexual selection and immunology has attempted to link variation in immune response with variation in sexual signals and/or mating success. To do this, many studies have measured immune response (Box 15.1) and looked for relationships between these measures and the quality of sexual signals. Table 15.1 summarizes such correlational studies

BOX 15.1: WHAT IS IMMUNITY AND HOW DO WE MEASURE IT?

Anyone who has studied immunity can agree at least on one point: the immune system is extremely complex. In fact, in vertebrates, immunity is often broken down into many parts, each corresponding to a certain "arm" of the immune system. When one takes into account the many parts of the immune system, defining what constitutes a strong immune system can become extremely difficult and tricky. Many scientists talk about an animal's "immunocompetence," or the overall ability of that animal to use its immune system to resist parasites. In practice, of course, measuring immunocompetence is next to impossible (Zuk and Stoehr, 2002). However, many studies try to be as thorough as possible by employing a variety of immunological measures to try to quantify an animal's overall immune response. The commonly used divisions of the immune system as well as methods used to quantify them are as follows:

- **Innate Immunity:** Innate immunity describes the actions of many cells or molecules that are always present and designed to fight off many different potential threats. These include natural killer cells, complement, macrophages, etc. Some immunologists also include physical barriers to infection, such as skin, in this category (Weir and Stewart, 1993), although the ecoimmunology literature does not usually consider such barriers. To quantify innate immunity, many researchers measure some part of the defenses already present, usually using a blood sample. One such popular method is to count the number and varieties of leukocytes in the blood (Skarstein and Folstad, 1996; Zuk and Johnsen, 1998), although such measures may be difficult to interpret (Sheldon and Verhulst, 1996; Norris and Evans, 2000). A more recent, and more easily interpreted, method involves measuring the ability of blood to kill bacteria (Millet et al., 2007).
- **Acquired Immunity:** Acquired immunity differs from innate immunity in that it is activated in response to a specific challenge. In vertebrates, it also produces immunological memory, whereby the immune system "remembers" specific pathogens that the organism has encountered in the past. This enables the mounting of a more efficient secondary response to that pathogen if it is ever encountered again. Methods to measure this type of immunity usually involve challenging the animal in some way with a novel antigen and measuring the response. Acquired immunity is further broken down into cell-mediated and humoral immunity (Weir and Stewart, 1993).
    - *Cell-mediated immunity* describes the response of the T lymphocytes. One very popular field method to measure this is the phytohemagglutinin (PHA) test, in which a mitogen is injected into the animal's body and the resulting swelling is measured (Møller and Petrie, 2002; Martín et al., 2007; Garvin et al., 2008; Griggio et al., 2010). PHA is used to stimulate T-lymphocyte proliferation in the lab, and thus this test has been thought to indicate cell-mediated immunity (Kennedy and Nager, 2006). However, while T lymphocytes are certainly involved in producing this swelling, other cells from the innate portion of the immune system may respond as well (Martin et al., 2006b; Tella et al., 2008). Thus, ecological immunologists must be cautious in their interpretations of this test (Kennedy and Nager, 2006).
    - *Humoral immunity* covers the actions of the B lymphocytes and the antibodies they produce. Popular tests to measure this involve challenges with various antigens designed to elicit an antibody response. The antigens commonly used include sheep red blood cells (Møller and Petrie, 2002; Bonneaud et al., 2005), tetanus–diphtheria vaccines (Råberg et al., 2000; Hanssen et al., 2008), and others (Aguilera and Amat, 2007). For more a complete discussion of methods to measure immunity, see reviews by Norris and Evans (2000), Grasman (2002), and Salvante (2006).

TABLE 15.1: Summary of Correlational Studies Examining the Relationship Between Immune Response and Sexual Signal Quality

Species	Measure of Immunity	Signal Considered	Relationship Between Signal Quality and Immune Response	Sources
Arthropods				
Hygrolycosa rubrofasciata (wolf spider)	Encapsulation response	Drumming rate	Positive	Ahtiainen et al., 2004
Hygrolycosa rubrofasciata (wolf spider)	Lytic activity	Male mobility	Positive	Ahtiainen et al., 2004
Calopteryx splendens xanthostoma (damselfly)	Increase in phenoloxidase levels after a challenge	Darkness and homogeneity of wing pigmentation	Negative	Siva-Jothy, 2000
Teleogryllus commodus (field cricket)	Encapsulation response	Calling song syllable duration	Positive	Simmons et al., 2005
Acheta domesticus (house cricket)	Hemocyte load	Number of syllables per chirp in calling song	Positive	Ryder and Siva-Jothy, 2000
Tenebrio molitor (mealworm beetle)	Phenoloxidase activity	Pheromone scent	Positive	Rantala et al., 2002
Tenebrio molitor (mealworm beetle)	Encapsulation response	Pheromone scent	Positive	Rantala et al., 2002
Fish				
Salvelinus alpines (Arctic charr)	Lymphocyte count	Degree of red coloration	Negative	Skarstein and Folstad, 1996
Reptiles				
Podarcis hispanica (Iberian wall lizard)	PHA* response	Chemical composition of femoral secretions	Positive	López and Martín, 2005
Psammodromus algirus (large Psammodromus)	PHA* response	Chemical composition of femoral secretions	Positive	Martín et al., 2007
Birds				
Gallus gallus (jungle fowl)	Lymphocyte count	Comb size	Negative	Zuk et al., 1995

TABLE 15.1: (Continued)

Species	Measure of Immunity	Signal Considered	Relationship Between Signal Quality and Immune Response	Sources
Phasianus colchicus (pheasant)	PHA* response	Spur length	Positive	Ohlsson et al., 2002
Lagopus lagopus scoticus (red grouse)	PHA* response	Comb color	Positive	Mougeot, 2008
Pavo cristatus (peafowl)	PHA* response	Male train length	Positive	Møller and Petrie, 2002
Pavo cristatus (peafowl)	Antibody response to sheep red blood cells	Male train length	Negative	Møller and Petrie, 2002
Melopsittacus undulates (budgerigar)	PHA* response	UV chroma of feathers	Positive	Griggio et al., 2010
Turdus merula (European blackbird)	PHA* response	Bill coloration	Positive	Faivre et al., 2003
Turdus merula (European blackbird)	Secondary antibody response	Bill coloration	Negative	Faivre et a., 2003
Sialia mexicana (Western bluebird)	Ability to survive an epidemic	Blue coloration	Negative	Keyser and Siefferman, 2005
Sternus vulgaris (European starling)	PHA* response	Song rate	Positive	Duffy and Ball, 2002
Sternus vulgaris (European starling)	Antibody response to keyhole limpet hemocyanin	Song rate	Positive	Duffy and Ball, 2002
Carpodacus mexicanus (house finch)	Ability to survive an epidemic	Red coloration	Positive	Nolan et al., 1998
Geothlypis trichas (common yellowthroat)	IgG levels	Size of facial mask	Positive	Garvin et al., 2008
Geothlypis trichas (common yellowthroat)	PHA* response	Size of facial mask	Negative	Garvin et al., 2008
Carduelis chloris (greenfinch)	Viral clearance rate	Male tail patch size	Positive	Lindström and Lundström, 2000
Carduelis chloris (greenfinch)	Antibody titers in response to challenge with *Brucella abortis*	Male breast coloration (chroma)	Negative	Aguilera and Amat, (2007)

*PHA = phytohemagglutinin response, a common assay of cell-mediated immunity in which phytohemagglutinin is injected and the swelling response is measured. Greater swelling is considered a sign of a stronger response (and therefore a stronger immune system).

and their results. Other studies have experimentally manipulated ornaments or challenged the immune system with an antigen. For instance, Saino and Møller (1996) manipulated the tail length of male barn swallows (a sexual signal in this species) and found that immune response decreased in males with lengthened tails, but that males that had had longer tails to begin with were better able to cope with the manipulation, implying that individual quality and immune response are linked. López et al. (2009) found that challenging the immune systems of male lizards caused them to develop duller ventral coloration.

Much research suggests that immunity and sexual signaling are linked, yet several questions about the nature of that relationship remain. First, we lack clearly defined predictions about the expected relationship between immune response and sexual signal quality. Studies that find opposite results may both be interpreted as supporting the same hypothesis, leading to a great deal of confusion. The mechanisms that link immunity and signaling remain controversial. Many hypotheses have been proposed (Fig. 15.1), and a considerable body of work has been devoted to trying to test these hypotheses.

However, we argue that most of the hypotheses generated to explain how sexual signaling relates to immunity are insufficient and do not explain the conflicting results obtained in the field. We use this chapter to address the following questions: (1) What is the predicted relationship between sexual signal quality and immune response? (2) What results have been obtained by studies looking for a relationship between sexual signals and immunity? (3) In cases where different studies have obtained conflicting results, how can we explain these discrepancies? We begin with a brief overview of the prominent mechanisms proposed to explain how immunity relates to sexual signal quality. We then explain why we think that these hypotheses by themselves are insufficient, and we draw on work done in other areas of the field of ecoimmunology in an attempt to explain some of the conflicting results obtained to date.

SEXUAL SELECTION AND IMMUNE RESPONSE: POPULAR MECHANISMS

IMMUNITY AS A LIFE-HISTORY TRAIT

Perhaps the simplest way of explaining the link between immune response and sexual signals is the one first described by Sheldon and Verhulst (1996). Under this scenario, immunity is viewed as a life-history trait, the costs of which must be balanced with competing costs in a way that maximizes reproductive success. If an animal allocates more energy to immunity, less energy will be available for breeding or other functions (see Chapter 8). The males best able to balance these costs or those that have more energy available to allocate will be the most successful. This implies that such energetic trade-offs are what determine the quality and honesty of sexual signals. Genetic resistance to parasites and the co-evolutionary cycles that are expected to result from that do not play a role in this mechanism, setting it apart from the original Hamilton-Zuk hypothesis (Westneat and Birkhead, 1998). The life-history explanation of immune response and sexual signaling is very general, and some of the more detailed physiological mechanisms proposed to explain why sexual signals may indicate immunity can be seen as extensions of this idea.

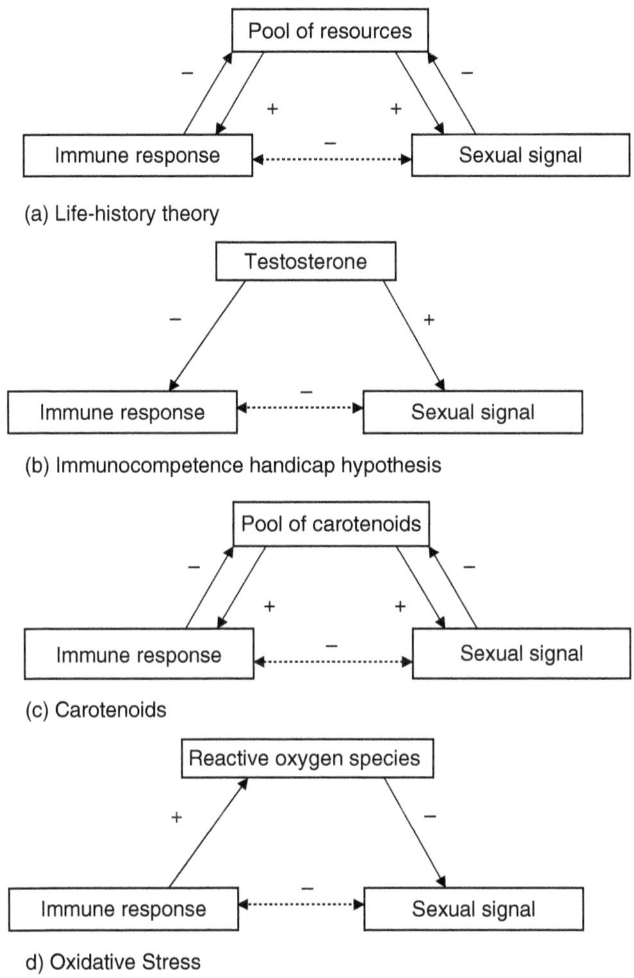

FIGURE 15.1: A summary of the mechanisms linking immunity and sexual signal quality. In all cases, there is a negative relationship predicted between sexual signaling and immune response. (a) The life-history theory of immunity predicts that the trade-off between immune response is due to limited energy or resources that can be allocated either to sexual signaling or to immune response. (b) The immunocompetence handicap hypothesis proposes that testosterone simultaneously increases sexual signaling while depressing immune response. (c) Carotenoids are thought to be a limited resource important both in signaling and immune function. Some have hypothesized that allocation of carotenoids either to sexual signals or to the immune system may be responsible for the trade-off between immunity and signaling. (d) The oxidative stress hypothesis predicts that immune responses generate reactive oxygen species, which then cause oxidative stress and lower the quality of sexual ornaments.

As mentioned earlier, under this hypothesis immune response must be costly to keep sexual signals honest. The costs of immunity have been demonstrated numerous times (reviewed in Zuk and Stoehr, 2002; Schmid-Hempel, 2003). Evidence for these costs comes both from studies that show a physiological and/or behavioral cost to immune activation (e.g., Råberg et al., 2000; Whitaker and Fair, 2002; Jacot et al., 2004; Marzal et al., 2007), and from selection experiments on laboratory populations. For instance, in domestic fowl, males selected for high immune response had smaller sexual ornaments (Verhulst et al., 1999), while in *Drosophila melanogaster*, males selected for increased ability to secure mates had lower immune responses (McKean and Nunney, 2008). The costs associated with immune response may ensure that only high-quality individuals are able to afford such costs and still produce a good sexual signal.

THE IMMUNOCOMPETENCE HANDICAP HYPOTHESIS

One of the first and most prominent ideas explaining the link between immunity and signaling is the immunocompetence handicap hypothesis (Folstad and Karter, 1992). This hypothesis suggests that testosterone provides the mechanistic link between signals and immune response. Because testosterone can be immunosuppressive and is necessary for the development of many sexual signals in vertebrates, it may create a system of trade-offs such that only males of high quality can afford to suppress their immune systems to the point necessary for signal development. Like the life-history explanation of immunity, the immunocompetence handicap hypothesis predicts that, within an individual, there should be a trade-off between immunity and sexual signal quality. However, the immunocompetence handicap hypothesis is novel in that it relies on hormones (and testosterone in particular) to create this trade-off and maintain signal honesty. This puts testosterone's regulation of immune response and sexual signals into a larger, evolutionary picture.

The immunocompetence handicap hypothesis as it was initially proposed differed greatly from the life-history explanation of immunity and signaling. Although some of their basic predictions are similar, the immunocompetence handicap hypothesis relied solely on testosterone as the factor mediating trade-offs between immunity and sexual signaling, without having to invoke any energy or resource-based explanations for why immune response should decrease as sexual signal quality increased. However, one could argue that if testosterone alone was responsible for this trade-off, nothing would prevent the spread of a mutant "cheater" whose immune system did not react to testosterone. Such a mutant would benefit from having highly exaggerated ornamentation without having to pay the costs in terms of immunosuppression. To explain why such a mutant could not arise, Wedekind and Folstad (1994) proposed that immunosuppression by testosterone was adaptive because it allowed for the reallocation of resources from the immune system to the sexual signal. Seen in this light, suppression of immunity by testosterone is merely a proximate explanation for the changes in immunity predicted by life-history theory. Thus, while the immunocompetence handicap hypothesis was initially proposed as a mechanism by which the Hamilton-Zuk hypothesis could operate, it deviates from the Hamilton-Zuk hypothesis by emphasizing the importance of trade-offs between immunity and signaling. Under this scenario, genetic resistance to parasites, which is crucial to the Hamilton-Zuk hypothesis, may become less important.

The immunocompetence handicap hypothesis has been extremely popular in the field of ecoimmunology and much work has focused on testing its predictions (Evans et al., 2000; Cox and John-Alder, 2007; Mougeot et al., 2009; Ros et al., 2009). However, it also suffers from several problems. One of these concerns the validity of a key assumption underlying the hypothesis, namely that testosterone is immunosuppressive. While Folstad and Karter (1992) cited numerous instances in which testosterone appeared to suppress immunity, a more recent review by

Roberts et al. (2004) found at best weak support for this assertion in most vertebrate taxa. Others have suggested that testosterone does not truly suppress immunity, but that cells of the immune system are redistributed from the blood to the tissues, making them harder to detect. This response is thought to promote wound healing and avoidance of infection after stress and competitive encounters (Braude et al., 1999). This redistribution effect was proposed to explain observed changes in only certain measures of immune response, namely lymphocyte numbers; other branches of the immune system, such as the humoral response, should not be affected by testosterone (Braude et al., 1999). Thus, the immunoredistribution hypothesis does not help to explain why immune response as a whole should appear to be lower. Also, the redistribution of immune response is thought to occur in response to stress, and there is reason to believe that stress may simply suppress immunity rather than redistributing it (Råberg et al., 1998; Adamo and Parsons, 2006).

While current evidence seems to indicate that testosterone does not always suppress immune response, this does not mean that the hypothesis as a whole is completely useless. Even if testosterone itself cannot suppress immunity, it may act indirectly through other means. Testosterone may influence behaviors such that individuals with high testosterone are more likely to become infected by parasites (Roberts et al., 2004). Alternatively, testosterone levels may be correlated with high levels of other hormones that influence immunity. Many researchers have suggested that corticosterone, a stress hormone that is often elevated in individuals with high testosterone, may provide the missing link (Evans et al., 2000; Owen-Ashley et al., 2004; Roberts et al., 2004), or that corticosterone and testosterone somehow interact to produce immunosuppression (Buchanan et al., 2003). For example, Evans et al. (2000) examined the effects of experimentally altered testosterone levels on corticosterone levels and antibody production in house sparrows. They found that increasing testosterone levels also increased corticosterone, and that once the effects of corticosterone were accounted for, testosterone appeared to improve immune response. Hormones such as corticosterone may also be involved in the adaptive suppression of immune response during times of stress (Råberg et al., 1998). Given the stress breeding males likely experience, this mechanism seems highly plausible (Zuk, 1990). However, as increased testosterone levels may be accompanied by increasing corticosterone levels (Evans et al., 2000), disentangling the effects of these two hormones may not be simple.

Other work suggests that testosterone may act by increasing damage done by oxidative stress, and that this damage reduces the quality of sexual signals (Alonso-Alvarez et al., 2007). This idea incorporates a newer mechanism, discussed below, which suggests that oxidative stress, and not testosterone *per se*, plays the main role in maintaining signal honesty. It should be pointed out, however, that regardless of the minute details, all of these ideas can be considered subsets of the immunocompetence handicap hypothesis, as they all suggest that testosterone is somehow related to signal development and immunosuppression, whether directly or indirectly. Thus, all of these modifications propose the same basic outcome, namely that immune response trades off against sexual signal quality.

Perhaps the greatest limitation of the immunocompetence handicap hypothesis arises from the fact that this mechanism, by its very nature, applies only to certain taxa, specifically vertebrates. Invertebrates lack testosterone, and thus the hypothesis has created a divide between work done on vertebrates and work done on invertebrates. Clearly one cannot expect to find testosterone-mediated immunosuppression or signal enhancement in invertebrates, and thus most researchers interested in the immunocompetence handicap hypothesis have focused on vertebrates and largely ignored work done on invertebrates. However, whether such a divide should really exist is questionable. Compared to vertebrates, the invertebrate immune system uses very different mechanisms to fight off invaders (see Chapter 13 and 17), yet despite these differences evidence suggests that invertebrate sexual signals may be related to immune response as well (Siva-Jothy, 2000; Ahtiainen et al., 2004; Simmons et al., 2005). Ahtiainen et al. (2004) found that male wolf spiders (*Hygrolycosa rubrofasciata*) that displayed faster drumming rates (a sexual signal) also had a stronger encapsulation response (a measure of immunity). These results are similar to results obtained in many vertebrate studies (Table 15.1), implying that sexual signals may convey information about immune response in invertebrates just as they are thought to in vertebrates (Lawniczak et al., 2007).

The similarity of the results obtained for vertebrate and invertebrate work raises the question of how signal honesty is maintained. The trade-off between immune response and sexual signaling seen in vertebrates also seems to be present in invertebrates (Jacot et al., 2004; McKean and Nunney, 2008; Leman et al., 2009), but if testosterone does not mediate such a trade-off, what does? Some researchers have proposed that the immunocompetence handicap hypothesis may still apply to invertebrates as long as some other substance performs the same function as testosterone by increasing signal quality while decreasing immune response (Rantala et al., 2003). Several candidate substances have been proposed, including juvenile hormone, which is thought to simultaneously promote traits such as dominance and sexual signaling while depressing immunity (Rantala et al., 2003; Contreras-Garduño et al., 2009). Thus, juvenile hormone is thought to act like testosterone in insects. However, juvenile hormone differs from testosterone in that it is present in both sexes, whereas testosterone is primarily associated with males (Folstad and Karter, 1992; Contreras-Garduño et al., 2009). Females not producing a sexual signal would presumably have no need to suppress immune response to promote signal expression. This hypothesis needs to focus more on how the effects of juvenile hormone differ between the sexes, if there is such a difference.

Others have suggested melanin as a substance that creates the immune-signaling trade-off. In some insect species, such as damselflies (*Calopteryx splendens xanthostoma*), melanin is used for pigmentation of sexual signals but also plays an important role in the insect immune response, making it a potentially limiting resource (Siva-Jothy, 2000; Cotter et al., 2008). This argument relies more on direct allocation of resources and not hormones, and thus does not closely resemble the immunocompetence handicap hypothesis (Lawniczak et al., 2007). Regardless of the substance proposed, all of these explanations predict a negative relationship between immunity and signal quality within an individual, just as the immunocompetence handicap hypothesis does in vertebrates. However, despite the potential similarities between patterns of immunity and sexual signaling across taxa, many of the studies focusing on vertebrates neglect the invertebrate literature. This divide is unfortunate, as invertebrate species can be excellent model systems in ecoimmunology (Lawniczak et al., 2007). It also highlights one of the major shortcomings of the immunocompetence handicap hypothesis, namely its lack of universal applicability. One might argue that this hypothesis can still be applied to invertebrates as long as researchers find some substance that acts in the same manner as testosterone to increase signal expression and reduce immune response in the more highly ornamented sex. Nevertheless, until researchers find more evidence that such a substance exists, it seems simpler and more universal to invoke the life-history explanation to explain trade-offs between immunity and sexual signals.

CAROTENOID-BASED SIGNALS

The immunocompetence handicap hypothesis is by no means the only hypothesis attempting to explain how sexual ornaments may reliably signal immunity. Some scientists have suggested that carotenoids provide the link. Carotenoids are molecules used by animals to produce red, orange, or yellow coloration (Olson and Owens, 1998; McGraw, 2006). They are also thought to play a role in immunity, possibly as immunostimulants or as antioxidants that mitigate damage caused by the free radicals generated during an immune response (Lozano, 1994; von Schantz et al., 1999). Animals lack the means to synthesize carotenoids themselves and must obtain them from their diet (Olson and Owens, 1998; McGraw, 2006). Such a restriction means that these molecules may be in short supply, and only healthy individuals would be able to divert enough carotenoids from their immune systems to display the bright colors often preferred in sexual signals. This can be viewed as a special case of the resource allocation issue described by Sheldon and Verhulst (1996), in which a limiting resource can be devoted either to the immune system or to other fitness-enhancing traits (Zuk and Stoehr, 2002). It should also be noted that the idea that allocation of carotenoids is important for signal honesty and the immunocompetence handicap hypothesis are not mutually exclusive ideas: testosterone may play a role in determining the amount of carotenoids circulating in the blood or moved to sexual signals (Alonso-Alvarez et al., 2009; Mougeot et al., 2009).

Carotenoid-based mechanisms for the operation of a link between sexual signals and parasites have received quite a bit of attention. A review of the Hamilton-Zuk hypothesis noted that many of the sexual signals that had been successfully linked to parasites were carotenoid-based (Zuk, 1992), and many studies have looked at the effects of carotenoid supplementation on immune response (Aguilera and Amat, 2007), sexual signaling (Hill, 1991), or both (Blount et al., 2003; Clotfelter et al., 2007; Biard et al., 2009). More recently, however, doubts about whether carotenoids deserve this special attention have been raised. First, doubt exists about the ability of carotenoids to combat oxidative damage, which is one of the assumptions often made about why carotenoids are important to immunity (Pérez-Rodríguez, 2009). Some have even suggested that high levels of carotenoids may be harmful to animals, although more evidence is needed on this subject (Olson and Owens, 1998; Vinkler and Albrecht, 2010). Furthermore, not all studies have found a relationship between carotenoid levels and sexual signals, even where carotenoids appear to influence immune response (Biard et al., 2009).

In some ways, the hypothesis that carotenoids determine signal honesty suffers from one of the same problems as the immunocompetence handicap hypothesis, namely that it cannot be universally applied. This mechanism by its very nature deals only with a limited number of signals. Carotenoid-based signals are common in many taxa, although most work has looked at birds (Olson and Owens, 1998; McGraw, 2006) or fishes (Olson and Owens, 1998; Clotfelter et al., 2007). However, many sexual and social signals do not seem to involve carotenoids. For example, much work has gone into studying the signaling honesty of badges of status, such as the throat patches in house sparrows, which are not carotenoid-based displays (Møller, 1987; Johnstone and Norris, 1993; Evans et al., 2000). Evidence of previous immune challenges may be reflected in bird plumage that is not colored by carotenoids (Hanssen et al., 2008). Moreover, Dunn et al. (2010) suggest that carotenoid-based and non–carotenoid-based ornaments may convey the same information about immunity to prospective mates. The authors examined two populations of the common yellowthroat warbler (*Geothlypis trichas*) and found that in one population, immune response was best correlated with the size and color of carotenoid-based bib plumage, but in the other population, immune response was best predicted by the size of a melanin-based facial mask. These results call into question whether carotenoid-based signals deserve special recognition among other sexual signals.

Given the many studies that have found an effect of carotenoids on immunity, it seems likely that some connection between carotenoids and parasites or immune response exists (Blount et al., 2003; Aguilera and Amat, 2007; Clotfelter et al., 2007). However, the exact mechanism by which carotenoids influence immune response and vice versa still requires verification (Pérez-Rodríguez, 2009; Vinkler and Albrecht, 2010). Until this relationship is resolved, researchers should be cautious about asserting that carotenoid-based ornaments are necessarily better honest indicators of immune function than other sexual signals.

REACTIVE OXYGEN SPECIES

One of the more recent mechanisms proposed to explain how signals may honestly indicate immunity suggests that oxidative stress from reactive oxygen species may create the trade-off between immunity and signal quality (von Schantz et al., 1999; Alonso-Alvarez et al., 2007; Dowling and Simmons, 2009). Reactive oxygen species are generated as a byproduct of metabolism, but they are also generated by immune responses (Constantini and Møller, 2009; Dowling and Simmons, 2009). These reactive oxygen species damage host tissues, and this damage may result in reduced quality of sexual signals (Dowling and Simmons, 2009). Thus, males that can mount a more efficient immune response may suffer less oxidative damage and thereby remain healthier and better able to produce flashy sexual signals (von Schantz et al., 1999). Originally, von Schantz et al. (1999) proposed that genes for parasite resistance played an important role in this process by increasing the efficiency of the immune response. Males lacking these genes would have to mount a costly general response to parasites and would suffer more oxidative damage. This reliance on specific genes for parasite resistance is similar to what Hamilton and Zuk (1982) proposed. However, more recent modifications of this idea have tied oxidative stress in more with the immunocompetence handicap hypothesis by suggesting that testosterone may also play an important role by increasing susceptibility to oxidative damage (Alonso-Alvarez et al., 2007). This newer approach has moved away from the focus on genetic resistance.

Given that this hypothesis is relatively new, fewer studies have tested it directly. However, Mougeot et al. (2009) found that red grouse with either testosterone supplementation or parasitic infection showed increased oxidative damage. In the males given testosterone, this oxidative damage occurred even though these males had higher levels of antioxidants circulating in the blood, suggesting either that testosterone makes individuals more susceptible to oxidative stress or that oxidative stress was increased to the point that even the higher levels of antioxidants were unable to prevent damage. Other studies have also found that testosterone increases susceptibility to oxidative stress (Alonso-Alvarez et al., 2007). However, our picture of the relationship between oxidative stress, immunity, and sexual signals is still incomplete. Many studies do not include any measure of an individual's ability to repair the damage caused by reactive oxygen species, which could represent an important component of fitness (Monaghan et al., 2009). The hypothesis set forth by von Schantz et al. (1999) emphasized the importance of underlying resistance genes in determining which males could mount more efficient immune response and thus suffer lower oxidative damage, yet this component of oxidative stress has been largely ignored (but see Kurtz et al., 2006). Thus, while the oxidative stress hypothesis seems plausible, more evidence is needed to determine whether this mechanism plays a prominent role in the honesty of sexual signaling.

THE PROBLEM: CONFLICTING PREDICTIONS AND RESULTS

While the mechanisms above suggest various ways that immunity may be linked to sexual signaling, they face a common problem when it comes to actually generating predictions about the relationship between immune response and signal quality. As mentioned above, all the mechanisms predict that there should be a trade-off between strength of the immune system and sexual signal quality, and at the level of the individual, the predictions are clear and straightforward. All of the mechanisms described predict that a greater investment in sexual signaling should result in a decrease in immune response and vice versa, and empirical data from selection experiments and experimental immune challenges support this assertion (Verhulst et al., 1999; Jacot et al., 2004; McKean and Nunney, 2008; Leman et al., 2009; López et al., 2009).

Problems arise, however, when scientists examine the correlation between immunity and sexual signaling across multiple individuals rather than within a single animal. If a negative relationship between immune response and signal quality exists within individuals, do we expect this same pattern to hold among different individuals? In other words, should males with the most preferred signals also have the weakest immune response, at least during the breeding season, as a consequence of having to pay the costs associated with signaling? If one considers only that immune response is costly, such a scenario makes sense. However, this view ignores the effects of differences in individual quality that could lead to differences among individuals in their ability to pay the costs associated with signaling. Obviously, as stated above, there should be a negative relationship within an individual between immune response and signal quality. However, if the costs of immune response act as a condition-dependent handicap, individuals in better condition should be better able to pay these costs and still produce a good sexual signal (Zahavi, 1977; Nur and Hasson, 1984; Maynard Smith, 1991; Getty, 2002). This pattern is sometimes referred to as the car/house paradox, after an example from economics. Within an individual, if money is spent on a car, less money will be available for buying a house. However, if one looked at the relationship across individuals, one tends to find that people with more expensive cars also have more expensive houses because they started off with more money to spend. This reasoning can also be applied to the trade-offs faced by individuals between investing in immune response and investing in sexual signals: individuals that had more resources to begin with should be better at both (Zuk et al., 1996). Many studies provide support for this idea by showing that higher-quality individuals seem to be better able to bear the costs associated with immune response and/or sexual signaling (Saino and Møller, 1996; Ardia, 2005; Mougeot et al., 2009).

At least some of the confusion can be blamed on the lack of distinction made between predictions *within* individuals and predictions *among* individuals (Getty, 2002). This has led to claims by some that a highly ornamented individual should have a weaker immune response, and by others that such individuals should have a stronger immune response. If researchers consider either outcome as support for the hypotheses described above, falsification of these hypotheses becomes difficult or impossible (Braude et al., 1999). Thus, it is important for researchers to take into account differences in individual quality and how these differences determine the relative costs individuals pay for signals and immunity (Nur and Hasson, 1984; Getty, 2002).

Compounding the confusion is that different studies looking for a correlation between immune response and sexual signal quality have obtained opposite results, shown summarized in Table 15.1. While more ornamented individuals often have better immunity, as one would predict, the opposite result has also been obtained (Table 15.1). For example, Skarstein and Folstad (1996) found that male arctic charr with brighter red coloration also tended to have lower lymphocyte counts, which presumably implied a weaker immune response. In contrast, López and Martín (2005) examined male scents in Iberian wall lizards and found that females preferred the scents of males able to mount a stronger immune response. They concluded that these scent marks were honest indicators of quality and immunity, and that females used these scents to select high-quality mates. Interestingly, both of these studies mention testosterone as a possible substance responsible for producing the observed relationship (Skarstein and Folstad, 1996; López and Martín, 2005). This is just one example of two studies testing the same idea, namely that sexual signal quality conveys information about immunity to prospective mates, and obtaining opposite results with regards to whether a preferred sexual signal indicates a strong or a weak immune response. Such conflicting results continue to plague the field. The breast coloration preferred by female greenfinches was associated with lower antibody response (Aguilera and Amat, 2007). In contrast, Griggio et al. (2010) found that male budgerigars whose breast plumage had higher UV chroma also were able to mount a stronger swelling response to phytohemagglutinin (PHA).

These contradictory results represent a largely unaddressed problem in sexual selection and ecoimmunology. Furthermore, all of the mechanisms described above predict essentially the same relationship between immunity and sexual signals, and yet none of them can explain why the relationship between immune response and sexual signal quality sometimes appears to be negative and sometimes appears to be positive.

The field has become bogged down by a focus on mechanisms that do not generate new predictions about the relationship between immune response and sexual signaling, while few efforts have been made to reconcile the different results that have been obtained. For example, several studies have attempted to determine whether testosterone suppresses immune response directly or indirectly via some other substance such as corticosterone (Evans et al., 2000; Owen-Ashley et al., 2004). This distinction is important to those interested in endocrine–immune interactions, but for researchers interested in explaining relationships between immune response and sexual signals, such a distinction is less helpful. Even if testosterone does suppress immune response by increasing corticosterone rather than having a direct effect, the predicted relationship between the quality of the sexual signal and immune response remains the same. The answer to the question of whether testosterone is directly immunosuppressive will not help us explain why some studies find a positive relationship between signal quality and immune response while others find a negative one.

If the hypotheses previously discussed are unable to explain the different results described above, where should we look for an answer? The first step may be to better understand the relationships between immunity and resistance as well as to understand the relationships and trade-offs that occur within the immune system itself. Below we review the information needed to generate testable predictions about the link between parasites, immune response, and sexual signaling, and discuss possible directions for future work.

IMMUNITY: WHAT WE KNOW AND WHAT WE NEED TO CONSIDER

SPECIFIC VERSUS GENERAL IMMUNITY

When trying to generate predictions about how immune response relates to signaling, it is important to distinguish between the strength of the immune system (as measured by some challenge that stimulates an immune response) and its ability to fight off particular pathogens of interest (Westneat and Birkhead, 1998; Adamo, 2004a; Viney et al., 2005). Many of the immune tests popular in the literature (PHA response, antibody response to sheep red blood cells, white cell counts, etc.) are designed to test "general immunocompetence" rather than an organism's ability to fight off any one pathogen. These methods have both strengths and weaknesses. On one hand, the ability of an organism to mount a response to a novel pathogen should reflect its ability to fight off the pathogens it already faces. However, a strong response to a generic immune challenge may not always translate into the ability to resist a given pathogen (Adamo, 2004a). Indeed, Westneat and Birkhead (1998) state that such immunological methods do not test the Hamilton-Zuk hypothesis. They argue that under the Hamilton-Zuk hypothesis, females select males with genes that confer resistance to a given pathogen or pathogens. This selection for genetic resistance against a single pathogen can produce an arms race between host and parasite that may be crucial for maintaining genetic variability in quality (Hamilton and Zuk, 1982). However, it may not lead to greater investment in the immune system because males with superior genetic resistance do not need to maintain as strong an immune system. Thus, such selection is less likely to be captured using general tests of immunity. Genetic resistance may even appear as a negative relationship between immunity and sexual signals, because males that are genetically resistant can afford to take more resources away from immunity and devote them to signaling (Zuk, 1994).

Some studies that have found a negative relationship between immune response and sexual signaling have attributed their results to this phenomenon, namely that individuals with genetic resistance can afford to suppress immunity without increasing disease susceptibility (Aguilera and Amat, 2007). However, few studies reporting such a negative relationship have measured resistance directly. Lindström and Lundström (2000) found that when male greenfinches (*Carduelis chloris*) were infected with a specific virus type, males with larger yellow tail patches had lower viral levels and cleared the infection faster; however, this clearance rate had no connection to antibody response, implying that such males had neither a higher nor a lower immune response than average. Whether genetically resistant males can suppress immunity and still have lower parasite loads and attract mates is unclear. A study by Zala et al. (2008) in mice found that females preferred males from a strain that was genetically resistant to infection by *Salmonella* even before such males were infected, implying that, at least in some circumstances, females can select directly for resistance genes. Males in the resistant strain were also able to maintain higher testosterone levels when infected, implying that such males could tolerate higher levels of immunosuppression.

If knowing the strength of an immune response is not enough to predict whether an animal can resist parasites, researchers must also examine pathogens more directly. Ideally, measures of immune response will correlate with parasite resistance, but this may not always be the case (Adamo, 2004a, 2004b; Viney et al., 2005). In cases where differences in immune response measurements do not seem to translate into differences in resistance, one must question what these differences in immune response actually mean. Is immune response simply correlated to condition and not to pathogen resistance, and if so, which is the better predictor of male ornament quality and mating success, immune response or resistance? It is also possible that different measures of immune response correlate to resistance to different pathogens. Different parts of the immune system may specialize in fighting off different types of pathogens (Weir and Stewart, 1993); thus, it may be important for researchers to consider which pathogens play the most important role in the lives of their study animal, and the kinds of immunological defenses needed to combat them. Which immune defenses are optimal may differ depending on the organism under study and the infectious agents that organism is likely to face (Schmid-Hempel and Ebert, 2003; Lee, 2006). For example, one measure of immune response, increase in lysozyme-like activity, was correlated with the ability of crickets to resist three different strains of pathogenic bacteria, but phenoloxidase activity did not influence resistance at all (Adamo, 2004b). Thus, if crickets consistently face challenges by pathogenic bacteria, one might predict that lysozyme-like activity would be important for survival and probably also mate choice.

In light of the distinction between immune response and pathogen resistance, researchers should consider the following questions when trying to predict the relationship between immune response and signal quality. (1) Are there genetic underpinnings to resistance, and how do they relate to immune response? To address this, researchers must first locate genes involved in parasite resistance, then look to see whether individuals possessing such genes have any measurable differences in the strength of their immune response. (2) Do the tests of immune response used actually reflect the organism's ability to fight off pathogens? Theoretically, if animals have a higher level of immune response, they should be better able to fight off pathogens. Experimental infections of individuals whose immune responses have already been measured may be helpful in answering this question. (3) If a more robust immune response is not related to greater resistance, which trait seems to be more important to female choice, specific resistance or a measure of overall immunity? This approach will doubtless prove challenging, as determining the many pathogens that animals may be exposed to under natural conditions is extremely difficult, and locating different genotypes that may confer resistance to these pathogens may prove more so.

While locating genes involved in resistance will doubtless be a long and difficult process, one possible starting point is to more closely examine genes in the major histocompatibility complex (MHC) (see Chapter 7). The MHC plays an important role in immune response by coding for cell surface antigens involved in the recognition and presentation of antigens to T cells (Milinski, 2006). Thus, it seems likely that these gene complexes are involved in pathogen resistance in both humans and non-human animals. Some empirical work supports this hypothesis; Westerdahl et al. (2005) found evidence suggesting that certain MHC alleles confer malaria resistance in songbirds, and certain MHC haplotypes may also play a role in resistance to ectoparasites in domestic chickens (Owen et al., 2008). Sticklebacks possessing lower MHC diversity displayed higher innate immunity (Kurtz et al., 2004), which the authors interpreted as compensating for a less-efficient adaptive immune system. Whether MHC alleles also influence the strength of adaptive immune response is less clear. In one study by Bonneaud et al. (2005), house sparrows possessing a particular MHC allele tended to have higher swelling responses to PHA and higher antibody responses when injected with sheep red blood cells. However, another study in nestlings of the same species failed to find any relationship between response to PHA and MHC alleles (Bonneaud et al., 2009).

While MHC alleles are not the only source of resistance to pathogens, they may be a promising starting point, and integrating studies looking at the role of MHC alleles in resistance and mate choice with those examining general immune response may prove fruitful (Zuk, 1994; Drury, 2010). Researchers pursuing this avenue should be cautious, however, to note that predicting that a female will choose a male because he possesses a specific MHC haplotype that confers resistance is different from predicting that a female will select a male that has a different MHC type than she does. Much of the work on the MHC and its role in mate choice has focused on this latter prediction, namely that females should select mates with dissimilar MHC alleles to increase the heterozygosity of the MHC in offspring. This increased heterozygosity is thought to increase the fitness of the offspring by allowing them to recognize a greater variety of pathogens (Milinski, 2006).

METHODS FOR MEASURING IMMUNITY

Given that the immune system consists of more than one branch, it is unlikely for any one test of immune response to accurately represent "immunocompetence" as a whole (Norris and Evans, 2000; Zuk and Stoehr, 2002; Adamo, 2004a). This realization has led scientists to employ multiple tests of immune response when trying to link immunity and signal quality. However, the results of these tests suggest that different components of the immune system may not all be linked to sexual signals in the same way. For instance, male peacock train length was related positively to PHA response, but negatively to the peacocks' ability to produce antibodies in response to stimulation by a novel antigen (in this case sheep red blood cells) (Møller and Petrie, 2002). Similar results, in which one measure of immune response increases with the quality of the sexual signal while another decreases, have been obtained for other species as well. Zuk and Johnsen (1998) found that, during the breeding season, male jungle fowl with larger combs had fewer lymphocytes in the blood but a higher response to PHA. Faivre et al. (2003) reported that blackbirds (*Turdus merula*) with preferred bill coloration had lower secondary antibody responses but higher swelling response to PHA. In contrast, Garvin et al. (2008) found exactly opposite results, namely that yellowthroat warblers (*Geothlypis trichas*) with larger facial masks had higher antibody levels but lower response to PHA.

Such variation obviously makes the results of these studies somewhat difficult to interpret. It also suggests the existence of more complex relationships between the different branches of the immune system. Some authors have suggested that trade-offs between different branches of the immune system can make it impossible to simultaneously produce a strong response in innate immunity, cell-mediated immunity, and humoral immunity (Møller and Petrie, 2002; Forsman et al., 2008; Garvin et al., 2008). Forsman et al. (2008) compared three different measures of immunity in house wren broods: response to PHA, ability to produce antibodies, and ability of the blood to kill bacteria. Broods of nestlings that produced a stronger antibody response tended to have a weaker PHA response, and vice versa (Forsman et al., 2008). These results imply that nestlings (and presumably other types of individuals) that have invested strongly in one branch of the immune system cannot invest as much in the other branches, creating trade-offs between different aspects of immunity. Other studies employing multiple measures of immunity have found similar trade-offs between different types of immune response in the presence of stressors such as parasites (Johnsen and Zuk, 1999; Arriero, 2009). These trade-offs within the immune system may explain why some studies have found a positive response between signal quality and immune response while others have found a negative one. In studies that measure one component of immunity, that component may or may not be the one in which the animal has invested the most. Thus, one would expect that the results of a study measuring only one type of immune response might differ from those of another study measuring a different type of immune response (Norris and Evans, 2000; Martin et al., 2007).

Having one branch of the immune system exhibit a strong response while another branch has a weak response makes things difficult for researchers hoping to find a clean, easily interpretable relationship between immune response and sexual signals. However, one should expect to find such "messy" results quite commonly due to the way animals allocate energy to the immune system. Life-history theory predicts that not all animals should invest in all arms of the immune system equally (Martin et al., 2006a; Lee, 2006; see also Chapter 2). Some attention has been focused on the relationship between an animal's "pace of life" and its investment in different components of immunity. "Slow-living" organisms, which have long periods of development and long life spans, are predicted to invest more in immune response (Martin et al., 2006b), and also to invest in specific, inducible defenses (e.g., antibody production). Such defenses are costly to develop but less costly to use, and for organisms that invest a great deal in development, the cost may be outweighed by the benefits later in life of improving survival, minimizing damage to self, and conserving energy to be invested in future reproduction. For "faster-living" species with lower survival rates, the cost of developing such defenses may not be worth the benefit, and thus they are expected to rely more on innate defenses (Martin et al., 2006a; Lee, 2006).

For a given species, then, some optimal combination of immune defenses exists. What defenses make up such an optimal response will depend on life-history as well as the pathogens to which an organism may be exposed (Schmid-Hempel and Ebert, 2003; Lee, 2006). This idea has been neglected in the sexual selection literature but may have profound implications for the role of immune response in mate choice. If females select mates based on the strength of the male's immune response, the types of response that are most important for that species should also be most important in the mating decision. For instance, females of long-lived species are expected to select males with a stronger humoral immune response, regardless of innate immunity or cell-mediated immunity. Such clear-cut results, of course, may be difficult to find, as the world cannot always be divided into fast-living and slow-living species. However, by looking at an organism's pace of life, researchers may be able to make predictions about the types of immune response that the organism should invest most heavily in, and these defenses may also play an important role in mate choice.

For researchers interested in how different branches of the immune system interact with one another as well as with sexual signaling, caution must be used when choosing how to measure immune response. One must consider what the immune tests chosen for a particular study are really measuring. Some of the immune tests that are popular in the literature do not measure the type of immune response they were originally thought to measure. The primary example of this is the PHA response, in which PHA is injected into the animal and the swelling response is measured (see Box 15.1). This swelling response was thought to indicate T-cell–mediated immunity. However, Martin et al. (2006c) found that other immune cells may take part in the reaction, and thus PHA response may indicate "cutaneous" immune response rather than a purely T-lymphocyte–mediated response. This is not to say that the PHA test is useless to researchers interested in immunity, especially as the response may be related to survival (Gonzalez et al., 1999). However, it does mean that researchers seeking to study the strength of cell-mediated immunity independent of other types of immunity should probably use a different test.

RESISTANCE VERSUS TOLERANCE

Virtually every model predicting the effect of parasites on sexual signals has assumed that females should select mates with the lowest parasite loads. The Hamilton-Zuk hypothesis, the transmission avoidance model, and the resource provisioning model all share this prediction (Clayton, 1991). However, as recent work by Råberg et al. (2007) points out, resistance is not the only method that animals may use to deal with parasites. Rather than fighting off a parasitic infection, animals may simply tolerate it by minimizing the damage done by the parasites (Råberg et al., 2007; Boots, 2008). This strategy would allow the animal to stay in better condition than its similarly infected fellows (Fig. 15.2).

The idea of parasite tolerance rather than resistance has long been recognized by plant biologists, but its application to animals has been neglected (Råberg et al., 2009). However, if tolerance plays a role in mate choice, its inclusion in models of sexual selection could drastically alter the expected outcome. Most studies assume a negative relationship between parasite load and the quality of sexual signals, but this may not be so if females can also choose for tolerance. Under such a scenario, higher-quality mates might not differ from the rest of the population in terms of overall parasite burden, but they would be better able to tolerate such infections. Individuals or groups that can better tolerate parasites also show reduced resistance (Råberg et al., 2007). Thus, under the scenario described above, preferred mates might show a lower immune response relative to the rest of the population because they are better able to tolerate parasites but less able to mount an effective response against them. These preferred mates might also have similar or higher parasite loads relative to the other males in the population. Some studies of wild populations have found results consistent with this idea (Clayton, 1991).

While the idea of tolerance opens up some intriguing new ideas about the nature of female preference and parasites, it is difficult to study. To demonstrate that tolerance exists, one must look at the effects of different parasite intensities on different host genotypes. If one genotype shows relatively less decline in condition as parasite intensity increases (i.e., if there is a significant

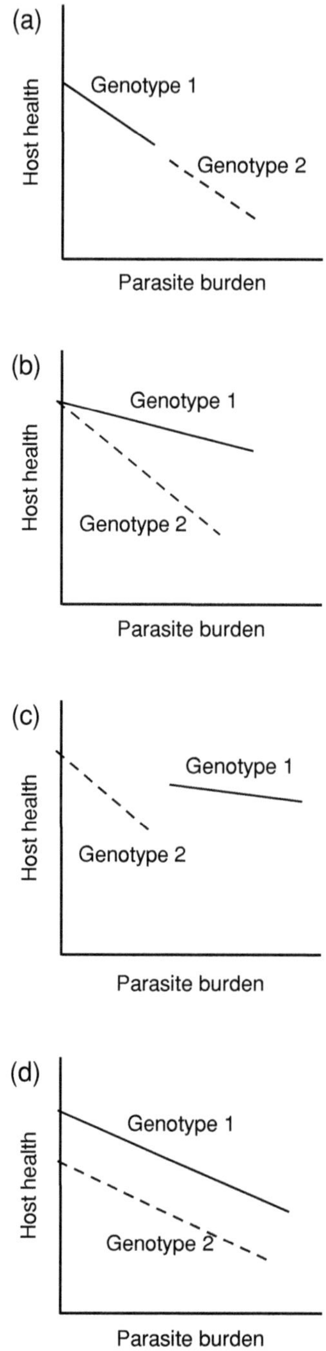

FIGURE 15.2: This diagram shows the relationship between host health and parasite burden for two different host genotypes. (a) Genotype 1 is more resistant than Genotype 2 and therefore maintains a lower parasite burden. There is no difference in tolerance. (b) Genotype 1 is more tolerant than Genotype 2, but there is no difference in resistance. (c) Genotype 1 is more tolerant than Genotype 2 but less resistant. (d) There are no differences in either resistance or tolerance between the two genotypes, but Genotype 1 has greater vigor overall. Figure adapted from Råberg et al. (2007).

interaction between genotype and parasite intensity), then one may conclude that that genotype is more tolerant than the others (Råberg et al., 2009). Given these requirements, demonstrating the effects of tolerance as it relates to mate choice in field studies would be extremely difficult. Laboratory experiments, however, may have more success at teasing apart these interactions. Thus, we encourage researchers with access to laboratory populations to explore this topic further. We second the appeal made by Boots (2008) for more research to demonstrate the existence of tolerance in animal populations, and we further encourage researchers to examine how tolerance may be shaped by sexual selection. Are males that can tolerate infection as attractive to females as those that can resist infection? Can females even distinguish between tolerant and resistant individuals, and would there be any benefit to doing so? Answering such questions may provide a better understanding of how sexual selection may help shape host–parasite interactions.

CONCLUSION AND FUTURE DIRECTIONS

While interest in the role of immunity in sexual selection has expanded greatly during the past two decades, progress has been hampered by a lack of clear predictions about the relationship between immunity and sexual signaling. Different studies have obtained opposite results but still interpreted them as supporting the same hypothesis, making the testing and falsification of such hypotheses difficult to impossible. The immunocompetence handicap hypothesis had been particularly prone to this problem (Braude et al., 1999). This difficulty may arise in part because none of the hypotheses currently popular in the field of sexual selection fully explains the complicated link between immune response and sexual signal quality.

We suggest that before scientists delve into the question of which hormones or antioxidants affect the immune system, we should first strive for a better understanding of how the immune system works, how differences in genetic backgrounds and life-history strategies can affect immune response, and how these differences may be important in mate choice. For example, it is not enough to simply claim that testosterone suppresses immunity when not all branches of the immune response may react in the same way. We need a better understanding of how the various components of the immune system relate to each other and to signaling before we focus on the detailed mechanisms controlling those interactions. Most importantly, we urge researchers to go in with clear predictions about what the relationship between immune response and sexual signaling will be based on a given model, rather than coming up with a *post hoc* explanation for why the results obtained support that model.

To derive accurate predictions from the models discussed in this review, we may first need to obtain further information. For example, if one studies the Hamilton-Zuk hypothesis, making predictions about the relationship between sexual signals and immune response may be extremely difficult without knowing more about the genes and/or mechanisms underlying resistance. In some cases, resistance may be conferred by differences in immune response (Adamo, 2004b). If these differences in immune response are heritable, females could select resistant mates with high immunity, and one would predict a positive relationship between immune response and sexual signaling. In other cases, however, having genes that confer parasite resistance may allow individuals to mount a weaker immune response, at least in some branches of the immune system (Kurtz et al., 2004). Under this scenario, one would predict a negative relationship between the quality of the sexual signal and the level of immune response.

Given this gap in our knowledge, researchers should explore the relationship between genetic resistance and facultative immune response. Theoretically, having genetic resistance may enable some individuals to get away with mounting a comparatively weaker immune response (Zuk, 1994; Westneat and Birkhead, 1998). However, depending on the genes involved, not all branches of the immune system may respond in the same manner. For instance, the MHC plays a role in recognition of antigens by T lymphocytes, and individuals that have MHC alleles that confer resistance to pathogens may be able to mount weaker innate immune responses (Kurtz et al., 2004). These alleles may facilitate the T-cell–mediated response, although this idea has received mixed support so far (Bonneaud et al., 2005, 2009). Researchers should also consider the possibility that tolerance, rather than resistance, could be important in mate choice. Possessing genes for tolerance may also influence the strength of an individual's immune response. Given that immune response is energetically costly, individuals that tolerate parasites rather than resisting them should not waste resources on mounting a strong immune response. Thus, we would expect that tolerance should be correlated with a lower-than-average level of immune response. Reduced immune response may even benefit such individuals by reducing the immunopathology that often occurs in the presence of parasites (Graham et al., 2005).

Researchers interested in how energetic trade-offs determine signal honesty may also profit from trying to disentangle the effects of condition on immunity and vice versa. Many studies suggest that immune response depends, at least in part, on condition (Gonzalez et al., 1999; Jacot et al., 2004; Aguilera and Amat, 2007; but see Ohlsson et al., 2002). This condition-dependence is in keeping with the view of immune response as a life-history trait. However, if the strength of an immune response and sexual signal quality both increase with condition, one may argue that the sexual signal itself evolved to convey information about condition to prospective mates, and that any information about immunity is simply secondary. Modeling suggests that females may benefit more by paying attention to indicators of condition rather than immune response itself (Adamo and Spiteri, 2009). However, immune response may also influence condition if it decreases parasite load. This feedback loop may be critical to the question of how important parasite resistance truly is in mate choice. If immune response is dependent on condition but the reverse is not true, females selecting highly ornamented males could be getting genes related to condition (e.g., genes for foraging ability) that have nothing to do with immune response. Any observed differences in immune response would simply be a product of differences in condition, and therefore not heritable. The Hamilton-Zuk hypothesis cannot operate under such a scenario.

In conclusion, the field of ecoimmunology and sexual selection has become tangled in a focus on mechanisms. To solve this problem, we propose that more work should focus on how different parts of the immune system interact with each other and how these interactions relate to disease resistance and/or tolerance. This does not mean that we believe that all researchers should become immunologists, but rather that it is important to have a clearer idea of what the immune system does, what our measurements of immunity actually tell us, and how immunity and resistance contribute to an animal's fitness. Such an understanding will help us determine what exactly females gain when they select their mates, and will prove invaluable to those studying the evolution of immunity as well as mate choice.

REFERENCES

Adamo, SA (2004a). How should behavioural ecologists interpret measurements of immunity? Anim. Behav. 68:1443–1449.

Adamo, SA (2004b). Estimating disease resistance in insects: phenoloxidase and lysozyme-like activity and disease resistance in the cricket *Gryllus texensis*. J. Insect Physiol. 50:209–216.

Adamo, SA and Parsons, NM (2006). The emergency life-history stage and immunity in the cricket *Gryllus texensis*. Anim. Behav. 72:235–244.

Adamo, SA and Spiteri, RJ (2009). He's healthy, but will he survive the plague? Possible constraints on mate choice for disease resistance. Anim. Behav. 77:67–78.

Aguilera, E and Amat, JA (2007). Carotenoids, immune response and the expression of sexual ornaments in male greenfinches (*Carduelis chlorisi*). Naturwissenschaften. 94:895–902.

Ahtiainen, JJ, Alatalo, RV, Kortet, R, and Rantala, MJ (2004). Sexual advertisement and immune function in an arachnid species (Lycosidae). Behav. Ecol. 15:602–606.

Alonso-Alvarez, C, Bertrand, S, Faivre, B, Chastel, O, and Sorci, G (2007). Testosterone and oxidative stress: the oxidation handicap hypothesis. Proc. R. Soc. Lond. B 274:819–825.

Alonso-Alvarez, C, Pérez-Rodríguez, L, Garcia, JT, and Viñuela, J (2009). Testosterone- mediated trade-offs in the old age: a new approach to the immunocompetence handicap hypothesis and carotenoid-based signalling. Proc. R. Soc. Lond. B 276:2093–2101.

Ardia, DR (2005). Individual quality mediates trade-offs between reproductive effort and immune function in tree swallows. J. Anim. Ecol. 74:517–524.

Arriero, E (2009). Rearing environment effects on immune defence in blue tit *Cyanistes caeruleus* nestlings. Oecologia. 159:697–704.

Biard, C, Hardy, C, Montreuil, S, and Moreau, J (2009). Dynamics of PHA-induced immune response and plasma carotenoids in birds: should we have a closer look? J. Exp. Biol. 212:1336–1343.

Blount, JD, Metcalfe, NB, Birkhead, TR, and Surai, PF (2003). Carotenoid modulation of immune function and sexual attractiveness in zebra finches. Science. 300:125–127.

Bonneaud, C, Richard, M, Faivre, B, Westerdahl, H, and Sorci, G (2005). An MHC class I allele associated to the expression of T-dependent immune response in the house sparrow. Immunogenetics. 57:782–789.

Bonneaud, C, Sinsheimer, JS, Richard, M, Chastel, O, and Sorci, G (2009). MHC polymorphisms fail to explain the heritability of phytohaemagglutinin-induced skin swelling in a wild passerine. Biol. Lett. 5:784–787.

Boots, M (2008). Fight or learn to live with the consequences? Trends Ecol. Evol. 23:248–250.

Braude, S, Tang-Martinez, Z, and Taylor, GT (1999). Stress, testosterone, and the immunoredistribution hypothesis. Behav. Ecol. 10:345–350.

Buchanan, KL, Evans, MR, and Goldsmith, AR (2003). Testosterone, dominance signaling and immunosuppression in the house sparrow, *Passer domesticus*. Behav. Ecol. Sociobiol. 55:50–59.

Clayton, DH (1991). The influence of parasites on host sexual selection. Parasitol. Today. 7:329–334.

Clotfelter, ED, Ardia, DR, and McGraw, KJ (2007). Red fish, blue fish: trade-offs between pigmentation and immunity in *Betta splendens*. Behav. Ecol. 18:1139–1145.

Constantini, D and Møller, AP (2009). Does immune response cause oxidative stress in birds? A meta-analysis. Comp. Biochem. Physiol. A. 153:339–344.

Contreras-Garduño, J, Córdoba-Aquilar, A, Lanz-Mendoza, H, and Rivera, AC (2009). Territorial behaviour and immunity are mediated by juvenile hormone: the physiological basis of honest signaling? Func. Ecol. 23:157–163.

Cotter, SC, Myatt, JP, Benskin, CMH, and Wilson, K (2008). Selection for cuticular melanism reveals immune function and life-history trade-offs in *Spodoptera littoralis*. J. Evol. Biol. 21:1744–1754.

Cox, RM, and John-Alder, HB (2007). Increased mite parasitism as a cost of testosterone in male striped plateau lizards *Sceloporus virgatus*. Func. Ecol. 21:327–334.

Dowling, DK and Simmons, LW (2009). Reactive oxygen species as universal constraints in life-history evolution. Proc. R. Soc. Lond. B 276:1737–1745.

Drury, JP (2010). Immunity and mate choice: a new outlook. Anim. Behav. 79:539–545.

Duffy, DL and Ball, GF (2002). Song predicts immunocompetence in male European starlings (*Sturnus vulgaris*). Proc. R. Soc. Lond. B. 269:847–852.

Dunn, PO, Garvin, JC, Whittingham, LA, Freeman-Gallant, CR, and Hasselquist, D (2010). Carotenoid and melanin-based ornaments signal similar aspects of male quality in two populations of the common yellowthroat. Func. Ecol. 24:149–158.

Evans, MR, Goldsmith, AR, and Norris, SRA (2000). The effects of testosterone on antibody production and plumage coloration in male house sparrows (*Passer domesticus*). Behav. Ecol. Sociobiol. 47:156–163.

Faivre, B, Préault, M, Salvadori, F, Théry, M, Gaillard, M, and Cézilly, F (2003). Bill colour and immunocompetence in the European blackbird. Anim. Behav. 65:1125–1131.

Folstad, I and Karter, AJ (1992). Parasites, bright males, and the immunocompetence handicap hypothesis. Am Nat. 139:603–622.

Forsman, AM, Vogel, LA, Sakaluk, SK, and Thompson, CF (2008). Immune-challenged house wren broods differ in the relative strengths of their responses among different axes of the immune system. J. Evol. Biol. 21:873–878.

French, SS, Moore, MC, and Demas, GE (2009) Ecological immunology: the organism in context. Integ. and Comp. Biol. 49:246–253.

Garvin, JC, Dunn, PO, Whittingham, LA, Steeber, DA, and Hasselquist, D. (2008). Do male ornaments signal immunity in the common yellowthroat? Behav. Ecol. 19:54–60.

Getty T (2002). Signaling health versus parasites. Am Nat. 159:363–371.

Gonzalez, G, Sorci, G, Møller, AP, Ninni, P, Haussy, C, and de Lope, F (1999). Immunocompetence and condition-dependent sexual advertisement in male house sparrows (*Passer domesticus*). J. Anim. Ecol. 68:1225–1234.

Graham, AL, Allen, JE, and Read, AF (2005). Evolutionary causes and consequences of immunopathology. Annu. Rev. Ecol. Evol. Syst. 36:373–397.

Grasman, KA (2002). Assessing immunological function in toxicological studies of avian wildlife. Integ. and Comp. Biol. 42:34–42.

Griggio, M, Zanollo, V, and Hoi, H (2010). UV plumage color is an honest signal of quality in male budgerigars. Ecol. Res. 25:77–82.

Hamilton, WD and Zuk, M (1982). Heritable true fitness and bright birds: a role for parasites? Science. 218:384–387.

Hanssen, SA, Hasselquist, D, Folstad, I, and Erikstad, KE (2008). A label of health: a previous immune challenge is reflected in the expression of a female plumage trait. Biol. Lett. 4:379–381.

Hill, GE (1991). Plumage coloration is a sexually selected indicator of male quality. Nature. 350: 337–339.

Jacot, A, Scheuber, H, and Brinkhof, MWG (2004) Costs of an induced immune response on sexual displays and longevity in field crickets. Evolution. 58:2280–2286.

Johnsen, TS and Zuk, M (1999). Parasites and tradeoffs in the immune response of female red jungle fowl. Oikos. 86:487–492.

Johnstone, RA and Norris, K (1993). Badges of status and the cost of aggression. Behav. Ecol. Sociobiol. 32:127–134.

Kennedy, MW and Nager, RG (2006). The perils and prospects of using phytohaemagglutinin in evolutionary ecology. Trends Ecol. Evol. 21:653–655.

Keyser, AJ and Siefferman, LM (2005). Viability selection against highly ornamented males. Evol. Ecol. Res. 14:855–861.

Kotiaho, JS (2001) Costs of sexual traits: a mismatch between theoretical considerations and empirical evidence. Biol. Rev. 76:365–376.

Kurtz, J, Kable, M, Aeschlimann, PB, Häberli, MA, Wegner, KM, Reusch, TBH, and Milinski, M (2004). Major histocompatibility complex diversity influences parasite resistance and innate immunity in sticklebacks. Proc. R. Soc. Lond. B. 271:197–204.

Kurtz, J, Wegner, KM, Kable, M, Reusch, TBH, Schaschl, H, Hasselquist, D, and Milinski, M (2006). MHC genes and oxidative stress in sticklebacks: an immuno-ecological approach. Proc. R. Soc. Lond. B. 273:1407–1414.

Lawniczak, MKN, Barnes, AI, Linklater, JR, Boone, JM, Wigby, S, and Chapman, T (2007). Mating and immunity in invertebrates. Trends Ecol. Evol. 22:48–55.

Lee, KA (2006). Linking immune defense and life history at the levels of the individual and the species. Integ. Comp. Biol. 46:1000–1015.

Leman, JC, Weddle, CB, Gershman, SN, Kerr, AM, Ower, GD, St. John, JM, Vogel, LA, and Sakaluk, SK (2009). Lovesick: immunological costs of mating to male sagebrush crickets. J. Evol. Biol. 22:163–171.

Lindström, K and Lundström, J (2000). Male greenfinches (*Carduelis chloris*) with brighter ornaments have higher virus infection clearance rate. Behav. Ecol. Sociobiol. 48:44–51.

López, P and Martín, J (2005). Female Iberian wall lizards prefer male scents that signal a better cell-mediated immune response. Biol. Lett. 1:404–406.

López, P, Gabirot, M, and Martín, J (2009) Immune challenge affects sexual coloration of male Iberian wall lizards. J. Exp. Zool. 311A: 96–104.

Lozano, GA (1994) Carotenoids, parasites, and sexual selection. Oikos. 70:309–311.

Martín, J, Civantos, E, Amo, L, and López, P (2007). Chemical ornaments of male lizards *Psammodromus algirus* may reveal their parasite load and health state to females. Behav. Ecol. Sociobiol. 62:173–179.

Martin, LB, Weil, ZM, and Nelson, RJ (2006a). Refining approaches and diversifying directions in ecoimmunology. Integ. Comp. Biol. 46:1030–1039.

Martin, LB, Hasselquist, D, and Wikelski, M (2006b). Investment in immune defense is linked to pace of life in house sparrows. Oecologia. 147:565–575.

Martin, LB, Han, P, Lewittes, J, Kuhlman, JR, Klasing, KC, and Wikelski, M (2006c). Phytohemagglutinin-induced skin swelling in birds: histological support for a classical immunoecological technique. Func. Ecol. 20:290–299.

Martin, LB, Weil, ZM, and Nelson, RJ (2007). Immune defense and reproductive pace of life in Peromyscus mice. Ecology. 88:2516–2528.

Marzal, A, Reviriego, M, de Lope, F, and Møller, AP (2007). Fitness cost of an immune response in the house martin (*Delichon urbica*). Behav. Ecol. Sociobiol. 61:1573–1580.

Maynard Smith, J (1991). Theories of sexual selection. Trends Ecol. Evol. 6:146–151.

McGraw, KJ (2006). Mechanics of carotenoid-based coloration. In: Hill, GE and McGraw, KJ, eds. Bird Coloration: Mechanisms and Measurements, pp. 177–243. Cambridge, MA: Harvard University Press.

McKean, KA and Nunney, L (2008). Sexual selection and immune function in *Drosophila melanogaster*. Evolution. 62:386–400.

Milinski, M (2006) The major histocompatibility complex, sexual selection, and mate choice. Annu. Rev. Ecol. Evol. Syst. 37:159–186.

Millet, S, Bennett, J, Lee, KA, Hau, M, and Klasing, KC (2007). Quantifying and comparing constitutive immunity across avian species. Dev. Comp. Immunol. 31:188–201.

Møller, AP (1987). Social control of deception among status signalling house sparrows *Passer domesticus*. Behav. Ecol. Sociobiol. 20:307–311.

Møller, AP (1990). Effects of a haematophagous mite on the barn swallow (*Hurundo rustica*). Evolution. 44:771–784.

Møller, AP (1994). Sexual Selection and the Barn Swallow. Oxford: Oxford University Press.

Møller, AP, Christe, P, and Lux, E (1999). Parasitism, host immune function, and sexual selection. Q. Rev. Biol. 74:3–20.

Møller, AP and Petrie, M (2002). Condition dependence, multiple sexual signals, and immunocompetence in peacocks. Behav. Ecol. 13:248–253.

Monaghan, P, Metcalfe, NB, and Torres, R (2009). Oxidative stress as a mediator of life history trade-offs: mechanisms, measurements and interpretation. Ecol. Lett. 12:75–92.

Mougeot, F (2008). Ornamental comb color predicts T-cell–mediated immune response in male red grouse *Lagopus lagopus scoticus*. Naturwissenschaften. 95:125–132.

Mougeot, F, Martinez-Padilla, J, Webster, LMI, Blount, JD, Pérez-Rodríguez, L, and Piertney, SB (2009) Honest sexual signaling mediated by parasite and testosterone effects on oxidative balance. Proc. R. Soc. Lond. B. 276:1093–1100.

Nolan, PM, Hill, GE, and Stoehr, AM (1998). Sex, size, and plumage redness predict house finch survival in an epidemic. Proc. R. Soc. Lond. B. 265:961–965.

Norris, K and Evans, MR (2000). Ecological immunology: life history trade-offs and immune defense in birds. Behav. Ecol. 11:19–26.

Nur, N and Hasson, O (1984). Phenotypic plasticity and the handicap principle. J. Theor. Biol. 110:275–297.

Ohlsson, T, Smith, HG, Råberg, L, and Hasselquist, D (2002). Pheasant sexual reflect nutritional conditions during early growth. Proc. R. Soc. Lond. B. 269:21–27.

Olson, VA and Owens, IPF (1998). Costly sexual signals: are carotenoids rare, risky or required? Trends. Ecol. Evol. 13:510–514.

Owen, JP, Delany, ME, and Mullens, BA (2008). MHC haplotype involvement in avian resistance to an ectoparasite. Immunogenetics. 60:621–631.

Owen-Ashley, NT, Hasselquist, D, and Wingfield, JC (2004). Androgens and the immunocompetence handicap hypothesis: unraveling direct and indirect pathways of immunosuppression in song sparrows. Am Nat. 164:490–505.

Pérez-Rodríguez, L (2009). Carotenoids in evolutionary ecology: re-evaluating the antioxidant role. BioEssays 31: 1116–1126.

Råberg, L, Grahn, M, Hasselquist, D, and Svensson, E (1998). On the adaptive significance of stress-induced immunosuppression. Proc. R. Soc. Lond. B. 265:1637–1641.

Råberg, L, Nilsson, J, Ilmonen, P, Stjernman, M, and Hasselquist, D (2000). The cost of an immune response: vaccination reduces parental effort. Ecol. Lett. 3:382–386.

Råberg, L, Sim, D, and Read, AF (2007). Disentangling genetic variation for resistance and tolerance to infectious diseases in animals. Science 318:812–814.

Råberg, L, Graham, AL, and Read, AF (2009). Decomposing health: tolerance and resistance to parasites in animals. Phil. Trans. R. Soc. Lond. B. 364:37–49.

Rantala, MJ, Jokinen, I, Kortet, R, Vainikka, A, and Suhonen, J (2002) Do pheromones reveal male immunocompetence? Proc. R. Soc. Lond. B. 269:1681–1685.

Rantala, MJ, Vainikka, A, and Kortet, R (2003) The role of juvenile hormone in immune function and pheromone production trade-offs: a test of the immunocompetence handicap hypothesis. Proc. R. Soc. Lond. B. 270:2257–2261.

Roberts, ML, Buchanan, KL, and Evans, MR (2004). Testing the immunocompetence handicap hypothesis: a review of the evidence. Anim. Behav. 68:227–239.

Ros, AFH, Correia, M, Wingfield, JC, and Oliveira, RF (2009) Mounting an immune response correlates with decreased androgen levels in male peafowl, *Pavo cristatus*. J. Ethol. 27:209–214.

Ryder, JJ and Siva-Jothy, MT (2000). Male calling song provides a reliable signal of immune function in a cricket. Proc. R. Soc. Lond. B. 267:1171–1175.

Saino, N and Møller, AP (1996). Sexual ornamentation and immunocompetence in the barn swallow. Behav. Ecol. 7:227–232.

Salvante, KG (2006). Techniques for studying integrated immune function in birds. Auk 123:575–586.

Schmid-Hempel, P (2003). Variation in immune defense as a question of evolutionary ecology. Proc. R. Soc. Lond. B. 270:357–366.

Schmid-Hempel, P and Ebert, D (2003). On the evolutionary ecology of specific immune defence. Trends Ecol. Evol. 18:27–32.

Sheldon, BC and Verhulst, S (1996). Ecological immunity: costly parasite defense and trade-offs in evolutionary ecology. Trends Ecol. Evol. 11:317–321.

Simmons, LW, Zuk, M, and Rotenberry, JT (2005). Immune function reflected in calling song characteristics in a natural population of the cricket *Teleogryllus commodus*. Anim. Behav. 69:1235–1241.

Siva-Jothy, MJ (2000). A mechanistic link between parasite resistance and expression of a sexually selected trait in a damselfly. Proc. R. Soc. Lond. B. 267:2523–2527.

Skarstein, F, and Folstad, I (1996). Sexual dichromatism and the immunocompetence handicap: an observational approach using arctic charr. Oikos 76:359–367.

Tella, JL, Lemus, JA, Carrete, M, and Blanco, G (2008). The PHA test reflects acquired T-cell–mediated immunocompetence in birds. PLoS One. 3:e3295.

Verhulst, S, Dieleman, SJ, and Parmentier, HK (1999) A trade-off between immunocompetence and sexual ornamentation in domestic fowl. Proc. Nat. Acad. Sci. USA. 96:4478–4481.

Viney, ME, Riley, EM, and Buchanan, KL (2005). Optimal immune responses: immunocompetence revisited. Trends Ecol. Evol. 20:665–669.

Vinkler, M and Albrecht, T (2010). Carotenoid maintenance handicap and the physiology of carotenoid-based signalization of health. Naturwissenschaften. 97:19–28.

von Schantz, T, Bensch, S, Grahn, M, Hasselquist, D, and Wittzell, H (1999). Good genes, oxidative stress and condition-dependent sexual signals. Proc. R. Soc. Lond. B. 266:1–12.

Wedekind, C and Folstad, I (1994). Adaptive or nonadaptive immunosuppression by sex hormones? Am. Nat. 143:936–938.

Weir, DM and Stewart, J (1993). Immunology, 7th ed. New York: Churchill Livingstone.

Westerdahl, H, Waldenström, J, Hansson, B, Hasselquist, D, von Schantz, T, and Bensch, S (2005) Associations between malaria and MHC genes in a migratory song bird. Proc. R. Soc. Lond. B. 272: 1511–1518.

Westneat, DF and Birkhead, TR (1998). Alternative hypotheses linking the immune system and mate choice for good genes. Proc. R. Soc. Lond. B. 265:1065–1073.

Whitaker, S and Fair, J (2002). The costs of immunological challenge to developing mountain chickadees, *Poecile gambeli*, in the wild. Oikos 99:161–165.

Zahavi, A (1975). Mate selection: a selection for a handicap. J. Theor. Biol. 53:205–214.

Zahavi, A (1977). The cost of honesty (further remarks on the handicap principle). J. Theor. Biol. 67: 603–605.

Zala, SM, Chan, BK, Bilbo, SD, Potts, WK, Nelson, RJ, and Penn, DJ (2008). Genetic resistance to infection influences a male's sexual attractiveness and modulation of testosterone. Brain Behav. Immun. 22: 381–387.

Zuk, M (1990). Reproductive strategies and disease susceptibility: an evolutionary viewpoint. Parasitol. Today. 6:231–233.

Zuk, M (1992). The role of parasites in sexual selection: current evidence and future directions. Adv. Study Behav. 21:39–68.

Zuk, M (1994). Immunology and the evolution of behavior. In: Real, LA, ed. Behavioral Mechanisms in Evolutionary Ecology. Chicago: Chicago University Press.

Zuk, M, Bryant, MJ, Kolluru, GT, and Mirmovitch, V (1996). Trade-offs in parasitology, evolution, and behavior. Parasitol. Today. 12:46–47.

Zuk, M and Johnsen, TS (1998). Seasonal changes in the relationship between ornamentation and immune response in red jungle fowl. Proc. R. Soc. Lond. B. 265:1631–1635.

Zuk, M, Johnsen, TS, and Maclarty, T (1995). Endocrine-immune interactions, ornaments and mate choice in red jungle fowl. Proc. R. Soc. Lond. B. 260:205–210.

Zuk, M and Stoehr, AM (2002). Immune defense and host life history. Am Nat. 160: S9–S22.

Zuk, M, Thornhill, R, Ligon, JD, and Johnson, K (1990). Parasites and mate choice in red jungle fowl. Am. Zool. 30:235–244.

CHAPTER 16

SEX DIFFERENCES IN IMMUNE RESPONSES TO VIRUSES

Dionne P. Robinson and Sabra L. Klein

INTRODUCTION

Males and females differ in their susceptibility to a variety of viral pathogens (Table 16.1). Although behavioral factors can influence exposure to viruses, several studies illustrate that physiological differences between males and females cause differential responses to infection. Females display reduced susceptibility to viral infections because they typically mount stronger immune responses than males. The innate recognition and response to viruses as well as downstream adaptive immune responses differ between males and females during viral infections. This results in sex differences in cytokine responses to infection that play a critical role in determining susceptibility to viruses. Immune responses to viruses can vary with changes in hormone concentrations naturally observed over the menstrual or estrous cycle, following contraception use, and during pregnancy. As a result of heightened immunity to viruses, both the intensity (i.e., viral load within an individual) and prevalence (i.e., number of infected individuals within a population) of viral infections are often lower for females than males (Table 16.1). There is growing awareness, however, that much of the disease attributed to viral infection results from aberrant host inflammatory responses (Theiler et al., 2008). Consequently, heightened antiviral, inflammatory, and cellular immune responses in females, though essential for virus clearance, may underlie increased development of symptoms of disease among females compared with males following infection (Table 16.1) (Anker, 2007; Ahmed et al., 2010).

The adaptive significance of immunological differences between males and females is multifaceted, possibly involving evolved mechanisms to reduce sickness behaviors in males as well as to reduce transmission of pathogens from mother to young. Alternatively, sex differences in the outcome of viral infections may be a side effect of hormonal mechanisms that evolved to promote reproduction and survival or reflect environmental perturbations that have significantly reduced the prevalence of pathogens in the environment at the cost of increasing the emergence of inflammatory diseases, including allergy and asthma, both of which are more severe in females than males (Schatz et al., 2006; Chen et al., 2008). The goals of this chapter are to: (1) evaluate the

TABLE 16.1: Sex Differences in the Intensity, Prevalence, and Severity of Disease or Mortality Following Viral Infections in Mammals

Virus	Host	Measure	Sex difference	Effect of hormone manipulation	Reference
Arboviruses (group A)	*Mus musculus*	I	M > F	T↑	Giron et al., 1973
Coxsackievirus	*Mus musculus*	I	M > F	T↑	Lyden et al., 1987; Huber and Pfaeffle, 1994; Huber et al., 1999
Cytomegalovirus	*Homo sapiens*	P	M < F		Villacres et al., 2004
Dengue virus	*Homo sapiens*	P	M > F		Guha-Sapir and Schimmer, 2005
Encephalomyocarditis virus	*Mus musculus*	I	M > F	T↑, C↓, E2↓	Friedman et al., 1972
Equine arteritis virus	*Equus caballus*	I	M > F	T↑, C↓	Little et al., 1992; McCollum et al., 1994
Epstein Barr virus	*Homo sapiens*	D	M > F		Murphy et al., 2009
Friend virus	*Mus musculus*	I	M > F	E2↓	Mirand et al., 1967; Bruland et al., 2003; Muehlenbein et al., 2006
Junin virus	*Calomys musculinus*	P	M > F		Mills et al.,1992, 1994
Hantaviruses (multiple species)	*Homo sapiens*	P	M > F		White et al., 1996; Williams et al., 1997; Ferrer et al., 1998; Armien et al., 2004
Puumala virus	*Myodes glareolus*	P	M > F		Bernshtein et al., 1999; Olsson et al., 2002
Bayou virus	*Oryzomys palustris*	P	M > F		McIntyre et al., 2005
Sin Nombre virus	*Peromyscus boylii*	P	M > F		Mills et al., 1997
Sin Nombre virus	*P. maniculatus*	P	M > F		Childs et al., 1994; Weigler et al., 1996; Mills et al., 1997; Pearce-Duvet et al., 2006; Douglass et al., 2007
Seoul virus	*Rattus norvegicus*	I	M > F	C↓,O↑	Klein et al., 2000, 2001, 2002; Easterbrook and Klein, 2008; Hannah et al., 2008
Sin Nombre virus	*Reithrodontomys megalotis*	P	M > F		Mills et al., 1997
Black Creek Canal virus	*Sigmodon hispidus*	P	M > F		Glass et al., 1998

Hepatitis B virus	*Homo sapiens*	I, P, D	M > F	T↑, C↓	Farza et al., 1987; DeLoia et al., 1989; Yu et al., 2000, 2001; Tsay et al., 2009; Wang et al., 2009b
Hepatitis C virus	*Homo sapiens*	P	M < F		Tsay et al., 2009
Herpes simplex virus type 1	*Mus musculus*	M	M > F		Han et al., 2001
Herpes simplex virus type 2	*Homo sapiens*	I, P	M < F	P4↑	Mertz, 1992; Fleming, 1997; Obasi, 1999; Langenberg, 1999; Gillgrass, 2003
	Mus musculus	I	M < F	P4↑	Teepe, 1990
Human immunodeficiency virus	*Homo sapiens*	I D	M > F M < F	E2↑	Farzadegan et al., 1998; Sterling et al., 2001; Gandhi et al., 2002; Napravnik et al., 2002; Katagiri et al., 2006
Human T-cell leukemia virus type 1	*Homo sapiens*	P	M < F		Eshima et al., 2009
Influenza A viruses	*Homo sapiens*	D, M	M < F		Noymer and Garenne, 2000; Sedyaningsih et al., 2007; Kandun et al., 2008; Chen et al., 2009; Fielding et al., 2009; Kumar et al., 2009; Oliveira et al., 2009; Fasina et al., 2010
Influenza A viruses	*Mus musculus*	M	M < F		Nohara et al., 2002
Measles	*Homo sapiens*	M	M < F		Garenne, 1994
Mousepox	*Mus musculus*	I	M > F	O↑	Brownstein et al., 1991; Brownstein and Gras, 1995
Polyoma virus	*Mus musculus*	I	M > F	T↑, C↓	Lamey et al., 1985
Theiler's virus	*Mus musculus*	I	M > F		Bihl et al., 1999; Butterfield et al., 2003; Fuller et al., 2005
Vesicular stomatitis virus	*Mus musculus*	I	M > F		Barna et al., 1996
West Nile virus	*Homo sapiens*	I	M > F		O'Leary et al., 2004; Jean et al., 2007
	Equus caballus	I,P	M > F	T↑, C↓	Porter et al., 2003; Epp et al., 2007

Key: I = intensity; P = prevalence; E2 = estradiol, P4 = progesterone; T = testosterone; C = castration of males; O = ovariectomy of females.

proximate mechanisms mediating sex differences in immune responses, (2) provide relevant examples of endocrine–immune interactions that modulate sex differences in response to viruses, and (3) hypothesize about the adaptive significance of why sex differences in responses to viruses may have evolved.

SEX DIFFERENCES IN IMMUNITY

SEX DIFFERENCES IN INNATE IMMUNITY

Males and females differ in their innate immune responses, suggesting that some sex differences may be germline-encoded (Fig. 16.1). For example, innate detection of nucleic acids by pattern recognition receptors (PRRs) differs between the sexes. The *Toll-like receptor 7* (*Tlr7*) gene is encoded on the X chromosome and may escape X inactivation, resulting in higher expression levels of *Tlr7* in females compared to males (Pisitkun et al., 2006). Exposure of peripheral mononuclear cells (PBMCs) to TLR7 ligands *in vitro* causes higher production of IFN-α in cells from women than from men (Berghofer et al., 2006; Meier et al., 2009). Conversely, stimulation of dendritic cells (DCs) with CpG, a TLR9 ligand, results in no sex bias in IFN-α production (Berghofer et al., 2006). Use of global analyses of sex differences in innate immunity reveals profound differences between the sexes in TLR pathways and induction of type I IFN responses.

For example, following a hantavirus challenge, expression of innate antiviral (e.g., *Tlr7, Myd88, Rigi, Visa, Irf7, Ifnβ, Ifnar1, Jak2, Stat3,* and *Mx2*) and pro-inflammatory (e.g., *Nfkβ, Ifnγ,* and *Tnfa*) genes is higher in female than male rats (Klein et al., 2004a; Easterbrook and Klein, 2008; Hannah et al., 2008). Putative androgen and estrogen response elements (AREs and EREs) are present in the promoters of several of these innate immunity genes, suggesting the propensity for sex steroids to directly affect dimorphic innate immune responses (Hannah et al., 2008).

Studies of both humans and rodents illustrate that inflammatory immune responses are generally higher in females than males, and this may explain why women are more likely to develop inflammatory autoimmune diseases, such as rheumatoid arthritis and systemic lupus erythematosus, than men (Da Silva, 1995; Fish, 2008). Female mice also exhibit stronger delayed-type hypersensitivity reactions than males (Boissier et al., 2003).

The number and activity of cells associated with innate immunity differ between the sexes. Phagocytic cells, including macrophages and neutrophils, can kill pathogens by generating reactive oxygen metabolites and nitric oxide, as well as by secreting enzymes. The phagocytic activity of neutrophils and macrophages is higher in females than males (Mondal and Rai, 1999; Spitzer, 1999). Following parasitic or antigenic stimulation, the production and release of prostaglandin E_2, thromboxane B_2, and nitric oxide are reportedly higher in females than males (Du et al., 1984; Barna et al., 1996; Spitzer, 1999). Other studies, however, demonstrate that plasma concentrations of several pro-inflammatory cytokines, including IL-6 and TNF-α, are higher in males than females (Diodato et al., 2001; Naugler et al., 2007). Macrophages derived from female, but not male, mice preferentially secrete IL-12 following nonspecific T-cell activation (Wilcoxen et al., 2000). Women with regular menstrual cycles as well as women tested during the luteal phase of their menstrual cycle have lower NK cell activity than men (Souza et al., 2001; Yovel et al., 2001).

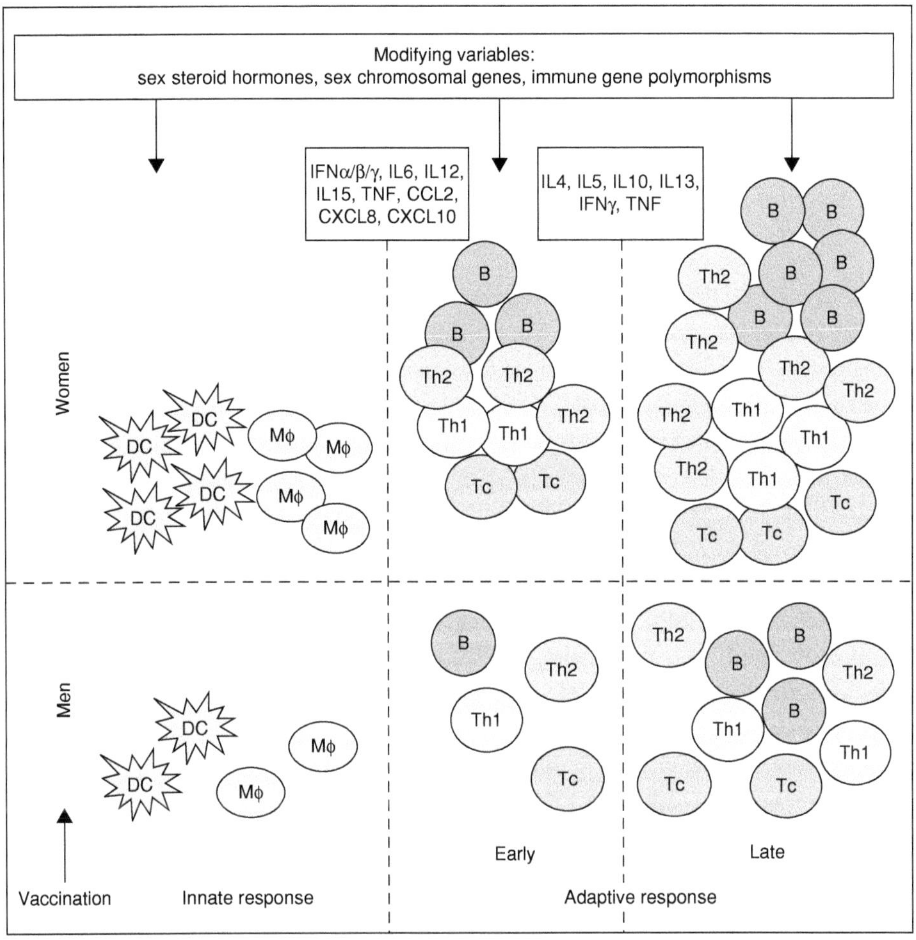

FIGURE 16.1. Sex-based differences in innate and adaptive immune responses after vaccination. After vaccination, the activity of innate immune cells, including dendritic cells (DCs) and macrophages (Mϕ), and the production of inflammatory cytokines (e.g., interferon [IFN] α/β/γ, interleukin [IL] 6, IL12, IL15, and tumor necrosis factor [TNF]-α) and chemokines (e.g., CC-chemokine ligand 2 [CCL2, also called MCP-1], CX-chemokine ligand [CXCL8, also called IL-8], and CXCL10 [also called IP-10]) are raised in women compared with men. The increased number and activity of innate immune cells in women drive heightened expansion and activity of B cells and T cells during the early adaptive immune response. Raised T-helper-2 (Th2) responses, including the production of IL4, IL5, IL10, and IL13, further expand B-cell responses and drive the raised humoral immune response in women during the late adaptive phase of the immune response. Several modifying variables, including sex steroid hormones (e.g., estradiol, progesterone, and testosterone), sex chromosomal genes (eg, *Il2rγ*, *Irak Tlr7, Tlr8, Foxp3*, and *Ikkγ*), and immune gene polymorphisms, are postulated to mediate dimorphic innate and adaptive immune responses to viral vaccines. Adapted from Klein et al. 2010, Lancet Infect Dis 10: 338–49.

Antigen-presenting cells (APCs) from females are more efficient at presenting peptides than are APCs from males (Weinstein et al., 1984). Following CNS infection, the expression of MHC class II on astrocytes, endothelial cells, and microglia is enhanced in female compared with male mice (Barna et al., 1996).

SEX DIFFERENCES IN ADAPTIVE IMMUNITY

Generally, females exhibit elevated humoral and cell-mediated immune responses to antigenic stimulation, vaccination, and infection than do males (Fig. 16.1) (Fish, 2008). Both basal levels of immunoglobulin (Ig) (Butterworth et al., 1967) as well as antibody responses to viruses and vaccines are consistently higher in females than males (Cook, 2008; Ahmed et al., 2010; Klein et al., 2010a). Cell-mediated immune responses also differ between males and females. Clinical studies reveal that men have lower absolute $CD3^+$ cell counts, absolute numbers of $CD4^+$ T cells, $CD4^+$:$CD8^+$ cell ratios, and helper T-cell type 1 (Th1) responses (Amadori et al., 1995; Villacres et al., 2004; Das et al., 2008; Wikby et al., 2008). Studies in mice further reveal that cytokine responses of $CD4^+$ T cells often differ between males and females (Araneo et al., 1991; Barrat et al., 1997; Roberts et al., 2001). For example, female mice produce higher Th2 cytokine responses (e.g., IL-4, IL-9, and IL-13) than males, which correlates with reduced worm burden following infection with helminth parasites (Hepworth et al., 2010). Female mice have higher proportions of regulatory T cells than males (Frisancho-Kiss et al., 2007), which is hypothesized to underlie tolerance of fetal antigens in women during pregnancy (Kallikourdis and Betz, 2007). Females exhibit higher cytotoxic T-cell activity than males, with recent analyses of PMA/ionomycin-stimulated T cells revealing upregulated expression of antiviral (e.g., *Ifnγ, Gzma, Ltb, Ifitm2, Ddx58* [*Rigi*], *Spink5, Oas3, Oas1,* and *Ifi6*) and pro-inflammatory (e.g., *Il12rβ2, Il1f5, Cxc3cl1, Cxcl2, Il16, Mist, Mbp*) genes in T cells isolated from women compared with T cells isolated from men (Hewagama et al., 2009). Following nonspecific T-cell stimulation, half of the genes (10/20) that are upregulated in T cells from women have EREs in their promoters (Hewagama et al., 2009). Several nonspecific indicators of cell-mediated immunity also are elevated in females, with females having higher mitogen-stimulated lymphocyte proliferation, faster wound healing, and increased immunological intolerance to foreign substances than males (Graff et al., 1969; Krzych et al., 1981; Blankenhorn et al., 2003).

SEX STEROIDS MODULATE SEX DIFFERENCES IN IMMUNITY

A majority of studies that have tested the hypothesis that adaptive immune responses differ between the sexes are conducted in the context of examining the direct effects of sex steroids on immune responses. The prevailing hypothesis for immunological differences between the sexes is that sex steroid hormones, particularly testosterone, 17β-estradiol (E2), and progesterone (P4) influence the functioning of immune cells. Receptors for sex steroids are expressed in various lymphoid tissue cells as well as in circulating lymphocytes, macrophages, and DCs (Cutolo et al., 1996; Roberts et al., 2001; Wunderlich et al., 2002; Sapino et al., 2003). The binding of sex steroids to their respective steroid receptors directly influences cell signaling pathways, including NF-κB, cJun, and interferon regulatory factor 1 (IRF1), resulting in differential production of cytokines and chemokines by cells of the immune system (McKay and Cidlowski, 1999; Pernis, 2007). Whereas cellular signaling through NF-κB, cJun, and IRF1 induces the expression of immune and inflammatory genes, steroid hormone, most notably androgen and corticosteroid, signaling antagonizes NF-κB–mediated responses, resulting in regulated communication between the endocrine and immune systems (McKay and Cidlowski, 1999).

ANDROGENS

Androgens are converted from progestins, are produced and released from Leydig cells in the testes, and occur in higher concentrations in postpubertal men than in women. Generally, androgens, including dihydrotestosterone (DHT) and testosterone, suppress activity of the immune cells (Olsen and Kovacs, 1996; Roberts et al., 2001). Exposure to testosterone *in vivo* reduces NK cell activity in mice (Hou and Zheng, 1988). Surface expression of TLR4 on macrophages is reduced by exposure to testosterone both *in vitro* and *in vivo*, which underlies increased susceptibility to endotoxic shock following gonadectomy of male mice (Rettew et al., 2008). Stimulation of murine macrophages with testosterone *in vitro* reduces the synthesis of pro-inflammatory products, including TNF-α and nitric oxide synthase (NOS) (D'Agostino et al., 1999). In contrast, testosterone and DHT increase synthesis of anti-inflammatory cytokines, such as IL-10, indicating that androgens affect anti-inflammatory responses by signaling through the androgen receptor (D'Agostino et al., 1999; Liva and Voskuhl, 2001). Men with androgen deficiency due to Klinefelter's syndrome or idiopathic hypogonadotropic hypogonadism have higher inflammatory cytokine (e.g., IL-1β, IL-2, TNF-α) concentrations, antibody titers, and $CD4^+$:$CD8^+$ T cell ratios than do healthy men (Kocar et al., 2000; Musabak et al., 2003; Malkin et al., 2004). Castrated male mice have increased numbers of $CD4^+$ and $CD8^+$ T cells than intact males (Roden et al., 2004). The immunosuppressive effects of androgens may reflect the inhibitory effects of androgen receptor signaling mechanisms on transcriptional factors that mediate the production of pro-inflammatory and antiviral cytokines (McKay and Cidlowski, 1999). Androgens also may enhance the expression of peroxisome proliferator-activated receptor-α in T cells, which can repress the activity of NF-κB and cJun to control inflammation in males (Dunn et al., 2007).

ESTROGENS

Androgens are the obligatory precursors for all estrogens, including estrone (E1), E2, and estriol (E3), with E2 occurring in highest concentrations in non-pregnant women and being responsible for a majority of the "classic" estrogenic effects in the reproductive and non-reproductive systems. Considerable research has focused on the effects of endogenous and synthetic estrogens on immune function (Ahmed et al., 2010). Estrogens affect both innate and adaptive immune function. Estrogen receptors (ERs) are expressed in various lymphoid tissue cells as well as in circulating lymphocytes, macrophages, and DCs (Danel et al., 1983; Cutolo et al., 1996; Curran et al., 2001; Roberts et al., 2001). There are two subtypes of the receptor for estrogens, ERα and ERβ, which exhibit differential expression in subsets of immune cells, with ERα being highly expressed in T cells and ERβ being upregulated in B cells (Phiel et al., 2005). Thus, the differential effects of estrogens on parameters of immunity may reflect not only the concentration of estrogen (i.e., whether physiological or pharmacological doses are used) but also the density distribution and type of ERs in immune cells. Exposure of NK cells to E2 *in vitro* can enhance NK cytotoxicity and production of IFN-γ (Sorachi et al., 1993; Nakaya et al., 2006). There are, however, reports that E2 downregulates the expression of NK cell surface activation markers, including CD69, NKp46, NKG2DL, and CD244, and secretion of granzyme B and FasL by splenic NK cells *in vivo* (Hao et al., 2007). Following exposure to endotoxin, the expression of TLR4 and CD14 on the surface of peritoneal macrophages as well as production of TNF-α is enhanced by treatment of female mice with E2 (Rettew et al., 2009). Estradiol can have bipotential effects on monocytes and macrophages, with low doses enhancing pro-inflammatory cytokine production (e.g., IL-1, IL-6, and TNF-α) and high concentrations reducing production of these cytokines (Bouman et al., 2005). High concentrations of E2 may suppress transcription of pro-inflammatory cytokine genes in macrophages through activation of peroxisome proliferator-activated receptor-γ (PPARγ) via competitive binding of activated ER to PPARγ response elements, in a manner similar to that demonstrated in adipocytes (Castrillo and Tontonoz, 2004; Bonofiglio et al., 2005).

Estrogens affect the differentiation and functioning of DCs, which influences the recognition and responses to viruses and possibly vaccines. *In vitro* exposure to E2 facilitates differentiation of bone marrow precursor cells, differentiated by the cytokine GM-CSF, into functional CD11c$^+$ DCs (Paharkova-Vatchkova et al., 2004). *In vitro* exposure to E2 also increases the synthesis of chemokines, including CXCL8 and CCL2 (MCP-1), by immature DCs (Bengtsson et al., 2004). Treatment of bone marrow precursor cells, differentiated by Flt3 ligand, with E2 *in vitro* inhibits the survival of these plasmacytoid DCs (Carreras et al., 2008). E2 treatment of human DCs differentiated from peripheral blood mononuclear cells (PMBCs) prior to infection with Newcastle disease virus downregulates antiviral responses, including production of IFN-α and CXCL10 (Escribese et al., 2008). Treatment of ovariectomized mice with E2 increases numbers of IFN-producing killer DCs, production of IFN-γ by CD11c$^+$ cells, and synthesis of pro-inflammatory cytokines, including IL-1, IL-6, and TNF-α (Miller and Hunt, 1996; Siracusa et al., 2008). E2 acts primarily through ERα, not ERβ, to regulate differentiation of DCs (Paharkova-Vatchkova et al., 2004; Carreras et al., 2008; Douin-Echinard et al., 2008).

Estrogens can enhance both cell-mediated and humoral immune responses (Straub, 2007). Generally, low E2 concentrations promote Th1 responses and cell-mediated immunity and high concentrations of E2 augment Th2 responses and humoral immunity (Straub, 2007). The binding of E2 to the ER increases *Ifnγ* transcription by interacting with EREs in the promoter region of the *Ifnγ* gene (Fox et al., 1991). Estrogens also upregulate mitogen activated protein kinase (MAPK), T-bet, and select microRNAs to increase production of IFN-γ by T cells, which can be reversed by treatment of cells with the ER antagonist ICI 1,82,780 (Karpuzoglu et al., 2007; Dai et al., 2008; Suzuki et al., 2008). Estradiol at physiological concentrations induces selective differentiation of antigen-specific CD4$^+$ T cells into Th1 cells and increases the expression of T-bet and production of IFN-γ by Th1 cells (Maret et al., 2003; Karpuzoglu et al., 2007). Conversely, at high doses, E2 reduces the expression of IRF1, which regulates IFN-γ production (Pernis, 2007) and may explain the dose-dependent effects of E2 on Th1 and Th2 immune responses. Estradiol augments the expression of chemokine receptors, including CCR5 and CCR1, on CD4$^+$ T cells (Mo et al., 2005).

Treatment of female multiple sclerosis patients or male mice with doses of E3 that produce pregnancy levels in circulation significantly reduce disease severity, suppresses pro-inflammatory cytokine production (e.g., TNF-α and IFN-γ), reduces numbers of $CD4^+$ and $CD8^+$ T cells, and increases proportions of $CD19^+$ B cells in circulation (Liu et al., 2003a; Soldan et al., 2003).

Estradiol enhances the expansion of $CD4^+CD25^+$ T cells in mice (Polanczyk et al., 2004). The number of regulatory T cells increases during proestrus and estrus in mice and during the follicular stage of the menstrual cycle in women (i.e., when E2 concentrations are highest) (Arruvito et al., 2007; Kallikourdis and Betz, 2007). The expression of several genes associated with activated regulatory T cells (McHugh et al., 2002), including *Tnfrsf4*, *Ltb*, *Ccl3*, and *Gadd45γ*, is higher among intact females than intact males and is reduced by gonadectomy of female mice (Cernetich et al., 2006). Th17 cells have recently been characterized as instrumental regulators of inflammation, and treatment of mice with high doses of E2 decreases production of IL-17 by Th17 cells (Wang et al., 2009a). The cellular and molecular mechanisms mediating estrogenic effects on inflammation have not been fully elucidated. The effects of estrogens on transcriptional factors, such as members of the NF-κB family and cJun, are cell-specific, in which estrogens can either enhance or inhibit these signaling pathways (Evans et al., 2001). Estradiol also preferentially increases the expression and nuclear binding of Bcl-3 and translocation of the NF-κB proteins p50 and p52 into the nuclei of splenocytes (Dai et al., 2007). Thus, there is growing evidence that estrogens regulate pro-inflammatory responses that are transcriptionally mediated by NF-κB. The effects of sex steroids on signaling through other transcriptional factors, including IRFs, to influence production of IFNs and expression of interferon-stimulated genes (ISGs) has not been well characterized.

Estrogens at physiological concentrations can stimulate antibody production by B cells, partially by inhibiting T-cell suppression of B-cell functions (Lu et al., 2002). Levels of Ig are highest prior to ovulation in women (Franklin and Kutteh, 1999). In female rhesus macaques, numbers of Ig-secreting cells are increased in the periovulatory period (Lu et al., 2002). Further, in mice, Ig plaque-forming cell responses are highest during proestrus (i.e., the stage of the estrous cycle prior to ovulation) (Krzych et al., 1981). Exposure of B cells to E2 increases expression of *Bcl2*, which may enhance survival of autoreactive B cells (Grimaldi et al., 2002). E2 also activates the expression of activation-induced deaminase (AID) through the binding of the E2–ER complex to the AID promoter, which enhances somatic hypermutation and class switch recombination in B cells (Pauklin et al., 2009). In contrast to physiological doses of E2, exposure to high doses of E2 suppresses numbers of B-cell lineage precursors in bone marrow and progression from the pro-B-cell to the early pre-B-cell stage (Medina and Kincade, 1994; Smithson et al., 1994; Medina et al., 2001). Although the impact of E2 on B-cell activity and development of antibody-mediated autoimmune diseases, such as systemic lupus erythematosus, has been characterized, considerably less is known about the role of E2 in modulating antibody responses to viruses and vaccines.

PROGESTERONE

In non-pregnant women, progestins, primarily progesterone, are produced by the corpora lutea in the ovaries and play a critical role in reproduction and immune function. Progesterone can have both stimulatory and suppressive effects on the immune system but is typically regarded as immunosuppressive. Progesterone receptors have been identified in epithelial cells, mast cells, granulocytes (e.g., eosinophils), macrophages, DCs, and lymphocytes (Miller and Hunt, 1996; Piccinni et al., 2000; Roberts et al., 2001; Butts et al., 2008). The expression of PRs is higher in DCs from females, which may explain why P4 is better able to suppress the activity (e.g., secretion of TNF-α) of DCs from female than male rats (Butts et al., 2008). Progesterone can bind to glucocorticoid receptors, which are more abundant in the immune system than are PRs, and may represent an alternative mechanism for progesterone-induced changes in immune function (Miller and Hunt, 1996; Jones et al., 2010).

Progesterone suppresses innate immune responses, including macrophage and NK cell activity as well as NF-κB signal transduction (Furukawa et al., 1984; Toder et al., 1984; Baley and Schacter, 1985; Miller and Hunt, 1996; Savita and Rai, 1998; McKay and Cidlowski, 1999). Progesterone can inhibit nitrite and nitric oxide production as well as *Tnfa* mRNA expression by murine macrophages (Miller and Hunt, 1996, 1998; Savita and Rai, 1998). Elevated concentrations of P4 during pregnancy inhibit the development of Th1 immune responses that can lead to fetal rejection and promote production of Th2 immune responses, including IL-4 and IL-5 production (Piccinni et al., 1995; Piccinni et al., 2000). In humans, elevated concentrations of P4 during the second trimester of pregnancy suppress the functional activity of regulatory T cells (Mjosberg et al., 2009). In contrast, in mice, the activity of regulatory T cells is increased at the maternal–fetal interface during pregnancy (Kallikourdis and Betz, 2007). Progesterone also suppresses antibody production, which may be caused by P4 inhibiting $CD8^+$ T-cell responses that in turn suppress antibody production by B cells (Lu et al., 2002). Although the immunomodulatory properties of P4 are well characterized in both mice and humans, the effects on responses to vaccines have not been adequately examined (Gillgrass et al., 2005b).

SEX DIFFERENCES IN RESPONSE TO VIRAL INFECTION AND VACCINATION

SEXUALLY TRANSMITTED VIRUSES

HUMAN IMMUNODEFICIENCY VIRUS

Sex differences are observed in response to a variety of viral agents that are transmitted sexually. Human immunodeficiency virus (HIV) replication exhibits a sexually dimorphic pattern. The amount of circulating HIV RNA in plasma is one marker of progression to acquired immunodeficiency syndrome (AIDS). HIV RNA levels are consistently lower in women than men (Farzadegan et al., 1998; Sterling et al., 2001; Napravnik et al., 2002). A meta-analysis of published studies revealed that women have approximately 41% less HIV RNA in circulation than do men, despite the fact that CD4+ T-cell counts and progression to AIDS are similar between the sexes (Napravnik et al., 2002). HIV loads in women often are below the cutoff value for initiation of antiretroviral therapy (Sterling et al., 2001). In addition to having lower HIV loads, women have higher antiviral responses to HIV than men. Specifically, plasmacytoid DCs from women produce more IFN-α in response to HIV-1 encoded TLR7 ligands than plasmacytoid DCs derived from males, resulting in higher levels of CD8+ T-cell activation in women (Meier et al., 2009). Progesterone can modulate the function of plasmacytoid DCs, and women with higher plasma P4 concentrations have greater numbers of IFN-α producing plasmacytoid DCs in response to the HIV TLR7 ligand than women with lower P4 concentrations (Meier et al., 2009). Use of P4-based hormone contraceptives also is associated with increased acquisition of HIV-1, loss of $CD4^+$ T cells, disease progression, and death rates among women (Hel et al., 2010).

In men, HIV infection causes hypogonadism (i.e., reduced androgen concentrations), which is associated with wasting syndrome, loss of bone mass, and depression (Grinspoon, 2005). Treatment of patients with anabolic steroids improves muscle mass, bone density, and quality of life in both men and women (Grinspoon, 2005); the immunological consequence of androgen treatment, however, has not been reported. In parallel with reduced androgen concentrations, E1 and E2 concentrations increase with the progression of HIV (Christeff et al., 1996; Teichmann et al., 2003). Consequently, E2 augments transcription of HIV *in vitro*, and this effect can be reversed by exposure to the ER antagonist ICI 1,82,780 (Katagiri et al., 2006).

In addition to hormones, X chromosomal variation alters the course of progression of AIDS differently in women than men. In HIV-1–infected patients, heterozygosity in the X chromosomal locus rs5968255 is a significant genetic determinant of progression to AIDS in women but not men (Siddiqui et al., 2009). Women who are carriers of the CT alleles have slower $CD4^+$ T-cell declines and lower HIV-1 loads and progress to AIDS approximately 8 years later than TT homozygous women (Siddiqui et al., 2009). Sex-specific differences in HIV/AIDS involve both genetic and hormonal mechanisms, but whether this affects responses to antiretroviral therapies has not been determined.

HERPES SIMPLEX VIRUSES

Herpes simplex virus type 2 (HSV-2), the causative agent of genital herpes infections and infection of the female reproductive tract, is influenced by ovarian sex hormones, including E2 and P4. The prevalence of HSV-2 typically is higher in women than men (Wald, 2004). In HSV-2–seropositive women, oral contraceptive use is associated with increased genital tract shedding of HSV-2 (Cherpes et al., 2005). *Ex vivo* E2 treatment of primary genital epithelial cells co-cultured with stromal cells increases HSV-2 shedding, whereas pretreatment of cells with P4 decreases HSV-2 shedding (MacDonald et al., 2007). In female mice, susceptibility to HSV-2 varies with stage of the estrous cycle (Gallichan and Rosenthal, 1996). High concentrations of P4 are associated with reduced survival, increased viral titers in the vagina, vaginal pathology, inflammation, infiltration of leukocytes (e.g., neutrophils), and the expression of chemokines (e.g., CCL5, CXCL2, and CXCL10) and chemokine receptors (e.g., CCR7) in vaginal tissue (Gillgrass et al., 2005a). Conversely, administration of E2 increases survival and reduces signs of inflammation and vaginal pathology during primary HSV-2 infection (Gillgrass et al., 2005a).

Mortality rates following exposure to HSV type 1 (HSV-1) also differ between the sexes. Male mice exhibit more severe pathology following corneal infection and are more likely to die from infection than are females; treatment of female mice with dihydrotestosterone (DHT) prior to infection significantly increases morbidity and mortality (Han et al., 2001). HSV-1 can infect the CNS, and in mice inoculated intraperitoneally with HSV-1 and examined during the first week after inoculation, HSV-1 loads are higher in the midbrain, ventricles, cortex, and cerebellum of female than male mice and are associated with worse outcome in females during the acute phase of infection (Burgos et al., 2005). In addition to having more virus in the CNS, females show greater dissemination of virus to peripheral tissues, including the gonads, spinal cord, and trigeminal ganglia than males (Burgos et al., 2005). Following ocular infection with HSV-1, treatment of latently infected ovariectomized female mice with E2 induces viral reactivation in the trigeminal ganglia, which is mediated by signaling through ERα (Vicetti Miguel et al., 2010).

Responses to vaccines against HSV-2 differ between the sexes. The vaccine provides protection against the development of symptoms associated with genital herpes in women but not in men (Stephenson, 2000). In ovariectomized female mice, immunization with an attenuated strain of HSV-2 protects against challenge with wild-type HSV-2 when females are treated with P4, but not E2. Progesterone, administered either alone or in combination with E2, reduces HSV-2 replication in the reproductive tract by increasing the number of DCs and T cells in the vaginal lamina propria and increasing titers of gB-specific vaginal IgA (Gillgrass et al., 2005b). Cessation of E2 treatment for 5 days, but not 1 to 3 days, prior to challenge with HSV-2 can increase protection in ovariectomized mice, suggesting that the effects of E2 can be reversed and are dependent on E2 clearance (Gillgrass et al., 2010). Immunization of female mice with regular estrous cycles with a recombinant adenovirus vector expressing HSV gB results in higher titers of gB-specific vaginal IgA during estrus than during either diestrus or proestrus, and exogenous administration of P4 to female mice at the time of immunization protects females from lethal intravaginal HSV-2 challenge (Gallichan and Rosenthal, 1996), indicating that sex steroids affect induction of protective immunity following vaccination against HSVs. Ocular infection with HSV-1 in HLA-DR*0101 and HLA-DR*0401 transgenic mice results in higher $CD4^+$ T-cell responses in females than males (Zhang et al., 2008).

CANCER-CAUSING VIRUSES

HEPATITIS B VIRUS

Hepatitis B virus (HBV) causes a chronic infection and is a major risk factor for the development of liver cancer and hepatocellular carcinoma. Worldwide, universal HBV vaccines have reduced the prevalence of HBV infection in children. Children infected neonatally, however, can become chronic HBV carriers who are at risk for developing liver cancer, especially in regions of Asia where HBV is endemic (Tsay et al., 2009). The prevalence of serum HBV surface antigen (HBsAg) is consistently higher in men than in women (Koulentaki et al., 1999; Robinson et al., 2005; Tsay et al., 2009). Male sex also is an independent factor associated with elevated HBV DNA titers (Chen et al., 2006). Further, development of hepatocellular carcinoma occurs at a 2:1 to 4:1 ratio of males to females (El-Serag and Rudolph, 2007). Increased rates of exposure to HBV in males do not completely explain why men are more likely to develop liver cancer than women. Studies show that among HBsAg-positive individuals, males are more than twice as likely to experience mortality from liver cancer as are females, suggesting that men may be more sensitive to the effect of HBV infection on the development of liver cancer (Wang et al., 2009b).

Hormones mediate sex differences in the development of liver cancer following infection with HBV. Among HBsAg-positive males, elevated concentrations of testosterone and expression of certain AR gene alleles (*SRD5A2* and *V89L*) correlate with an increased risk of hepatocellular carcinoma (Yu et al., 2000, 2001). Furthermore, the development of chemically induced hepatocellular carcinoma is delayed in AR male knockout mice compared with wild-type male mice (Ma et al., 2008). In HBV transgenic mice, castration of males reduces, whereas replacement of testosterone in castrated males increases, serum HBsAg concentrations (Farza et al., 1987; DeLoia et al., 1989). The effect of androgens on HBV is mediated by the AR because male *Tfm* mice (i.e., mice with a mutation in the AR) do not show elevated concentrations of HBsAg as do wild-type males following HBV infection (Breidbart et al., 1993). One mechanism by which androgens affect HBV replication is through direct binding to AREs that have been identified in the enhancer I of HBV (Wang et al., 2009b). In addition to direct modulation of virus transcription, hormones can alter host immune responses to infection. For example, chemically induced hepatocellular carcinoma is more severe in male than female mice, which is mediated by increased IL-6 production by Kupffer cells in the livers of male mice (Naugler et al., 2007). These studies further reveal that E2 reduces the synthesis of IL-6 by Kupffer cells through inhibition of Myd88-dependent induction of NF-κB (Naugler et al., 2007). Thus, sex steroids modulate sex differences in the prevalence of HBV and development of liver cancer through effects on the transcription of virus and host immune responses to HBV.

Following vaccination against HBV, among both children and adults, anti-HBV antibody titers are higher in females than males (Jilg et al., 1984; Morris et al., 1989; Hess et al., 1992; Fang et al., 1994; Bock et al., 1996). Immunization with the combined HAV/HBV vaccine also results in higher antibody titers among adult women than men (Van der Wielen et al., 2006; Hohler et al., 2007). In multivariate analyses, being male is a significant predictor of being "non-responsive" to the HBV vaccine; thus, adult females show higher rates of seroconversion following exposure to the HBV vaccine than do males (Zeeshan et al., 2007). Greater efficacy of the HBV vaccine in females also may contribute to reduced prevalence of HBV and development of liver cancer among females compared with males.

RESPIRATORY VIRUSES

INFLUENZA A VIRUSES

Sex differences in the incidence of influenza A viruses have been documented in humans (Wingard, 1984; Noymer and Garenne, 2000; Chen et al., 2007). Although exposure rates are often higher in men (Noymer and Garenne, 2000), fatality following exposure to pathogenic influenza A viruses is reportedly higher in women (Sedyaningsih et al., 2007; Kandun et al., 2008; Fasina et al., 2010). Hospitalization with severe disease from 2009 H1N1 was reportedly higher in young women (Chen et al., 2009; Fielding et al., 2009; Kumar et al., 2009; Oliveira et al., 2009). The reason for the greater proportion of hospitalized women than men is not known. In addition to pregnancy, which is a critical factor associated with increased severity of disease (Jamieson et al., 2009; Louie et al., 2010), many cases involve comorbid conditions, including chronic lung disease (e.g., asthma), which are typically more severe in females than males (Singh et al., 1999; Schatz and Camargo, 2003; Schatz et al., 2006; Moorman et al., 2007).

Although sex differences in the incidence of influenza virus infection may reflect differences in exposure to these viruses, differential disease severity between the sexes may involve biological differences in response to infection. The extent to which immune responses differ between males and females during influenza virus infection requires assessment as this may contribute to differential severity of disease between the sexes. Disease associated with highly pathogenic influenza viruses and the clinical manifestations that ensue in humans, including fever, viral pneumonia, encephalitis, and acute respiratory distress syndrome, is hypothesized to be mediated by the profound pro-inflammatory cytokine and chemokine response (referred to as the "cytokine storm") initiated by the host in response to infection (Guan et al., 2004; de Jong et al., 2006). Humans, macaques, and mice infected with highly pathogenic strains of influenza virus, including the 1918 H1N1 or avian H5N1, produce excessively high concentrations of pro-inflammatory cytokines and chemokines, which correlate with elevated mortality (Cook et al., 1995; Guan et al., 2004; de Jong et al., 2006; Kash et al., 2006; Kobasa et al., 2007; Szretter et al., 2007). Whether the duration and magnitude of the cytokine storm initiated during influenza virus infection differ between males and females has not been reported. Preliminary studies using mouse-adapted pathogenic H1N1 influenza reveal that females mount higher inflammatory responses and are more likely to die following inoculation than are males (Robinson et al. 2011). Thus, elevated immunity in females against influenza A viruses may represent a delicate balance between immune responses conferring protection through clearance of virus or causing pathology through increased production of proteins and an influx of immune cells into the lungs.

In addition to influenza virus pathogenesis, males and females differ in response to influenza virus vaccines. Rates of immunization against influenza viruses are similar between the sexes (Qureshi et al., 2004; Jimenez-Garcia et al., 2008; Opstelten et al., 2008), but antibody responses to influenza A virus vaccines are consistently higher in women than men (Cook et al., 2006; Edwards et al., 2007; Cook, 2008; Engler et al., 2008). In response to the trivalent inactivated influenza vaccine (TIV), for example, women generate a more robust antibody response than men (Engler et al., 2008). Importantly, the antibody response of women to a half-dose of the TIV vaccine is equivalent to the antibody response of men to the full dose (Engler et al., 2008). Women also consistently report more severe local and systemic reactions to influenza A virus vaccines than men (Beyer et al., 1996; Nichol et al., 1996; Engler et al., 2008).

ZOONOTIC VIRUSES

HANTAVIRUSES

Globally, hantaviruses are one of the most widely distributed zoonotic pathogens that are maintained in the environment by rodents. Spillover of hantaviruses from rodents to humans causes hantavirus cardiopulmonary syndrome (HCPS) or hemorrhagic fever with renal syndrome (HFRS), depending on the species of virus (Klein and Calisher, 2007). Reported human hantavirus infections in the Americas and Europe, as well as field observations of several rodent-virus systems, indicate that more males than females are infected with hantaviruses (Childs et al., 1994; Weigler et al., 1996; White et al., 1996; Mills et al., 1997; Williams et al., 1997; Glass et al., 1998; Mills and Childs, 1998; Bernshtein et al., 1999). Sex differences in hantavirus infection become apparent only after puberty, suggesting that sex steroid hormones may underlie the dimorphism in infection in rodents (Mills et al., 1997; Childs et al., 1988). Sex steroids can modulate sex differences in infection through effects on the immune system or on the expression of behaviors (e.g., aggression) that increase the likelihood of being exposed to hantaviruses (Zuk and McKean, 1996; Root et al., 1999; Klein et al., 2000, 2004b). Recent data from HFRS patients reveal sex-specific patterns in the acute immune response to Puumala virus (PUUV) infection, in which women produce higher levels of IL-9, FGF-2, and GM-CSF and lower levels of CXCL8 and CXCL10 than men (Klingstrom et al., 2008). Although a similar proportion of men and women have antibodies against PUUV (Ahlm et al., 1994), men are more likely to develop symptoms of disease (i.e., be hospitalized) during PUUV infection than are women (Vapalahti et al., 2003; Klingstrom et al., 2008). Whether sexually dimorphic immune responses during hantavirus infection cause differences in severity of disease between men and women requires further investigation.

Laboratory studies of Norway rats inoculated with Seoul virus (i.e., the hantavirus that naturally infects Norway rats) reveal that when given the same challenge, male and female rats are equally likely to become infected (Klein et al., 2000). After inoculation, however, male rats have more copies of Seoul virus RNA in the lungs for a longer duration of time and shed significantly more virus in saliva than females (Klein et al., 2000, 2001, 2002; Hannah et al., 2008). Additionally, large-scale genomic analyses reveal that the expression of genes that encode for immunological proteins associated with innate antiviral defenses, pro-inflammatory responses, T-cell responsiveness, and antibody production is higher in females than males (Klein et al., 2004a). The induction of PRRs (*Tlr7* and *Rig-I*), expression of antiviral genes (*Myd88, Visa, Jun, Irf7, Ifnβ, Ifnar1, Jak2, Stat3,* and *Mx2*), and production of Mx protein also are elevated in the lungs of intact females compared with intact males (Easterbrook and Klein, 2008; Hannah et al., 2008). Similarly, immunocompetence, as measured by swelling in response to PHA, is higher in female than male deer mice during Sin Nombre virus infection (Lehmer et al., 2007). Conversely, the production of regulatory factors, including *FoxP3* and TGF-β, is elevated in the lungs of Seoul virus-infected males compared with females (Easterbrook and Klein, 2008). These sexually dimorphic immune responses may be dependent on E2 in females and testosterone in males, as gonadectomy reverses these differences. Elevated antiviral immune responses in females may contribute to less efficient Seoul virus replication and shedding in females than males.

Steroid hormones can bind to their respective receptors, which then translocate to the nucleus and bind to HREs in the promoter region of hormone-responsive genes, thereby influencing gene transcription. To test the hypothesis that genes associated with antiviral defenses against hantaviruses are transcriptionally regulated by steroids, computational analyses were employed to identify HREs (i.e., estrogen response elements [EREs], androgen response elements [AREs], progesterone response elements [PREs], and glucocorticoid response elements [GREs]) in promoters of antiviral genes in rats. Putative AREs and EREs have been identified in the promoters of several of antiviral genes, including *Tlr3, Tlr7, Myd88, Irf7, Jun, Hsp70, Ifnar1,* and *Mx2*, suggesting that sex steroids may directly affect dimorphic antiviral responses against Seoul virus infection (Hannah et al., 2008).

Hantaviruses are transmitted through the passage of virus in saliva during aggressive encounters (Glass et al., 1988; Hinson et al., 2004), and male rodents shed more virus in saliva and are more likely to be infected with hantaviruses than females (Hinson et al., 2004; Klein et al., 2004b). Wild-caught male rats that engage in elevated levels of aggression are more likely to be infected with Seoul virus and have higher circulating testosterone concentrations than males that engage in less aggression (Easterbrook et al., 2007). Suppressed antiviral defenses combined with the propensity to engage in aggression, both of which are mediated by testosterone, may make males better transmitters of hantaviruses than females.

VIRUS-INDUCED AUTOIMMUNITY

COXSACKIEVIRUS

Coxsackievirus B-3 (CVB3) is a major etiologic agent of cardiovascular disease in humans (Liu et al., 2003b; Frisancho-Kiss et al., 2006). Clinical myocarditis occurs two times more frequently in men than women (Woodruff, 1980). Mortality rates following inoculation of BALB/c mice with CVB3 are elevated in male (60%) compared with female (25%) mice (Huber and Pfaeffle, 1994). Inoculation of male and female mice with varying doses of CVB3 consistently results in more severe myocarditis in males than females at all doses examined (Lyden et al., 1987). Whether CVB3 virus titers differ between the sexes is debated (Huber et al., 1999; Frisancho-Kiss et al., 2006). Female mice infected during estrus have greater CVB3 titers in the cardiac muscle than females infected during diestrus or proestrus; CVB3 infection during diestrus and proestrus, however, results in more severe myocarditis, suggesting that E2 and P4 concentrations may be more important determinants of the outcome of CVB3 infection than virus titers alone (Huber, 2008a). The precise cause of CVB3-induced cardiovascular disease is not known but likely involves the host immune response to the virus, which can be altered by sex steroid hormones.

Infection of mice with CVB3 results in a skewing of the Th1/Th2 $CD4^+$ T-cell response, in which males exhibit a predominant Th1 (e.g., IFN-γ production) response and females present with a Th2 phenotypic (e.g., IL-4 production) response (Huber and Pfaeffle, 1994; Huber et al., 1999; Frisancho-Kiss et al., 2006). Injection of female mice with high doses of testosterone increases, whereas injection of male mice with high doses of E2 decreases, inflammatory responses (e.g., IFN-γ production) and CVB3 titers in the heart (Lyden et al., 1987; Huber et al., 1999). CVB3 infection in female mice during diestrus (i.e., when E2 levels are lowest) results in greater activation of the Th1 response as a consequence of greater activation and expansion of $CD4^+$ T cells compared with females infected during other stages of the estrous cycle (Huber, 2008b). Conversely, infection during estrus (i.e., when E2 levels are elevated) results in the activation of regulatory T cells (Huber, 2008b). Exogenous administration of E2 to female mice reduces myocarditis regardless of estrous cycle stage at the time of CVB3 inoculation (Huber, 2008b). The protective effect of E2 correlates with the expansion of regulatory T cells within the myocardium (Huber, 2008b). In mice, testosterone administration to males and females results in the activation and expansion of $\gamma\delta$ T cells in response to CVB3, which may augment Th1 responses and contribute to greater myocarditis (Huber et al., 1999). The dominant Th1 cytokine profile in male mice infected with CVB3 is dependent on IL-12Rβ1 signaling, but not on signaling through STAT4 (signal transducers and activators of transcription 4); in contrast, the absence of STAT4 reverses Th2 responses in females (Frisancho-Kiss et al., 2006). Immune cells that infiltrate the myocardium following CVB3 also differ between the sexes: male mice show a greater recruitment of monocytes, macrophages, granulocytes, and DCs and females have a greater influx of B cells (Frisancho-Kiss et al., 2007). Following inoculation with CVB3, activation of TLR4-mediated signaling of inflammatory responses is higher in macrophages from males, which contribute to heightened Th1 responses by increasing the production of IL-18, a critical signal for induction of IFN-γ production from CD4⁺ T cells (Barbulescu et al., 1998; Frisancho-Kiss et al., 2007). Conversely, the innate immune response to CVB3 in female mice is characterized by an increase in Tim-3 and CD80 expression on macrophages, which, through interactions with CTLA-4 on CD4⁺ T cells, results in the expansion of regulatory T cells (Sanchez-Fueyo et al., 2003; Frisancho-Kiss et al., 2007). The extent to which sex hormones affect the innate recognition and response to CVB3 requires further consideration.

MOSQUITO-BORNE VIRUSES

WEST NILE VIRUS

West Nile virus (WNV), a mosquito-borne virus of the flavivirus family, can cause neuroinvasive disease in humans, horses, and birds. Since its emergence in the United States in 1999, WNV has been responsible for the fastest-spreading neuroinvasive disease in recorded history (O'Leary et al., 2004; Hayes et al., 2005; Warner et al., 2006). Infection with WNV results in a range of symptoms from mild fever to more severe symptoms that include meningitis and encephalitis (Campbell et al., 2002). The majority of deaths following WNV infection in healthy individuals are attributed to the progression of severe neurological symptoms, which are commonly referred to as West Nile neuroinvasive disease (WNND) (Campbell et al., 2002). WNV incidence and mortality increase with age (Nash et al., 2001; Weinberger et al., 2001). Although males and females are equally susceptible to WNV, males are at greater risk than females for developing WNND following WNV infection (Campbell et al., 2002; O'Leary et al., 2004; Warner et al., 2006; Theiler et al., 2008). In 2005, California identified nearly 900 cases of WNV in a seven-month time span; 56% of confirmed cases were males (Jean et al., 2007). Further examination of WNND cases revealed that males (62%) were significantly more likely to develop WNND than females (Jean et al., 2007). The incidence of WNND also increased with age; males, however, consistently exhibited a greater risk for developing WNND and mortality than females across all age groups (i.e., 5–17, 18–64, and >65 years of age) (Warner et al., 2006). The male-biased incidence of WNND was most pronounced among individuals over 65 years age, which may reflect sex differences in comorbid conditions, including hypertension and diabetes (Warner et al., 2006).

Studies of horses infected with WNV have been instrumental for addressing how males and females differ in response to infection. Male horses are more likely than females to be diagnosed with WNV, to present with clinical signs of CNS dysfunction, and to develop WNND (Porter et al., 2003). The fatality rate for male horses following WNV infection is 82%, whereas the fatality rates for castrated males and females are 45% and 36%, respectively (Epp et al., 2007). Overall, gonadally intact male horses are 8.5 times more likely to die from WNV infection than castrated males or females, suggesting that testosterone may negatively influence the outcome of WNV infection in males (Epp et al., 2007). Several studies using mouse models of WNV have identified key components of the immune response that are critical for control of WNV infection, including B cells (Diamond et al., 2003; Brien et al., 2009). In the absence of B cells, which produce IgM following WNV infection, mice rapidly progress to CNS abnormalities as a result of WNV invasion of the CNS (Diamond et al., 2003). Whether females exhibit greater B-cell activity in response to WNV should be considered.

EVOLUTION OF SEX DIFFERENCES IN IMMUNE RESPONSES AGAINST VIRUSES

A majority of this chapter has addressed how males and females differ in their responses to viruses, providing extensive documentation of hormonal effects on immune responses to viral infections and vaccines. It is equally important to consider why males and females have evolved such different strategies for responding to viruses. One prevailing hypothesis is that sexual selection drives differences in immune function between males and females. In other words, biological factors associated with differential reproduction, successful competition for mates, and increased survival may affect immune function. Males and females rely on different strategies for maximizing reproductive success, with males generally competing for access to females and females selecting the highest-quality males with whom to mate. Presumably the mechanisms, including androgens, that are required for successful competition and acquisition of mates may also contribute to differential susceptibility to infection (Zuk, 2009). Thus, sex differences in the outcome of infection may reflect differential selection pressures acting on each sex to maximize reproductive output (Zuk and McKean, 1996).

In addition to sexual selection, life-history strategies may result in constraints on responses to infection. Presumably, engaging in behaviors that limit exposure to pathogens or mounting immune responses to reduce susceptibility to pathogens requires metabolic resources that might otherwise be used for other biological processes, such as growth, maintenance of secondary sex characteristics, and reproduction (Barnard and Behnke, 2001; Sheldon and Verhulst, 1996; Zuk and Stoehr, 2002). Thus, trade-offs exist among life-history strategies that influence investment in survival and reproduction. Studies in mice illustrate that trade-offs exist between engaging in competitive behaviors (which presumably increase mating opportunities and territorial resources) and susceptibility to infection in males. Specifically, dominant male mice that are more likely to engage in successful aggressive encounters are more susceptible to infection with nematodes, such as *Heligmosomoides polygyrus*, protozoa, including *Babesia microti*, and viruses, such as herpes simplex, than lower-rank, less aggressive males (Barnard et al., 1993, 1998; Padgett et al., 1998). This trade-off between competitive abilities and susceptibility to infection may be mediated by changes in steroid hormone concentrations or available metabolic resources (Folstad and Karter, 1992; Sheldon and Verhulst, 1996). An assumption underlying these theories is that increased virus load and reduced immune function among males is merely an adverse side effect of positive selection for other traits or characteristics that increase reproductive success and survival. These theories, however, do not account for why the intensity and prevalence of some pathogens is higher in females than males or why females can experience a worse disease outcome following infection (Escobedo et al., 2010; Klein et al., 2010b) (see Table 16.1).

In females, the selective pressure of successful pregnancy may affect hormone-mediated alterations in immunity. During pregnancy, the immune system shifts away from inflammatory responses associated with Th1-mediated immunity, while maintaining or augmenting the anti-inflammatory Th2 and regulatory T-cell responses, which are modulated by hormones, including P4 (Szekeres-Bartho et al., 1996). This immunological shift serves to prevent immune-mediated rejection of the fetus, but also reduces the capacity of pregnant females to mount a strong antiviral immune response, which is needed to control a primary viral infection (Gaunt and Ramin, 2001; Szekeres-Bartho, 2002). This hypothesis, however, does not explain why non-pregnant women generally mount higher innate and adaptive immune responses than males.

Innate and adaptive immune responses have evolved to protect hosts through effective control and clearance of pathogens. If these immune responses become too high or remain elevated for too long, then this can cause pathology (Katze et al., 2008). As a result of elevated immunity among females, many inflammatory-mediated diseases, including autoimmunity, allergy, and asthma, are more frequent and more severe in females than males (Schatz et al., 2006; Chen et al., 2008; Libert et al., 2010). The hygiene hypothesis contends that reduced exposure to microorganisms in industrialized countries, such as the United States and Europe, has contributed to dysregulation of the immune system and the emergence of inflammatory diseases (Rook, 2009). The extent to which this hypothesis can be applied to better understand sex-dependent differences in the ability of the immune system to clear viruses versus cause immunopathology has not been considered.

Heightened immunity among females may underlie the development of behaviors that could "advertise" immune status to other individuals in a population. Several studies illustrate that elevated immune responses, including the production of pro-inflammatory cytokines, underlie the expression of sickness behaviors and development of depression (Dantzer et al., 2008; Solomon and Herman, 2009). Consequently, females experience greater sickness behavior, including suppression of sexual activity, than males, which reflects hypothalamic-pituitary-adrenal activity as well as increased cytokine production following challenge (Avitsur and Yirmiya, 1999; Yirmiya et al., 1995; Spinedi et al., 2002). Females also are more likely to be diagnosed with depression than males (Solomon and Herman, 2009). From an evolutionary perspective, if males are more likely than females to compete for mates, territories, and resources, then selection may favor physiological mechanisms that limit the expression of sickness behaviors and depression in males.

Taken together, sex differences in response to viruses are evolutionarily well conserved and are present across diverse host and pathogen species. Hormones play a fundamental role in modulating the immune responses to viruses and contribute significantly to sex differences in the outcome of infection. The mechanistic interactions between hormones and the immune system likely evolved through the process of sexual selection and contribute significantly to survival and reproductive success. Because males and females differ in the selective pressures that maximize reproductive success, the influence of these pressures on the immune system and outcome of infection differs as well. Future research must continue to uncover the mechanisms mediating sex differences in response to viruses and consider why the mechanistic interactions have evolved.

Despite experimental, clinical, and observational data illustrating sex differences the prevalence, incidence, and susceptibility to infectious diseases, a majority of studies do not appropriately consider sex in either the experimental design or analyses of data (Beery and Zucker, 2011). Furthermore, a majority of biomedical animal studies are conducted using rats or mice and rarely consider using alternative animal models (Beery and Zucker, 2011). Future studies must continue to examine sex differences in response to diverse viruses to more rigorously test the hypothesis that sex differences in response to viruses are evolutionarily well conserved. More stringent consideration also should be given to the selection of animal models because some viruses (e.g., hantaviruses) have distinct patterns of infection in rodent species other than laboratory mice. From a public health perspective, understanding both how and why males and females differ in their responses to viral infections and vaccines may lead to the development and optimization of treatments that provide equal protection for both sexes, thereby forming a fundamental basis for personalized medicine.

REFERENCES

Ahlm C, Linderholm M, Juto P, Stegmayr B, Settergren B, 1994. Prevalence of serum IgG antibodies to Puumala virus (haemorrhagic fever with renal syndrome) in northern Sweden. Epidemiol Infect 113:129–136.

Ahmed SA, Karpuzoglu E, Khan D, 2010. Effects of sex steroids on innate and adaptive immunity. In: Klein SL, Roberts CW, eds. Sex Hormones and Immunity to Infection. Berlin: Springer-Verlag, pp. 19–53.

Amadori A, Zamarchi R, De Silvestro G, Forza G, Cavatton G, Danieli GA, Clementi M, Chieco-Bianchi L, 1995. Genetic control of the CD4/CD8 T-cell ratio in humans. Nat Med 1:1279–1283.

Anker M, 2007. Addressing Sex and Gender in Epidemic-Prone Infectious Diseases. Geneva: World Health Organization Press, pp. 1–40.

Araneo BA, Dowell T, Diegel M, Daynes RA, 1991. Dihydrotestosterone exerts a depressive influence on the production of interleukin-4 (IL-4), IL-5, and gamma-interferon, but not IL-2 by activated murine T cells. Blood 78:688–699.

Armien B, Pascale JM, Bayard V, Munoz C, Mosca I, Guerrero G, Armien A, Quiroz E, Castillo Z, Zaldivar Y, Gracia F, Hjelle B, Koster F, 2004. High seroprevalence of hantavirus infection on the Azuero peninsula of Panama. Am J Trop Med Hyg 70:682–687.

Arruvito L, Sanz M, Banham AH, Fainboim L, 2007. Expansion of CD4+CD25+and FOXP3+ regulatory T cells during the follicular phase of the menstrual cycle: implications for human reproduction. J Immunol 178:2572–2578.

Avitsur R, Yirmiya R, 1999. The immunobiology of sexual behavior: gender differences in the suppression of sexual activity during illness. Pharmacol Biochem Behav 64:787–796.

Baley JE, Schacter BZ, 1985. Mechanisms of diminished natural killer cell activity in pregnant women and neonates. J Immunol 134:3042–3048.

Barbulescu K, Becker C, Schlaak JF, Schmitt E, Meyer zum Buschenfelde KH, Neurath MF, 1998. IL-12 and IL-18 differentially regulate the transcriptional activity of the human IFN-gamma promoter in primary CD4+ T lymphocytes. J Immunol 160:3642–3647.

Barna M, Komatsu T, Bi Z, Reiss CS, 1996. Sex differences in susceptibility to viral infection of the central nervous system. J Neuroimmunol 67:31–39.

Barnard CJ, Behnke JM, 2001. From psychoneuroimmunology to ecological immunology: life history strategies and immunity trade-offs. In: Ader R, Felten DL, Cohen N, eds, Psychoneuroimmunology. San Diego: Academic Press, p. 35–47.

Barnard CJ, Behnke JM, Gage AR, Brown H, Smithurst PR, 1998. The role of parasite-induced immunodepression, rank and social environment in the modulation of behaviour and hormone concentration in male laboratory mice (Mus musculus). Proc R Soc Lond B Biol Sci 265:693–701.

Barnard CJ, Behnke JM, Sewell J, 1993. Social behaviour, stress and susceptibility to infection in house mice (Mus musculus): effects of duration of grouping and aggressive behaviour prior to infection on susceptibilty to Babesia microti. Parasitology 107:183–192.

Barrat F, Lesourd B, Boulouis HJ, Thibault D, Vincent-Naulleau S, Gjata B, Louise A, Neway T, Pilet C, 1997. Sex and parity modulate cytokine production during murine ageing. Clin Exp Immunol 109:562–568.

Beery AK, Zucker I, 2011. Sex bias in neuroscience and biomedical research. Neurosci Biobehav Rev 35: 565–572.

Bengtsson AK, Ryan EJ, Giordano D, Magaletti DM, Clark EA, 2004. 17beta-estradiol (E2) modulates cytokine and chemokine expression in human monocyte-derived dendritic cells. Blood 104:1404–1410.

Berghofer B, Frommer T, Haley G, Fink L, Bein G, Hackstein H, 2006. TLR7 ligands induce higher IFN-alpha production in females. J Immunol 177:2088–2096.

Bernshtein AD, Apekina NS, Mikhailova TV, Myasnikov YA, Khlyap LA, Korotkov YS, Gavrilovskaya IN, 1999. Dynamics of Puumala hantavirus infection in naturally infected bank voles (*Clethrinomys glareolus*). Arch Virol 144:2415–2428.

Beyer WE, Palache AM, Kerstens R, Masurel N, 1996. Gender differences in local and systemic reactions to inactivated influenza vaccine, established by a meta-analysis of fourteen independent studies. Eur J Clin Microbiol Infect Dis 15:65–70.

Bihl F, Brahic M, Bureau JF, 1999. Two loci, Tmevp2 and Tmevp3, located on the telomeric region of chromosome 10, control the persistence of Theiler's virus in the central nervous system of mice. Genetics 152:385–392.

Blankenhorn EP, Troutman S, Clark LD, Zhang XM, Chen P, Heber-Katz E, 2003. Sexually dimorphic genes regulate healing and regeneration in MRL mice. Mamm Genome 14:250–260.

Bock HL, Kruppenbacher J, Sanger R, Hobel W, Clemens R, Jilg W, 1996. Immunogenicity of a recombinant hepatitis B vaccine in adults. Arch Intern Med 156:2226–2231.

Boissier J, Chlichlia K, Digon Y, Ruppel A, Mone H, 2003. Preliminary study on sex-related inflammatory reactions in mice infected with *Schistosoma mansoni*. Parasitol Res 91:144–150.

Bonofiglio D, Gabriele S, Aquila S, Catalano S, Gentile M, Middea E, Giordano F, Ando S, 2005. Estrogen receptor alpha binds to peroxisome proliferator-activated receptor response element and negatively interferes with peroxisome proliferator-activated receptor gamma signaling in breast cancer cells. Clin Cancer Res 11:6139–6147.

Bouman A, Heineman MJ, Faas MM, 2005. Sex hormones and the immune response in humans. Hum Reprod Update 11:411–423.

Breidbart S, Burk RD, Saenger P, 1993. Hormonal regulation of hepatitis B virus gene expression: influence of androgen receptor. Pediatr Res 34:300–302.

Brien JD, Uhrlaub JL, Hirsch A, Wiley CA, Nikolich-Zugich J, 2009. Key role of T cell defects in age-related vulnerability to West Nile virus. J Exp Med 206:2735–2745.

Brownstein DG, Bhatt PN, Gras L, Jacoby RO, 1991. Chromosomal locations and gonadal dependence of genes that mediate resistance to ectromelia (mousepox) virus-induced mortality. J Virol 65:1946–1951.

Brownstein DG, Gras L, 1995. Chromosome mapping of Rmp-4, a gonad-dependent gene encoding host resistance to mousepox. J Virol 69:6958–6964.

Bruland T, Lavik LA, Dai HY, Dalen A, 2003. A glucocorticoid response element in the LTR U3 region of Friend murine leukaemia virus variant FIS-2 enhances virus production in vitro and is a major determinant for sex differences in susceptibility to FIS-2 infection in vivo. J Gen Virol 84:907–916.

Burgos JS, Ramirez C, Sastre I, Alfaro JM, Valdivieso F, 2005. Herpes simplex virus type 1 infection via the bloodstream with apolipoprotein E dependence in the gonads is influenced by gender. J Virol 79: 1605–1612.

Butterfield RJ, Roper RJ, Rhein DM, Melvold RW, Haynes L, Ma RZ, Doerge RW, Teuscher C, 2003. Sex-specific quantitative trait loci govern susceptibility to Theiler's murine encephalomyelitis virus-induced demyelination. Genetics 163:1041–1046.

Butterworth M, McClellan B, Allansmith M, 1967. Influence of sex in immunoglobulin levels. Nature 214:1224–1225.

Butts CL, Bowers E, Horn JC, Shukair SA, Belyavskaya E, Tonelli L, Sternberg EM, 2008. Inhibitory effects of progesterone differ in dendritic cells from female and male rodents. Gend Med 5:434–447.

Campbell GL, Marfin AA, Lanciotti RS, Gubler DJ, 2002. West Nile virus. Lancet Infect Dis 2:519–529.

Carreras E, Turner S, Paharkova-Vatchkova V, Mao A, Dascher C, Kovats S, 2008. Estradiol acts directly on bone marrow myeloid progenitors to differentially regulate GM-CSF or Flt3 ligand-mediated dendritic cell differentiation. J Immunol 180:727–738.

Castrillo A, Tontonoz P, 2004. Nuclear receptors in macrophage biology: at the crossroads of lipid metabolism and inflammation. Annu Rev Cell Dev Biol 20:455–480.

Cernetich A, Garver LS, Jedlicka AE, Klein PW, Kumar N, Scott AL, Klein SL, 2006. Involvement of gonadal steroids and gamma interferon in sex differences in response to blood-stage malaria infection. Infect Immun 74:3190–3203.

Chen CJ, Yang HI, Su J, Jen CL, You SL, Lu SN, Huang GT, Iloeje UH, 2006. Risk of hepatocellular carcinoma across a biological gradient of serum hepatitis B virus DNA level. JAMA 295:65–73.

Chen JM, Chen JW, Dai JJ, Sun YX, 2007. A survey of human cases of H5N1 avian influenza reported by the WHO before June 2006 for infection control. Am J Infect Control 35:467–469.

Chen KF, Gaydos C, Rothman RE, 2009. Update on emerging infections: news from the Centers for Disease Control and Prevention. Hospitalized patients with novel influenza A (H1N1) virus infection—California, April-May, 2009. Ann Emerg Med 54:732–736.

Chen W, Mempel M, Schober W, Behrendt H, Ring J, 2008. Gender difference, sex hormones, and immediate type hypersensitivity reactions. Allergy 63:1418–1427.

Cherpes TL, Melan MA, Kant JA, Cosentino LA, Meyn LA, Hillier SL, 2005. Genital tract shedding of herpes simplex virus type 2 in women: effects of hormonal contraception, bacterial vaginosis, and vaginal group B Streptococcus colonization. Clin Infect Dis 40:1422–1428.

Childs JE, Glass GE, Korch GW, LeDuc JW, 1988. The ecology and epizootiology of hantaviral infections in small mammal communities of Baltimore: A review and synthesis. Bulletin of the Society of Vector Biology 13:113–122.

Childs JE, Ksiazek TG, Spiropoulou CF, Krebs JW, Morzunov S, Maupin GO, Gage KL, Rollin PE, Sarisky J, Enscore RE, Frey JK, Peters CJ, Nichol ST, 1994. Serologic and genetic identification of *Peromyscus maniculatus* as the primary rodent reservoir for a new hantavirus in the Southwestern United States. J Infect Dis 169:1271–1280.

Christeff N, Lortholary O, Casassus P, Thobie N, Veyssier P, Torri O, Guillevin L, Nunez EA, 1996. Relationship between sex steroid hormone levels and CD4 lymphocytes in HIV infected men. Exp Clin Endocrinol Diabetes 104:130–136.

Cook DN, Beck MA, Coffman TM, Kirby SL, Sheridan JF, Pragnell IB, Smithies O, 1995. Requirement of MIP-1 alpha for an inflammatory response to viral infection. Science 269:1583–1585.

Cook IF, 2008. Sexual dimorphism of humoral immunity with human vaccines. Vaccine 26:3551–3555.

Cook IF, Barr I, Hartel G, Pond D, Hampson AW, 2006. Reactogenicity and immunogenicity of an inactivated influenza vaccine administered by intramuscular or subcutaneous injection in elderly adults. Vaccine 24:2395–2402.

Curran EM, Berghaus LJ, Vernetti NJ, Saporita AJ, Lubahn DB, Estes DM, 2001. Natural killer cells express estrogen receptor-alpha and estrogen receptor-beta and can respond to estrogen via a non-estrogen receptor-alpha-mediated pathway. Cell Immunol 214:12–20.

Cutolo M, Accardo S, Villaggio B, Barone A, Sulli A, Coviello DA, Carabbio C, Felli L, Miceli D, Farruggio R, Carruba G, Castagnetta L, 1996. Androgen and estrogen receptors are present in primary cultures of human synovial macrophages. J Clin Endocrinol Metab 81:820–827.

D'Agostino P, Milano S, Barbera C, Di Bella G, La Rosa M, Ferlazzo V, Farruggio R, Miceli DM, Miele M, Castagnetta L, Cillari E, 1999. Sex hormones modulate inflammatory mediators produced by macrophages. Ann N Y Acad Sci 876:426–429.

Da Silva JA, 1995. Sex hormones, glucocorticoids and autoimmunity: facts and hypotheses. Ann Rheum Dis 54:6–16.

Dai R, Phillips RA, Ahmed SA, 2007. Despite inhibition of nuclear localization of NF-kappa B p65, c-Rel, and RelB, 17-beta estradiol up-regulates NF-kappa B signaling in mouse splenocytes: the potential role of Bcl-3. J Immunol 179:1776–1783.

Dai R, Phillips RA, Zhang Y, Khan D, Crasta O, Ahmed SA, 2008. Suppression of LPS-induced Interferon-gamma and nitric oxide in splenic lymphocytes by select estrogen-regulated microRNAs: a novel mechanism of immune modulation. Blood 112:4591–4597.

Danel L, Souweine G, Monier JC, Saez S, 1983. Specific estrogen binding sites in human lymphoid cells and thymic cells. J Steroid Biochem 18:559–563.

Dantzer R, O'Connor JC, Freund GG, Johnson RW, Kelley KW, 2008. From inflammation to sickness and depression: when the immune system subjugates the brain. Nat Rev Neurosci 9:46–56.

Das BR, Bhanushali AA, Khadapkar R, Jeswani KD, Bhavsar M, Dasgupta A, 2008. Reference ranges for lymphocyte subsets in adults from western India: influence of sex, age and method of enumeration. Indian J Med Sci 62:397–406.

de Jong MD, Simmons CP, Thanh TT, Hien VM, Smith GJ, Chau TN, Hoang DM, Chau NV, Khanh TH, Dong VC, Qui PT, Cam BV, Ha do Q, Guan Y, Peiris JS, Chinh NT, Hien TT, Farrar J, 2006. Fatal outcome of human influenza A (H5N1) is associated with high viral load and hypercytokinemia. Nat Med 12: 1203–1207.

DeLoia JA, Burk RD, Gearhart JD, 1989. Developmental regulation of hepatitis B surface antigen expression in two lines of hepatitis B virus transgenic mice. J Virol 63:4069–4073.

Diamond MS, Shrestha B, Mehlhop E, Sitati E, Engle M, 2003. Innate and adaptive immune responses determine protection against disseminated infection by West Nile encephalitis virus. Viral Immunol 16:259–278.

Diodato MD, Knoferl MW, Schwacha MG, Bland KI, Chaudry IH, 2001. Gender differences in the inflammatory response and survival following haemorrhage and subsequent sepsis. Cytokine 14:162–169.

Douglass RJ, Calisher CH, Wagoner KD, Mills JN, 2007. Sin Nombre virus infection of deer mice in Montana: characteristics of newly infected mice, incidence, and temporal pattern of infection. J Wildl Dis 43:12–22.

Douin-Echinard V, Laffont S, Seillet C, Delpy L, Krust A, Chambon P, Gourdy P, Arnal JF, Guery JC, 2008. Estrogen receptor alpha, but not beta, is required for optimal dendritic cell differentiation and [corrected] CD40-induced cytokine production. J Immunol 180:3661–3669.

Du JT, Vennos E, Ramey E, Ramwell PW, 1984. Sex differences in arachidonate cyclo-oxygenase products in elicited rat peritoneal macrophages. Biochim Biophys Acta 794:256–260.

Dunn SE, Ousman SS, Sobel RA, Zuniga L, Baranzini SE, Youssef S, Crowell A, Loh J, Oksenberg J, Steinman L, 2007. Peroxisome proliferator-activated receptor (PPAR)alpha expression in T cells mediates gender differences in development of T cell-mediated autoimmunity. J Exp Med 204:321–330.

Easterbrook JD, Kaplan JB, Glass GE, Pletnikov MV, Klein SL, 2007. Elevated testosterone and reduced 5-HIAA concentrations are associated with wounding and hantavirus infection in male Norway rats. Horm Behav 52:474–481.

Easterbrook JD, Klein SL, 2008. Corticosteroids modulate Seoul virus infection, regulatory T-cell responses and matrix metalloprotease 9 expression in male, but not female, Norway rats. J Gen Virol 89:2723–2730.

Edwards KM, Burns VE, Allen LM, McPhee JS, Bosch JA, Carroll D, Drayson M, Ring C, 2007. Eccentric exercise as an adjuvant to influenza vaccination in humans. Brain Behav Immun 21:209–217.

El-Serag HB, Rudolph KL, 2007. Hepatocellular carcinoma: epidemiology and molecular carcinogenesis. Gastroenterology 132:2557–2576.

Engler RJ, Nelson MR, Klote MM, VanRaden MJ, Huang CY, Cox NJ, Klimov A, Keitel WA, Nichol KL, Carr WW, Treanor JJ, 2008. Half- vs full-dose trivalent inactivated influenza vaccine (2004–2005): age, dose, and sex effects on immune responses. Arch Intern Med 168:2405–2414.

Epp T, Waldner C, West K, Townsend H, 2007. Factors associated with West Nile virus disease fatalities in horses. Can Vet J 48:1137–1145.

Escobedo G, De Leon MA, Morales-Montor J, 2010. Sex differences in parasitic infections: Beyond the dogma of female-biased resistance. In: Klein SL, Roberts C, eds. Sex Hormones and Immunity to Infection. Berlin: Springer-Verlag, pp. 187–205.

Escribese MM, Kraus T, Rhee E, Fernandez-Sesma A, Lopez CB, Moran TM, 2008. Estrogen inhibits dendritic cell maturation to RNA viruses. Blood 112:4574–4584.

Eshima N, Iwata O, Iwata S, Tabata M, Higuchi Y, Matsuishi T, Karukaya S, 2009. Age and gender specific prevalence of HTLV-1. J Clin Virol 45:135–138.

Evans MJ, Eckert A, Lai K, Adelman SJ, Harnish DC, 2001. Reciprocal antagonism between estrogen receptor and NF-kappaB activity in vivo. Circ Res 89:823–830.

Fang JW, Lai CL, Chung HT, Wu PC, Lau JY, 1994. Female children respond to recombinant hepatitis B vaccine with a higher titre than male. J Trop Pediatr 40:104–107.

Farza H, Salmon AM, Hadchouel M, Moreau JL, Babinet C, Tiollais P, Pourcel C, 1987. Hepatitis B surface antigen gene expression is regulated by sex steroids and glucocorticoids in transgenic mice. Proc Natl Acad Sci USA 84:1187–1191.

Farzadegan H, Hoover DR, Astemborski J, Lyles CM, Margolick JB, Markham RB, Quinn TC, Vlahov D, 1998. Sex differences in HIV-1 viral load and progression to AIDS. Lancet 352:1510–1514.

Fasina F, Ifende V, Ajibade A, 2010. Avian influenza A(H5N1) in humans: lessons from Egypt. Euro Surveill 15.

Ferrer JF, Jonsson CB, Esteban E, Galligan D, Basombrio MA, Peralta-Ramos M, Bharadwaj M, Torrez-Martinez N, Callahan J, Segovia A, Hjelle B, 1998. High prevalence of hantavirus infection in Indian communities of the Paraguayan and Argentinean Gran Chaco. Am J Trop Med Hyg 59:438–444.

Fielding J, Higgins N, Gregory J, Grant K, Catton M, Bergeri I, Lester R, Kelly H, 2009. Pandemic H1N1 influenza surveillance in Victoria, Australia, April–September, 2009. Euro Surveill 14.

Fish EN, 2008. The X-files in immunity: sex-based differences predispose immune responses. Nat Rev Immunol 8:737–744.

Folstad I, Karter AJ, 1992. Parasites, bright males, and the immunocompetence handicap. Am Nat 139: 603–622.

Fox HS, Bond BL, Parslow TG, 1991. Estrogen regulates the IFN-gamma promoter. J Immunol 146: 4362–4367.

Franklin RD, Kutteh WH, 1999. Characterization of immunoglobulins and cytokines in human cervical mucus: influence of exogenous and endogenous hormones. J Reprod Immunol 42:93–106.

Friedman SB, Grota LJ, Glasgow LA, 1972. Differential susceptibility of male and female mice to encephalomyocarditis virus: effects of castration, adrenalectomy, and the administration of sex hormones. Infect Immun 5:637–644.

Frisancho-Kiss S, Davis SE, Nyland JF, Frisancho JA, Cihakova D, Barrett MA, Rose NR, Fairweather D, 2007. Cutting edge: cross-regulation by TLR4 and T cell Ig mucin-3 determines sex differences in inflammatory heart disease. J Immunol 178:6710–6714.

Frisancho-Kiss S, Nyland JF, Davis SE, Frisancho JA, Barrett MA, Rose NR, Fairweather D, 2006. Sex differences in coxsackievirus B3-induced myocarditis: IL-12Rbeta1 signaling and IFN-gamma increase inflammation in males independent from STAT4. Brain Res 1126:139–147.

Fuller AC, Kang B, Kang HK, Yahikozowa H, Dal Canto MC, Kim BS, 2005. Gender bias in Theiler's virus-induced demyelinating disease correlates with the level of antiviral immune responses. J Immunol 175:3955–3963.

Furukawa K, Itoh K, Okamura K, Kumagai K, Suzuki M, 1984. Changes in NK cell activity during the estrous cycle and pregnancy in mice. J Reprod Immunol 6:353–363.

Gallichan WS, Rosenthal KL, 1996. Effects of the estrous cycle on local humoral immune responses and protection of intranasally immunized female mice against herpes simplex virus type 2 infection in the genital tract. Virology 224:487–497.

Gandhi M, Bacchetti P, Miotti P, Quinn TC, Veronese F, Greenblatt RM, 2002. Does patient sex affect human immunodeficiency virus levels? Clin Infect Dis 35:313–322.

Garenne M, 1994. Sex differences in measles mortality: a world review. Int J Epidemiol 23:632–642.

Gaunt G, Ramin K, 2001. Immunological tolerance of the human fetus. American Journal of Perinatology 18:299–312.

Gillgrass A, Chege D, Bhavanam S, Kaushic C, 2010. Estradiol limits viral replication following intravaginal immunization leading to diminished mucosal IgG response and non-sterile protection against genital herpes challenge. Am J Reprod Immunol 63:299–309.

Gillgrass AE, Fernandez SA, Rosenthal KL, Kaushic C, 2005a. Estradiol regulates susceptibility following primary exposure to genital herpes simplex virus type 2, while progesterone induces inflammation. J Virol 79:3107–3116.

Gillgrass AE, Tang VA, Towarnicki KM, Rosenthal KL, Kaushic C, 2005b. Protection against genital herpes infection in mice immunized under different hormonal conditions correlates with induction of vagina-associated lymphoid tissue. J Virol 79:3117–3126.

Giron DJ, Allen PT, Pindak FF, Schmidt JP, 1973. Further studies on the influence of steroids on viral infection in mice. Infect Immun 8:151–155.

Glass GE, Childs JE, Korch GW, LeDuc JW, 1988. Association of intraspecific wounding with hantaviral infection in wild rats (*Rattus norvegicus*). Epidemiol Infect 101:459–472.

Glass GE, Livingstone W, Mills JN, Hlady WG, Fine JB, Biggler W, Coke T, Frazier D, Atherley S, Rollin PE, Ksiazek TG, Peters CJ, Childs JE, 1998. Black Creek Canal Virus infection in *Sigmodon hispidus* in southern Florida. Am J Trop Med Hyg 59:699–703.

Graff RJ, Lappe MA, Snell GD, 1969. The influence of the gonads and adrenal glands on the immune response to skin grafts. Transplantation 7:105–111.

Grimaldi CM, Cleary J, Dagtas AS, Moussai D, Diamond B, 2002. Estrogen alters thresholds for B cell apoptosis and activation. J Clin Invest 109:1625–1633.

Grinspoon S, 2005. Androgen deficiency and HIV infection. Clin Infect Dis 41:1804–1805.

Guan Y, Poon LL, Cheung CY, Ellis TM, Lim W, Lipatov AS, Chan KH, Sturm-Ramirez KM, Cheung CL, Leung YH, Yuen KY, Webster RG, Peiris JS, 2004. H5N1 influenza: a protean pandemic threat. Proc Natl Acad Sci USA 101:8156–8161.

Guha-Sapir D, Schimmer B, 2005. Dengue fever: new paradigms for a changing epidemiology. Emerg Themes Epidemiol 2:1.

Han X, Lundberg P, Tanamachi B, Openshaw H, Longmate J, Cantin E, 2001. Gender influences herpes simplex virus type 1 infection in normal and gamma interferon-mutant mice. J Virol 75:3048–3052.

Hannah MF, Bajic VB, Klein SL, 2008. Sex differences in the recognition of and innate antiviral responses to Seoul virus in Norway rats. Brain Behav Immun 22:503–516.

Hao S, Zhao J, Zhou J, Zhao S, Hu Y, Hou Y, 2007. Modulation of 17beta-estradiol on the number and cytotoxicity of NK cells in vivo related to MCM and activating receptors. Int Immunopharmacol 7: 1765–1775.

Hayes EB, Komar N, Nasci RS, Montgomery SP, O'Leary DR, Campbell GL, 2005. Epidemiology and transmission dynamics of West Nile virus disease. Emerg Infect Dis 11:1167–1173.

Hel Z, Stringer E, Mestecky J, 2010. Sex steroid hormones, hormonal contraception, and the immunobiology of human immunodeficiency virus-1 infection. Endocr Rev 31:79–97.

Hepworth MR, Hardman MJ, Grencis RK, 2010. The role of sex hormones in the development of Th2 immunity in a gender-biased model of Trichuris muris infection. Eur J Immunol 40:406–416.

Hess G, Hingst V, Cseke J, Bock HL, Clemens R, 1992. Influence of vaccination schedules and host factors on antibody response following hepatitis B vaccination. Eur J Clin Microbiol Infect Dis 11:334–340.

Hewagama A, Patel D, Yarlagadda S, Strickland FM, Richardson BC, 2009. Stronger inflammatory/cytotoxic T-cell response in women identified by microarray analysis. Genes Immun 10:509–516.

Hinson ER, Shone SM, Zink MC, Glass GE, Klein SL, 2004. Wounding: the primary mode of Seoul virus transmission among male Norway rats. Am J Trop Med Hyg 70:310–317.

Hohler T, Groeger-Bicanic G, Hoet B, Stoffel M, 2007. Antibody persistence and immune memory elicited by combined hepatitis A and B vaccination in older adults. Vaccine 25:1503–1508.

Hou J, Zheng WF, 1988. Effect of sex hormones on NK and ADCC activity of mice. Int J Immunopharmacol 10:15–22.

Huber S, 2008a. Host immune responses to coxsackievirus B3. Curr Top Microbiol Immunol 323:199–221.

Huber SA, 2008b. Coxsackievirus B3-induced myocarditis: infection of females during the estrus phase of the ovarian cycle leads to activation of T regulatory cells. Virology 378:292–298.

Huber SA, Kupperman J, Newell MK, 1999. Hormonal regulation of CD4(+) T-cell responses in coxsackievirus B3-induced myocarditis in mice. J Virol 73:4689–4695.

Huber SA, Pfaeffle B, 1994. Differential Th1 and Th2 cell responses in male and female BALB/c mice infected with coxsackievirus group B type 3. J Virol 68:5126–5132.

Jamieson DJ, Honein MA, Rasmussen SA, Williams JL, Swerdlow DL, Biggerstaff MS, Lindstrom S, Louie JK, Christ CM, Bohm SR, Fonseca VP, Ritger KA, Kuhles DJ, Eggers P, Bruce H, Davidson HA, Lutterloh E, Harris ML, Burke C, Cocoros N, Finelli L, MacFarlane KF, Shu B, Olsen SJ, 2009. H1N1 2009 influenza virus infection during pregnancy in the USA. Lancet 374:451–458.

Jean CM, Honarmand S, Louie JK, Glaser CA, 2007. Risk factors for West Nile virus neuroinvasive disease, California, 2005. Emerg Infect Dis 13:1918–1920.

Jilg W, Lorbeer B, Schmidt M, Wilske B, Zoulek G, Deinhardt F, 1984. Clinical evaluation of a recombinant hepatitis B vaccine. Lancet 2:1174–1175.

Jimenez-Garcia R, Hernandez-Barrera V, Carrasco-Garrido P, Lopez de Andres A, Perez N, de Miguel AG, 2008. Influenza vaccination coverages among children, adults, health care workers and immigrants in Spain: related factors and trends, 2003–2006. J Infect 57:472–480.

Jones LA, Kreem S, Shweash M, Paul A, Alexander J, Roberts CW, 2010. Differential modulation of TLR3- and TLR4-mediated dendritic cell maturation and function by progesterone. J Immunol 185:4525–4534.

Kallikourdis M, Betz AG, 2007. Periodic accumulation of regulatory T cells in the uterus: preparation for the implantation of a semi-allogeneic fetus? PLoS ONE 2:e382.

Kandun IN, Tresnaningsih E, Purba WH, Lee V, Samaan G, Harun S, Soni E, Septiawati C, Setiawati T, Sariwati E, Wandra T, 2008. Factors associated with case fatality of human H5N1 virus infections in Indonesia: a case series. Lancet 372:744–749.

Karpuzoglu E, Phillips RA, Gogal RM, Jr., Ansar Ahmed S, 2007. IFN-gamma-inducing transcription factor, T-bet is upregulated by estrogen in murine splenocytes: role of IL-27 but not IL-12. Mol Immunol 44:1808–1814.

Kash JC, Tumpey TM, Proll SC, Carter V, Perwitasari O, Thomas MJ, Basler CF, Palese P, Taubenberger JK, Garcia-Sastre A, Swayne DE, Katze MG, 2006. Genomic analysis of increased host immune and cell death responses induced by 1918 influenza virus. Nature 443:578–581.

Katagiri D, Hayashi H, Victoriano AF, Okamoto T, Onozaki K, 2006. Estrogen stimulates transcription of human immunodeficiency virus type 1 (HIV-1). Int Immunopharmacol 6:170–181.

Katze MG, Fornek JL, Palermo RE, Walters KA, Korth MJ, 2008. Innate immune modulation by RNA viruses: emerging insights from functional genomics. Nat Rev Immunol 8:644–654.

Klein SL, Bird BH, Glass GE, 2000. Sex differences in Seoul virus infection are not related to adult sex steroid concentrations in Norway rats. J Virol 74:8213–8217.

Klein SL, Bird BH, Glass GE, 2001. Sex differences in immune responses and viral shedding following Seoul virus infection in Norway rats. Am J Trop Med Hygiene 65:57–63.

Klein SL, Calisher CH, 2007. Emergence and persistence of hantaviruses. Current Topics in Microbiology and Immunology 315:217–252.

Klein SL, Cernetich A, Hilmer S, Hoffman EP, Scott AL, Glass GE, 2004a. Differential expression of immunoregulatory genes in male and female Norway rats following infection with Seoul virus. J Med Virol 74:180–190.

Klein SL, Jedlicka A, Pekosz A, 2010a. The Xs and Y of immune responses to viral vaccines. Lancet Infect Dis 10:338–349.

Klein SL, Pekosz A, Passaretti C, Anker M, Olukoya P, 2010b. Sex, gender and influenza. Geneva: World Health Organization, pp. 1–58.

Klein SL, Wisniewski AB, Marson AL, Glass GE, Gearhart JP, 2002. Early exposure to genistein exerts long-lasting effects on the endocrine and immune systems in rats. Mol Med 8:742–749.

Klein SL, Zink MC, Glass GE, 2004b. Seoul virus infection increases aggressive behaviour in male Norway rats. Animal Behaviour 67:421–429.

Klingstrom J, Lindgren T, Ahlm C, 2008. Sex-dependent differences in plasma cytokine responses to hantavirus infection. Clin Vaccine Immunol 15:885–887.

Kobasa D, Jones SM, Shinya K, Kash JC, Copps J, Ebihara H, Hatta Y, Kim JH, Halfmann P, Hatta M, Feldmann F, Alimonti JB, Fernando L, Li Y, Katze MG, Feldmann H, Kawaoka Y, 2007. Aberrant innate immune response in lethal infection of macaques with the 1918 influenza virus. Nature 445:319–323.

Kocar IH, Yesilova Z, Ozata M, Turan M, Sengul A, Ozdemir I, 2000. The effect of testosterone replacement treatment on immunological features of patients with Klinefelter's syndrome. Clin Exp Immunol 121: 448–452.

Koulentaki M, Spanoudakis S, Kantidaki E, Drandakis P, Tzagarakis N, Biziagos E, Moschandrea J, Kouroumalis EA, 1999. Prevalence of hepatitis B and C markers in volunteer blood donors in Crete. A 5-year study. J Viral Hepat 6:243–248.

Krzych U, Strausser HR, Bressler JP, Goldstein AL, 1981. Effects of sex hormones on some T and B cell functions as evidenced by differential immune expression between male and female mice and cyclic pattern of immune responsiveness during the estrous cycle. Prog Clin Biol Res 70:145–150.

Kumar A, Zarychanski R, Pinto R, Cook DJ, Marshall J, Lacroix J, Stelfox T, Bagshaw S, Choong K, Lamontagne F, Turgeon AF, Lapinsky S, Ahern SP, Smith O, Siddiqui F, Jouvet P, Khwaja K, McIntyre L, Menon K, Hutchison J, Hornstein D, Joffe A, Lauzier F, Singh J, Karachi T, Wiebe K, Olafson K, Ramsey C, Sharma S, Dodek P, Meade M, Hall R, Fowler RA, 2009. Critically ill patients with 2009 influenza A(H1N1) infection in Canada. JAMA 302:1872–1879.

Lamey PJ, Ferguson MM, Marshall W, 1985. Sex hormone involvement in the development of experimental virally induced murine salivary gland tumors. J Oral Pathol 14:414–421.

Lehmer EM, Clay CA, Wilson E, St Jeor S, Dearing MD, 2007. Differential resource allocation in deer mice exposed to sin nombre virus. Physiol Biochem Zool 80:514–521.

Libert C, Dejager L, Pinheiro I, 2010. The X chromosome in immune functions: when a chromosome makes the difference. Nat Rev Immunol 10:594–604.

Little TV, Holyoak GR, McCollum WH, Timoney PJ, 1992. Output of equine arteritis virus from persistently infected stallions is testosterone-dependent. Proceedings of the Sixth International Conference on Equine Infectious Diseases, pp. 225–229.

Liu HB, Loo KK, Palaszynski K, Ashouri J, Lubahn DB, Voskuhl RR, 2003a. Estrogen receptor alpha mediates estrogen's immune protection in autoimmune disease. J Immunol 171:6936–6940.

Liu PY, Death AK, Handelsman DJ, 2003b. Androgens and cardiovascular disease. Endocr Rev 24:313–340.

Liva SM, Voskuhl RR, 2001. Testosterone acts directly on CD4+ T lymphocytes to increase IL-10 production. J Immunol 167:2060–2067.

Louie JK, Acosta M, Jamieson DJ, Honein MA, 2010. Severe 2009 H1N1 influenza in pregnant and postpartum women in California. N Engl J Med 362:27–35.

Lu FX, Abel K, Ma Z, Rourke T, Lu D, Torten J, McChesney M, Miller CJ, 2002. The strength of B cell immunity in female rhesus macaques is controlled by CD8+ T cells under the influence of ovarian steroid hormones. Clin Exp Immunol 128:10–20.

Lyden DC, Olszewski J, Feran M, Job LP, Huber SA, 1987. Coxsackievirus B-3-induced myocarditis. Effect of sex steroids on viremia and infectivity of cardiocytes. Am J Pathol 126:432–438.

Ma WL, Hsu CL, Wu MH, Wu CT, Wu CC, Lai JJ, Jou YS, Chen CW, Yeh S, Chang C, 2008. Androgen receptor is a new potential therapeutic target for the treatment of hepatocellular carcinoma. Gastroenterology 135:947–955, 955 e941–945.

MacDonald EM, Savoy A, Gillgrass A, Fernandez S, Smieja M, Rosenthal KL, Ashkar AA, Kaushic C, 2007. Susceptibility of human female primary genital epithelial cells to herpes simplex virus, type-2 and the effect of TLR3 ligand and sex hormones on infection. Biol Reprod 77:1049–1059.

Malkin CJ, Pugh PJ, Jones RD, Kapoor D, Channer KS, Jones TH, 2004. The effect of testosterone replacement on endogenous inflammatory cytokines and lipid profiles in hypogonadal men. J Clin Endocrinol Metab 89:3313–3318.

Maret A, Coudert JD, Garidou L, Foucras G, Gourdy P, Krust A, Dupont S, Chambon P, Druet P, Bayard F, Guery JC, 2003. Estradiol enhances primary antigen-specific CD4 T cell responses and Th1 development in vivo. Essential role of estrogen receptor alpha expression in hematopoietic cells. Eur J Immunol 33:512–521.

McCollum WH, Little TV, Timoney PJ, Swerczek TW, 1994. Resistance of castrated male horses to attempted establishment of the carrier state with equine arteritis virus. J Comp Pathol 111:383–388.

McHugh RS, Whitters MJ, Piccirillo CA, Young DA, Shevach EM, Collins M, Byrne MC, 2002. CD4(+) CD25(+) immunoregulatory T cells: gene expression analysis reveals a functional role for the glucocorticoid-induced TNF receptor. Immunity 16:311–323.

McIntyre NE, Chu YK, Owen RD, Abuzeineh A, De la Sancha N, Dick CW, Holsomback T, Nisbett RA, Jonsson C, 2005. A longitudinal study of Bayou virus, hosts, and habitat. Am J Trop Med Hyg 73:1043–1049.

McKay LI, Cidlowski JA, 1999. Molecular control of immune/inflammatory responses: interactions between nuclear factor-kappa B and steroid receptor-signaling pathways. Endocr Rev 20:435–459.

Medina KL, Garrett KP, Thompson LF, Rossi MI, Payne KJ, Kincade PW, 2001. Identification of very early lymphoid precursors in bone marrow and their regulation by estrogen. Nat Immunol 2:718–724.

Medina KL, Kincade PW, 1994. Pregnancy-related steroids are potential negative regulators of B lymphopoiesis. Proc Natl Acad Sci USA 91:5382–5386.

Meier A, Chang JJ, Chan ES, Pollard RB, Sidhu HK, Kulkarni S, Wen TF, Lindsay RJ, Orellana L, Mildvan D, Bazner S, Streeck H, Alter G, Lifson JD, Carrington M, Bosch RJ, Robbins GK, Altfeld M, 2009. Sex differences in the Toll-like receptor-mediated response of plasmacytoid dendritic cells to HIV-1. Nat Med 15:955–959.

Miller L, Hunt JS, 1996. Sex steroid hormones and macrophage function. Life Sci 59:1–14.

Miller L, Hunt JS, 1998. Regulation of TNF-alpha production in activated mouse macrophages by progesterone. J Immunol 160:5098–5104.

Mills JN, Childs JE, 1998. Ecologic studies of rodent reservoirs: their relevance for human health. Emerg Infect Dis 4:529–537.

Mills JN, Ellis BA, Childs JE, McKee KT, Jr., Maiztegui JI, Peters CJ, Ksiazek TG, Jahrling PB, 1994. Prevalence of infection with Junin virus in rodent populations in the epidemic area of Argentine hemorrhagic fever. Am J Trop Med Hyg 51:554–562.

Mills JN, Ellis BA, McKee KT, Jr., Calderon GE, Maiztegui JI, Nelson GO, Ksiazek TG, Peters CJ, Childs JE, 1992. A longitudinal study of Junin virus activity in the rodent reservoir of Argentine hemorrhagic fever. Am J Trop Med Hyg 47:749–763.

Mills JN, Ksiazek TG, Ellis BA, Rollin PE, Nichol ST, Yates TL, Gannon WL, Levy CE, Engelthaler DM, Davis T, Tanda DT, Frampton JW, Nichols CR, Peters CJ, Childs JE, 1997. Patterns of association with host and habitat: antibody reactive with Sin Nombre virus in small mammals in the major biotic communities of the southwestern United States. Am J Trop Med Hyg 56:273–284.

Mirand EA, Back N, Grace JT, Jr., Buffet R, 1967. Effect of pituitary and gonadal hormones on Friend Virus Disease in mice. Proc Soc Exp Biol Med 124:1055–1059.

Mjosberg J, Svensson J, Johansson E, Hellstrom L, Casas R, Jenmalm MC, Boij R, Matthiesen L, Jonsson JI, Berg G, Ernerudh J, 2009. Systemic reduction of functionally suppressive CD4dimCD25highFoxp3+ Tregs in human second trimester pregnancy is induced by progesterone and 17beta-estradiol. J Immunol 183:759–769.

Mo R, Chen J, Grolleau-Julius A, Murphy HS, Richardson BC, Yung RL, 2005. Estrogen regulates CCR gene expression and function in T lymphocytes. J Immunol 174:6023–6029.

Mondal S, Rai U, 1999. Sexual dimorphism in phagocytic activity of wall lizard's splenic macrophage and its control by sex steroids. Gen Comp Endocr 116:291–298.

Moorman JE, Rudd RA, Johnson CA, King M, Minor P, Bailey C, Scalia MR, Akinbami LJ, Centers for Disease Control and Prevention, 2007. National surveillance for asthma—United States, 1980–2004. Morbidity and Mortality Weekly Report Surveillance Summaries/CDC 56:1–54.

Morris CA, Oliver PR, Reynolds F, Selkon JB, 1989. Intradermal hepatitis B immunization with yeast-derived vaccine: serological response by sex and age. Epidemiol Infect 103:387–394.

Muehlenbein MP, Cogswell FB, James MA, Koterski J, Ludwig GV, 2006. Testosterone correlates with Venezuelan equine encephalitis virus infection in macaques. Virol J 3:19.

Murphy G, Pfeiffer R, Camargo MC, Rabkin CS, 2009. Meta-analysis shows that prevalence of Epstein-Barr virus-positive gastric cancer differs based on sex and anatomic location. Gastroenterology 137:824–833.

Musabak U, Bolu E, Ozata M, Oktenli C, Sengul A, Inal A, Yesilova Z, Kilciler G, Ozdemir IC, Kocar IH, 2003. Gonadotropin treatment restores in vitro interleukin-1beta and tumour necrosis factor-alpha production by stimulated peripheral blood mononuclear cells from patients with idiopathic hypogonadotropic hypogonadism. Clin Exp Immunol 132:265–270.

Nakaya M, Tachibana H, Yamada K, 2006. Effect of estrogens on the interferon-gamma producing cell population of mouse splenocytes. Biosci Biotechnol Biochem 70:47–53.

Napravnik S, Poole C, Thomas JC, Eron JJ, Jr., 2002. Gender difference in HIV RNA levels: a meta-analysis of published studies. J Acquir Immune Defic Syndr 31:11–19.

Nash D, Mostashari F, Fine A, Miller J, O'Leary D, Murray K, Huang A, Rosenberg A, Greenberg A, Sherman M, Wong S, Layton M, 2001. The outbreak of West Nile virus infection in the New York City area in 1999. N Engl J Med 344:1807–1814.

Naugler WE, Sakurai T, Kim S, Maeda S, Kim K, Elsharkawy AM, Karin M, 2007. Gender disparity in liver cancer due to sex differences in MyD88-dependent IL-6 production. Science 317: 121–124.

Nichol KL, Margolis KL, Lind A, Murdoch M, McFadden R, Hauge M, Magnan S, Drake M, 1996. Side effects associated with influenza vaccination in healthy working adults. A randomized, placebo-controlled trial. Arch Intern Med 156:1546–1550.

Nohara K, Izumi H, Tamura S, Nagata R, Tohyama C, 2002. Effect of low-dose 2,3,7,8-tetrachlorodibenzo-p-dioxin (TCDD) on influenza A virus-induced mortality in mice. Toxicology 170:131–138.

Noymer A, Garenne M, 2000. The 1918 influenza epidemic's effects on sex differentials in mortality in the United States. Population and Development Review 26:565–581.

O'Leary DR, Marfin AA, Montgomery SP, Kipp AM, Lehman JA, Biggerstaff BJ, Elko VL, Collins PD, Jones JE, Campbell GL, 2004. The epidemic of West Nile virus in the United States, 2002. Vector Borne Zoonotic Dis 4:61–70.

Oliveira W, Carmo E, Penna G, Kuchenbecker R, Santos H, Araujo W, Malaguti R, Duncan B, Schmidt M, 2009. Pandemic H1N1 influenza in Brazil: analysis of the first 34,506 notified cases of influenza-like illness with severe acute respiratory infection (SARI). Euro Surveill 14.

Olsen NJ, Kovacs WJ, 1996. Gonadal steroids and immunity. Endocr Rev 17:369–384.

Olsson GE, White N, Ahlm C, Elgh F, Verlemyr AC, Juto P, Palo RT, 2002. Demographic factors associated with hantavirus infection in bank voles (*Clethrionomys glareolus*). Emerg Infect Dis 8:924–929.

Opstelten W, van Essen GA, Ballieux MJ, Goudswaard AN, 2008. Influenza immunization of Dutch general practitioners: vaccination rate and attitudes towards vaccination. Vaccine 26:5918–5921.

Padgett DA, Sheridan JF, Dorne J, Berntson GG, Candelora J, Glaser R, 1998. Social stress and the reactivation of latent herpes simplex virus type 1. Proc Natl Acad Sci USA 95:7231–7235.

Paharkova-Vatchkova V, Maldonado R, Kovats S, 2004. Estrogen preferentially promotes the differentiation of CD11c+ CD11b(intermediate) dendritic cells from bone marrow precursors. J Immunol 172: 1426–1436.

Pauklin S, Sernandez IV, Bachmann G, Ramiro AR, Petersen-Mahrt SK, 2009. Estrogen directly activates AID transcription and function. J Exp Med 206:99–111.

Pearce-Duvet JM, St Jeor SC, Boone JD, Dearing MD, 2006. Changes in sin nombre virus antibody prevalence in deer mice across seasons: the interaction between habitat, sex, and infection in deer mice. J Wildl Dis 42:819–824.

Pernis AB, 2007. Estrogen and CD4+ T cells. Curr Opin Rheumatol 19:414–420.

Phiel KL, Henderson RA, Adelman SJ, Elloso MM, 2005. Differential estrogen receptor gene expression in human peripheral blood mononuclear cell populations. Immunol Lett 97:107–113.

Piccinni MP, Giudizi MG, Biagiotti R, Beloni L, Giannarini L, Sampognaro S, Parronchi P, Manetti R, Annunziato F, Livi C, et al., 1995. Progesterone favors the development of human T helper cells producing Th2-type cytokines and promotes both IL-4 production and membrane CD30 expression in established Th1 cell clones. J Immunol 155:128–133.

Piccinni MP, Scaletti C, Maggi E, Romagnani S, 2000. Role of hormone-controlled Th1- and Th2-type cytokines in successful pregnancy. J Neuroimmunol 109:30–33.

Pisitkun P, Deane JA, Difilippantonio MJ, Tarasenko T, Satterthwaite AB, Bolland S, 2006. Autoreactive B cell responses to RNA-related antigens due to TLR7 gene duplication. Science 312:1669–1672.

Polanczyk M, Yellayi S, Zamora A, Subramanian S, Tovey M, Vandenbark AA, Offner H, Zachary JF, Fillmore PD, Blankenhorn EP, Gustafsson JA, Teuscher C, 2004. Estrogen receptor-1 (Esr1) and -2 (Esr2) regulate the severity of clinical experimental allergic encephalomyelitis in male mice. Am J Pathol 164:1915–1924.

Porter MB, Long MT, Getman LM, Giguere S, MacKay RJ, Lester GD, Alleman AR, Wamsley HL, Franklin RP, Jacks S, Buergelt CD, Detrisac CJ, 2003. West Nile virus encephalomyelitis in horses: 46 cases (2001). J Am Vet Med Assoc 222:1241–1247.

Qureshi AM, Hughes NJ, Murphy E, Primrose WR, 2004. Factors influencing uptake of influenza vaccination among hospital-based health care workers. Occup Med (Lond) 54:197–201.

Rettew JA, Huet YM, Marriott I, 2009. Estrogens augment cell surface TLR4 expression on murine macrophages and regulate sepsis susceptibility in vivo. Endocrinology 150:3877–3884.

Rettew JA, Huet-Hudson YM, Marriott I, 2008. Testosterone reduces macrophage expression in the mouse of toll-like receptor 4, a trigger for inflammation and innate immunity. Biol Reprod 78:432–437.

Roberts CW, Walker W, Alexander J, 2001. Sex-associated hormones and immunity to protozoan parasites. Clin Microbiol Rev 14:476–488.

Robinson DP, Lorenzo ME, Jian W, Klein SL. 2011. Elevated 17beta-estradiol protects females from influenza A virus pathogenesis by suppressing inflammatory responses. PLoS Pathog 7:e1002149. doi:10.1371/journal.ppat.1002149.

Robinson T, Bullen C, Humphries W, Hornell J, Moyes C, 2005. The New Zealand Hepatitis B Screening Programme: screening coverage and prevalence of chronic hepatitis B infection. N Z Med J 118:U1345.

Roden AC, Moser MT, Tri SD, Mercader M, Kuntz SM, Dong H, Hurwitz AA, McKean DJ, Celis E, Leibovich BC, Allison JP, Kwon ED, 2004. Augmentation of T cell levels and responses induced by androgen deprivation. J Immunol 173:6098–6108.

Rook GAW, 2009. Review series on helminths, immune modulation and the hygiene hypothesis: The broader implications of the hygiene hypothesis. Immunology 126:3–11.

Root JJ, Calisher CH, Beaty BJ, 1999. Relationships of deer mouse movement, vegetative structure, and prevalence of infection with Sin Nombre virus. J Wildl Dis 35:311–318.

Sanchez-Fueyo A, Tian J, Picarella D, Domenig C, Zheng XX, Sabatos CA, Manlongat N, Bender O, Kamradt T, Kuchroo VK, Gutierrez-Ramos JC, Coyle AJ, Strom TB, 2003. Tim-3 inhibits T helper type 1-mediated auto- and alloimmune responses and promotes immunological tolerance. Nat Immunol 4:1093–1101.

Sapino A, Cassoni P, Ferrero E, Bongiovanni M, Righi L, Fortunati N, Crafa P, Chiarle R, Bussolati G, 2003. Estrogen receptor alpha is a novel marker expressed by follicular dendritic cells in lymph nodes and tumor-associated lymphoid infiltrates. Am J Pathol 163:1313–1320.

Savita, Rai U, 1998. Sex steroid hormones modulate the activation of murine peritoneal macrophages: receptor mediated modulation. Comp Biochem Physiol C Pharmacol Toxicol Endocrinol 119: 199–204.

Schatz M, Camargo CA, Jr., 2003. The relationship of sex to asthma prevalence, health care utilization, and medications in a large managed care organization. Ann Allergy Asthma Immunol 91:553–558.

Schatz M, Clark S, Camargo CA, Jr., 2006. Sex differences in the presentation and course of asthma hospitalizations. Chest 129:50–55.

Sedyaningsih ER, Isfandari S, Setiawaty V, Rifati L, Harun S, Purba W, Imari S, Giriputra S, Blair PJ, Putnam SD, Uyeki TM, Soendoro T, 2007. Epidemiology of cases of H5N1 virus infection in Indonesia, July 2005–June 2006. J Infect Dis 196:522–527.

Sheldon BC, Verhulst S, 1996. Ecological immunology: costly parasite defences and trade-offs in evolutionary ecology. Trends Ecol Evol 11:317–321.

Siddiqui RA, Sauermann U, Altmuller J, Fritzer E, Nothnagel M, Dalibor N, Fellay J, Kaup FJ, Stahl-Hennig C, Nurnberg P, Krawczak M, Platzer M, 2009. X chromosomal variation is associated with slow progression to AIDS in HIV-1-infected women. Am J Hum Genet 85:228–239.

Singh AK, Cydulka RK, Stahmer SA, Woodruff PG, Camargo CA, Jr., 1999. Sex differences among adults presenting to the emergency department with acute asthma. Multicenter Asthma Research Collaboration Investigators. Arch Intern Med 159:1237–1243.

Siracusa MC, Overstreet MG, Housseau F, Scott AL, Klein SL, 2008. 17beta-Estradiol Alters the Activity of Conventional and IFN-Producing Killer Dendritic Cells. J Immunol 180:1423–1431.

Smithson G, Beamer WG, Shultz KL, Christianson SW, Shultz LD, Kincade PW, 1994. Increased B lymphopoiesis in genetically sex steroid-deficient hypogonadal (hpg) mice. J Exp Med 180:717–720.

Soldan SS, Alvarez Retuerto AI, Sicotte NL, Voskuhl RR, 2003. Immune modulation in multiple sclerosis patients treated with the pregnancy hormone estriol. J Immunol 171:6267–6274.

Solomon MB, Herman JP, 2009. Sex differences in psychopathology: of gonads, adrenals and mental illness. Physiol Behav 97:250–258.

Sorachi K, Kumagai S, Sugita M, Yodoi J, Imura H, 1993. Enhancing effect of 17 beta-estradiol on human NK cell activity. Immunol Lett 36:31–35.

Souza SS, Castro FA, Mendonca HC, Palma PV, Morais FR, Ferriani RA, Voltarelli JC, 2001. Influence of menstrual cycle on NK activity. J Reprod Immunol 50:151–159.

Spinedi E, Gaillard RC, Chisari A, 2002. Sexual dimorphism of neuroendocrine-immune interactions. Front Horm Res 29:91–107.

Spitzer JA, 1999. Gender differences in some host defense mechanisms. Lupus 8:380–383.

Stephenson J, 2000. Genital herpes vaccine shows limited promise. JAMA 284:1913–1914.

Sterling TR, Vlahov D, Astemborski J, Hoover DR, Margolick JB, Quinn TC, 2001. Initial plasma HIV-1 RNA levels and progression to AIDS in women and men. N Engl J Med 344:720–725.

Straub RH, 2007. The complex role of estrogens in inflammation. Endocr Rev 28:521–574.

Suzuki T, Yu HP, Hsieh YC, Choudhry MA, Bland KI, Chaudry IH, 2008. Mitogen activated protein kinase (MAPK) mediates non-genomic pathway of estrogen on T cell cytokine production following trauma-hemorrhage. Cytokine 42:32–38.

Szekeres-Bartho J, 2002. Immunological relationship between the mother and the fetus. Int Rev Immunol 21:471–495.

Szekeres-Bartho J, Faust Z, Varga P, Szereday L, Kelemen K, 1996. The immunological pregnancy protective effect of progesterone is manifested via controlling cytokine production. Am J Reprod Immunol 35: 348–351.

Szretter KJ, Gangappa S, Lu X, Smith C, Shieh WJ, Zaki SR, Sambhara S, Tumpey TM, Katz JM, 2007. Role of host cytokine responses in the pathogenesis of avian H5N1 influenza viruses in mice. J Virol 81:2736–2744.

Teichmann J, Schmidt A, Lange U, Stracke H, Discher T, Friese G, Lohmeyer J, Bretzel RG, 2003. Longitudinal evaluation of serum estradiol and estrone in male patients infected with the human immunodeficiency virus. Eur J Med Res 8:77–80.

Theiler RN, Rasmussen SA, Treadwell TA, Jamieson DJ, 2008. Emerging and zoonotic infections in women. Infect Dis Clin North Am 22:755–772.

Toder V, Nebel L, Elrad H, Blank M, Durdana A, Gleicher N, 1984. Studies of natural killer cells in pregnancy. II. The immunoregulatory effect of pregnancy substances. J Clin Lab Immunol 14:129–133.

Tsay PK, Tai DI, Chen YM, Yu CP, Wan SY, Shen YJ, Lin DY, 2009. Impact of gender, viral transmission and aging in the prevalence of hepatitis B surface antigen. Chang Gung Med J 32:155–164.

Van der Wielen M, Van Damme P, Chlibek R, Smetana J, von Sonnenburg F, 2006. Hepatitis A/B vaccination of adults over 40 years old: comparison of three vaccine regimens and effect of influencing factors. Vaccine 24:5509–5515.

Vapalahti O, Mustonen J, Lundkvist A, Henttonen H, Plyusnin A, Vaheri A, 2003. Hantavirus infections in Europe. Lancet Infect Dis 3:653–661.

Vicetti Miguel RD, Sheridan BS, Harvey SA, Schreiner RS, Hendricks RL, Cherpes TL, 2010. 17-beta estradiol promotion of herpes simplex virus type 1 reactivation is estrogen receptor dependent. J Virol 84:565–572.

Villacres MC, Longmate J, Auge C, Diamond DJ, 2004. Predominant type 1 CMV-specific memory T-helper response in humans: evidence for gender differences in cytokine secretion. Hum Immunol 65:476–485.

Wald A, 2004. Herpes simplex virus type 2 transmission: risk factors and virus shedding. Herpes 11 Suppl 3:130A–137A.

Wang C, Dehghani B, Li Y, Kaler LJ, Vandenbark AA, Offner H, 2009a. Oestrogen modulates experimental autoimmune encephalomyelitis and interleukin-17 production via programmed death 1. Immunology 126:329–335.

Wang SH, Yeh SH, Lin WH, Wang HY, Chen DS, Chen PJ, 2009b. Identification of androgen response elements in the enhancer I of hepatitis B virus: a mechanism for sex disparity in chronic hepatitis B. Hepatology 50:1392–1402.

Warner RD, Kimbrough RC, Alexander JL, Rush Pierce J, Jr., Ward T, Martinelli LP, 2006. Human West Nile Virus neuroinvasive disease in Texas, 2003 epidemic: regional differences. Ann Epidemiol 16:749–755.

Weigler BJ, Ksiazek TG, Vandenbergh JG, Levin M, Sullivan WT, 1996. Serological evidence for zoonotic hantaviruses in North Carolina rodents. J Wildl Dis 32:354–357.

Weinberger M, Pitlik SD, Gandacu D, Lang R, Nassar F, Ben David D, Rubinstein E, Izthaki A, Mishal J, Kitzes R, Siegman-Igra Y, Giladi M, Pick N, Mendelson E, Bin H, Shohat T, 2001. West Nile fever outbreak, Israel, 2000: epidemiologic aspects. Emerg Infect Dis 7:686–691.

Weinstein Y, Ran S, Segal S, 1984. Sex-associated differences in the regulation of immune responses controlled by the MHC of the mouse. J Immunol 132:656–661.

White DJ, Means RG, Birkhead GS, Bosler EM, Grady LJ, Chatterjee N, Woodall J, Hjelle B, Rollin PE, Ksiazek TG, Morse DL, 1996. Human and rodent hantavirus infection in New York State: public health significance of an emerging infectious disease. Arch Intern Med 156:722–726.

Wikby A, Mansson IA, Johansson B, Strindhall J, Nilsson SE, 2008. The immune risk profile is associated with age and gender: findings from three Swedish population studies of individuals 20–100 years of age. Biogerontology 9:299–308.

Wilcoxen SC, Kirkman E, Dowdell KC, Stohlman SA, 2000. Gender-dependent IL-12 secretion by APC is regulated by IL-10. J Immunol 164:6237–6243.

Williams RJ, Bryan RT, Mills JN, Palma RE, Vera I, De Velasquez F, Baez E, Schmidt WE, Figueroa RE, Peters CJ, Zaki SR, Khan AS, Ksiazek TG, 1997. An outbreak of hantavirus pulmonary syndrome in western Paraguay. Am J Trop Med Hyg 57:274–282.

Wingard DL, 1984. The sex differential in morbidity, mortality, and lifestyle. Annu Rev Public Health 5: 433–458.

Woodruff JF, 1980. Viral myocarditis. A review. Am J Pathol 101:425–484.

Wunderlich F, Benten WP, Lieberherr M, Guo Z, Stamm O, Wrehlke C, Sekeris CE, Mossmann H, 2002. Testosterone signaling in T cells and macrophages. Steroids 67:535–538.

Yirmiya R, Avitsur R, Donchin O, Cohen E, 1995. Interleukin-1 inhibits sexual behavior in female but not in male rats. Brain Behav Immun 9:220–233.

Yovel G, Shakhar K, Ben-Eliyahu S, 2001. The effects of sex, menstrual cycle, and oral contraceptives on the number and activity of natural killer cells. Gynecol Oncol 81:254–262.

Yu MW, Cheng SW, Lin MW, Yang SY, Liaw YF, Chang HC, Hsiao TJ, Lin SM, Lee SD, Chen PJ, Liu CJ, Chen CJ, 2000. Androgen-receptor gene CAG repeats, plasma testosterone levels, and risk of hepatitis B-related hepatocellular carcinoma. J Natl Cancer Inst 92:2023–2028.

Yu MW, Yang YC, Yang SY, Cheng SW, Liaw YF, Lin SM, Chen CJ, 2001. Hormonal markers and hepatitis B virus-related hepatocellular carcinoma risk: a nested case-control study among men. J Natl Cancer Inst 93:1644–1651.

Zeeshan M, Jabeen K, Ali AN, Ali AW, Farooqui SZ, Mehraj V, Zafar A, 2007. Evaluation of immune response to Hepatitis B vaccine in health care workers at a tertiary care hospital in Pakistan: an observational prospective study. BMC Infect Dis 7:120.

Zhang X, Castelli FA, Zhu X, Wu M, Maillere B, BenMohamed L, 2008. Gender-dependent HLA-DR-restricted epitopes identified from herpes simplex virus type 1 glycoprotein D. Clin Vaccine Immunol 15: 1436–1449.

Zuk M, 2009. The sicker sex. PLoS Pathog 5:e1000267.

Zuk M, McKean KA, 1996. Sex differences in parasite infections: patterns and processes. Int J Parasitol 26:1009–1023.

Zuk M, Stoehr AM, 2002. Immune defense and host life history. Am Nat 160:S9–S22.

CHAPTER 17

IMMUNOPATHOLOGY IN ECOLOGICAL IMMUNOLOGY

E. Rhiannon Pursall and Jens Rolff

INTRODUCTION

Immunopathology, the damage to the host caused by the immune response *per se* (Graham et al., 2005), is well established to cause the severity of many of the world's major diseases. However, its role in the wild has hardly been investigated. In this chapter, we argue that immunopathology is an important factor in an ecological immunology context: it contributes to the cost of an immune response and hence shapes the evolution and plasticity of immune responses. Moreover, it has also been proposed that immunopathology influences virulence evolution of pathogens (Day et al., 2007). Here we will briefly review immunopathology in humans and in other organisms, with a focus on insects. We will then consider the importance of immunopathology in an evolutionary ecology context. Finally, we will discuss the relationship between immunopathology and tolerance—that is, the ability to reduce the harm caused by pathogens without attacking them directly.

The type of immunopathology considered here is that which is associated with the immune response and occurs as a result of the activation of the immune system. This form of immunopathology is not the only source of damage that can be inflicted on the host by its own immune system. Autoimmunity also causes self-harm, but unlike the type of immunopathology considered here does so in the absence of a pathogenic immune challenge (Bertrand et al., 2006). Autoimmunity describes a situation where the host system mistakes self-tissue for a foreign invader and mounts an immune response (Goldsby et al., 2000), and has so far been described only for vertebrates. This notion of "mistaking" is crucial in the distinction between autoimmunity and immunopathology. Immunopathology does not "make a mistake," because a real immune threat is present, but immunopathology results in the immune response being inappropriate or poorly regulated or targeted. For example, rheumatoid arthritis apparently occurs without infection, but is still considered "immunopathology" because the damage is essentially caused by the immune system

(Day et al., 2007). Immunopathology can also be distinguished from the other costs of immunity that are frequently investigated in ecoimmunology studies. These costs include trade-offs between immunity and life-history traits such as reproduction, during which resources relating to energy, physiology, and time are invested in immunity to the expense of other life-history traits (Schmid-Hempel, 2003). In the framework of costs and trade-offs, immunopathology is distinct from autoimmunity. The latter is a cost of maintaining an immune system, while immunopathology is part of the cost of mounting an immune response (Rolff & Siva-Jothy, 2003).

Immunopathology is potentially very important in an ecological setting, as it very likely influences a suite of life-history traits. The degree of immunopathology suffered is essentially a measure of the adaptiveness of the stress response by an organism. Stress imposes a very important evolutionary force in natural populations, as organisms are frequently exposed to stressful situations (Costantini, 2008), including potential pathogen attack. Evidence shows that certain factors, such as the duration of stress, determine whether a stressor will cause the direction of response to favor an enhancing or suppressive direction (Dhabhar, 2009). Life-history is affected to a large extent by stress because an organism's response can either reinstate homeostasis and increase survival and fitness, or it can be maladaptive, for example through immunopathology.

The benefits of integrating medical and biochemical approaches to immunopathology with those of evolutionary ecology and immunology were recently highlighted (Graham, 2002). Graham (2002) argued that ecological analysis could help to understand how co-infection can exacerbate immunopathology, and how an evolutionary ecological approach could generate testable predictions about the optimal strategy for dealing with co-infection. In the 20th-anniversary edition of *Trends in Ecology and Evolution* it was suggested that the next 20 years of ecology and evolution needed to pay attention to how and why immune genes harm organisms, due to the immense human suffering caused by immunopathology (Read and Clark, 2006).

INFLAMMATION AS A MECHANISM OF IMMUNOPATHOLOGY

The inflammatory process was first characterized in a traditional pathology context by Celsius around AD 40 by four distinct features: *rubor, calor, dolor*, and *tumor* (redness, heat, pain, and swelling) (Nathan, 2002). A fifth sign of inflammation, *functio laesa* (disturbance of function) was added to the definition by Galen in the second century AD (Rather, 1971). Inflammation broadly describes the response by the tissues to trauma or infection. This response is the innate first line of defense against pathogen invasion and is therefore a crucial component of immunity (Janeway and Medzhitov, 2002; Nathan, 2002; Sorci and Faivre, 2009; in insects: Stramer et al., 2005). In the short term, inflammation is an adaptive response and constitutes the vital facilitation of a return to physiological homeostasis and tissue repair following injury or infection (Vasto et al., 2007). However, exposure to inflammation has been linked to many diseases (reviewed in Hotamisligil, 2009), and in these cases the inflammatory process can cause more harm than the initial immune insult (Nathan, 2002). The major benefits of inflammation are that it is rapid and fairly nonspecific, and therefore pathogens cannot easily evolve features to escape

(Sorci and Faivre, 2009). However, this adaptive nature of inflammation is also what causes it to be immunopathological, as the rapid response cannot easily distinguish between cells and tissues of the invader or the host, leading to host "self-harm." The way in which inflammation can cause disease is through the development of the initial acute inflammatory response into a chronic condition. This occurs when tissue health is not restored, or if a low-grade irritation to the immune system is maintained (Licastro et al., 2005). Therefore, diseases caused by inflammation tend to be associated with aging, and often symptoms are not expressed until many years after the initial inflammatory response (e.g., Bengtsson and Lindstrom, 2003).

Considerable research effort has focused on elucidating the particular mechanisms that cause inflammation to inflict immunopathological damage. An important consequence of inflammation, oxidative damage, has been investigated in a range of species. In an initial move towards incorporating immunopathology into ecological research, the effect of oxidative stress on life-history strategies has been studied. Although this research is in its infancy, it appears that parameters to reduce oxidative stress are associated with life-history variation, and that these parameters are highly conserved across animal taxa (Costantini, 2008). The importance of oxidative stress in ecology was given attention by von Schantz and colleagues in 1999, who suggested that sexual ornaments may represent traits that are particularly sensitive to oxidative stress, and therefore this stress may play an important role in sexual selection (von Schantz et al., 1999). Oxidative stress is potentially important for life-history theory because it is induced by key ecologically important factors: reproduction, immune response, and physical activity (Costantini, 2008). In the following section, we will focus on the production of oxidative stress from immune responses, as this represents a major source of immunopathology in many organisms.

At the biochemical level, during the inflammatory immune response the infiltration of tissue with immune cells such as monocytes and macrophages can cause immunopathological tissue damage (Lambeth et al., 2008). Micromolecules such as reactive oxygen species (ROS) are released by phagocytotic cells during pathogen attack, and are generically cytotoxic because they do not differentiate between cells and tissues of the host and infective agent (Cherry and Silverman, 2006; Sorci and Faivre, 2009). Oxidative stress resulting from ROS is regularly cited as a major cause of immunopathology in humans and non-human animals, and has sparked huge interest in investigating how organisms regulate the balance between pro-oxidants and antioxidants (Costantini, 2008). An excess of pro-oxidants and a reduction in antioxidants is characterized by oxidative stress, which can damage DNA, proteins, and lipids (Sies, 1997; Lesser, 2006; De Block and Stocks, 2008). This has detrimental consequences for survival and fitness; for example, mouse macrophages show a higher mutation rate when continuously exposed to nitric oxide, suggesting that oxidative stress could cause cancer (Zhuang et al., 1998). In an experiment into avian immunopathology, activation of the inflammatory response led to the production of cytotoxic compounds that damaged host tissues—that is, oxidative stress (Bertrand et al., 2006). Bertrand et al. (2006) could therefore have demonstrated the mechanism by which inflammation causes immunopathology through oxidative stress.

There is evidence for immunopathology caused by ROS in invertebrate immune systems, by similar mechanisms as in human immunity. In response to pathogen invasion, the nematode *Caenorhabditis elegans* produces ROS, leading to cell damage (Chavez et al., 2009). In fruit flies (*Drosophila melanogaster*), pathogen invasion triggers an increased production of ROS, and if ROS-production regulators are inhibited, fly lethality is elevated (Lemaitre and Hoffman, 2007).

Insects indeed show inflammation (Stramer et al., 2005), and the mechanisms of immunopathology caused by oxidative stress arising from inflammatory responses could be conserved between vertebrates and invertebrates.

In summary, experimental studies conducted *in vivo* and *in vitro* have demonstrated that inflammation can alter host homeostasis through immunopathology, which can ultimately lead to survival costs (Sorci and Faivre, 2009). At the cellular level, oxidative stress may be an important mediator of inflammation-induced immunopathology. At the population level, there is a hypothesis that a decrease in childhood inflammation has led to an increase in life span in modern Western societies (Finch and Crimmins, 2004; Crimmins and Finch, 2006). This hypothesis will be discussed in more detail in following sections. Although studies of oxidative stress from an ecological viewpoint are currently scarce, avian model systems have provided an interesting insight into how this immunopathological stress may have evolutionary implications. Bird life-history is "paradoxical" to that of mammals in many ways. For example, birds have relatively higher metabolic rate, body temperature, and blood glucose levels compared to mammals, which should result in lower life span as a result of tissue damage and physiological senescence (Costantini, 2008). However, birds exhibit considerably enhanced longevity compared to mammals of the equivalent body mass, and a hypothesis to explain this proposes that birds have evolved superior mechanisms for coping with, and reducing, oxidative stress (Costantini, 2008). In 2006, Kurtz and colleagues applied a medical method of assessing tissue damage to an ecological immunology study for the first time in an experiment using sticklebacks (Apeltes, *Gasterosteus aculeatus*). One important outcome was that on exposure to parasitic infection, oxidative damage was correlated positively with gene expression at the major histocompatibility complex (MHC), genes crucial to specific parasite defense. Increased MHC gene expression also correlated negatively with body condition, suggesting that oxidative damage was a key mediator in the damage caused during an immune response to pathogen attack (Kurtz et al., 2006).

IMMUNOPATHOLOGY IN HUMANS

As previously discussed, immunopathology in humans has been found to account for many diseases. First, age-related diseases that present themselves later in adult life are considered "immunopathological" conditions and are attributed to inflammation. In these cases, inflammation can develop into a chronic condition, either because tissue repair and physiological homeostasis are not initially achieved by the inflammatory response, or because the immune injury is sustained. Inflammation enters a chronic state if maintained at a low level for a long period of time, and sometimes this can be carried through many years of life. Ultimately, chronic inflammation can lead to age-related diseases, thereby outlining the mechanism by which immunopathology can cause immune responses to have long-term costs.

For example, a persistent elevation in inflammation has been attributed to the development of cancer, because the generation of highly reactive intermediates during inflammation increases mutation rates (Ewald, 2010). Figure 17.1 demonstrates the wealth and nature of the so-called age-related human diseases that have been directly attributed to chronic inflammation, and how

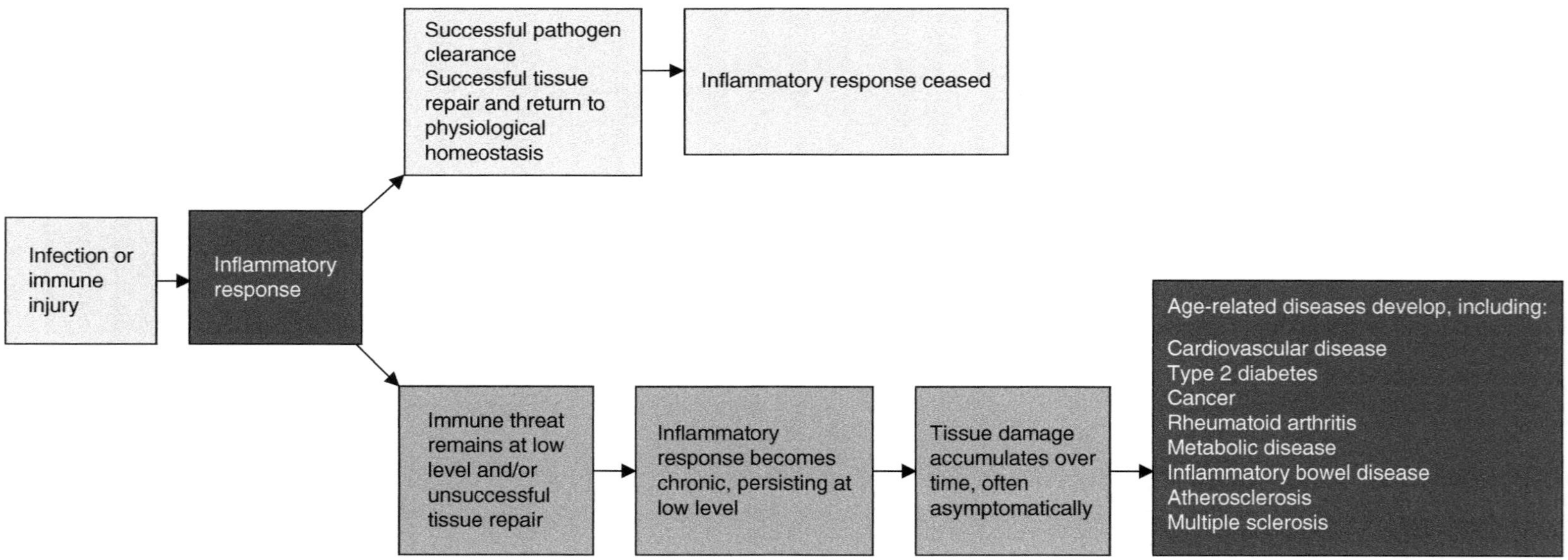

FIGURE 17.1: The age-related diseases caused by inflammation. Sources of information: Licastro et al. (2005); Lin and Karin (2007); Ridker et al. (1997); Libby (2002); Sin and Man (2003); Balkwill and Mantovani (2001); Wellen and Hotamlisligil (2005); Hotamisligil (2006); Tak and Firestein (2001); Barnes and Karin (1997)

they develop. These diseases are incredibly important to many modern societies in terms of shaping longevity patterns and affecting a range of social and economic issues. For example, heart disease, which has been well studied in the context of immunopathology, has recently emerged as the leading cause of mortality in Western adults and is the leading single cause of death worldwide (Mathers and Loncar, 2006; McPherson et al., 2007).

In addition, many of the world's most important diseases, such as malaria and tuberculosis, would not be as severe without their large immunopathological element. In tuberculosis, both a suite of *Mycobacterium tuberculosis* virulence factors and an inappropriate immune response by the host cause tissue damage in the host (Kaushal et al., 2002). This tissue damage ultimately leads to the high death rates (1,577,000 worldwide deaths in 2005, World Health Organisation Tuberculosis fact sheet No. 104) of this disease. In malaria, both the inability of the immune system to combat *Plasmodium* replication and the inability of the host to regulate an excessive immune response have been established as contributors to severity (Scholzen et al., 2010).

Epidemic-causing viruses that have been of huge importance in human populations in the past century are thought to owe a large amount of their lethality to immunopathology. For example, the 1918 influenza and Ebola viruses cause a nonspecific and abnormally sustained innate immune response that contributes significantly to symptoms associated with death (Zampieri et al., 2007).

STUDY OF IMMUNOPATHOLOGY IN NON-HUMAN ORGANISMS

Inflammation is not confined to humans: the inflammatory response has been characterized in both vertebrates and invertebrates (Libert et al., 2006; Sorci and Faivre, 2009). Furthermore, along with the medical evidence of immunopathology as a direct result of inflammation in humans, this damaging response has also been found in invertebrates. In *Drosophila*, a disease equivalent to the human metabolic syndrome is caused by immunopathology of inflammation (Schilder and Marden, 2006).

Immunopathology in organisms lacking an acquired immune system (i.e., invertebrates) is extremely understudied. This most likely results from the relative deficiency of research into immunopathology in an ecological immunology context compared to that in a medical one. However, invertebrates offer a very valuable model system to study immunity from an ecological perspective, as population consequences of immunopathology such as mortality phenotypes can be relatively easily investigated. This is due to qualities of invertebrate study systems not present for their vertebrate counterparts, including comparatively short generation time and relative ease of generating large sample sizes. Moreover, insects, as the most species-rich metazoan taxon, are particularly interesting in the context of immunopathology. The insect immune response faces the problem of an open body cavity, the hemocoel. This requires fine-tuned localized immune responses (Siva-Jothy et al., 2005; Bidla et al., 2009) to avoid widespread damage caused by cytotoxic immune reactions such as the phenoloxidase cascade (Cerenius et al., 2008).

IMMUNOPATHOLOGY AND INFLAMMATION IN INSECTS

Immunopathology has been shown to occur in insects, which adds to their suitability as model invertebrate systems for studying the ecological consequences of immunopathology. In the mealworm beetle (*Tenebrio molitor*), the production of melanin to melanize and restrict the growth and movement of an invading pathogen inflicted detrimental effects on host tissue (Sadd and Siva-Jothy, 2006). Along with the foreign body being encapsulated, melanin was also deposited on tissue adjacent to the infection site. An organ complex that was found to be particularly susceptible to this collateral damage was the Malpighian tubule system, which functions similarly to mammalian kidneys. The reason that the tubules are highly susceptible to damage comes partly from the insect morphology: tubules run through the insect body and are therefore often close to a wound site. In addition, the tubules are primarily responsible for regulating two very important processes, osmoregulation and toxic waste removal, and have a microstructure prone to damage by melanin deposition (Spring et al., 2007). Therefore, any harm suffered would presumably have a large impact upon insect fitness. Melanization of this tissue in an experimental setting caused the reduction of functioning of the organ, measured as fluid pumping rate (Sadd and Siva-Jothy, 2006).

In a further example of the presence of immunopathology in insects, Brandt et al. (2004) showed that a parasite (*Salmonella enterica serovar typhimurium*) elicits an immune response in its host (*Drosophila melanogaster*) that is damaging to both parasite and host. Using eiger mutants, which exhibit reduced tumor necrosis factor (TNF)-like signaling, they proposed that the flies died from a metabolic collapse analogous to shock syndromes observed in vertebrates.

The evidence for immunopathology in insects is relatively sparse and represents only a few insect species. This highlights the need for more experimental investigation into the mechanisms and outcomes of immunopathology in this group of animals. Yet we would like to highlight that the few examples studied are likely to represent a common phenomenon. The melanization response, which has been proposed to underlie the findings by Sadd and Siva-Jothty (2006), is highly conserved within arthropods (Cerenius et al., 2008), and similar results can therefore be predicted. Further indirect support for this notion comes from a recent study by Bidla et al. (2009) demonstrating the localized melanization in responses in *D. melanogaster* and in the wax moth (*Galleria mellonella*) after wounding. While wound closure is pivotal to maintain homeostasis, a systemic response could be detrimental as it would release cytotoxic substances into the open hemocoel.

Inflammation in insects has been studied in great detail in the context of wound repair. This might actually be of great importance given the frequency of wounding during mating (Siva-Jothy, 2009) and the loss of limbs in many insect species (Stoks et al., 2008, and references therein). Stramer et al. (2005) have revealed that the recruitment of hemocytes to wounds resembles the inflammation responses in vertebrates.

Many ectoparasites cause wounds as well, and while speculation at the moment, ectoparasites in insects might impose natural selection on the reduction of immunopathology. Ectoparasitic water mites of the genus *Arrenurus*, for example, seem to have a 100% success rate in exploiting their damselfly (Coenagrionidae, *Coenagrion puella*) hosts (Rolff and Martens, 1997; Rolff, 2000; Rolff, personal observations) and reduce host fitness (Braune and Rolff, 2001). The damselflies mount an immune response against these mites (Åbro, 1992), but this immune response is completely inefficient. In other words, the hosts pay the energetic and immunopathological costs of the immune response without any gain. Ectoparasitic mites are widespread in insects (Knell and Webberley, 2004).

IMMUNOPATHOLOGY AND AGING

Due to the ongoing increase in human life span throughout the developed world (Finch, 2010), mechanisms that control aging are gaining in their research importance. In the past 160 years, for example, female life expectancy has increased by an average of 3 months per year in certain counties (Oeppen and Vaupel, 2002). Additionally, life spans of companions of humans such as dogs and cats are also on the rise (Day, 2010). Thus, a comprehensive understanding of how and why animals age would greatly benefit areas such as drug development in the medical and veterinary sciences.

A critical link between immunopathology and aging in vertebrates is that the balance between an adaptive immune defense and immunopathology is disturbed in older organisms compared to younger ones (Bird, 2010). Evidence suggests that aging organisms produce elevated levels of pro-inflammatory cytokines, which in humans leads to so-called age-related diseases, including diabetes, osteoporosis, and atherosclerosis (Aw et al., 2007).

Aging is accompanied by a decline in immunity, characterized by increased morbidity and mortality caused by infectious disease (insects: Ramsden et al., 2008; vertebrates: Finch and Crimmins, 2004). Alongside immune decline, chronic inflammation sets in as an organism reaches old age, in a seemingly maladaptive action by the innate immune system. This chronic inflammation develops into illness termed age-related disease, such as heart disease, cancer, and diabetes (Fig. 17.1). Given that age-related disease is a major measurable cost of immunopathology in humans, it is possible that the effects of immunopathology on aging could be the main mechanism behind any ecological consequences of immune-mediated self-harm in the wild. Moreover, since aging is associated with immunopathological chronic inflammation in many organisms (Licastro et al., 2005; Vasto et al., 2007), the potentially huge ecological impact of age-related immunopathology could extend across the whole animal kingdom.

A long life span and a robust immune system have been argued to require high quality of the same damage-control mechanism (Cichon, 1997; Chavez et al., 2009). Accumulation of ROS is a prominent theory of aging (Lin and Beal, 2006; Jang and Remmen, 2009; Sesti et al., 2010) due to the tissue damage caused. It is this tissue damage that is also the measurable effect of immunopathology. Genes that are involved in the process of senescence are also critical to ROS regulation in host–pathogen interactions, resulting in immunopathology being intrinsically coupled with aging. Reduced insulin-like peptide signaling has been found to reduce the life span of nematodes, flies, and mammals (Tatar et al., 2003).

Aging, immunopathology, and inflammation are intrinsically linked because aging is characterized by an imbalance between the pro- and anti-inflammatory status, leading to prolonged pro-inflammatory status. This has been demonstrated in insects (Pletcher et al., 2007). Immunopathology could be the mechanism behind the hypothesis proposed by Finch and

Crimmins (2004) that childhood inflammation translates directly into adult morbidity and mortality costs. In other words, early-life inflammation could have long-term or delayed immunopathological costs. In support of this idea, many human diseases associated with aging are linked with inflammatory events earlier in life (Licastro et al., 2005).

The link between aging, inflammation, and immunopathology may help us to understand how the apparently maladaptive nature of inflammation has not been removed by selection in humans. During most of human evolution, infectious diseases were widespread and have been proposed as the major cause of death over this evolutionary time. In the pre-industrial age, most people died of infectious diseases between the ages of 35 and 40 (Oeppen and Vaupel, 2002; Crimmins and Finch, 2006). Since the age of antibiotics and advanced medical treatment during the past two centuries, mortality due to infectious diseases has been greatly reduced (Finch and Crimmins, 2004; Crimmins and Finch, 2006; Gurven et al., 2008). In parallel, life expectancy has also increased; reaching old age is much more common since the beginning of the 20th century (Riley, 2001). Although all ages have benefited from increased life expectancy during the past 250 years, the increase among the elderly began many years after the increases at younger ages (Crimmins and Finch, 2006).

A strong inflammatory response, which was adaptive in ancestral populations when the environmental pathogen load was so high and medical treatment was so low, could now be considered maladaptive, owing to the increases in life expectancy. Indeed, low inflammatory responses have actually been found to benefit longevity. In a study of inflammatory molecules, individuals that were genetically predisposed to a weak inflammation response had an increased chance of living to an older age, provided they did not catch any serious infectious diseases (Caruso et al., 2005). The effect of the changing environment in westernized countries compared to that of our ancestors with regards to inflammation can be demonstrated in a particularly extreme form: because of the absence of the gut infections that our immune systems evolved alongside, the balance between the normal bacterial content of the gut and the immune system has been upset, so that we can now suffer inflammatory diseases without any initial infection (MacDonald and Monteleone, 2005).

Although this explanation for the immunopathological nature of inflammation is fairly neat, the fact that it is demonstrated in many other organisms that have not undergone dramatic life span extension in recent years means that further explanation is required. This highlights once again the need for further investigation into immunopathology from an evolutionary perspective. A more general explanation for the existence of immunopathology is that the primary mechanism behind it, inflammation, needs to be a strong response in order to have the desired effect of pathogen clearance. The price paid for a strong immune response in the short term could be a decline in health in the long-term—that is, an antagonistic pleiotropic effect. This explanation is plausible, since a reduction in post-reproductive age could understandably be a small price to pay for an increased chance of surviving an early-life disease.

An immune response experimentally elicited in early life in the mealworm beetle (*Tenebrio molitor*) caused a reduction in adult life span (Pursall and Rolff, 2011). This provided a proof-of-principle that an immune insult in early life translates directly into adult survival costs, a hypothesis that had been proposed previously (Finch and Crimmins, 2004). The experimental immune insult was deliberately non-pathogenic to ensure that the effects of immune response alone were being tested in the absence of direct pathogenic effects (Fig. 17.2). This provided an insight into

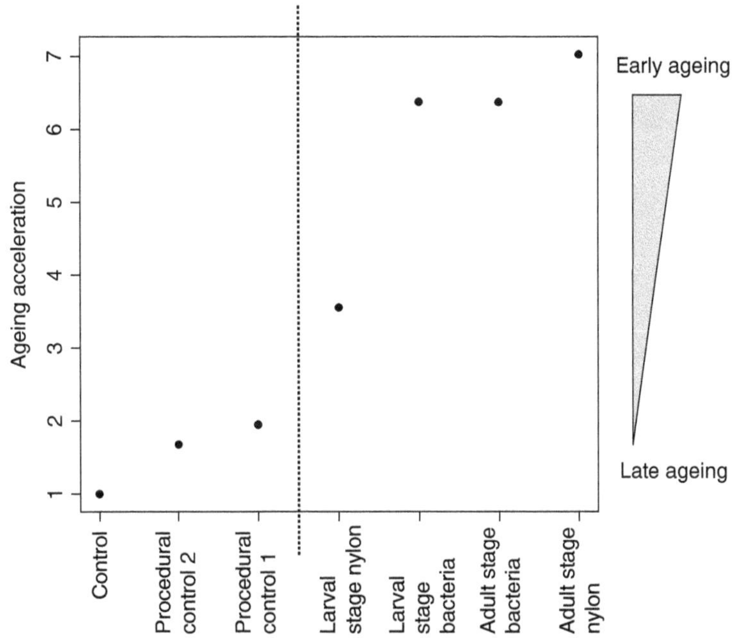

FIGURE 17.2: Early-life immune challenge accelerates ageing 'Ageing acceleration' on the y-axis represents the reduction in maximum lifespan caused by the treatments (the percentage of survivors to the 90th percentile in the control group/ the percentage of survivors to the 90th percentile in the treatment group). The dashed line represents the point at which the difference in aging acceleration between Control and experimental groups becomes statistically significant. Control n = 133; Procedural Control 1 n = 69; Procedural Control 2 n = 62; Nylon Larval Stage n = 97; Bacteria Adult Stage n = 29; Bacteria Larval Stage n = 58; Nylon Adult Stage n = 32. Experimental groups were as follows: Control = no experimental manipulation in larval or adult stage; Procedural Control 2 = injection with sterile insect ringer in larval stage only; Procedural Control 1 = injection with sterile insect ringer in both larval and adult stage; Larval stage nylon = insertion of nylon filament in larval stage, and injection of sterile insect ringer in adult stage; Larval stage bacteria = injection of dead bacteria in larval stage, and injection of sterile insect ringer in adult stage; Adult stage bacteria = injection of sterile insect ringer at larval stage and injection of dead bacteria in adult stage; Adult stage nylon = injection of sterile insect ringer in larval stage and insertion of nylon filament in adult stage. Pursall and Rolff, 2011.

whether immunopathology could be a potential mechanism behind any long-term costs. Maximum life span was significantly reduced following an early-life experimental immune challenge with either dead bacteria or another non-pathogenic antigen, nylon filament. Maximum life span measures the proportion of a cohort reaching the 90th percentile of overall survival time, and therefore specifically concentrates on the aging pattern of the population. In contrast, median life span was not significantly affected by early-life immune challenge.

The fact that long-term costs of early infection were apparently carried through beetle metamorphosis from the larval stage (in which insects were infected) to the adult stage (in which individuals died) suggested that a mechanism could be immunopathology. This is because immunopathology has been shown to occur in this study organism in the form of damage to malpighian tubules (Sadd and Siva-Jothy, 2006), an organ complex that is carried in a "constant" state (i.e., without being rearranged) from the larval stage to the adult stage (Truman and Riddiford, 1999). We speculate that collateral damage to self-tissue during an early-stage immune response caused an acceleration of aging in the mealworm beetle. This mechanism could explain the correlation between the decreased incidence of childhood disease and enhanced longevity observed in recent decades in Western societies (Finch and Crimmins, 2004; Crimmins and Finch, 2006).

IMMUNOPATHOLOGY AND TOLERANCE VERSUS RESISTANCE STRATEGIES

We propose that immunopathology could determine whether a host adopts a tolerance or resistance strategy in response to pathogen attack (Fig. 17.3). Therefore, in a wider context, the extent of immunopathological damage suffered from immune responses by different species could determine the evolution of different tolerance and resistance patterns in nature. A resistance strategy involves limiting parasite burden, whereas a tolerance strategy is concerned with limiting the damage caused by the parasite burden (Caldwell et al., 1958; Råberg et al., 2007; Ayres and Schneider, 2008). Both of these mechanisms of immunity demonstrate genetic variation (Råberg et al., 2007).

Tolerance relative to resistance is considerably understudied in animals compared to plants, and a greater understanding of tolerance in general would greatly benefit the study of enemy–victim interactions (Svensson and Råberg, 2010). A greater understanding of how tolerance strategies evolve would also help to investigate immunopathology at the population level. If our

FIGURE 17.3: Proposed interactions between immunopathology and the tolerance-resistance strategies of the immune system. As risk from immunopathological damage decreases, the host shifts its immune response along the "gradient" from a resistance to a tolerance approach. However, if the risk from pathogen-induced damage accelerates past the risk of immunopathological-induced harm, the host would invest more in resistance strategies (the shape of the curves is not known and just chosen for illustration purposes).

predictions were supported by experimental evidence, and hosts were found to adopt a tolerance strategy depending on whether damage risk is greater from the pathogen or host, this would confirm the importance of tolerance and immunopathology to be incorporated into ecological immunology.

A rare example of the study of tolerance from an ecological immunology perspective was presented by Corby-Harris et al. (2007). They reported that in *Drosophila melanogaster*, the ability to clear pathogens is not correlated with survival after infection (Fig. 17.4), though both traits demonstrated ample genetic variation. They concluded the most likely explanation to be the existence of a tolerance mechanism in *D. melanogaster*. To investigate our prediction (see Fig. 17.2), a useful study could be to investigate the extent of immunopathological compared to pathogen-induced damage suffered by the host in the system studied by Corby-Harris et al. (2007). If immunopathological damage was low relative to pathogen-induced damage, this could lend support to our theory.

Ayres and Schneider (2008) showed that manipulating the level of "tolerance" to a pathogen caused a complex set of immunity phenotypes, which affect the host's future levels of resistance and tolerance to various pathogens (Fig. 17.5). In this experiment, a reduction in immune response-imposed self-harm was considered to be a method of shifting the phenotype in a more tolerant direction. Specifically, the self-harm considered here was melanization, which was shown in another insect (*Tenebrio molitor*) to have immunopathological costs (Sadd and Siva-Jothy, 2006). Alterations in melanization intensities could be a common mechanism throughout insect taxa to mediate the levels of tolerance and resistance to pathogens. Therefore, an interesting area

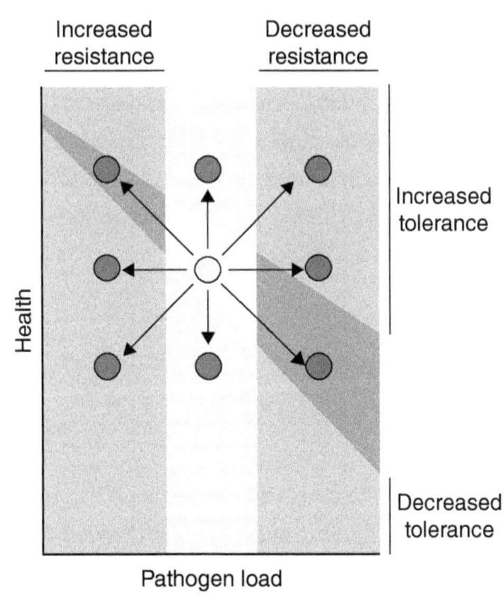

FIGURE 17.4: Mutations in a single gene, CG3066, a component of the melanization cascade, generated almost all combinations of resistance and tolerance phenotypes in *Drosophila melanogaster* against a variety of bacterial pathogens (Ayres and Schneider, 2008, copyright David Schneider), as depicted in the graph. The yellow bar depicts changes in tolerance, the blue bar depicts changes in resistance, and the green bars depict changes in both traits.

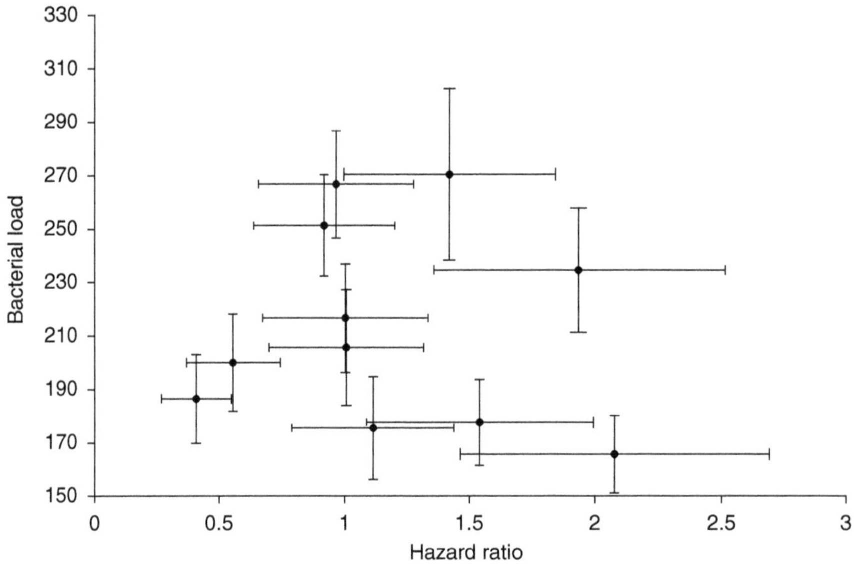

FIGURE 17.5: The relationship between tolerance, measured as survival (hazard ratio), and immunity, measured as bacterial clearance in *Drosophila melanogaster*. The authors did not find a genetic correlation between survival and immunity across the families (Corby-Harris et al., 2007).

of research would be to assess melanization intensities in different species under different pathogen levels, and investigate whether changing melanization causes similar patterns in the adoption of tolerance versus resistance patterns between species. This would help build up the picture of the ecological consequences of immunopathology.

Resisting a pathogen by eliciting a large immune response may be more harmful than succumbing to infection (Little and Killick, 2007). In their experiment, Little and Killick (2007) found that a higher immune response correlated with an elevated level of mortality, compared to a lower immune response. This provides evidence for the costs of immunity, and one important cost could be immunopathology (Little and Killick, 2007).

FINAL CONSIDERATIONS

Models of virulence are often parasite-centric, and therefore do not consider immunopathology to a satisfactory degree (Day et al., 2007). Day et al. (2007) showed that although this is the case, trade-off models of the evolution of virulence could be expanded to incorporate immunopathology. Such models are an important component in evolutionary ecology-based theory of disease and therefore represent one area in which immunopathology should be incorporated. One of the main model predictions was that if immunopathology is taken into account, parasites actually evolve to become less virulent. This was recently demonstrated in an elegant experiment by Long et al. (2008). They increased immunopathology in rodents by neutralizing interleukin, a regulator of inflammation in mammals. This elevation in immunopathology had the potential to select for reduced virulence in *Plamodium chabaudi* (rodent malaria). There are likely many other topics in evolutionary ecology that require the incorporation of immunopathology in order to present more realistic theories of immunity, because immunopathology is apparently so widespread and influential on fitness. For example, it will be interesting to see how important a cost immunopathology is in the context of genital wounding during copulation, which seems to be widespread (Siva-Jothy, 2009). Decomposing the costs of using an immune response into immunopathology and energetic costs such as resource allocations has the potential to improve our understanding of host–parasite interactions in the wild. Lee et al. (2006) demonstrated the role of nutrient composition in shaping immune responses. Certain immune effectors depend not only on energetic resources, but also for example on the environmental availability of essential amino acids.

Ecological immunology usually addresses two categories of costs of immune activation: life-history consequences and energetic costs (e.g., Bertrand et al., 2006). We argue that immunopathology should be included in the list of "costs" that evolutionary ecologists consider when studying infection and immunity.

Studies similar to that by Bertrand et al. (2006) need to be done that present a potential immunopathological threat that is "realistic"—that is, could be imagined to occur in the wild. Bertrand et al. (2006) used a relatively small lipopolysaccharide challenge, which resulted in a comparably small measurable immunopathological consequence. However, a real-life immunopathological threat is perhaps more likely to be long-lasting or chronic, for example through a persistent low-level environmental pathogen, or long-term immune regulation dysfunction.

In 1998, Råberg et al. published a review outlining two hypotheses on how adaptive immunosuppression could arise: the resource-limitation hypothesis and the immunopathology-avoidance hypothesis (Råberg et al., 1998). A decade later, there remains insubstantial experimental evidence that could help to distinguish which mechanism is acting in different circumstances (Bourgeon et al., 2009). Results from a recent study by Bougeon et al. (2009) concluded that the two hypotheses are not mutually exclusive. This further emphasizes that immunopathology should be routinely considered by evolutionary ecologists, because there is a good chance that it is a mechanism that has occurred in many studies into immunity costs, alongside other mechanisms more readily identified (such as resource trade-offs).

ACKNOWLEDGMENTS

We would like to thank Oliver Otti for thorough feedback on a draft of this chapter. We would also like to thank David Schneider and Vanessa Corby-Harris for providing figures.

REFERENCES

Åbro, A. 1992. Attachment and feeding devices of water mite larvae (*Arrenurus* spp.) parasitic on damselflies (Odonata, Zygoptera). Zool. Scripta 8: 221–234.

Aw, D, Silva, AB, and Palmer, DB (2007) Immunosenescence: emerging challenges for an ageing population. Immunol. 120(4): 435–446.

Ayres, JS, and Schneider, DS (2008) Two ways to survive infection: what resistance and tolerance can teach us about treating infectious diseases. Nat. Rev. Immunol. 8: 889–895.

Balkwill F and Mantovani A (2001) Inflammation and cancer: back to Virchow? *Lancet* 357: 539–545.

Barnes, PJ, Karin, M (1997) Nuclear factor-kB – A pivotal transcription factor in chronic inflammatory diseases. New Eng. J. Med. 336: 1066–1071.

Bengtsson, T, and Lindstrom, M (2003) Airborne infectious diseases during infancy and mortality in later life in southern Sweden, 1766–1894. Int. J. Epidemiol. 32: 286–294.

Bertrand, B, Alonso-Alvarez, C, Devevey, G, Faivre, B, Prost, J, and Sorci, G (2006) Carotenoids modulate the trade-off between egg production and resistance to oxidative stress in zebra finches. Oecologia 147(4): 576–584.

Bidla, G, Hauling, T, Dushay, MS, Theopold, U (2009) Activation of insect phenoloxidase after injury: Endogenous versus foreign elicitors. Journal of Innate Immunity 1: 301–308.

Bird, L (2010) Antiviral immunity: Ageing immune systems work too hard. Nat. Rev. Immunol. 10: 6–7.

Bourgeon, S, Le Maho, Y, and Raclot, T (2009). Proximate and ultimate mechanisms underlying immunosuppression during the incubation fast in female eiders: Roles of triiodothyronine and corticosterone. Gen. Comp. Endocrinol. 163(1–2): 77–82.

Brandt, SM, Dionne, MS, Khush, RS, Pham, LN, Vigdal, TJ, and Schneider, DS (2004) Secreted bacterial effectors and host-produced eiger/TNF drive death in a salmonella-infected fruit fly. Plos. Biol. 2: e418.

Braune, P, and Rolff, J (2001) Parasitism and survival in a damselfly: does host sex matter? Proc. R. Soc. B: Biol. Sci. 268(1472): 1133–1137.

Caldwell, RM, Schafer, JF, Compton, LE, and Patterson, FL (1958) Tolerance to cereal leaf rusts. Science 26(3326): 714–715.

Caruso, C, Candore, G, Colonna-Romano, G, Lio, D and Franceschi, C (2005) Inflammation and life-span. Science 307: 208.

Cerenius, L, Lee, BL, and Soderhall, K (2008) The proPO-system: pros and cons for its role in invertebrate immunity. Trends Immunol. 29(6): 263–271.

Chavez, V, Mohri-Shiomi, A, and Garsin, DA (2009) Ce-Duox1/BLI-3 generates reactive oxygen species as a protective innate immune mechanism in *Caenorhabditis elegans*. Infection. Immun. 77(11): 4983–4989.

Cherry, S, and Silverman, N (2006) Host–pathogen interactions in drosophila: new tricks from an old friend. Nat. Immunol. 7: 911–917.

Cichon, M (1997) Evolution of longevity through optimal resource allocation. Proc. R. Soc. B 264: 1383–1388.

Corby-Harris, V, Habel, KE, Ali, FG, and Promislow, DE (2007) Alternative measures of response to *Pseudomonas aeruginosa* infection in *Drosophila melanogaster*. J. Evol. Biol. 20: 526–533.

Costantini, D (2008) Oxidative stress in ecology and evolution: lessons from avian studies. Ecol. Letters 11 (11): 1238–1251.

Crimmins, EM, and Finch, CE (2006) Infection, inflammation, height, and longevity. PNAS 103: 498–503.

Day, M (2010) Ageing, immunosenescence and inflammageing in the dog and cat. J. Comp. Pathol. 142: S60–S69.

Day, T, Graham, AL, and Read, AF (2007) Evolution of parasite virulence when host responses cause disease. Proc. R. Soc. B: Biol. Sci. 274(1626): 2685–2692.

De Block, M, and Stocks, R (2008) Compensatory growth and oxidative stress in a damselfly. Proc. R. Soc. B: Biol. Sci. 275(1636): 781–785.

Dhabhar, FS (2009) Enhancing versus suppressive effects of stress on immune function: implications for immunoprotection and immunopathology. NeuroImmunoMod. 16 (5): 300–317.

Ewald, PW (2010) 99th Dahlem Conference on Infection, Inflammation and Chronic Inflammatory Disorders: Symbionts and immunopathology in chronic diseases: insights from evolution. Clin. Exp. Immunol. 160(1): 27–34.

Finch, C (2010) Evolution of the human lifespan and diseases of aging: Roles of infection, inflammation, and nutrition. PNAS 107: 1718–1724.

Finch, CE, and Crimmins, EM (2004) Inflammatory exposure and historical changes in human life-spans. Science 305: 1736–1739.

Graham, A (2002) When T-helper cells don't help: immunopathology during concomitant infection. Quart. Rev. Biol. 77(4) 409–434.

Graham, AL, Allen, JE, and Read, AF (2005) Evolutionary causes and consequences of immunopathology. Ann. Rev. Ecol., Evol., System. 36: 373–397.

Goldsby RA, Kindt TJ Osborne BA (2000) Kuby immunology. W. H. Freeman, New York, NY.

Gurven, M, Kaplan, H, Winking, J, Finch, C, and Crimmins, EM (2008) Aging and inflammation in two epidemiological worlds. J. Gerontol. Med. Sci. 63A: 196–199.

Hotamisligil, GS (2006) Inflammation and metabolic disorders. Nature 444: 860–867.

Janeway, CA, Jr., and Medzhitov, R (2002) Innate immune recognition. Ann. Rev. Immunol. 20: 197–216.

Jang, YC, and Van Remmen, H (2009) The mitochondrial theory of aging: Insight from transgenic and knockout mouse models. Exp. Gerontol. 44(4): 256–260.

Kaushal, D, Schroeder, BG, Tyagi, S, Yoshiatsu, T, Scott, C, Ko, C, Carpenter, LM, Mehrotra, J, Manabe, YC, Fleischmann, RD, and Bishai, WR (2002) Reduced immunopathology and mortality despite tissue persistence in a *Mycobacterium tuberculosis* mutant lacking alternative factor, SigH. PNAS 99(12): 8330–8335.

Kurtz, J, Wegner, MW, Kalbe, M, Reusch, TBH, Schaschi, H, Hasselquist, D, and Milinski, M (2006) MHC genes and oxidative stress in sticklebacks: an immuno-ecological approach. Proc. R. Soc. B. Biol. Sci. 273(1592): 1407–1414.

Lambeth, JD, Krause, KH, and Clark, RA (2008) NOX enzymes as novel targets for drug development. Seminars. Immunopathol. 30(3): 339–363.

Lee, KP, Cory, JS, Wilson, K, Raubenheimer, D, and Simpson, SJ (2006) Flexible diet choice offsets protein costs of pathogen resistance in a caterpillar. Proc. R. Soc. B Biol. Sci. 273(1588): 823–829.

Lemaitre, B, and Hoffman, J (2007) The host defense of *Drosophila melanogaster*. Ann. Rev. Immunol. 25: 697–743.

Lesser, M (2006) Oxidative stress in marine environments: biochemistry and physiological ecology. Ann. Rev. Physiol. 68: 253–278.

Libert, S, Chao, Y, Chu, X, and Pletcher, SD (2006) Trade-offs between longevity and pathogen resistance in *Drosophila melanogaster* are mediated by NFκB signaling. Aging Cell 5 (6): 533–543.

Libby (2002) Inflammation and atherosclerosis. *Nature* 420: 868–874.

Licastro, F, Candore, G, Lio, D, Porcellini, E, Colonna-Romano, G, Franceschi, C, and Caruso, C (2005) Innate immunity and inflammation in aging: a key for understanding age-related diseases. Imm. Ageing 2: 8–22.

Lin, MT, and Beal, MF, (2006) Mitochondrial dysfunction and oxidative stress in neurodegenerative diseases. Nature 443: 787–795.

Lin WW and Karin M (2007) A cytokine-mediated link between innate immunity, inflammation, and cancer. *J. Clin. Invest.* 117: 1175–1183.

Little, TJ, and Killick, SC (2007) Evidence for a cost of immunity when the crustacean *Daphnia magna* is exposed to the bacterial pathogen *Pasteuria ramose*. J. Anim. Ecol. 76: 1202–1207.

Long, GH, Chan, BHK, Allen, JE, Read, AF, and Graham, AL (2008) Experimental manipulation of immune-mediated disease and its fitness costs for rodent malaria parasites. BMC Evol. Biol. 8: 128–139.

MacDonald, TT, and Monteleone, G (2005) Immunity, inflammation and allergy in the gut. Science 307: 1920–1925.

Mathers, CD, and Loncar, D (2006) Projections of global mortality and burden of disease from 2002 to 2030. PloS Med. 3: e442.

McPherson, R, Pertsemilidis, A, Kavaslar, N, Sterwart, A, Roberts, R, Cox, DR, Hinds, DA, Pennecchio, LA, Tybjaerg-Hansen, A, Folsom, AR, Boerwinkle, E, Hobbs, HH, Cohen, JC (2007) A common allele on chromosome 9 associated with coronary heart disease. Science 316(5830): 1488–1491.

Nathan, C (2002) Points of control in inflammation. Nature 420: 846–852.

Oeppen, J, and Vaupel, JW (2002) Broken limits to life expectancy. Science 296: 1029–1031.

Pletcher, SD, Kabil, H, and Partridge, L (2007) Chemical complexity and the genetics of aging. Ann. Rev. Ecol. Evol. Systemat. 38: 299–326.

Pursall, ER and Rolff, J (2011) Immune Responses Accelerate Ageing: Proof-of-Principle in an Insect Model. PloS One 6(5): e19972.

Råberg, L, Grahn, M, Hasselquist, D, and Svensson, E (1998) On the adaptive significance of stress-induced immunosuppression. Proc. R. Soc. B Biol. Sci. 265(1406): 1638–1641.

Råberg, L, Sim, D, and Read, AF (2007) Disentagling genetic variation for resistance and tolerance to infectious diseases in animals. Science 318: 812–814.

Ramsden, S, Cheung, YY, and Seroude, L (2008) Functional analysis of the Drosophila immune response during aging. Ag. Cell 7: 225–236.

Rather, LJ (1971) Disturbance of function (functio laesa): the legendary fifth cardinal sign of inflammation, added by Galen to the four cardinal signs of Celsus. Bull. New York Acad. Med. 47(3): 303–322.

Read, AF, and Clark, JS (2006) The past 20 years of ecology and evolution. TREE 21(6): 287.

Ridker PM, Cushman M, Stampfer MJ, Tracy RP, Hennekens CH (1997) Inflammation, aspirin, and the risk of cardiovascular disease in apparently healthy men. *N. Engl. J. Med.* 337:422–424.

Riley, JC (2001) Rising Life Expectancy: A Global History. Cambridge University Press, Cambridge, UK.

Rolff, J (2000) Water mite parasitism in damselflies during emergence: two hosts, one pattern. Ecography 23: 273–282.

Rolff, J, and Martens, A (1997) Completing the life cycle: detachment of an aquatic parasite (*Arrenurus cuspidator*, Hydrachnellae) from an aerial host (*Coenagriom puella*, Odonata). Can. J. Zool. 75(4): 655–659.

Rolff, J, and Siva-Jothy, MT (2003) Invertebrate ecological immunology. Science 301(5632): 472–475.

Sadd, BM, and Siva-Jothy, MT (2006) Self-harm caused by an insect's innate immunity. Proc. R. Soc. B 273: 2571–2574.

Schilder, RJ, and Marden, JH (2006) Metabolic syndrome and obesity in an insect. Proc. Nat. Acad. Sci. 103: 18805–18809.

Scholzen, A, Minigo, G, and Plebanski, M (2010) Heroes or villains? T regulatory cells in malaria infection. Trends Parasitol. 26(1): 16–25.

Schmid-Hempel, P (2003) Variation in immune defences as a question of evolutionary ecology. Proc. Roy. Soc. B Biol. Sci. 270 (1513): 357–366.

Sesti, F, Liu, S and Cai, SQ (2010) Oxidation of potassium channels by ROS: a general mechanism of aging and neurodegeneration? Trends Cell Biol. 20 (1): 45–51.

Sies, H (1997) Oxidative stress: oxidants and antioxidants. Exp. Physiol. 82: 291–295.

Sin DD and Man P (2003) Why are patients with chronic obstructive pulmonary disease at increased risk of cardiovasculr diseases? *Circulation* 107:1514–1519.

Siva-Jothy, MT (2009). Reproductive immunity. In: Rolff, J, and Reynolds, S, eds. Insect Infection and Immunity. Oxford University Press, Oxford, pp. 241–250.

Siva-Jothy, MT, Rolff, J, and Moret, Y (2005) Insect immunity: an ecological immunology perspective. In Simpson SJ. Advances in Insect Physiology, Vol. 32. Elsevier Ltd., Oxford, UK.

Sorci, G, and Faivre, B (2009) Inflammation and oxidative stress in vertebrate host–parasite systems. Phil. Trans. R. Soc. B Biol. Sci. 364(1513): 71–83.

Spring, JH, Robichaux, SR, Kaufmann, N, Brodsky, JL (2007) Localization of a Drosophila DRIP-like aquaporin in the malpighian tubules of the house cricket, *Acheta domesticus.* Comp. Biochem. Physiol. A. 148(1): 92–100.

Stramer, B, Wood, W, Galko, MJ, and Redd, MJ (2005) Live imaging of wound inflammation in Drosophila embryos reveals key roles for small GTPases during in vivo cell migration J. Cell Biol. 168(4): 567–573.

Stoks R, Johansson F, De Block M (2008) Life-history plasticity under time stress in damselfly larvae. In Dragonflies and Damselflies - Model organisms for ecological and evolutionary research, pp 39–51. Oxford University Press, Oxford, UK.

Svensson, EI, and Raberg, L (2010) Resistance and tolerance in animal enemy–victim coevolution. TREE 25: 267–274.

Tak PP and Firestein GS (2001) NF-kB: a key role in inflammatory diseases. *J. Clin. Invest.* 107: 7–11.

Tatar, M, Bartke, A, and Antebi, A (2003) The endocrine regulation of aging by insulin-like signals. Science 299(5611): 1346–1351.

Truman, JW, and Riddiford, LM (1999) The origins of insect metamorphosis. Nature 401: 447–450.

Vasto, S, Candore, G, Balistreri, C, and Caruso, M (2007) Inflammatory networks in ageing, age-related diseases and longevity. Mech. Ag. Devel. 128 (1): 83–91.

Von Schantz, T, Bensch, S, and Grahn, M (1999). Good genes, oxidative stress and condition-dependent sexual signals. Proc. R. Soc. B Biol. Sci. 266(1414): 1–12.

Knell, RJ, and Webberley, M (2004) Sexually transmitted diseases of insects: distribution, evolution, ecology and host behaviour. Biological Reviews 79: 557–581.

Wellen, KE, and Hotalisligil, GS (2005) Inflammation, stress and diabetes. *J. Clin. Invest.* 115: 1111–1119.

Zampieri, CA, Sullivan, NJ, and Nabel, GJ (2007). Immunopathology of highly virulent pathogens: insights from Ebola virus. Nat. Immunol. 8: 1159–1164.

Zhuang, J, Lin, C, and Lin, D (1998). Mutagenesis associated with nitric oxide production in macrophages. PNAS 95(14): 8286–8291.

CHAPTER 18

THE EVOLUTIONARY ECOLOGY OF INFECTIOUS DISEASE VIRULENCE

Lars Råberg and Martin Stjernman

INTRODUCTION

Infectious diseases are a common problem for most organisms. From our perspective as potential hosts for parasites and pathogens, the arguably most interesting aspect of an infectious disease is its virulence—that is, the harm that an infection causes the host (see Box 18.1 for various definitions of virulence). The degree of virulence differs considerably between diseases, and these differences are due to variation both among host and parasite species. For instance, the severity of human malaria depends to a large extent on which parasite species is causing the infection (Garamszegi, 2006). Thus, there is interspecific variation among malaria parasites for virulence. Correspondingly, the virulence of a given parasite often differs between host species. For example, the virulence of simian immunodeficiency virus (SIV) differs greatly depending on which primate species it infects (Mandl et al., 2008). What causes this diversity in virulence and how has it arisen?

One explanation for this diversity is that virulence is shaped by natural selection and that different levels of virulence are favored under different environmental conditions. There are several examples of evolution of virulence in ecological time, mainly from so-called serial passage experiments in the laboratory, but also in natural host–parasite systems. In serial passage experiments, a parasite is successively transferred between host individuals, typically in a novel host species. Such experiments generally result in increased virulence in the new host and reduced virulence in the original host (Ebert, 1998). When it comes to virulence evolution in natural populations, myxomatosis in rabbits is still the most famous and well-documented case. The Myxoma virus was introduced into the Australian rabbit population in 1950, and later also in Europe, as a means of pest control. Virulence was initially extremely high, with a case mortality close to 100%. However, within about a decade, virulence was reduced considerably (to an average case mortality of about 70%), and similar patterns were observed in the United Kingdom and France (Fenner and Fantini, 1999). Other examples of evolution of virulence concern human immunodeficiency virus (HIV)

BOX 18.1: WHAT IS VIRULENCE?

Virulence is generally defined as the harm that an infection causes the host. A more strict ecological/evolutionary definition is the reduction in host fitness caused by infection. Thus, it includes effects of parasitism on both host survival and fecundity. However, parasite-centered models of virulence evolution typically focus on host survival, because this is what affects the fitness of the parasite (unless a parasite is transmitted vertically [i.e., from mother to offspring]). Specifically, theoretical models usually define virulence as "parasite-induced mortality rate." In contrast, in empirical studies, effects of parasitism on host survival are more often measured as case mortality (i.e., the probability of dying from a disease once infected). It is important to realize that these different measures are not necessarily well correlated: a disease with a relatively high case mortality may or may not kill its host quickly (Day, 2002). As an example, infections by HIV and Ebola virus differ dramatically in virulence in the sense of parasite-induced mortality rate, even though the case mortality is about the same.

We also want to point out that "virulence" sometimes has quite different meanings. In particular, in the plant pathology literature it is usually taken to mean the ability of a pathogen to infect a particular host, something we would refer to as "infectivity."

infections in humans in Italy, where virulence has increased between 1984 and 2006 (Müller et al., 2009); diphtheria in humans, where virulence has decreased since the introduction of toxoid vaccine in 1958 (Read and Mackinnon, 2008); and Marek's disease in poultry, where virulence has increased continuously during the past 60 years (Read and Mackinnon, 2008). Thus, it is clear that virulence is evolutionary labile, but what are the selective pressures that mold this trait?

The virulence of an infection is jointly determined by the host and the parasite; it depends both on the host's ability to defend itself and on the parasite's ability to grow and damage the host. Thus, the evolution of virulence observed in, for example, the Myxoma-rabbit system described above could be a result of parasite evolution, host evolution, or both. To understand the evolution of virulence it is therefore necessary to disentangle host and parasite sources of genetic variation for virulence, and to elucidate how natural selection acts on each of these components. For this purpose, we propose a simple path model where variation in virulence is decomposed into different host and parasite components (Fig. 18.1). According to this model, the virulence of an infection depends both on the infection intensity (number of parasites per host or unit host tissue) and the damage that each one of these parasites (bacterial cells, virus particles, etc.) causes. Moreover, both infection intensity and damage per parasite are chimeric traits, influenced by both parasite and host. In other words, both parasite and host genes can affect virulence via two different paths.

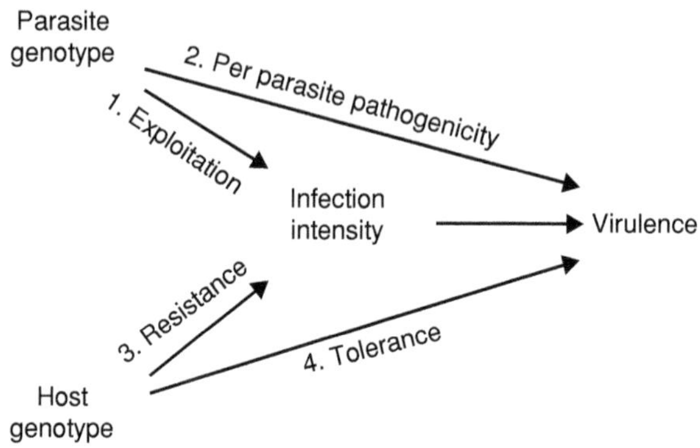

FIGURE 18.1: The sources of genetic variation for infectious disease virulence. Both parasite and host genes can in principle contribute to variation in virulence in two conceptually different ways: by affecting the intensity of infection (parasite load), or by affecting the virulence of a given infection intensity. Parasite "exploitation" (path 1) and host "resistance" (path 3) affect the growth of the parasite and thereby the virulence of the infection. "Per parasite pathogenicity" (path 2) and host "tolerance" (path 4) affect the harm caused per parasite.

First, there can be parasite or host genes affecting infection intensity (paths 1 and 3, respectively, in Fig. 18.1) and thereby virulence. Second, there can be parasite or host genes affecting the virulence of a given infection intensity (paths 2 and 4, respectively). These different sources of variation can be distinguished by means of a fully factorial experiment where a number of parasite genotypes are used to infect a number of host genotypes (see Fig. 18.2 for an example). In this chapter, we will use the model in Figure 18.1 as a framework to help understand virulence evolution. For each of the paths, we first examine the empirical evidence for genetic variability. We then review theoretical models of how natural selection should act on such variation, and critically evaluate the available empirical evidence for the assumptions and predictions of these models. Throughout, we strive to point out areas where there is a lack of empirical evidence, or a mismatch between theory and empirical data. We focus on animal studies, but with occasional references to the plant literature for comparison and completeness.

PARASITE EVOLUTION

Most studies of virulence evolution have taken a parasite-centered perspective and considered virulence as a property of the parasite (reviewed in Frank, 1996; Alizon et al., 2009; but see Ebert and Hamilton, 1996; Day and Burns, 2003; Restif and Koella, 2003). Thus, this field of research focuses on how parasite genetic variation for virulence affects the Darwinian fitness of the parasite. The fitness of a parasite is determined by its ability to transmit to and infect new hosts, and is usually described by the "basic reproductive ratio" (Anderson and May, 1982):

$$R_0 = \frac{\beta S}{b + v + \rho} \qquad (1)$$

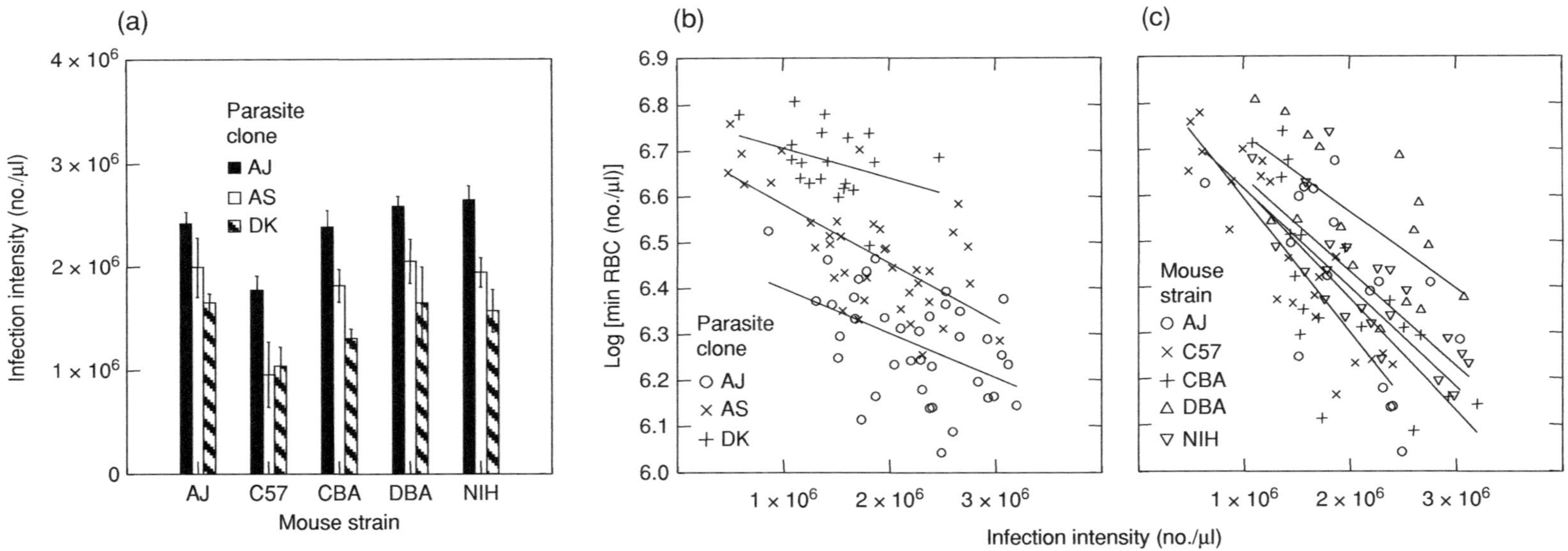

FIGURE 18.2: Disentangling different host and parasite sources of variation in virulence. The different paths or components of genetic variation outlined in Figure 18.1 can be disentangled by an infection experiment where a number of host genotypes are infected by a number of parasite genotypes in a fully factorial design. To illustrate this we here reanalyze the data in Råberg et al. (2007). In this experiment, five different inbred strains of laboratory mice were infected by three different clones of the rodent malaria parasite *Plasmodium chabaudi*. Infection intensity was measured as density of parasites (no/μL blood) at the peak of infection, whereas virulence was measured as the minimum level of red blood cells (RBC) density during infection while controlling for the pre-infection density of RBCs. To test for genetic variation for parasite "exploitation" and host "resistance," we performed an ANOVA with infection intensity as dependent variable and mouse strain and parasite clone as factors; both factors were highly significant (panel a). The main effect of clone shows that there is variation for exploitation, and the effect of strain shows that there is variation for resistance. To test for variation in "per parasite pathogenicity" and host "tolerance," we performed an ANCOVA with minimum RBC density (adjusted for pre-infection density) as dependent variable, parasite clone and mouse strain as factors, and infection intensity as covariate; the effects of both clone and strain were significant (panels b and c), showing that there is variation for per parasite pathogenicity and tolerance, respectively. (Note that the operational definition of tolerance used here differs from the one used in Råberg et al. [2007]. In the previous analysis we used minimum RBC density as the dependent variable and defined tolerance as the slope of a regression of minimum RBC density against infection intensity; here we use minimum RBC density adjusted for pre-infection density of RBC as dependent variable and define tolerance as the adjusted minimum RBC density at a given infection intensity. See Råberg et al. [2009] for further discussion of statistical issues in analysis of tolerance.)

where β is the transmission rate, S is the number of susceptible hosts in the population, b is the background mortality rate of the host (i.e., when uninfected), v is the infection-induced mortality rate (i.e., virulence), and ρ is the recovery rate (i.e., the rate at which infected hosts clear the parasite). Thus, R_0 is the number of new infections arising from one infection. The numerator describes the rate at which new infections are produced, while the denominator is the mortality rate of infections. In other words, the numerator and denominator correspond to the traditional life-history traits fecundity and survival, respectively. The level of R_0 determines the long-term success of a parasite; if $R_0 \geq 1$ the parasite will persist, whereas if $R_0 < 1$ it will go extinct. R_0 can be calculated for a parasite population/species as a whole and can be used as a tool to study the epidemiology of a disease. Alternatively, it can be calculated separately for different genotypes of a parasite to help predict evolution; this is how it has been used in studies of virulence evolution.

PATH 1: EXPLOITATION

To date, parasite-centered studies of virulence evolution have focused on parasite variation for infection intensity (Frank, 1996; Alizon et al., 2009). These studies typically assume that a genetically homogenous host population is infected by a number of parasite strains that differ in the infection intensity they induce, and that virulence is a function of infection intensity. Virulence (v in terms of Eq. (1)) can then be represented by

$$v = cn_p \tag{2}$$

where c is a constant, n is the infection intensity, and the subscript p indicates that the trait is determined by the parasite (i.e., the parasite exhibits genetic variation for the trait while the host is genetically homogenous). We will refer to parasite genetic variation in n as variation in "exploitation." This focus on infection intensity is well justified as there is ample evidence that parasites generally exhibit considerable genetic variation for this trait. For example, in animal models of various diseases, infection intensities usually differ considerably among isolates (e.g., Dunn and North, 1995; Pennington et al., 1997; Mackinnon and Read, 1999; Tumpey et al., 2005). Genetic variation for exploitation has also been shown in a variety of natural host–parasite systems (e.g., Ebert, 1994; Ben-Ami et al., 2008; De Roode and Altizer, 2010). In fact, it seems such variation is found whenever it is looked for.

How, then, does variation in exploitation affect the Darwinian fitness of the parasite? Models of virulence evolution typically assume that exploitation, via its effects on infection intensity, affects several components of parasite fitness. First, exploitation is assumed to be positively correlated with the rate of transmission to new hosts; the higher the infection intensity, the more new hosts are infected per unit time. Second, exploitation is also assumed to be negatively correlated with the recovery rate such that parasites that reach higher infection intensities take longer to clear. Thus, exploitation has positive effects on two components of parasite fitness. However, exploitation should also be positively correlated with virulence such that parasites that reach higher infection intensities kill their hosts, and thereby terminate the infection, faster. Thus, under the assumption that transmission rate (β), virulence (v), and recovery rate (ρ) are all functions of infection intensity (n), there will be trade-offs between transmission rate and recovery rate on the

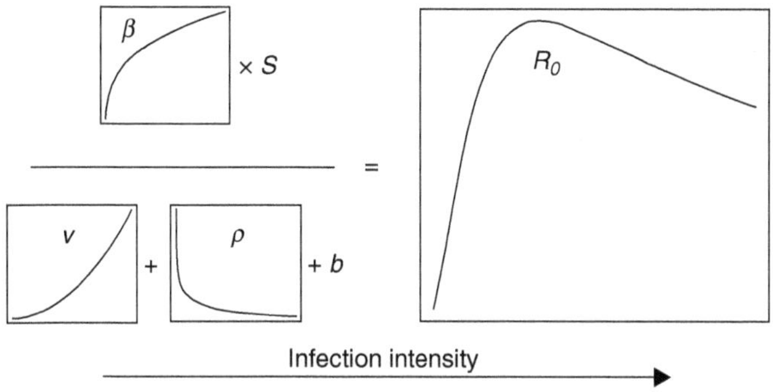

FIGURE 18.3. The trade-off hypothesis for the evolution of virulence. The basic reproductive ratio (R_0; i.e., the number of secondary infections resulting from one infected host individual) is a function of transmission rate (β), number of susceptible host individuals (S), host mortality rate (b), virulence (v), and clearance rate (ρ). Given that either β or v is a nonlinear function of infection intensity (as shown in the figure), R_0 will have an intermediate optimum. Adapted from Mackinnon and Read, 2004.

one hand, and virulence on the other. Eq. (1) is therefore known as the "trade-off hypothesis" for virulence evolution. Given that either $\beta(n)$ or $v(n)$ is a nonlinear function (decelerating and accelerating, respectively), parasite fitness is maximized at some intermediate level of infection intensity (Fig. 18.3). In contrast, if $\beta(n)$ and $v(n)$ are both linear, selection favors maximal exploitation and virulence.

What is the empirical evidence for these basic assumptions of the trade-off hypothesis? One of the best studied host–parasite systems with respect to virulence evolution is the rodent malaria parasite *Plasmodium chabaudi* in laboratory mice. Read and coworkers have performed a series of experiment with clones of *P. chabaudi* isolated from thicket rats (*Thamnomys rutilans*) collected in the Central African Republic in 1969 and 1970, and demonstrated that, as assumed by the trade-off hypothesis, infection intensity is positively correlated with virulence and transmission, but negatively correlated with clearance rate (Mackinnon and Read, 1999; Mackinnon and Read, 2004). A number of studies of other host–parasite systems support the assumptions that infection intensity is positively correlated with virulence and negatively correlated with clearance rate, and there are also several studies demonstrating a correlation between infection intensity and transmission (see Alizon et al., 2009, for a recent review). However, very few have investigated the shape of these relationships, something that is critical for the trade-off hypothesis (see above). A notable exception is a recent study on HIV in humans that showed that transmission rate was a saturating function of viral load (Fraser et al., 2007). Moreover, a study of monarch butterflies and a protozoan parasite directly quantified overall transmission success as a function of infection intensity and found that it was indeed maximized at an intermediate level (De Roode et al., 2008). Similarly, an experiment with the freshwater crustacean *Daphnia magna* and its bacterial parasite *Pasteuria ramosa* showed that overall production of bacterial spores was maximized at intermediate levels of virulence (measured as life span of infected hosts) (Jensen et al., 2006). Thus, even though several authors have recently questioned the validity of the trade-off hypothesis (e.g., Ebert and Bull, 2003), we conclude that there are currently a number of studies supporting its assumptions. Nevertheless, more data, in particular as regards the shape of the relationships (whether they are linear or not) between β, v, and n, are clearly desirable to determine the generality of the trade-off hypothesis.

How, then, can the trade-off hypothesis help us understand the diversity in virulence observed in nature? A number of environmental factors that can affect the strength of selection on the different components of parasite fitness, and hence the position of the optimal level of exploitation, have been identified. We will focus on three factors—parasite prevalence, host population structure, and host heterogeneity—because we feel these are likely to be particularly important for explaining patterns of virulence evolution within and across parasite species. Moreover, there has recently been considerable empirical and/or theoretical progress in the study of these factors.

PARASITE PREVALENCE AND WITHIN-HOST COMPETITION

The basic version of the trade-off hypothesis represented by Eq. (1) assumes that each host is infected by a single parasite genotype—that is, parasite genetic variation (e.g., for exploitation) occurs only among hosts. Since natural selection requires variation to act on, selection will under this scenario therefore act only between hosts. However, it is common that a proportion of hosts are infected by two or more parasite genotypes simultaneously (so-called "mixed" or "multiple infections") such that there is genetic variation for exploitation not only among but also within hosts (Read and Taylor, 2000). Under these circumstances natural selection will act at two levels: between and within hosts. Between-host selection should still follow the trade-off hypothesis and select for an intermediate optimal level of exploitation, but how does selection act on within-host variation in exploitation? Theoretical models addressing this question have outlined two different scenarios for how ecological interactions between parasite genotypes within hosts generate selection on exploitation. First, co-infecting parasite genotypes may interact via their effects on host mortality. Imagine a host that is simultaneously infected by two genotypes that differ in exploitation, but where the two genotypes do not affect each others' infection intensities and transmission rates. The virulence of the infection will, however, be determined by the most virulent genotype, or by the combined effects of the two genotypes. In any case, the less virulent genotype will not have time to realize its total transmission potential before the more virulent genotype has killed the host (van Baalen and Sabelis, 1995; Frank, 1996) (Fig. 18.4a and b). Thus, in the presence of other genotypes, it always pays to exploit the host as fast as possible, a form of "the tragedy of the commons" (Hardin, 1968). This is often referred to as the "co-infection" scenario following May and Nowak (1995).

Second, parasite genotypes may interact via their effects on each others' infection intensities and transmission rates, a form of contest competition (Fig. 18.4c and d). Such competition can occur if the growth of the parasite is dependent on some limited resource. Alternatively, contest competition can be mediated by the host's immune response (a form of "apparent competition") or allelopathic substances ("interference competition") (Read and Taylor, 2001; Råberg et al., 2006). It is usually assumed that parasite genotypes with a higher level of exploitation have an advantage in contest competition. The extreme scenario, where a more virulent genotype immediately outcompetes a less virulent genotype from a host individual, so that only the most virulent genotype in a mixed infection is transmitted, is often termed "superinfection" (Nowak and

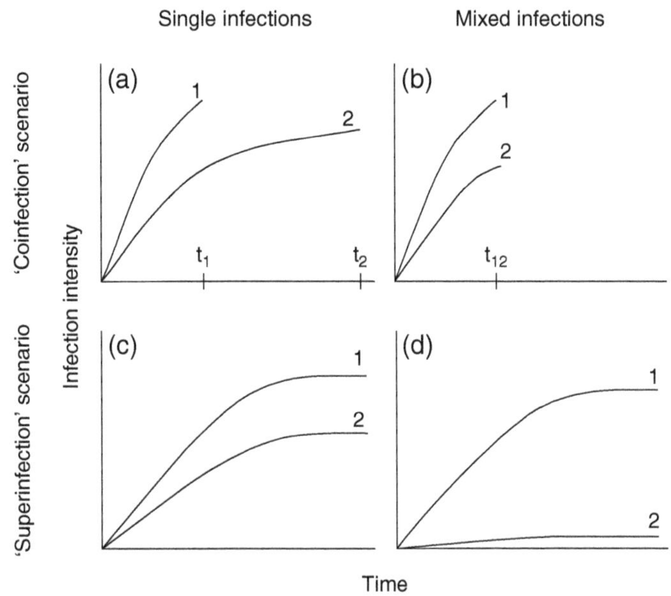

FIGURE 18.4: Within-host interactions and selection on virulence. Graphs show infection intensities over time for infections with a relatively virulent (1) and a relatively avirulent parasite genotype (2). The transmission of each genotype is proportional to the area under the infection intensity curve. a) and b) The "co-infection" scenario. In single infections (a), the less virulent genotype grows more slowly and kills the host later (at t_2) and thereby achieves higher transmission than the more virulent genotype (which grows more quickly but kills the host already at t_1). In mixed infections (b), host life span is determined by the more virulent genotype (host is killed at t_{12}), so that the less virulent genotype does not have time to realize its potential transmission. c) and d) The "superinfection" or contest competition scenario. The more virulent genotype suppresses the infection intensity of the avirulent genotype in mixed infections, and thereby achieves disproportionably high transmission in mixed compared to single infections.

May, 1994). One could also imagine less extreme forms of contest competition where co-infecting genotypes affect the growth and infection intensities, and thereby transmission rates, of each other, but without necessarily outcompeting each other.

It is important to distinguish between the co-infection and superinfection scenarios because they yield different predictions regarding the variability of exploitation in the parasite population. Whereas the co-infection model predicts a largely monomorphic parasite population, the superinfection model predicts that considerable genetic variation for exploitation and virulence will be maintained (Nowak and May, 1994; May and Nowak, 1995). However, as outlined above, both predict selection for higher levels of exploitation in multiple as compared to single infections. Hence, within-host selection can generally be expected to shift the overall optimal level of exploitation and virulence upwards relative to the single-infection optimum (which is determined by Eq. (1); but see Chao et al., 2000, for a case where within-host selection favors reduced virulence). The degree to which the optimum is shifted depends on the proportion of infections that consist of multiple parasite genotypes, which in turn should depend on the prevalence of the parasite. In other words, a higher prevalence of parasites in the host population should result in more multiple infections, which in turn will intensify within-host selection for higher virulence. Thus, variation in prevalence of parasites among host populations and species may potentially explain some of the variation in virulence observed in nature.

What is the empirical evidence for these different scenarios, then? Under the superinfection scenario, one would expect there to be a deficit of multiple infections in host populations. However, the most common finding is that the distribution of parasite genotypes is aggregated with more multiple infections than expected by chance (Lord et al., 1999), a pattern indicating that the superinfection scenario is uncommon in nature. Still, there is growing evidence for the less extreme form of contest competition where co-infecting genotypes affect the growth and infection intensities, and thereby transmission rates, of each other, but without necessarily outcompeting each other (Read and Taylor, 2000). For example, in an infection experiment with *Plasmodium chabaudi* in laboratory mice, the infection intensity of an avirulent parasite clone was severely reduced in the presence of a more virulent competitor (De Roode et al., 2004). In fact, there are apparently no *in vivo* studies that have demonstrated that infection intensities of co-infecting genotypes are completely unaffected by each other (Read and Taylor 2001). Hence, the pure co-infection scenario (*sensu* May and Nowak, 1995) seems to be as uncommon as the pure superinfection scenario in the real world. Instead, more or less moderate forms of contest competition seem to be the most common attribute of multiple infections. The crucial question—how such within-host competition affects virulence evolution—has, however, been difficult to address. The results of serial passage experiments, which typically result in evolution of higher virulence, provide indirect evidence that within-host competition selects for enhanced virulence (Ebert, 1998), but studies directly testing how exploitation relates to competitive ability have been lacking, mainly because it has been difficult to measure infection intensities of different parasite genotypes in multiple infections. The advent of quantitative PCR, which allows one to measure individual infection intensities of parasite genotypes that differ in as little as one or a few base pairs at the target gene, has however made it possible to quantify parasite loads of each genotype.

De Roode et al. (2005) took advantage of this technique and tested for relationships between exploitation, virulence, and competitiveness in *Plasmodium chabaudi* in laboratory mice. They allowed six different but closely related genotypes (derived from the same original isolate, "AS") differing in exploitation and virulence to compete against an unrelated and more virulent reference strain ("AJ"). The relative success of the different AS strains in mixed infections with AJ was positively correlated with their infection intensity and virulence in single infections. In a follow-up experiment, five different unrelated strains were competed against each other in pairs, and in this set-up there also was a correlation between exploitation and competitiveness (Bell et al., 2006). Similar results were obtained in a study of single and mixed infections of one virulent and two avirulent isolates of the bacterium *Pasteuria ramosa* in *Daphnia magna* (Ben-Ami et al., 2008). Hosts infected with one of the two relatively avirulent clones survived longer than hosts infected by the virulent clone, and as a consequence, the avirulent clones produced a higher total number of spores than the virulent clone in single infections. However, in multiple infections with the virulent and one of the avirulent clones, the spore production of the avirulent clones was suppressed almost to zero, whereas the virulent clone was unaffected. Thus, within-host selection favored higher virulence, although in this case the relationship between exploitation and competitiveness is not entirely clear as spore production was measured only as the accumulated spore load at the time of host death (which differed between isolates).

In contrast to the above two studies, an experiment with single and mixed infections of a virulent and an avirulent strain of the trematode parasite *Schistosoma mansoni* in its intermediate snail host *Biomphalaria glabrata* showed that the less virulent strain was more competitive (Gower and Webster, 2005). In addition, in this system virulence (measured as host survival) was negatively correlated with infection intensity. Thus, the results of experiments testing for associations between exploitation, virulence, and competitiveness are somewhat mixed. While the studies of rodent malaria unequivocally support the assumptions and predictions of theoretical models, the experiment with schistosomes in snails contradict them. Clearly, more of these types of experiments are desirable before it is possible to draw any general conclusions about how multiple infections generally affect the optimal level of virulence.

HOST POPULATION STRUCTURE

The basic version of the trade-off hypothesis (Eq. (1)) also assumes (beside that infections consist of a single parasite genotype) that the host population consists of a single, completely mixed deme, so that the transmission probability to all hosts is the same. This is probably rarely the case. Instead, host populations are typically geographically structured, consisting of a number of more or less well-connected subpopulations where transmission is higher within than between subpopulations. How does such spatial structuring affect the optimal levels of exploitation and virulence?

Lipsitch et al. (1995) addressed this problem by modeling a situation where the host population consists of pairs of host individuals, and investigated how the contact rate within pairs affects the optimal level of virulence, assuming that more structured populations have a higher pair-wise contact rate. Their model showed that optimal virulence decreases with increasing structure (increasing contact rate within host pairs). Boots and Sasaki (1999) modeled the effect of transmission distance on optimal virulence more explicitly. In their model, a situation with local transmission (between nearest neighbors) selected for lower exploitation than more random transmission (both local and global). Hence, also this model predicts a negative relationship between degree of structuring and optimal virulence. There are also several other models showing that structured host populations generally favor lower levels of virulence than unstructured ones (reviewed by Messinger and Ostling, 2009). There has been some confusion as to the exact reason why population structure should favor lower virulence, but a recent model showed it can be explained as a result of several processes working in combination (Wild et al., 2009): Because host population structure reduces the dispersal distance of the parasite, there are two types of costs of increased exploitation and transmission in a structured population that are absent in an unstructured population. First, the more structured the population, the more a parasite genotype must compete with itself for uninfected hosts. Second, population structure also increases the average relatedness among parasites within subpopulations, and hence competition with relatives for uninfected hosts. The lower optimal level of virulence in structured versus unstructured populations can therefore be explained by the fact that a parasite that exploits its subpopulation of hosts more prudently achieves higher inclusive fitness.

Boots et al. (2004) extended these models by also incorporating acquired immunity. Under this scenario, both high and low virulence is evolutionarily stable over a wide range of population structures, and the outcome depends on initial levels of virulence. The reason that high virulence can be favored in structured populations is that when hosts can recover and become immune it is an evolutionarily stable strategy to kill hosts (i.e., be more virulent) because this frees space for new susceptible hosts (whereas surviving hosts become immune and thereby block transmission).

These results have important implications for managing diseases of humans and domestic animals. Specifically, the models predict that, under some circumstances, increasing movement of people and animals could select for more virulent parasites. So far, there are, however, very few empirical tests of these theories, although the ones that exist are largely supportive. Thus, an experimental evolution study with *Escherichia coli* and its phage T4 showed that local transmission favored prudent, relatively avirulent phage, whereas random transmission across bacteria populations favored higher exploitation (Kerr et al., 2006). Similarly, an elegant experimental evolution study of the moth *Plodia interpunctella* and its granulosis virus showed that the dispersal ability of the host had evolutionary consequences for the pathogen, although in this case the virus evolved lower infectivity, rather than virulence, under the low dispersal regime (mimicking a structured host population) (Boots and Mealor, 2007). Apart from these, there are, however, no empirical studies of the effect of spatial structure on the optimal level of virulence. Thus, although theoretical models are fairly concordant, more empirical studies are badly needed.

HOST HETEROGENEITY

Until now we have assumed that the host population is completely homogenous, but that is of course rarely the case in nature. Host heterogeneity can occur at different scales. First, as reviewed in the next section of this chapter, there is typically genetic variation within host species for traits that affect parasite growth—that is, resistance. Second, many, if not most, parasites are capable of infecting several related host species, for example different mammals (Woolhouse et al., 2001). How do these kinds of heterogeneity affect virulence evolution?

Regoes et al. (2000) investigated the effect of host heterogeneity on virulence with a model that incorporated a virulence trade-off between two host types, such that a generalist parasite genotype that can infect both host types is relatively virulent in host type 1 and relatively avirulent in host type 2, and vice versa (i.e., a parasite genotype-by-host type interaction with crossing reaction norms). However, the model did not include a link between virulence and transmission, as is commonly assumed in other models of virulence evolution (cf. Eq. (1)). Under these assumptions, the model predicted that generalist parasites should have lower virulence than specialists (that infect only one of the host types). This result can be explained by the fact that in the absence of a virulence–transmission trade-off, there is nothing that prevents a specialist parasite from evolving ever-increasing virulence. Although this scenario is in our view a bit unrealistic, there is some support for the prediction that specialist parasites should evolve higher virulence than generalists: a comparative study of primate malaria parasites showed that the level of exploitation was negatively related to number of host species (Garamszegi, 2006).

Ganusov et al. (2002) addressed this issue with a model incorporating variation in lethal infection intensity (effectively variation in "tolerance"; see further the section on host defense below) among hosts, but no trade-off. In contrast to Regoes et al., Ganusov et al. predicted that virulence should increase with increasing heterogeneity. Yet an alternative scenario is provided in a model developed by Gandon (2004), where host types differ in resistance so that the optimal level of exploitation differs between host types (the more resistant the host, the higher the optimal level of exploitation). Thus, this model resembles the one by Ganusov et al. (2002) in the sense that it does not assume any trade-off between host types. However, unlike Ganusov et al., Gandon's model predicts that the parasite will simply evolve a level of exploitation and virulence that is sub-optimal in the most resistant host type and super-optimal in the least resistant, and where the resulting trait values depend on the relative abundance of the two host types.

Although different model formulations give radically different predictions, one message from these studies is that the nature of the effect of host heterogeneity on virulence is likely to be important. As indicated in the review of the models above, host heterogeneity can affect virulence in two qualitatively different ways. First, there can be a general difference in resistance or tolerance, and thereby infection intensity and/or virulence, among host types (where host type could refer to genotype, species, etc.). Second, there can be parasite genotype-by-host type interactions for infection intensity and/or virulence. There are as yet only a few empirical studies of how host type and parasite genotype interact to determine the expression of infection intensity and virulence. An infection experiment where different strains of laboratory mice were infected with different clones of *Plasmodium chabaudi* in a fully factorial design showed host genotype-by-parasite genotype interactions for infection intensity and virulence, but the different parasite genotypes ranked in largely the same way in different mouse strains. Hence, the interactions were due to differences in variance rather than crossing of reaction norms (Grech et al., 2006). However, a recent study of monarch butterflies (*Danaus plexippus*) and the protozoan parasite *Ophryocystis elektroscirrha* showed that there can also be host genotype-by-parasite genotype interactions in the form of crossing reaction norms (De Roode and Altizer, 2010). We are not aware of any studies that have quantified parasite genotype-by-host species interactions for virulence, but the results of serial passage experiments, where adaptation to and increased virulence in a new host typically result in lower virulence in the original host, suggest that such interactions can be important (Ebert, 1998). We conclude that, based on theoretical models, it seems that host heterogeneity can have important effects on the evolution of virulence, but it is difficult to say in what way until more empirical studies of the nature of parasite genotype-by-host type interactions are available.

To summarize the section on exploitation, we conclude that factors like parasite prevalence, host population structure, and host heterogeneity can all affect the optimal level of exploitation and thus result in population differentiation or interspecific variation in virulence. In addition to these, several other potentially important factors have been identified. For instance, it has been argued that increased host background mortality selects for higher exploitation (Lenski and May, 1994; but see Day and Gandon, 2007), that horizontal transmission selects for higher exploitation than vertical transmission (Bull et al., 1991, but see Day and Gandon, 2007), that vector-borne transmission selects for higher exploitation than direct transmission (Ewald, 1994, but see Day, 2001), that expanding host populations select for higher exploitation (Lenski and May, 1994), and that improved host resistance, for example through vaccination, selects for higher exploitation (Gandon et al., 2001; Mackinnon et al., 2008).

PATH 2: PER PARASITE PATHOGENICITY

As reviewed above, studies of virulence evolution from the parasite's perspective have traditionally assumed that the only parasite source of variation for virulence is exploitation, so that in a homogenous host population, virulence is determined by parasite-induced variation in infection intensity (n_p) times some constant (c) (see Eq. (2)). However, in principle it is possible that there could be parasite genetic variation not only for n but also for c. In other words, Eq. (2) could be written $v = c_p n_p$, where the subscript p again indicates that the parameter is determined by the parasite. We refer to parasite-induced variation in c as variation in "per parasite pathogenicity" (PPP). We use this term instead of "per parasite virulence" to imply that it is determined by the parasite, rather than jointly by host and parasite. Although this kind of variation has to our knowledge never been explicitly looked for, there are studies indicating that it might be important. As shown in Figure 18.2, there is evidence for such variation in rodent malaria. There is also evidence for genetic variation in PPP from several natural host–parasite systems. For example, in a study of *Daphnia* and its bacterial parasite *Pasteuria ramosa*, Little et al. (2008) infected *Daphnia* from a single clone with one of two different isolates of the bacterium differing in virulence (measured as the rate at which the *Daphnia* was killed). Following the assumptions of the trade-off hypothesis, one would expect the more virulent (early killing) isolate to have higher infection intensity, but, surprisingly, Little et al. found the directly opposite pattern, with the more virulent parasite isolate having lower infection intensity. This pattern is best explained by the fact that the two different isolates differed not only in the infection intensity they reached ("exploitation"), but also in the harm they induced per parasite (PPP).

Another study indicating the existence of variation in PPP is the study of evolution of HIV virulence in Italy mentioned earlier (Müller et al., 2009). In this study, infection intensity was measured as the viral load during the chronic phase of the infection (i.e., after the resolution of the primary infection, which is characterized by higher viral load), while virulence was measured as the decline in concentration of CD4+ cells during the infection, with steeper slope taken to mean higher virulence. Müller et al. found that both virulence and infection intensity increased during the study period (1984–2006). Under the trade-off hypothesis, one would expect the increase in virulence to be fully explained by increasing viral load. However, the increase in viral load over time explained only 22% of the increase in virulence, and the factor "time" (year) was significant while controlling for infection intensity. This is consistent with the fact that not only exploitation but also PPP evolved, although in this case it is difficult to rule out alternative explanations, for example environmental changes in host tolerance (i.e., path 4; see below) over time.

Perhaps the most compelling evidence that variation in PPP can be important comes from a recent study of monarch butterflies and their protozoan parasite *Ophrocystis elektroscirrha*. In an experiment where 18 parasite isolates were used to infect 10 monarch families in a fully factorial design, parasite isolates differed in the infection intensity (spore load) they reached, and spore load was strongly negatively correlated with host life span. However, when controlling for spore load, there was still a significant effect of parasite isolate on longevity, indicating that isolates differed not only in the infection intensity they induced, but also in PPP (De Roode and Altizer, 2010). There are also studies of plant–parasite systems indicating the presence of genetic variation for PPP (Stewart et al., 2005; Pagàn et al., 2007). Thus, there is good evidence that parasite genetic variation in virulence can be a result of not only exploitation but also PPP, although at the moment we can only speculate about the relative importance of these two sources. We note, however, that it would be relatively easy to disentangle them in many systems; indeed, the necessary data should already be available in several cases.

How, then, should natural selection act on variation in PPP? Clearly, all else equal, one would expect natural selection to favor reduced PPP because this would prolong the infection period without reducing transmission rate (i.e., the trade-off in Eq. (1) would be set aside). The fact that parasites are generally not completely avirulent, even though there is at least sometimes standing genetic variation for PPP, suggests that the evolution of PPP is usually constrained in some way. In other words, it seems likely that there should be trade-offs between PPP and other components of parasite fitness. Identifying the nature of such trade-offs seems to us an important next step in the study of virulence evolution. Similar arguments have been put forward by Ganusov and Antia (2003), who advocated a focus on detailed understanding of the quantitative relationship between intensity, transmission, and pathogenicity. Frank and Schmid-Hempel (2007) argued that trade-offs (like Eq. (1)) are difficult to measure directly and that a thorough understanding of virulence evolution requires the identification of the timing and nature of components of parasite life history that are affected by pathogenicity. Such understanding is best achieved through detailed studies of the mechanisms of pathogenesis, further highlighting the importance of a per-parasite–centered view of virulence. Under any circumstance, we suggest that variation in PPP could contribute to intra- and interspecific variation in virulence, although it is difficult to say how until we know more about how selection acts on this source of variation.

HOST EVOLUTION

Studies of evolution of host defense against parasites have generally not been framed in the context of virulence evolution (but see Ebert and Hamilton, 1996; Day and Burns, 2003). Yet it is clear that the ability of the host to defend itself against disease should have significant effects on the virulence of an infection. Thus, a complete picture of virulence evolution can be obtained only by considering not only parasite but also host evolution. For this reason we have chosen to devote a large part of this chapter to host defense. To help define the different host traits we discuss in this section, and relate them to the parasite traits discussed in the previous section, we will use a function known as the "basic depression ratio of the host" (D_0), developed by Bowers (2001). We chose this function because it is directly analogous to R_0 and involves largely the same parameters, something that makes comparison easy.

The basic depression ratio is defined as "the number of host individuals per infected by which the infected population is depressed below its uninfected level" (Bowers, 2001). Thus, natural selection should act to minimize D_0. The basic depression ratio can be written in the following way to reveal its relationship with R_0 (Bowers, 2001):

$$D_0 = \frac{\beta S(v+b)}{(a-b)(b+v+\rho)} = \frac{R_0(v+b)}{(a-b)} \tag{3}$$

where a is the intrinsic birth rate while β, S, b, v, ρ, and R_0 are defined as in Eq. (1). Note that R_0 and D_0 contain exactly the same parameters, with the exception that D_0 also includes a. We will not provide any detailed analysis of D_0 here. Rather, besides using this function to define the different host traits we discuss, the point we want to convey by including Eq. (3) is that the performance of a host genotype can be described with the same parameters as the performance of a parasite genotype, and that both the host and the parasite can contribute to the variation in these parameters (with the exception of a and b, which are, naturally, determined only by the host). Thus, in nature, the traits β, v, and ρ are all shared between host and parasite—that is, they are traits of the infection rather than the parasite or host alone. In the present section we will, however, focus solely on host evolution and consider the parasite as a constant. For the moment, we therefore assume that a number of different host genotypes are infected by a genetically homogenous parasite population so that all genetic variation in β, v, and ρ (and, of course, a and b) stem from the host, in contrast to the situation in the section on parasite evolution above, where we assumed that the parasite was the source of variation in β, v, and ρ. Thus, D_0 describes how the extent to which a parasite reduces the population size of the host depends on variation in different host traits. Since the topic of this chapter is virulence, we will in the remainder of this section discuss how the host contributes to variation in v, and how natural selection acts on such variation.

PATH 3: RESISTANCE

Studies of the effect of host defense on virulence have focused on the ability of the host to control the growth and infection intensity of the parasite. The effect of such host defense on the virulence of an infection can be described in a way analogous to the equation we used to describe the parasite's effect on virulence (Eq. (2)), namely:

$$v = cn_h \tag{4}$$

where c is a constant, n is the infection intensity, and the subscript h indicates that the trait is determined by the host. We refer to variation in n_h as variation in resistance. Thus, resistance influences the virulence of an infection in the same way as exploitation, namely through its effects on infection intensity.

To avoid confusion, we want to point out that the term "resistance" has been used in several different ways in the animal host–parasite literature. Besides the definition given above (i.e., the ability of the host to limit the growth of a parasite once infection has occurred), "resistance" has also been used to mean the ability to prevent infection in the first place (often referred to as "avoidance resistance" [Boots et al., 2009], implying that it is a behavioral trait, but it could as well be a physiological or morphological trait). Thus, while the former type of resistance influences virulence, the latter affects infectivity (β in Eq. (3)). It is often useful to distinguish between these two traits, because their evolution is likely governed by different selection pressures (cf. Dybdahl and Storfer, 2003). Here we are interested in how host defenses influence virulence and will therefore use "resistance" to mean the ability of the host to limit the growth of the parasite, unless otherwise stated.

Resistance is typically measured by challenging animals with a given dose of a parasite and then measuring the infection intensity after a specified period of time. If the average infection intensity differs between host genotypes, there is genetic variation for resistance. There is accumulating evidence that animals commonly display extensive genetic variation for resistance.

For example, different strains of inbred laboratory mice typically differ in resistance to all sorts of parasites (e.g., Stevenson et al., 1982; Ma et al., 1998). Likewise, domestic animals often show high heritability for resistance (Stear et al., 2001; Nejsum et al., 2009). When it comes to more natural populations, a number of candidate gene and QTL mapping studies of humans and other mammals have demonstrated effects of particular loci on resistance (e.g., Paterson et al., 1998; Timmann et al., 2007; Kaslow et al., 2008; Kloch et al., 2010). Nevertheless, quantitative estimates of the overall magnitude of genetic variation for resistance in natural animal populations are still surprisingly few, although such variation has been demonstrated in both vertebrates and invertebrates (Smith et al., 1999; Lazzaro et al., 2004). Taken together, the available studies leave little room for doubt that genetic variation for resistance is pervasive.

How, then, should we expect natural selection to act on genetic variation for resistance? A common idea in studies of the evolutionary ecology of resistance is that resistance confers not only fitness benefits in the form of reduced virulence but also some kind of costs, which constrains the magnitude of resistance so that there should be an intermediate optimum (Simms and Rausher, 1987; Behnke et al., 1992; Schmid-Hempel, 2003). While this scenario is plausible, more detailed theoretical modeling shows that it is not the only possible outcome. A number of theoretical models of evolution of resistance have been developed, but the majority of these have focused on resistance in the form of preventing infection in the first place. In terms of Eq. (3), these studies have focused on variation in β. Here we are more interested in how the host controls n in Eq. (4) and thereby v in Eq. (3), because this is the kind of resistance that affects virulence (see Fig. 18.1). What does theory tell us about the nature of selection on such variation?

In our view, the most relevant theoretical models in this context are the ones by Miller et al. (2005), Fornoni et al. (2004), and Day and Burns (2003). We briefly review the assumptions and predictions of these three models here. The models by Miller et al. (2005) and Fornoni et al. (2004) considered two types of host defense: resistance and tolerance (note that Miller et al. use "resistance" as the umbrella term and "control" to specify the type of defense that limits infection intensity). For the moment we will focus on the predictions regarding resistance, but we will come back to tolerance (path 4) in the next section. All three models assume that resistance comes at a cost. The costs of resistance to a particular parasite can be divided into two types based on whether they are a result of maintaining resistance mechanisms in the absence of infection, or inducing a response to fight a parasite once infection has occurred. Both Miller et al. and Fornoni et al. assumed a trade-off between resistance and intrinsic growth rate (in the absence of infection)—that is, a maintenance cost. Importantly, they considered different shapes of this cost function, including decelerating, linear, and accelerating costs (i.e., decreasing, constant, and increasing marginal costs of resistance, respectively). Miller et al. used the approach of "adaptive dynamics" to analyze their model and found that a decelerating cost function results in selection for both maximal and minimal resistance, which in turn results in a dimorphic host population. In contrast, accelerating costs leads to an intermediate optimum for resistance. Similarly, Fornoni et al., using a more traditional optimality analysis, found that linear and decelerating cost functions lead to selection for maximal or minimal levels of resistance, whereas accelerating costs results in an intermediate optimum. (Actually, Fornoni et al. modeled the relative shapes of the cost and benefit functions, but for simplicity and easier comparison with other models we here assume that the benefit function is linear and focus on the effects of varying the shape of the cost function.) In contrast to the other two models, Day and Burns (2003) included a response cost rather than a maintenance cost, such that the strength of the induced response is negatively correlated with fecundity. Also in this case, the evolutionarily stable strategy was dependent on the shape of the cost function, with nonlinear costs resulting in an intermediate optimum.

Thus, it is clear that the nature of selection on resistance predicted by these theoretical models is heavily dependent on model details, in particular the shape of the cost function. An intermediate optimal level of resistance, as is commonly assumed in the less mathematically oriented literature, is largely restricted to scenarios with accelerating cost functions. Thus, these theoretical excursions show that the presence of costs of resistance does not necessarily lead to the commonly expected intermediate optimum. We therefore conclude that even though accelerating costs and an intermediate optimum is perhaps the most likely scenario (see Boots and Haraguchi, 1999) it should not be taken for granted.

What do empirical studies tell us about costs of resistance? Costs have been measured in many different currencies, including immunopathology (Graham et al., 2005), energy and other resources (Demas et al., 1997; Svensson et al. 1998), and Darwinian fitness or components thereof (e.g., fecundity or survival). From an evolutionary perspective, it does not matter whether the proximate cost is energy, immunopathology or something else; the crucial question is whether resistance ultimately has negative effects on fitness under some circumstances. Here, we will therefore focus on studies that have measured costs of resistance in some component of fitness.

MAINTENANCE COSTS

Costs of maintaining resistance can be estimated by measuring the fitness of more or less resistant host genotypes in the presence and absence of the parasite under consideration. If resistant genotypes have higher fitness than non-resistant genotypes in the presence of parasites, but lower fitness in their absence, this demonstrates that resistance is costly. In terms of Eqs. (3) and (4), this involves testing for a trade-off between the intrinsic rate of growth r (i.e., a-b in Eq. (3)) in the absence of parasites and n_h. This is the kind of cost that is most commonly assumed in theoretical models.

Evidence for maintenance costs comes from two recent experiments with fruit flies. Vijendravarma et al. (2008) selected *Drosophila melanogaster* for resistance to the microsporidian parasite *Tubulinosema kingi* for 61 weeks. At the end of the experiment, selected lines had on average 27% lower spore loads than unselected controls; hence, selected lines had become more resistant and were better able to limit infection intensities. Moreover, in the presence of the parasite, selected lines had 30% longer survival and 15% higher fecundity than unselected, showing that the enhanced resistance actually translated into reduced virulence. What was the cost of this resistance then? In the absence of the parasite, there was no difference in survival, but selected lines had 9% lower fecundity than controls. Thus, this study nicely demonstrated a cost of maintaining resistance to this parasite.

In another *Drosophila* study, McKean et al. (2008) tested for correlations between fecundity and resistance to a naturally occurring bacterial pathogen (*Providenca rettgeri*) under limited and unlimited food conditions. Using a panel of 40 "hemiclone" families isolated from a single population of *Drosophila melanogaster*, they found a heritability of about 0.1 for resistance (measured as bacterial load 28 hours after infection). There was also a strong negative genetic correlation ($r = -0.4$) between fecundity (in the absence of the pathogen) and resistance, but only under limited food conditions. Thus, this study shows that the ability to limit infection intensities comes at a cost, but that the expression of this trade-off is environment dependent.

We conclude that there is currently good evidence that resistance can be costly to maintain, at least in invertebrates, although it should be noted that there are also well-designed studies that have failed to demonstrate maintenance costs of resistance, so such costs are apparently not omnipresent (Little et al., 2002). What is critically lacking at the moment is information about the shape of cost functions; we are not aware of such data in any animal host–parasite system. The situation is somewhat better in the plant literature, where accelerating cost functions have indeed been demonstrated in several cases (Bergelson et al., 2001). Since the predictions of theoretical models are heavily dependent on the shape of cost functions, it is currently a bit unclear whether the observed costs actually yield an intermediate optimum for resistance. Thus, determining the shape of costs functions is clearly an area for future research in animal systems. It should in principle be relatively straightforward to address this issue using a setup like the one employed by McKean et al. (2008), although it may be logistically demanding to obtain the sample size required to test for curvature of the fitness–resistance relationship.

RESPONSE COSTS

Costs of inducing a response are more difficult to measure, because the reduction in fitness of infected hosts is a result of both direct effects of parasite replication and indirect effects due to induced host responses. Response costs have therefore instead been identified by inducing responses to non-pathogenic antigens, for example killed bacteria, vaccines, or other immunostimulatory compounds. This approach assumes that the response to non-pathogenic antigens correlates with resistance to real parasites, an assumption that is supported by infection experiments with laboratory mice (Biozzi et al., 1984; Luster et al., 1993). Moret and Schmid-Hempel (2000) used this approach to test for survival costs of mounting an immune response in bumblebees. The bumblebees were kept under starvation and immunized with lipopolysaccharide (LPS) and/or micro-latex beads, or injected with physiological saline (PBS) as a control. Immunization reduced the survival by as much as 50% to 70%, showing that response costs can be significant under stressful conditions. When it comes to vertebrates there is some evidence of response costs in the form of reduced fecundity or traits related to fecundity. For example, Ilmonen et al. (2000) and Råberg et al. (2000) immunized female pied flycatchers and blue tits, respectively, with diphtheria–tetanus vaccine during the nestling feeding period and compared the level of parental effort (nestling feeding rate) and reproductive success with that of controls (injected with PBS). Vaccination reduced parental effort in the blue tits and reproductive success in the flycatchers. Similarly, a study of lactating cows found that immunization with LPS reduced milk production (Waldron et al., 2003). In contrast, Williams et al. (1999) found no effect of immunization with sheep red blood cells on reproductive success in starlings. Thus, even if the results are somewhat mixed, the majority of these studies indicate that mounting an immune response is costly. Under the reasonable assumption that mounting an appropriate immune response to an infecting parasite improves resistance, these studies indicate the presence of response costs of resistance. This type of cost is usually not included in theoretical models (but see Day and Burns, 2003), but could potentially constrain resistance in the same way as maintenance costs.

Nevertheless, some objections against the evidence for the importance of response costs can be raised. First, it has been argued that the reduction in fitness observed in some of the studies above could be due not to the induced immune response, but rather to some unwanted side effect of immunization—for example, if the adjuvant used in vaccines is toxic (Staszewski and Boulinier, 2004). Second, even if the above studies are taken as evidence of response costs, it is hard to know if the magnitude is representative for costs of responses to real pathogens, because not much is known about the strength of responses to non-pathogenic antigens as compared to responses to real parasites. Third, as in the case of maintenance costs (see above), little is as yet known about the shape of response cost functions, making it difficult to predict if and how response costs affect the optimal level of resistance. Indeed, in the case of response costs there is, as far as we are aware, not even evidence of a positive correlation between cost and strength of response. Actually, the currently available data suggest a lack of such a correlation. For example, in the bumblebee study (Moret and Schmid-Hempel, 2000), the magnitude of the fitness cost did not vary with the dose of LPS, although a higher dose presumably induces a stronger response (see also Nilsson et al., 2007). A positive correlation between response and cost is a very basic first assumption behind the idea of optimal resistance (cf., Day and Burns, 2003). Thus, in our view, there is clearly some way to go to demonstrate that response costs actually constrain the magnitude of resistance and produce an intermediate optimum.

SELECTION ON RESISTANCE

An alternative approach to investigate whether there is an intermediate optimum for resistance is to directly estimate the pattern of natural selection acting on resistance. If there is an intermediate optimum, one would expect stabilizing selection on resistance in the presence of parasites. In an effort to elucidate the shape of natural selection on resistance, Stjernman et al. (2008) studied a nest box breeding population of blue tits (*Cyanistes caeruleus*) and investigated the relationship between infection intensity with the avian malaria parasite *Haemoproteus majoris* and the probability of survival from one breeding season to the next. Like other studies of natural animal host–parasite systems (e.g., Smith et al., 1999; Coltman et al., 2001), they assumed that infection intensity reflected the resistance of the host. Based on data collected during three years and a total sample size of 334 infected birds, there was a parabolic relationship between infection intensity and host survival. While this result indicates that there is indeed an intermediate optimal level of resistance to this parasite, it is not informative about whether resistance is constrained by maintenance or response costs. Admittedly, this is probably a rather weak approach to test for an intermediate optimum because it is easy to imagine factors that could obscure the expected parabolic relationship between resistance and fitness. For example, if variation in infection intensity is a result of variation in the parasite rather than the host (i.e., n_p instead of n_h), one would—following the assumptions of the trade-off hypothesis (Eq. (1))—expect a negative correlation between infection intensity and host fitness. Similarly, if resistance is highly condition-dependent, one would also expect a negative correlation between infection intensity and host fitness (Råberg and Stjernman, 2003). Both these problems could potentially be circumvented by measuring selection on "breeding values" rather phenotypic values (Rausher, 1992), although this would clearly be a formidable task in most natural animal host–parasite systems. However, given that a parabolic fitness function is actually found based on just phenotypic values, it is difficult to come up with sensible alternative explanations other than that it should be a result of some kind of costs constraining the magnitude of resistance.

WHY VARIATION IN RESISTANCE?

Assuming that there is, after all, in general an intermediate optimum for resistance, the obvious follow-up questions are: what determines the position of this optimum, and what can make the position of the optimum vary among populations of a host species, and between host species? In other words, what factors can lead to population differentiation and ultimately interspecific variation for resistance? Obviously, the position of the optimal level of resistance will ultimately be determined by the relative magnitudes of the costs and benefits of resistance, but what determines the magnitude of costs and benefits then? One factor that could be expected to have an impact on the benefit function is the abundance of the parasite—that is, the "force of infection" (Day and Burns, 2003). A reasonable assumption is that a higher abundance of a parasite should increase the benefits of resistance against that parasite such that there should be a positive correlation between abundance and optimal level of resistance. This assumption underlies studies in ecological immunology that have looked for inter- and interspecific variation in immune responsiveness to non-pathogenic antigens, and there is some evidence for this scenario. For example, a study of four different populations of small ground finches (*Geospiza fuliginosa*) in the Galapagos archipelago showed a positive correlation between parasite abundance and investment in immune defense across populations (Lindström et al., 2004). However, as usual, more detailed modeling has shown that there are also other possible scenarios. In a model where host resistance affects recovery rate (and thereby also virulence in the sense of case mortality), the optimal level of resistance was highest at intermediate parasite abundance (van Baalen, 1998). The rationale behind this result is that when the abundance of the parasite is sufficiently high, there is no reason to clear infections because the host will immediately be re-infected; instead, it is better to avoid the costs of resistance and reproduce as fast as possible.

Another factor that can affect the benefits of resistance is host life span. It is sometimes argued that species with a high background mortality rate (b in Eq. (3)) should invest little in resistance because they die quickly anyway, regardless of whether they are able to defend themselves against parasites (e.g., Medzhitov and Janeway, 1997). However, once again, modeling complicates the picture. For example, a model with acquired immunity and maintenance costs showed that the evolutionarily stable strategy for short-lived species was to invest as little as possible in resistance, whereas long-lived species should invest either a little or a lot depending on initial conditions (van Boven and Weissing, 2004). Thus, although current models do not give a general prediction, it is clear that host life expectancy can affect the optimal degree of resistance.

It is also conceivable that the magnitude of costs can vary with the environment. In general, trade-offs are often more pronounced under adverse environmental conditions, and there is some evidence that this also applies to various forms of resistance. For example, lines of *Drosophila melanogaster* selected for enhanced resistance to an endoparasitoid (enhanced ability to encapsulate and thereby kill parasitoid eggs) had reduced competitive ability, but only under limited food conditions (Kraaijeveld and Godfray, 1997). Thus, according to this reasoning, poor environments should select for lower levels of resistance. To the best of our knowledge, there are, however, as yet no direct empirical tests of this idea.

PATH 4: TOLERANCE

Just like parasite-centered studies of virulence evolution have focused on variation for "exploitation" (i.e., path 1 in Fig. 18.1) and ignored "PPP" (path 2), so have host-centered studies focused on "resistance" (path 3) and largely ignored other potential host sources of variation in virulence. As outlined in Figure 18.1, hosts can in principle reduce the virulence of an infection not only by reducing the infection intensity, but also by reducing the fitness effects per parasite. In terms of Eq. (4), this means reducing c rather than n. Thus, Eq. (4) can be written $v = c_h n_h$, with the subscript h indicating that the component is determined by the host. We refer to c_h as "tolerance." Thus tolerance is analogous to PPP, but a result of variation in host rather that parasite genes.

The distinction between resistance and tolerance has a long history in plant pathology, where numerous studies have demonstrated genetic variation for both traits (Caldwell et al., 1958; Schafer, 1971; Clarke, 1986; Simms and Triplett, 1994), but this distinction has only recently been applied in animal host–parasite systems. In an infection experiment with *Plasmodium chabaudi* in laboratory mice, Råberg et al. (2007) found that not only infection intensities but also the virulence (measured as anemia and weight loss) at a given infection intensity differed between mouse strains, indicating variation in both resistance and tolerance among strains. Perhaps the best evidence for tolerance in a natural host–parasite system comes from a recent study of monarch butterflies and a protozoan parasite. In the experiment described above (path 2) where a number of butterfly families were infected with a number of parasite isolates in a fully factorial design, the virulence of a given infection intensity (spore load) differed strongly between families, demonstrating genetic variation for tolerance to this parasite (De Roode and Altizer, 2010). There are also results from a number of studies of other host–parasite systems that are best interpreted as tolerance, although more formal analyses are still lacking in these cases (reviewed in Råberg et al., 2009). Thus, on current evidence, we suggest that animals often exhibit genetic variation not only for resistance but also for tolerance.

How, then, should natural selection act on this variation? All else equal, selection should of course favor enhanced tolerance. However, tolerance, like resistance, is likely to confer not only benefits but also costs. Two types of costs have been proposed. First, higher tolerance could reduce fitness in the absence of parasites. This is often referred to as an allocation cost, and is directly analogous to a maintenance cost of resistance. Second, there could be a trade-off between resistance and tolerance, so that tolerance is a cost of resistance and vice versa. Both these types of costs could potentially constrain evolution of tolerance. Moreover, the strength of selection on tolerance should also be influenced by the level of resistance, and vice versa; a fully resistant host has no need for tolerance, and a fully tolerant host has no need for resistance. Thus, one should expect correlational selection on resistance and tolerance—that is, selection for particular combinations of trait values (Sinervo and Svensson, 2002). Fornoni et al. (2004) modeled the joint evolution of resistance and tolerance and found that the shape of the fitness surface depends on the shape of

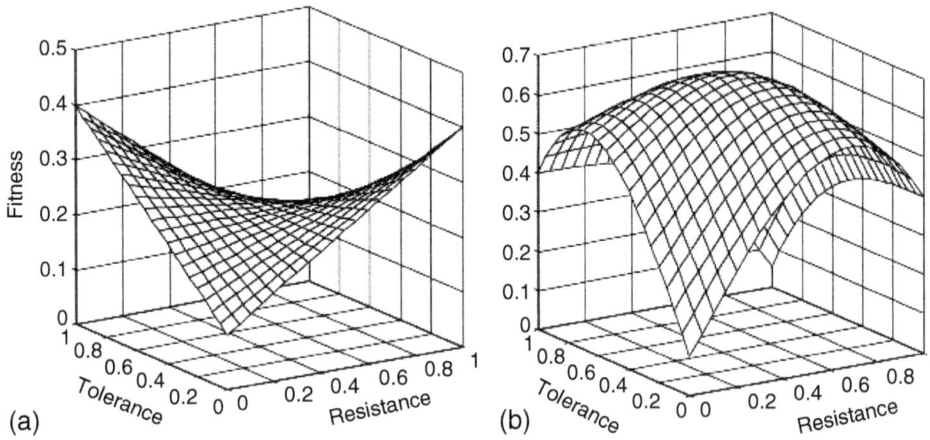

FIGURE 18.5: Fitness surfaces for resistance and tolerance. a) Linear cost functions for resistance and tolerance yield adaptive peak at maximum resistance and minimum tolerance, and minimum resistance and maximal tolerance. b) Accelerating cost functions yield adaptive peak at intermediate resistance and tolerance. Fitness surfaces are drawn using Eq. (3) in Fornoni et al. (2004).

the cost functions of tolerance and resistance. Specifically, if cost functions are linear, natural selection favors either maximum resistance and minimum tolerance, or minimum resistance and maximum tolerance (Fig. 18.5a). In contrast, if cost functions are accelerating, selection favors intermediate levels of both resistance and tolerance (Fig. 18.5b). Thus, as for resistance (see path 3 above), investigating not only the presence and magnitude of costs but also the shape of cost functions is key to understand the evolution of tolerance.

In the plant literature, both allocation costs and trade-offs between tolerance and resistance have been demonstrated, but when it comes to animals, our understanding of tolerance costs is currently very limited. In the rodent malaria study, there was a negative correlation between tolerance and resistance across mouse strains, indicating that tolerance can come at a cost in terms of reduced resistance (Råberg et al., 2007). In contrast, the study of monarchs found evidence for neither an allocation cost nor a tolerance–resistance trade-off (De Roode and Altizer, 2010).

We conclude that although a number of studies indicate the presence of genetic variation for tolerance, more studies of natural host–parasite systems are definitely needed before it is possible to draw conclusions about the relative importance of tolerance and resistance for the expression of virulence. Moreover, the understanding of how natural selection acts on tolerance is hampered by a lack of studies of the nature of costs of tolerance.

CO-EVOLUTION

Until now we have treated host and parasite evolution separately. However, hosts and parasites obviously do not evolve in isolation; rather, evolution of the host is likely to impose selection on the parasite, and vice versa. In other words, hosts and parasites can be expected to co-evolve.

The emerging message from the previous sections of this chapter is that both theory and empirical studies suggest that there should often be an intermediate optimal level of virulence for both parasites and hosts (although, as we have pointed out, the evidence for this is not yet entirely convincing). However, the position of this optimum is presumably not the same for the two parties. Because parasites usually have larger population sizes and shorter generation times than their hosts, parasites are often assumed to have a higher rate of adaptation (although the clonal genetic structure of many parasites may constrain their evolutionary potential; Hellgren et al., 2010). Consequently, it seems likely that the level of virulence is closer to the optimum of the parasite than the host in most systems. Evidence that parasites are generally ahead in the co-evolutionary race is provided by local adaptation studies. For example, a study of *Daphina magna* and its protozoan parasite *Pleistophora intestinalis* showed that sympatric host–parasite combinations were more virulent than allopatric ones (Ebert, 1994). Moreover, two recent studies have shown that the average virulence in human-HIV and butterfly-protozoan infections is close to the parasite's optimum (Fraser et al., 2007; De Roode et al., 2008). Thus, one would generally expect strong selection on hosts for reduced virulence. As outlined above, such selection can act on two different traits: resistance and tolerance. Interestingly, it has been argued that the co-evolutionary consequences of these two types of defense should differ (Rausher, 2001; Kover, 2006). The reasons for this are as follows.

If hosts evolve resistance, this will reduce the infection intensity below the optimum for the parasite. This should lead to selection for counter-adaptations in the parasite that enhance infection intensity—that is, selection for increased exploitation, something that in turn should select for further increase in resistance, and so on. This kind of antagonistic dynamics is often referred to as "arms race co-evolution" (Woolhouse et al., 2002). It seems unlikely that resistance and exploitation can escalate indefinitely; instead, some form of cyclical dynamics with escalation followed by reduction of trait values and then escalation again is perhaps more plausible (Bergelson et al., 2001). In any case, the antagonistic nature of evolution of resistance and exploitation is likely to result in open-ended non-equilibrium evolutionary dynamics; a process that would maintain genetic variation in both host and parasite.

If hosts evolve tolerance instead, this will not have any direct negative effects on the parasite. It has therefore been argued that tolerance should not select for counter-adaptations in the parasite, and consequently not result in antagonistic co-evolution (Rausher, 2001; Kover, 2006). Evolution of tolerance may, however, still prompt parasite evolution in cases where infection intensity is correlated with transmission rate, as assumed by the trade-off hypothesis (Eq. (1)). Since tolerance reduces the mortality cost for the parasite of a given infection intensity, the optimal level of exploitation should increase with increasing host tolerance (cf., Miller et al., 2006). The dynamics of such co-evolution remains to be modeled, but we suspect it will result in a new stable equilibrium instead of the non-equilibrium dynamics typical of antagonistic co-evolution. Thus, according to this reasoning, resistance leads to antagonistic co-evolution that maintains genetic variation, while tolerance should not.

It is currently unclear what determines whether hosts evolve resistance or tolerance to reduce virulence towards their own optimum. Which evolutionary trajectory is taken should primarily be influenced by the relative amounts of genetic variation for the two types of defense, but the relative costs and benefits of the traits are also likely to be important. In addition, one could imagine a number of other factors that could have an effect, including host life span, the level of parasite exploitation, force of infection, and so forth. This issue is clearly ripe for some modeling.

How do host–parasite interactions unfold in nature, then? Are hosts and parasites really co-evolving in response to each other, and, if so, which of the components outlined in Figure 18.1 are actually involved? Perhaps the detailed longitudinal study of myxomatosis in rabbits provides some clues. Fenner et al. followed the evolution of both virus and rabbit carefully (Fenner and Fantini, 1999). To track the evolution of the virus, they collected isolates from the field and measured their virulence in standard laboratory rabbit strains. Similarly, they followed the evolution of the host by assessing the virulence of reference strains of the virus in rabbits collected in the wild. Interestingly, the virulence of reference virus strains in wild rabbits decreased continuously from the introduction of the virus in 1950 up to the end of the 1970s, indicating that the host evolved improved defense. The virulence of field isolates of the virus in laboratory rabbit strains decreased during the first decade or so, but increased slightly during the next two decades, presumably as a response to the improved defense of the hosts (Fenner and Fantini, 1999; Woolhouse et al., 2002). Thus, this unique study nicely demonstrates that both host and parasite evolved. However, quantitative analyses of temporal changes in infection intensities have, as far as we are aware, not been published, so it is not entirely clear which of the host and parasite components of virulence described in Figure 18.1 actually evolved. In particular, the reduced virulence of reference virus strains in wild collected rabbits could be a result of either enhanced resistance or tolerance, or both.

Another approach to study host–parasite co-evolution is to look for signatures of selection in host and parasite genes likely to be involved in expression of virulence, for example by comparing DNA sequences from closely related species. Such studies are based on the rationale that in a neutrally evolving gene, the rate of non-synonymous and synonymous substitutions (often referred to as dN and dS; i.e., the rate of substitutions that do or do not change the amino acid sequence, respectively) should be the same, whereas in a gene that is subject to positive selection (i.e., selection for change), dN should be higher than dS (Nielsen, 2005). Genes involved in antagonistic co-evolution can be expected to be subject to positive selection, and studies of both host immune genes and various parasite genes (e.g., genes encoding surface proteins acting as antigens) have indeed found elevated dN/dS ratios in such genes (Aguileta et al., 2009; Obbard et al., 2009). Because resistance, but not tolerance, is expected to result in antagonistic co-evolution and the repeated selective sweeps that are detected by dN/dS analyses, the evidence for positive selection on host immune genes and parasite surface antigens indicates that continuous evolution of resistance and exploitation is an important process.

We conclude that while the studies that have demonstrated host–parasite co-evolution of virulence in ecological time are very few (which is not very surprising, given the logistical difficulties of such studies), there is strong evidence from molecular genetic data for co-evolution. Moreover, the latter type of studies indicate that antagonistic co-evolution involving resistance is an important process, whereas the extent to which tolerance is also involved in host–parasite co-evolution remains to be elucidated.

CONCLUSION

We have proposed that virulence is best understood as a trait determined by both parasite and host, and that both partners can influence virulence in two different ways: by affecting the infection intensity, and by affecting the harm of a given intensity. Moreover, in line with numerous other authors, we have taken a Panglossian approach (*sensu* Gould and Lewontin, 1979) and argued that variation in virulence within and among host–parasite systems is best understood as a result of natural selection on the different parasite and host traits outlined in Figure 18.1.

Many theoretical models have been developed to address how natural selection should act on some of the components in Figure 18.1, in particular paths 1 and 3, but also to some extent path 4. In some cases, different models addressing a particular problem yield largely concordant predictions. For example, the result that within-host competition selects for increased exploitation and virulence is relatively insensitive to model details, with some well-founded exceptions (Chao et al., 2000; Brown et al., 2002). In other cases, models with different assumptions yield contrasting predictions, for example as regards the effect of host heterogeneity on optimal virulence (see above). Unfortunately, empirical tests of model assumptions and predictions are currently lagging severely behind model development in most areas. The most extreme case is probably the issue of within-host competition and virulence evolution, which has been addressed by more than a dozen different models, while there are so far only a few empirical tests of the relationship between virulence and competitiveness. Moreover, to date, empirical studies have generally tested models only in a rough qualitative way, but ignored more quantitative but nevertheless important details. For example, while there is accumulating evidence that resistance can entail fitness costs, there is as yet no information about the shape of this cost function in animals, although this has key effects on model predictions. Similarly, quite a few studies have shown a positive correlation between infection intensity and transmission rate—as assumed by the trade-off hypothesis—but very few have tested whether this relationship is saturating, something that is critical for the prediction of an intermediate optimum. Clearly there is a need for well-designed, theory-informed empirical studies on nearly all aspects of virulence evolution.

ACKNOWLEDGEMENTS

We thank Barbara Tschirren for comments on the manuscript.

REFERENCES

Aguileta, G, Refrégier, G, Yockteng, R, Fournier, E and Giraud, T (2009). Rapidly evolving genes in pathogens: Methods for detecting positive selection and examples among fungi, bacteria, viruses and protists. Infection, Genetics and Evolution 9:656–670.

Alizon, S, Hurford, A, Mideo, N and van Baalen, M (2009). Virulence evolution and the trade-off hypothesis: history, current state of affairs and the future. Journal of Evolutionary Biology 22:245–259.

Anderson, RM and May, RM (1982). Coevolution of hosts and parasites. Parasitology 85:411–426.

Behnke, JM, Barnard, CJ and Wakelin, D (1992). Understanding chronic nematode infections-evolutionary considerations, current hypotheses and the way forward. International Journal for Parasitology 22: 861–907.

Bell, AS, De Roode, JC, Sim, D and Read, AF (2006). Within-host competition in genetically diverse malaria infections: parasite virulence and competitive success. Evolution 60:1358–1371.

Ben-Ami, F, Mouton, L and Ebert, D (2008). The effects of multiple infections on the expression and evolution of virulence in a *Daphnia*-endoparasite system. Evolution 62:1700–1711.

Bergelson, J, Dwyer, G and Emerson, J (2001). Models and data on plant-enemy coevolution. Annual Review of Genetics 35:469–499.

Biozzi, G, Mouton, D, Stiffel, C and Boutjillier, Y (1984). A major role of the macrophage in quantitative genetic regulation of immunoresponsiveness and antitinfectious immunity. Advances in Immunology 36:189–234.

Boots, M, Best, A, Miller, MR and White, A (2009). The role of ecological feedbacks in the evolution of host defence: what does theory tell us? Philosophical Transactions of the Royal Society B Biological Sciences 364:27–36.

Boots, M and Haraguchi, Y (1999). The evolution of costly resistance in host-parasite systems. American Naturalist 153:359–370.

Boots, M, Hudson, P and Sasaki, A (2004). Large shifts in pathogen virulence relate to host population structure. Science 303:842–844.

Boots, M and Mealor, M (2007). Local interactions select for lower pathogen infectivity. Science 315: 1284–1287.

Boots, M and Sasaki, A (1999). "Small worlds" and the evolution of virulence: infection occurs locally and at a distance. Proceedings of the Royal Society of London B 266:1933–1938.

Bowers, R (2001). The basic reproduction ratio of the host: the evolution of host resistance to microparasites. Proceedings of the Royal Society of London B 268:243–250.

Brown, SP, Hochberg, ME and Grenfell, BT (2002). Does multiple infection select for raised virulence? Trends in Microbiology 10:401–405.

Bull, JJ, Molineux, I and Rice, W (1991). Selection of benevolence in a host-parasite system. Evolution 45: 875–882.

Caldwell, R, Schafer, J, Compton, L and Patterson, F (1958). Tolerance to cereal leaf rusts. Science 128: 714–715.

Chao, L, Hanley, KA, Burch, CL, Dahlberg, C and Turner, PE (2000). Kin selection and parasite evolution: Higher and lower virulence with hard and soft selection. Quarterly Review of Biology 75:261–275.

Clarke, D (1986). Tolerance of parasites and disease in plants and its significance in host-parasite interactions. Advances in Plant Pathology 5:161–197.

Coltman, DW, Pilkington, J, Kruuk, LEB, Wilson, K and Pemberton, JM (2001). Positive genetic correlation between parasite resistance and body size in a free-living ungulate population. Evolution 55:2116–2125.

Day, T (2001). Parasite transmission modes and the evolution of virulence. Evolution 55:2389–2400.

Day, T (2002). On the evolution of virulence and the relationship between various measures of mortality. Proceedings of the Royal Society of London B 269:1317–1323.

Day, T and Burns, JG (2003). A consideration of patterns of virulence arising from host- parasite coevolution. Evolution 57:671–676.

Day, T and Gandon, S (2007). Applying population-genetic models in theoretical evolutionary epidemiology. Ecology Letters 10:876–888.

De Roode, JC and Altizer, S (2010). Host-parasite genetic interactions and virulence- transmission relationships innatural populations of monarch butterflies. Evolution 64:502–514.

De Roode, JC, Culleton, R, Cheesman, SJ, Carter, R and Read, AF (2004). Host heterogeneity is a determinant of competitive exclusion or coexistence in genetically diverse malaria infections. Proceedings of the Royal Society of London B 271:1073–1080.

De Roode, JC, Pansini, R, Cheesman, SJ, Helinski, MEH, Huijben, S, Wargo, A, Bell, AS, Chan, BHK, Walliker, D and Read, AF (2005). Virulence and competitive ability in genetically diverse malaria infections. Proceedings of the National Academy of Sciences of the United States of America 102:7624–7628.

De Roode, JC, Yates, AJ and Altizer, S (2008). Virulence-transmission trade-offs and population divergence in virulence in a naturally occurring butterfly parasite. Proceedings of the National Academy of Sciences of the United States of America 105:7489–7494.

Demas, GE, Chefer, V, Talan, MI and Nelson, RJ (1997). Metabolic costs of mounting an antigen-stimulated immune response in adult and aged C57BL/6J mice. American Journal of Physiology–Regulatory Integrative and Comparative Physiology 42:R1631–R1637.

Dunn, P and North, RJ (1995). Virulence ranking of some *Mycobacterium tuberculosis* and *Mycobacterium bovis* strains according to their ability to multiply in the lungs, induce lung pathology, and cause mortality in mice. Infection and Immunity 63:3428–3437.

Dybdahl, MF and Storfer, A (2003). Parasite local adaptation: Red Queen versus Suicide King. Trends in Ecology and Evolution 18:523–530.

Ebert, D (1994). Virulence and local adaptation of a horizontally transmitted parasite. Science 265: 1084–1086.

Ebert, D (1998). Evolution: Experimental evolution of parasites. Science 282:1432–1435.

Ebert, D and Bull, JJ (2003). Challenging the trade-off model for the evolution of virulence: is virulence management feasible? Trends in Microbiology 11:15–20.

Ebert, D and Hamilton, WD (1996). Sex against virulence: The coevolution of parasitic diseases. Trends in Ecology and Evolution 11:79–82.

Ewald, P (1994). Evolution of Infectious Disease. Oxford University Press, Oxford.

Fenner, F and Fantini, B (1999). Biological Control of Vertebrate Pests: The History of Myxomatosis, an Experiment in Evolution. CABI Publishing, Wallingford.

Fornoni, J, Nuñez-Farfan, J, Valverde, PL and Rausher, M (2004). Evolution of mixed strategies of plant defense allocation against natural enemies. Evolution 58:1685–1695.

Frank, SA (1996). Models of parasite virulence. Quarterly Review of Biology 71:37–78.

Frank, SA and Schmid-Hempel, P (2007). Mechanisms of pathogenesis and the evolution of parasite virulence. Journal of Evolutionary Biology 21:396–404.

Fraser, C, Hollingsworth, TD, Chapman, R, de Wolf, F and Hanage, WP (2007). Variation in HIV-1 set-point viral load: Epidemiological analysis and an evolutionary hypothesis. Proceedings of the National Academy of Sciences of the United States of America 104:17441–17446.

Gandon, S (2004). Evolution of multihost parasites. Evolution 58:455–469.

Gandon, S, Mackinnon, MJ, Nee, S and Read, AF (2001). Imperfect vaccines and the evolution of pathogen virulence. Nature 414:751–756.

Ganusov, VV and Antia, R (2003). Trade-offs and the evolution of virulence of microparasites: do details matter? Theoretical Population Biology 64:211–220.

Ganusov, VV, Bergstrom, CT and Antia, R (2002). Within-host population dynamics and the evolution of microparasites in a heterogenous host population. Evolution 56:213–223.

Garamszegi, LZ (2006). The evolution of virulence and host specialization in malaria parasites of primates. Ecology Letters 9:933–940.

Gould, S and Lewontin, R (1979). The spandrels of San Marco and the Panglossian paradigm: a critique of the adaptationist programme. Proceedings of the Royal Society of London B 205:581–598.

Gower, CM and Webster, JP (2005). Intraspecific competition and the evolution of virulence in a parasitic trematode. Evolution 59:544–553.

Graham, AL, Allen, JE and Read, AF (2005). Evolutionary causes and consequences of immunopathology. Annual Review of Ecology and Systematics 36:373–397.

Grech, K, Watt, K and Read, AF (2006). Host-by-parasite interactions for virulence and resistance in a malaria model system. Journal of Evolutionary Biology 19:1620–1630.

Hardin, G (1968). The tragedy of the commons. Science 162:1243–1248.

Hellgren, O, Andersson, M and Råberg, L (2011). The genetic structure of *Borrelia afzelii* varies with geographic but not ecological sampling scale. Journal of Evolutionary Biology 24:159–167.

Ilmonen, P, Taarna, T and Hasselquist, D (2000). Experimentally activated immune defence in female pied flycatchers results in reduced breeding success. Proceedings of the Royal Society of London B 267:665–670.

Jensen, K, Little, TJ, Skorping, A and Ebert, D (2006). Empirical support for optimal virulence in a castrating parasite. Plos Biology 4:e197.

Kaslow, R, Mcnicholl, J and Hill, AVS (2008). Genetic Susceptibility to Infectious Diseases. Oxford University Press.

Kerr, B, Neuhauser, C, Bohannan, B and Dean, A (2006). Local migration promotes competitive restraint in a host-pathogen "tragedy of the commons". Nature 442:75–78.

Kloch, A, Babik, W, Bajer, A, Sinski, E and Radwan, J (2010). Effects of an MHC-DRB genotype and allele number on the load of gut parasites in the bank vole *Myodes glareolus*. Molecular Ecology 19:255–265.

Kover, PX (2006). Evolutionary genetics of host-parasite interactions. In: Fox C, Wolf J (eds). Evolutionary Genetics: Concepts and Case Studies. Oxford University Press, Oxford, pp. 447–463.

Kraaijeveld, AR and Godfray, HCJ (1997). Trade-off between parasitoid resistance and larval competitive ability in *Drosophila melanogaster*. Nature 389:278–280.

Lazzaro, B, Sceurman, B and Clarke, A (2004). Genetic basis of natural variation in *D. melanogaster* antimicrobial immunity. Science 303:1873–1876.

Lenski, R and May, RM (1994). The evolution of virulence in parasites and pathogens: reconciliation of two competing hypotheses. Journal of Theoretical Biology 169:253–265.

Lindström, K, Foufopulos, J, Pärn, H and Wikelski, M (2004). Immunological investment reflect parasite abundance in island population of Darwin's finches. Proceedings of the Royal Society of London B 271:1513–1519.

Lipsitch, M, Herre, E and Nowak, MA (1995). Host population structure and the evolution of virulence: a "law of diminishing returns". Evolution 49:743–748.

Little, TJ, Carius, HJ, Sakwinska, O and Ebert, D (2002). Competitiveness and life-history characteristics of Daphnia with respect to susceptibility to a bacterial pathogen. Journal of Evolutionary Biology 15: 796–802.

Little, TJ, Chadwick, W and Watt, K (2008). Parasite variation and the evolution of virulence in a *Daphnia*-microparasite system. Parasitology 135:308.

Lord, CC, Barnard, B, Day, K, Hargrove, JW, McNamara, JJ, Paul, REL, Trenholme, K and Woolhouse, MEJ (1999). Aggregation and distribution of strains in microparasites. Philosophical Transactions of the Royal Society of London B 354:799–807.

Luster, M, Portier, C, Pait, D, Rosenthal, G, Geromec, D, Corsini, E, Blaylock, B, Pollock, P, Kouchi, Y, Craig, W, White, K, Munson, A and Comment, C (1993). Risk assessment in immunotoxicology II. Relationships between immune and host resistance tests. Fundamental and Applied Toxicology 21:71–82.

Ma, Y, Seiler, KP, Eichwald, EJ, Weis, JH, Teuscher, C and Weis, JJ (1998). Distinct characteristics of resistance to *Borrelia burgdorferi*-induced arthritis in C57BL/6N mice. Infection and Immunity 66:161–168.

Mackinnon, MJ, Gandon, S and Read, AF (2008). Virulence evolution in response to vaccination: the case of malaria. Vaccine 26:C42–52.

Mackinnon, MJ and Read, AF (1999). Genetic relationships between parasite virulence and transmission in the rodent malaria *Plasmodium chabaudi*. Evolution 53:689–703.

Mackinnon, MJ and Read, AF (2004). Virulence in malaria: an evolutionary viewpoint. Philosophical Transactions of the Royal Society of London B 359:965–986.

Mandl, JN, Barry, AP, Vanderford, TH, Kozyr, N, Chavan, R, Mucking, S, Barrat, FJ, Coffman, RL, Staprans, SI and Feinberg, MB (2008). Divergent TLR7 and TLR9 signaling and type I interferon production distinguish pathogenic and nonpathogenic AIDS virus infections. Nature Medicine 14:1077–1087.

May, RM and Nowak, MA (1995). Coinfection and the evolution of parasite virulence. Proceedings of the Royal Society of London B 261:209–215.

McKean, K, Yourth, C, Lazzaro, B and Clark, A (2008). The evolutionary costs of immunological maintenance and deployment. BMC Evolutionary Biology 8:76.

Medzhitov, R and Janeway, CA (1997). Innate immunity: The virtues of a nonclonal system of recognition. Cell 91:295–298.

Messinger, S and Ostling, A (2009). The consequences of spatial structure for the evolution of pathogen transmission rate and virulence. American Naturalist 174:441–454.

Miller, MR, White, A and Boots, M (2005). The evolution of host resistance: Tolerance and control as distinct strategies. Journal of Theoretical Biology 236:198–207.

Miller, MR, White, A and Boots, M (2006). The evolution of parasites in response to tolerance in their hosts: The good, the bad, and apparent commensalism. Evolution 60:945–956.

Moret, Y and Schmid-Hempel, P (2000). Survival for immunity: The price of immune system activation for bumblebee workers. Science 290:1166–1168.

Müller, V, Maggiolo, F, Suter, F, Ladisa, F, De Luca, A, Antinori, A, Sighinolfi, L, Quiros- Roldan, E, Carosi, G and Torti, C (2009). Increasing clinical virulence in two decades of the Italian HIV epidemic. Plos Pathogens 5:e1000454.

Nejsum, P, Roepstorff, A, Jørgensen, C, Fredholm, M, Göring, H and Anderson, T (2009). High heritability for *Ascaris* and *Trichuris* infection levels in pigs. Heredity 102:357–364.

Nielsen, R (2005). Molecular signatures of natural selection. Annual Review of Genetics 39:197–218.

Nilsson, J-Å, Granbom, M and Råberg, L (2007). Does the strength of an immune response reflect its energetic cost? Journal of Avian Biology 38:488–494.

Nowak, MA and May, RM (1994). Superinfection and the evolution of parasite virulence. Proceedings of the Royal Society of London B 255:81–89.

Obbard, D, Welch, J, Kim, K-W and Jiggins, F (2009). Quantifying adaptive evolution in the *Drosophila* immune system. Plos Genetics 5:e1000698.

Pagàn, I, Alonso-Blanco, C and García-Arenal, F (2007). The relationship of within-host multiplication and virulence in a plant-virus system. Plos ONE 2:e786.

Paterson, S, Wilson, K and Pemberton, JM (1998). Major histocompatibility complex variation associated with juvenile survival and parasite resistance in a large unmanaged ungulate population *(Ovis aries L.)*. Proceedings of the National Academy of Sciences USA 95:3714–3719.

Pennington, PM, Allred, CD, West, CS, Alvarez, R and Barbour, AG (1997). Arthritis severity and spirochete burden are determined by serotype in the *Borrelia turicatae*-Mouse model of Lyme disease. Infection and Immunity 65:285–292.

Råberg, L, De Roode, JC, Bell, AS, Stamou, P, Gray, D and Read, AF (2006). The role of immune-mediated apparent competition in genetically diverse malaria infections. American Naturalist 168:41–53.

Råberg, L, Nilsson, JA, Ilmonen, P, Stjernman, M and Hasselquist, D (2000). The cost of an immune response: vaccination reduces parental effort. Ecology Letters 3:382–386.

Råberg, L, Sim, D and Read, AF (2007). Disentangling genetic variation for resistance and tolerance to infectious diseases in animals. Science 318:812–814.

Råberg, L and Stjernman, M (2003). Natural selection on immune responsiveness in blue tits *Parus caeruleus*. Evolution 57:1670–1678.

Råberg, L, Graham, AL and Read, AF (2009). Decomposing health: tolerance and resistance to parasites in animals. Philosophical Transactions of the Royal Society B Biological Sciences 364:37–49.

Rausher, M (1992). The measurement of selection on quantitative traits: biases due to environmental covariances between traits and fitness. Evolution 46:616–626.

Rausher, MD (2001). Co-evolution and plant resistance to natural enemies. Nature 411:857–864.

Read, AF and Mackinnon, MJ (2008). Pathogen evolution in a vaccinated world. In: Stearns S, Koella JC (eds). Evolution in Health and Disease. Oxford University Press, Oxford, pp. 139–152.

Read, AF and Taylor, LH (2000). Within-host ecology of infectious diseases: patterns and consequences. In: Thompson RCA (ed). Molecular Epidemiology of Infectious Diseases. Arnold, London, pp. 59–75.

Read, AF and Taylor, LH (2001). The ecology of genetically diverse infections. Science 292:1099–1102.

Regoes, RR, Nowak, MA and Bonhoeffer, S (2000). Evolution of virulence in a heterogenous host population. Evolution 54:64–71.

Restif, O and Koella, JC (2003). Shared control of epidemiological traits in a coevolutionary model of host-parasite interactions. American Naturalist 161:827–836.

Schafer, J (1971). Tolerance to plant disease. Annual Review of Phytopathology 9:235–252.

Schmid-Hempel, P (2003). Variation in immune defence as a question of evolutionary ecology. Proceedings of the Royal Society of London B 270:357–366.

Simms, EL and Rausher, MD (1987). Costs and benefits in plant resistance to herbivory. American Naturalist 130:570–581.

Simms, EL and Triplett, J (1994). Costs and benefits of plant responses to disease: resistance and tolerance. Evolution 48:1973–1985.

Sinervo, B and Svensson, E (2002). Correlational selection and the evolution of genomic architecture. Heredity 89:329–338.

Smith, JA, Wilson, K, Pilkington, JG and Pemberton, JM (1999). Heritable variation in resistance to gastro-intestinal nematodes in an unmanaged mammal population. Proceedings of the Royal Society of London B 266:1283–1290.

Staszewski, V and Boulinier, T (2004). Vaccination: a way to address questions in behavioural and population ecology. Trends in Parasitology 20:17–22.

Stear, MJ, Bishop, S, Mallard, B and Raadsma, H (2001). The sustainability, feasability and desirability of breeding livestock for disease resistance. Research in Veterinary Science 71:1–7.

Stevenson, MM, Lyanga, JJ and Skamene, E (1982). Murine malaria: genetic control of resistance to *Plasmodium chabaudi*. Infection and Immunity 38:80–88.

Stewart, A, Logdson, J and Kelley, S (2005). An empirical study of the evolution of virulence under both horizontal and vertical transmission. Evolution 59:730–739.

Stjernman, M, Råberg, L and Nilsson, J-Å (2008). Maximum host survival at intermediate parasite infection intensities. Plos ONE 3:e2463.

Svensson, E, Råberg, L, Koch, C and Hasselquist, D (1998). Energetic stress, immunosuppression and the costs of an antibody response. Functional Ecology 12:912–919.

Timmann, C, Evans, J, König, I, Kleensang, A, Rüschendorf, F, Lenzen, J, Sievertsen, J, Becker, C, Enuameh, Y, Osei Kwakye, K, Opuku, E, Browne, E, Ziegler, A, Nürnberg, P and Horstmann, R (2007). Genome-wide linkage analysis of malaria infection intensity and mild disease. PLoS Genetics 3:e48.

Tumpey, T, Basler, C, Aguilar, P, Zeng, H, Solórzano, A, Swayne, D, Cox, N, Katz, J, Taubenberger, J, Palese, P and García-Sastre, A (2005). Characterization of the reconstructed 1918 Spanish influenza pandemic virus. Science 310:77–80.

van Baalen, M (1998). Coevolution of recovery ability and virulence. Proceedings of the Royal Society of London B 265:317–325.

van Baalen, M and Sabelis, MW (1995). The dynamics of multiple infection and the evolution of virulence. American Naturalist 146:881–910.

van Boven, M and Weissing, F (2004). The evolutionary economics of immunity. American Naturalist 163:277–294.

Vijendravarma, R, Kraaijeveld, AR and Godfray, HCJ (2008). Experimental evolution shows *Drosophila melanogaster* resistance to a microsporidian pathogen has fitness costs. Evolution 63:104–114.

Waldron, MR, Nishida, T, Nonnecke, BJ and Overton, TR (2003). Effect of lipopolysaccharide on indices of peripheral and hepatic metabolism in lactating cows. Journal of Dairy Science 86:3447–3459.

Wild, G, Gardner, A and West, SA (2009). Adaptation and the evolution of parasite virulence in a connected world. Nature 459:983–986.

Williams, TD, Christians, JK, Aiken, JJ and Evanson, M (1999). Enhanced immune function does not depress reproductive output. Proceedings of the Royal Society of London B 266:753–757.

Woolhouse, MEJ, Taylor, LH and Haydon, DT (2001). Population biology of multihost pathogens. Science 292:1109–1112.

Woolhouse, MEJ, Webster, JP, Domingo, E, Charlesworth, B and Levin, B (2002). Biological and biomedical implications of the co-evolution of pathogens and their hosts. Nature Genetics 32:569–577.

CHAPTER 19

EVOLUTIONARY GENETICS OF INFECTIOUS DISEASE

Paul Schmid-Hempel

INTERACTION BETWEEN GENOTYPES

When scrutinizing the variation within host and parasite species, genotypes play an important role. It is a common observation that only certain parasite genotypes can affect a given host genotype and, vice versa, only some host genotypes are resistant against a variety of parasite genotypes. For example, when cross-infecting six isolates of rust (*Melamspora lini*) in 67 different plants of flax (*Linum marginale*) taken from the wild, infection success varied consistently and systematically with genotype such that ten different resistance classes could be identified, and one flax genotype even was entirely resistant (Burdon and Jarosz, 1991). Similar observations are made in a range of different systems (Table 19.1). In fact, genotype-associated variation in host resistance and parasite infectivity or virulence is the most interesting and relevant kind of variation in host–parasite systems for many questions of evolutionary parasitology. For example, such variation is suspected to maintain genetic diversity in host (and parasite) populations through the effects of rapid antagonistic co-evolution that generates negative frequency-dependent selection on host genotypes (Haldane, 1949; Berenos et al., 2010). Similarly, specific host–parasite interactions are a requirement so that this process also can select for the evolution and maintenance of sex and meiotic recombination (Salathé et al., 2008).

Interestingly, when the scope of the analysis is widened to the specificity of host and parasite species as a whole, various patterns emerge. Parasite specificity at this level varies considerably, too, but host choice often seems opportunistic (Poulin, 2007). However, it is often observed that only a few parasite species manage to infect a wide range of host species, whereas most parasites are restricted to a relatively small set of hosts. Factoring in the observation that many more parasite species are and will be uncovered with the use of molecular tools (rather than by morphology or biochemistry), where previously only one species was recognized, it appears likely that many more parasite species are fairly host-specific as a whole (Poulin and Keeney, 2008). Because parasite (and host) speciation is an extension of the underlying microevolutionary processes, the existence of a good deal of within-species genetic variation in parasite infectivity (and host resistance) is

TABLE 19.1: Examples of Genetic Interactions in Host–Parasite Systems

Host	Parasite	Observation	Reference
Flax (*Linum* sp.).	Flax rust (*Melamspora lini*)	Resistance associated with resistance alleles. Infection success can be classified into resistance types.	Burdon and Jarosz, 1991; Ellis et al., 1999
Plants (*Glycine canescens*)	Soybean rust (*Phakopsora pachyrhizi)*	Variation of resistance based on single dominant gene with many alleles	Burdon, 1987
Wheat	Fungus (*Mycosphaerella graminicola*)	Genetic variance found for host resistance and parasite infectivity. Based on 47 host cultivars, 80 parasite isolates.	Kema et al., 1996a, 1996b
Snail (*Biomphalaria glabrata*)	Trematode (*Schistosoma mansoni*)	Resistance is heritable, selectable, and specific to certain trematode lines.	Webster and Woolhouse, 1998a, 1998b
Snail (*Potamopyrgus antipodarium*)	Trematodes (*Microphallus* sp.)	Snail clones vary in resistance. Rare clones less infected.	Dybdahl and Krist, 2004
Drosophila melanogaster	Parasitoid (*Leptopilina boulardi*)	Cellular immune response against parasitoid varies according to host and parasite genotypes.	Carton and Nappi, 2001
Drosophila melanogaster	Bacterium (*Serratia marescens*)	Host lines vary in susceptibility; candidate genes identified.	Lazzaro et al., 2004
Pig (*Sus domesticus*)	Bacterium (*Escherichia coli*)	Resistance depends on combination of host and parasite genotypes.	Edfors-Lilja, 1991

obviously expected to relate to specificity at the species level. But how exactly these quantities relate to each other is not well understood. For example, it is not necessarily the case that antagonistic host–parasite co-evolution drives population differentiation, even though it can maintain genetic diversity in host populations (Berenos et al., 2010).

Host–parasite interactions within populations are typically analyzed by cross-infecting a range of different parasite strains or isolates into a range of different host lines. The specific interaction is then retrieved by an ANOVA type of statistics that can identify an interaction term and assess its significance. For example, different clones of *Daphnia magna* that are infected by different isolates of the bacterium *Pasteuria ramosa* varied in their susceptibility. Overall, a clear statistical sign of interactions was observed, even as isolates can represent a mixture of a few parasite strains rather than just a single genotype. The precise nature and strength of this effect also depended on dosage (Carius et al., 2001), but the basic nature of an interaction persisted (Fig. 19.1).

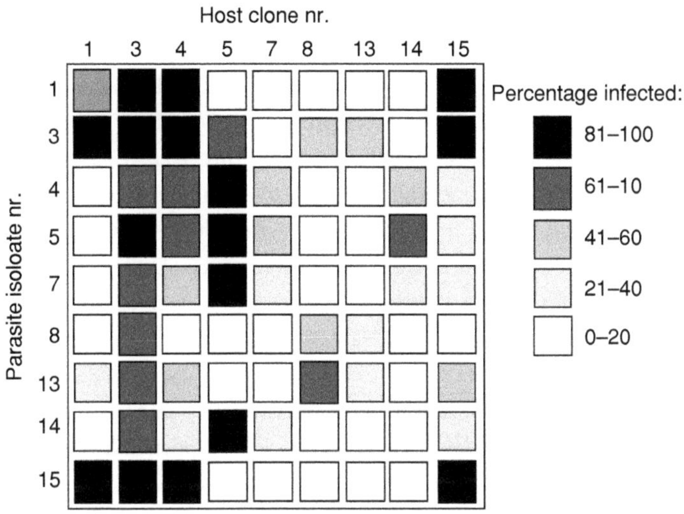

FIGURE 19.1: Genetic interactions. A range of clones of the host *Daphnia magna* (numbers 1–15) were exposed to a range of parasite isolates of the parasite *Pasteuria ramosa* (numbers 1–15) and the infection success was measured (generally with nine replicates). The shading indicates the percentage of host clones infected. In all, the host clone × parasite isolate interaction effect was statistically significant ($p < 0.001$). (From Carius et al., 2001)

Because of the widespread occurrence of interaction effects and their significance for several important questions of host–parasite co-evolution, interactions have been classified into different types. Two of these models have been widely cited and used.

GENE-FOR-GENE MODEL

Originally formulated for the interaction of a plant pathogen (flax rust) with its host plant (flax) (Flor, 1956), the model suggests that a genetically encoded factor from the plant (the susceptible/resistance gene) interacts with a genetically encoded factor from the parasite (the virulence/avirulence gene). At the biochemical level, the parasite produces an elicitor (coded by an avirulence gene) that the plant can recognize (coded by a resistance gene); the outcome in such a case is no infection. Infection occurs when the elicitor is not recognizable and/or when the plant lacks a corresponding recognition gene. Hence, the model is "asymmetric" as it proposes a universally susceptible host type and a super-parasite that can infect all hosts. Despite this shortcoming, the gene-for-gene model is a useful approximation for plant–parasite interactions because it captures some of the underlying biochemistry and genetics.

MATCHING ALLELES MODEL

This model is motivated by the observation of recognition systems in hosts that respond to different allele combinations on the side of the parasite. This idea is modeled after a key-and-lock mechanism. For example, if hosts are able to recognize specific patterns on the surface of the parasite—the epitopes—then host defense is activated and infection blocked. The respective "matching" between host and parasite genotypes that underlies this recognition thus yields the probability that the infection takes place or is prevented. Matching allele models are a good approximation in many cases of animal–pathogen interactions. To take a more elaborate example, *Drosophila melanogaster* has major genes that affect resistance against two different kinds of parasitoid wasps, whereas the parasitoids have virus-like particles that aid in undermining the host's defense responses. These particles show Mendelian inheritance (Dupas et al., 1998; Poirie et al., 2000). A number of additional models have been formulated, such as the inverse gene-for-gene model, or models based on linkage (Schmid-Hempel, 2011).

Whereas these models are simple and intuitive and have some justification from the underlying biochemical and molecular processes, they remain models—that is, these models are abstractions, and any real case is likely to deviate in one or several aspects. Not the least, these models implicitly convey the idea that host–parasite interactions are somehow ordered, for example, as a symmetric matrix of interactions. Nothing could be further from the truth. Rather, host and parasite genotypes posses something like "non-linear identities." In other words, a given parasite type can infect very different host types, but a closely related one (e.g., as given by its average genetic distance) might show a completely different pattern. Similarly, different parasite types might primarily vary in the range of hosts they can infect rather than in which of a constant number of host types they can attack. The same is true for host resistance. Hence, rather than having an ordered interaction space, real host–parasite interactions show a very rugged interaction landscape. A likely reason for this ruggedness is the prevalence of epistatic fitness components, as will be discussed below. Only in some cases will the major peaks in this interaction landscape be dominant and ordered enough to fit a gene-for-gene model. In addition, genetic polymorphism for the gene-for-gene model, for instance, does not always reveal itself from real data, and inferred polymorphisms might be an artifact of how things can be observed (Frank, 1994, 1996).

GENETIC ARCHITECTURE OF HOST RESISTANCE AND PARASITE VIRULENCE

HOST RESISTANCE

How many genes affect host resistance, what is their effect, how do they interact, and where in the genome are they located? Such are the questions that define the genetic architecture of host resistance. Several methods are available to study this question. These include quantitative trait loci (QTL) analyses, the direct study of genomic sequences, or the use of comparative genomics. All of these methods have some advantages and disadvantages. QTL studies, for example, are powerful tools for many cases, especially when no genomic sequence is yet available. On the other hand, QTL studies tend to detect genes of major effect and are not sensitive enough for many small effect genes, or their interactions. Nevertheless, the accumulating evidence and the application of several methods provide valid insights into the genetic architecture of host resistance.

QTL studies have often been applied to plant–parasite systems, often driven by the search for marker-assisted breeding programs (Klover and Caicedo, 2001). But also for many animal systems, QTL studies have proven to be useful (Wilfert and Schmid-Hempel, 2008). Such studies suggest that a limited number of additive QTL loci are involved, typically in the order of a few. These additive loci are main effects in statistical terms. For example, five main effect loci dominate the resistance of maize (*Zea mays*) to stalk rot (caused by fungi, *Fusarium* spp.), and four to nine loci affect the resistance of oat (*Avena sativa*) to mildew (Jones, 1986). Similar numbers are reported for invertebrate hosts (Carton et al., 2005). These main effect loci seem typically dispersed across the chromosomes—at least, it is often difficult to find a statistical association among loci because either too few loci are involved or too few chromosomes are available for a powerful test (Wilfert and Schmid-Hempel, 2008).

Most remarkably, QTLs change according to the different sets of loci. Hence, if parasite strain A infects a host, different loci will give a signal than if parasite strain B infects, even if the same host line is compared. The chances to recover the same QTL locus is significantly lower when the involved host and parasite genotypes are different than when environmental conditions change (Wilfert and Schmid-Hempel, 2008). Therefore, the genetic architecture of host resistance—at least as revealed by QTL studies—is not fixed but rather fluid and changes according to the infecting strain. This is likely to be a corollary of the observation that host–parasite interactions happen in a rugged interaction landscape. Looking at the genetics once again, epistatic interactions could be a relevant issue.

Epistasis occurs when alleles at different loci do not act independently—that is, when fitness, contingent on alleles at different loci, is not simply additive but shows an interaction component. There has been a lot of discussion as to whether epistasis is widespread and important enough to, for example, select for the evolution of sex and recombination (which depends on the existence of linkage disequilibrium) (Otto and Nuismer, 2004). On one hand, it is important to recall the difference between physiological epistasis, which refers to the effects of gene combinations for individual fitness, and population-wide epistasis, which refers to the epistatic component of average fitness in the population. The latter is relevant for the dynamics of the co-evolutionary process and is connected to the existence of linkage disequilibria—that is, a statistical association between alleles at different loci. The distinction is also important as main effect loci can generate physiological epistasis but no such effect at the population level. Regardless, the survey of QTL-based studies showed that epistasis at this level is very common and important. Across animals and plants, the main effect loci explained some 30% to 50% of the phenotypic variance (i.e., resistance), but in almost 80% of the studies epistatic effects were identified that included, on average, two pairs of loci. These loci often were not the same ones that produced a main effect. Hence, there is a considerable fraction of phenotypic variance that is determined by epistasis in addition to the main effects of other loci. Almost every chromosome (or linkage group) has at least one locus involved in resistance, but most epistatic interactions between resistance loci seem to occur between loci on different chromosomes rather than between loci within linkage groups (Wilfert and Schmid-Hempel, 2008). A part of the explanation is certainly due to an artifact of the method, as QTL studies tend to pick up major effects. On the other hand, the role of epistasis is almost certainly underestimated, since its detection as a statistical interaction term is notoriously difficult. Nevertheless, epistasis relevant for host resistance has been found in many different systems. Examples include viruses (Bonhoeffer et al., 2004), bacteria (Lenski, 1988), fungi (Cortesi et al.,

2001), plants (Coego et al., 2005), arthropods (Salathé and Ebert, 2003; Lazzaro et al., 2004), mice (Desrosiers et al., 2005), and humans (Moreno et al., 2004).

Studies in the flour beetle (*Tribolium castaneum*), for example, have furthermore shown how different elements of genetic variance add to resistance against a typical microparasite, the microsporidium *Nosema whitei*. Measures of host resistance, such as survival after infection and spore load of the parasite, revealed strong maternal effects, but also a major fraction of phenotypic variance that is explained by epistasis (Wegner et al., 2008). The relative contribution furthermore varied depending on whether the host was infected by a single isolate or several isolates, where epistatic effects became less important (but still explained the largest part of the variance) (Wegner et al., 2009). These results confirm a general pattern of a rather fluid genetic architecture, at least as far as the main components are concerned.

By all standards the expressed genes, that is, the product of host resistance or parasite virulence genes, such as proteins or cytotoxic molecules, are much closer to the actual host–parasite interaction than the genes themselves. The most relevant and accessible set is the transcriptome, the set of mRNAs expressed by a host as a result of a parasitic challenge. This "gene expression profile" varies among individuals, populations, and infections and can be statistically analyzed (Oleksiak et al., 2002; Gibson 2003). Of course, variation in expression is ultimately based on genetic variation in the regulatory genes, and the heritable component of gene expression can be studied, too, or mapped onto a genomic map (the eQTL map) (Gibson and Weir, 2005; Rockman and Kruglyak, 2006). Technically, such regulation is often characterized as cis- or trans-regulation. The latter refers to regulatory genes somewhere in the genome, whereas the former (cis) is achieved by the promoter sequence immediately upstream of the gene.

Transcriptome studies of infected hosts, and of infecting parasites, are relatively recent because the technology has become available for larger-scale studies and within reasonable cost limits only in the past few years. Vectors of human diseases, such as mosquitoes, were among the early examples of hosts to be studied. Often these studies show that several hundreds or thousands of genes are up- or down-regulated after an infection. Of those, however, only a few dozen are characteristic of a particular infection. Typically, the respective genes are those involved in immune defenses such as antimicrobial peptides, or general stress-response genes (Dana et al., 2005; Vododvar et al., 2005; Xu et al., 2005; Altincicek et al., 2008; Navajas et al., 2008). Not surprisingly, some genes also alter their expression depending on the genetic background they find themselves in (an epistatic effect) (Dirlewanger et al., 1994).

Expression studies are bound to become more relevant for host–parasite studies, precisely because host–parasite interactions vary according to who infects whom and according to environmental conditions. For example, the expressed mixture of different antimicrobial peptides might be the key to understanding specific host–parasite interactions rather than the variation in the genes themselves (Riddell et al., 2009) (Fig. 19.2).

GENETICS OF PARASITE VIRULENCE

The most detailed work on this question comes from viruses and, notably, from bacteria. Any gene that contributes to infection, establishment, replication, and transmission of the parasite is therefore under scrutiny. Such genes have been collectively called "virulence genes" by microbiologists,

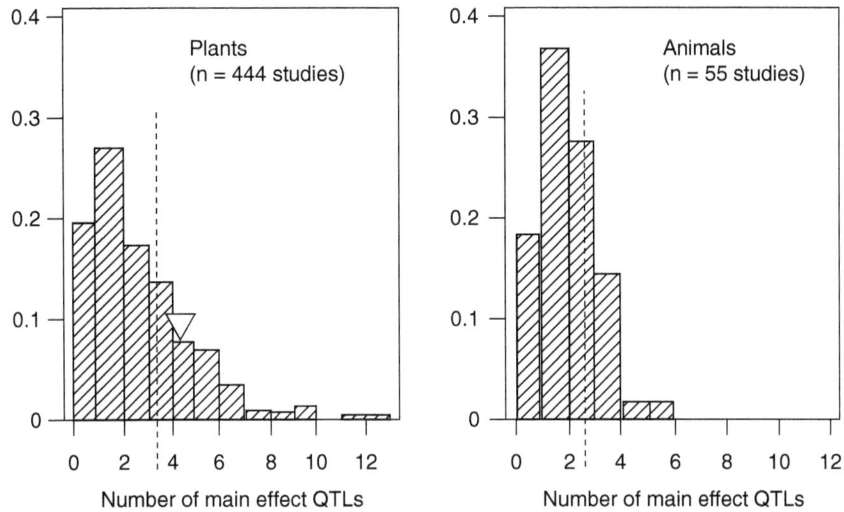

FIGURE 19.2: Number of resistance genes. Shown is the frequency distribution of main effect (additive) QTL loci reported for studies of plants (left) and animals (right). The vertical lines indicate the mean for plants (3.24 ± 2.22 S.E. loci, n = 444 studies) and animals (2.47 ± 1.18, n = 55), respectively. The triangle refers to the value reported for plants by Klover and Caicedo, 2001. (From Wilfert and Schmid-Hempel, 2008.)

even though they might not directly contribute to fitness loss for the host (which is the more usual meaning of virulence for evolutionary ecologists).

Even though bacterial genomes are small and simple compared to those of eukaryotes, they are quite dynamic over evolutionary time—which can mean only a few generations in this case. Genes are added to a bacterial genome, modified, duplicated, degraded, and excised again after some while. The most important class of such genes are the "virulence factors," genes that code for traits that allow the bacterium to become "virulent" (i.e., in this general sense to be a successful pathogen). In bacterial populations, virulence factors are often transferred horizontally, between lines, by transposons, phages, transformation (uptake of plasmids or DNA), or conjugation (between bacterial cells, with transfer of plasmids). Such horizontal gene transfer is a major force shaping microbial pathogen evolution (Dobrindt et al., 2004).

Microbial virulence genes are often clustered together on the so-called pathogenicity islands. These show characteristic elements such as a deviant GC content of the genomic sequence, flanking of repeat sequences, functional or pseudogenes that affect gene mobility, and so forth (Hacker and Kaper, 2000). These pathogenicity islands have an evolutionary dynamics of their own, as they become transferred, adapt to their new genomic background, and become excised and transferred again somewhat later. Many important pathogens have emerged because pathogenicity islands have been transferred to a previously harmless line. A classical example is the agent of cholera, the bacterium *Vibrio*, or *Salmonella* spp., where the emergence of pathogenic strains and several epidemic waves can be traced back to such islands (Cotter and diRita, 2000).

Also for the parasite, the regulation of virulence genes is of the essence. A well-studied case is the bacterial complex, *Bordetella*. *B. pertussis* (the agent of whooping cough), *B. parapertussis*, and

B. bronchiseptica infect the respiratory tract of mammals and differ in their pathogenicity. All of these forms use virtually identical virulence genes regulated by similar regulatory complexes. The degree of pathogenicity or whether the infection is acute or chronic depends on how the genes are expressed (Cotter and diRita, 2000; Beier and Gross, 2006). Different usage of a similar set of genes thus determines the phenotypically important characteristics of the pathogen rather than the genetic differences *per se* (but recall that regulatory genes are under selection themselves). The mechanisms of expression are the relevant background against which a study of parasite strategies or virulence can unfold (Biron et al., 2005). The dynamics of a host–parasite system whose variation is primarily based on expression differences is likely to evolve differently from one that is based on variation in the sequence of effector genes. Whether this leads to faster or slower evolution, or towards more or less diversity in the host–parasite system, remains to be explored (Carroll, 2005).

HETEROZYGOSITY IN HOST–PARASITE INTERACTIONS

In small populations, or when related individuals preferentially mate with one another, the average degree of heterozygosity decreases. A decrease of heterozygosity means a loss of genetic diversity; hence, the question is whether this also relates to a decreased level of defense against parasites and increased susceptibility to infections. The problem emerges at two levels: individual heterozygosity and population-wide average heterozygosity.

There is indeed evidence that population-wide heterozygosity correlates with measures of parasitism. For example, when different island populations of the Galapagos hawk (*Buteo galapagoensis*) were compared, low levels of heterozygosity related to increased infection levels (and to lower levels of constitutively expressed antibodies) (Whiteman et al., 2005). Similarly, infection levels increase with the degree of inbreeding in ungulates (Cassinello et al., 2001). Heterozygosity correlates negatively with parasite prevalence in deer mouse populations, even when controlled for population density (Meagher, 1999). Comparing stocks of salmon (*Oncorhynchus tshawytscha*) with either homozygous or heterozygous MHC class II genes shows that homozygous stocks are more susceptible to viral disease and *Myxobolus* causing whirling disease (Arkush et al., 2002). The relationship is not always that simple, however. For example, among 182 plant species, the outcrossing rate (associated with higher levels of heterozygosity) relates positively to the number of fungal species reported for the host. This is considered as evidence that sexual reproduction (i.e., outcrossing) is selected for by the presence of parasites. Hence, genetic diversity is still considered to be "protective" within species (Busch et al., 2004). Therefore, helpful evidence comes from experimental breeding of lines. In *Drosophila melanogaster*, for example, more inbred lines suffered more severe damage when exposed to bacteria (*Bacillus thuringiensis, Serrratia marescens*) (Spielman et al., 2004).

Social insects are a particularly interesting case. This is because the mother (the colony queen, or queens) can "decide" about the level of genetic diversity among her offspring by varying either the number of males she mates with (polyandry) or by varying the recombination rate in her eggs. In social hymenoptera (ants, wasps, bees) the males are haploid and therefore only the mother has a recombination effect. The biology of social insects implies that many of the consequences of this "decision" are immediately relevant to the mother because she depends on the functioning of her colony (and thus her offspring that stay at home to help) that will eventually allow them to actually reproduce—that is, to produce sexual offspring (daughter queens and drones) that leave the colony and start the next generation. Several studies have shown that increasing genetic diversity among offspring does indeed reduce parasite load and increases the fitness of the mother (as measured by the colony's output in terms of sexuals). For example, colonies of the bumblebee, *Bombus terrestris*, headed by mothers that were artificially inseminated with sperm of drones representing high genetic diversity (unrelated males) had lower parasite loads and higher fitness than those headed by mothers inseminated by sperm of low genetic diversity (Baer and Schmid-Hempel, 1999). Similar findings have been made for honeybees, where colonies of queens inseminated by high-diversity sperm had lower variance of infection levels and higher growth rates compared to low-diversity colonies (Tarpy, 2003). Across 119 ant species, the degree of colony genetic diversity correlated negatively with parasite load even when corrected for phylogenetic effects (Schmid-Hempel and Crozier, 1999). Also in termites (*Zootermopsis angusticollis*), which have a "standard" diplo-diploid genetic system, experimental lines that are more inbred show higher mortality rates and higher parasite loads when infected by bacteria or fungal spores (Calleri et al., 2006). Hence, the effect seems to be widespread across a range of taxa and must therefore have general relevance. Social hymenoptera, in particular, are also remarkable because they not only show some of the highest genome-wide recombination rates among eukaryotes (Wilfert et al., 2007) but are also renowned for having excessive degrees of polyandry. For example, among the honeybees (*Apis* spp.), the European honeybee (*A. mellifera*) mates with around 15 males (Oldroyd et al. 1998), the giant honeybee (*A. dorsata*) with around 30 males (Moritz et al., 1995), *A. andreniformis* with 10 to 20 males (Oldroyd et al., 1997), and *A. cerana* with an average of 18 males (Oldroyd et al. 1998). Yet, at the same time, the honeybee (*A. mellifera*) has the highest known genome-wide recombination rate of all animals (Gadau et al., 2000; Wilfert et al., 2007). It is not known whether all of this diversity is selected for by defense against parasites, but given the examples where diversity has been shown to be protective for a colony as a whole, it might well be the case. At any rate, social insects provide an excellent testing ground for the relationship of genetic diversity and parasitism, since the effects are played out every generation in a well-defined selection arena, the colony.

There is also considerable evidence that heterozygosity and genetic diversity at the individual level is favorable under parasitism. Examples include studies from fisheries, where less homozygous fish (typically measured by MHC-allelic diversity) are generally more susceptible to infections; this is the case for rainbow trout (*Oncorhynchus mykiss*) infected by bacteria (Ferguson and Drahushchak, 1990), or Chinook salmon infected by bacteria and fungi (Ferguson and Drahushchak, 1990). Other examples include lower parasite loads, less severe symptoms, or higher survival rates in less inbred/more heterozygous individuals of the California sea lion (*Zalophus californicus*; Acevedo-Whitehouse et al., 2003), bighorn sheep (*Ovis canadensis*) infected by lungworms (*Protostrongylus* spp.; Luikart et al., 2008), Soay sheep (*Ovies aries*) infected by nematodes (Coltman et al., 1999), or house finches (*Carpodacus mexicanus*) experimentally infected by bacteria (*Mycoplasma gallisepticum*; Hawley et al., 2005). Hence, there is good evidence that more heterozygous individuals are better protected against parasitism.

Again, the pattern is not all that easy. In many studies, no relationship was discovered. For example, no effect of outcrossing rate on infection levels was reported for freshwater snails

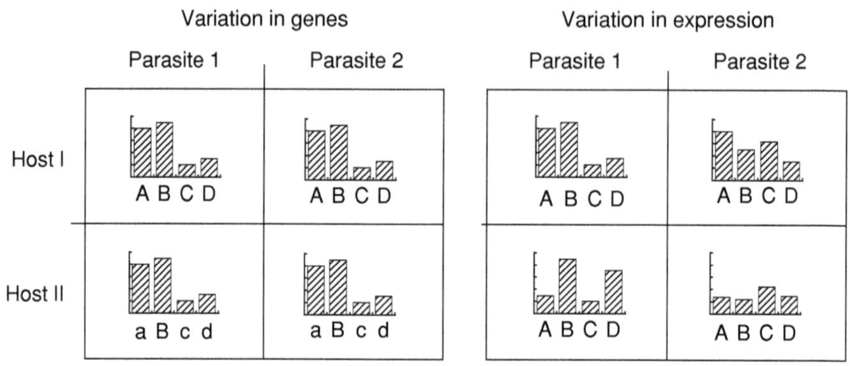

FIGURE 19.3: Two simplified models of how specific interactions between host (host genotypes I, II) and parasite (parasite strains 1, 2) come about, viewed from the host perspective. *Left:* The genomic sequences relevant for products affecting the interaction with the parasite vary among host lines, but their expression remains the same. *Right:* The genomic sequences are the same but the relative expression of the genes varies. Shown are four hypothetical host genes (A, B, C, D) and their variants (a, c, d). Parasite strains are assumed to vary with respect to the relevant properties without further details. The bars indicate the respective level of expression. Any real case is likely to have elements of both schemes, but it matters whether specific interactions are closer to one or the other model.

(*Lymnaea truncatula*; Trouve et al., 2003), no effect of heterozygosity of MHC alleles on infection rate with bacteria (*Aeromonas salmonicia*) in Atlantic salmon (*Salmo salar*; Lohm et al., 2002), and no relationship of infection by two species of gastrointestinal nematodes among several hundred females of reindeer (*Rangifer tarandus*) on the island of Svalbard (Côté et al., 2005). The study on sticklebacks is especially illuminating. The minimum parasite load in free-living fish was found at an intermediate level of heterozygosity (measured as number of MHC alleles that an individual harbors) rather than at a maximum level of heterozygosity (Fig. 19.3). The interpretation is that more diversity in MHC alleles would lower fitness due to an increased risk of self-reactivity of the MHC-based immune response (Reusch et al., 2001). Regardless of whether this is the case, the pattern is clear: some processes limit the usefulness of an ever-increased degree of heterozygosity, which otherwise seems to generally benefit the individual in the face of parasitism.

To the extent that individual or population-wide heterozygosity is protective against parasitism, this becomes important for issues in conservation biology. There are several examples where parasitic diseases have led to severe population crashes of endangered species. Examples include African wild dogs (*Lycaon pictus*) affected by an epidemic of canine distemper virus (Creel and Creel, 2002), frogs affected by fungi (Pounds et al., 2006), or the recent epidemic of a virus-induced facial tumor in Tasmanian devils (*Sarcophilus harrisii*; Bostanci, 2005), among many other cases.

SIGNATURES OF SELECTION

As host and parasite co-evolve, gene frequencies change, genes are lost or become dysfunctional (pseudogenes), and new genes emerge by mutation, duplication, or horizontal transfer (such as with pathogenicity islands). In any case, the co-evolutionary process should leave traces in the genomes of both parties. At the larger scale, such traces can be used to reconstruct the phylogenetic relationships among lineages of host and parasite species, or of host types and parasite strains within a given species. The latter is often used to trace the origin of epidemics, such as those caused by flu viruses (Ferguson et al., 2003).

In many cases, however, the question is what genes or genetic regions are affected by the co-evolution and in which ways. A classical textbook example is human sickle-cell anemia, where the underlying genetics can be simplified to one locus and two alleles (S, P). The polymorphism of these alleles is maintained by overdominance (heterosis) because the genotype SP has a selective advantage over the homozygous genotypes SS (which is highly anemic due to dysfunctional red blood cells) and PP (which is susceptible to *Plasmodium* infections, causing malaria). In areas where malaria has disappeared, the polymorphism still lingers and shows a signature of past selection pressures. For most other cases, the situation is more complex and the genetic signs of past selection are more subtle. As far as humans are concerned, such signs are of great interest to medicine because they can reveal what genotypes might protect or be susceptible against a certain parasitic disease (Sabeti et al., 2002; Nielsen 2005; Nielsen et al., 2005).

A major problem in identifying signatures of selection by parasites is to separate the effects of population history and mutations from those of selection. These different processes can sometimes generate very similar effects. For example, the degree of genetic variation between species increases by mutation accumulation and by positive directional selection, and genetic variation within species is reduced by both negative directional selection and by selective sweeps (Nielsen, 2005). Hence, it is not always easy or possible to separate processes from observing effects. Fortunately, though, the history of a population—for example, the occurrence of bottlenecks—affects the entire genome, whereas selection by parasites usually affects only a limited number of loci. Such differences, in combination with other elements, can be exploited to test for genetic signs that selection has occurred. Here, we can list a few of these approaches.

HOW TO TEST FOR SELECTION SIGNATURES

SELECTION AFFECTS NON-SYNONYMOUS MUTATIONS

Because the genetic code has some redundancies, the third position of a coding triplet is often redundant—that is, different nucleotides code for the same amino acid. An example is arginine, where the third position can be either A, G, C, or T. For histidine the third position is either C or T. Hence, a change in this third position does not affect the phenotype and, consequently, no signature of selection is to be expected. By contrast, selection affects the first two nucleotides and, hence, a signature of selection is can be seen. This difference is exploited by the MacDonald-Kreitman test (Box 19.1).

BOX 19.1: MACDONALD-KREITMANTEST

Not all changes in a nucleotide sequence of a coding region have the same effect. Due to the redundancy in the genetic code, some changes are synonymous (S)—that is, have no effect on the amino acid that is coded for by the nucleotide triplet. This is typically the case for the third position in the triplet. By contrast, some changes lead to a different amino acid—that is, are non-synonymous (N). This is typically the case for the first and second position of a triplet. Because a non-synonymous change leads to a different amino acid and then to a different protein, such a change can potentially be selected for or against, while the synonymous change is neutral in terms of selection. Once the reading frame for a genomic sequence (defining the first, second, and third positions for each triplet) is known, this difference can be exploited to detect signals of selection from a genomic sequence. The basic principle is that the neutral synonymous changes will reflect the effects of drift while the non-synonymous changes reflect the effects of selection. Hence, the comparison of changes at synonymous and non-synonymous sites reveals whether selection has happened. This test therefore is a convenient tool to study protein evolution. The standard procedure is the MacDonald-Kreitman test (MacDonald and Kreitman, 1991). It is a very robust test, for example, against the effects of population history.

The test is run by comparing pairs of sequences from different species or populations, as in the hypothetical example of Box 19.1, Figure 1. For each pairwise comparison, the number of non-synonymous (d_N) and synonymous (d_S) changes and the respective ratio (d_N/d_S) are calculated. Under neutral evolution, the two kinds of substitutions should have the same neutral consequences and so behave similarly. Hence, under neutral evolution we expect $d_N/d_S = 1$. If the sequences are under positive selection, the non-synonymous changes should outweigh the synonymous changes and hence $d_N/d_S > 1$. The opposite is true for negative selection, and hence $d_N/d_S < 1$. Whether any observed ratio d_N/d_S is indeed different from 1 can be tested statistically with the MacDonald-Kreitman procedure. A potential problem with the test is that negative selection on newly arising mutations dominates in natural populations, since the majority of new mutations are deleterious. Hence, it is much easier to see a signal of negative than of positive selection (see Box 19.1, Fig. 1). If positive selection is found, this is therefore a conservative estimate.

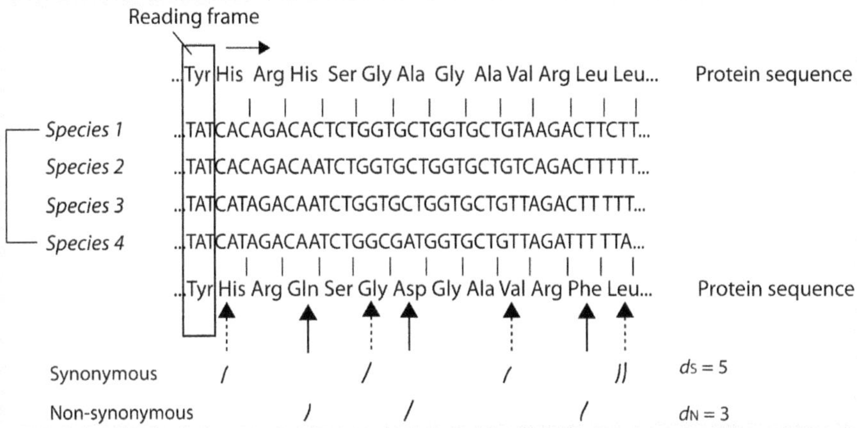

BOX 19.1, FIGURE 1: Hypothetical example of the MacDonald-Kreitman test. Here, the aligned nucleotide sequences of four species are shown. The reading frame defines the corresponding protein (amino acid) sequence and so identifies synonymous and non-synonymous changes. When the sequences of species 1 and species 4 are compared, this example yields five changes that code for the same amino acid (synonymous, $d_S = 5$) and three changes where the meaning of the codon changes and a different amino acid is coded for (non-synonymous changes, $d_N = 3$). The ratio of non-synonymous to synonymous changes is therefore $d_N/d_S = 3/5 = 0.6$. If this were a statistically supported value, the sequence would be under negative selection.

POPULATION DIVERGENCE

Selection by parasites can lead to the genetic separation of different populations. Such genetic divergence can be analyzed by testing for nucleotide diversity from genomic sequences (Box 19.2). Tajima's D is often used for this purpose. It starts from the observation that without selection, the allele frequencies should be inversely proportional to the rank of their occurrence, and that deviations indicate the effects of selection. The HKA (Hudson-Kreitman-Aguadé) test is another approach used for the same purpose. It considers changes at loci between pairs of species or populations (Box 19.2).

PHYLOGENY OF HAPLOTYPES

Haplotypes are very useful tools to study signs of selection, too. For example, positive directional selection reduces nucleotide diversity among haplotypes and leads to few and common haplotypes. The phylogenetic tree furthermore shows how deep the branches are that connect a pair of given haplotypes. Such information can yield an insight into the population history and the selective forces (Box 19.3).

BOX 19.2: SEPARATION OF POPULATIONS

ANALYSIS OF ALLELE FREQUENCY SPECTRA

When a population is screened for the different alleles at a locus, then we might find that allele A occurs two times ($i = 2$), allele B occurs 6 times ($i = 6$), and so forth. The distribution of these frequencies is itself a distribution, $f(i)$. The distribution $f(i)$ is the frequency spectrum of alleles. Its shape is affected by the selective processes that have happened in this population (or species). In particular, under neutrality the distribution is expected to follow $f(i): \frac{1}{i}$ —that is, the frequencies should be inversely proportional to the occurrence of the alleles (Box 19.2, Fig. 1). The different effects of the selective process on the distribution of alleles can be used to test for signatures of selection.

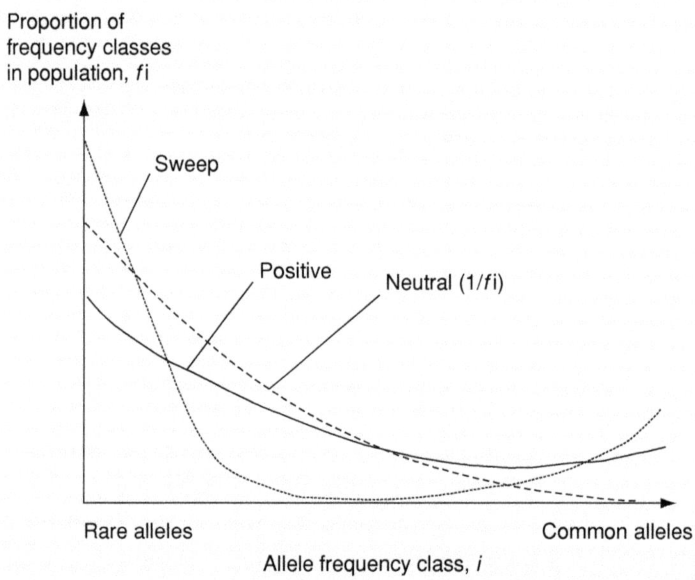

BOX 19.2, FIGURE 1: Frequency spectra of alleles under different assumptions: neutrality (broken line), positive selection (solid line), and a selective sweep (dotted line).

Tajima's D

This is the most widely used test exploiting this principle. It is based on sequence data. As with the divergence tests, the test uses the observed nucleotide diversity, π, within a population or between species, together with the number of segregating sites, S (i.e., number of positions in the sequence that have at least one different nucleotide across this sample). Under neutral evolution, the expected value is $E(\pi)=\theta=4N\mu$, where N is the effective population size and μ is the (neutral) mutation rate. On the other hand, given the number of segregating sites, divergence θ can also be estimated from $\theta_S=\frac{S}{\sum_{i=1}^{n-1}\frac{1}{i}}$

If the difference $d=\pi-\Theta_s$ is too large, then neutrality is rejected and (positive) selection in this population is inferred. The test statistic is calculated as:

$$\text{Tajima's } D=\frac{d}{\sqrt{Var(d)}}$$

The test statistic, D, follows a beta distribution. A modification of this test uses phylogenetic information to give polarity to the character states (Fu and Li, 1993).

GENETIC DIVERGENCE

A locus under (positive) selection will diverge more in different species or populations than a locus that is neutral (Box 19.2, Fig. 2). The genetic divergence (i.e., the average per-site substitution rate) within a population or species under neutral evolution is expected to be $\theta=4N\mu$, where N is the effective population size and μ the (neutral) mutation rate. The parameter Θ is a fundamental quantity in population genetics with an infinite site model (i.e., with an infinite number of loci where every new mutation occurs at another site). With neutral evolution the observed average per-site substitution rate ("nucleotide diversity") is expected to correspond to $E(\pi)=\theta$. Selective processes will change this value and, hence, the difference between Θ and π is indicative for deviations from neutrality. By contrast, the value of N is the same for all loci in the same population or species.

To calculate the nucleotide diversity, π, one has to average the number of pairwise differences between the n aligned sequences in the sample as follows:

$$\pi=\frac{n}{n-1}\sum_{ij}f_if_jp_{ij}$$

where f_i and f_j are the relative frequencies of sequences i and j, respectively, in the sample of n sequences; p_{ij} is the number of nucleotide substitutions in the sequence pair ij. Finally, S is the fraction of segregating sites—that is, the number of positions in the sequence out of all positions that have at least two different nucleotides in the comparison. Under neutral evolution the expected number of segregating sites is $E(S)=\theta\sum_{i=1}^{n-1}\frac{1}{i}$, where i is the running index for the sequences under investigation.

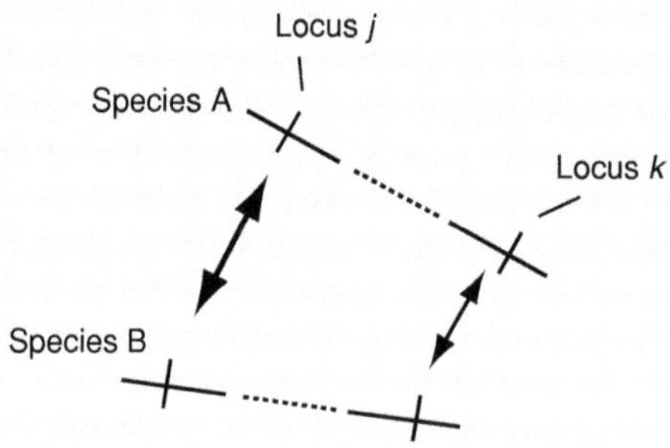

BOX 19.2, FIGURE 2. The genetic divergence, π between loci (j, k) in two different species or populations (A, B) might differ due to selection. For example, locus j might be under positive selection, in which case the divergence is higher than at a neutral locus k.

HKA (Hudson-Kreitman-Aguadé) test

In this test (Hudson et al., 1987), a sample from m loci is taken. At each of these m loci, the sequences are compared and the differences in nucleotides counted. From these counts, it is calculated across all loci i:

- K_{ij} = the number of loci that are different between a pair of species (j = 1, 2) (i.e., the number of segregating sites)
- D_i = the average number of nucleotide differences between the two species (or sequences)

From these quantities, the test statistic is calculated; it is approximately distributed as X^2 with $df = 2m - 2$. The general assumptions of the HKA test include constant population sizes, no recombination within a locus but recombination between loci.

Lewontin-Krakauer test

Based on the same principles, this test identifies whether a given locus of interest is under selection when compared with other loci. This test does not require sequence data but can be run with di-allelic loci. The idea is that the selected locus should diverge more than expected by chance. In the test, for each locus, a measure of divergence is calculated across populations as:

$$F = \frac{\sigma_p^2}{\bar{p}(1-\bar{p})}$$

where $\bar{p}$ and σ_p^2 are the mean and variance of the allele frequencies at this locus. If the value of F is too large among loci compared to a random expectation, neutrality is rejected and (positive) selection can be inferred. A major problem is to define what "too large" means, because the value of F depends in sensitive ways on population demography. Variants of this test use the F_{ST} values for SNPs as a measure of population divergence (Akey et al., 2002) or are based on whole genome scans (Beaumont and Balding, 2004).

SELECTIVE SWEEPS

A sweep occurs when an allele at a given locus increases relatively rapidly under the effect of selection. The allele will eventually go to fixation and, in our context, the suspected selective factor would be resistance to parasitism. Because an allele at a locus (or perhaps alleles at several loci) is favored by selection, the locus itself must show reduced diversity. At the same time, sites adjacent to the selected locus will be dragged along by the hitchhiking effect (Maynard Smith and Haigh, 1974). Such patterns can be used to identify signatures of selection (Box 19.4).

EXAMPLES OF SELECTION SIGNATURES

BALANCING SELECTION

As an example, defense against bacteria has a genetic basis in the cabbage plant (*Arabidopsis thaliana*) in form of the so-called R genes that are involved in pathogen recognition. Plant populations vary considerably in the R-gene loci. Looking at the sequences flanking the Rpm1 gene that confers the ability to recognize *Pseudomonas* suggested that these R genes were under balancing selection. Furthermore, the R genes must have maintained genetic polymorphism for millions of years (Stahl et al., 1999). This study, for instance, used Tajima's D as well as checking nucleotide

BOX 19.3: ANALYZING HAPLOTYPES

Haplotypes are immensely useful tools for genetic analyses, notably for the search of selected sites. Haplotypes arise naturally with gametes and may be statistically inferred from the genomic sequences in diploid organisms. Any given population usually contains several haplotypes for a given gene (Box 19.3, Fig. 1).

Comparing haplotypes can give information about their frequency spectra and therefore lead to tests on signatures of selection. Another use of this information is to construct haplotype networks. In this network, haplotypes are connected according to their similarity, with those differing only by one nucleotide change immediately adjacent to each other. The network structure under neutral evolution is characterized by a certain mix of short and long connecting lines, as well as by a certain branching pattern, if the haplotypes have been under positive selection. Box 19.3, Figure 2, shows an example of a haplotype network.

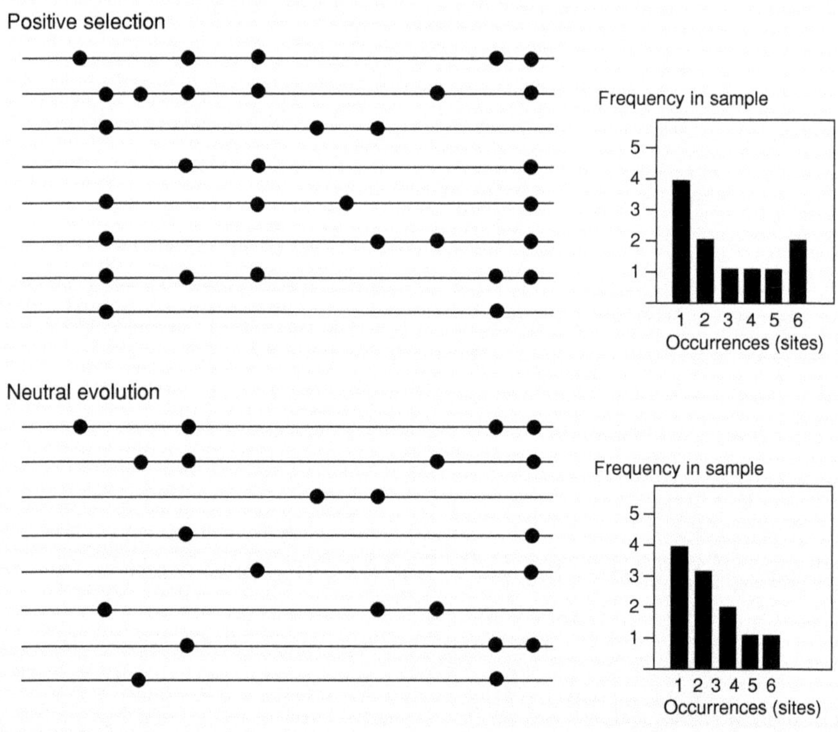

BOX 19.3, FIGURE 1: Haplotypes under positive selection (top panel) and under neutral evolution (lower panel). Each line represents a sequence and each dot is a mutation (e.g., a nucleotide substitution). Sequences that carry the same pattern of substitutions belong to the same haplotype. The frequency of substitutions in the sample is indicated in the histograms at right (frequency spectrum). These show the frequency of affected sequences that carry a substitution at the same site.

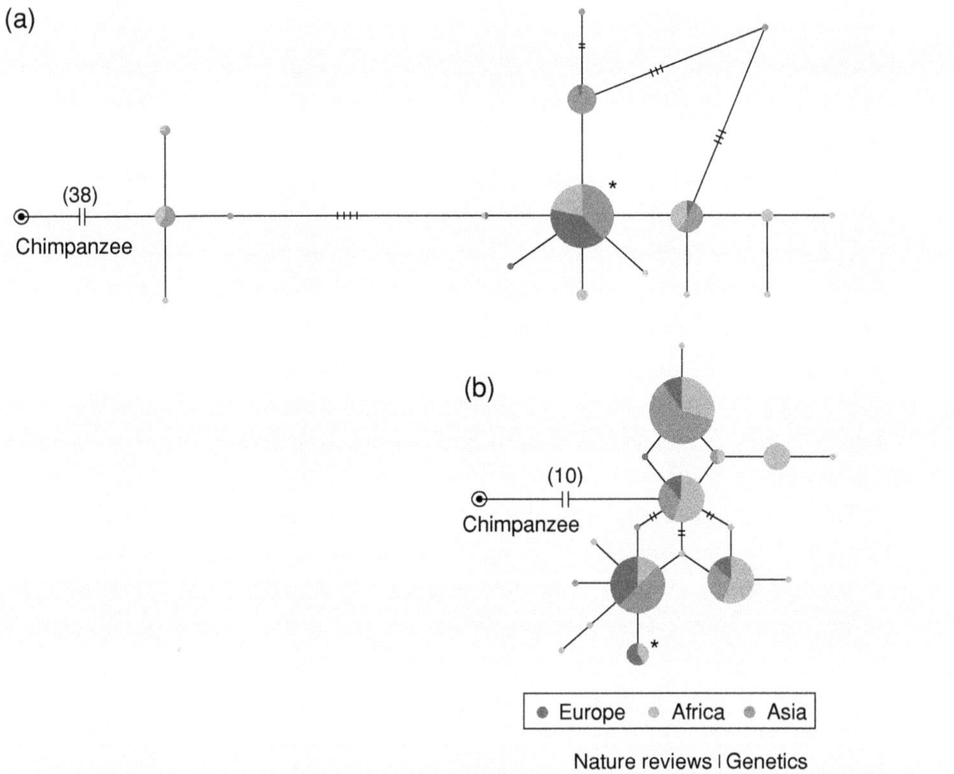

BOX 19.3, FIGURE 2: Haplotype network for a housekeeping gene (cytochrome P450; top panel) and a gene involved in immune defense (chemokine receptor CCR5; lower panel). The network connects an outgroup (chimpanzee) with samples from human populations in Europe, Africa, and Asia (from Bamshad et al., 2002). The upper panel suggest positive selection, since the derived haplotypes (with chimpanzees as the ancestral type) are more common than expected under neutral evolution. In the lower panel, balancing selection is likely, since there are two main clusters of haplotypes separated by long branches.

BOX 19.4: SELECTIVE SWEEPS AND LINKAGE

A genomic region (a locus) that is under selection will have a changed nucleotide diversity. For example, under positive selection, the diversity of nucleotides is reduced. Furthermore, this effect will leave a trace of the same kind on adjacent sites because of the "hitchhiking effect" (Maynard Smith and Haigh, 1974). Hitchhiking of adjacent sites occurs because genomic sites located nearby are to a certain degree linked to the locus under selection. Even though such linkage is eroded over time by recombination events, the trace of selection on a locus in the neighborhood persists for some time and for some distance from the locus (Box 19.4, Fig. 1).

The effect on a selected locus and the nearby sites can be used to test for a signal of selection. For example, the frequency spectrum of alleles also changes at a selected site in predictable ways. The change in the frequency spectrum furthermore extends to nearby sites due to the hitchhiking effect. The expectations for the frequency spectrum at a given site as a function of its distance from the selected locus (Box 19.4, Fig. 1) can be calculated. The comparison with the observed frequency spectra at sites in various distances from a putatively selected locus indicates whether, for example, a selective sweep has indeed occurred (Kim and Stephan, 2002).

For the same reasons, selection will leave a trace on local linkage disequilibria (LD) along the genome (Box 19.4, Fig. 1) as well as on haplotype structure. This is understood for the effect of balancing selection, but analyses for selective sweeps and positive selection are still not fully resolved. Under balancing selection, the local LD is reduced when this selection regime has had sufficient time to establish the polymorphism. If balancing selection is recent, LD might transiently be increased. Similarly, haplotype structure is affected by selective sweeps. This effect can be used to test for such events (Sabeti et al., 2002).

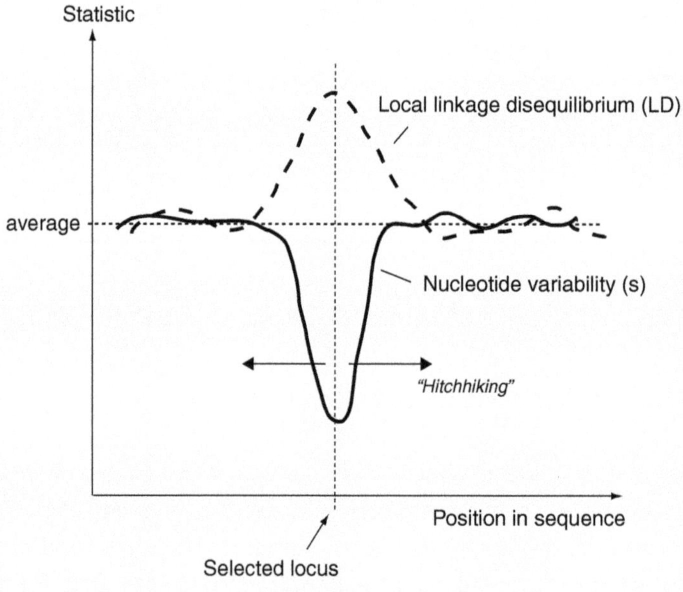

BOX 19.4, FIGURE 1: A selected locus shows reduced nucleotide variability and increased local linkage. The effect extends to nearby sites due to the hitchhiking effect.

divergence at synonymous and non-synonymous sites. Simulating different genetic scenarios added to the identification of the most probable selection regime. A similar conclusion of balancing selection was also reached when considering fungal parasites instead of bacteria (Bittner-Eddy et al., 2000).

DIRECTIONAL SELECTION

It appears that directional selection has been important to diversify host defense genes with their gene families (e.g., Stotz et al., 2000). But directional selection is also suspected for RNAi genes that are involved in antiviral defenses (Obbard et al. 2006; Obbard et al. 2009) or antibacterial peptides in *Drosophila* (Lazzaro and Clark, 2003). Similarly, Relish, a transcription factor downstream of the pathway that is, for example, triggered by gram-negative bacteria infecting insects, and the corresponding recognition molecules, gram-negative binding protein (GNBP), show a strong sign of directional selection (Bulmer and Crozier, 2005).

SELECTIVE SWEEPS

Immune genes are known to have been subject to selective sweeps. TEP1 is such an example; it is an important defense gene in the mosquito *Anopheles gambiae* when infected by Plasmodium, which uses the insect as its vector (Obbard et al., 2008). In fact, immune genes might undergo selective sweeps more often than non-immune genes (Hurst and Smith, 1999; Schlenke and Begun, 2003). Also many regions of the human genome show signs of selective sweeps, especially also loci known to be involved in host defenses or associated with disease risks (Nielsen et al., 2005).

Hence, the evidence suggests that all forms of selection can occur. The differences among different groups of genes are the most interesting information in this context. For example, gene sequences of the core signal transduction pathways evolve rapidly (Jiggins and Hurst, 2003), but receptors and opsonization molecules are either conserved or highly divergent (Jiggins and Kim, 2006; Lazzaro, 2008). Strong selection is also found for key genes of the antiviral defense as mentioned above (Obbard et al., 2009), or peptidoglycan recognition proteins (PGRPs) of ants (Viljakainen et al., 2009). Among the immune system's effectors, antimicrobial peptides (AMPs) play an important role. Together with other components, AMPs regulate and control parasite development (Boulanger et al., 2006). In fact, AMPs are found everywhere in the innate immune systems of plants, invertebrates, and vertebrates, and are fundamentally important for organisms. AMPs appear to rapidly evolve in vertebrates (e.g., Duda et al., 2002; Lynn et al., 2004), their coding sequence and specificity varying among closely related species. Although the final jury is still out, rapid evolution of AMPs seems not to be the general rule in insects (Viljakainen et al., 2009; Yoon et al., 2009). Comparative genomics suggests that while recognition molecules rapidly diverge, the effector molecule families are more conserved, even though they can diversify by duplication or the emergence of new families (Waterhouse et al., 2007; Kafatos et al., 2009). Also, different scenarios suggest different things. For example, the Red Queen scenario of sustained antagonistic co-evolution between host and parasites would suggest that balancing selection should be the dominant pattern. Recurrent selective sweeps might produce a pattern that is similar to some degree, but there is no requirement of negative frequency-dependent selection as in the Red Queen model. With the accumulation of large data sets from sequencing and genomic projects, the signatures of selection will become easier to study and the comparative aspect, which eventually will be the most telling one, can be studied with unprecedented detail.

GENETIC STRUCTURE OF PROTOZOAN POPULATIONS

We here take protozoa as an example for parasites more generally, but protozoa such as trypanosomes, apicomplexa, or sarcodina (amoebae) are important groups of parasites to start with. Indeed, some of the most dangerous and medically important parasites are found in this group. Examples are tropical malaria (*Plasmodium falciparum*), leishmaniasis (*Leishmania*), Chagas' disease (*Trypanosoma cruzi*), and sleeping sickness (*Trypanosoma brucei*). Evolutionary change in the parasites, the effect of control measures, or strategies of prevention must obviously take into account their populations. Therefore, it is of considerable interest to understand how the populations of protozoan parasites are structured. And there is a longstanding controversy about how these populations are structured, notably whether protozoa exchange genes with one another (sexual reproduction) or whether they are clonal.

The so-called clonal theory of parasitic protozoa suggests that protozoa have clonal populations, with perhaps only a few exceptions (Tibayrenc and Ayala, 2002). Evidence for this view is mastered from the observation of genetic variation in protozoan parasite populations, such as measures of heterozygosity, linkage disequilibria, or deviations from Hardy-Weinberg frequencies. The latter is often observed in protozoan populations and would indicate a deviation from random mating among individuals, which might result from the lack of sexual reproduction. However, the structure of parasite populations more generally, and of protozoans in particular, is highly structured anyway because parasites infect individual hosts that represent "islands" in a sea of non-suitable habitat. Furthermore, there is a certain degree of "matching" between host and parasite genotypes, which adds to the effect of separation. As trivial as this might seem, the implications have only more recently begun to be addressed. For example, under these conditions a population of protozoan parasites is unlikely to be in Hardy-Weinberg equilibrium even if there were perfect panmixia otherwise (Heitman, 2006; McKenzie et al., 2008). The general problem has also been labeled the "iceberg effect," since only the tips of an underlying population structure can be typically analyzed when genetically typing the isolates from individual hosts (Tibayrenc, 1999). Hence, whether or not protozoa are clonal or sexual must be established by checking for the occurrence of genetic exchange. With this basic knowledge the frequency of exchange can be estimated and the observed population structure (Table 19.2) can be interpreted better.

The exchange of genetic material between protozoan cells is now established in a few cases. Such exchange must occur in a host, and some researchers suggest that this requirement explains why multiple infections are an advantage for a parasite even though it must thereby share its host with others. Regardless, the classical case is *Trypanosoma brucei,* which uses the tsetse fly as its

TABLE 19.2: Genetic Structure of Some Eukaryote Parasite Populations[a]

Species	Structure[b]
Fungi:	
Candida albicans	No evidence for clonality
Candida tropicalis/paratropicalis	Clonality hypothetical
Cryptococcus neoformans	Clonality hypothetical
Sacharomyces cerevisiae	Clonality hypothetical
Protozoa:	
Entamoeba histolytica	Some evidence for clonality
Giardia spp.	Some evidence for clonality
Leishmania spp.	Some evidence for clonality
Leishmania spp.	Clonality hypothetical
Crithidia fasciolata	No evidence for sex found[c]
Naegleria spp.	Some evidence for clonality
Plasmodium falciparum	Clonality hypothetical
Toxoplasma gondii	Clonality hypothetical
Trichomonas spp.	Clonality hypothetical
Trypanosoma brucei, West and East Africa	Clonality ensured
Trypanosoma brucei, East Africa (wild)	Clonality hypothetical
Trypanosoma brucei rhodensiense	Clonality hypothetical
Trypanosoma cruzei	Clonality ensured
Trypanosoma vivax	Clonality ensured

[a] based on Tibayrenc and Ayala, 2002; [c] Votypka et al., 2001.
[b] Clonality hypothetical: uncertainty due to small sample sizes studied.
[c] Some evidence for clonality: based on too few markers studied to firmly decide.

vector (Jenni et al., 1986). The exchange includes meiosis and self-fertilization (Heitman, 2006). Similarly, genetic exchange has been confirmed for *T. cruzi* and *Leishmania major* (Heitman, 2006; Akopyants et al., 2009). The agent of malaria, *Plasmodium*, has a sexual cycle with the gametocytes fusing into a zygote, even if this frequently results in self-fertilization (Hall et al., 2005). Some apicomplexan parasites are known to have gametocysts and a sexual phase in their life cycle. Examples are *Hepatozoon*, *Toxoplasma*, *Eimeria*, and *Cryptosporidium* (West et al., 2000). It is not always clear, however, whether such genetic exchange includes a regular meiosis or whether alternative mechanisms are used. For example, in trypanosomes it is also possible that the normal 2n cells fuse into a 4n cell that subsequently undergoes fission to reconstitute the normal diploid state (Gibson and Stevens, 1999).

The evidence thus suggests that genetic exchange among protozoa does occur but is probably not very widespread. Alternatively, the potential for exchange might be present in a wide range of taxa, given the evidence that conserved genes, which are involved in regulating meiosis, can be found in basal groups (such as in *Giardia*; Heitman, 2006). The most pertinent question in this context is then to understand how often genetic exchange occurs in the natural situation, given that the basic set of genes is present, or given that it can be observed under controlled conditions. One possibility is that ecological conditions, such as the frequency of multiple infections, might dominate the rate of exchange. In the case of malaria, for example, where there is still a lot of controversy about what exactly populations look like, the population structure is closer to clonal in areas of low transmission, but even in areas of high transmission the evidence for panmixia is ambiguous, perhaps due to a high rate of self-fertilization (Razakandrainibe et al., 2005).

It is therefore certainly a good description that protozoan parasites show "epidemic clonality," whereby longer episodes of clonal reproduction are interrupted by occasional genetic exchange (Maynard Smith et al., 1993). The frequency of exchange must in turn be related to the frequency of multiple infections and, perhaps, to the probability that two co-infecting strains "match" each other to some degree. It is also possible that some strains might be more prone to engage in exchange than others. With epidemic clonality, a parasite strain does exist for a certain time and then disappears by exchange. How long such strains persist is a crucial question and an important parameter for any disease dynamics. We are still a long way from understanding the situation. However, modern methods of high-throughput genotyping and methods of analyzing such data give justified hope that we will see much progress in the area in the years to come.

REFERENCES

Acevedo-Whitehouse K, Gulland F, Greig D, Amos W (2003) Inbreeding-dependent pathogen susceptibility in California sea lions. Nature 422:35

Akey JM, Zhang G, Zhang K, Jin L, Shriver MD (2002) Interrogating a high-density SNP map for signatures of natural selection. Genome Research 12:1805–1814

Akopyants NS, Kimbllin N, Secundino N, Patrick R, Peters N, Lawyer P, Dobson DE, Beverley SM, Sacks DL (2009) Demonstration of genetic exchange during cyclical development of *Leishmania* in the sand fly vector. Science 324:265–268

Altincicek B, Knorr E, Vilcinskas A (2008) Beetle immunity: Identification of immune-inducible genes from the model insect *Tribolium castaneum*. Developmental and Comparative Immunology 32: 585–595

Arkush KD, Giese AR, Mendonca HL, McBride AM, Marty GD, Hedrick PW (2002) Resistance to three pathogens in the endangered winter-run chinook salmon (*Oncorhynchus tshawytscha*): effects of inbreeding and major histocompatibility complex genotypes. Canadian Journal of Fisheries and Aquatic Sciences 59:966–975

Baer B, Schmid-Hempel P (1999) Experimental variation in polyandry affects parasite loads and fitness in a bumble-bee. Nature 397:151–154

Bamshad MJ, Mummidi S, Gonzalez E, Ahuja SS, Dunn DM, Watkins WS, Wooding S, Stone AC, Jorde LB, Weiss RB, Ahua SK (2002) A strong signature of balancing selection in the 5' cis-regulatory region of CCR5. Proceedings of the National Academy of Sciences USA 99:10539–10544

Beaumont MA, Balding DJ (2004) Identifying adaptive genetic divergence among populations from genome scans. Molecular Ecology 13:969–980

Beier D, Gross R (2006) Regulation of bacterial virulence by two-component systems. Current Opinion in Microbiology 9:143–152

Berenos C, Wegner MK, Schmid-Hempel P (2010) Antagonistic co-evolution with parasites maintains host genetic diversity: an experimental test. Proceedings of the Royal Society London B 278: 218–224

Biron DG, Moura H, Marché L, Hughes AL, Thomas F (2005) Towards a new conceptual approach to "parasitoproteomics." Trends In Ecology and Evolution 21:162–168

Bittner-Eddy PD, Crute IR, Holub EB, Beynon JL (2000) RPP13 is a simple locus in *Arabidopsis thaliana* for alleles that specify downy mildew resistance to different avirulence determinants in *Peronospora parasitica*. Plant Journal 21:177–188

Bonhoeffer S, Chappey C, Parkin NT, Whitcomb JM, Petropoulo CJ (2004) Evidence for positive epistasis in HIV-1. Science 206:1547–1550

Bostanci A (2005) A devil of a disease. Science 307:1035

Boulanger N, Bulet P, Lowenberger CA (2006) Antimicrobial peptides in the interactions between insects and flagellate parasites. Trends in Parasitology 22:262–268

Bulmer MS, Crozier RH (2005) Variation in positive selection in termite GNBPs and Relish. Molecular Biology and Evolution 23:317–326

Burdon JJ (1987) Diseases and Plant Population Biology. Cambridge University Press, Cambridge

Burdon JJ, Jarosz AM (1991) Host-pathogen interactions in natural populations of *Linum marginale* and *Melampsora lini*: 1. Patterns of resistance and racial variation in a large host population. Evolution 45: 205–217

Busch JW, Neiman M, Koslow JM (2004) Evidence for maintenance of sex by pathogens in plants. Evolution 58:2584–2590

Calleri DV, Reid EM, Rosengaus RB, Vargo EL, Traniello JFA (2006) Inbreeding and disease resistance in a social insect: effects of heterozygosity on immnnocompetence in the termite *Zootermopsis angusticollis*. Proceedings of the Royal Society London B 273:2633–2640

Carius HJ, Little TJ, Ebert D (2001) Genetic variation in a host-parasite association: potential for coevolution and frequency-dependent selection. Evolution 55:1136–1145

Carroll SB (2005) Evolution at two levels: on genes and form. PLoS Biology 3:1159–1116

Carton Y, Nappi A, Poirie M (2005) The genetics of anti-parasite resistance in invertebrates. Developmental and Comparative Immunology 29:9–32

Carton Y, Nappi AJ (2001) Immunogenetic aspects of the cellular immune response of *Drosophila* against parasitoids. Immunogenetics 52:157–164

Cassinello J, Gomendino M, Roldan ERS (2001) Relationship between coefficient of inbreeding and parasite burden in endangered gazelles. Conservation Genetics 15:1171–1174

Coego A, Ramirez V, Gil M, Flors V, Mauch-Mani B, Vera P (2005) An *Arabidopsis* homeodomain transcription factor, OVEREXPRESSOR OFCATIONIC PEROXIDASE 3, mediates resistance to infection by necrotrophic pathogens. Plant Cell 17:2123–2137

Coltman DW, al. e (1999) Parasite-mediated selection against inbred Soay sheep in a free-living, island population. Evolution 53:1259–1267

Cortesi P, McCulloch C, Song H, Lin H, Milgroom M (2001) Genetic control of horizontal virus transmission in the chestnut blight fungus, *Cryphonectria parasitica*. Genetics 159:107–118

Côté S, D., Stien A, Irvine RJ, Dallas J, F., Marshall F, Halvorsen O, Langvatn R, Albon SD (2005) Resistance to abomasal nematodes and individual genetic variability in reindeer. Molecular Ecology 14:4159–4168

Cotter PA, diRita VJ (2000) Bacterial virulence gene regulation: an evolutionary perspective. Annual Review of Microbiology 54:519–565

Creel S, Creel NM (2002) The African Wild Dog. Behavior, Ecology and Conservation. Princeton University Press, Princeton

Dana AN, Hong YS, Kern MK, Hillenmeyer ME, Harker BW, Lobo NF, Hogan JR, Romans P, Collins FH (2005) Gene expression patterns associated with blood-feeding in the malaria mosquito *Anopheles gambiae*. BMC Genomics 6:5

Desrosiers M, Kielczewska A, Loredo-Osti J, Adam S, Makrigiannis A, Lemieux S, Pham T, Lodoen M, Morgan K, Lanier L, Vidal S (2005) Epistasis between mouse *Klra* and major histocompatibility complex class I loci is associated with a new mechanism of natural killer cell-mediated innate resistance to cytomegalovirus infection. Nature Genetics 37:593–599

Dirlewanger E, Isaac PG, Ranade S, Belajouza M, Cousin R, Vienne HD (1994) Restriction fragment length polymoprhisms analysis of loci associated with disease resistance genes and developmental traits in *Pisum sativum*. Theoretical and Applied Genetics 88:17–27

Dobrindt U, Hochhut B, Hentschel U, Hacker J (2004) Genomic islands in pathogenic and environmental microorganisms. Nature Reviews Microbiology 2:414–424

Duda TFJ, Vanhoye D, Nicolas P (2002) Roles of diversifying selection and coordinated evolution in the evolution of antimicrobial peptides. Molecular Biology and Evolution 19:858–864

Dupas S, Frey F, Carton Y (1998) A single parasitoid segregating factor controls immune suppression in *Drosophila*. Heredity 89:306–311

Dybdahl MF, Krist AC (2004) Genotypic vs. condition effects on parasite-driven rare advantage. Journal of Evolutionary Biology 17:967–973

Edfors-Lilja I (1991) *Escherichia coli* resistance in pigs. In: Owen JB, Axford RFE (eds) Breeding for Resistance in Farm Animals. CABI Publishing, Wallingford, U.K., pp. 424–436

Ellis JG, Lawrence GJ, Luck JE, Dodds PN (1999) Identification of regions in alleles of the flax rust resistance gene L that determine differences in gene-for-gene specificity. Plant Cell 11:495–506

Ferguson MM, Drahushchak LR (1990) Disease resistance and enzyme heterozygosity in rainbow trout. Heredity 64:413–417

Ferguson NM, Glavani AP, Bush RM (2003) Ecological and immunological determinants of influenza evolution. Nature 422:428–433

Flor HH (1956) The complementary genic systems in flax and flax rust. Advances in Genetics 8:29–54

Frank SA (1994) Recognition and polymorphism in host-parasite genetics. Philosophical Transactions of the Royal Society London B 346:283–293

Frank SA (1996) Problems inferring the specificity of plant-pathogen genetics. Evolutionary Ecology 10: 319–322

Fu YX, Li WH (1993) Statistical tests of neutrality of mutations. Genetics 133:693–709

Gadau J, Page REJ, Werren JH, Schmid-Hempel P (2000) Genome organization and social evolution in Hymenoptera. Naturwissenschaften 87:87–89

Gibson G (2003) Population genomics: celebrating individual expression. Heredity 90:1–5

Gibson G, Weir B (2005) The quantitative genetics of transcription. Trends in Genetics 21:616–623

Gibson W, Stevens J (1999) Genetic exchange in the Trypanosomatidae. Advances in Parasitology 43: 1–46

Hacker J, Kaper JB (2000) Pathogenicity islands and the evolution of microbes. Annual Review of Microbiology 54:641–679

Haldane JBS (1949) Disease and evolution. La Ricerca Scientifica 19, Suppl. :68–76

Hall N, Karras M, Raine JD, Carlton J, Kooij TW, Berriman M, Florens L, Janssen CS, Pain A, Christophides GK, James K, Rutherford K, Harris B, Harris D, Chrucher C, Quail MA, Ormond D, Doggett J, Trueman HE, Mendoza J, Bidwell SL, Rajandream M-A, Carucci DJ, Yates III JR, Kafatos FC, Janse C, Barrell B, Turner CMR, Waters AP, Sinden RE (2005) A comprehensive survey of the *Plasmodium* life cycle by genomic, transcriptomic, and proteomic analyses. Science 307:82–86

Hawley DM, Sydenstricker KV, Kollias GV, Dhondt AA (2005) Genetic diversity predicts pathogen resistance and cell-mediated immunocompetence in house finches. Biology Letters 1:326–329

Heitman J (2006) Sexual reproduction and the evolution of microbial pathogens. Current Biology 16: R711–R725

Hudson RR, Kreitman RR, Aguadé M (1987) A test of neutral molecular evolution based on nucleotide data. Genetics 116:153–159

Hurst LD, Smith NG (1999) Do essential genes evolve slowly? Current Biology 9:747–750

Jenni L, Marti S, Schweitzer J, Betschart B, LePage RWF, Wells JM, Tait A, Paindavoine P, Pays E, Steinert M (1986) Hybrid formation between African trypanosomes during cyclical transmission. Nature 322: 173–175

Jiggins FM, Hurst GDD (2003) The evolution of parasite recognition genes in the innate immune system: purifying selection on *Drosophila melanogaster* peptidoglycan recognition proteins. Journal of Molecular Evolution 57:598–605

Jiggins FM, Kim K-W (2006) Contrasting evolutionary patterns in *Drosophila* immune receptors. Journal of Molecular Evolution 63:769–780

Jones IT (1986) Inheritance of adult-plant resistance to mildew in oats. Annals of Applied Biology 109: 187–192

Kafatos F, Waterhouse R, Zdobnov E, Christophides G (2009) Comparative genomics of insect immunity. In: Rolff J, Reynolds SE (eds) Insect Infection and Immunity. Oxford University Press, Oxford, pp. 86–105

Kema GHJ, Sayoud R, Annone JG, Van SCH, Van GM, de Bree J (1996a) Genetic variation for virulence and resistance in the wheat-*Mycosphaerella graminicola* pathosystem I. Interactions between pathogen isolates and host cultivars. Phytopathology 86:200–212

Kema GHJ, Sayoud R, Annone JG, Van Silfhout CH (1996b) Genetic variation for virulence and resistance in the wheat-*Mycosphaerella graminicola* pathosystem: II. Analysis of interaction between pathogen isolates and host cultivars. Phytopathology 86:213–220

Kim YJ, Stephan W (2002) Detecting a local signature of genetic hitchhiking along a recombining chromosome. Genetics 160:765–777

Klover PX, Caicedo AL (2001) The genetic architecture of disease resistance in plants and the maintenance of recombination by parasites. Molecular Ecology 10:1–16

Lazzaro B, Sceurman BK, Clark AG (2004) Genetic basis of natural variation in *D. melanogaster* antibacterial immunity. Science 303:1873–1876

Lazzaro BP (2008) Natural selection on the *Drosophila* antimicrobial immune system. Current Opinion in Microbiology 11:284–289

Lazzaro BP, Clark AG (2003) Molecular population genetics of inducible antibacterial peptide genes in *Drosophila*. Molecular Biology and Evolution 20:914–923

Lenski RE (1988) Experimental studies of pleiotropy and epistasis in *Escherichia coli*. I. Variation in competitive fitness among mutants resistant to virus T4. Evolution 42:425–432

Lohm J, Grahn M, Langefors A, Andersen O, Storset A, Von Schantz T (2002) Experimental evidence for major histocompatibility complex-allele-specific resistance to a bacterial infection. Proceedings of the Royal Society London B 269:2029–2033

Luikart G, Pilgrim K, Visty J, Ezenwa VO, Schwartz MK (2008) Candidate gene microsatellite variation is associated with parasitism in wild bighorn sheep. Biology Letters 4:228–231

Lynn DJ, Lloyd AT, Fares MA, O'Farrelly C (2004) Evidence of positively selected sites in mammalian alphadefensins. Molecular Biology and Evolution 21:819–827

MacDonald JH, Kreitman M (1991) Adaptive protein evolution in the Adh locus in *Drosophila*. Nature 351:652–654

Maynard Smith J, Haigh J (1974) The hitchhiking effect of a favorable gene. Genetical Research 23: 23–35

Maynard Smith J, Smith NH, O'Rourke M, Spratt BG (1993) How clonal are bacteria? Proceedings of the National Academy of Sciences USA 90:4384–4388

McKenzie FE, Smith DL, O'Meara WP, Riley EM (2008) Strain theory of malaria: The first 50 years. Advances in Parasitology 66:1–46

Meagher S (1999) Genetic diversity and *Capillaria hepatica* (nematoda) prevalence in Michigan deer mouse populations. Evolution 53:1318–1324

Moreno M, Silva E, Ramirez L, Palacio L, Rivera D, Arcos-Burgos M (2004) Chagas' disease susceptibility/resistance: linkage disequilibrium analysis suggests epistasis between major histocompatibility complex and interleukin-10. Tissue Antigens 64:18–24

Moritz RFA, Kryger P, Koeniger G, Koeniger N, Estoup A, Tingek S (1995) High degree of polyandry in *Apis dorsata* queens detected by DNA microsatellite variability. Behavioral Ecology & Sociobiology 37: 357–363

Navajas M, Migeon A, Alaux C, Martin-Magniette ML, Robinson GE, Ebvans JD, Cros-Arteil S, Crauser D, Le Conte Y (2008) Differential gene expression of the honey bee *Apis mellifera* associated with *Varroa destructor* infection. BMC Genomics 9:301

Nielsen R (2005) Molecular signatures of natural selection. Annual Review of Genetics 39:197–218

Nielsen R, Williamson S, Kim Y, Hubisz MJ, Clark AG, Bustamante C (2005) Genomic scans for selective sweeps using SNP data. Genome 15:1566–1575

Obbard DJ, Callister DM, Jiggins FM, Soares DC, Yan G, Little TJ (2008) The evolution of TEP1, an exceptionally polymorphic immunity gene in *Anopheles gambiae*. BMC Evolutionary Biology 8:274

Obbard DJ, Gordon KHJ, Buck AH, Jiggins FM (2009) The evolution of RNAi as a defence against viruses and transposable elements. Philosophical Transactions of the Royal Society B 364:99–115

Obbard DJ, Jiggins FM, Halligan DL, Little TJ (2006) Natural selection drives extremely rapid evolution in antiviral RNAi genes. Current Biology 16:580–585

Oldroyd BP, Clifton MJ, Parker K, Wongsiri S, Rinderer TE, Crozier RH (1998) Evolution of mating behavior in the genus *Apis* and an estimate of mating frequency in *Apis cerana* (Hymenoptera: Apidae). Annals of the Entomological Society of America 85:335–343

Oldroyd BP, Clifton MJ, Womgsiri S, Rinderer TE, Sylvester HA, Crozier RH (1997) Polyandry in the genus *Apis*, particularly *Apis andreniformis*. Behavioral Ecology and Sociobiology 40:17–26

Oleksiak MF, Churchill GA, Crawford DL (2002) Variation in gene expression within and among natural populations. Nature Genetics 32:261–266

Otto SP, Nuismer SL (2004) Species interaction and the evolution of sex. Science 304:1018–1021

Poirie M, Frey F, Hita M, Huguet E, Lemeunier F, Periquet G, Carton Y (2000) Drosophila resistance genes against parasitoids: chromosomal location and linkage analysis. Proceedings of the Royal Society London B 267:1417–1421

Poulin R (2007) Evolutionary Ecology of Parasites, 2nd ed. Princeton University Press, Princeton

Poulin R, Keeney DB (2008) Host specificity under molecular and experimental scrutiny. Trends in Parasitology 24:24–28

Pounds JA, Bustamante MR, Coloma LA, Consuegra JA, Fogden MP, Foster PN, La Marca E, Masters KL, Merino-Viteri A, Puschendorf R, Ron SR, Sánchez-Azofeifa GA, Still CJ, Young BE (2006) Widespread amphibian extinctions from epidemic disease driven by global warming. Nature 439:161–167

Razakandrainibe FG, Duran P, Koella JC, De Meeus T, Rousset F, Ayala FJ, Renaud Fc (2005) "Clonal" population structure of malaria agent, *Plasmodium falciparum*, in high-infection regions. Proceedings of the National Academy of Sciences USA 102:17388–17399

Reusch TBH, Häberli ME, Aeschlimann PB, Milinski M (2001) Female sticklebacks count alleles in a strategy of sexual selection explaining MHC polymorphism. Nature 414:300–302

Riddell CR, Adams S, Schmid-Hempel P, Mallon EB (2009) Differential expression of immune defences is associated with specific host-parasite interactions in insects. PLoS One 4: e7621

Rockman MV, Kruglyak L (2006) Genetics of global gene expression. Nature 7:862–872

Sabeti PC, Reich DE, Higgins JM, Levine HZP, Richter DJ, Schaffner SF, Gabriel SB, Platko JV, Patterson NJ, McDonald GJ, Ackerman HC, Campbell SJ, Altshuler D, Cooper R, Kwiatkowski D, Ward R, Lander ES (2002) Detecting recent positive selection in the human genome from haplotype structure. Nature 419:832–837

Salathé M, Kouyos RD, Bonhoeffer S (2008) The state of affairs in the kingdom of the Red Queen. Trends in Ecology and Evolution 23:439–445

Salathé P, Ebert D (2003) The effects of parasitism and inbreeding on the competitive ability in *Daphnia magna*: evidence for synergistic epistasis: evidence for synergistic epistasis. Journal of Evolutionary Biology 16:976–986

Schlenke TA, Begun DJ (2003) Natural selection drives *Drosophila* immune system evolution. Genetics 164:1471–1480

Schmid-Hempel P (2011) Evolutionary Parasitology: The Integrated Study of Infections, Immunology, Ecology, and Genetics. Oxford University Press, Oxford

Schmid-Hempel P, Crozier RH (1999) Polyandry versus polygyny versus parasites. Philosophical Transactions of the Royal Society B 354:507–515

Spielman D, Brook BW, Briscoe DA, Frankha R (2004) Does inbreeding and loss of genetic diversity decrease disease resistance? Conservation Genetics 5:439–448

Stahl EA, Dwyer G, Mauricio R, Kreitmann M, Bergelson J (1999) Dynamics of disease resistance polymorphism at the Rpm1 locus of *Arabidopsis*. Nature 400:667–671

Stotz HU, Bishop JG, Bergmann CW, Koch M, Albersheim P, Darvill AG, Labavitch JM (2000) Identification of target amino acids that affect interactions of fungal polygalacturonases and their plant inhibitors. Physiological and Molecular Plant Pathology 56:117–130

Tarpy DR (2003) Genetic diversity within honeybee colonies prevents severe infections and promotes colony growth. Proceedings of the Royal Society B 270:99–103

Tibayrenc M (1999) Toward an integrated genetic epidemiology of parasitic protozoa and other pathogens. Annual Review of Genetics 33:449–477

Tibayrenc M, Ayala FJ (2002) The clonal theory of parasitic protozoa: 12 years on. Trends in Parasitology 18:405–410

Trouve S, Degen L, Renaud F, Goudet J (2003) Evolutionary implications of a high selfing rate in the freshwater snail *Lymnaea truncatula*. Evolution 57:2303–2314

Viljakainen L, Evans JD, Hasselmann M, Rueppell O, Tingek S, Pamilo P (2009) Rapid evolution of immune proteins in social insects. Molecular Biology and Evolution 26:1791–1801

Vododvar N, Vinals M, Liehl P, Basset A, Degrouard J, Spellmann P, Boccard F, Lemaitre B (2005) *Drosophila* host defense after oral infection by an entomopathogenic *Pseudomonas* species. Proceedings of the National Academy of Sciences USA 102:11414–11419

Votypka J, Ray DS, Lukes J (2001) *Crithidia fasciculata*: A test for genetic exchange. Experimental Parasitology 99:104–107

Waterhouse RM, Kriventseva EV, Meister S, YX Z, Alvarez KS, Bartholomay LC, Barillas. Mury C, Bian G, Blandin S, Christensen BM, Dong Y, Jiang H, Kanost MR, Koutsos AC, Levashina EA, Li J, Ligoxygakis P, MacCallum RM, Mayhew GF, Mendes A, Michel K, Osta MA, Paskewitz S, Whin SW, Vlachou D, Raikhle AS, Kafatos FC, Dimopoulos G, Zdobnov EM, Christophides GK (2007) Evolutionary dynamics of immune-related genes and pathways in disease-vector mosquitoes. Science 316:1738–1743

Webster JP, Woolhouse ME (1998a) Cost of resistance: relationship between reduced fertility and increased resistance in a snail-schistosome host-parasite system. Proceedings of the Royal Society B 266:391–396

Webster JP, Woolhouse ME (1998b) Selection and strain specificity of compatibility between snail intermediate hosts and their parasitic schistosomes. Evolution 52:1627–1634

Wegner MK, Berenos C, Schmid-Hempel P (2008) Non-additive genetic components in resistance of the red flour beetle against parasitic infection. Evolution 62:2381–2392

Wegner MK, Berenos C, Schmid-Hempel P (2009) Host genetic architecture in single and multiple infections. Journal of Evolutionary Biology 22:396–404

West SA, Smith TG, Read AF (2000) Sex allocation and population structure in apicomplexan (protozoa) parasites. Proceedings of the Royal Society London B 267:257–263

Whiteman NK, Matson KD, Bollmer JL, Parker PG (2005) Disease ecology in the Galapagos hawk (*Buteo galapagoensis*): host genetic diversity, parasite load and natural antibodies. Proceedings of the Royal Society London B 273:797–804

Wilfert L, Gadau J, Schmid-Hempel P (2007) Variation in genomic recombination rates among animal taxa and the case of social insects. Heredity 98:189–197

Wilfert L, Schmid-Hempel P (2008) The genetic architecture of susceptibility to parasites. BMC Evolutionary Biology 8:187

Xu X, Dong Y, Abraham EG, Kocan A, Srinivasan P, Ghosh AK, Sinden RE, Ribeiro JMC, Jacobs-Lorena M, Kafatos FC, Dimopoulos G (2005) Transcriptome analysis of *Anopheles stephensi–Plasmodium berghei* interactions. Molecular and Biochemical Parasitology 142:76–87

Yoon HJ, Sohn MR, Choo YM, Li J, Sohn HD, Jin BR (2009) Defensin gene sequences of three different bumblebees, *Bombus* spp. Journal of Asian-Pacific Entomology 12:27–31

INDEX

www.ingramcontent.com/pod-product-compliance
Ingram Content Group UK Ltd.
Pitfield, Milton Keynes, MK11 3LW, UK
UKHW051224160926
475595UK00022B/166

9 780199 737345